Stochastic Processes

Theory for Applications

This definitive textbook provides a solid introduction to discrete and continuous stochastic processes, tackling a complex field in a way that instills a deep understanding of the relevant mathematical principles, and develops an intuitive grasp of the way these principles can be applied to modeling real-world systems.

- Basic underlying principles and axioms are made clear from the start, and new topics are developed as needed, encouraging and enabling students to develop an instinctive grasp of the fundamentals.
- Mathematical proofs are made easy for students to understand and remember, helping them quickly learn how to choose and apply the best possible models to real-world situations.
- Includes a review of elementary probability theory; detailed coverage of Poisson, Gaussian and Markov processes; the basic elements of queueing theory; and theory and applications of inference, hypothesis testing, detection and estimation, in addition to more advanced topics.

Written by one of the world's leading information theorists, based on his 20 years' experience of teaching stochastic processes to graduate students, this is an exceptional resource for anyone looking to develop their understanding of stochastic processes.

Robert G. Gallager is a Professor Emeritus at MIT, and one of the world's leading information theorists. He is a Member of the US National Academy of Engineering, and the US National Academy of Sciences, and his numerous awards and honors include the IEEE Medal of Honour (1990) and the Marconi Prize (2003). He was awarded the MIT Graduate Student Teaching Award in 1993, and this book is based on his 20 years' of experience of teaching this subject to students.

"The book is a wonderful exposition of the key ideas, models, and results in stochastic processes most useful for diverse applications in communications, signal processing, analysis of computer and information systems, and beyond. The text provides excellent intuition, with numerous beautifully crafted examples, and exercises. Foundations are included in a natural way that enhances clarity and the reader's ability to apply the results."

Bruce Hajek, *University of Illinois, Urbana-Champaign*

"This book provides a beautiful treatment of the fundamentals of stochastic process. Gallager's clear exposition conveys a deep and intuitive understanding of this important subject. Graduate students and researchers alike will benefit from this text, which will soon be a classic."

Randall Berry, *Northwestern University*

"In *Stochastic Processes: Theory for Applications*, Robert Gallager has produced another in his series of outstanding texts. Using a style that is very intuitive and approachable, but without sacrificing the underlying rigor of the subject matter, he has focused his treatment exactly at the level that engineers and applied scientists need to understand in order to have a working knowledge of this field. The breadth and sequencing of the coverage are also excellent. This book will be a useful resource both for students entering the field and for practitioners seeking to deepen their understanding of stochastic methods."

H. Vincent Poor, *Princeton University*

"Professor Gallager's book is the first of a plethora of textbooks on stochastic processes for engineers that strike the perfect balance between broad coverage, rigor, and motivation for applications. With a wealth of illustrative examples and challenging exercises, this book is the ideal text for graduate students in any field that applies stochastic processes."

Abbas El Gamal, *Stanford University*

Stochastic Processes

Theory for Applications

Robert G. Gallager
MIT

CAMBRIDGE
UNIVERSITY PRESS

University Printing House, Cambridge CB2 8BS, United Kingdom

One Liberty Plaza, 20th Floor, New York, NY 10006, USA

477 Williamstown Road, Port Melbourne, VIC 3207, Australia

314-321, 3rd Floor, Plot 3, Splendor Forum, Jasola District Centre, New Delhi-110025, India

79 Anson Road, #06-04/06, Singapore 079906

Cambridge University Press is part of the University of Cambridge.

It furthers the University's mission by disseminating knowledge in the pursuit of education, learning and research at the highest international levels of excellence.

www.cambridge.org
Information on this title: www.cambridge.org/9781107039759

First published 2013
Reprinted 2017

A catalogue record for this publication is available from the British Library

Library of Congress Cataloging in Publication data
Gallager, Robert G.
Stochastic processes: theory for applications / Robert G. Gallager, MIT.
pages cm
ISBN 978-1-107-03975-9 (hardback)
1. Stochastic processes – Textbooks. I. Title.
QA274.G344 2013
519.2′3–dc23

2013005146

ISBN 978-1-107-03975-9 Hardback

Additional resources for this publication at www.cambridge.org/stochasticprocesses

To Marie, with thanks for her love and encouragement while I finished this book

Contents

Preface *page* xv
Suggestions for instructors and self study xix
Acknowledgements xxi

1 **Introduction and review of probability** 1
1.1 Probability models 1
1.1.1 The sample space of a probability model 3
1.1.2 Assigning probabilities for finite sample spaces 4
1.2 The axioms of probability theory 5
1.2.1 Axioms for events 7
1.2.2 Axioms of probability 8
1.3 Probability review 9
1.3.1 Conditional probabilities and statistical independence 9
1.3.2 Repeated idealized experiments 11
1.3.3 Random variables 12
1.3.4 Multiple random variables and conditional probabilities 14
1.4 Stochastic processes 16
1.4.1 The Bernoulli process 17
1.5 Expectations and more probability review 19
1.5.1 Random variables as functions of other random variables 23
1.5.2 Conditional expectations 25
1.5.3 Typical values of random variables; mean and median 28
1.5.4 Indicator random variables 29
1.5.5 Moment generating functions and other transforms 29
1.6 Basic inequalities 31
1.6.1 The Markov inequality 32
1.6.2 The Chebyshev inequality 32
1.6.3 Chernoff bounds 33
1.7 The laws of large numbers 36
1.7.1 Weak law of large numbers with a finite variance 36
1.7.2 Relative frequency 39
1.7.3 The central limit theorem (CLT) 39
1.7.4 Weak law with an infinite variance 44
1.7.5 Convergence of random variables 45
1.7.6 Convergence with probability 1 48

1.8 Relation of probability models to the real world 51
1.8.1 Relative frequencies in a probability model 52
1.8.2 Relative frequencies in the real world 52
1.8.3 Statistical independence of real-world experiments 55
1.8.4 Limitations of relative frequencies 56
1.8.5 Subjective probability 57
1.9 Summary 57
1.10 Exercises 58

2 Poisson processes 72
2.1 Introduction 72
2.1.1 Arrival processes 72
2.2 Definition and properties of a Poisson process 74
2.2.1 Memoryless property 75
2.2.2 Probability density of S_n and joint density of $S_1, \dots, S_n$ 78
2.2.3 The probability mass function (PMF) for $N(t)$ 79
2.2.4 Alternative definitions of Poisson processes 80
2.2.5 The Poisson process as a limit of shrinking Bernoulli processes 82
2.3 Combining and splitting Poisson processes 84
2.3.1 Subdividing a Poisson process 86
2.3.2 Examples using independent Poisson processes 87
2.4 Non-homogeneous Poisson processes 89
2.5 Conditional arrival densities and order statistics 92
2.6 Summary 96
2.7 Exercises 97

3 Gaussian random vectors and processes 105
3.1 Introduction 105
3.2 Gaussian random variables 105
3.3 Gaussian random vectors 107
3.3.1 Generating functions of Gaussian random vectors 108
3.3.2 IID normalized Gaussian random vectors 108
3.3.3 Jointly-Gaussian random vectors 109
3.3.4 Joint probability density for Gaussian n-rv s (special case) 112
3.4 Properties of covariance matrices 114
3.4.1 Symmetric matrices 114
3.4.2 Positive definite matrices and covariance matrices 115
3.4.3 Joint probability density for Gaussian n-rv s (general case) 117
3.4.4 Geometry and principal axes for Gaussian densities 118
3.5 Conditional PDFs for Gaussian random vectors 120
3.6 Gaussian processes 124
3.6.1 Stationarity and related concepts 126
3.6.2 Orthonormal expansions 128
3.6.3 Continuous-time Gaussian processes 130
3.6.4 Gaussian sinc processes 132

3.6.5 Filtered Gaussian sinc processes 134
3.6.6 Filtered continuous-time stochastic processes 136
3.6.7 Interpretation of spectral density and covariance 138
3.6.8 White Gaussian noise 139
3.6.9 The Wiener process/Brownian motion 142
3.7 Circularly-symmetric complex random vectors 144
3.7.1 Circular symmetry and complex Gaussian random variables 145
3.7.2 Covariance and pseudo-covariance of complex n-dimensional random vectors 146
3.7.3 Covariance matrices of complex n-dimensional random vectors 148
3.7.4 Linear transformations of $\boldsymbol{W} \sim \mathcal{CN}(0, [I_\ell])$ 149
3.7.5 Linear transformations of $\boldsymbol{Z} \sim \mathcal{CN}(0, [K])$ 150
3.7.6 The PDF of circularly-symmetric Gaussian n-dimensional random vectors 150
3.7.7 Conditional PDFs for circularly-symmetric Gaussian random vectors 153
3.7.8 Circularly-symmetric Gaussian processes 154
3.8 Summary 155
3.9 Exercises 156

4 Finite-state Markov chains 161
4.1 Introduction 161
4.2 Classification of states 163
4.3 The matrix representation 168
4.3.1 Steady state and $[P^n]$ for large n 168
4.3.2 Steady state assuming $[P] > 0$ 171
4.3.3 Ergodic Markov chains 172
4.3.4 Ergodic unichains 173
4.3.5 Arbitrary finite-state Markov chains 175
4.4 The eigenvalues and eigenvectors of stochastic matrices 176
4.4.1 Eigenvalues and eigenvectors for $\mathsf{M} = 2$ states 176
4.4.2 Eigenvalues and eigenvectors for $\mathsf{M} > 2$ states 177
4.5 Markov chains with rewards 180
4.5.1 Expected first-passage times 181
4.5.2 The expected aggregate reward over multiple transitions 185
4.5.3 The expected aggregate reward with an additional final reward 188
4.6 Markov decision theory and dynamic programming 189
4.6.1 Dynamic programming algorithm 190
4.6.2 Optimal stationary policies 194
4.6.3 Policy improvement and the search for optimal stationary policies 197
4.7 Summary 201
4.8 Exercises 202

5 Renewal processes 214

5.1 Introduction 214
5.2 The strong law of large numbers and convergence with probability 1 217
5.2.1 Convergence with probability 1 (WP1) 217
5.2.2 Strong law of large numbers 219
5.3 Strong law for renewal processes 221
5.4 Renewal–reward processes; time averages 226
5.4.1 General renewal–reward processes 230
5.5 Stopping times for repeated experiments 233
5.5.1 Wald's equality 236
5.5.2 Applying Wald's equality to $\mathsf{E}[N(t)]$ 238
5.5.3 Generalized stopping trials, embedded renewals, and G/G/1 queues 239
5.5.4 Little's theorem 242
5.5.5 M/G/1 queues 245
5.6 Expected number of renewals; ensemble averages 249
5.6.1 Laplace transform approach 250
5.6.2 The elementary renewal theorem 251
5.7 Renewal–reward processes; ensemble averages 254
5.7.1 Age and duration for arithmetic processes 255
5.7.2 Joint age and duration: non-arithmetic case 258
5.7.3 Age $Z(t)$ for finite t: non-arithmetic case 259
5.7.4 Age $Z(t)$ as $t \to \infty$: non-arithmetic case 262
5.7.5 Arbitrary renewal–reward functions: non-arithmetic case 264
5.8 Delayed renewal processes 266
5.8.1 Delayed renewal–reward processes 268
5.8.2 Transient behavior of delayed renewal processes 269
5.8.3 The equilibrium process 270
5.9 Summary 270
5.10 Exercises 271

6 Countable-state Markov chains 287

6.1 Introductory examples 287
6.2 First-passage times and recurrent states 289
6.3 Renewal theory applied to Markov chains 294
6.3.1 Renewal theory and positive recurrence 294
6.3.2 Steady state 296
6.3.3 Blackwell's theorem applied to Markov chains 299
6.3.4 Age of an arithmetic renewal process 300
6.4 Birth–death Markov chains 302
6.5 Reversible Markov chains 303
6.6 The M/M/1 sampled-time Markov chain 307
6.7 Branching processes 309
6.8 Round-robin service and processor sharing 312

6.9 Summary 317
6.10 Exercises 319

7 Markov processes with countable-state spaces 324
7.1 Introduction 324
7.1.1 The sampled-time approximation to a Markov process 328
7.2 Steady-state behavior of irreducible Markov processes 329
7.2.1 Renewals on successive entries to a given state 330
7.2.2 The limiting fraction of time in each state 331
7.2.3 Finding $\{p_j(i); j \geq 0\}$ in terms of $\{\pi_j; j \geq 0\}$ 332
7.2.4 Solving for the steady-state process probabilities directly 335
7.2.5 The sampled-time approximation again 336
7.2.6 Pathological cases 336
7.3 The Kolmogorov differential equations 337
7.4 Uniformization 341
7.5 Birth–death processes 342
7.5.1 The M/M/1 queue again 342
7.5.2 Other birth–death systems 343
7.6 Reversibility for Markov processes 344
7.7 Jackson networks 350
7.7.1 Closed Jackson networks 355
7.8 Semi-Markov processes 357
7.8.1 Example – the M/G/1 queue 360
7.9 Summary 361
7.10 Exercises 363

8 Detection, decisions, and hypothesis testing 375
8.1 Decision criteria and the maximum a posteriori probability (MAP) criterion 376
8.2 Binary MAP detection 379
8.2.1 Sufficient statistics I 381
8.2.2 Binary detection with a one-dimensional observation 381
8.2.3 Binary MAP detection with vector observations 385
8.2.4 Sufficient statistics II 391
8.3 Binary detection with a minimum-cost criterion 395
8.4 The error curve and the Neyman–Pearson rule 396
8.4.1 The Neyman–Pearson detection rule 402
8.4.2 The min–max detection rule 403
8.5 Finitely many hypotheses 403
8.5.1 Sufficient statistics with $\mathsf{M} \geq 2$ hypotheses 407
8.5.2 More general minimum-cost tests 409
8.6 Summary 409
8.7 Exercises 410

9 Random walks, large deviations, and martingales 417

9.1 Introduction 417

9.1.1 Simple random walks 418

9.1.2 Integer-valued random walks 419

9.1.3 Renewal processes as special cases of random walks 419

9.2 The queueing delay in a G/G/1 queue 420

9.3 Threshold crossing probabilities in random walks 423

9.3.1 The Chernoff bound 423

9.3.2 Tilted probabilities 425

9.3.3 Large deviations and compositions 428

9.3.4 Back to threshold crossings 431

9.4 Thresholds, stopping rules, and Wald's identity 433

9.4.1 Wald's identity for two thresholds 434

9.4.2 The relationship of Wald's identity to Wald's equality 435

9.4.3 Zero-mean random walks 435

9.4.4 Exponential bounds on the probability of threshold crossing 436

9.5 Binary hypotheses with IID observations 438

9.5.1 Binary hypotheses with a fixed number of observations 438

9.5.2 Sequential decisions for binary hypotheses 442

9.6 Martingales 444

9.6.1 Simple examples of martingales 444

9.6.2 Scaled branching processes 446

9.6.3 Partial isolation of past and future in martingales 446

9.7 Submartingales and supermartingales 447

9.8 Stopped processes and stopping trials 450

9.8.1 The Wald identity 452

9.9 The Kolmogorov inequalities 453

9.9.1 The SLLN 455

9.9.2 The martingale convergence theorem 456

9.10 A simple model for investments 458

9.10.1 Portfolios with constant fractional allocations 461

9.10.2 Portfolios with time-varying allocations 465

9.11 Markov modulated random walks 468

9.11.1 Generating functions for Markov random walks 469

9.11.2 Stopping trials for martingales relative to a process 471

9.11.3 Markov modulated random walks with thresholds 471

9.12 Summary 472

9.13 Exercises 473

10 Estimation 488

10.1 Introduction 488

10.1.1 The squared-cost function 489

10.1.2 Other cost functions 490

10.2 MMSE estimation for Gaussian random vectors 491

10.2.1 Scalar iterative estimation 494
10.2.2 Scalar Kalman filter 496
10.3 LLSE estimation 498
10.4 Filtered vector signal plus noise 500
10.4.1 Estimate of a single random variable in IID vector noise 502
10.4.2 Estimate of a single random variable in arbitrary vector noise 502
10.4.3 Vector iterative estimation 503
10.4.4 Vector Kalman filter 504
10.5 Estimation for circularly-symmetric Gaussian rv s 505
10.6 The vector space of random variables; orthogonality 507
10.7 MAP estimation and sufficient statistics 512
10.8 Parameter estimation 513
10.8.1 Fisher information and the Cramer–Rao bound 516
10.8.2 Vector observations 518
10.8.3 Information 519
10.9 Summary 521
10.10 Exercises 523

References 528
Index 530

Preface

This text has evolved over some 20 years, starting as lecture notes for two first-year graduate subjects at MIT, namely, *Discrete Stochastic Processes* (6.262) and *Random Processes, Detection, and Estimation* (6.432). The two sets of notes are closely related and have been integrated into one text. Instructors and students can pick and choose the topics that meet their needs, and suggestions for doing this follow this preface.

These subjects originally had an application emphasis, the first on queueing and congestion in data networks and the second on modulation and detection of signals in the presence of noise. As the notes have evolved, it has become increasingly clear that the mathematical development (with minor enhancements) is applicable to a much broader set of applications in engineering, operations research, physics, biology, economics, finance, statistics, etc.

The field of stochastic processes is essentially a branch of probability theory, treating probabilistic models that evolve in time. It is best viewed as a branch of mathematics, starting with the axioms of probability and containing a rich and fascinating set of results following from those axioms. Although the results are applicable to many areas, they are best understood initially in terms of their mathematical structure and interrelationships.

Applying axiomatic probability results to a real-world area requires creating a probability model for the given area. Mathematically precise results can then be derived within the model and translated back to the real world. If the model fits the area sufficiently well, real problems can be solved by analysis within the model. However, since models are almost always simplified approximations of reality, precise results within the model become approximations in the real world.

Choosing an appropriate probability model is an essential part of this process. Sometimes an application area will have customary choices of models, or at least structured ways of selecting them. For example, there is a well-developed taxonomy of queueing models. A sound knowledge of the application area, combined with a sound knowledge of the behavior of these queueing models, often lets one choose a suitable model for a given issue within the application area. In other cases, one can start with a particularly simple model and use the behavior of that model to gain insight about the application, and use this to iteratively guide the selection of more general models.

An important aspect of choosing a probability model for a real-world area is that a prospective choice depends heavily on prior understanding, at both an intuitive and mathematical level, of results from the range of mathematical models that might be involved. This partly explains the title of the text – *Theory for Applications*. The aim is to guide the reader in both the mathematical and intuitive understanding necessary in developing and using stochastic process models in studying application areas.

Application-oriented students often ask why it is important to understand axioms, theorems, and proofs in mathematical models when the precise results in the model become approximations in the real-world system being modeled. One answer is that a deeper understanding of the mathematics leads to the required intuition for understanding the differences between model and reality. Another answer is that theorems are transferable between applications, and thus enable insights from one application area to be transferred to another.

Given the need for precision in the theory, however, why is an axiomatic approach needed? Engineering and science students learn to use calculus, linear algebra, and undergraduate probability effectively without axioms or rigor. Why does this not work for more advanced probability and stochastic processes?

Probability theory has more than its share of apparent paradoxes, and these show up in very elementary arguments. Undergraduates are content with this, since they can postpone these questions to later study. For the more complex issues in graduate work, however, reasoning without a foundation becomes increasingly frustrating, and the axioms provide the foundation needed for sound reasoning without paradoxes.

I have tried to avoid the concise and formal proofs of pure mathematics, and instead use explanations that are longer but more intuitive while still being precise. This is partly to help students with limited exposure to pure mathematics, and partly because intuition is vital when going back and forth between a mathematical model and a real-world problem. In doing research, we grope toward results, and successful groping requires both a strong intuition and precise reasoning.

The text neither uses nor develops measure theory. Measure theory is undoubtedly important in understanding probability at a deep level, but most of the topics useful in many applications can be understood without measure theory. I believe that the level of precision here provides a good background for a later study of measure theory.

The text does require some background in probability at an undergraduate level. Chapter 1 presents this background material as a review, but it is too concentrated and deep for most students without prior background. Some exposure to linear algebra and analysis (especially concrete topics like vectors, matrices, and limits) is helpful, but the text develops the necessary results. The most important prerequisite is the mathematical maturity and patience to couple precise reasoning with intuition.

The organization of the text, after the review in Chapter 1 is as follows: Chapters 2, 3, and 4 treat three of the simplest and most important classes of stochastic processes, first Poisson processes, next Gaussian processes, and finally finite-state Markov chains. These are beautiful processes where almost everything is known, and they contribute insights, examples, and initial approaches for almost all other processes. Chapter 5 then treats renewal processes, which generalize Poisson processes and provide the foundation for the rest of the text.

Chapters 6 and 7 use renewal theory to generalize Markov chains to countable state spaces and continuous time. Chapters 8 and 10 then study decision making and estimation, which in a sense gets us out of the world of theory and back to using the theory.

Chapter 9 treats random walks, large deviations, and martingales and illustrates many of their applications.

Most results here are quite old and well established, so I have not made any effort to attribute results to investigators. My treatment of the material is indebted to the texts by Bertsekas and Tsitsiklis [2], Sheldon Ross [22] and William Feller [8] and [9].

Suggestions for instructors and self study

The subject of stochastic processes contains many beautiful and surprising results at a relatively simple level. These results should be savored and contemplated rather than rushed. The urge to go too quickly, to sacrifice understanding for shallow bottom lines, and to cover all the most important topics should be resisted. This text covers all the material in two full term graduate subjects at MIT, plus many other topics added for enrichment, so it cannot be 'covered' in one term.

My conviction is that if a student acquires a deep understanding of any, say, 20% of the material, then that student will be able to read and understand the rest with relative ease at a later time. Better still, a full appreciation of that 20% will make most students eager to learn more. In other words, instructors have a good deal of freedom, subject to a prerequisite structure, to choose topics of interest to them and their students to cover in a one term course.

One of the two MIT courses leading to this text covers Chapters 1, 2, 4, 5, 6, 7, and 9, skipping many of the more detailed parts of the latter five chapters. The other course covers Chapters 1, 3, 8, and 10, again omitting many topics. The first course is largely discrete and the second largely continuous, and a different mix is probably more appropriate for a student taking only one subject.

The topics in Chapter 1 are largely covered in good elementary probability subjects, but students are usually better at doing plug and chug exercises on these topics than having the depth of understanding required by the subsequent topics. Thus instructors should spend some time reviewing these topics.

It is difficult to be precise about the extent to which one topic is a prerequisites of another. The table below lists the prerequisites of each section. Most sections have only one prerequisite, but that recursively includes the prerequisites of the prerequisite. Instructors and students are encouraged to use their own judgement here.

Sect.	Prereq.	Sect.	Prereq.	Sect.	Prereq.	Sect.	Prereq.	Sect.	Prereq.
1.2	1.1	3.4	3.3	5.6	5.5	7.5	7.2	9.7	9.6
1.3	1.2	3.5	3.4	5.7	5.6	7.6	7.5	9.8	9.7
1.4	1.3	3.6	3.5	5.8	5.7	7.7	7.6	9.9	9.8
1.5	1.4	3.7	3.5	6.1	5.5, 4.3	7.8	7.2	9.10	9.9
1.6	1.5	4.1	1.5	6.2	6.1	8.1	1.8	9.11	9.9
1.7	1.6	4.2	4.1	6.3	6.2	8.2	8.1, 3.3	10.1	1.8, 3.5
1.8	1.7	4.3	4.2	6.4	6.3	8.3	8.2	10.2	10.1
2.1	1.5	4.4	4.3	6.5	6.4	8.4	8.3	10.3	10.2
2.2	2.1	4.5	4.4	6.6	6.5	8.5	8.4	10.4	10.3
2.3	2.2	4.6	4.5	6.7	6.3	9.1	5.5	10.5	10.4, 3.7
2.4	2.3	5.1	1.7, 2.2	6.8	6.3	9.2	9.1	10.6	10.3
2.5	1.2	5.2	5.1	7.1	6.3	9.3	9.2	10.7	10.3
3.1	1.7	5.3	5.2	7.2	7.1	9.4	9.3	10.8	10.3
3.2	3.1	5.4	5.3	7.3	7.2	9.5	9.4, 8.2		
3.3	3.2	5.5	5.4	7.4	7.2	9.6	9.5		

Acknowledgements

This book has its roots in a book called *Discrete Stochastic Processes* that I wrote back in 1996, some lecture notes on continuous random processes from about the same time, and lecture notes that I have been writing at MIT from 2007 to 2012 for a subject also entitled Discrete Stochastic Processes.

I am deeply grateful to Professors John Wyatt, John Tsitsiklis, and Lizhong Zheng who have used these notes in teaching the Discrete Stochastic Process course in recent years. Their many general observations about the value and teachability of various topics and the suggestions of alternative approaches have been invaluable. Their ability (particularly in John Tsitsiklis' case) to catch minor flaws in proofs and suggest cleaner approaches has saved me from many errors. They are also responsible for a number of excellent new exercises.

Natasha Blitvic, Mina Karzand, and Fabian Kozynski who were teaching assistants in the course for the last five years, were also very helpful both in creating and improving the wording in a number of exercises, but also in explaining why students were having difficulty and how to improve the presentation.

I am also indebted to a number of people in the MIT community who have been helpful in the evolution of this book. Professors Dimitri Berksekas and Yuri Polyansky have helped in discussing topics for the book and in reading various sections. Shan-Yuan Ho has been enormously helpful in reading the entire manuscript and catching many things, ranging from typos to poorly presented concepts. She also wrote the solution manual for the old book and has been helpful through the whole evolution of this project.

A number of friends, students, and even people who found the text on the web have also been helpful in catching errors and inconsistencies and suggesting better approaches to various topics. Murat Azizoglu and Baris Nakiboglu have been particularly helpful in this regard. Others are Ivan Bersenco, Dimitris Bisias, Nathan Jones, Tarek Lahlou, Vahid Montazerhodju, Emre Teletar, Roy Yates, and Andrew Young. Finally, I am grateful to the many students who have taken the course in the last five years and who look puzzled (or sleepy) when something is explained badly, and who ask questions when it is almost clear.

1 Introduction and review of probability

1.1 Probability models

Stochastic processes constitute a branch of probability theory treating probabilistic systems that evolve in time. There seems to be no very good reason for trying to define stochastic processes precisely, but as we hope will become evident in this chapter, there is a very good reason for trying to be precise about probability itself. Those particular topics in which evolution in time is important will then unfold naturally. Section 1.5 gives a brief introduction to one of the very simplest stochastic processes, the Bernoulli process, and then Chapters 2, 3, and 4 develop three basic stochastic process models which serve as simple examples and starting points for the other processes to be discussed later.

Probability theory is a central field of mathematics, widely applicable to scientific, technological, and human situations involving uncertainty. The most obvious applications are to situations, such as games of chance, in which repeated trials of essentially the same procedure lead to differing outcomes. For example, when we flip a coin, roll a die, pick a card from a shuffled deck, or spin a ball onto a roulette wheel, the procedure is the same from one trial to the next, but the outcome (heads (H) or tails (T) in the case of a coin, 1 to 6 in the case of a die, etc.) varies from one trial to another in a seemingly random fashion.

For the case of flipping a coin, the outcome of the flip could be predicted from the initial position, velocity, and angular momentum of the coin and from the nature of the surface on which it lands. Thus, in one sense, a coin flip is deterministic rather than random and the same can be said for the other examples above. When these initial conditions are unspecified, however, as when playing these games, the outcome can again be viewed as random in some intuitive sense.

Many scientific experiments are similar to games of chance in the sense that multiple trials of apparently the *same* procedure lead to results that *vary* from one trial to another. In some cases, this variation is due to slight variations in the experimental procedure, in some it is due to noise, and in some, such as in quantum mechanics, the randomness is generally believed to be fundamental. Similar situations occur in many types of systems, especially those in which noise and random delays are important. Some of these systems, rather than being repetitions of a common basic procedure, are systems that evolve over time while still containing a sequence of underlying similar random occurrences.

This intuitive notion of randomness, as described above, is a very special kind of uncertainty. Rather than involving a lack of understanding, it involves a type of uncertainty that can lead to probabilistic models with precise results. As in any scientific field, the models might or might not correspond to reality very well, but when they do correspond to reality, there is the sense that the situation is completely understood, while still being random.

For example, we all feel that we understand flipping a coin or rolling a die, but still accept randomness in each outcome. The theory of probability was initially developed particularly to give precise and quantitative meaning to these types of situations. The remainder of this section introduces this relationship between the precise view of probability theory and the intuitive view as used in applications and everyday language.

After this introduction, the following sections of this chapter review probability theory as a mathematical discipline, with a special emphasis on the laws of large numbers. In the final section, we use the theory and the laws of large numbers to obtain a fuller understanding of the relationship between theory and the real world.[1]

Probability theory, as a mathematical discipline, started to evolve in the seventeenth century and was initially focused on games of chance. The importance of the theory grew rapidly, particularly in the twentieth century, and it now plays a central role in risk assessment, statistics, data networks, operations research, information theory, control theory, theoretical computer science, quantum theory, game theory, neurophysiology, and many other fields.

The core concept in probability theory is that of a *probability model*. Given the extent of the theory, both in mathematics and in applications, the simplicity of probability models is surprising. The first component of a probability model is a *sample space*, which is a *set* whose elements are called *sample points* or *outcomes*. Probability models are particularly simple in the special case where the sample space is finite, and we consider only this case in the remainder of this section. The second component of a probability model is a class of *events*, which can be considered for now simply as the class of all subsets of the sample space. The third component is a *probability measure*, which can be regarded for now as the assignment of a non-negative number to each outcome, with the restriction that these numbers must sum to 1 over the sample space. The probability of an event is the sum of the probabilities of the outcomes comprising that event.

These probability models play a dual role. In the first, the many known results about various classes of models, and the many known relationships between models, constitute the essence of probability theory. Thus one often studies a model not because of any relationship to the real world, but simply because the model provides a building block or example useful for the theory and thus ultimately useful for other models. In the other role, when probability theory is applied to some game, experiment, or other situation

[1] It would be appealing to show how probability theory evolved from real-world random situations, but probability theory, like most mathematical theories, has evolved from complex interactions between theoretical developments and initially oversimplified models of real situations. The successes and flaws of such models lead to refinements of the models and the theory, which in turn suggest applications to totally different fields.

involving randomness, a probability model is used to represent the experiment (in what follows, we refer to all of these random situations as experiments).

For example, the standard probability model for rolling a die uses $\{1, 2, 3, 4, 5, 6\}$ as the sample space, with each possible outcome having probability 1/6. An *odd* result, i.e., the subset $\{1, 3, 5\}$, is an example of an event in this sample space, and this event has probability 1/2. The correspondence between model and actual experiment seems straightforward here. Both have the same set of outcomes and, given the symmetry between faces of the die, the choice of equal probabilities seems natural. Closer inspection, however, reveals an important difference between the model and the actual rolling of a die.

The model above corresponds to a single roll of a die, with a probability defined for each possible outcome. In a real-world experiment where a single die is rolled, one of the six faces, say face k comes up, but there is no *observable* probability for k.

Our intuitive notion of rolling dice, however, involves an experiment with n consecutive rolls of a die. There are then 6^n possible outcomes, one for each possible n-tuple of individual die outcomes. As reviewed in subsequent sections, the standard probability model for this repeated-roll experiment is to assign probability 6^{-n} to each possible n-tuple, which leads to a probability $\binom{n}{m}(1/6)^m(5/6)^{n-m}$ that the face k comes up on m of the n rolls, i.e., that the relative frequency of face k is m/n. The distribution of these relative frequencies is increasingly clustered around $1/6$ as n is increased. Thus if a real-world experiment for tossing n dice is reasonably modeled by this probability model, we would also expect the relative frequency to be close to 1/6 for large n. This relationship through relative frequencies in a repeated experiment helps overcome the non-observable nature of probabilities in the real world.

1.1.1 The sample space of a probability model

An *outcome* or *sample point* in a probability model corresponds to a complete result (with all detail specified) of the experiment being modeled. For example, a game of cards is often appropriately modeled by the arrangement of cards within a shuffled 52-card deck, thus giving rise to a set of 52! outcomes (incredibly detailed, but trivially simple in structure), even though the entire deck might not be played in one trial of the game. A poker hand with four aces is an *event* rather than an *outcome* in this model, since many arrangements of the cards can give rise to four aces in a given hand. The possible outcomes in a probability model (and in the experiment being modeled) are mutually exclusive and collectively constitute the entire sample space (space of possible outcomes). An outcome ω is often called a *finest grain* result of the model in the sense that a singleton event $\{\omega\}$ containing only ω clearly contains no proper subsets. Thus events (other than singleton events) typically give only partial information about the result of the experiment, whereas an outcome fully specifies the result.

In choosing the sample space for a probability model of an experiment, we often omit details that appear irrelevant for the purpose at hand. Thus in modeling the set of outcomes for a coin toss as $\{H, T\}$, we ignore the type of coin, the initial velocity and angular momentum of the toss, etc. We also omit the rare possibility that the coin comes to rest on its edge. Sometimes, conversely, the sample space is enlarged beyond what is relevant in the interest of structural simplicity. An example is the above use of a shuffled deck of 52 cards.

The choice of the sample space in a probability model is similar to the choice of a mathematical model in any branch of science. That is, one simplifies the physical situation by eliminating detail of little apparent relevance. One often does this in an iterative way, using a very simple model to acquire initial understanding, and then successively choosing more detailed models based on the understanding from earlier models.

The mathematical theory of probability views the sample space simply as an abstract set of elements, and from a strictly mathematical point of view, the idea of doing an experiment and getting an outcome is a distraction. For visualizing the correspondence between the theory and applications, however, it is better to view the abstract set of elements as the set of possible outcomes of an idealized experiment in which, when the idealized experiment is performed, one and only one of those outcomes occurs. The two views are mathematically identical, but it will be helpful to refer to the first view as a probability model and the second as an idealized experiment. In applied probability texts and technical articles, these idealized experiments, rather than real-world situations, are often the primary topic of discussion.[2]

1.1.2 Assigning probabilities for finite sample spaces

The word *probability* is widely used in everyday language, and most of us attach various intuitive meanings[3] to the word. For example, everyone would agree that something virtually impossible should be assigned a probability close to 0 and something virtually certain should be assigned a probability close to 1. For these special cases, this provides a good rationale for choosing probabilities. The meaning of *virtually* and *close to* are slightly unclear at the moment, but if there is some implied limiting process, we would all agree that, in the limit, certainty and impossibility correspond to probabilities 1 and 0 respectively.

Between virtual impossibility and certainty, if one outcome appears to be closer to certainty than another, its probability should be correspondingly greater. This intuitive notion is imprecise and highly subjective; it provides little rationale for choosing numerical probabilities for different outcomes, and, even worse, little rationale justifying that probability models bear any precise relation to real-world situations.

Symmetry can often provide a better rationale for choosing probabilities. For example, the symmetry between H and T for a coin, or the symmetry between the six faces of a die, motivates assigning equal probabilities, 1/2 each for H and T and 1/6 each for the six faces of a die. This is reasonable and extremely useful, but there is no completely convincing reason for choosing probabilities based on symmetry.

[2] This is not intended as criticism, since we will see that there are good reasons to concentrate initially on such idealized experiments. However, readers should always be aware that modeling errors are the major cause of misleading results in applications of probability, and thus modeling must be seriously considered before using the results.

[3] It is popular to try to define probability by likelihood, but this is unhelpful since the words are essentially synonyms.

Another approach is to perform the experiment many times and choose the probability of each outcome as the relative frequency of that outcome (i.e., the number of occurrences of that outcome divided by the total number of trials). Experience shows that the relative frequency of an outcome often approaches a limiting value with an increasing number of trials. Associating the probability of an outcome with that limiting relative frequency is certainly close to our intuition and also appears to provide a testable criterion between model and real world. This criterion is discussed in Sections 1.8.1 and 1.8.2 and provides a very concrete way to use probabilities, since it suggests that the randomness in a single trial tends to disappear in the aggregate of many trials. Other approaches to choosing probability models will be discussed later.

1.2 The axioms of probability theory

As the applications of probability theory became increasingly varied and complex during the twentieth century, the need arose to put the theory on a firm mathematical footing. This was accomplished by an axiomatization of the theory, successfully carried out by the great Russian mathematician A. N. Kolmogorov [18] in 1932. Before stating and explaining these axioms of probability theory, the following two examples explain why the simple approach of the last section, assigning a probability to each sample point, often fails with infinite sample spaces.

Example 1.2.1 Suppose we want to model the phase of a sine wave, where the phase is viewed as being 'uniformly distributed' between 0 and 2π. If this phase is the only quantity of interest, it is reasonable to choose a sample space consisting of the set of real numbers between 0 and 2π. There are uncountably[4] many possible phases between 0 and 2π, and with any reasonable interpretation of uniform distribution, one must conclude that each sample point has probability 0. Thus, the simple approach of the last section leads us to conclude that any event in this space with a finite or countably infinite set of sample points should have probability 0. That simple approach does not help in finding the probability, say, of the interval $(0, \pi)$.

For this example, the appropriate view is the one taken in all elementary probability texts, namely to assign a *probability density* $1/(2\pi)$ to the phase. The probability of an event can then usually be found by integrating the density over that event. Useful as densities are, however, they do not lead to a general approach over arbitrary sample spaces.[5]

---

[4] A set is uncountably infinite if it is infinite and its members cannot be put into one-to-one correspondence with the positive integers. For example, the set of real numbers over some interval such as $(0, 2\pi)$ is uncountably infinite. The Wikipedia article on countable sets provides a friendly introduction to the concepts of countability and uncountability.

[5] It is possible to avoid the consideration of infinite sample spaces here by quantizing the possible phases. This is analogous to avoiding calculus by working only with discrete functions. Both usually result in both artificiality and added complexity.

Example 1.2.2 Consider an infinite sequence of coin tosses. The usual probability model is to assign probability 2^{-n} to each possible initial n-tuple of individual outcomes. Then in the limit $n \to \infty$, the probability of any given sequence is 0. Again, expressing the probability of an event involving infinitely many tosses as a sum of individual sample-point probabilities does not work. The obvious approach (which we often adopt for this and similar situations) is to evaluate the probability of any given event as an appropriate limit, as $n \to \infty$, of the outcome from the first n tosses.

We will later find a number of situations, even for this almost trivial example, where working with a finite number of elementary experiments and then going to the limit is very awkward. One example, to be discussed in detail later, is the strong law of large numbers (SLLN). This law looks directly at events consisting of infinite length sequences and is best considered in the context of the axioms to follow.

Although appropriate probability models can be generated for simple examples such as those above, there is a need for a consistent and general approach. In such an approach, rather than assigning probabilities to sample points, which are then used to assign probabilities to events, *probabilities must be associated directly with events.* The axioms to follow establish consistency requirements between the probabilities of different events. The axioms, and the corollaries derived from them, are consistent with one's intuition, and, for finite sample spaces, are consistent with our earlier approach. Dealing with the countable unions of events in the axioms will be unfamiliar to some students, but will soon become both familiar and consistent with intuition.

The strange part of the axioms comes from the fact that defining the class of events as the collection of *all* subsets of the sample space is usually inappropriate when the sample space is uncountably infinite. What is needed is a class of events that is large enough that we can almost forget that some very strange subsets are excluded. This is accomplished by having two simple sets of axioms, one defining the class of events,[6] and the other defining the relations between the probabilities assigned to these events. In this theory, all events have probabilities, but those truly weird subsets that are not events do not have probabilities. This will be discussed more after giving the axioms for events.

The axioms for events use the standard notation of set theory. Let Ω be the sample space, i.e., the set of all sample points for a given experiment. It is assumed throughout that Ω is non-empty. The events are subsets of the sample space. The union of n subsets (events) $A_1, A_2, \dots, A_n$ is denoted by either $\bigcup_{i=1}^{n} A_i$ or $A_1 \bigcup \cdots \bigcup A_n$, and consists of all points in at least one of $A_1, A_2, \dots, A_n$. Similarly, the intersection of these subsets is denoted by either $\bigcap_{i=1}^{n} A_i$ or[7] $A_1A_2 \cdots A_n$ and consists of all points in all of $A_1, A_2, \dots, A_n$.

A *sequence* of events is a collection of events in one-to-one correspondence with the positive integers, i.e., $A_1, A_2, \dots$ ad infinitum. A countable union, $\bigcup_{i=1}^{\infty} A_i$ is the set of

[6] A class of elements satisfying these axioms is called a σ-algebra or, less commonly, a σ-field.

[7] Intersection is also sometimes denoted as $A_1 \bigcap \cdots \bigcap A_n$, but is usually abbreviated as $A_1A_2 \cdots A_n$.

points in one or more of $A_1, A_2, \ldots$. Similarly, a countable intersection $\bigcap_{i=1}^{\infty} A_i$ is the set of points in all of $A_1, A_2, \ldots$. Finally, the complement A^c of a subset (event) A is the set of points in Ω but not A.

1.2.1 Axioms for events

Given a sample space Ω, the class of subsets of Ω that constitute the set of events satisfies the following axioms:

1. Ω is an event.
2. For every sequence of events $A_1, A_2, \ldots$, the union $\bigcup_{n=1}^{\infty} A_n$ is an event.
3. For every event A, the complement A^c is an event.

There are a number of important corollaries of these axioms. First, the empty set $\emptyset$ is an event. This follows from Axioms 1 and 3, since $\emptyset = \Omega^c$. The empty set does not correspond to our intuition about events, but the theory would be extremely awkward if it were omitted. Second, every finite union of events is an event. This follows by expressing $A_1 \bigcup \cdots \bigcup A_n$ as $\bigcup_{i=1}^{\infty} A_i$, where $A_i = \emptyset$ for all $i > n$. Third, every finite or countable intersection of events is an event. This follows from De Morgan's law,

$$\left[\bigcup\nolimits_n A_n\right]^c = \bigcap\nolimits_n A_n^c.$$

Although we will not make a big fuss about these axioms in the rest of the text, we will be careful to use only complements and countable unions and intersections in our analysis. Thus subsets that are not events will not arise.

Note that the axioms do not say that all subsets of Ω are events. In fact, there are many rather silly ways to define classes of events that obey the axioms. For example, the axioms are satisfied by choosing only the universal set Ω and the empty set $\emptyset$ to be events. We shall avoid such trivialities by assuming that for each sample point ω, the singleton subset $\{\omega\}$ is an event. For finite sample spaces, this assumption, plus the axioms above, imply that all subsets are events.

For uncountably infinite sample spaces, such as the sinusoidal phase above, this assumption, plus the axioms above, still leaves considerable freedom in choosing a class of events. As an example, the class of all subsets of Ω satisfies the axioms but surprisingly does not allow the probability axioms to be satisfied in any sensible way. How to choose an appropriate class of events requires an understanding of measure theory which would take us too far afield for our purposes. Thus we neither assume nor develop measure theory here.[8]

From a pragmatic standpoint, we start with the class of events of interest, such as those required to define the random variables (rv s) needed in the problem. That class is then extended so as to be closed under complementation and countable unions. Measure theory shows that this extension is possible.

[8] There is no doubt that measure theory is useful in probability theory, and serious students of probability should certainly learn measure theory at some point. For application-oriented people, however, it seems advisable to acquire more insight and understanding of probability, at a graduate level, before concentrating on the abstractions and subtleties of measure theory.

1.2.2 Axioms of probability

Given any sample space Ω and any class of events $\mathtt{E}$ satisfying the axioms of events, a probability rule is a function $\Pr\{\cdot\}$ mapping each $A \in \mathtt{E}$ to a (finite[9]) real number in such a way that the following three probability axioms[10] hold:

1. $\Pr\{\Omega\} = 1$.
2. For every event A, $\Pr\{A\} \geq 0$.
3. The probability of the union of any sequence $A_1, A_2, \ldots$ of disjoint[11] events is given by

$$\Pr\left\{\bigcup_{n=1}^{\infty} A_n\right\} = \sum_{n=1}^{\infty} \Pr\{A_n\}, \tag{1.1}$$

where $\sum_{n=1}^{\infty} \Pr\{A_n\}$ is shorthand for $\lim_{m\to\infty} \sum_{n=1}^{m} \Pr\{A_n\}$.

The axioms imply the following useful corollaries:

$$\Pr\{\emptyset\} = 0. \tag{1.2}$$

$$\Pr\left\{\bigcup_{n=1}^{m} A_n\right\} = \sum_{n=1}^{m} \Pr\{A_n\} \qquad \text{for } A_1, \ldots, A_m \text{ disjoint.} \tag{1.3}$$

$$\Pr\left\{A^c\right\} = 1 - \Pr\{A\} \qquad \text{for all } A. \tag{1.4}$$

$$\Pr\{A\} \leq \Pr\{B\} \qquad \text{for all } A \subseteq B. \tag{1.5}$$

$$\Pr\{A\} \leq 1 \qquad \text{for all } A. \tag{1.6}$$

$$\sum_{n} \Pr\{A_n\} \leq 1 \qquad \text{for } A_1, A_2, \ldots \text{ disjoint.} \tag{1.7}$$

$$\Pr\left\{\bigcup_{n=1}^{\infty} A_n\right\} = \lim_{m\to\infty} \Pr\left\{\bigcup_{n=1}^{m} A_n\right\}. \tag{1.8}$$

$$\Pr\left\{\bigcup_{n=1}^{\infty} A_n\right\} = \lim_{n\to\infty} \Pr\{A_n\} \qquad \text{for } A_1 \subseteq A_2 \subseteq \cdots. \tag{1.9}$$

$$\Pr\left\{\bigcap_{n=1}^{\infty} A_n\right\} = \lim_{n\to\infty} \Pr\{A_n\} \qquad \text{for } A_1 \supseteq A_2 \supseteq \cdots. \tag{1.10}$$

To verify (1.2), consider a sequence of events, $A_1, A_2, \ldots$ for which $A_n = \emptyset$ for each n. These events are disjoint since $\emptyset$ contains no outcomes, and thus has no outcomes in common with itself or any other event. Also, $\bigcup_n A_n = \emptyset$ since this union contains no outcomes. Axiom 3 then says that

$$\Pr\{\emptyset\} = \lim_{m\to\infty} \sum_{n=1}^{m} \Pr\{A_n\} = \lim_{m\to\infty} m\Pr\{\emptyset\}.$$

Since $\Pr\{\emptyset\}$ is a real number, this implies that $\Pr\{\emptyset\} = 0$.

To verify (1.3), apply Axiom 3 to the disjoint sequence $A_1, \ldots, A_m, \emptyset, \emptyset, \ldots$.

To verify (1.4), note that $\Omega = A \bigcup A^c$. Then apply (1.3) to the disjoint sets A and A^c.

[9] The word *finite* is redundant here, since the set of real numbers, by definition, does not include $\pm\infty$. The set of real numbers with $\pm\infty$ appended, is called the *extended* set of real numbers.

[10] Sometimes finite additivity, (1.3), is included as an additional axiom. This inclusion is quite intuitive and avoids the technical and somewhat peculiar proofs given for (1.2) and (1.3).

[11] Two sets or events A_1, A_2 are disjoint if they contain no common events, i.e., if $A_1A_2 = \emptyset$. A collection of sets or events are disjoint if all pairs are disjoint.

To verify (1.5), note that if $A \subseteq B$, then $B = A \bigcup (B-A)$, where $B-A$ is an alternative way to write $B \bigcap A^c$. We see then that A and $B-A$ are disjoint, so from (1.3),

$$\Pr\{B\} = \Pr\left\{A \bigcup (B-A)\right\} = \Pr\{A\} + \Pr\{B-A\} \geq \Pr\{A\},$$

where we have used Axiom 2 in the last step.

To verify (1.6) and (1.7), first substitute Ω for B in (1.5) and then substitute $\bigcup_n A_n$ for A.

Finally, (1.8) is established in Exercise 1.2(e), and (1.9) and (1.10) are simple consequences of (1.8).

The axioms specify the probability of any *disjoint* union of events in terms of the individual event probabilities, but what about a finite or countable union of arbitrary events? Exercise 1.2(c) shows that in this case, (1.3) can be generalized to

$$\Pr\left\{\bigcup_{n=1}^{m} A_n\right\} = \sum_{n=1}^{m} \Pr\{B_n\}, \tag{1.11}$$

where $B_1 = A_1$ and for each $n > 1$, $B_n = A_n - \bigcup_{m=1}^{n-1} A_m$ is the set of points in A_n but not in any of the sets $A_1, \ldots, A_{n-1}$. That is, the sets B_n are disjoint. The probability of a countable union of disjoint sets is then given by (1.8). In order to use this, one must know not only the event probabilities for $A_1, A_2 \ldots$, but also the probabilities of their intersections. The union bound, which is derived in Exercise 1.2(d), depends only on the individual event probabilities, and gives the following frequently useful upper bound on the union probability.

$$\Pr\left\{\bigcup_n A_n\right\} \leq \sum_n \Pr\{A_n\} \qquad \text{(union bound).} \tag{1.12}$$

1.3 Probability review

1.3.1 Conditional probabilities and statistical independence

Definition 1.3.1 *For any two events A and B in a probability model, the* ***conditional probability*** *of A, conditional on B, is defined if* $\Pr\{B\} > 0$ *by*

$$\Pr\{A|B\} = \Pr\{AB\}/\Pr\{B\}. \tag{1.13}$$

To motivate this definition, consider a discrete experiment in which we make a partial observation B (such as the result of a given medical test on a patient) but do not observe the complete outcome (such as whether the patient is sick and the outcome of other tests). The event B consists of all the sample points with the given outcome of the given test. Now let A be an arbitrary event (such as the event that the patient is sick). The conditional probability, $\Pr\{A|B\}$ is intended to represent the probability of A from the observer's viewpoint.

For the observer, the sample space can now be viewed as the set of sample points in B, since only those sample points are now possible. For any event A, only the event AB, i.e., the original set of sample points in A that are also in B, is relevant, but the probability of

A in this new sample space should be scaled up from $\Pr\{AB\}$ to $\Pr\{AB\}/\Pr\{B\}$, i.e., to $\Pr\{A|B\}$.

With this scaling, the set of events conditional on B becomes a probability space, and it is easily verified that all the axioms of probability theory are satisfied for this conditional probability space. Thus all known results about probability can also be applied to such conditional probability spaces.

Another important aspect of the definition in (1.13) is that it maintains consistency between the original probability space and this new conditional space in the sense that for any disjoint events, $A_1, A_2, \ldots$, and any event B with $\Pr\{B\} > 0$,

$$\Pr\left\{\left(\bigcup_n A_n\right) \mid B\right\} = \sum_n \Pr\{A_n \mid B\}.$$

This means that we can easily move back and forth between unconditional and conditional probability spaces.

The intuitive statements about partial observations and probabilities from the standpoint of an observer are helpful in reasoning probabilistically, but sometimes cause confusion. For example, Bayes' law, in the form

$$\Pr\{A|B\} \Pr\{B\} = \Pr\{B|A\} \Pr\{A\},$$

is an immediate consequence of the definition of conditional probability in (1.13). However, if we can only interpret $\Pr\{A|B\}$ when B is 'observed' or occurs 'before' A, then we cannot interpret $\Pr\{B|A\}$ and $\Pr\{A|B\}$ together. This caused immense confusion in probabilistic arguments before the axiomatic theory and clean definitions based on axioms were developed.

Definition 1.3.2 *Two events, A and B, are **statistically independent** (or, more briefly, independent) if*

$$\Pr\{AB\} = \Pr\{A\} \Pr\{B\}.$$

For $\Pr\{B\} > 0$, this is equivalent to $\Pr\{A|B\} = \Pr\{A\}$. This latter form often corresponds to a more intuitive view of independence, since it says that A and B are independent if the observation of B does not change the observer's probability of A.

The notion of independence is of vital importance in defining, and reasoning about, probability models. We will see many examples where very complex systems become very simple, both in terms of intuition and analysis, when appropriate quantities are modeled as statistically independent. An example will be given in the next subsection where repeated independent experiments are used to understand arguments about relative frequencies.

Often, when the assumption of independence is unreasonable, it is reasonable to assume conditional independence, where A and B are said to be *conditionally independent* given C if $\Pr\{AB|C\} = \Pr\{A|C\} \Pr\{B|C\}$. Most of the stochastic processes to be studied here are characterized by various forms of independence or conditional independence.

For more than two events, the definition of statistical independence is a little more complicated.

Definition 1.3.3 *The events $A_1, \ldots, A_n$, $n > 2$ are **statistically independent** if for each collection S of two or more of the integers 1 to n.*

$$\Pr\Big\{\bigcap\nolimits_{i\in S} A_i\Big\} = \prod\nolimits_{i\in S} \Pr\{A_i\}. \tag{1.14}$$

This includes the entire collection $\{1, \ldots, n\}$, so a necessary condition for independence is that

$$\Pr\Big\{\bigcap\nolimits_{i=1}^{n} A_i\Big\} = \prod\nolimits_{i=1}^{n} \Pr\{A_i\}. \tag{1.15}$$

It might be surprising that (1.15) does not imply (1.14), but the example in Exercise 1.4 will help clarify this. This definition will become clearer (and simpler) when we see how to view independence of events as a special case of independence of random variables.

1.3.2 Repeated idealized experiments

Much of our intuitive understanding of probability comes from the notion of repeating the same idealized experiment many times (i.e., performing multiple trials of the same experiment). However, the axioms of probability contain no explicit recognition of such repetitions. The appropriate way to handle n repetitions of an idealized experiment is through an extended experiment whose sample points are n-tuples of sample points from the original experiment. Such an extended experiment is viewed as *n trials* of the original experiment. The notion of multiple trials of a given experiment is so common that one sometimes fails to distinguish between the original experiment and an extended experiment with multiple trials of the original experiment.

To be more specific, given an original sample space Ω, the sample space of an n-repetition model is the Cartesian product

$$\Omega^n = \{(\omega_1, \omega_2, \ldots, \omega_n) : \omega_i \in \Omega \text{ for each } i, 1 \le i \le n\}, \tag{1.16}$$

i.e., the set of all n-tuples for which each of the n components of the n-tuple is an element of the original sample space Ω. Since each sample point in the n-repetition model is an n-tuple of points from the original Ω, it follows that an event in the n-repetition model is a subset of Ω^n, i.e., a collection of n-tuples $(\omega_1, \ldots, \omega_n)$, where each ω_i is a sample point from Ω. This class of events in Ω^n should include each event of the form $\{(A_1 A_2 \cdots A_n)\}$, where $\{(A_1 A_2 \cdots A_n)\}$ denotes the collection of n-tuples $(\omega_1, \ldots, \omega_n)$ where $\omega_i \in A_i$ for $1 \le i \le n$. The set of events (for n repetitions) must also be extended to be closed under complementation and countable unions and intersections.

The simplest and most natural way of creating a probability model for this extended sample space and class of events is through the assumption that the n trials are statistically independent. More precisely, we assume that for each extended event $\{A_1 \times A_2 \times \cdots \times A_n\}$ contained in Ω^n, we have

$$\Pr\{A_1 \times A_2 \times \cdots \times A_n\} = \prod\nolimits_{i=1}^{n} \Pr\{A_i\}, \tag{1.17}$$

where $\Pr\{A_i\}$ is the probability of event A_i in the original model. Note that since Ω can be substituted for any A_i in this formula, the subset condition of (1.14) is automatically satisfied. In fact, the Kolmogorov extension theorem asserts that for any probability model, there is an extended independent n-repetition model for which the events in each trial are independent of those in the other trials. In what follows, we refer to this as the probability model for n independent identically distributed (IID) trials of a given experiment.

The niceties of how to create this model for n IID arbitrary experiments depend on measure theory, but we simply rely on the existence of such a model and the independence of events in different repetitions. What we have done here is very important conceptually. A probability model for an experiment does not say anything directly about repeated experiments. However, questions about independent repeated experiments can be handled directly within this extended model of n IID repetitions. This can also be extended to a countable number of IID trials.

1.3.3 Random variables

The outcome of a probabilistic experiment often specifies a collection of numerical values such as temperatures, voltages, numbers of arrivals or departures in various time intervals, etc. Each such numerical value varies, depending on the particular outcome of the experiment, and thus can be viewed as a mapping from the set Ω of sample points to the set $\mathbb{R}$ of real numbers (note that $\mathbb{R}$ does not include $\pm\infty$). These mappings from sample points to real numbers are called random variables (rv s).

Definition 1.3.4 *A **random variable** (rv) is essentially a function X from the sample space Ω of a probability model to the set of real numbers $\mathbb{R}$. Three modifications are needed to make this precise. First, X might be undefined or infinite for a subset of Ω that has 0 probability.*[12] *Second, the mapping $X(\omega)$ must have the property that $\{\omega \in \Omega : X(\omega) \le x\}$ is an event*[13] *for each $x \in \mathbb{R}$. Third, every finite set of rv s $X_1, \dots, X_n$ has the property that for each $x_1 \in \mathbb{R}, \dots, x_n \in \mathbb{R}$, the set $\{\omega : X_1(\omega) \le x_1, \dots, X_n(\omega) \le x_n\}$ is an event.*

As with any function, there is often confusion between the function itself, which is called X in the definition above, and the value $X(\omega)$ taken on for a sample point ω. This is particularly prevalent with rv s since we intuitively associate a rv with its sample value when an experiment is performed. We try to control that confusion here by using X, $X(\omega)$, and x, respectively, to refer to the rv, the sample value taken for a given sample point ω, and a generic sample value.

Definition 1.3.5 *The **cumulative distribution function** (CDF)*[14] *of a rv X is a function $\mathsf{F}_X(x)$ mapping each $x \in \mathbb{R}$ into $\mathsf{F}_X(x) = \Pr\{\omega \in \Omega : X(\omega) \le x\}$. The argument ω is usually omitted for brevity, so $\mathsf{F}_X(x) = \Pr\{X \le x\}$.*

[12] For example, consider a probability model in which Ω is the closed interval $[0, 1]$ and the probability is uniformly distributed over Ω. If $X(\omega) = 1/\omega$, then the sample point 0 maps to ∞ but X is still regarded as a rv. These subsets of 0 probability are usually ignored, both by engineers and mathematicians.

[13] These last two modifications are technical limitations connected with measure theory. They can usually be ignored, since they are satisfied in all but the most bizarre conditions. However, just as it is important to know that not all subsets in a probability space are events, one should know that not all functions from Ω to $\mathbb{R}$ are rv s.

[14] The CDF is often referred to simply as the distribution function. We will use the word distribution in a generic sense to refer to any probabilistic characterization from which the CDF can in principle be found.

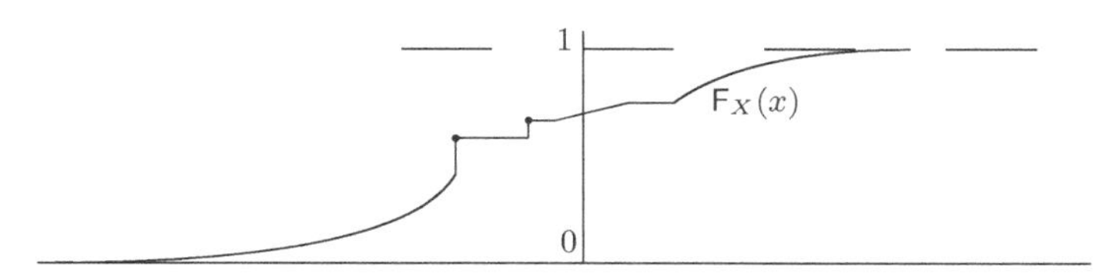

Figure 1.1 Example of a CDF for a rv that is neither continuous nor discrete. If $F_X(x)$ has a discontinuity at some x_o, it means that there is a discrete probability at x_o equal to the magnitude of the discontinuity. In this case $F_X(x_o)$ is given by the height of the upper point at the discontinuity.

Note that x is the argument of $F_X(x)$ and the subscript X denotes the particular rv under consideration. As illustrated in Figure 1.1, the CDF $F_X(x)$ is non-decreasing with x and must satisfy $\lim_{x\to-\infty} F_X(x) = 0$ and $\lim_{x\to\infty} F_X(x) = 1$. Exercise 1.5 proves that $F_X(x)$ is continuous from the right (i.e., that for every $x \in \mathbb{R}$, $\lim_{\epsilon\downarrow 0} F_X(x+\epsilon) = F_X(x)$).

Because of the definition of a rv, the set $\{X \leq x\}$ for any rv X and any real number x must be an event, and thus $\Pr\{X \leq x\}$ must be defined for all real x.

The concept of a rv is often extended to complex rv s and vector rv s. A *complex rv* is a mapping from the sample space to the set of finite complex numbers, and a *vector rv* is a mapping from the sample space to the finite vectors in some finite-dimensional vector space. Another extension is that of defective rv s. A *defective* rv X is a mapping from the sample space to the extended real numbers which satisfies the conditions of a rv except that the set of sample points mapped into $\pm\infty$ has positive probability.

When rv s are referred to (without any modifier such as complex, vector, or defective), the original definition, i.e., a function from Ω to $\mathbb{R}$, is intended.

If X has only a finite or countable number of possible sample values, say $x_1, x_2, \ldots,$ the probability $\Pr\{X = x_i\}$ of each sample value x_i is called the probability mass function (PMF) at x_i and denoted by $p_X(x_i)$; such a rv is called *discrete*. The CDF of a discrete rv is a 'staircase function,' staying constant between the possible sample values and having a jump of magnitude $p_X(x_i)$ at each sample value x_i. Thus the PMF and the CDF each specify the other for discrete rv s.

If the CDF $F_X(x)$ of a rv X has a (finite) derivative at x, the derivative is called the *density*, or more precisely the *probability density function* (PDF) of X at x and denoted by $f_X(x)$; for $\delta > 0$ sufficiently small, $f_X(x)\delta$ then approximates the probability that X is mapped to a value between x and $x + \delta$. A rv is said to be *continuous* if there is a function $f_X(x)$ such that, for each $x \in \mathbb{R}$, the CDF satisfies $F_X(x) = \int_{-\infty}^{x} f_X(y)\, dy$. If such a $f_X(x)$ exists, it is called the PDF. Essentially this means that $f_X(x)$ is the derivative of $F_X(x)$, but it is slightly more general in that it permits $f_X(x)$ to be discontinuous.

Elementary probability courses work primarily with the PMF and the PDF, since they are convenient for computational exercises. We will work more often with the CDF here. This is partly because it is always defined, partly to avoid saying everything thrice, for discrete, continuous, and other rv s, and partly because the CDF is often most important in limiting arguments such as steady-state time-average arguments. For CDFs, PDFs and PMFs, the subscript denoting the rv is often omitted if the rv is clear from the context. The same convention is used for complex or vector rv s.

Table 1.1 The PDF, mean, variance and MGF for some common continuous rv s

Name	PDF $f_X(x)$	Mean	Variance	MGF $g_X(r)$
Exponential:	$\lambda \exp(-\lambda x); \quad x{\ge}0$	$\frac{1}{\lambda}$	$\frac{1}{\lambda^2}$	$\frac{\lambda}{\lambda - r}$; for $r < \lambda$
Erlang:	$\frac{\lambda^n x^{n-1} \exp(-\lambda x)}{(n-1)!}; \quad x{\ge}0$	$\frac{n}{\lambda}$	$\frac{n}{\lambda^2}$	$\left(\frac{\lambda}{\lambda - r}\right)^n$; for $r < \lambda$
Gaussian:	$\frac{1}{\sigma\sqrt{2\pi}} \exp\left(\frac{-(x-a)^2}{2\sigma^2}\right)$	a	σ^2	$\exp(ra + r^2\sigma^2/2)$
Uniform:	$\frac{1}{a}; \quad 0{\le}x{\le}a$	$\frac{a}{2}$	$\frac{a^2}{12}$	$\frac{\exp(ra) - 1}{ra}$

Table 1.2 The PMF, mean, variance and MGF for some common discrete rv s

Name	PMF $p_M(m)$	Mean	Variance	MGF $g_M(r)$
Binary:	$p_M(1) = p;\ p_M(0) = 1 - p$	p	$p(1-p)$	$1 - p + pe^r$
Binomial:	$\binom{n}{m} p^m (1-p)^{n-m};\ 0{\le}m{\le}n$	np	$np(1-p)$	$[1 - p + pe^r]^n$
Geometric:	$p(1-p)^{m-1};\ m{\ge}1$	$\frac{1}{p}$	$\frac{1-p}{p^2}$	$\frac{pe^r}{1-(1-p)e^r}$; for $r < \ln \frac{1}{1-p}$
Poisson:	$\frac{\lambda^n \exp(-\lambda)}{n!};\ n{\ge}0$	λ	λ	$\exp[\lambda(e^r - 1)]$

Tables 1.1 and 1.2 list some widely used rv s. If the PDF or PMF is given only in a limited region, it is zero outside of that region. The mean, variance, and moment generating function (MGF) are defined in Section 1.5.

1.3.4 Multiple random variables and conditional probabilities

Often we must deal with multiple rv s in a single probability experiment. If $X_1, X_2, \ldots, X_n$ are rv s or the components of a vector rv, their joint CDF is defined by

$$F_{X_1 \cdots X_n}(x_1, \ldots, x_n) = \Pr\{\omega \in \Omega : X_1(\omega) \le x_1,\ X_2(\omega) \le x_2, \ldots,\ X_n(\omega) \le x_n\}. \quad (1.18)$$

This definition goes a long way toward explaining why we need the notion of a sample space Ω when all we want to talk about is a set of rv s. The CDF of a rv fully describes the individual behavior of that rv (and gives rise to its name, as in Tables 1.1 and 1.2), but Ω and the above mappings are needed to describe how the rv s interact.

For a vector rv $\boldsymbol{X}$ with components $X_1, \ldots, X_n$, or a complex rv X with real and imaginary parts X_1, X_2, the CDF is also defined by (1.18). Note that $\{X_1 \le x_1,\ X_2 \le x_2, \ldots,\ X_n \le x_n\}$ is an event and the corresponding probability is non-decreasing in each argument x_i. Also the CDF of any subset of rv s is obtained by setting the other arguments to $+\infty$. For example, the CDF of a single rv (called a *marginal* CDF for a given joint CDF) is given by

$$\mathsf{F}_{X_i}(x_i) = \mathsf{F}_{X_1 \cdots X_{i-1} X_i X_{i+1} \cdots X_n}(\infty, \ldots, \infty, x_i, \infty, \ldots, \infty).$$

If the rv s are all discrete, there is a joint PMF which specifies and is specified by the joint CDF. It is given by

$$\mathsf{p}_{X_1 \ldots X_n}(x_1, \ldots, x_n) = \Pr\{X_1 = x_1, \ldots, X_n = x_n\}.$$

Similarly, if the joint CDF can be differentiated as below, then it specifies and is specified by the joint PDF,

$$\mathsf{f}_{X_1 \cdots X_n}(x_1, \ldots, x_n) = \frac{\partial^n \mathsf{F}(x_1, \ldots, x_n)}{\partial x_1 \partial x_2 \cdots \partial x_n}.$$

Two rv s, say X and Y, are *statistically independent* (or, more briefly, *independent*) if

$$\mathsf{F}_{XY}(x, y) = \mathsf{F}_X(x)\mathsf{F}_Y(y) \qquad \text{for each } x \in \mathbb{R}, y \in \mathbb{R}. \tag{1.19}$$

If X and Y are discrete rv s, then the definition of independence in (1.19) is equivalent to the corresponding statement for PMFs,

$$\mathsf{p}_{XY}(x_i, y_j) = \mathsf{p}_X(x_i)\mathsf{p}_Y(y_j) \qquad \text{for each value } x_i \text{ of } X \text{ and } y_j \text{ of } Y.$$

Since $\{X = x_i\}$ and $\{Y = y_j\}$ are events, the conditional probability of $\{X = x_i\}$ conditional on $\{Y = y_j\}$ (assuming $\mathsf{p}_Y(y_j) > 0$) is given by (1.13) to be

$$\mathsf{p}_{X|Y}(x_i \mid y_j) = \frac{\mathsf{p}_{XY}(x_i, y_j)}{\mathsf{p}_Y(y_j)}.$$

If $\mathsf{p}_{X|Y}(x_i \mid y_j) = \mathsf{p}_X(x_i)$ for all i, j, then it is seen that X and Y are independent. This captures the intuitive notion of independence better than (1.19) for discrete rv s, since it can be viewed as saying that the PMF of X is not affected by the sample value of Y.

If X and Y have a joint density, then (1.19) is equivalent to

$$\mathsf{f}_{XY}(x, y) = \mathsf{f}_X(x)\mathsf{f}_Y(y) \qquad \text{for each } x \in \mathbb{R}, y \in \mathbb{R}. \tag{1.20}$$

If the joint density exists and the marginal density $\mathsf{f}_Y(y)$ is positive, the conditional density can be defined as $\mathsf{f}_{X|Y}(x|y) = \mathsf{f}_{XY}(x, y)/\mathsf{f}_Y(y)$. In essence, $\mathsf{f}_{X|Y}(x|y)$ is the density of X conditional on $Y = y$, but, being more precise, it is a limiting conditional density as $\delta \to 0$ of X conditional on $Y \in [y, y + \delta)$.

If X and Y have a joint density, then statistical independence can also be expressed as

$$\mathsf{f}_{X|Y}(x|y) = \mathsf{f}_X(x) \qquad \text{for each } x \in \mathbb{R},\ y \in \mathbb{R} \text{ such that } \mathsf{f}_Y(y) > 0. \tag{1.21}$$

This often captures the intuitive notion of statistical independence for continuous rv s better than (1.20)

More generally, the probability of an arbitrary event A, conditional on a given value of a continuous rv Y, is given by

$$\Pr\{A \mid Y = y\} = \lim_{\delta \to 0} \frac{\Pr\{A, Y \in [y, y + \delta]\}}{\Pr\{Y \in [y, y + \delta]\}}.$$

We next generalize the above results about two rv s to the case of n rv s $\boldsymbol{X} = X_1, \ldots, X_n$. Statistical independence is then defined by the equation

$$\mathsf{F}_{\boldsymbol{X}}(x_1, \ldots, x_n) = \prod_{i=1}^{n} \Pr\{X_i \le x_i\} = \prod_{i=1}^{n} \mathsf{F}_{X_i}(x_i) \qquad \text{for all } x_1, \ldots, x_n \in \mathbb{R}. \tag{1.22}$$

In other words, $X_1, \ldots, X_n$ are independent if the events $X_i \le x_i$ for $1 \le i \le n$ are independent for all choices of $x_1, \ldots, x_n$. If the density or PMF exists, (1.22) is equivalent to a product form for the density or mass function. A set of rv s is said to be pairwise independent if each pair of rv s in the set is independent. As shown in Exercise 1.23, pairwise independence does not imply that the entire set is independent.

Independent rv s are very often also identically distributed, i.e., they all have the same CDF. These cases arise so often that we abbreviate 'independent identically distributed' by IID. For the IID case (1.22) becomes

$$\mathsf{F}_{\boldsymbol{X}}(x_1, \ldots, x_n) = \prod_{i=1}^{n} \mathsf{F}_{X}(x_i).$$

1.4 Stochastic processes

A stochastic process (or random process[15]) is an infinite collection of rv s defined on a common probability model. These rv s are usually indexed by an integer or a real number often interpreted as time. Thus each sample point of the probability model maps to an infinite collection of sample values of rv s. If the index is regarded as time, then each sample point maps to a function of time called a sample path or sample function. These sample paths might vary continuously with time or might vary only at discrete times, and if they vary at discrete times, those times might be deterministic or random.

In many cases, this collection of rv s comprising the stochastic process is the only thing of interest. In this case, the sample points of the probability model can be taken to be the sample paths of the process. Conceptually, then, each event is a collection of sample paths. Often the most important of these events can be defined in terms of a finite set of rv s.

As an example of sample paths that change at only discrete times, we might be concerned with the times at which customers arrive at some facility. These 'customers' might be customers entering a store, incoming jobs for a computer system, arriving packets in a communication system, or orders for a merchandising warehouse.

The Bernoulli process is an example of how such customers could be modeled and is perhaps the simplest non-trivial stochastic process. We now define this process and develop a few of its many properties. We will return to it frequently as an example.

[15] Stochastic and random are synonyms, but *random* has become more popular for random variables and *stochastic* for stochastic processes. The reason for the author's choice is that the common-sense intuition associated with randomness appears more important than mathematical precision in reasoning about rv s, whereas for stochastic processes, common-sense intuition causes confusion much more frequently than with rv s. The less familiar word *stochastic* warns the reader to be more careful.

1.4.1 The Bernoulli process

A *Bernoulli process* is a sequence, $Z_1, Z_2, \ldots$, of IID binary rv s.[16] Let $p = \Pr\{Z_i = 1\}$ and $q = 1-p = \Pr\{Z_i = 0\}$. We often visualize a Bernoulli process as evolving in discrete time with the event $\{Z_i = 1\}$ representing an arriving customer at time i and $\{Z_i = 0\}$ representing no arrival. Thus at most one arrival occurs at each integer time.

When viewed as arrivals in time, it is interesting to understand something about the intervals between successive arrivals and about the aggregate number of arrivals up to any given time (see Figure 1.2). By convention, the first interarrival time is the time of the first arrival. These interarrival times and aggregate numbers of arrivals are rv s that are functions of the underlying sequence $Z_1, Z_2, \ldots$. The topic of rv s that are defined as functions of other rv s (i.e., whose sample values are functions of the sample values of the other rv s) is taken up in more generality in Section 1.5.1, but the interarrival times and aggregate arrivals for Bernoulli processes are so specialized and simple that it is better to treat them from first principles.

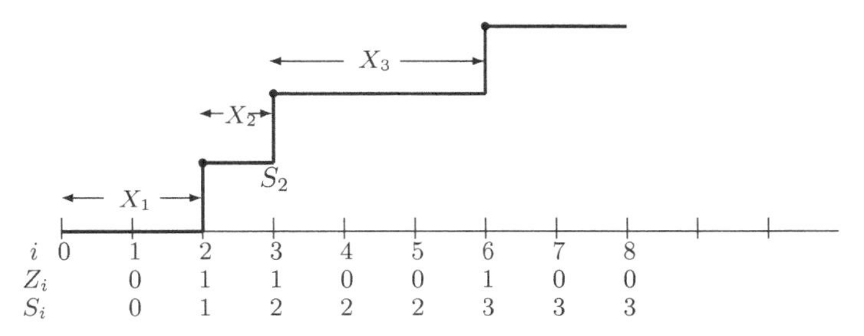

Figure 1.2 Illustration of a sample path for a Bernoulli process: the sample values of the binary rv s Z_i are shown below the time instants. The sample value of the aggregate number of arrivals, $S_n = \sum_{i=1}^{n} Z_i$, is the illustrated step function, and the interarrival intervals are the intervals between steps.

The first interarrival time, X_1, is 1 if and only if the first binary rv Z_1 is 1, and thus $\mathsf{p}_{X_1}(1) = p$. Next, $X_1 = 2$ if and only if $Z_1 = 0$ and $Z_2 = 1$, so $\mathsf{p}_{X_1}(2) = p(1-p)$. Continuing, we see that X_1 has the *geometric* PMF,

$$\mathsf{p}_{X_1}(j) = p(1-p)^{j-1}, \qquad \text{where } j \geq 1.$$

Each subsequent interarrival time, $X_2, X_3, \ldots$, can be found in this same way.[17] Each X_R has the same geometric PMF and is statistically independent of $X_1, \ldots, X_{k-1}$. Thus the sequence of interarrival times is an IID sequence of geometric rv s.

It can be seen from Figure 1.2 that a sample path of interarrival times also determines a sample path of the binary arrival rv s, $\{Z_i;\ i \geq 1\}$. Thus the Bernoulli process can also be characterized in terms of a sequence of IID geometric rv s.

[16] We say that a sequence $Z_1, Z_2, \ldots$, of rv s are IID if for each integer n, the rv s $Z_1, \ldots, Z_n$ are IID. There are some subtleties in going to the limit $n \to \infty$, but we can avoid most such subtleties by working with finite n-tuples and going to the limit at the end.

[17] This is one of those maddening arguments that, while intuitively obvious, requires some careful reasoning to be completely convincing. We go through several similar arguments with great care in Chapter 2, and suggest that skeptical readers wait until then to think this through carefully.

For our present purposes, the most important rv s in a Bernoulli process are the partial sums $S_n = \sum_{i=1}^n Z_i$. Each rv S_n is the number of arrivals up to and including time n, i.e., S_n is simply the sum of n binary IID rv s and its distribution is known as the binomial distribution. That is, $\mathsf{p}_{S_n}(k)$ is the probability that k out of n of the Z_i's have the value 1. There are $\binom{n}{k} = n!/[k!(n-k)!]$ arrangements of a binary n-tuple with k 1s, and each has probability $p^k q^{n-k}$. Thus

$$\mathsf{p}_{S_n}(k) = \binom{n}{k} p^k q^{n-k}. \tag{1.23}$$

We will use the binomial PMF extensively as an example in explaining the laws of large numbers and the central limit theorem (CLT) later in this chapter, and will often use it in later chapters as an example of a sum of IID rv s. For these examples, we need to know how $\mathsf{p}_{S_n}(k)$ behaves asymptotically as $n \to \infty$ and $k \to \infty$. We make a short digression here to state and develop tight upper and lower bounds to the binomial PMF that makes this asymptotic behavior clear. The ratio of k to n in these bounds is denoted $\widetilde{p} = k/n$ and the bounds are given in terms of a quantity $D(\widetilde{p}\|p)$ called the binary Kullback–Liebler divergence (or relative entropy) and defined by

$$\mathsf{D}(\widetilde{p}\|p) = \widetilde{p} \ln\left(\frac{\widetilde{p}}{p}\right) + (1-\widetilde{p}) \ln\left(\frac{1-\widetilde{p}}{1-p}\right) \geq 0. \tag{1.24}$$

Theorem 1.4.1 *Let $\mathsf{p}_{S_n}(k)$ be the PMF of the binomial distribution for an underlying binary PMF $\mathsf{p}_Z(1) = p > 0$, $\mathsf{p}_Z(0) = 1 - p > 0$ and let $\widetilde{p} = k/n$. Then for each integer $\widetilde{p}n$, $1 \leq \widetilde{p}n \leq n-1$,*

$$\mathsf{p}_{S_n}(\widetilde{p}n) < \sqrt{\frac{1}{2\pi n\widetilde{p}(1-\widetilde{p})}} \exp\big[-n\mathsf{D}(\widetilde{p}\|p)\big], \tag{1.25}$$

$$\mathsf{p}_{S_n}(\widetilde{p}n) > \left(1 - \frac{1}{12n\widetilde{p}(1-\widetilde{p})}\right) \sqrt{\frac{1}{2\pi n\widetilde{p}(1-\widetilde{p})}} \exp\big[-n\mathsf{D}(\widetilde{p}\|p)\big]. \tag{1.26}$$

Also, $\mathsf{D}(\widetilde{p}\|p) \geq 0$ with strict inequality for all $\widetilde{p} \neq p$.

Discussion The parameter $\widetilde{p} = k/n$ is the relative frequency of 1s in the n-tuple $Z_1, \ldots, Z_n$. The upper bound in (1.25), as a function of n for fixed p and $\widetilde{p}$, decreases exponentially with n (aside from a slowly varying coefficient proportional to $1/\sqrt{n}$).

The lower bound in (1.26) is the same as the upper bound in (1.25) except for the initial term of (1.26). For fixed $\widetilde{p}$ satisfying $0 < \widetilde{p} < 1$, this initial term approaches 1 with increasing n, so that the upper and lower bounds are asymptotically the same in the sense that their ratio goes to 1 as $n \to \infty$. An upper and lower bound of this type is said to be asymptotically tight, and the result is denoted as

$$\mathsf{p}_{S_n}(\widetilde{p}n)) \sim \sqrt{\frac{1}{2\pi n\widetilde{p}(1-\widetilde{p})}} \exp\big[-n\mathsf{D}(\widetilde{p}\|p)\big] \qquad \text{for } 0 < \widetilde{p} < 1, \tag{1.27}$$

where the symbol $\sim$ here means that the ratio of the upper to lower bound approaches 1 uniformly in p and $\widetilde{p}$ as $n \to \infty$ for $\widetilde{p}$ and p in any interval bounded away from 0 and 1.

The reason for fussing about uniform convergence here is that $\widetilde{p}n$ must be an integer, so that one cannot simply pick any fixed $\widetilde{p}$ and go to the limit $n \to \infty$. The reason for requiring $\widetilde{p}$ to be bounded away from 0 and 1 for this asymptotic result can be seen by looking at $\mathsf{p}_{S_n}(1) = np(1-p)^{n-1}$. This corresponds to $\widetilde{p} = 1/n$. The extra factor in the lower bound $(1 - 1/(12n\widetilde{p}(1-\widetilde{p})))$ then does not quite approach 1 with increasing n.

The divergence, $\mathsf{D}(\widetilde{p}\|p)$, provides a measure of the difference between $\widetilde{p}$ and p that is also useful in detection and the study of large deviations. This same type of quantity appears for arbitrary discrete rv s in Chapter 9.

A guided proof of the theorem is given in Exercise 1.9.

We saw earlier that the Bernoulli process can also be characterized as a sequence of IID geometrically distributed interarrival intervals. An interesting generalization of this arises by allowing the interarrival intervals to be arbitrary discrete or continuous non-negative IID rv s rather than geometric rv s. These processes are known as *renewal processes* and are the topic of Chapter 5. Poisson processes are special cases of renewal processes in which the interarrival intervals have an exponential PDF. These are treated in Chapter 2 and have many connections to Bernoulli processes.

Renewal processes are examples of *discrete stochastic processes*. The distinguishing characteristic of such processes is that interesting things (arrivals, departures, changes of state) occur at discrete instants of time separated by deterministic or random intervals. Discrete stochastic processes are to be distinguished from noise-like stochastic processes in which changes are continuously occurring and the sample paths are continuously varying functions of time. The description of discrete stochastic processes above is not intended to be precise, but Chapters 2, 4, and 5 are restricted to discrete stochastic processes in this sense, whereas Chapter 3 is restricted to continuous processes.

1.5 Expectations and more probability review

The *expected value* $\mathsf{E}[X]$ of a rv X is also called the *expectation* or the *mean* and is frequently denoted as $\overline{X}$. Before giving a general definition, we discuss several special cases. First consider non-negative discrete rv s. The expected value $\mathsf{E}[X]$ is then given by

$$\mathsf{E}[X] = \sum_x x\,\mathsf{p}_X(x). \tag{1.28}$$

If X has a finite number of possible sample values, the above sum must be finite since each sample value must be finite. On the other hand, if X has a countable number of non-negative sample values, the sum in (1.28) might be either finite or infinite. Example 1.5.1 illustrates a case in which the sum is infinite. The expectation is said to *exist* only if the sum is finite (i.e., if the sum converges to a real number), and in this case $\mathsf{E}[X]$ is given by (1.28). If the sum is infinite, we say that $\mathsf{E}[X]$ does not exist, but also say[18] that

[18] It seems metaphysical to say that something does not exist but has infinite value. However, the word *exist* here is shorthand for *exist as a real number*, which makes it quite reasonable to also consider the value in the extended real number system, which includes $\pm\infty$.

$\mathsf{E}[X] = \infty$. In other words, (1.28) can be used in both cases, but $\mathsf{E}[X]$ is said to *exist* only if the sum is finite.

Example 1.5.1 This example will be useful frequently in illustrating rv s that have an infinite expectation. Let N be a positive integer-valued rv with the CDF $\mathsf{F}_N(n) = n/(n+1)$ for each integer $n \geq 1$. Then N is clearly a positive rv since $\mathsf{F}_N(0) = 0$ and $\lim_{n\to\infty} \mathsf{F}_N(n) = 1$. For each $n \geq 1$, the PMF is given by

$$\mathsf{p}_N(n) = \mathsf{F}_N(n) - \mathsf{F}_N(n-1) = \frac{n}{n+1} - \frac{n-1}{n} = \frac{1}{n(n+1)}. \tag{1.29}$$

Since $\mathsf{p}_N(n)$ is a PMF, we see that $\sum_{n=1}^{\infty} 1/[n(n+1)] = 1$, which is a frequently useful fact. The following equation, however, shows that $\mathsf{E}[N]$ does not exist and has infinite value.

$$\mathsf{E}[N] = \sum_{n=1}^{\infty} n\,\mathsf{p}_N(n) = \sum_{n=1}^{\infty} \frac{n}{n(n+1)} = \sum_{n=1}^{\infty} \frac{1}{n+1} = \infty,$$

where we have used the fact that the harmonic series diverges.

We next derive an alternative expression for the expected value of a non-negative discrete rv. This new expression is given directly in terms of the CDF. We then use this new expression as a general definition of expectation which applies to all non-negative rv s, whether discrete, continuous, or arbitrary. It contains none of the convergence questions that could cause confusion for arbitrary rv s or for continuous rv s with wild densities.

For a non-negative discrete rv X, Figure 1.3 illustrates that (1.28) is simply the integral of the complementary CDF, where the *complementary CDF* F^{c} of a rv is defined as $\mathsf{F}^{\mathsf{c}}_X(x) = \Pr\{X > x\} = 1 - \mathsf{F}_X(x)$:

$$\mathsf{E}[X] = \int_0^{\infty} \mathsf{F}^{\mathsf{c}}_X \, dx = \int_0^{\infty} \Pr\{X > x\} \, dx. \tag{1.30}$$

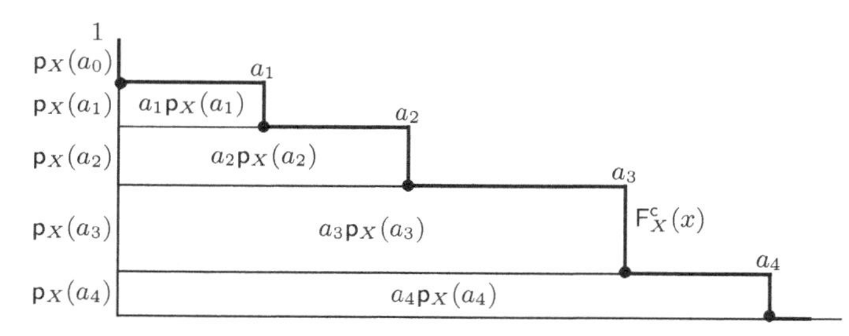

Figure 1.3 The figure shows the complementary CDF $\mathsf{F}^{\mathsf{c}}_X$ of a non-negative discrete rv X. For this example, X takes on five possible values, $0 = a_0 < a_1 < a_2 < a_3 < a_4$. Thus $\mathsf{F}^{\mathsf{c}}_X(x) = \Pr\{X > x\} = 1 - p_X(a_0)$ for $x < a_1$. For $a_1 \leq x < a_2$, $\Pr\{X > x\} = 1 - \mathsf{p}_X(a_0) - \mathsf{p}_X(a_1)$. Similar drops occur in $\Pr\{X > x\}$ as x reaches a_2, a_3, and a_4. From (1.28), $\mathsf{E}(X)$ is $\sum_i a_i \mathsf{p}_X(a_i)$, which is the sum of the rectangles in the figure. This is also the area under the curve $\mathsf{F}^{\mathsf{c}}_X(x)$, i.e., $\int_0^{\infty} \mathsf{F}^{\mathsf{c}}_X(x)\,dx$. It can be seen that this argument applies to any non-negative rv, thus verifying (1.30).

Although Figure 1.3 only illustrates the equality of (1.28) and (1.30) for one special case, one easily sees that the argument applies to any non-negative discrete rv, including those with countably many values, by equating the sum of the indicated rectangles with the integral.

For a non-negative integer-valued rv X, (1.30) reduces to a simpler form that is often convenient when X has a countable set of sample values:

$$\mathsf{E}[X] = \sum_{n=0}^{\infty} \Pr\{X > n\} = \sum_{n=1}^{\infty} \Pr\{X \geq n\}. \tag{1.31}$$

For an arbitrary non-negative rv X, we can visualize quantizing the rv, finding the expected value of the quantized rv, and then going to the limit of arbitrarily fine quantizations. Each quantized rv is discrete, so its expectation is given by (1.30) applied to the quantized rv. Each such expectation can be viewed as a Riemann sum for the integral $\int_0^\infty F_X^c(x)\,dx$ of the original rv.

There are no mathematical subtleties in integrating an arbitrary non-negative non-increasing function, so $\int_0^\infty F_X^c(x)\,dx$ must have either a finite or infinite limit. This leads us to the following fundamental definition of expectation for arbitrary non-negative rv s.

Definition 1.5.2 *The **expectation** $\mathsf{E}[X]$ of a non-negative rv X is defined by (1.30). The expectation is said **to exist** if and only if the integral is finite. Otherwise the expectation is said to not exist and is also said to be infinite.*

Exercise 1.6 shows that this definition is consistent with the conventional definition of expectation for the case of continuous rv s, i.e.,

$$\mathsf{E}[X] = \lim_{b \to \infty} \int_0^b x \mathsf{f}_X(x)\,dx. \tag{1.32}$$

This can also be seen using integration by parts.

Next consider rv s with both positive and negative sample values. If X has a finite number of positive and negative sample values, say $a_1, a_2, \ldots, a_n$, the expectation $\mathsf{E}[X]$ is given by

$$\begin{aligned} \mathsf{E}[X] &= \sum_i a_i \mathsf{p}_X(a_i) \\ &= \sum_{a_i \leq 0} a_i\, \mathsf{p}_X(a_i) + \sum_{a_i > 0} a_i\, \mathsf{p}_X(a_i). \end{aligned} \tag{1.33}$$

If X has a countably infinite set of sample values, then (1.33) can still be used if each of the sums in (1.33) converges to a finite value, and otherwise the expectation does not exist (as a real number). It can be seen that each sum in (1.33) converges to a finite value if and only if $\mathsf{E}[|X|]$ exists (i.e., converges to a finite value) for the non-negative rv $|X|$.

If $\mathsf{E}[X]$ does not exist (as a real number), it still might have the value ∞ if the first sum converges and the second does not, or the value $-\infty$ if the second sum converges and the first does not. If both sums diverge, then $\mathsf{E}[X]$ is undefined, even as $\pm\infty$. In this latter case, the partial sums can be arbitrarily small or large depending on the order in which the terms of (1.33) are summed (see Exercise 1.8).

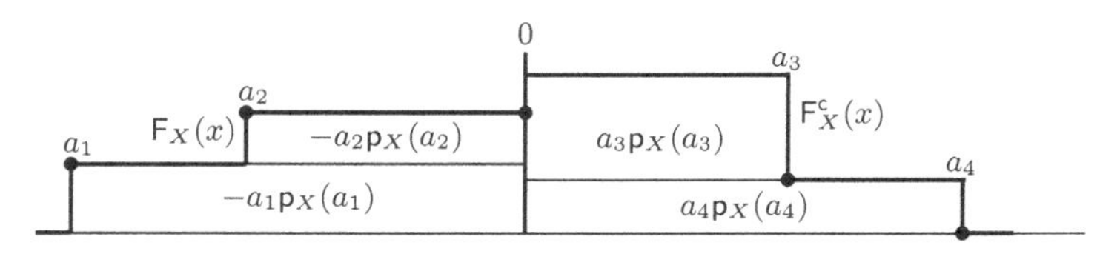

Figure 1.4 For this example, X takes on four possible sample values, $a_1 < a_2 < 0 < a_3 < a_4$. The figure plots $\mathsf{F}_X(x)$ for $x \le 0$ and $\mathsf{F}^{\mathsf{c}}_X(x)$ for $x > 0$. As in Figure 1.3, $\int_{x\ge 0} \mathsf{F}^{\mathsf{c}}_X(x)\,dx = a_3\mathsf{p}_X(a_3) + a_4\mathsf{p}_X(a_4)$. Similarly, $\int_{x<0} \mathsf{F}_X(x)\,dx = -a_1\mathsf{p}_X(a_1) - a_2\mathsf{p}_X(a_2)$.

As illustrated for a finite number of sample values in Figure 1.4, the expression in (1.33) can also be expressed directly in terms of the CDF and complementary CDF as

$$\mathsf{E}[X] = -\int_{-\infty}^{0} \mathsf{F}_X(x)\,dx + \int_{0}^{\infty} \mathsf{F}^{\mathsf{c}}_X(x)\,dx. \tag{1.34}$$

Since $\mathsf{F}^{\mathsf{c}}_X(x) = 1 - \mathsf{F}_X(x)$, this can also be expressed as

$$\mathsf{E}[X] = \int_{-\infty}^{\infty} \left[u(x) - \mathsf{F}_X(x)\right] dx,$$

where $u(x)$ is the unit step, $u(x) = 1$ for $x \ge 0$ and $u(x) = 0$ otherwise.

The first integral in (1.34) corresponds to the negative sample values and the second to the positive sample values, and $\mathsf{E}[X]$ exists if and only if both integrals are finite (i.e., if $\mathsf{E}[|X|]$ is finite).

For continuous-valued rv s with positive and negative sample values, the conventional definition of expectation (assuming that $\mathsf{E}[|X|]$ exists) is given by

$$\mathsf{E}[X] = \int_{-\infty}^{\infty} x\,\mathsf{f}_X(x)\,dx. \tag{1.35}$$

This is equal to (1.34) by the same argument as with non-negative rv s. Also, as with nonnegative rv s, (1.34) also applies to arbitrary rv s. We thus have the following fundamental definition of expectation.

Definition 1.5.3 *The* ***expectation*** $\mathsf{E}[X]$ *of a rv X* ***exists****, with the value given in (1.34), if each of the two terms in (1.34) is finite. The expectation does not exist, but has value* ∞ *(*$-\infty$*), if the first term is finite (infinite) and the second infinite (finite). The expectation does not exist and is undefined if both terms are infinite.*

We should not view the general expression in (1.34) for expectation as replacing the need for the conventional expressions in (1.35) and (1.33). We will use all of these expressions frequently, using whichever is most convenient. The main advantages of (1.34) are that it applies equally to all rv s, it poses no questions about convergence, and it is frequently useful, especially in limiting arguments.

Example 1.5.4 The *Cauchy* rv X is the classic example of a rv whose expectation does not exist and is undefined. The probability density is $\mathsf{f}_X(x) = 1/[\pi(1+x^2)]$. Thus $x\mathsf{f}_X(x)$

is proportional to $1/x$ both as $x \to \infty$ and as $x \to -\infty$. It follows that $\int_0^\infty x\mathsf{f}_X(x)\,dx$ and $\int_{-\infty}^0 -x\mathsf{f}_X(x)\,dx$ are both infinite. On the other hand, we see from symmetry that the Cauchy principal value of the integral in (1.35) is given by

$$\lim_{A\to\infty} \int_{-A}^{A} \frac{x}{\pi(1+x^2)}\,dx = 0.$$

There is usually little motivation for considering the upper and lower limits of the integration to have the same magnitude, and the Cauchy principal value usually has little significance for expectations.

1.5.1 Random variables as functions of other random variables

Random variables are often defined in terms of each other. For example, if h is a function from $\mathbb{R}$ to $\mathbb{R}$ and X is a rv, then $Y = h(X)$ is the rv that maps each sample point ω to the composite function $h(X(\omega))$. The CDF of Y can be found from this, and the expected value of Y can then be evaluated by (1.34).

It is often more convenient to find $\mathsf{E}[Y]$ directly using the CDF of X. Exercise 1.19 indicates that $\mathsf{E}[Y]$ is given by $\int h(x)\mathsf{f}_X(x)\,dx$ for continuous rv s and by $\sum_x h(x)\mathsf{p}_X(x)$ for discrete rv s. In order to avoid continuing to use separate expressions for continuous and discrete rv s, we express both of these relations by

$$\mathsf{E}[Y] = \int_{-\infty}^{\infty} h(x)\,d\mathsf{F}_X(x). \tag{1.36}$$

This is known as a Stieltjes integral, which can be used as a generalization of both the continuous and discrete cases. For most purposes, we use Stieltjes integrals[19] as a notational shorthand for either $\int h(x)\mathsf{f}_X(x)\,dx$ or $\sum_x h(x)\mathsf{p}_X(x)$.

The existence of $\mathsf{E}[X]$ does not guarantee the existence of $\mathsf{E}[Y]$, but we will treat the question of existence as it arises rather than attempting to establish any general rules.

Particularly important examples of such expected values are the moments $\mathsf{E}[X^n]$ of a rv X and the central moments $\mathsf{E}\left[(X-\overline{X})^n\right]$ of X, where $\overline{X}$ is the mean $\mathsf{E}[X]$. The second central moment is called the *variance*, denoted by σ_X^2 or $\mathsf{VAR}[X]$. It is given by

$$\sigma_X^2 = \mathsf{E}\left[(X-\overline{X})^2\right] = \mathsf{E}\left[X^2\right] - \overline{X}^2. \tag{1.37}$$

The *standard deviation* σ_X of X is the square root of the variance and provides a measure of dispersion of the rv around the mean. Thus the mean is often viewed as a 'typical value' for the outcome of the rv (see Section 1.5.3) and σ_X is similarly viewed as a typical difference between X and $\overline{X}$. An important connection between the mean and

[19] More specifically, the Riemann–Stieltjes integral, abbreviated here as the Stieltjes integral, is denoted as $\int_a^b h(x)dF_X(x)$. This integral is defined as the limit of a generalized Riemann sum, $\lim_{\delta\to 0}\sum_n h(x_n)[\mathsf{F}(y_n) - \mathsf{F}(y_{n-1})]$, where $\{y_n; n \ge 1\}$ is a sequence of increasing numbers from a to b satisfying $y_n - y_{n-1} \le \delta$ and $y_{n-1} < x_n \le y_n$ for all n. The Stieltjes integral is defined to exist over finite limits if the limit exists and is independent of the choices of $\{y_n\}$ and $\{x_n\}$ as $\delta \to 0$. It exists over infinite limits if it exists over finite lengths and a limit over the integration limits can be taken. See Rudin [24] for an excellent elementary treatment of Stieltjes integration, and see Exercise 1.15 for some examples.

standard deviation is that $\mathsf{E}\left[(X-x)^2\right]$ is minimized over x by choosing x to be $\mathsf{E}[X]$ (see Exercise 1.24).

Next suppose X and Y are rv s and consider the rv[20] $Z = X + Y$. If we assume that X and Y are independent, then the CDF of $Z = X + Y$ is given by[21]

$$\mathsf{F}_Z(z) = \int_{-\infty}^{\infty} \mathsf{F}_X(z-y)\, d\mathsf{F}_Y(y) = \int_{-\infty}^{\infty} \mathsf{F}_Y(z-x)\, d\mathsf{F}_X(x). \tag{1.38}$$

If X and Y both have densities, this can be rewritten as

$$\mathsf{f}_Z(z) = \int_{-\infty}^{\infty} \mathsf{f}_X(z-y)\mathsf{f}_Y(y)\, dy = \int_{-\infty}^{\infty} \mathsf{f}_Y(z-x)\mathsf{f}_X(x)\, dx. \tag{1.39}$$

Equation (1.39) is the familiar convolution equation from linear systems, and we similarly refer to (1.38) as the convolution of CDFs (although it has a different functional form from (1.39)). If X and Y are non-negative rv s, then the integrands in (1.38) and (1.39) are non-zero only between 0 and z, so we often use 0 and z as the limits in (1.38) and (1.39).

If $X_1, X_2, \ldots, X_n$ are independent rv s, then the distribution of the rv $S_n = X_1 + X_2 + \cdots + X_n$ can be found by first convolving the CDFs of X_1 and X_2 to get the CDF of S_2 and then, for each $n \geq 2$, convolving the CDF of S_n and X_{n+1} to get the CDF of S_{n+1}. The CDFs can be convolved in any order to get the same resulting CDF.

Whether or not $X_1, X_2, \ldots, X_n$ are independent, the expected value of $S_n = X_1 + X_2 + \cdots + X_n$ satisfies

$$\mathsf{E}[S_n] = \mathsf{E}[X_1 + X_2 + \cdots + X_n] = \mathsf{E}[X_1] + \mathsf{E}[X_2] + \cdots + \mathsf{E}[X_n]. \tag{1.40}$$

This says that the expected value of a sum is equal to the sum of the expected values, whether or not the rv s are independent (see Exercise 1.14). The following example shows how this can be a valuable problem-solving aid with an appropriate choice of rv s.

Example 1.5.5 Consider a switch with n input nodes and n output nodes. Suppose each input is randomly connected to a single output in such a way that each output is also connected to a single input. That is, each output is connected to input 1 with probability $1/n$. Given this connection, each of the remaining outputs is connected to input 2 with probability $1/(n-1)$, and so forth.

An input node is said to be *matched* if it is connected to the output of the same number. We want to show that the expected number of matches (for any given n) is 1. Note that the first node is matched with probability $1/n$, and therefore the expectation of a match for node 1 is $1/n$. Whether or not the second input node is matched depends on the choice of output for the first input node, but it can be seen from symmetry that

[20] The question whether a real-valued function of a rv is itself a rv is usually addressed by the use of measure theory, and since we neither use nor develop measure theory in this text, we usually simply assume (within the limits of common sense) that any such function is itself a rv. However, the sum $X + Y$ of rv s is so important that Exercise 1.13 provides a guided derivation of this result for $X + Y$. In the same way, the sum $S_n = X_1 + \cdots + X_n$ of any finite collection of rv s is also a rv.

[21] See Exercise 1.15 for some peculiarities about this definition.

the *marginal CDF* for the output node connected to input 2 is $1/n$ for each output. Thus the expectation of a match for node 2 is also $1/n$. In the same way, the expectation of a match for each input node is $1/n$. From (1.40), the expected total number of matches is the sum over the expected number for each input, and is thus equal to 1. This exercise would be much more difficult without the use of (1.40).

If the rv s $X_1, \ldots, X_n$ are independent, then, as shown in Exercises 1.14 and 1.21, the variance of $S_n = X_1 + \cdots + X_n$ is given by

$$\sigma^2_{S_n} = \sum_{i=1}^{n} \sigma^2_{X_i}. \tag{1.41}$$

If $X_1, \ldots, X_n$ are also identically distributed (i.e., $X_1, \ldots, X_n$ are IID) with variance σ^2_X, then $\sigma^2_{S_n} = n\sigma^2_X$. Thus the standard deviation of S_n is $\sigma_{S_n} = \sqrt{n}\sigma_X$. Sums of IID rv s appear everywhere in probability theory and play an especially central role in the laws of large numbers. It is important to remember that the mean of S_n is linear in n but the standard deviation increases only with the square root of n. Figure 1.5 illustrates this behavior.

1.5.2 Conditional expectations

Just as the *conditional CDF* of one rv conditioned on a sample value of another rv is important, the *conditional expectation* of one rv based on the sample value of another is equally important. Initially let X be a positive discrete rv and let y be a sample value of another discrete rv Y such that $\mathsf{p}_Y(y) > 0$. Then the conditional expectation of X given $Y = y$ is defined to be

$$\mathsf{E}[X \mid Y{=}y] = \sum_x x\,\mathsf{p}_{X|Y}(x \mid y). \tag{1.42}$$

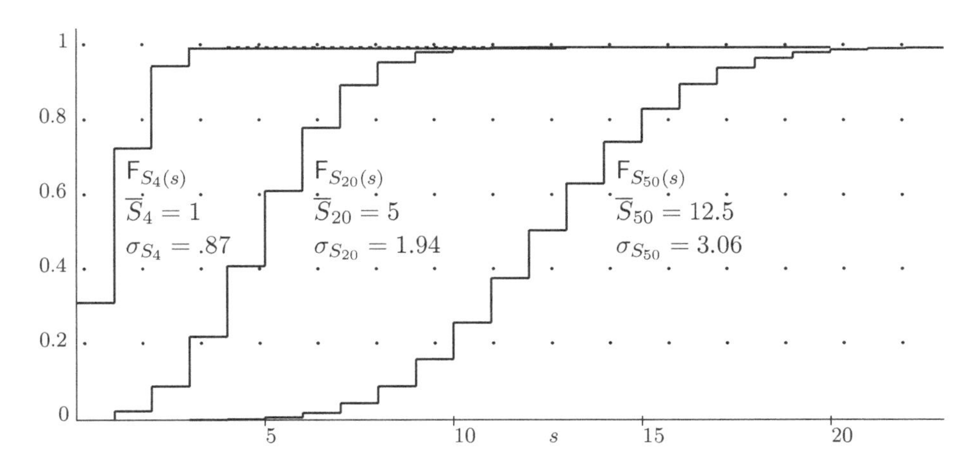

Figure 1.5 The CDF $\mathsf{F}_{S_n}(s)$ of $S_n = X_1 + \cdots + X_n$, where $X_1, \ldots, X_n$ are typical IID rv s and n takes the values 4, 20, and 50. The particular rv X in the figure is binary with $\mathsf{p}_X(1) = 1/4$, $\mathsf{p}_X(0) = 3/4$. Note that the mean of S_n is proportional to n and the standard deviation to $\sqrt{n}$.

This is simply the ordinary expected value of X using the conditional probabilities in the reduced sample space corresponding to $Y = y$. This value can be finite or infinite as before. More generally, if X can take on positive or negative values, then there is the possibility that the conditional expectation is undefined. In other words, for discrete rv s, the conditional expectation is determined in exactly the same way as the ordinary expectation, except that it is taken using conditional probabilities over the reduced sample space.

More generally yet, let X be an arbitrary rv and let y be a sample value of a discrete rv Y with $\mathsf{p}_Y(y) > 0$. The conditional CDF of X conditional on $Y = y$ is defined as

$$\mathsf{F}_{X|Y}(x \mid y) = \frac{\Pr\{X \leq x, Y = y\}}{\Pr\{Y = y\}}.$$

Since this is an ordinary CDF in the reduced sample space where $Y = y$, (1.34) expresses the expectation of X conditional on $Y = y$ as

$$\mathsf{E}\big[X \mid Y = y\big] = -\int_{-\infty}^{0} \mathsf{F}_{X|Y}(x \mid y)\,dx + \int_{0}^{\infty} \mathsf{F}^{\mathsf{c}}_{X|Y}(x \mid y)\,dx. \tag{1.43}$$

The forms of conditional expectation in (1.42) and (1.43) are given for individual sample values of Y for which $\mathsf{p}_Y(y) > 0$.

We next show that the conditional expectation of X conditional on a discrete rv Y can also be viewed as a rv. With the possible exception of a set of zero probability, each $\omega \in \Omega$ maps to $\{Y = y\}$ for some y with $\mathsf{p}_Y(y) > 0$ and $\mathsf{E}\big[X \mid Y = y\big]$ is defined for that y. Thus we can define $\mathsf{E}[X \mid Y]$ as[22] a rv that is a function of Y, mapping ω to a sample value, say y of Y, and mapping that y to $\mathsf{E}\big[X \mid Y = y\big]$. Regarding a conditional expectation as a rv that is a function of the conditioning rv is a powerful tool both in problem solving and in advanced work. We now explain how this can be used to express the unconditional mean of X as

$$\mathsf{E}[X] = \mathsf{E}\big[\mathsf{E}[X \mid Y]\big], \tag{1.44}$$

where the inner expectation is over X for each sample value of Y and the outer expectation is over the rv $\mathsf{E}[X \mid Y]$, which is a function of Y.

Example 1.5.6 Consider rolling two dice, say a red die and a black die. Let X_1 be the number on the top face of the red die, and X_2 that for the black die. Let $S = X_1 + X_2$. Thus X_1 and X_2 are IID integer rv s, each uniformly distributed from 1 to 6. Conditional on $S = j$, X_1 is uniformly distributed between 1 and $j - 1$ for $j \leq 7$ and between $j - 6$ and 6 for $j \geq 7$. For each $j \leq 7$, it follows that $\mathsf{E}\big[X_1 \mid S = j\big] = j/2$. Similarly, for $j \geq 7$, $\mathsf{E}\big[X_1 \mid S = j\big] = j/2$. This can also be seen by the symmetry between X_1 and X_2.

[22] This assumes that $\mathsf{E}\big[X \mid Y = y\big]$ is finite for each y, which is one of the reasons that expectations are said to exist only if they are finite.

The rv $\mathsf{E}[X_1 \mid S]$ is thus a discrete rv taking on values from 1 to 6 in steps of 1/2 as the sample value of S goes from 2 to 12. The PMF of $\mathsf{E}[X_1 \mid S]$ is given by $\mathsf{p}_{\mathsf{E}[X_1|S]}(j/2) = \mathsf{p}_S(j)$. Using (1.44), we can then calculate $\mathsf{E}[X_1]$ as

$$\mathsf{E}[X_1] = \mathsf{E}\big[\mathsf{E}[X_1 \mid S]\big] = \sum_{j=2}^{12} \frac{j}{2}\,\mathsf{p}_S(j) = \frac{\mathsf{E}[S]}{2} = \frac{7}{2}.$$

This example is not intended to show the value of (1.44) in calculating expectation, since $\mathsf{E}[X_1] = 7/2$ is initially obvious from the uniform integer distribution of X_1. The purpose is simply to illustrate what the rv $\mathsf{E}[X_1 \mid S]$ means.

To illustrate (1.44) in a more general way, while still assuming X to be discrete, we can write out this expectation by using (1.42) for $\mathsf{E}\big[X \mid Y = y\big]$:

$$\begin{aligned}\mathsf{E}[X] = \mathsf{E}\big[\mathsf{E}[X \mid Y]\big] &= \sum_y \mathsf{p}_Y(y)\mathsf{E}\big[X \mid Y = y\big] \\ &= \sum_y \mathsf{p}_Y(y) \sum_x x\,\mathsf{p}_{X|Y}(x|y). \end{aligned} \tag{1.45}$$

Operationally, there is nothing very fancy in the example or in (1.44). Combining the sums, (1.45) simply says that $\mathsf{E}[X] = \sum_{y,x} x\,\mathsf{p}_{YX}(y,x)$. As a concept, however, viewing the conditional expectation $\mathsf{E}[X \mid Y]$ as a rv based on the conditioning rv Y is often a useful theoretical tool. This approach is equally useful as a tool in problem solving, since there are many problems where it is easy to find conditional expectations, and then to find the total expectation by averaging over the conditioning variable. For this reason, this result is sometimes called either the total expectation theorem or the iterated expectation theorem. Exercise 1.20 illustrates the advantages of this approach, particularly where it is initially clear that the expectation is finite. The following cautionary example, however, shows that this approach can sometimes hide convergence questions and give the wrong answer.

Example 1.5.7 Let Y be a geometric rv with the PMF $\mathsf{p}_Y(y) = 2^{-y}$ for integer $y \geq 1$. Let X be an integer rv that, conditional on Y, is binary with equiprobable values $\pm 2^y$ given $Y = y$. We then see that $\mathsf{E}\big[X \mid Y = y\big] = 0$ for all y, and thus, (1.45) indicates that $\mathsf{E}[X] = 0$. On the other hand, it is easy to see that $\mathsf{p}_X(2^k) = \mathsf{p}_X(-2^k) = 2^{-k-1}$ for each integer $k \geq 1$. Thus the expectation over positive values of X is ∞ and that over negative values is $-\infty$. In other words, the expected value of X is undefined and (1.45) is incorrect in this case.

The difficulty in the above example cannot occur if X is a non-negative rv. Then (1.45) is simply a sum of a countable number of non-negative terms, and thus it either converges to a finite sum independent of the order of summation, or it diverges to ∞, again independent of the order of summation.

If X has both positive and negative components, we can separate it into $X = X^+ + X^-$, where $X^+ = \max(0, X)$ and $X^- = \min(X, 0)$. Then (1.45) applies to X^+ and $-X^-$ separately. If at most one is infinite, then (1.45) applies to X, and otherwise X is undefined. This is summarized in the following theorem.

Theorem 1.5.8 (Total expectation) *Let X and Y be discrete rv s. If X is non-negative, then $\mathsf{E}[X] = \mathsf{E}\big[\mathsf{E}[X \mid Y]\big] = \sum_y \mathsf{p}_Y(y)\mathsf{E}\big[X \mid Y = y\big]$. If X has both positive and negative values, and if at most one of $\mathsf{E}\big[X^+\big]$ and $\mathsf{E}\big[-X^-\big]$ is infinite, then $\mathsf{E}[X] = \mathsf{E}\big[\mathsf{E}[X \mid Y]\big] = \sum_y \mathsf{p}_Y(y)\mathsf{E}\big[X \mid Y = y\big]$.*

We have seen above that if Y is a discrete rv, then the conditional expectation $\mathsf{E}\big[X|Y = y\big]$ is only a little more complicated than the unconditional expectation, and this is true whether X is discrete, continuous, or arbitrary. If X and Y are continuous, we can essentially extend these results to probability densities. In particular, defining $\mathsf{E}\big[X \mid Y = y\big]$ as

$$\mathsf{E}\big[X \mid Y = y\big] = \int_{-\infty}^{\infty} x\,\mathsf{f}_{X|Y}(x \mid y)\,dx, \tag{1.46}$$

we have

$$\mathsf{E}[X] = \int_{-\infty}^{\infty} \mathsf{f}_Y(y)\mathsf{E}\big[X \mid Y{=}y\big]\,dy = \int_{-\infty}^{\infty} \mathsf{f}_Y(y) \int_{-\infty}^{\infty} x\,\mathsf{f}_{X|Y}(x \mid y)\,dx\,dy. \tag{1.47}$$

We do not state this as a theorem because the details about the integration do not seem necessary for the places where it is useful.

1.5.3 Typical values of random variables; mean and median

The CDF of a rv often has more detail than we are interested in, and the mean is often taken as a ‘typical value.’ Similarly, in statistics, the average of a set of numerical data values is often taken to be representative of the entire set. For example, students always want to know the average of the scores in an exam, and investors always want to know the Dow–Jones average. Economists are also interested, for example, in such averages as the average annual household income over various geographical regions. These averages often take on an importance and a life of their own, particlarly in terms of how they vary in time.

The median of a rv (or set of data values) is often an alternative choice of a single number to serve as a typical value. We say that α is a median of X if $\Pr\{X \le \alpha\} \ge 1/2$ and $\Pr\{X \ge \alpha\} \ge 1/2$. It is possible for the median to be non-unique, with all values in an interval satisfying the definition. Exercise 1.10 illustrates what this definition means. In addition, Exercise 1.11 shows that if the mean exists, then the median is an x that minimizes $\mathsf{E}[|X - x|]$.

Another interesting property of the median, suggested in Exercise 1.34, is that in essence a median of a large number of IID sample values of a rv is close to a median of the distribution with high probability. Another property, relating the median α to the mean $\overline{X}$ of a rv with standard deviation σ, is (see Exercise 1.33)

$$|\,\overline{X} - \alpha\,| \le \sigma. \tag{1.48}$$

The question now arises whether the mean or the median is preferable as a single number describing a rv. The question is too vague to be answered in any generality, but the answer depends heavily on the intended use of the single number. To illustrate this, consider a rv whose sample values are the yearly household incomes of a large society (or, almost equivalently, consider a large data set consisting of these yearly household incomes).

For the mean, the probability of each sample value is weighted by the household income, so that a household income of $\$10^9$ is weighted the same as 100 000 household incomes of $\$10^4$ each. For the median, this weighting disappears, and if our billionaire has a truly awful year with only $\$10^6$ income, the median is unchanged. If one is interested in the total purchasing power of the society, then the mean might be the more appropriate value. On the other hand, if one is interested in the well-being of the society, the median is the more appropriate value.[23]

1.5.4 Indicator random variables

For any event A, the *indicator rv* of A, denoted $\mathbb{I}_A$, is a binary rv that has the value 1 for all $\omega \in A$ and the value 0 otherwise. It then has the PMF $\mathsf{p}_{\mathbb{I}_A}(1) = \Pr\{A\}$ and $\mathsf{p}_{\mathbb{I}_A}(0) = 1 - \Pr\{A\}$. The corresponding CDF $\mathsf{F}_{\mathbb{I}_A}$ is then illustrated in Figure 1.6. It is easily seen that $\mathsf{E}[\mathbb{I}_A] = \Pr\{A\}$.

Figure 1.6 The CDF $\mathsf{F}_{\mathbb{I}_A}$ of an indicator random variable $\mathbb{I}_A$.

Indicator rv s are useful because they allow us to apply the many known results about rv s and particularly binary rv s to events. For example, the laws of large numbers are expressed in terms of sums of rv s. If these rv s are taken to be the indicator functions for the occurencees of an event over successive trials, then the law of large numbers applies to the relative frequency of that event.

1.5.5 Moment generating functions and other transforms

The *MGF* for a rv X is given by

$$\mathsf{g}_X(r) = \mathsf{E}\left[e^{rX}\right] = \int_{-\infty}^{\infty} e^{rx}\, d\mathsf{F}_X(x), \tag{1.49}$$

[23] Unfortunately, the choice between median and mean (and many similar choices) is often made for commercial or political expediency rather than scientific or common-sense appropriateness.

where r is a real variable. The integrand is non-negative, and we can study where the integral exists (i.e., where it is finite) by separating it as follows:

$$\mathsf{g}_X(r) = \int_0^{\infty} e^{rx}\, d\mathsf{F}_X(x) + \int_{-\infty}^{0} e^{rx}\, d\mathsf{F}_X(x). \tag{1.50}$$

Both integrals exist for $r = 0$, since the first is $\Pr\{X > 0\}$ and the second is $\Pr\{X \le 0\}$. The first integral is increasing in r, and thus if it exists for one value of r, it also exists for all smaller values. For example, if X is a non-negative exponential rv with the density $\mathsf{f}_X(x) = e^{-x}$, then the first integral exists if and only if $r < 1$, and it then has the value $1/(1-r)$. As another example, if X satisfies $\Pr\{X > A\} = 0$ for some finite A, then the first integral is at most e^{rA}, which is finite for all real r.

Let $r_+(X)$ be the supremum of values of r for which the first integral exists. Then $0 \le r_+(X) \le \infty$ and the first integral exists for all $r < r_+(X)$. In the same way, let $r_-(X)$ be the infimum of values of r for which the the second integral exists. Then $0 \ge r_-(X) \ge -\infty$ and the second integral exists for all $r > r_-(X)$.

Combining the two integrals, the region of r over which $\mathsf{g}_X(r)$ exists is an interval $I(X)$ from $r_-(X) \le 0$ to $r_+(X) \ge 0$. Either or both of the endpoints, $r_-(X)$ and $r_+(X)$, might be included in $I(X)$; i.e., each end of the interval $I(X)$ can be open or closed. We denote these quantities as I, r_-, and r_+ when the rv X is clear from the context. Exercise 1.25 illustrates r_- and r_+ further.

If $\mathsf{g}_X(r)$ exists in an open region of r around 0 (i.e., if $r_- < 0 < r_+$), then derivatives[24] of all orders exist in that region. They are given by

$$\frac{d^k \mathsf{g}_X(r)}{dr^k} = \int_{-\infty}^{\infty} x^k e^{rx} d\mathsf{F}_X(x) \quad ; \quad \left.\frac{d^k \mathsf{g}_X(r)}{dr^k}\right|_{r=0} = \mathsf{E}\left[X^k\right]. \tag{1.51}$$

This shows that finding the MGF often provides a convenient way to calculate the moments of a rv (see Exercise 3.2 for an example). If any moment of a rv fails to exist, however, then the MGF is infinite throughout either $(0,\infty)$ or $(\infty, 0)$ (see Exercise 1.37).

Another important feature of MGFs is their usefulness in treating sums of independent rv s. For example, let $S_n = X_1 + X_2 + \cdots + X_n$. Then

$$\begin{aligned} \mathsf{g}_{S_n}(r) = \mathsf{E}\left[e^{rS_n}\right] &= \mathsf{E}\left[\exp\left(\sum\nolimits_{i=1}^{n} rX_i\right)\right] \\ = \mathsf{E}\left[\prod\nolimits_{i=1}^{n} \exp(rX_i)\right] &= \prod\nolimits_{i=1}^{n} \mathsf{g}_{X_i}(r). \end{aligned} \tag{1.52}$$

In the last step, we have used a result of Exercise 1.14, which shows that for independent rv s, the mean of the product is equal to the product of the means. If $X_1, \ldots, X_n$ are also IID, then

$$\mathsf{g}_{S_n}(r) = [\mathsf{g}_X(r)]^n. \tag{1.53}$$

[24] This result depends on interchanging the order of differentiation (with respect to r) and integration (with respect to x). This can be shown to be permissible because $\mathsf{g}_X(r)$ exists for r both greater and smaller than 0, which in turn implies, first, that $1 - \mathsf{F}_X(x)$ must approach 0 at least exponentially as $x \to \infty$ and, second, that $\mathsf{F}_X(x)$ must approach 0 at least exponentially as $x \to -\infty$.

We will use this property frequently in treating sums of IID rv s. Note that this also implies that the region over which the MGF s of S_n and X exist is the same, i.e., $I(S_n) = I(X)$.

The real variable r in the MGF can be replaced by a complex variable, giving rise to a number of other transforms. A particularly important case is to view r as a pure imaginary variable, say $i\theta$ where $i = \sqrt{-1}$ and θ is real. Then[25] $\mathsf{g}_X(i\theta) = \mathsf{E}\left[e^{i\theta x}\right]$ is called the *characteristic function* of X. Since $|e^{i\theta x}|$ is 1 for all x, $\mathsf{g}_X(i\theta)$ exists for all rv s X and all real θ, and its magnitude is at most 1.

A minor but important variation on the characteristic function of X is the Fourier transform of the probability density of X. If X has a density $\mathsf{f}_X(x)$, then the Fourier transform of $\mathsf{f}_X(x)$ is given by

$$\mathsf{g}_X(-i2\pi\theta) = \int_{-\infty}^{\infty} \mathsf{f}_X(x) \exp(-i2\pi\theta x)\, dx. \tag{1.54}$$

The major advantage of the Fourier transform (aside from its familiarity) is that $\mathsf{f}_X(x)$ can usually be found from $\mathsf{g}_X(-i2\pi\theta)$ as the inverse Fourier transform,[26]

$$\mathsf{f}_X(x) = \int_{-\infty}^{\infty} \mathsf{g}_X(-i2\pi\theta) \exp(i2\pi\theta x)\, d\theta, \tag{1.55}$$

The Z-transform is the result of replacing e^r with z in $\mathsf{g}_X(r)$. This is useful primarily for integer-valued rv s, but if one transform can be evaluated, the other can be found immediately. Finally, if we use $-s$, viewed as a complex variable, in place of r, we get the two-sided Laplace transform of the density of the rv. Note that for all of these transforms, multiplication in the transform domain corresponds to convolution of the CDFs or densities, and summation of independent rv s. The simplicity of taking products of transforms is a major reason that transforms are so useful in probability theory.

1.6 Basic inequalities

Inequalities play a particularly fundamental role in probability, partly because many important models are too complex to find exact answers, and partly because many of the most useful theorems establish limiting rather than exact results. In this section, we study three related inequalities: the Markov, Chebyshev, and Chernoff bounds. These are used repeatedly both in the next section and in the remainder of the text.

[25] The notation here can be slightly dangerous, since one cannot necessarily take an expression for $\mathsf{g}_X(r)$, valid for real r, and replace r by $i\theta$ with real θ to get the characteristic function.

[26] This integral does not necessarily converge, particularly if X does not have a PDF. However, it can be shown (see [27] Chap. 2.12, or [9] Chap. 15) that the characteristic function/Fourier transform of an arbitrary rv does uniquely specify the CDF.

1.6.1 The Markov inequality

This is the simplest and most basic of these inequalities. It states that if a non-negative rv Y has a mean $\mathsf{E}[Y]$, then, for every $y > 0$, $\Pr\{Y \geq y\}$ satisfies[27]

$$\Pr\{Y \geq y\} \leq \frac{\mathsf{E}[Y]}{y} \qquad \text{for every } y > 0 \quad \text{(Markov inequality).} \tag{1.56}$$

Figure 1.7 derives this result using the fact (see Figure 1.3) that the mean of a non-negative rv is the integral of its complementary CDF, i.e., of the area under the curve $\Pr\{Y > z\}$. Exercise 1.30 gives another simple proof using an indicator random variable.

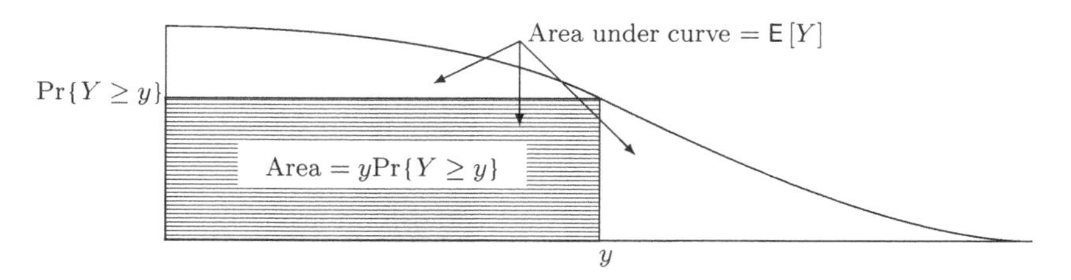

Figure 1.7 Demonstration that $y\Pr\{Y \geq y\} \leq \mathsf{E}[Y]$. By letting $y \to \infty$, it can also be seen that the shaded area becomes a negligible portion of the area $\mathsf{E}[Y]$, so that $\lim_{y\to\infty} y\Pr\{Y > y\} = 0$ if $\mathsf{E}[Y] \leq \infty$.

As an example of this inequality, assume that the average height of a population of people is 1.6 meters. Then the Markov inequality states that at most half of the population have a height exceeding 3.2 meters. We see from this example that the Markov inequality is often very weak. However, for any $y > 0$, we can consider a rv that takes on the value y with probability ϵ and the value 0 with probability $1 - \epsilon$; this rv satisfies the Markov inequality at the point y with equality. Figure 1.7 (as elaborated in Exercise 1.45) also shows that, for any non-negative rv Y with a finite mean,

$$\lim_{y\to\infty} y\Pr\{Y \geq y\} = 0. \tag{1.57}$$

This will be useful shortly in the proof of Theorem 1.7.4.

1.6.2 The Chebyshev inequality

We now use the Markov inequality to establish the well-known Chebyshev inequality. Let Z be an arbitrary rv with finite mean $\mathsf{E}[Z]$ and finite variance σ_Z^2, and define Y as the non-negative rv $Y = (Z - \mathsf{E}[Z])^2$. Thus $\mathsf{E}[Y] = \sigma_Z^2$. Applying (1.56),

$$\Pr\left\{(Z - \mathsf{E}[Z])^2 \geq y\right\} \leq \frac{\sigma_Z^2}{y} \qquad \text{for every } y > 0.$$

[27] The CDF of any given rv Y is known (at least in principle), and thus one might question why an upper bound is ever preferable to the exact value. One answer is that Y might be given as a function of many other rv s and that the parameters (such as the mean) used in a bound are often much easier to find than the CDF. Another answer is that such inequalities are often used in theorems which state results in terms of simple statistics such as the mean rather than the entire CDF. This will be evident as we use these bounds.

Replacing y with ϵ^2 and noting that the event $\{(Z - \mathsf{E}[Z])^2 \geq \epsilon^2\}$ is the same as $|Z - \mathsf{E}[Z]| \geq \epsilon$, this becomes

$$\Pr\{|Z - \mathsf{E}[Z]| \geq \epsilon\} \leq \frac{\sigma_Z^2}{\epsilon^2} \qquad \text{for every } \epsilon > 0 \quad \text{(Chebyshev inequality).} \tag{1.58}$$

Note that the Markov inequality bounds just the upper tail of the CDF and applies only to non-negative rv s, whereas the Chebyshev inequality bounds both tails of the CDF. The more important differences, however, are that the Chebyshev bound requires a finite variance and approaches zero inversely with the square of the distance from the mean, whereas the Markov bound does not require a finite variance and goes to zero inversely with the distance from 0 (and thus asymptotically with distance from the mean).

The Chebyshev inequality is particularly useful when Z is the sample average, $(X_1 + X_2 + \cdots + X_n)/n$, of a set of IID rv s. This will be used shortly in proving the weak law of large numbers (WLLN).

1.6.3 Chernoff bounds

We saw that the Chebyshev inequality on a rv Z results from applying the Markov inequality to $(Z - \overline{Z})^2$. In the same way, the Chernoff (or exponential) bound on a rv Z results from applying the Markov inequality to e^{rZ} for a given r. The Chernoff bound, when it applies, approaches 0 exponentially with increasing distance from the mean.

For any given rv Z, let $I(Z)$ be the interval over which the MGF $\mathsf{g}_Z(r) = \mathsf{E}\left[e^{rZ}\right]$ exists. Letting $Y = e^{rZ}$ for any $r \in I(Z)$, the Markov inequality (1.56) applied to Y is

$$\Pr\{\exp(rZ) \geq y\} \leq \frac{\mathsf{g}_Z(r)}{y} \qquad \text{for every } y > 0 \quad \text{(Chernoff bound).}$$

This takes on a more meaningful form if y is replaced by e^{rb}. Note that $\exp(rZ) \geq \exp(rb)$ is equivalent to $Z \geq b$ for $r > 0$ and to $Z \leq b$ for $r < 0$. Thus, for any real b, we get the following two bounds, one for $r > 0$ and the other for $r < 0$:

$$\Pr\{Z{\geq}b\} \leq \mathsf{g}_Z(r)\exp(-rb) \qquad \text{(Chernoff bound for } r > 0,\ r \in I(Z)\text{);} \tag{1.59}$$

$$\Pr\{Z{\leq}b\} \leq \mathsf{g}_Z(r)\exp(-rb) \qquad \text{(Chernoff bound for } r < 0,\ r \in I(Z)\text{).} \tag{1.60}$$

Note that $g_Z(r) = 1$ for $r = 0$, so each bound has the value 1 for $r = 0$ and is thus valid (but uninteresting) for $r = 0$. It is somewhat simpler, however, when optimizing over r to include $r = 0$ in each bound. We will soon see that for $b \geq \overline{Z}$, the bound in (1.60) is at least 1 over the range of r where it is valid, and thus only (1.59) can be useful for $b \geq \overline{Z}$. Similarly, only (1.60) can be useful for $b \leq \overline{Z}$ (and thus neither is useful for $b = \overline{Z}$).

The most important application of Chernoff bounds is to sums of IID rv s. Let $S_n = X_1 + \cdots + X_n$ where $X_1, \ldots, X_n$ are IID with the MGF $\mathsf{g}_X(r)$. Then $\mathsf{g}_{S_n}(r) = [\mathsf{g}_X(r)]^n$, so (1.59) and (1.60) (with b replaced by na) become

$$\Pr\{S_n{\geq}na\} \leq [\mathsf{g}_X(r)]^n \exp(-rna) \qquad \text{for } r > 0,\ r \in I(X); \tag{1.61}$$

$$\Pr\{S_n{\leq}na\} \leq [\mathsf{g}_X(r)]^n \exp(-rna) \qquad \text{for } r < 0,\ r \in I(X). \tag{1.62}$$

These equations are easier to understand if we define the *semi-invariant MGF*, $\gamma_X(r)$, as

$$\gamma_X(r) = \ln \mathsf{g}_X(r). \tag{1.63}$$

The semi-invariant MGF for a typical rv X is sketched in Figure 1.8. Important features of the figure, as explained in the caption, are, first, that $\gamma'_X(0) = \mathsf{E}[X]$ and, second, that $\gamma''_X(r) \geq 0$ for r in the interior of $I(X)$.

In terms of $\gamma_X(r)$, (1.61) and (1.62) become

$$\Pr\{S_n{\geq}na\} \leq \exp(n[\gamma_X(r) - ra]) \qquad \text{for } r \geq 0, \ r \in I(X); \tag{1.64}$$

$$\Pr\{S_n{\leq}na\} \leq \exp(n[\gamma_X(r) - ra]) \qquad \text{for } r \leq 0, \ r \in I(X). \tag{1.65}$$

These bounds are exponential in n for fixed a and r, and for any given a, we can optimize (1.64), simultaneously for all n, by minimizing $\gamma(r) - ra$ over $r \geq 0$, $r \in I(X)$. Similarly, (1.65) can be minimized over $r \leq 0$. We will explore these minimizations in detail in Section 9.3, but for now, consider the case where $a > \overline{X}$. Note that $[\gamma(r){-}ra]_{r=0} = 0$ and $d[\gamma(r){-}ra]/dr = \overline{X}{-}a < 0$. This means that $\gamma_X(r){-}ra$ must be negative for sufficiently small $r > 0$ and thus means that the bound in (1.64) is decreasing exponentially in n for small enough $r > 0$.

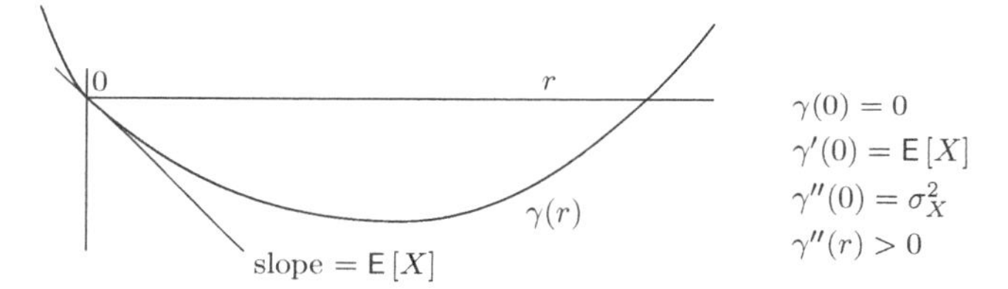

Figure 1.8 Semi-invariant MGF $\gamma(r)$ for a typical rv X assuming that 0 is in the interior of $I(X)$. Since $\gamma(r) = \ln g(r)$, we see that $d\gamma(r)/dr = [g(r)]^{-1} dg(r)/dr$. Thus $\gamma'(0) = \mathsf{E}[X]$. Also, for r in the interior of $I(X)$, Exercise 1.26 shows that $\gamma''(r) \geq 0$ and, in fact, $\gamma''(r)$ is strictly positive except in the uninteresting case where X is deterministic (takes on a single value with probability 1). As indicated in the figure, the straight line of slope $\mathsf{E}[X]$ through the origin is tangent to $\gamma(r)$ at $r = 0$.

For $r < 0$, $\gamma(r) - ra$ is positive for r close to 0, so the bound in (1.65) exceeds 1 and is worthless. Since $\gamma''(r) > 0$ over $I(X)$, $d[\gamma(r)]/dr < 0$ for all negative $r \in I(X)$, so the bound in (1.65) is worthless for all r in its range when $a > \overline{X}$. This is not surprising, since $\Pr\{S_n \geq na\} = 1 - \Pr\{S_n \leq na\}$, so both bounds cannot be exponentially decreasing in n for the same a.

The same argument can be reversed for $a < \overline{X}$, and in this case the bound (1.65) is exponentially decreasing in n for negative r sufficiently close to 0 and (1.64) is worthless. We also see that both bounds are worthless for $a = \overline{X}$, and this will be more obvious after discussing the central limit theorem.

The roles of (1.64) and (1.65) can be reversed by replacing $\{X_n;\, n \geq 1\}$ with $\{-X_n;\, n \geq 1\}$ and therefore we consider only (1.64) with $a > \overline{X}$ as the Chernoff bound in what follows.

The results of this section can be summarized in the following lemma.

Lemma 1.6.1 *Let $\{X_i;\ i \geq 1\}$ be IID rv s and let $S_n = X_1 + \cdots + X_n$ for each $n \geq 1$. Let $I(X)$ be the interval over which the semi-invariant MGF $\gamma_X(r)$ is finite and assume 0 is in the interior of $I(X)$. Then, for each $n \geq 1$ and each real number a,*

$$\Pr\{S_n \geq na\} \leq \exp(n\mu_X(a)) \qquad \text{where } \mu_X(a) = \inf_{r\geq 0,\, r\in I(X)} \gamma_X(r) - ra. \tag{1.66}$$

Furthermore, $\mu_X(a) < 0$ for $a > \overline{X}$ and $\mu_X(a) = 0$ for $a \leq \overline{X}$.

These Chernoff bounds will be used in the next section to help understand several laws of large numbers. They will also be used and further developed in Chapter 9. They are useful for detection, random walks, information theory, and other areas where large deviations are important.

The following example evaluates these bounds for the case where the IID rv s are binary. We will see that in this case the bounds are exponentially tight in a sense to be described.

Example 1.6.2 Let X be binary with $\mathsf{p}_X(1) = p$ and $\mathsf{p}_X(0) = q = 1 - p$. Then $\mathsf{g}_X(r) = q + pe^r$ for $-\infty < r < \infty$. Also, $\gamma_X(r) = \ln(q + pe^r)$. To be consistent with the expression for the binomial PMF in (1.25), we will find the Chernoff bound on $\Pr\{S_n \geq \tilde{p}n\}$ for $\tilde{p} > p$. Thus, according to Lemma 1.6.1, we first evaluate

$$\mu_X(\tilde{p}) = \inf_{r\geq 0}[\gamma_X(r) - \tilde{p}r].$$

The minimum over all r occurs at that r for which $\gamma_X'(r) = \tilde{p}$, i.e., at

$$\frac{pe^r}{q + pe^r} = \tilde{p}.$$

Rearranging terms,

$$e^r = \frac{\tilde{p}q}{p\tilde{q}}, \qquad \text{where } \tilde{q} = 1 - \tilde{p}. \tag{1.67}$$

For $\tilde{p} > p$, this r is positive, achieving the minimization over $r \geq 0$. Substituting this r into $\ln(q + pe^r) - r\tilde{p}$ and rearranging terms,

$$\mu_X(\tilde{p}) = \tilde{p}\ln\frac{p}{\tilde{p}} + \tilde{q}\ln\frac{\tilde{q}}{q}. \tag{1.68}$$

Substituting this into (1.66),

$$\Pr\{S_n \geq n\tilde{p}\} \leq \exp\left\{n\left[\tilde{p}\ln\frac{p}{\tilde{p}} + \tilde{q}\ln\frac{q}{\tilde{q}}\right]\right\} \qquad \text{for } \tilde{p} > p. \tag{1.69}$$

So far, it seems that we have simply developed another upper bound on the tails of the CDF for the binomial. It will then perhaps be surprising to compare this bound with the asymptotically correct value (repeated below) for the binomial PMF in (1.27) with $\tilde{p} = k/n$:

$$\mathsf{p}_{S_n}(k) \sim \sqrt{\frac{1}{2\pi n\tilde{p}\tilde{q}}}\exp\{-n\mathsf{D}(\tilde{p}\|p)\}, \qquad \text{where } \mathsf{D}(\tilde{p}\|p) = \left[\tilde{p}\ln\frac{\tilde{p}}{p} + \tilde{q}\ln\frac{\tilde{q}}{q}\right]. \tag{1.70}$$

For any integer value of $n\widetilde{p}$ with $\widetilde{p} > p$, we can lower bound $\Pr\{S_n \geq n\widetilde{p}\}$ by the single term $\mathsf{p}_{S_n}(n\widetilde{p})$. Thus $\Pr\{S_n \geq n\widetilde{p}\}$ is both upper and lower bounded by quantities that decrease exponentially with n at the same rate. In a sense, then, the upper bound and the lower bound are essentially the same for large n. We can express the sense in which they are the same by considering the log of the upper bound in (1.69) and the lower bound arising from (1.70).

$$\lim_{n\to\infty} \frac{\ln \Pr\{S_n \geq n\widetilde{p}\}}{n} = -\mathsf{D}(\widetilde{p}\|p), \qquad \text{where } \widetilde{p} > p. \tag{1.71}$$

In other words, the Chernoff bound when optimized over r is not only an upper bound, but is also exponentially tight in the sense of (1.71).

In Section 9.3 we will show that this property is typical for sums of IID rv s. Thus we see that the Chernoff bounds are not 'just bounds,' but rather are bounds that when optimized provide the correct asymptotic exponents for the tails of the distribution of sums of IID rv s. In this sense these bounds are quite different from the Markov and Chebyshev bounds.

1.7 The laws of large numbers

The laws of large numbers are a collection of results in probability theory that describe the behavior of the arithmetic average of n rv s for large n. For any n rv s, $X_1, \dots, X_n$, the *arithmetic average* is the rv $(1/n)\sum_{i=1}^{n} X_i$. Since in any outcome of the experiment, the sample value of this rv is the arithmetic average of the sample values of $X_1, \dots, X_n$, this rv is usually called the *sample average*. If $X_1, \dots, X_n$ are viewed as successive variables in time, this sample average is called the time average. Under fairly general assumptions, the standard deviation of the sample average goes to 0 with increasing n, and, in various ways depending on the assumptions, the sample average approaches the mean.

These results are central to the study of stochastic processes because they allow us to relate time averages (i.e., the average over time of individual sample paths) to ensemble averages (i.e., the mean of the value of the process at a given time). In this section, we develop and discuss one of these results, the weak law of large numbers for IID rv s. We also briefly discuss another of these results, the strong law of large numbers. The strong law requires considerable patience to understand, and its derivation and fuller discussion are postponed to Chapter 5 where it is first needed. We also discuss the central limit theorem, partly because it enhances our understanding of the weak law, and partly because of its importance in its own right.

1.7.1 Weak law of large numbers with a finite variance

Let $X_1, X_2, \dots, X_n$ be IID rv s with a finite mean $\overline{X}$ and finite variance σ^2. Let $S_n = X_1 + \cdots + X_n$, and consider the sample average S_n/n. We saw in (1.41) that $\sigma^2_{S_n} = n\sigma^2$. Thus the variance of S_n/n is

$$\mathsf{VAR}\left[\frac{S_n}{n}\right] = \mathsf{E}\left[\left(\frac{S_n - n\overline{X}}{n}\right)^2\right] = \frac{1}{n^2}\,\mathsf{E}\left[\left(S_n - n\overline{X}\right)^2\right] = \frac{\sigma^2}{n}. \tag{1.72}$$

This says that the variance of the sample average S_n/n is σ^2/n, which approaches 0 as n increases. Figure 1.9 illustrates this decrease in the variance of S_n/n with increasing n. In contrast, recall that Figure 1.5 illustrated how the variance of S_n increases with n. From (1.72), we see that

$$\lim_{n\to\infty} \mathsf{E}\left[\left(\frac{S_n}{n} - \overline{X}\right)^2\right] = 0. \tag{1.73}$$

As a result, we say that S_n/n *converges in mean square* to $\overline{X}$.

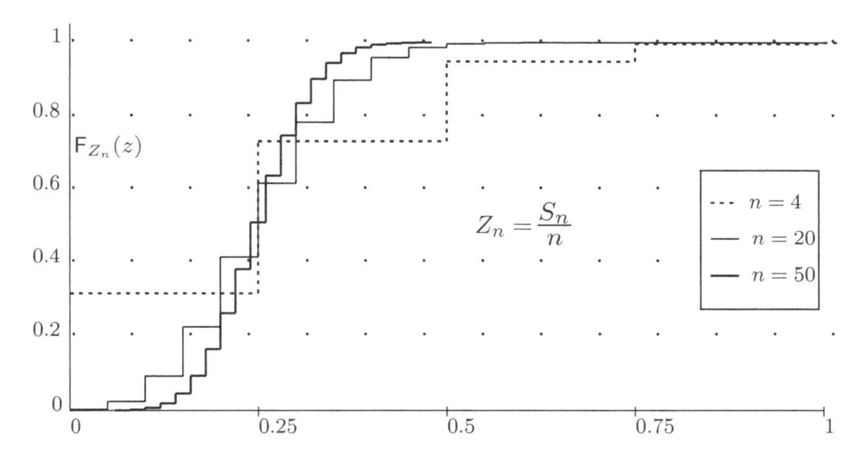

Figure 1.9 The same CDF as Figure 1.5, scaled differently to give the CDF of the sample average Z_n. It can be visualized that as n increases, the CDF of Z_n becomes increasingly close to a unit step at the mean, 0.25, of the variables X being summed.

This convergence in mean square says that the sample average, $Z_n = S_n/n$, differs from the mean, $\overline{X}$, by a random variable whose variance approaches 0 with increasing n. This convergence in mean square is one sense in which S_n/n approaches $\overline{X}$, but the idea of a sequence of rv s (i.e., a sequence of functions) approaching a constant is clearly much more involved than a sequence of numbers approaching a constant. The laws of large numbers bring out this central idea in a more fundamental, and usually more useful, way. We start the development by applying the Chebyshev inequality (1.58) to the sample average,

$$\Pr\left\{\left|\frac{S_n}{n} - \overline{X}\right| > \epsilon\right\} \le \frac{\sigma^2}{n\epsilon^2}. \tag{1.74}$$

This is an upper bound on the probability that S_n/n differs by more than ϵ from its mean, $\overline{X}$. This is illustrated in Figure 1.9 which shows the CDF of S_n/n for various n. The figure suggests that $\lim_{n\to\infty} \mathsf{F}_{S_n/n}(z) = 0$ for all $z < \overline{X}$ and $\lim_{n\to\infty} \mathsf{F}_{S_n/n}(z) = 1$

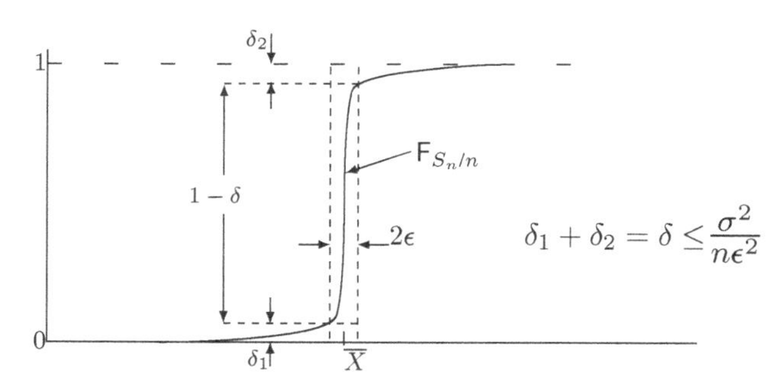

Figure 1.10 Approximation of the CDF $\mathsf{F}_{S_n/n}$ of a sample average by a step function at the mean. From (1.74), the probability δ that S_n/n differs from $\overline{X}$ by more than ϵ (i.e., $\Pr\{|S_n/n - \overline{X}| \geq \epsilon\}$) is at most $\sigma^2/n\epsilon^2$. The complementary event, where $|S_n/n - \overline{X}| < \epsilon$, has probability $1 - \delta \geq 1 - \sigma^2/n\epsilon^2$. This means that we can construct a rectangle of width 2ϵ centered on $\overline{X}$ and of height $1-\delta$ such that $\mathsf{F}_{S_n/n}$ enters the rectangle at the lower left (say at $(\overline{X} - \epsilon, \delta_1)$) and exits at the upper right, say at $(\overline{X} + \epsilon, 1 - \delta_2)$). Now visualize increasing n while holding ϵ fixed. In the limit, $1 - \delta \to 1$ so $\Pr\{|S_n/n - \overline{X}| \geq \epsilon\} \to 0$. Since this is true for every $\epsilon > 0$ (usually with slower convergence as ϵ gets smaller), $\mathsf{F}_{S_n/n}(z)$ approaches 0 for every $z < \overline{X}$ and approaches 1 for every $z > \overline{X}$, i.e., $\mathsf{F}_{S_n/n}$ approaches a unit step at $\overline{X}$. Note that there are two 'fudge factors' here, ϵ and δ and, since we are approximating an entire CDF, neither can be omitted, except by directly going to a limit as $n \to \infty$.

for all $z > \overline{X}$. This is stated more cleanly in the following weak law of large numbers, abbreviated WLLN.

Theorem 1.7.1 (WLLN with finite variance) *For each integer $n \geq 1$, let $S_n = X_1 + \cdots + X_n$ be the sum of n IID rv s with a finite variance. Then*

$$\lim_{n\to\infty} \Pr\left\{\left|\frac{S_n}{n} - \overline{X}\right| > \epsilon\right\} = 0 \qquad \text{for every } \epsilon > 0. \tag{1.75}$$

Proof For every $\epsilon > 0$, $\Pr\{|S_n/n - \overline{X}| > \epsilon\}$ is bounded between 0 and $\sigma^2/n\epsilon^2$. Since the upper bound goes to 0 with increasing n, the theorem is proved. □

Discussion The algebraic proof above is both simple and rigorous. However, the graphical description in Figure 1.10 probably provides more intuition about how the limit takes place. It is important to understand both.

We refer to (1.75) as saying that S_n/n converges to $\overline{X}$ in probability. To make sense out of this, we should view $\overline{X}$ as a deterministic rv, i.e., a rv that takes the value $\overline{X}$ for each sample point of the space. Then (1.75) says that the probability that the absolute difference, $|S_n/n - \overline{X}|$, exceeds any given $\epsilon > 0$ goes to 0 as[28] $n \to \infty$.

One should ask at this point what (1.75) adds to the more specific bound in (1.74), which in fact provides an upper bound on the rate of convergence for the limit in (1.75). The answer is that (1.75) remains valid when the theorem is generalized. For variables that are not IID or have an infinite variance, (1.74) is no longer necessarily valid. In some

[28] Saying this in words gives one added respect for mathematical notation, and perhaps in this case, it is preferable to simply understand the mathematical statement (1.75).

situations, as we see later, it is valuable to know that (1.75) holds, even if the rate of convergence is extremely slow or unknown.

One difficulty with the bound in (1.74) is that it is extremely loose in most cases. If S_n/n actually approached $\overline{X}$ this slowly, the WLLN would often be more a mathematical curiosity than a highly useful result. If we assume that the MGF of X exists in an open interval around 0, then (1.74) can be strengthened considerably. Note that the event $\{|S_n/n - \overline{X}| \geq \epsilon\}$ can be expressed as the union of the events $\{S_n/n \geq \overline{X} + \epsilon\}$ and $\{-S_n/n \geq \mathsf{E}[-X] + \epsilon\}$. Recall from Lemma 1.6.1 that for any $\epsilon > 0$

$$\Pr\{S_n/n \geq \mathsf{E}[X] + \epsilon\} \leq \exp(n\mu_X(\mathsf{E}[X] + \epsilon)), \tag{1.76}$$

$$\Pr\{-S_n/n \geq \mathsf{E}[-X] + \epsilon\} \leq \exp(n\mu_{(-X)}(\mathsf{E}[-X] + \epsilon)), \tag{1.77}$$

where both these bounds are exponentially decreasing in n. The two probabilities being bounded here are the quantities marked δ_2 and δ_1 in Figure 1.10. We see then that when the MGF exists in a region around $\overline{X}$, the convergence in δ for fixed ϵ is exponential in n rather than harmonic as suggested in Figure 1.10.

Each of these upper bounds is exponentially decreasing in n where, from Lemma 1.6.1, $\mu_X(a) = \inf_r\{\gamma_X(r) - ra\} < 0$ for $a \neq \overline{X}$. Thus, for any $\epsilon > 0$,

$$\Pr\left\{|S_n/n - \overline{X}| \geq \epsilon\right\} \leq \exp[n\mu_X(\overline{X} + \epsilon)] + \exp[n\mu_X(\overline{X} - \epsilon)]. \tag{1.78}$$

The bound here, for any fixed $\epsilon > 0$, decreases geometrically in n rather than harmonically. In terms of Figure 1.10, the height of the rectangle must approach 1 at least geometrically in n.

1.7.2 Relative frequency

We next show that the WLLN applies to the relative frequency of an event as well as to the sample average of a rv. Suppose that A is some event in a single experiment, and that the experiment is independently repeated n times. Then, in the probability model for the n repetitions, let A_i be the event that A occurs at the ith trial, $1 \leq i \leq n$. The events $A_1, A_2, \ldots, A_n$ are then IID.

If we let $\mathbb{I}_{A_i}$ be the indicator rv for A on the ith trial, then the rv $S_n = \mathbb{I}_{A_1} + \mathbb{I}_{A_2} + \cdots + \mathbb{I}_{A_n}$ is the number of occurrences of A over the n trials. It follows that

$$\text{relative frequency of } A = \frac{S_n}{n} = \frac{\sum_{i=1}^{n} \mathbb{I}_{A_i}}{n}. \tag{1.79}$$

Thus the relative frequency of A is the sample average of the binary rv s $\mathbb{I}_{A_i}$, and everything we know about the sum of IID rv s applies equally to the relative frequency of an event. In fact, everything we know about sums of IID *binary* rv s applies to relative frequency.

1.7.3 The central limit theorem (CLT)

The WLLN says that with high probability, S_n/n is close to $\overline{X}$ for large n, but it establishes this via an upper bound on the tail probabilities rather than an estimate of what

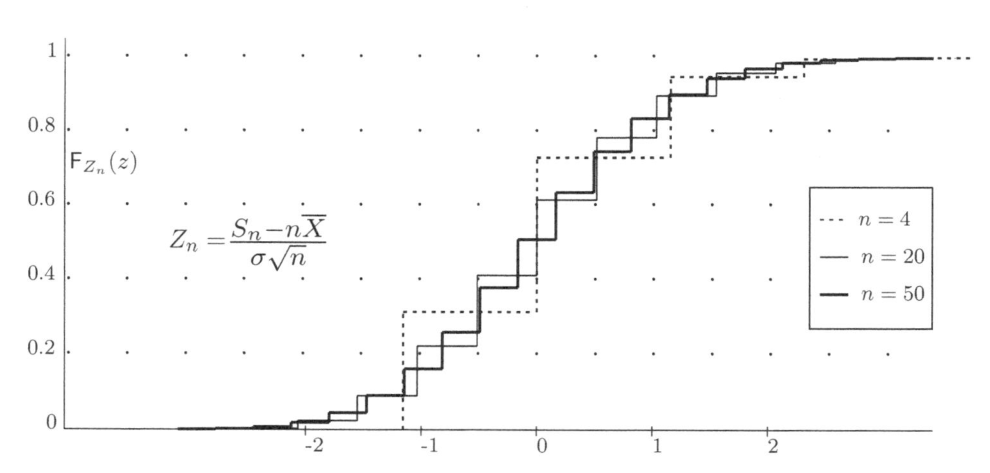

Figure 1.11 The same CDFs as Figure 1.5 normalized to 0 mean and unit standard deviation, i.e., the CDFs of $Z_n = (S_n/n - \overline{X})\sqrt{n}/\sigma$ for $n = 4, 20, 50$. Note that as n increases, the CDF of Z_n slowly starts to resemble the normal CDF.

$\mathsf{F}_{S_n/n}$ looks like. If we look at the shape of $\mathsf{F}_{S_n/n}$ for various values of n in the example of Figure 1.9, we see that the function $\mathsf{F}_{S_n/n}$ becomes increasingly compressed around $\overline{X}$ as n increases (in fact, this is the essence of what the WLLN is saying). If we normalize the rv S_n/n to 0 mean and unit variance, we get a normalized rv, $Z_n = (S_n/n - \overline{X})\sqrt{n}\,/\sigma$. The CDF of Z_n is illustrated in Figure 1.11 for the same underlying X as used for S_n/n in Figure 1.9. The curves in the two figures are the same except that each curve has been horizontally scaled by $\sqrt{n}$ in Figure 1.11.

Inspection of Figure 1.11 shows that the normalized CDFs there seem to be approaching a limiting distribution. The critically important central limit theorem (CLT) states that there is indeed such a limit, and it is the normalized Gaussian CDF.

Theorem 1.7.2 (CLT) *Let $X_1, X_2, \ldots$ be IID rv s with finite mean $\overline{X}$ and finite variance σ^2. Then for every real number z,*

$$\lim_{n\to\infty} \Pr\left\{\frac{S_n - n\overline{X}}{\sigma\sqrt{n}} \leq z\right\} = \Phi(z), \tag{1.80}$$

where $\Phi(z)$ is the normal CDF, i.e., the Gaussian distribution with mean 0 and variance 1,

$$\Phi(z) = \int_{-\infty}^{z} \frac{1}{\sqrt{2\pi}} \exp\left(-\frac{y^2}{2}\right) dy.$$

Discussion The rv $Z_n = (S_n - n\overline{X})/(\sigma\sqrt{n})$, for each $n \geq 1$ on the left-hand side of (1.80), has mean 0 and variance 1. The CLT, as expressed in (1.80), says that the sequence of CDFs $\mathsf{F}_{Z_1}(z), \mathsf{F}_{Z_2}(z), \ldots$ converges at each value of z to $\Phi(z)$ as $n \to \infty$. In other words, $\lim_{n\to\infty} \mathsf{F}_{Z_n}(z) = \Phi(z)$ for each $z \in \mathbb{R}$. This is called *convergence in* *distribution*, since it is the sequence of CDFs, rather than the sequence of rv s that is converging. The theorem is illustrated by Figure 1.11.

The CLT tells us quite a bit about how $\mathsf{F}_{S_n/n}$ converges to a step function at $\overline{X}$. To see this, rewrite (1.80) in the form

$$\lim_{n\to\infty} \Pr\left\{\frac{S_n}{n} - \overline{X} \le \frac{\sigma z}{\sqrt{n}}\right\} = \Phi(z). \tag{1.81}$$

This is illustrated in Figure 1.12 where we have used the Gaussian distribution as an approximation for the probability on the left.

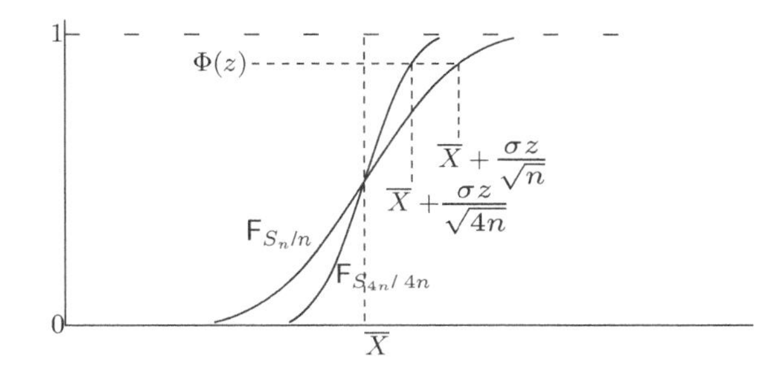

Figure 1.12 Approximation of the CDF $\mathsf{F}_{S_n/n}$ of a sample average by a Gaussian distribution of the same mean and variance. The WLLN says that $\mathsf{F}_{S_n/n}$ approaches a step function at $\overline{X}$ as $n \to \infty$. The CLT says that the shape of S_n/n approximates a Gaussian distribution $\mathcal{N}(\overline{X}, \sigma^2/n)$. If n is increased by a factor of 4, the Gaussian distribution curve is horizontally scaled inward toward $\overline{X}$ by a factor of 2. The CLT says both that the CDF of S_n/n is scaled horizontally as $1/\sqrt{n}$ and also that it is better approximated by the Gaussian of the given mean and variance as n increases.

The reason why the word *central* appears in the CLT can also be seen from (1.81). Asymptotically, we are looking at a limit (as $n \to \infty$) of the probability that the sample average differs from the mean by at most a quantity going to 0 as $1/\sqrt{n}$. This should be contrasted with the corresponding optimized Chernoff bound in (1.78) which looks at the limit of the probability that the sample average differs from the mean by at most a constant amount. Those latter results are exponentially decreasing in n and are known as large deviation results.

Theorem 1.7.2 says nothing about the rate of convergence to the normal distribution. The Berry–Esseen theorem (see, for example, Feller, [9]) provides some guidance about this for cases in which the third central moment $\mathsf{E}\left[|X - \overline{X}|^3\right]$ exists. This theorem states that

$$\left|\Pr\left\{\frac{(S_n - n\overline{X})}{\sigma\sqrt{n}} \le z\right\} - \Phi(z)\right| \le \frac{C\mathsf{E}\left[|X - \overline{X}|^3\right]}{\sigma^3\sqrt{n}}, \tag{1.82}$$

where C can be upper bounded by 0.766 (later improved to 0.4784). We will come back shortly to discuss convergence in greater detail.

The CLT helps explain why Gaussian rv s play such a central role in probability theory. In fact, many of the cookbook formulas of elementary statistics are based on the tacit assumption that the underlying variables are Gaussian, and the CLT helps explain why these formulas often give reasonable results.

One should be careful to avoid reading more into the CLT than it says. For example, the normalized sum $(S_n - n\overline{X})/\sigma\sqrt{n}$ need not have a density that is approximately Gaussian. In fact, if the underlying variables are discrete, the normalized sum is discrete and has no density. The PMF of the normalized sum might have very detailed and wild fine structure; this does not disappear as n increases, but becomes 'integrated out' in the CDF.

A proof of the CLT requires mathematical tools that will not be needed subsequently.[29] Thus we give a proof only for the binomial case. Before doing this, however, we will show that the PMF for S_n in the binomial approaches a sampled form of the Gaussian density. This detailed form of the PMF does not follow from the CLT and is often valuable in its own right.

Theorem 1.7.3 *Let $\{X_i;\, i \geq 1\}$ be a sequence of IID binary rv s with $p = \mathsf{p}_X(1) > 0$ and $q = 1 - p = \mathsf{p}_X(0) > 0$. Let $S_n = X_1 + \cdots + X_n$ for each $n \geq 1$ and let α be a fixed constant satisfying $1/2 < \alpha < 2/3$. Then constants C and n_o exist such that for all integers k $|k - np| \leq n^{\alpha}$,*

$$p_{S_n}(k) \;=\; \frac{1 \pm Cn^{3\alpha-2}}{\sqrt{2\pi npq}} \exp\left(\frac{-(k-np)^2}{2npq}\right) \qquad \text{for } n \geq n_o, \tag{1.83}$$

where this 'equation' is to be interpreted as an upper bound when the $\pm$ sign is replaced with $+$ and a lower bound when $\pm$ is replaced with $-$.

Note that $n^{3\alpha-2}$ goes to 0 with increasing n for $\alpha < 2/3$, so the ratio between the upper and lower bound in (1.83) approaches 1 at the rate $n^{3\alpha-2}$. This is independent of k within the range $|k - n| \leq n^{\alpha}$. We will see why this fussiness is required when we go from the PMF to the CDF.

Proof Let $\widetilde{p} = k/n$ and $\widetilde{q} = 1 - \widetilde{p}$. From (1.27) we have

$$\mathsf{p}_{S_n}(\widetilde{p}n) \sim \frac{1}{\sqrt{2\pi n\widetilde{p}\widetilde{q}}} \exp -n\mathsf{D}(\widetilde{p}\|p)\,,$$

where $\mathsf{D}(\widetilde{p}\|p) = \widetilde{p}\ln(\widetilde{p}/p) + \widetilde{q}\ln(\widetilde{q}/q)$. The ratio between $\mathsf{p}_{S_n}(k)$ and its asymptotic value is bounded in (1.25) and (1.26) by $1 - 1/(12n\widetilde{p}(1-\widetilde{p}))$. Since $0 < p < 1$ and $\widetilde{p}$ approaches p with increasing n, we can replace this asymptotic form by

$$\mathsf{p}_{S_n}(\widetilde{p}n) = \frac{1 \pm C_1 n^{-1}}{\sqrt{2\pi n\widetilde{p}\widetilde{q}}} \exp -n\mathsf{D}(\widetilde{p}\|p) \qquad \text{for large enough } C_1 \text{ and } n. \tag{1.84}$$

[29] Many elementary texts provide 'simple proofs,' using transform techniques. These proofs can be correct if they quote powerful enough transform results, but many of them indicate that the normalized sum has a density that approaches the Gaussian density; this is incorrect for all discrete rv s. A direct proof is given by Feller ([8] and [9]).

The remainder of the proof shows that $\mathsf{D}(\widetilde{p}\|p)$ approaches the exponential term in (1.83) as $n \to \infty$. Let $\epsilon = \widetilde{p} - p$. Then

$$\mathsf{D}(\widetilde{p}\|p) = (p+\epsilon)\ln\left(1+\frac{\epsilon}{p}\right) + (q-\epsilon)\ln\left(1-\frac{\epsilon}{q}\right)$$

$$= \frac{\epsilon^2}{2p} - \frac{\epsilon^3}{6p^2} + \cdots + \frac{\epsilon^2}{2q} + \frac{\epsilon^3}{6q^2}\cdots \tag{1.85}$$

$$= \frac{\epsilon^2}{2pq} - \frac{\epsilon^3}{6p^2} + \cdots + \frac{\epsilon^3}{6q^2}\cdots, \tag{1.86}$$

where in (1.85) we have used the power series expansion, $\ln(1+u) = u - u^2/2 + u^3/3 - \cdots$. In (1.86), we used the fact that $1/p + 1/q = 1/pq$. Substituting this into the exponential term of (1.84),

$$\exp -n\mathsf{D}(\widetilde{p}\|p) = \exp\left(-\frac{n\epsilon^2}{2pq}\right)\exp\left(\frac{n\epsilon^3}{6p^2} + \cdots - \frac{n\epsilon^3}{6q^2} + \cdots\right) \tag{1.87}$$

$$= \exp\left(-\frac{n\epsilon^2}{2pq}\right)(1 \pm C_2 n^{-3\alpha-2}), \tag{1.88}$$

where in (1.88) we used, first, the fact that the neglected terms in (1.87) are decreasing at least geometrically with ϵ, second, the condition in the theorem that $|\epsilon| = |k-np|/n \le n^{\alpha-1}$, and third, the expansion $e^x = 1 + x + x^2/2 + \cdots$. Substituting this into (1.84), we get

$$\mathsf{p}_{S_n}(\widetilde{p}n) = \frac{(1 \pm C_1 n^{-1})(1 \pm C_2 n^{-3\alpha+2})}{\sqrt{2\pi n \widetilde{p}\widetilde{q}}}\exp\left(-\frac{n\epsilon^2}{2pq}\right)$$

Finally, in the square root term in the denominator, we can represent $\widetilde{p}\widetilde{q}$ as $pq(1 \pm n^{-1+\alpha})$. Then, combining all the approximation terms, we get (1.83). □

Proof of Theorem 1.7.2 (binomial case) The CLT (for the binary case) in the form of Theorem 1.7.2 simply converts the PMF of (1.83) into a CDF. To do this carefully, we first choose arbitrary real numbers $z' < z$ and show that

$$\lim_{n\to\infty} \Pr\left\{z' \le \frac{S_n - n\overline{X}}{\sigma\sqrt{n}} \le z\right\} = \int_{z'}^{z} \frac{1}{\sqrt{2\pi}}\exp\frac{-y^2}{2}\,dy. \tag{1.89}$$

Choose $\alpha \in (\frac{1}{2}, \frac{2}{3})$ and for that α, choose n_o and C to satisfy (1.83). Then choose m to satisfy $m \ge n_o$ and $-m^\alpha \le z'\sqrt{mpq} < z\sqrt{mpq} \le m^\alpha$. Since $\alpha > 1/2$, this guarantees that

$$-n^\alpha \le z'\sqrt{npq} < z\sqrt{npq} \le n^\alpha \qquad \text{for } n \ge m. \tag{1.90}$$

For $n \geq m$, we then have

$$\Pr\left\{z' \leq \frac{S_n - np}{\sqrt{npq}} \leq z\right\} = \sum_{k=\lceil np+z'\sqrt{npq}\,\rceil}^{\lfloor np+z\sqrt{npq}\rfloor} \mathsf{p}_{S_n}(k) \tag{1.91}$$

$$= \left(1 \pm Cn^{3\alpha-2}\right)\left(\sum_{k=\lceil np+z'\sqrt{npq}\,\rceil}^{\lfloor np+z\sqrt{npq}\rfloor} \frac{1}{\sqrt{2\pi npq}} \exp\left(\frac{-(k-np)^2}{2npq}\right)\right).$$

As seen from (1.90), each term in (1.91) satisfies $|k - np| \leq n^{\alpha}$, which justifies the bounds in the following sum. That following sum can be viewed as a Riemann sum for the integral in (1.80). Thus the sum approaches the integral as $n^{-1/2}$. Taking the limit as $n \to \infty$ in (1.91), the term $Cn^{3\alpha-2}$ approaches 0, justifying (1.89). The theorem follows by taking the limit $z' \to -\infty$. □

Since the CLT provides such explicit information about the convergence of S_n/n to $\overline{X}$, it is reasonable to ask why the WLLN is so important. The first reason is that the WLLN is so simple that it can be used to give clear insights into situations where the CLT could confuse the issue. A second reason is that the CLT requires a variance, where, as we see next, the WLLN does not. A third reason is that the WLLN can be extended to many situations in which the variables are not independent and/or not identically distributed.[30] A final reason is that the WLLN provides an upper bound on the tails of $\mathsf{F}_{S_n/n}$, whereas the CLT provides only an approximation.

1.7.4 Weak law with an infinite variance

We now establish the WLLN without assuming a finite variance.

Theorem 1.7.4 (WLLN) *For each integer $n \geq 1$, let $S_n = X_1 + \cdots + X_n$, where $X_1, X_2, \ldots$ are IID rv s satisfying $\mathsf{E}[|X|] < \infty$. Then for any $\epsilon > 0$,*

$$\lim_{n\to\infty} \Pr\left\{\left|\frac{S_n}{n} - \mathsf{E}[X]\right| > \epsilon\right\} = 0. \tag{1.92}$$

Proof[31] We use a truncation argument; such arguments are used frequently in dealing with rv s that have infinite variance. The underlying idea in these arguments is important, but some less important details are treated in Exercise 1.41. Let b be a positive number (which we later take to be increasing with n), and for each variable X_i, define a new rv $\breve{X}_i$ (see Figure 1.13) by

$$\breve{X}_i = \begin{cases} X_i & \text{for } \mathsf{E}[X] - b \leq X_i \leq \mathsf{E}[X] + b, \\ \mathsf{E}[X] + b & \text{for } X_i > \mathsf{E}[X] + b, \\ \mathsf{E}[X] - b & \text{for } X_i < \mathsf{E}[X] - b. \end{cases} \tag{1.93}$$

[30] Central limit theorems also hold in many of these more general situations, but they do not hold as widely as the WLLN.

[31] The details of this proof can be omitted without loss of continuity. However, truncation arguments are important in many places and should be understood at some point.

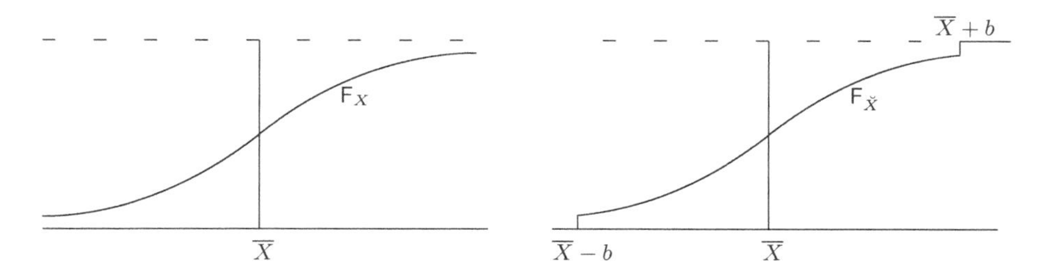

Figure 1.13 The truncated rv $\breve{X}$ for a given rv X has a CDF which is truncated at $\overline{X} \pm b$.

The truncated variables $\breve{X}_i$ are IID and, because of the truncation, must have a finite second moment. Thus the WLLN applies to the sample average $\breve{S}_n = \breve{X}_1 + \cdots + \breve{X}_n$. More particularly, using the Chebyshev inequality in the form of (1.74) on $\breve{S}_n/n$, we get

$$\Pr\left\{\left|\frac{\breve{S}_n}{n} - \mathsf{E}\big[\breve{X}\big]\right| > \frac{\epsilon}{2}\right\} \leq \frac{4\sigma^2_{\breve{X}}}{n\epsilon^2} \leq \frac{8b\mathsf{E}\left[|X|\right]}{n\epsilon^2},$$

where Exercise 1.41 demonstrates the final inequality. Exercise 1.41 also shows that $\mathsf{E}\big[\breve{X}\big]$ approaches $\mathsf{E}[X]$ as $b \to \infty$ and thus that

$$\Pr\left\{\left|\frac{\breve{S}_n}{n} - \mathsf{E}[X]\right| > \epsilon\right\} \leq \frac{8b\mathsf{E}\left[|X|\right]}{n\epsilon^2}, \tag{1.94}$$

for all sufficiently large b. This bound also applies to S_n/n in the case where $S_n = \breve{S}_n$, so we have the following bound (see Exercise 1.41 for further details):

$$\Pr\left\{\left|\frac{S_n}{n} - \mathsf{E}[X]\right| > \epsilon\right\} \leq \Pr\left\{\left|\frac{\breve{S}_n}{n} - \mathsf{E}[X]\right| > \epsilon\right\} + \Pr\left\{S_n \neq \breve{S}_n\right\}. \tag{1.95}$$

The original sum S_n is the same as $\breve{S}_n$ unless one of the X_i has an outage, i.e., $|X_i - \overline{X}| > b$. Thus, using the union bound, $\Pr\{S_n \neq \breve{S}_n\} \leq n\Pr\left\{|X_i - \overline{X}| > b\right\}$. Substituting this and (1.94) into (1.95),

$$\Pr\left\{\left|\frac{S_n}{n} - \mathsf{E}[X]\right| > \epsilon\right\} \leq \frac{8b\mathsf{E}\left[|X|\right]}{n\epsilon^2} + \frac{n}{b}\left[b\Pr\{\,|X - \mathsf{E}[X]\,| > b\}\right]. \tag{1.96}$$

We now show that for any $\epsilon > 0$ and $\delta > 0$, $\Pr\left\{|S_n/n - \overline{X}| \geq \epsilon\right\} \leq \delta$ for all sufficiently large n. We do this, for given ϵ, δ, by choosing $b(n)$ for each n so that the first term in (1.96) is equal to $\delta/2$. Thus $b(n) = n\delta\epsilon^2/16\mathsf{E}\left[|X|\right]$. This means that $n/b(n)$ in the second term is independent of n. Now from (1.57), $\lim_{b\to\infty} b\Pr\left\{|X - \overline{X}| > b\right\} = 0$, so by choosing $b(n)$ sufficiently large (and thus n sufficiently large), the second term in (1.96) is also at most $\delta/2$. □

1.7.5 Convergence of random variables

This section has developed a number of results about how the sequence of sample averages, $\{S_n/n;\ n \geq 1\}$, for a sequence of IID rv s $\{X_i;\ i \geq 1\}$, approaches the mean $\overline{X}$.

In the case of the CLT, the limiting distribution around the mean is also specified to be Gaussian. At the outermost intuitive level, i.e., at the level most useful when first looking at a very complicated set of issues, viewing the limit of the sample averages as being essentially equal to the mean is highly appropriate.

At the next intuitive level down, the meaning of the word *essentially* becomes important and thus involves the details of the above laws. All of the results involve how the rv s S_n/n change with n and become better and better approximated by $\overline{X}$. When we talk about a sequence of rv s (namely a sequence of functions on the sample space) being approximated by a rv or numerical constant, we are talking about some kind of *convergence*, but it clearly is not as simple as a sequence of real numbers (such as $1/n$ for example) converging to some given number (0 for example).

The purpose of this subsection is to give names and definitions to these various forms of convergence. This will give us increased understanding of the laws of large numbers already developed, but, equally important, it will allow us to develop another law of large numbers called the *strong law of large numbers* (SLLN). Finally, it will put us in a position to use these convergence results later for sequences of rv s other than the sample averages of IID rv s.

We discuss four types of convergence in what follows, convergence in distribution, in probability, in mean square, and with probability 1. For the first three, we first recall the type of large-number result with that type of convergence and then give the general definition.

For convergence with probability 1 (WP1), we will define this type of convergence and then provide some understanding of what it means. This will be used in Chapter 5 to state and prove the SLLN.

We start with the CLT, which, from (1.80) says

$$\lim_{n\to\infty} \Pr\left\{\frac{S_n - n\overline{X}}{\sqrt{n}\,\sigma} \le z\right\} = \int_{-\infty}^{z} \frac{1}{\sqrt{2\pi}} \exp\left(\frac{-x^2}{2}\right) dx \qquad \text{for every } z \in \mathbb{R}.$$

This is illustrated in Figure 1.11 and says that the sequence (in n) of CDF's $\Pr\left\{[S_n - n\overline{X}]/\sqrt{n\sigma^2} \le z\right\}$ converges at every z to the normal CDF at z. This is an example of *convergence in distribution*.

Definition 1.7.5 *A sequence of rv s, $Z_1, Z_2, \dots$,* ***converges in distribution*** *to a rv Z if $\lim_{n\to\infty} \mathsf{F}_{Z_n}(z) = \mathsf{F}_Z(z)$ at each z for which $\mathsf{F}_Z(z)$ is continuous.*

For the CLT example, the rv s that converge in distribution are $\{[S_n - n\overline{X}]/\sqrt{n\sigma^2};\ n \ge 1\}$, and they converge in distribution to the normal Gaussian rv.

Convergence in distribution does not say that the rv s themselves converge in any reasonable sense, but only that their CDF's converge. For example, let $Y_1, Y_2, \dots$ be IID rv s with the CDF F_Y. For each $n \ge 1$, if we let $Z_n = Y_n + 1/n$, then it is easy to see that $\{Z_n;\ n \ge 1\}$ converges in distribution to Y. However (assuming Y has variance σ_Y^2 and is independent of each Z_n), we see that $Z_n - Y$ has variance $2\sigma_Y^2$. Thus Z_n does not get close to Y as $n \to \infty$ in any reasonable sense, and $Z_n - Z_m$ does not get small as n

and m both get large.[32] As an even more trivial example, the IID sequence $\{Y_n;\ n \geq 1\}$ converges in distribution to Y.

For the CLT, it is the rv s $[S_n - n\overline{X}]/\sqrt{n\sigma^2}$ that converge in distribution to the normal. As shown in Exercise 1.44, however, the rv $[S_n - n\overline{X}]/\sqrt{n\sigma^2} - [S_{2n} - 2n\overline{X}]/\sqrt{2n\sigma^2}$ is not close to 0 in any reasonable sense, even though the two terms have CDFs that are very close for large n.

For the next type of convergence of rv s, the WLLN, in the form of (1.92), says that

$$\lim_{n\to\infty} \Pr\left\{\left|\frac{S_n}{n} - \overline{X}\right| > \epsilon\right\} = 0 \qquad \text{for every } \epsilon > 0.$$

This is an example of *convergence in probability,* as defined below.

Definition 1.7.6 *A sequence of rv s $Z_1, Z_2, \ldots$* ***converges in probability*** *to a rv Z if $\lim_{n\to\infty} \Pr\{|Z_n - Z| > \epsilon\} = 0$ for every $\epsilon > 0$.*

For the WLLN example, Z_n in the definition is the sample average S_n/n and Z is the constant rv $\overline{X}$. It is probably simpler and more intuitive in thinking about convergence of rv s to think of the sequence of rv s $\{Y_n = Z_n - Z;\ n \geq 1\}$ as converging to 0 in some sense.[33] As illustrated in Figure 1.9, convergence in probability means that $\{Y_n;\ n \geq 1\}$ converges in distribution to a unit step function at 0.

An equivalent statement, as illustrated in Figure 1.10, is that $\{Y_n;\ n \geq 1\}$ converges in probability to 0 if $\lim_{n\to\infty} \mathsf{F}_{Y_n}(y) = 0$ for all $y < 0$ and $\lim_{n\to\infty} \mathsf{F}_{Y_n}(y) = 1$ for all $y > 0$. This shows that convergence in probability to a constant is a special case of convergence in distribution, since with convergence in probability, the sequence F_{Y_n} of CDF's converges to a unit step at 0. Note that $\lim_{n\to\infty} \mathsf{F}_{Y_n}(y)$ is not specified at $y = 0$. However, the step function is not continuous at 0, so the limit there need not be specified for convergence in distribution.

Convergence in probability says quite a bit more than convergence in distribution. As an important example of this, consider the difference $Y_n - Y_m$ for n and m both large. If $\{Y_n;\ n \geq 1\}$ converges in probability to 0, then Y_n and Y_m are both close to 0 with high probability for large n and m, and thus close to each other. More precisely, $\lim_{m\to\infty, n\to\infty} \Pr\{|Y_n - Y_m| > \epsilon\} = 0$ for every $\epsilon > 0$. If the sequence $\{Y_n;\ n \geq 1\}$ merely converges in distribution to some arbitrary distribution, then, as we saw, $Y_n - Y_m$ can be large with high probability, even when n and m are large. Another example of this is given in Exercise 1.44.

It appears paradoxical that the CLT is more explicit about the convergence of S_n/n to $\overline{X}$ than the WLLN, but it corresponds to a weaker type of convergence. The resolution of this paradox is that the sequence of rv s in the CLT is $\{[S_n - n\overline{X}]/\sqrt{n\sigma^2};\ n \geq 1\}$. The presence of $\sqrt{n}$ in the denominator of this sequence provides much more detailed

[32] In fact, saying that a sequence of rv s converges in distribution is unfortunate but standard terminology. It would be just as concise, and far less confusing, to say that a sequence of CDFs converges rather than saying that a sequence of rv s converges in distribution.

[33] Definition 1.7.6 gives the impression that convergence to a rv Z is more general than convergence to a constant or convergence to 0, but converting the rv s to $Y_n = Z_n - Z$ makes it clear that this added generality is quite superficial.

information about how S_n/n approaches $\overline{X}$ with increasing n than the limiting unit step of $\mathsf{F}_{S_n/n}$ itself. For example, it is easy to see from the CLT that $\lim_{n\to\infty} \mathsf{F}_{S_n/n}(\overline{X}) = 1/2$, which cannot be derived directly from the weak law.

Yet another kind of convergence is *convergence in mean square*. An example of this, for the sample average S_n/n of IID rv s with a variance, is given in (1.73), repeated below:

$$\lim_{n\to\infty} \mathsf{E}\left[\left(\frac{S_n}{n} - \overline{X}\right)^2\right] = 0.$$

The general definition is as follows.

Definition 1.7.7 *A sequence of rv s $Z_1, Z_2, \ldots$* ***converges in mean square*** *to a rv Z if* $\lim_{n\to\infty} \mathsf{E}\left[(Z_n - Z)^2\right] = 0$.

Our derivation of the WLLN (Theorem 1.7.1) was essentially based on the mean-square convergence of (1.73). Using the same approach, Exercise 1.43 shows in general that convergence in mean square implies convergence in probability. Convergence in probability does not imply mean-square convergence, since as shown in Theorem 1.7.4, the WLLN holds without the need for a variance.

Figure 1.14 illustrates the relationship between these forms of convergence, i.e., mean-square convergence implies convergence in probability, which in turn implies convergence in distribution. The figure also shows convergence with probability 1 (WP1), which is the next form of convergence to be discussed.

1.7.6 Convergence with probability 1

Convergence with probability 1, abbreviated as convergence WP1, is often referred to as convergence a.s. (almost surely) and convergence a.e. (almost everywhere). The SLLN, which is discussed briefly in this section and further discussed and proven in various forms in Chapters 5 and 9, provides an extremely important example of convergence WP1. The general definition is as follows.

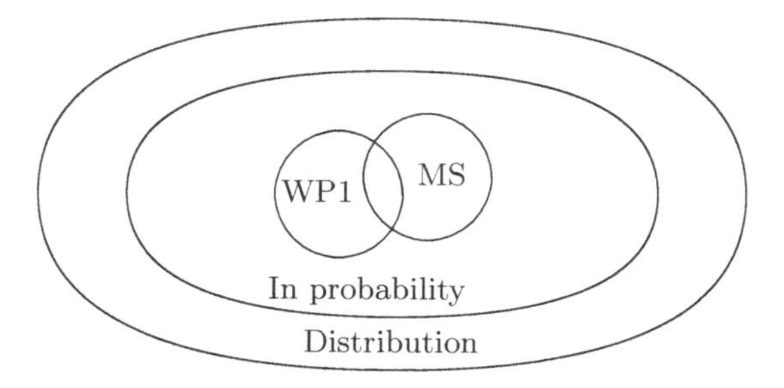

Figure 1.14 Relationship between different kinds of convergence: Convergence in distribution is the most general and is implied by all the others. Convergence in probability is the next most general and is implied both by convergence with probability 1 (WP1) and by mean-square (MS) convergence, neither of which implies the other.

Definition 1.7.8 *Let $Z_1, Z_2, \dots$ be a sequence of rv s in a sample space Ω and let Z be another rv in Ω. Then $\{Z_n;\ n \geq 1\}$* ***converges*** *to Z* ***with probability*** *1 (WP1) if*

$$\Pr\left\{\omega \in \Omega : \lim_{n\to\infty} Z_n(\omega) = Z(\omega)\right\} = 1. \tag{1.97}$$

The condition $\Pr\{\omega \in \Omega : \lim_{n\to\infty} Z_n(\omega) = Z(\omega)\} = 1$ is often stated more compactly as $\Pr\{\lim_n Z_n = Z\} = 1$, and even more compactly as $\lim_n Z_n = Z$ WP1, but the form here is the simplest for initial understanding. As discussed in Chapter 5, the SLLN says that if $X_1, X_2, \dots$ are IID with $\mathsf{E}[|X|] < \infty$, then the sequence of sample averages, $\{S_n/n; n \geq 1\}$ converges WP1 to $\overline{X}$.

In trying to understand (1.97), note that each sample point ω of the underlying sample space Ω maps to a sample value $Z_n(\omega)$ of each rv Z_n, and thus maps to a sample path $\{Z_n(\omega);\ n \geq 1\}$. For any given ω, such a sample path is simply a sequence of real numbers. That sequence of real numbers might converge to $Z(\omega)$ (which is a real number for the given ω), it might converge to something else, or it might not converge at all. Thus a set of ω exists for which the corresponding sample path $\{Z_n(\omega);\ n \geq 1\}$ converges to $Z(\omega)$, and a second set for which the sample path converges to something else or does not converge at all. Convergence WP1 of the sequence of rv s is thus defined to occur when the first set of sample paths above is an event that has probability 1.

For each ω, the sequence $\{Z_n(\omega);\ n \geq 1\}$ is simply a sequence of real numbers, so we briefly review what the limit of such a sequence is. A sequence of real numbers $b_1, b_2, \dots$ is said to have a limit b if, for every $\epsilon > 0$, there is an integer m_ϵ such that $|b_n - b| \leq \epsilon$ for all $n \geq m_\epsilon$. An equivalent statement is that $b_1, b_2, \dots$ has a limit b if, for every integer $k \geq 1$, there is an integer $m(k)$ such that $|b_n - b| \leq 1/k$ for all $n \geq m(k)$.

Figure 1.15 illustrates this definition for those, like the author, whose eyes blur on the second or third 'there exists', 'such that', etc. in a statement. As illustrated, an important

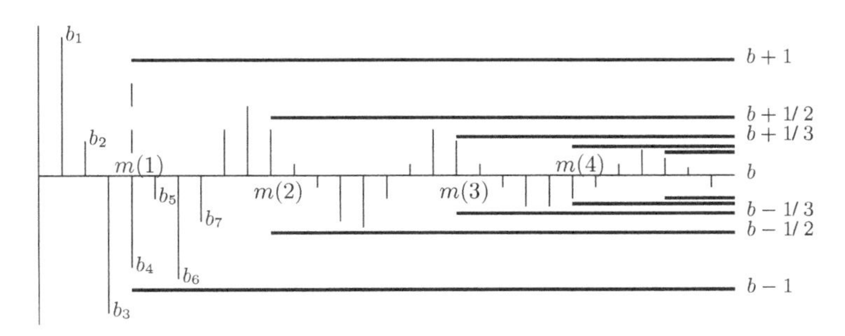

Figure 1.15 Illustration of a sequence of real numbers $b_1, b_2, \dots$ that converge to a number b. The figure illustrates an integer $m(1)$ such that for all $n \geq m(1)$, b_n lies in the interval $b \pm 1$. Similarly, for each $k \geq 1$, there is an integer $m(k)$ such that b_n lies in $b \pm 1/k$ for all $n \geq m(k)$. Thus $\lim_{n\to\infty} b_n = b$ means that for a sequence of ever tighter constraints, the kth constraint can be met for all sufficiently large n, (i.e., all $n \geq m(k)$). Intuitively, convergence means that the elements $b_1, b_2, \dots$ get close to b *and stay close*. The sequence of positive integers $m(1), m(2), \dots$ is non-decreasing, but otherwise arbitrary, depending only on the sequence $\{b_n;\ n \geq 1\}$. For sequences that converge very slowly, the integers $m(1), m(2), \dots$ are simply correspondingly larger.

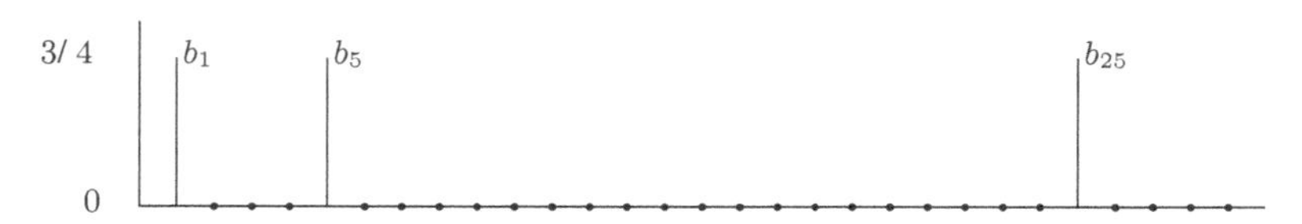

Figure 1.16 Illustration of a non-convergent sequence of real numbers $b_1, b_2, \ldots$. The sequence is defined by $b_n = 3/4$ for $n = 1, 5, 25, \ldots, 5^j, \ldots$ for all integer $j \geq 0$. For all other n, $b_n = 0$. The terms for which $b_n \neq 0$ become increasingly rare as $n \to \infty$. Note that $b_n \in [-1, 1]$ for all n, but there is no $m(2)$ such that $b_n \in [-\frac{1}{2}, \frac{1}{2}]$ for all $n \geq m(2)$. Thus the sequence does not converge.

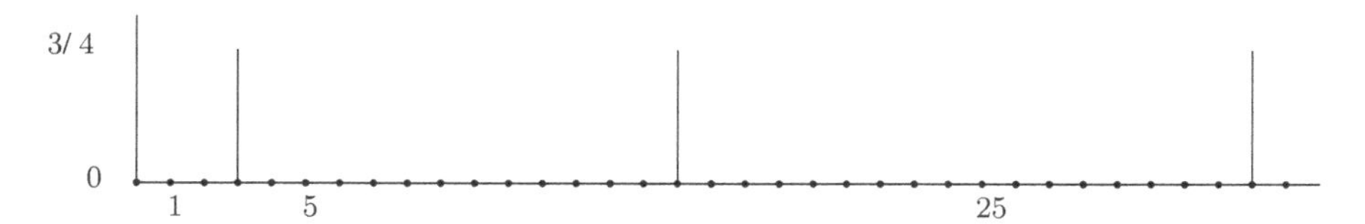

Figure 1.17 Illustration of a sample path of a sequence of rv s $\{Y_n; n \geq 0\}$ where, for each $j \geq 0$, $Y_n = 1$ for an equiprobable choice of $n \in [5^j, 5^{j+1})$ and $Y_n = 0$ otherwise.

aspect of convergence of a sequence $\{b_n;\ n \geq 1\}$ of real numbers is that b_n becomes close to b for large n and stays close for all sufficiently large values of n.

Figure 1.16 gives an example of a sequence of real numbers that does not converge. Intuitively, this sequence is close to 0 (and in fact identically equal to 0) for most large n, but it does not stay close, because of ever more rare outages.

The following example illustrates how a sequence of rv s can converge in probability but not converge WP1. The example also provides some clues as to why convergence WP1 is important.

Example 1.7.9 Consider a sequence $\{Y_n;\ n \geq 1\}$ of rv s for which the sample paths constitute the following slight variation of the sequence of real numbers in Figure 1.16. In particular, as illustrated in Figure 1.17, the non-zero term at $n = 5^j$ in Figure 1.16 is replaced by a non-zero term at a randomly chosen n in the interval[34] $[5^j,\ 5^{j+1})$.

Since each sample path contains a single one in each segment $[5^j,\ 5^{j+1})$, and contains zeroes elsewhere, none of the sample paths converges. In other words, $\Pr\{\omega : \lim Y_n(\omega) = 0\} = 0$. On the other hand, $\Pr\{Y_n = 0\} = 1 - 5^{-j}$ for $5^j \leq n < 5^{j+1}$, so $\lim_{n\to\infty} \Pr\{Y_n = 0\} = 1$.

Thus this sequence of rv s converges to 0 in probability, but does not converge to 0 WP1. This sequence also converges in mean square and (since it converges in probability) in distribution. Thus we have shown (by example) that convergence WP1 is not implied by any of the other types of convergence we have discussed. We will show in Section 5.2 that convergence WP1 does imply convergence in probability and in distribution but not in mean square (as illustrated in Figure 1.14).

[34] There is no special significance to the number 5 here other than making the figure easy to visualize. We could replace 5 by 2 or 3 etc.

The interesting point in this example is that this sequence of rv s is not bizarre (although it is somewhat specialized to make the analysis simple). Another important point is that this definition of convergence has a long history of being accepted as the 'useful,' 'natural,' and 'correct' way to define convergence for a sequence of real numbers. Thus it is not surprising that convergence WP1 will turn out to be similarly useful for sequences of rv s.

There is a price to be paid in using the concept of convergence WP1. We must look at the entire sequence of rv s and can no longer analyze finite n-tuples and then go to the limit as $n \to \infty$. This requires a significant additional layer of abstraction, which involves additional mathematical precision and initial loss of intuition. For this reason we put off further discussion of convergence WP1 and the SLLN until Chapter 5 where it is needed.

1.8 Relation of probability models to the real world

Whenever experienced and competent engineers or scientists construct a probability model to represent aspects of some system that either exists or is being designed for some application, they must acquire a deep knowledge of the system and its surrounding circumstances, and concurrently consider various types of probability models used in probabilistic analyses of the same or similar systems. Usually very simple probability models help in understanding the real-world system, and knowledge about the real-world system helps in understanding what aspects of the system are well modeled by a given probability model. For a text such as this, there is insufficient space to understand the real-world aspects of each system that might be of interest. We must use the language of various canonical real-world systems for motivation and insight when studying probability models for various classes of systems, but such models must necessarily be chosen more for their tutorial than practical value.

There is a danger, then, that readers will come away with the impression that analysis is more challenging and important than modeling. To the contrary, for work on real-world systems, modeling is almost always more difficult, more challenging, and more important than analysis. The objective here is to provide the necessary knowledge and insight about probabilistic models so that the reader can later combine this with a deep understanding of particular real application areas. This will result in a useful interactive use of models, analysis, and experimentation.

In this section, our purpose is not to learn how to model real-world problems, since, as said above, this requires deep and specialized knowledge of whatever application area is of interest. Rather it is to understand the following conceptual problem that was posed in Section 1.1. Suppose we have a probability model of some real-world experiment involving randomness in the sense expressed there. When the real-world experiment being modeled is performed, there is an outcome, which presumably is one of the outcomes of the probability model, but there is no observable probability.

It appears to be intuitively natural, for experiments that can be carried out repeatedly under essentially the same conditions, to associate the probability of a given event with the relative frequency of that event over many repetitions. We now have the background to understand this approach. We first look at relative frequencies within the probability model, and then within the real world.

1.8.1 Relative frequencies in a probability model

We have seen that, for any probability model, an extended probability model exists for n IID idealized experiments of the original model. For any event A in the original model, the indicator function $\mathbb{I}_A$ is a random variable, and the relative frequency of A over n IID experiments is the sample average of n IID rv s each with the distribution of $\mathbb{I}_A$. From the WLLN, this relative frequency converges in probability to $\mathsf{E}[\mathbb{I}_A] = \Pr\{A\}$. By taking the limit $n \to \infty$, the SLLN says that the relative frequency of A converges with probability 1 to $\Pr\{A\}$.

In plain English, this says that for large n, the relative frequency of an event (in the n-repetition IID model) is essentially the same as the probability of that event. The word *essentially* is carrying a great deal of hidden baggage. For the WLLN, for any $\epsilon, \delta > 0$, the relative frequency is within some ϵ of $\Pr\{A\}$ with a confidence level $1 - \delta$ whenever n is sufficiently large. For the SLLN, the ϵ and δ are avoided, but only by looking directly at the limit $n \to \infty$. Despite the hidden baggage, though, relative frequency and probability are related as indicated.

1.8.2 Relative frequencies in the real world

In trying to sort out if and when the laws of large numbers have much to do with real-world experiments, we should ignore the mathematical details for the moment and agree that for large n, the relative frequency of an event A over n IID trials of an idealized experiment is essentially $\Pr\{A\}$. We can certainly visualize a real-world experiment that has the same set of possible outcomes as the idealized experiment and we can visualize evaluating the relative frequency of A over n repetitions with large n. If that real-world relative frequency is essentially equal to $\Pr\{A\}$, and this is true for the various events A of greatest interest, then it is reasonable to hypothesize that the idealized experiment is a reasonable model for the real-world experiment, at least so far as those given events of interest are concerned.

One problem with this comparison of relative frequencies is that we have carefully specified a model for n IID repetitions of the idealized experiment, but have said nothing about how the real-world experiments are repeated. The IID idealized experiments specify that the conditional probability of A at one trial is the same no matter what the results of the other trials are. Intuitively, we would then try to isolate the n real-world trials so they do not affect each other, but this is a little vague. The following examples help explain this problem and several others in comparing idealized and real-world relative frequencies.

Example 1.8.1 (Coin tossing) Tossing coins is widely used as a way to choose the first player in various games, and is also sometimes used as a primitive form of gambling. Its importance, however, and the reason for its frequent use, is its simplicity. When tossing a coin, we would argue from the symmetry between the two sides of the coin that each should be equally probable (since any procedure for evaluating the probability of one side should apply equally to the other). Thus since H and T are the only outcomes (the remote possibility of the coin balancing on its edge is omitted from the model), the reasonable and universally accepted model for coin tossing is that H and T each have probability 1/2.

On the other hand, the two sides of a coin are embossed in different ways, so that the mass is not uniformly distributed. Also the two sides do not behave in quite the same way when bouncing off a surface. Each denomination of each currency behaves slightly differently in this respect. Thus, not only do coins violate symmetry in small ways, but different coins violate it in different ways.

How do we test whether this effect is significant? If we assume for the moment that successive tosses of the coin are well modeled by the idealized experiment of n IID trials, we can essentially find the probability of H for a particular coin as the relative frequency of H in a sufficiently large number of independent tosses of that coin. This gives us slightly different relative frequencies for different coins, and thus slightly different probability models for different coins.

The assumption of independent tosses is also questionable. Consider building a carefully engineered machine for tossing coins and using it in a vibration-free environment. A standard coin is inserted into the machine in the same way for each toss and we count the number of heads and tails. Since the machine has essentially eliminated the randomness, we would expect all the coins, or almost all the coins, to come up the same way – the more precise the machine, the less independent the results. By inserting the original coin in a random way, a single trial might have equiprobable results, but successive tosses are certainly not independent. The successive trials would be closer to independent if the tosses were done by a slightly inebriated individual who tossed the coins high in the air.

The point of this example is that there are many different coins and many ways of tossing them, and the idea that one model fits all is reasonable under some conditions and not under others. Rather than retreating into the comfortable world of theory, however, note that we can now find the relative frequency of heads for any given coin and essentially for any given way of tossing that coin.[35]

---

[35] We are not suggesting that distinguishing different coins for the sake of coin tossing is an important problem. Rather, we are illustrating that even in such a simple situation, the assumption of identically prepared experiments is questionable and the assumption of independent experiments is questionable. The extension to n repetitions of IID experiments is not necessarily a good model for coin tossing. In other words, one has to question not only the original model but also the n-repetition model.

Example 1.8.2 (Binary data) Consider the stream of bits transmitted over a communication link or stored in a data facility. The bit stream is often a mixture of encoded voice, video, graphics, text, etc., with relatively long runs of each, interspersed with various protocols for retrieving the original non-binary data.

The simplest (and most common) model for this is to assume that each binary digit is 0 or 1 with equal probability and that successive digits are statistically independent. This is the same as the model for coin tossing after the trivial modification of converting $\{H, T\}$ into $\{0, 1\}$. This is also a rather appropriate model for designing a communication or storage facility, since all n-tuples are then equiprobable (in the model) for each n, and thus the facilities need not rely on any special characteristics of the data. On the other hand, if one wants to compress the data, reducing the required number of transmitted or stored bits per incoming bit, then a more elaborate model is needed.

Developing such an improved model would require finding out more about the data source – a naive application of calculating relative frequencies of n-tuples would probably not be the best choice. On the other hand, there are well-known data compression schemes that in essence track dependencies in the data and use them for compression in a coordinated way. These schemes are called *universal data-compression* schemes since they do not rely on a probability model. At the same time, they are best analyzed by looking at how they perform for various idealized probability models.

The point of this example is that choosing probability models often depends heavily on how the model is to be used. Models more complex than IID binary digits are usually based on what is known about the input processes. Measuring relative frequencies and associating them with probabilities is the basic underlying conceptual connection between real-world and models, but in practice this is essentially the relationship of last resort. For most of the applications we will consider, there is a long history of modeling to build on, with experiments as needed.

Example 1.8.3 (Fable) In the year 2008, the financial structure of the USA failed and the world economy was brought to its knees. Much has been written about the role of greed on Wall Street and incompetence in Washington. Another aspect of the collapse, however, was a widespread faith in stochastic models for limiting risk. These models encouraged people to engage in investments that turned out to be far riskier than the models predicted. These models were created by some of the brightest PhDs from the best universities, but they failed miserably because they modeled everyday events very well, but modeled the rare events and the interconnection of events poorly. They failed badly by not understanding their application, and in particular, by trying to extrapolate typical behavior when their primary goal was to protect against highly atypical situations. The moral of the fable is that brilliant analysis is not helpful when the modeling is poor; as computer engineers say, ‘garbage in, garbage out.’

The examples above show that the problems of modeling a real-world experiment are often connected with the question of creating a model for a set of experiments that are not exactly the same and do not necessarily correspond to the notion of independent repetitions within the model. In other words, the question is not only whether the probability model is reasonable for a single experiment, but also whether the IID repetition model is appropriate for multiple copies of the real-world experiment.

At least we have seen, however, that if a real-world experiment can be performed many times with a physical isolation between performances that is well modeled by the IID repetition model, then the relative frequencies of events in the real-world experiment correspond to relative frequencies in the idealized IID repetition model, which correspond to probabilities in the original model. In other words, under appropriate circumstances, the probabilities in a model become essentially observable over many repetitions.

We will see later that our emphasis on IID repetitions was done for simplicity. There are other models for repetitions of a basic model, such as Markov models, that we study later. These will also lead to relative frequencies approaching probabilities within the repetition model. Thus, for repeated real-world experiments that are well modeled by these repetition models, the real-world relative frequencies approximate the probabilities in the model.

1.8.3 Statistical independence of real-world experiments

We have been discussing the use of relative frequencies of an event A in a repeated real-world experiment to test $\Pr\{A\}$ in a probability model of that experiment. This can be done essentially successfully if the repeated trials correpond to IID trials in the idealized experiment. However, the statement about IID trials in the idealized experiment is a statement about probabilities in the extended n-trial model. Thus, just as we tested $\Pr\{A\}$ by repeated real-world trials of a single experiment, we should be able to test $\Pr\{A_1, \dots, A_n\}$ in the n-repetition model by a much larger number of real-world repetitions of n-tuples rather than single trials.

To be more specific, choose two large integers, m and n, and perform the underlying real-world experiment mn times. Partition the mn trials into m runs of n trials each. For any given n-tuple $A_1, \dots, A_n$ of successive events, find the relative frequency (over m trials of n tuples) of the n-tuple event $A_1, \dots, A_n$. This can then be used essentially to test the probability $\Pr\{A_1, \dots, A_n\}$ in the model for n IID trials. The individual event probabilities can also be tested, so the condition for independence can be tested.

The observant reader will note that there is a tacit assumption above that successive n tuples can be modeled as independent, so it seems that we are simply replacing a big problem with a bigger problem. This is not quite true, since if the trials are dependent with some given probability model for dependent trials, then this test for independence will essentially reject the independence hypothesis for large enough n. In other words, we cannot completely verify the correctness of an independence hypothesis for the n-trial model, although in principle we could eventually falsify it if it is false.

Choosing models for real-world experiments is primarily a subject for statistics, and we will not pursue it further except for brief discussions when treating particular application areas. The purpose here has been to treat a fundamental issue in probability theory. As stated before, probabilities are non-observables – they exist in the theory but are not directly measurable in real-world experiments. We have shown that probabilities essentially become observable in the real-world via relative frequencies over repeated trials.

1.8.4 Limitations of relative frequencies

Most real-world applications that are modeled by probability models have such a large sample space that it is impractical to conduct enough trials to choose probabilities from relative frequencies. Even a shuffled deck of 52 cards would require many more than $52! \approx 8 \times 10^{67}$ trials for most of the outcomes to appear even once. Thus relative frequencies can be used to test the probability of given individual events of importance, but are usually impractical for choosing the entire model and even more impractical for choosing a model for repeated trials.

Since relative frequencies give us a concrete interpretation of what probability means, however, we can now rely on other approaches, such as symmetry, for modeling. From symmetry, for example, it is clear that all 52! possible arrangements of a card deck should be equiprobable after shuffling. This leads, for example, to the ability to calculate probabilities of different poker hands, etc., which are such popular exercises in elementary probability classes.

Another valuable modeling procedure is that of constructing a probability model where the possible outcomes are independently chosen n-tuples of outcomes in a simpler model. More generally, most of the random processes to be studied in this text are defined as various ways of combining simpler idealized experiments.

What is really happening as we look at modeling increasingly sophisticated systems and studying increasingly sophisticated models is that we are developing mathematical results for simple idealized models and relating those results to real-world results (such as relating idealized statistically independent trials to real-world independent trials). The association of relative frequencies to probabilities forms the basis for this, but is usually exercised only in the simplest cases.

The way one selects probability models of real-world experiments in practice is to use scientific knowledge and experience, plus simple experiments, to choose a reasonable model. The results from the model (such as the laws of large numbers) are then used both to hypothesize results about the real-world experiment and to provisionally reject the model when further experiments show it to be highly questionable. Although the results about the model are mathematically precise, the corresponding results about the real-world are at best insightful hypotheses whose most important aspects must be validated in practice.

1.8.5 Subjective probability

There are many useful applications of probability theory to situations other than repeated trials of a given experiment. When designing a new system in which randomness (of the type used in probability models) is hypothesized, one would like to analyze the system before actually building it. In such cases, the real-world system does not exist, so indirect means must be used to construct a probability model. Often some sources of randomness, such as noise, can be modeled in the absence of the system. Often similar systems or simulation can be used to help understand the system and help in formulating appropriate probability models. However, the choice of probabilities is to a certain extent subjective.

Another type of situation (such as risk analysis for nuclear reactors) deals with a large number of very unlikely outcomes, each catastrophic in nature. Experimentation clearly cannot be used to establish probabilities, and it is not clear that probabilities have any real meaning here. It can be helpful, however, to choose a probability model on the basis of subjective beliefs which can be used as a basis for reasoning about the problem.

Choosing a probability model on the basis of subjective beliefs can be helpful here if used as a basis for reasoning about the problem. When handled well, this can at least make the subjective biases clear, leading to a more rational approach. When handled poorly (as for example in some risk analyses of large financial systems) it can hide both the real risks and the arbitrary nature of possibly poor decisions.

We will not discuss the various, often ingenious, methods for choosing subjective probabilities. The reason is that subjective beliefs should be based on intensive and long-term exposure to the particular problem involved; discussing these problems in abstract probability terms weakens this link. We will focus instead on the analysis of idealized models. These can be used to provide insights for subjective models, and more refined and precise results for objective models.

1.9 Summary

This chapter started with an introduction to the relation between probability theory and real-world experiments involving randomness. While almost all work in probability theory works with established probability models, it is important to think through what these probabilities mean in the real world, and elementary subjects rarely address these questions seriously.

The next section discussed the axioms of probability theory, along with some insights about why these particular axioms were chosen. This was followed by a review of conditional probabilities, statistical independence, and rv s. Stochastic processes in the guise of the Bernoulli process were then introduced with an emphasis on asymptotic behavior. The review of probability was then resumed with an emphasis on the underlying structure of the field rather than reviewing details and problem solving techniques.

Next the laws of large numbers were developed and discussed at a somewhat deeper level than most elementary courses. This involved a fair amount of abstraction, combined with mathematical analysis. The central idea is that the sample average of n IID rv s approaches the mean with increasing n. As a special case, the relative frequency of an event A approaches $\Pr\{A\}$. What the word *approaches* means here is both tricky and vital in understanding probability theory. The SLLN and convergence WP1 require mathematical maturity, and are postponed to Chapter 5 where they are first used.

The final section came back to the fundamental problem of understanding the relation between probability theory and randomness in the real-world. It was shown, via the laws of large numbers, that probabilities become essentially observable via relative frequencies calculated over repeated experiments.

There are too many texts on elementary probability to mention here, and most of them serve to give added understanding and background to the material in this chapter. We recommend Bertsekas and Tsitsiklis [2], both for a careful statement of the fundamentals and for a wealth of well-chosen and carefully explained examples.

Texts that cover similar material to that here are [22] and [14]. Kolmogorov [18] is readable for the mathematically mature and is also of historical interest as the translation of the 1933 book that first put probability on a firm mathematical basis. Feller [8] is the classic extended and elegant treatment of elementary material from a mature point of view. Rudin [23] is an excellent text on measure theory for those with advanced mathematical preparation.

1.10 Exercises

Exercise 1.1 Let A_1 and A_2 be arbitrary events and show that $\Pr\{A_1 \bigcup A_2\} + \Pr\{A_1A_2\} = \Pr\{A_1\} + \Pr\{A_2\}$. Explain which parts of the sample space are being double counted on both sides of this equation and which parts are being counted once.

Exercise 1.2 This exercise derives the probability of an arbitrary (non-disjoint) union of events, derives the union bound, and derives some useful limit expressions.

(a) For two arbitrary events A_1 and A_2, show that

$$A_1 \bigcup A_2 = A_1 \bigcup (A_2 - A_1), \qquad \text{where } A_2 - A_1 = A_2 A_1^c.$$

Show that A_1 and $A_2 - A_1$ are disjoint. Hint: Venn diagrams were invented to help understand expressions like these.

(b) For an arbitrary sequence of events, $\{A_n;\, n \geq 1\}$, let $B_1 = A_1$ and for each $n \geq 2$ define $B_n = A_n - \bigcup_{m=1}^{n-1} A_m$. Show that $B_1, B_2, \ldots,$ are disjoint events and show that for each $n \geq 2$, $\bigcup_{m=1}^{n} A_m = \bigcup_{m=1}^{n} B_m$. Hint: Use induction.

(c) Show that

$$\Pr\left\{\bigcup_{n=1}^{\infty} A_n\right\} = \Pr\left\{\bigcup_{n=1}^{\infty} B_n\right\} = \sum_{n=1}^{\infty} \Pr\{B_n\}.$$

Hint: Use the axioms of probability for the second equality.

(d) Show that for each n, $\Pr\{B_n\} \leq \Pr\{A_n\}$. Use this to show that

$$\Pr\left\{\bigcup_{n=1}^{\infty} A_n\right\} \leq \sum_{n=1}^{\infty} \Pr\{A_n\}.$$

(e) Show that $\Pr\{\bigcup_{n=1}^{\infty} A_n\} = \lim_{m\to\infty} \Pr\{\bigcup_{n=1}^{m} A_n\}$. Hint: Combine (c) and (b). Note that this says that the probability of a limit of unions is equal to the limit of the probabilities. This might well appear to be obvious without a proof, but you will see situations later where similar appearing interchanges cannot be made.

(f) Show that $\Pr\{\bigcap_{n=1}^{\infty} A_n\} = \lim_{n\to\infty} \Pr\{\bigcap_{i=1}^{n} A_i\}$. Hint: Remember De Morgan's equalities.

Exercise 1.3 Find the probability that a five-card poker hand, chosen randomly from a 52-card deck, contains four aces. That is, if all 52! arrangements of a deck of cards are equally likely, what is the probability that all four aces are in the first five cards of the deck.

Exercise 1.4 Consider a sample space of eight equiprobable sample points and let A_1, A_2, A_3 be three events each of probability 1/2 such that $\Pr\{A_1A_2A_3\} = \Pr\{A_1\}\Pr\{A_2\}\Pr\{A_3\}$.

(a) Create an example where $\Pr\{A_1A_2\} = \Pr\{A_1A_3\} = \frac{1}{4}$ but $\Pr\{A_2A_3\} = \frac{1}{8}$. Hint: Make a table with a row for each sample point and a column for each event and try different ways of assigning sample points to events (the answer is not unique).

(b) Show that, for your example, A_2 and A_3 are not independent. Note that the definition of statistical independence would be very strange if it allowed A_1, A_2, A_3 to be independent while A_2 and A_3 are dependent. This illustrates why the definition of independence requires (1.14) rather than just (1.15).

Exercise 1.5 This exercise shows that for all rv s X, $\mathsf{F}_X(x)$ is continuous from the right.

(a) For any given rv X, any real number x, and each integer $n \geq 1$, let $A_n = \{\omega : X > x + 1/n\}$, and show that $A_1 \subseteq A_2 \subseteq \cdots$. Use this and the corollaries to the axioms of probability to show that $\Pr\{\bigcup_{n\geq 1} A_n\} = \lim_{n\to\infty} \Pr\{A_n\}$.

(b) Show that $\Pr\{\bigcup_{n\geq 1} A_n\} = \Pr\{X > x\}$ and that $\Pr\{X > x\} = \lim_{n\to\infty}\{X > x + 1/n\}$.

(c) Show that for $\epsilon > 0$, $\lim_{\epsilon\to 0} \Pr\{X \leq x + \epsilon\} = \Pr\{X \leq x\}$.

(d) Define $\widetilde{\mathsf{F}}_X(x) = \Pr\{X < x\}$. Show that $\widetilde{\mathsf{F}}_X(x)$ is continuous from the left. In other words, the continuity from the right for the CDF arises from the almost arbitrary (but universally accepted) choice in defining the CDF as $\Pr\{X \leq x\}$ rather than $\Pr\{X < x\}$.

Exercise 1.6 Show that for a continuous non-negative rv X,

$$\int_0^{\infty} \Pr\{X > x\}\, dx = \int_0^{\infty} x\mathsf{f}_X(x)\, dx. \tag{1.98}$$

Hint 1: First rewrite $\Pr\{X > x\}$ on the left-hand side of (1.98) as $\int_x^{\infty} \mathsf{f}_X(y)\, dy$. Then think through, to your level of comfort, how and why the order of integration can be interchanged in the resulting expression.

Hint 2: As an alternative approach, derive (1.98) using integration by parts.

Exercise 1.7 Suppose X and Y are discrete rv s with the PMF $\mathsf{p}_{XY}(x_i, y_j)$. Show (a picture will help) that this is related to the joint CDF by

$$\mathsf{p}_{XY}(x_i, y_j) = \lim_{\delta>0,\delta\to 0} \left[\mathsf{F}(x_i, y_j) - \mathsf{F}(x_i - \delta, y_j) - \mathsf{F}(x_i, y_j - \delta) + \mathsf{F}(x_i - \delta, y_j - \delta)\right].$$

Exercise 1.8 A variation of Example 1.5.1 is to let M be a rv that takes on both positive and negative values with the PMF

$$\mathsf{p}_M(m) = \frac{1}{2|m|\,(|m|+1)}.$$

In other words, M is symmetric around 0 and $|M|$ has the same PMF as the non-negative rv N of Example 1.5.1.

(a) Show that $\sum_{m\geq 0} m\mathsf{p}_M(m) = \infty$ and $\sum_{m<0} m\mathsf{p}_M(m) = -\infty$. (Thus show that the expectation of M not only does not exist but is undefined even as an extended real number.)

(b) Suppose that the terms in $\sum_{m=-\infty}^{\infty} m\mathsf{p}_M(m)$ are summed in the order of two positive terms for each negative term (i.e., in the order $1, 2, -1, 3, 4, -2, 5, \cdots$). Find the limiting value of the partial sums in this series. Hint: You may find it helpful to know that

$$\lim_{n\to\infty}\left[\sum_{i=1}^{n}\frac{1}{i} - \int_1^n \frac{1}{x}\,dx\right] = \gamma,$$

where γ is the Euler–Mascheroni constant, $\gamma = 0.57721\cdots$.

(c) Repeat (b) where, for any given integer $k > 0$, the order of summation is k positive terms for each negative term.

Exercise 1.9 (Proof of Theorem 1.4.1) The bounds on the binomial in this theorem are based on the *Stirling bounds*. These say that for all $n \geq 1$, $n!$ is upper and lower bounded by

$$\sqrt{2\pi n}\left(\frac{n}{e}\right)^n < n! < \sqrt{2\pi n}\left(\frac{n}{e}\right)^n e^{1/12n}. \tag{1.99}$$

The ratio $\sqrt{2\pi n}(n/e)^n/n!$ of the first two terms is monotonically increasing with n toward the limit 1, and the ratio $\sqrt{2\pi n}(n/e)^n \exp(1/12n)/n!$ is monotonically decreasing toward 1. The upper bound is more accurate, but the lower bound is simpler and known as the Stirling approximation. See [8] for proofs and further discussion of the above facts.

(a) Show from (1.99) and from the above monotone property that

$$\binom{n}{k} < \sqrt{\frac{n}{2\pi k(n-k)}}\;\frac{n^n}{k^k(n-k)^{n-k}}.$$

Hint: First show that $n!/k! < \sqrt{n/k}\,n^n k^{-k} e^{-n+k}$ for $k < n$.

(b) Use the result of (a) to upper bound $\mathsf{p}_{S_n}(k)$ by

$$\mathsf{p}_{S_n}(k) < \sqrt{\frac{n}{2\pi k(n-k)}}\;\frac{p^k(1-p)^{n-k}n^n}{k^k(n-k)^{n-k}}.$$

Show that this is equivalent to the upper bound in Theorem 1.4.1.

(c) Show that

$$\binom{n}{k} > \sqrt{\frac{n}{2\pi k(n-k)}}\;\frac{n^n}{k^k(n-k)^{n-k}}\left[1 - \frac{n}{12k(n-k)}\right].$$

(d) Derive the lower bound in Theorem 1.4.1.

(e) Show that $\mathsf{D}(\widetilde{p}\|p) = \widetilde{p}\ln(\widetilde{p}/p) + (1-\widetilde{p})\ln[(1-\widetilde{p})/(1-p)]$ is 0 at $\widetilde{p} = p$ and non-negative elsewhere.

Exercise 1.10 Let X be a ternary rv taking on the three values 0, 1, 2 with probabilities p_0, p_1, p_2 respectively. Find the median of X for each of the cases below.

(a) $p_0 = 0.2,\ p_1 = 0.4,\ p_2 = 0.4$.

(b) $p_0 = 0.2,\ p_1 = 0.2,\ p_2 = 0.6$.

(c) $p_0 = 0.2,\ p_1 = 0.3,\ p_2 = 0.5$.

Note 1: The median is not unique in (c). Find the interval of values that are medians. Note 2: Some people force the median to be distinct by defining it as the midpoint of the interval satisfying the definition given here.

(d) Now suppose that X is non-negative and continuous with the density $\mathsf{f}_X(x) = 1$ for $0 \le x \le 0.5$ and $\mathsf{f}_X(x) = 0$ for $0.5 < x \le 1$. We know that $\mathsf{f}_X(x)$ is positive for all $x > 1$, but it is otherwise unknown. Find the median or interval of medians.

The median is sometimes (incorrectly) defined as that α for which $\Pr\{X > \alpha\} = \Pr\{X < \alpha\}$. Show that it is possible for no such α to exist. Hint: Look at the examples above.

Exercise 1.11 **(a)** For any given rv Y, express $\mathsf{E}[|Y|]$ in terms of $\int_{y<0} \mathsf{F}_Y(y)\,dy$ and $\int_{y\ge 0} \mathsf{F}^{\mathsf{c}}_Y(y)\,dy$. Hint: Review the argument in Figure 1.4.

(b) For some given rv X with $\mathsf{E}[|X|] < \infty$, let $Y = X - \alpha$. Using (a), show that

$$\mathsf{E}[|X-\alpha|] = \int_{-\infty}^{\alpha} \mathsf{F}_X(x)\,dx + \int_{\alpha}^{\infty} \mathsf{F}^{\mathsf{c}}_X(x)\,dx.$$

(c) Show that $\mathsf{E}[|X-\alpha|]$ is minimized over α by choosing α to be a median of X. Hint: Both the easy way and the most instructive way to do this is to use a graphical argument involving shifting Figure 1.4. Be careful to show that when the median is an interval, all points in this interval achieve the minimum.

Exercise 1.12 Let X be a rv with CDF $\mathsf{F}_X(x)$. Find the CDF of the following rv s.

(a) The maximum of n IID rv s, each with CDF $\mathsf{F}_X(x)$.

(b) The minimum of n IID rv s, each with CDF $\mathsf{F}_X(x)$.

(c) The difference of the rv s defined in (a) and (b); assume X has a density $\mathsf{f}_X(x)$. Hint: Let M_+ and M_- be the maximum and minimum respectively of the X_i. First, conditional on $X_1 = x$, find the joint probability that $X_1 = M_+$ and $M_+ - M_- \le r$.

Exercise 1.13 Let X and Y be rv s in some sample space Ω and let $Z = X + Y$, i.e., for each $\omega \in \Omega$, $Z(\omega) = X(\omega) + Y(\omega)$.

(a) Show that the set of ω for which $Z(\omega) = \pm\infty$ has probability 0.

(b) To show that $Z = X + Y$ is a rv, we must show that for each real number α, the set $\{\omega \in \Omega : X(\omega) + Y(\omega) \le \alpha\}$ is an event. We proceed indirectly. For an arbitrary positive integer n and an arbitrary integer $k > 0$, let $B(n,k) = \{\omega : X(\omega) \le k/n\} \bigcap \{Y(\omega) \le \alpha + (1-k)/n\}$. Let $D(n) = \bigcup_k B(n,k)$ and show that $D(n)$ is an event.

(c) On a two-dimensional sketch for a given α, show the values of $X(\omega)$ and $Y(\omega)$ for which $\omega \in D(n)$. Hint: This set of values should be bounded by a staircase function.

(d) Show that

$$\{\omega : X(\omega) + Y(\omega) \le \alpha\} = \bigcap_n D(n). \tag{1.100}$$

Explain why this shows that $Z = X + Y$ is a rv.

(e) Explain why (d) implies that if $Y = X_1 + X_2 + \cdots + X_n$ and if $X_1, X_2, \ldots, X_n$ are rv s, then Y is a rv. Hint: Only one or two lines of explanation are needed.

Exercise 1.14 **(a)** Let $X_1, X_2, \ldots, X_n$ be rv s with expected values $\overline{X}_1, \ldots, \overline{X}_n$. Show that $\mathsf{E}[X_1 + \cdots + X_n] = \overline{X}_1 + \cdots + \overline{X}_n$. You may assume that the rv s have a joint density function, but do not assume that the rv s are independent.

(b) Now assume that $X_1, \ldots, X_n$ are statistically independent and show that the expected value of the product is equal to the product of the expected values.

(c) Again assuming that $X_1, \ldots, X_n$ are statistically independent, show that the variance of the sum is equal to the sum of the variances.

Exercise 1.15 (Stieltjes integration) **(a)** Let $h(x) = u(x)$ and $\mathsf{F}_X(x) = u(x)$, where $u(x)$ is the unit step, i.e., $u(x) = 0$ for $-\infty < x < 0$ and $u(x) = 1$ for $x \ge 0$. Using the definition of the Stieltjes integral in Footnote 19 on p. 23, show that $\int_{-1}^{1} h(x)\, d\mathsf{F}_X(x)$ does not exist. Hint: Look at the term in the Riemann sum including $x = 0$ and look at the range of choices for $h(x)$ in that interval. Intuitively, it might help initially to view $d\mathsf{F}_X(x)$ as a unit impulse at $x = 0$.

(b) Let $h(x) = u(x - a)$ and $\mathsf{F}_X(x) = u(x - b)$, where a and b are in $(-1, +1)$. Show that $\int_{-1}^{1} h(x)\, d\mathsf{F}_X(x)$ exists if and only if $a \ne b$. Show that the integral has the value 1 for $a < b$ and the value 0 for $a > b$. Argue that this result is still valid in the limit of integration over $(-\infty, \infty)$.

(c) Let X and Y be independent discrete rv s, each with a finite set of possible values. Show that $\int_{-\infty}^{\infty} \mathsf{F}_X(z - y) d\mathsf{F}_Y(y)$, defined as a Stieltjes integral, is equal to the distribution of $Z = X + Y$ at each z other than the possible sample values of Z, and is undefined at each sample value of Z. Hint: Express F_X and F_Y as sums of unit steps. Note: This failure of Stieltjes integration is not a serious problem; $\mathsf{F}_Z(z)$ is a step function, and the integral is undefined at its points of discontinuity. We automatically define $\mathsf{F}_Z(z)$ at those step values so that F_Z is a CDF (i.e., is continuous from the right). This problem does not arise if either X or Y is continuous.

Exercise 1.16 Let $X_1, X_2, \ldots, X_n, \ldots$ be a sequence of IID continuous rv s with the common PDF $\mathsf{f}_X(x)$; note that $\Pr\{X{=}\alpha\} = 0$ for all α and that $\Pr\{X_i{=}X_j\} = 0$ for all $i \ne j$. For $n \ge 2$, define X_n as a *record-to-date* of the sequence if $X_n > X_i$ for all $i < n$.

(a) Find the probability that X_2 is a record-to-date. Use symmetry to obtain a numerical answer without computation. A one- or two-line explanation should be adequate.

(b) Find the probability that X_n is a record-to-date, as a function of $n \ge 1$. Again use symmetry.

(c) Find a simple expression for the expected number of records-to-date that occur over the first m trials for any given integer m. Hint: Use indicator functions. Show that this expected number is infinite in the limit $m \to \infty$.

Exercise 1.17 (Continuation of Exercise 1.16) **(a)** Let N_1 be the index of the *first* record-to-date in the sequence. Find $\Pr\{N_1 > n\}$ for each $n \geq 2$. Hint: There is a far simpler way to do this than working from Exercise 1.16(b).

(b) Show that N_1 is a rv.

(c) Show that $\mathsf{E}[N_1] = \infty$.

(d) Let N_2 be the index of the *second* record-to-date in the sequence. Show that N_2 is a rv. Hint: You need not find the CDF of N_2 here.

(e) Contrast your result in (c) to the result from Exercise 1.16(c) saying that the expected number of records-to-date is infinite over an infinite number of trials. Note: This should be a shock to your intuition – there is an infinite expected wait for the first of an infinite sequence of occurrences, each of which must eventually occur.

Exercise 1.18 (Another direction from Exercise 1.16) **(a)** For any given $n \geq 2$, find the probability that X_n and X_{n+1} are both records-to-date. Hint: The idea in Exercise 1.16(b) is helpful here, but the result is not.

(b) Is the event that X_n is a record-to-date statistically independent of the event that X_{n+1} is a record-to-date?

(c) Find the expected number of adjacent pairs of records-to-date over the sequence $X_1, X_2, \ldots$. Hint: A helpful fact here is that $1/[n(n+1)] = 1/n - 1/(n+1)$.

Exercise 1.19 **(a)** Assume that X is a non-negative discrete rv taking on values $a_1, a_2, \ldots$, and let $Y = h(X)$ for some non-negative function h. Let $b_i = h(a_i)$, $i \geq 1$ be the ith value taken on by Y. Show that $\mathsf{E}[Y] = \sum_i b_i \mathsf{p}_Y(b_i) = \sum_i h(a_i)\mathsf{p}_X(a_i)$. Find an example where $\mathsf{E}[X]$ exists but $\mathsf{E}[Y] = \infty$.

(b) Let X be a non-negative continuous rv with density $\mathsf{f}_X(x)$ and let $h(x)$ be differentiable, non-negative, and strictly increasing in x. Let $A(\delta) = \sum_{n\geq 1} h(n\delta)[\mathsf{F}(n\delta) - \mathsf{F}(n\delta - \delta)]$, i.e., $A(\delta)$ is a δth-order approximation to the Stieltjes integral $\int h(x)\,d\mathsf{F}(x)$. Show that if $A(1) < \infty$, then $A(2^{-k}) \leq A(2^{-(k-1)}) < \infty$. Show from this that $\int h(x)\,d\mathsf{F}(x)$ converges to a finite value. Note: This is a very special case, but it can be extended to many cases of interest. It seems better to consider these convergence questions as required rather than to consider them in general.

Exercise 1.20 **(a)** Consider a positive, integer-valued rv whose CDF is given at integer values by

$$\mathsf{F}_Y(y) = 1 - \frac{2}{(y+1)(y+2)} \qquad \text{for integer } y \geq 0.$$

Use (1.30) to show that $\mathsf{E}[Y] = 2$. Hint: Note the PMF given in (1.29).

(b) Find the PMF of Y and use it to check the value of $\mathsf{E}[Y]$.

(c) Let X be another positive, integer-valued rv. Assume its conditional PMF is given by

$$\mathsf{p}_{X|Y}(x|y) = \frac{1}{y} \qquad \text{for } 1 \leq x \leq y.$$

Find $\mathsf{E}\left[X \mid Y = y\right]$ and show that $\mathsf{E}[X] = 3/2$. Explore finding $\mathsf{p}_X(x)$ until you are convinced that using the conditional expectation to calculate $\mathsf{E}[X]$ is considerably easier than using $\mathsf{p}_X(x)$.

(d) Let Z be another integer-valued rv with the conditional PMF

$$\mathsf{p}_{Z|Y}(z|y) = \frac{1}{y^2} \qquad \text{for } 1 \le z \le y^2.$$

Find $\mathsf{E}\left[Z \mid Y = y\right]$ for each integer $y \ge 1$ and find $\mathsf{E}[Z]$.

Exercise 1.21 **(a)** Show that, for uncorrelated rv s, the expected value of the product is equal to the product of the expected values (by definition, X and Y are uncorrelated if $\mathsf{E}\left[(X - \overline{X})(Y - \overline{Y})\right] = 0$).

(b) Show that if X and Y are uncorrelated, then the variance of $X + Y$ is equal to the variance of X plus the variance of Y.

(c) Show that if $X_1, \dots, X_n$ are uncorrelated, then the variance of the sum is equal to the sum of the variances.

(d) Show that independent rv s are uncorrelated.

(e) Let X, Y be identically distributed ternary valued rv s with the PMF $\mathsf{p}_X(-1) = \mathsf{p}_X(1) = 1/4$; $\mathsf{p}_X(0) = 1/2$. Find a simple joint probability assignment such that X and Y are uncorrelated but dependent.

(f) You have seen that the MGF of a sum of independent rv s is equal to the product of the individual MGFs. Give an example where this is false if the variables are uncorrelated but dependent.

Exercise 1.22 Suppose X has the Poisson PMF, $\mathsf{p}_X(n) = \lambda^n \exp(-\lambda)/n!$ for $n \ge 0$ and Y has the Poisson PMF, $\mathsf{p}_Y(m) = \mu^n \exp(-\mu)/n!$ for $n \ge 0$. Assume that X and Y are independent. Show that $Z = X + Y$ is a Poisson rv and find the conditional distribution of Y conditional on $Z = n$.

Exercise 1.23 **(a)** Suppose X, Y and Z are binary rv s, each taking on the value 0 with probability 1/2 and the value 1 with probability 1/2. Find a simple example in which X, Y, Z are statistically *dependent* but are *pairwise* statistically *independent* (i.e., X, Y are statistically independent, X, Z are statistically independent, and Y, Z are statistically independent). Give $\mathsf{p}_{XYZ}(x, y, z)$ for your example. Hint: In the simplest example, there are four joint values for x, y, z that have probability 1/4 each.

(b) Is pairwise statistical independence enough to ensure that

$$\mathsf{E}\left[\prod_{i=1}^{n} X_i\right] = \prod_{i=1}^{n} \mathsf{E}[X_i]$$

for a set of rv s $X_1, \dots, X_n$?

Exercise 1.24 Show that $\mathsf{E}[X]$ is the value of α that minimizes $\mathsf{E}\left[(X - \alpha)^2\right]$.

Exercise 1.25 For each of the following rv s, find the endpoints r_- and r_+ of the interval for which the MGF $g(r)$ exists. Determine in each case whether $\mathsf{g}_X(r)$ exists at r_- and r_+. For (a) and (b) you should also find and sketch $g(r)$. For (c) and (d), $g(r)$ has no closed form.

(a) Let λ, θ, be positive numbers and let X have the density

$$\mathsf{f}_X(x) = \frac{1}{2}\lambda \exp(-\lambda x);\ x \ge 0; \quad \mathsf{f}_X(x) = \frac{1}{2}\theta \exp(\theta x);\ x < 0.$$

(b) Let Y be a Gaussian rv with mean m and variance σ^2.

(c) Let Z be a non-negative rv with density

$$\mathsf{f}_Z(z) = k(1+z)^{-2} \exp(-\lambda z) \quad z \geq 0,$$

where $\lambda > 0$ and $k = [\int_{z \geq 0} (1+z)^{-2} \exp(-\lambda z)\, dz]^{-1}$. Hint: Do not try to evaluate $\mathsf{g}_Z(r)$. Instead, investigate values of r for which the integral is finite and infinite.

(d) For the Z of (c), find the limit of $\gamma'(r)$ as r approaches λ from below. Then replace $(1+z)^2$ with $|1+z|^3$ in the definition of $\mathsf{f}_Z(z)$ and k and show whether the above limit is then finite or not. Hint: Use tilted rv s; no integration is required.

Exercise 1.26 **(a)** Assume that the MGF of the rv X exists (i.e., is finite) in the interval (r_-, r_+), $r_- < 0 < r_+$, and assume $r_- < r < r_+$ throughout. For any finite constant c, express the MGF of $X - c$, i.e., $\mathsf{g}_{(X-c)}(r)$, in terms of $\mathsf{g}_X(r)$ and show that $\mathsf{g}_{(X-c)}(r)$ exists for all r in (r_-, r_+). Explain why $\mathsf{g}''_{(X-c)}(r) \geq 0$.

(b) Show that $\mathsf{g}''_{(X-c)}(r) = [\mathsf{g}''_X(r) - 2c\mathsf{g}'_X(r) + c^2\mathsf{g}_X(r)]e^{-rc}$.

(c) Use (a) and (b) to show that $\mathsf{g}''_X(r)\mathsf{g}_X(r) - [\mathsf{g}'_X(r)]^2 \geq 0$. Let $\gamma_X(r) = \ln \mathsf{g}_X(r)$ and show that $\gamma''_X(r) \geq 0$. Hint: Choose $c = \mathsf{g}'_X(r)/\mathsf{g}_X(r)$.

(d) Assume that X is non-deterministic, i.e., there is no value of α such that $\Pr\{X = \alpha\} = 1$. Show that the inequality sign '$\geq$' may be replaced by '$>$' everywhere in (a), (b) and (c).

Exercise 1.27 A computer system has n users, each with a unique name and password. Due to a software error, the n passwords are randomly permuted internally (i.e., each of the $n!$ possible permutations is equally likely). Only those users lucky enough to have had their passwords unchanged in the permutation are able to continue using the system.

(a) What is the probability that a particular user, say user 1, is able to continue using the system?

(b) What is the expected number of users able to continue using the system? Hint: Let X_i be a rv with the value 1 if user i can use the system and 0 otherwise.

Exercise 1.28 Suppose the rv X is continuous and has the CDF $\mathsf{F}_X(x)$. Consider another rv $Y = \mathsf{F}_X(X)$. That is, for each sample point ω such that $X(\omega) = x$, we have $Y(\omega) = \mathsf{F}_X(x)$. Show that Y is uniformly distributed in the interval 0 to 1.

Exercise 1.29 Let Z be an integer-valued rv with the PMF $\mathsf{p}_Z(n) = 1/k$ for $0 \leq n \leq k-1$. Find the mean, variance, and MGF of Z. Hint: An elegant way to do this is to let U be a uniformly distributed continuous rv over $(0, 1]$ that is independent of Z. Then $U + Z$ is uniform over $(0, k]$. Use the known results about U and $U + Z$ to find the mean, variance, and MGF for Z.

Exercise 1.30 (Alternative approach 1 to the Markov inequality) **(a)** Let Y be a non-negative rv and $y > 0$ be some fixed number. Let A be the event that $Y \geq y$. Show that $y\,\mathbb{I}_A \leq Y$ (i.e., that this inequality is satisfied for every $\omega \in \Omega$).

(b) Use your result in (a) to prove the Markov inequality.

Exercise 1.31 (Alternative approach 2 to the Markov inequality) **(a)** Minimize $\mathsf{E}[Y]$ over all non-negative rv s such that $\Pr\{Y \geq b\} \geq \beta$ for some given $b > 0$ and $0 < \beta < 1$. Hint: Use a graphical argument similar to that in Figure 1.7. What is the rv that achieves the minimum. Hint: It is binary.

(b) Use (a) to prove the Markov inequality and also point out the distribution that meets the inequality with equality.

Exercise 1.32 (The one-sided Chebyshev inequality) This inequality states that if a zero-mean rv X has a variance σ^2, then it satisfies the inequality

$$\Pr\{X \geq b\} \leq \frac{\sigma^2}{\sigma^2 + b^2} \qquad \text{for every } b > 0, \tag{1.101}$$

with equality for some b only if X is binary and $\Pr\{X = b\} = \sigma^2/(\sigma^2 + b^2)$. We prove this here using the same approach as in Exercise 1.31. Let X be a zero-mean rv that satisfies $\Pr\{X \geq b\} = \beta$ for some $b > 0$ and $0 < \beta < 1$. The variance σ^2 of X can be expressed as

$$\sigma^2 = \int_{-\infty}^{b^-} x^2 \mathsf{f}_X(x)\, dx + \int_b^\infty x^2\, \mathsf{f}_X(x)\, dx. \tag{1.102}$$

We will first minimize σ^2 over all zero-mean X satisfying $\Pr\{X \geq b\} = \beta$.

(a) Show that the second integral in (1.102) satisfies $\int_b^\infty x^2 \mathsf{f}_X(x)\, dx \geq b^2\beta$.

(b) Show that the first integral in (1.102) is constrained by

$$\int_{-\infty}^{b^-} \mathsf{f}_X(x)\, dx = 1 - \beta \qquad \text{and} \qquad \int_{-\infty}^{b^-} x\mathsf{f}_X(x)\, dx \leq -b\beta.$$

(c) Minimize the first integral in (1.102) subject to the constraints in (b). Hint: If you scale $f_X(x)$ up by $1/(1 - \beta)$, it integrates to 1 over $(-\infty, b)$ and the second constraint becomes an expectation. You can then minimize the first integral in (1.102) by inspection.

(d) Combine the results in (a) and (c) to show that $\sigma^2 \geq b^2\beta/(1 - \beta)$. Find the minimizing distribution. Hint: It is binary.

(e) Use (d) to establish (1.101). Also show (trivially) that if Y has a mean $\overline{Y}$ and variance σ^2, then $\Pr\{Y - \overline{Y} \geq b\} \leq \sigma^2/(\sigma^2 + b^2)$

Exercise 1.33 (Proof of (1.48)) Here we show that if X is a zero-mean rv with a variance σ^2, then the median α satisfies $|\alpha| \leq \sigma$.

(a) First show that $|\alpha| \leq \sigma$ for the special case where X is binary with equiprobable values at $\pm\sigma$.

(b) For all zero-mean rv s X with variance σ^2 other than the special case in (a), show that

$$\Pr\{X \geq \sigma\} < 0.5.$$

Hint: Use the one-sided Chebyshev inequality of Exercise 1.32.

(c) Show that $\Pr\{X \geq \alpha\} \geq 0.5$. Other than the special case in (a), show that this implies that $\alpha < \sigma$.

(d) Other than the special case in (a), show that $|\alpha| < \sigma$. Hint: Repeat (b) and (c) for the rv $-X$. You have then shown that $|\alpha| \leq \sigma$ with equality only for the binary case with values $\pm\sigma$. For rv s Y with a non-zero mean, this shows that $|\alpha - \overline{Y}| \leq \sigma$.

Exercise 1.34 We stressed the importance of the mean of a rv X in terms of its association with the sample average via the WLLN. Here we show that in essence the WLLN allows us to evaluate the entire CDF, say $\mathsf{F}_X(x)$ of X via sufficiently many independent sample values of X.

(a) For any given y, let $\mathbb{I}_j(y)$ be the indicator function of the event $\{X_j \leq y\}$, where $X_1, X_2, \ldots, X_j, \ldots$ are IID rv s with the CDF $\mathsf{F}_X(x)$. State the WLLN for the IID rv s $\{\mathbb{I}_1(y), \mathbb{I}_2(y), \ldots\}$.

(b) Does the answer to (a) require X to have a mean or variance?

(c) Suggest a procedure for evaluating the median of X from the sample values of $X_1, X_2, \ldots$. Assume that X is a continuous rv and that its PDF is positive in an open interval around the median. You need not be precise, but try to think the issue through carefully.

What you have seen here, without stating it precisely or proving it is that the median has a law of large numbers associated with it, saying that the sample median of n IID samples of a rv is close to the true median with high probability.

Exercise 1.35 **(a)** Show that for any $0 < k < n$

$$\binom{n}{k+1} \leq \binom{n}{k}\frac{n-k}{k}.$$

(b) Extend (a) to show that, for all $\ell \leq n-k$,

$$\binom{n}{k+\ell} \leq \binom{n}{k}\left[\frac{n-k}{k}\right]^{\ell}.$$

(c) Let $\widetilde{p} = k/n$ and $\widetilde{q} = 1 - \widetilde{p}$. Let S_n be the sum of n binary IID rv s with $\mathsf{p}_X(0) = q$ and $\mathsf{p}_X(1) = p$. Show that for all $\ell \leq n-k$,

$$\mathsf{p}_{S_n}(k+\ell) \leq \mathsf{p}_{S_n}(k)\left(\frac{\widetilde{q}p}{\widetilde{p}q}\right)^{\ell}.$$

(d) For $k/n > p$, show that

$$\Pr\{S_n \geq kn\} \leq \frac{\widetilde{p}q}{\widetilde{p}-p}\,\mathsf{p}_{S_n}(k).$$

(e) Now let ℓ be fixed and $k = \lceil n\widetilde{p} \rceil$ for fixed $\widetilde{p}$ such that $1 > \widetilde{p} > p$. Argue that as $n \to \infty$,

$$\mathsf{p}_{S_n}(k+\ell) \sim \mathsf{p}_{S_n}(k) \left(\frac{\widetilde{q}p}{\widetilde{p}q} \right)^{\ell} \qquad \text{and} \qquad \Pr\{S_n \geq kn\} \sim \frac{\widetilde{p}q}{\widetilde{p}-p}\, \mathsf{p}_{S_n}(k),$$

where $a(n) \sim b(n)$ means that $\lim_{n\to\infty} a(n)/b(n) = 1$.

Exercise 1.36 A sequence $\{a_n;\ n \geq 1\}$ of real numbers has the limit 0 if, for all $\epsilon > 0$, there is an $m(\epsilon)$ such that $|a_n| \leq \epsilon$ for all $n \geq m(\epsilon)$. Show that the sequences in (a) and (b) below satisfy $\lim_{n\to\infty} a_n = 0$ but the sequence in (c) does not have a limit.

(a) $a_n = \dfrac{1}{\ln(\ln(n+1))}$

(b) $a_n = n^{10}\exp(-n)$

(c) $a_n = 1$ for $n = 10^{\ell}$ for each positive integer ℓ and $a_n = 0$ otherwise.

(d) Show that the definition can be changed (with no change in meaning) by replacing ϵ with either $1/k$ or 2^{-k} for every positive integer k.

Exercise 1.37 Represent the MGF of a rv X by

$$\mathsf{g}_X(r) = \int_{-\infty}^{0} e^{rx}\,d\mathsf{F}(x) + \int_{0}^{\infty} e^{rx}\,d\mathsf{F}(x).$$

In each of the following parts, you are welcome to restrict X to be either discrete or continuous.

(a) Show that the first integral always exists (i.e., is finite) for $r \geq 0$ and that the second integral always exists for $r \leq 0$.

(b) Show that if the second integral exists for a given $r_1 > 0$, then it also exists for all r in the range $0 \leq r \leq r_1$.

(c) Show that if the first integral exists for a given $r_2 < 0$, then it also exists for all r in the range $r_2 \leq r \leq 0$.

(d) Show that the range of r over which $\mathsf{g}_X(r)$ exists is an interval from some $r_- \leq 0$ to some $r_+ \geq 0$ (the interval might or might not include each endpoint, and the magnitude of either or both endpoints might be 0, ∞, or any point between).

(e) Find an example where $r_+ = 1$ and the MGF does not exist for $r = 1$. Find another example where $r_+ = 1$ and the MGF does exist for $r = 1$. Hint: Consider $\mathsf{f}_X(x) = e^{-x}$ for $x \geq 0$ and figure out how to modify it to $\mathsf{f}_Y(y)$ so that $\int_0^\infty e^{y}\mathsf{f}_Y(y)\,dy < \infty$ but $\int_0^\infty e^{y+\epsilon y}\mathsf{f}_Y(y) = \infty$ for all $\epsilon > 0$.

Exercise 1.38 Let $\{X_n;\ n \geq 1\}$ be a sequence of independent but not identically distributed rv s. We say that the WLLN holds for this sequence if for all $\epsilon > 0$

$$\lim_{n\to\infty} \Pr\left\{ \left| \frac{S_n}{n} - \frac{\mathsf{E}[S_n]}{n} \right| \geq \epsilon \right\} = 0, \quad \text{where } S_n = X_1 + X_2 + \cdots + X_n. \qquad \text{(WLLN)}.$$

(a) Show that the WLLN holds if there is some constant A such that $\sigma^2_{X_n} \leq A$ for all n.

(b) Suppose that $\sigma^2_{X_n} \leq A\, n^{1-\alpha}$ for some α, $0 < \alpha < 1$ and for all n. Show that the WLLN holds in this case.

Exercise 1.39 Let $\{X_i;\ i \geq 1\}$ be IID binary rv s. Let $\Pr\{X_i = 1\} = \delta$, $\Pr\{X_i = 0\} = 1 - \delta$. Let $S_n = X_1 + \cdots + X_n$. Let m be an arbitrary but fixed positive integer. Think! Then evaluate the following and explain your answers:

(a) $\lim_{n\to\infty} \sum_{i:n\delta-m\leq i\leq n\delta+m} \Pr\{S_n = i\}$;

(b) $\lim_{n\to\infty} \sum_{i:0\leq i\leq n\delta+m} \Pr\{S_n = i\}$;

(c) $\lim_{n\to\infty} \sum_{i:n(\delta-1/m)\leq i\leq n(\delta+1/m)} \Pr\{S_n = i\}$.

Exercise 1.40 The WLLN is said to hold for a zero-mean sequence $\{S_n;\ n \geq 1\}$ if

$$\lim_{n\to\infty} \Pr\left\{\left|\frac{S_n}{n}\right| > \epsilon\right\} = 0 \qquad \text{for every } \epsilon > 0.$$

The CLT is said to hold for $\{S_n;\ n \geq 1\}$ if for some $\sigma > 0$ and all $z \in \mathbb{R}$,

$$\lim_{n\to\infty} \Pr\left\{\frac{S_n}{\sigma\sqrt{n}} \leq z\right\} = \Phi(z),$$

where $\Phi(z)$ is the normal CDF. Show that if the CLT holds, then the WLLN also holds. Note 1: If you hate ϵ, δ arguments, you will hate this. Note 2: It will probably ease the pain if you convert the WLLN statement to: For every $\epsilon > 0,\ \delta > 0$ there exists an $n(\epsilon, \delta)$ such that for every $n \geq n(\epsilon,\ \delta)$,

$$\Pr\left\{\frac{S_n}{n} < -\epsilon\right\} \leq \delta \qquad \text{and } \Pr\left\{\frac{S_n}{n} > 1 - \epsilon\right\} \leq \delta. \tag{i}$$

Exercise 1.41 (Details in the proof of Theorem 1.7.4) **(a)** Show that if $X_1, X_2, \ldots$ are IID, then the truncated versions $\breve{X}_1, \breve{X}_2, \ldots$ are also IID.

(b) Show that each $\breve{X}_i$ has a finite mean $\mathsf{E}[\breve{X}]$ and finite variance $\sigma^2_{\breve{X}}$. Show that the variance is upper bounded by the second moment around the original mean $\overline{X}$, i.e., show that $\sigma^2_{\breve{X}} \leq \mathsf{E}[|\ \breve{X} - \mathsf{E}\,[X]\ |^2]$.

(c) Assume that $\breve{X}_i$ is X_i truncated to $\overline{X} \pm b$. Show that $|\breve{X} - \overline{X}| \leq b$ and that $|\breve{X} - \overline{X}| \leq |X - \overline{X}|$. Use this to show that $\sigma^2_{\breve{X}} \leq b\mathsf{E}[|\breve{X} - \overline{X}|] \leq 2b\mathsf{E}\,[|X|]$.

(d) Let $\breve{S}_n = \breve{X}_1 + \cdots + \breve{X}_n$ and show that for any $\epsilon > 0$,

$$\Pr\left\{\left|\frac{\breve{S}_n}{n} - \mathsf{E}\left[\breve{X}\right]\right| \geq \frac{\epsilon}{2}\right\} \leq \frac{8b\mathsf{E}\,[|X|]}{n\epsilon^2}.$$

(e) Sketch the form of $\mathsf{F}_{\breve{X}-\overline{X}}(x)$ and use this, along with (1.34), to show that for all sufficiently large b, $\left|\mathsf{E}\left[\breve{X} - \overline{X}\right]\right| \leq \epsilon/2$. Use this to show that

$$\Pr\left\{\left|\frac{\breve{S}_n}{n} - \mathsf{E}\,[X]\right| \geq \epsilon\right\} \leq \frac{8b\mathsf{E}\,[|X|]}{n\epsilon^2} \qquad \text{for all large enough } b.$$

(f) Use the following equation to justify (1.96).

$$\begin{aligned}\Pr\left\{\left|\frac{S_n}{n} - \mathsf{E}\,[X]\right| > \epsilon\right\} &= \Pr\left\{\left|\frac{S_n}{n} - \mathsf{E}\,[X]\right| > \epsilon \bigcap S_n = \breve{S}_n\right\} \\ &\quad + \Pr\left\{\left|\frac{S_n}{n} - \mathsf{E}\,[X]\right| > \epsilon \bigcap S_n \neq \breve{S}_n\right\}.\end{aligned}$$

Exercise 1.42 Let $\{X_i;\ i \geq 1\}$ be IID rv s with mean 0 and infinite variance. Assume that $\mathsf{E}\left[|X_i|^{1+h}\right] = \beta$ for some given h, $0 < h < 1$ and some finite β. Let $S_n = X_1 + \cdots + X_n$.

(a) Show that $\Pr\{|X_i| \geq y\} \leq \beta\, y^{-1-h}$.

(b) Let $\{\breve{X}_i; i \geq 1\}$ be truncated variables defined as

$$\breve{X}_i = \begin{cases} b & : \ X_i \geq b, \\ X_i & : \ -b \leq X_i \leq b, \\ -b & : \ X_i \leq -b. \end{cases}$$

Show that $\mathsf{E}\left[\breve{X}^2\right] \leq 2\beta b^{1-h}/(1-h)$. Hint: For a non-negative rv Z, $\mathsf{E}\left[X^2\right] = \int_0^\infty 2\,z\Pr\{Z \geq z\}\, dz$ (you can establish this, if you wish, by integration by parts).

(c) Let $\breve{S}_n = \breve{X}_1 + \cdots + \breve{X}_n$. Show that $\Pr\left\{S_n \neq \breve{S}_n\right\} \leq n\beta b^{-1-h}$.

(d) Show that

$$\Pr\left\{\left|\frac{S_n}{n}\right| \geq \epsilon\right\} \leq \beta\left[\frac{2b^{1-h}}{(1-h)n\epsilon^2} + \frac{n}{b^{1+h}}\right].$$

(e) Optimize your bound with respect to b. How fast does this optimized bound approach 0 with increasing n?

Exercise 1.43 (MS convergence $\Longrightarrow$ convergence in probability) Assume that $\{Z_n;\ n \geq 1\}$ is a sequence of rv s and α is a number with the property that $\lim_{n\to\infty} \mathsf{E}\left[(Z_n - \alpha)^2\right] = 0$.

(a) Let $\epsilon > 0$ be arbitrary and show that for each $n \geq 0$,

$$\Pr\{|Z_n - \alpha| \geq \epsilon\} \leq \frac{\mathsf{E}\left[(Z_n - \alpha)^2\right]}{\epsilon^2}.$$

(b) For the ϵ above, let $\delta > 0$ be arbitrary. Show that there is an integer m such that $\mathsf{E}\left[(Z_n - \alpha)^2\right] \leq \epsilon^2\delta$ for all $n \geq m$.

(c) Show that this implies convergence in probability.

Exercise 1.44 Let $X_1, X_2 \ldots$ be a sequence of IID rv s each with mean 0 and variance σ^2. Let $S_n = X_1 + \cdots + X_n$ for all n and consider the random variable $S_n/\sigma\sqrt{n} - S_{2n}/\sigma\sqrt{2n}$. Find the limiting CDF for this sequence of rv s as $n \to \infty$. The point of this exercise is to see clearly that the CDF of $S_n/\sigma\sqrt{n}$ is converging but that the sequence of rv s is not converging.

Exercise 1.45 Use Figure 1.7 to verify (1.57). Hint: Show that $y\Pr\{Y \geq y\} \leq \int_{z \geq y} z\, d\mathsf{F}_Y(z)$ and show that $\lim_{y\to\infty} \int_{z\geq y} z\, d\mathsf{F}_Y(z) = 0$ if $\mathsf{E}[Y]$ is finite.

Exercise 1.46 Show that $\prod_{m \geq n}(1 - 1/m) = 0$. Hint: Show that

$$\left(1 - \frac{1}{m}\right) = \exp\left(\ln\left(1 - \frac{1}{m}\right)\right) \leq \exp\left(-\frac{1}{m}\right).$$

Exercise 1.47 Consider a discrete rv X with the PMF

$$\mathsf{p}_X(-1) = (1 - 10^{-10})/2,$$
$$\mathsf{p}_X(1) = (1 - 10^{-10})/2,$$
$$\mathsf{p}_X(10^{12}) = 10^{-10}.$$

(a) Find the mean and variance of X. Assuming that $\{X_m;\ m \geq 1\}$ is an IID sequence with the distribution of X and that $S_n = X_1 + \cdots + X_n$ for each n, find the mean and variance of S_n. (No explanations needed.)

(b) Let $n = 10^6$ and describe the event $\{S_n \leq 10^6\}$ in words. Find an exact expression for $\Pr\{S_n \leq 10^6\} = \mathsf{F}_{S_n}(10^6)$.

(c) Find a way to use the union bound to get a simple upper bound and approximation of $1 - \mathsf{F}_{S_n}(10^6)$.

(d) Sketch the CDF of S_n for $n = 10^6$. You can choose the horizontal axis for your sketch to go from -1 to $+1$ or from -3×10^3 to 3×10^3 or from -10^6 to 10^6 or from 0 to 10^{12}, whichever you think will best describe this CDF.

(e) Now let $n = 10^{10}$. Give an exact expression for $\Pr\{S_n \leq 10^{10}\}$ and show that this can be approximated by e^{-1}. Sketch the CDF of S_n for $n = 10^{10}$, using a horizontal axis going from slightly below 0 to slightly more than 2×10^{12}. Hint: First view S_n as conditioned on an appropriate rv.

(f) Can you make a qualitative statement about how the CDF of a rv X affects the required size of n before the WLLN and the CLT provide much of an indication about S_n.

Exercise 1.48 Let $\{Y_n;\ n \geq 1\}$ be a sequence of rv s and assume that $\lim_{n\to\infty} \mathsf{E}[|Y_n|] = 0$. Show that $\{Y_n;\ n \geq 1\}$ converges to 0 in probability. Hint 1: Look for the easy way. Hint 2: The easy way uses the Markov inequality.

2 Poisson processes

2.1 Introduction

A Poisson process is a simple and widely used stochastic process for modeling the times at which arrivals enter a system. It is in many ways the continuous-time version of the Bernoulli process. Section 1.4.1 characterized the Bernoulli process by a sequence of independent identically distributed (IID) binary random variables (rv s), $Y_1, Y_2, \dots,$ where $Y_i = 1$ indicates an arrival at increment i and $Y_i = 0$ otherwise. We observed (without any careful proof) that the process could also be characterized by a sequence of geometrically distributed interarrival times.

For the Poisson process, arrivals may occur at arbitrary positive times, and the probability of an arrival at any particular instant is 0. This means that there is no very clean way of describing a Poisson process in terms of the probability of an arrival at any given instant. It is more convenient to define a Poisson process in terms of the sequence of interarrival times, $X_1, X_2, \dots,$ which are defined to be IID. Before doing this, we describe arrival processes in a little more detail.

2.1.1 Arrival processes

An *arrival process* is a sequence of increasing rv s, $0 < S_1 < S_2 < \cdots$, where $S_i < S_{i+1}$ means that $S_{i+1} - S_i$ is a positive rv, i.e., a rv X such that $\mathsf{F}_X(0) = 0$. The rv s $S_1, S_2, \dots$ are called arrival epochs[1] and represent the successive times at which some random repeating phenomenon occurs. Note that the process starts at time 0 and that multiple arrivals cannot occur simultaneously (the phenomenon of bulk arrivals can be handled by the simple extension of associating a positive integer rv to each arrival). We will sometimes permit simultaneous arrivals or arrivals at time 0 as events of zero probability, but these can be ignored. In order to fully specify the process by the sequence $S_1, S_2, \dots$ of rv s, it is necessary to specify the joint distribution of the subsequences $S_1, \dots, S_n$ for all $n > 1$.

Although we refer to these processes as arrival processes, they could equally well model departures from a system, or any other sequence of incidents. Although it is quite common, especially in the simulation field, to refer to incidents or arrivals as events, we

[1] Epoch is a synonym for time. We use it here to avoid overusing the word time in overlapping ways.

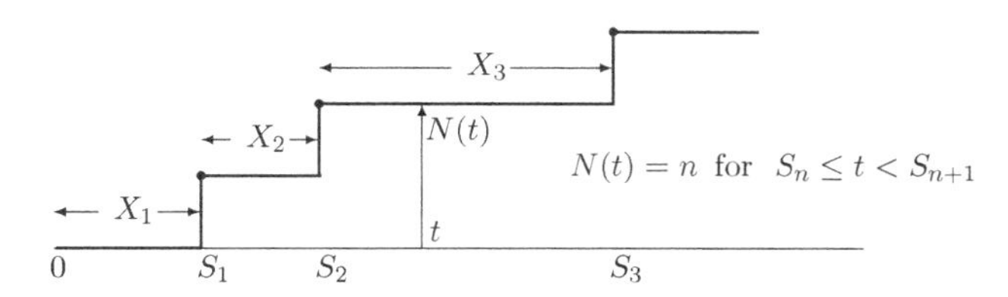

Figure 2.1 A sample function of an arrival process and its arrival epochs $\{S_1, S_2, \ldots\}$, its interarrival times $\{X_1, X_2, \ldots\}$, and its counting process $\{N(t);\ t > 0\}$

shall avoid that here. The nth arrival epoch S_n is a rv and $\{S_n \le t\}$, for example, is an event. This would make it confusing to refer to the nth arrival itself as an event.

As illustrated in Figure 2.1, any arrival process $\{S_n;\ n \ge 1\}$ can also be specified by either of two alternative stochastic processes. The first alternative is the sequence of interarrival times, $\{X_i;\ i \ge 1\}$. The X_i here are positive rv s defined in terms of the arrival epochs by $X_1 = S_1$ and $X_i = S_i - S_{i-1}$ for $i > 1$. Similarly, each arrival epoch S_n is specified by $\{X_i;\ i \ge 1\}$ as

$$S_n = \sum\nolimits_{i=1}^{n} X_i. \tag{2.1}$$

Thus the joint distribution of $X_1, \ldots, X_n$ for all $n > 1$ is sufficient (in principle) to specify the arrival process. Since the interarrival times are IID in most cases of interest, it is usually simpler to specify the joint distribution of the X_i than of the S_i.

The second alternative is the counting process $\{N(t);\ t > 0\}$, where for each $t > 0$ the rv $N(t)$ is the aggregate number of arrivals[2] up to and including time t.

The counting process $\{N(t);\ t > 0\}$, illustrated in Figure 2.1, is an uncountably infinite family of rv s where the *counting rv* $N(t)$, for each $t > 0$, is the aggregate number of arrivals in the interval $(0, t]$, i.e., the number of arrivals in the set $\{\tau : 0 < \tau \le t\}$. By convention, a parenthesis is used to indicate that the corresponding endpoint of an interval is not contained in the interval and a bracket is used if it is. Thus, as another example, $(a, b) = \{\tau : a < \tau < b\}$. The rv $N(0)$ is defined to be 0 with probability 1, which means that we are considering only arrivals at strictly positive times.

The counting process $\{N(t);\ t > 0\}$ for any arrival process has the property that $N(\tau) \ge N(t)$ for all $\tau \ge t > 0$ (i.e., $N(\tau) - N(t)$ is a non-negative rv for $\tau \ge t > 0$).

For any given integer $n \ge 1$ and time $t > 0$, the nth arrival epoch, S_n, and the counting rv, $N(t)$, are related by

$$\{S_n \le t\} = \{N(t) \ge n\}. \tag{2.2}$$

To see this, note that $\{S_n \le t\}$ is the event that the nth arrival occurs at some epoch $\tau \le t$. This event implies that $N(\tau) = n$, and thus that $\{N(t) \ge n\}$. Similarly, $\{N(t) = m\}$ for some $m \ge n$ implies $\{S_m \le t\}$, and thus that $\{S_n \le t\}$. This equation is essentially obvious from Figure 2.1, but is one of those peculiar obvious things that is often difficult

[2] Thus, for the Bernoulli process with an increment size of 1, $N(n)$ is the rv denoted as S_n in Section 1.4.1.

to see. An alternative form, which is occasionally more transparent, comes from taking the complement of both sides of (2.2), getting

$$\{S_n > t\} = \{N(t) < n\}. \tag{2.3}$$

For example, the event $\{S_1 > t\}$ means that the first arrival occurs after t, which means $\{N(t) < 1\}$ (i.e., $\{N(t) = 0\}$). These relations will be used constantly in going back and forth between arrival epochs and counting rv s. In principle, (2.2) or (2.3) can be used to specify the joint cumulative distribution function (CDF) of arrival epochs in terms of joint CDFs of counting variables and vice versa, so either characterization can be used to specify an arrival process.

In summary, then, an arrival process can be specified by the joint distributions of the arrival epochs, or of the interarrival times, or of the counting rv s. In principle, specifying any one of these specifies the others also.[3]

2.2 Definition and properties of a Poisson process

The most elegantly simple, and most important, arrival processes are those for which the interarrival times are IID. These arrival processes are called *renewal processes* and form the subject of Chapter 5.

Poisson processes constitute the simplest and most widely used class of renewal processes. They are characterized by having exponentially distributed interarrival times. We will soon see why these exponentially distributed interarrival times simplify these processes and make it possible to look at them in so many simple ways. We start by being explicit about the definitions of renewal and Poisson processes.

Definition 2.2.1 *A **renewal process** is an arrival process for which the sequence of interarrival times is a sequence of positive IID rv s.*

Definition 2.2.2 *A **Poisson process** is a renewal process in which the interarrival times have an exponential CDF; i.e., for some real $\lambda > 0$, each X_i has the density*[4] $\mathsf{f}_X(x) = \lambda \exp(-\lambda x)$ *for* $x \geq 0$.

The parameter λ is called the rate of the process. We shall see later that for any interval of size t, λt is the expected number of arrivals in that interval, motivating the designation of λ as a rate.

[3] By definition, a stochastic process is a collection of rv s, so one might ask whether an arrival process (as a stochastic process) is 'really' the arrival epoch process $\{S_n;\ n \geq 1$ or the interarrival process $X_i;\ i > 0\}$ or the counting process $\{N(t);\ t > 0\}$. The arrival epoch process comes to grips with the actual arrivals, the interarrival process is often the simplest, and the counting process 'looks' most like a stochastic process in time since $N(t)$ is a rv for each $t \geq 0$. It seems preferable, since the descriptions are so clearly equivalent, to view arrival processes in terms of whichever description is most convenient.

[4] With this density, $\Pr\{X_i{>}0\} = 1$, so that we can regard X_i as a positive rv. Since events of probability 0 can be ignored, the density $\lambda \exp(-\lambda x)$ for $x \geq 0$ and zero for $x < 0$ is effectively the same as the density $\lambda \exp(-\lambda x)$ for $x > 0$ and zero for $x \leq 0$.

2.2.1 Memoryless property

What makes the Poisson process unique among renewal processes is the memoryless property of the exponential distribution.

Definition 2.2.3 (Memoryless random variables) *A rv X possesses the* ***memoryless property*** *if X is a positive rv (i.e.,* $\Pr\{X > 0\} = 1$*) for which*

$$\Pr\{X > t + x\} = \Pr\{X > x\}\Pr\{X > t\} \qquad \textit{for all } x, t \geq 0. \tag{2.4}$$

Note that (2.4) is a statement about the complementary CDF of X. There is no intimation that the *event* $\{X > t+x\}$ in the equation has any particular relation to the events $\{X > t\}$ or $\{X > x\}$.

For an exponential rv X of rate $\lambda > 0$, $\Pr\{X > x\} = e^{-\lambda x}$ for $x \geq 0$. This satisfies (2.4) for all $x \geq 0,\ t \geq 0$, so X is memoryless. Conversely, an arbitrary rv X is memoryless only if it is exponential. To see this, let $h(x) = \ln[\Pr\{X > x\}]$ and observe that since $\Pr\{X > x\}$ is non-increasing in x, $h(x)$ is also. In addition, (2.4) says that $h(t + x) = h(x) + h(t)$ for all $x, t \geq 0$. These two statements (see Exercise 2.6) imply that $h(x)$ must be linear in x, and thus $\Pr\{X > x\}$ must be exponential in x.

Since a memoryless rv X must be exponential, there must be some $\lambda > 0$ such that $\Pr\{X > t\} = e^{-\lambda t} > 0$ for all $t \geq 0$. This means that we can rewrite (2.4) as

$$\Pr\{X > t + x \mid X > t\} = \Pr\{X > x\}\,. \tag{2.5}$$

If X is interpreted as the waiting time until some given arrival, then (2.5) states that, given that the arrival has not occurred by time t, the distribution of the remaining waiting time (given by x on the left-hand side of (2.5)) is the same as the original waiting time distribution (given on the right-hand side of (2.5)), i.e., the remaining waiting time has no 'memory' of previous waiting.

Example 2.2.4 Suppose X is the waiting time, starting at time 0, for a bus to arrive, and suppose X is memoryless. After waiting from 0 to t, the distribution of the remaining waiting time from t is the same as the original distribution starting from 0. The still waiting customer is, in a sense, no better off at time t than at time 0. On the other hand, if the bus is known to arrive regularly every 16 minutes, then it will certainly arrive within a minute for a person who has already waited 15 minutes. Thus regular arrivals are not memoryless. The opposite situation is also possible. If the bus frequently breaks down, then a 15-minute wait can indicate that the remaining wait is probably very long, so again X is not memoryless. We study these non-memoryless situations when we study renewal processes in Chapter 5.

Although memoryless distributions must be exponential, it can be seen that if the definition of memoryless is restricted to integer times, then the geometric distribution becomes memoryless, and it can be seen as before that this is the only memoryless integer-time distribution. In this respect, the Bernoulli process (which has geometric interarrival times) is like a discrete-time version of the Poisson process (which has exponential interarrival times).

We now use the memoryless property of exponential rv s to find the distribution of the first arrival in a Poisson process after an arbitrary given time $t > 0$. We not only find this distribution, but also show that this first arrival after t is independent of all arrivals up to and including t. More precisely, we prove the following theorem.

Theorem 2.2.5 *For a Poisson process of rate λ, and any given $t > 0$, the length of the interval from t until the first arrival after t is a positive rv Z with the CDF $1 - \exp[-\lambda z]$ for $z \geq 0$. This rv is independent of both $N(t)$ and the $N(t)$ arrival epochs before time t. It is also independent of the set of rv s $\{N(\tau); \tau \leq t\}$.*

The basic idea behind this theorem is that Z, conditional on the epoch τ of the last arrival before t, is simply the remaining time until the next arrival. Since the interarrival time starting at τ is exponential and thus memoryless, Z is independent of $\tau \leq t$, and of all earlier arrivals. The following proof carries this idea out in detail.

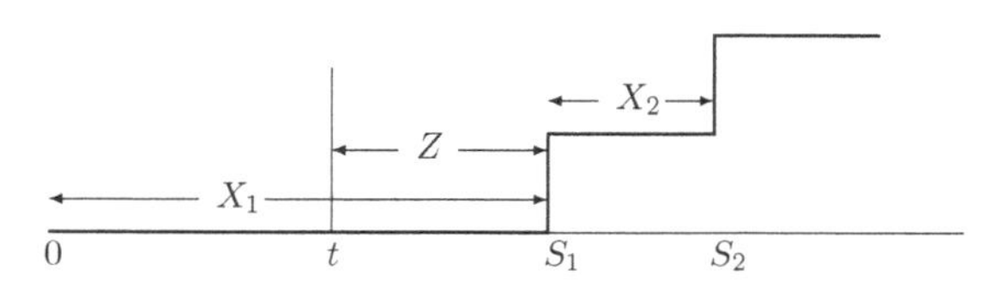

Figure 2.2 For arbitrary fixed $t > 0$, consider the event $\{N(t) = 0\}$. Conditional on this event, Z is the distance from t to S_1, i.e., $Z = X_1 - t$.

Proof Let Z be the distance from t until the first arrival after t. We first condition on $N(t) = 0$ (see Figure 2.2). Given $N(t) = 0$, we see that $X_1 > t$ and $Z = X_1 - t$. Thus,

$$\begin{aligned}\Pr\{Z > z \mid N(t){=}0\} &= \Pr\{X_1 > z + t \mid N(t){=}0\} \\ &= \Pr\{X_1 > z + t \mid X_1 > t\} &(2.6) \\ &= \Pr\{X_1 > z\} \;=\; e^{-\lambda z}. &(2.7)\end{aligned}$$

In (2.6), we used the fact that $\{N(t) = 0\} = \{X_1 > t\}$, which is clear from Figure 2.1 (and also from (2.3)). In (2.7) we used the memoryless condition in (2.5) and the fact that X_1 is exponential.

Next consider the conditions that $N(t) = n$ (for arbitrary $n \geq 1$) and $S_n = \tau$ (for arbitrary $\tau \leq t$). The argument here is basically the same as that with $N(t) = 0$, with a few extra details (see Figure 2.3).

Conditional on $N(t) = n$ and $S_n = \tau$, the first arrival after t is the first arrival after the arrival at S_n, i.e., $Z = z$ corresponds to $X_{n+1} = z + (t - \tau)$.

$$\begin{aligned}\Pr\{Z > z \mid N(t){=}n, S_n{=}\tau\} &= \Pr\{X_{n+1} > z{+}t{-}\tau \mid N(t){=}n,\ S_n{=}\tau\} &(2.8) \\ &= \Pr\{X_{n+1} > z{+}t{-}\tau \mid X_{n+1}{>}t{-}\tau,\ S_n{=}\tau\} &(2.9) \\ &= \Pr\{X_{n+1} > z{+}t{-}\tau \mid X_{n+1}{>}t{-}\tau\} &(2.10) \\ &= \Pr\{X_{n+1} > z\} \;=\; e^{-\lambda z}, &(2.11)\end{aligned}$$

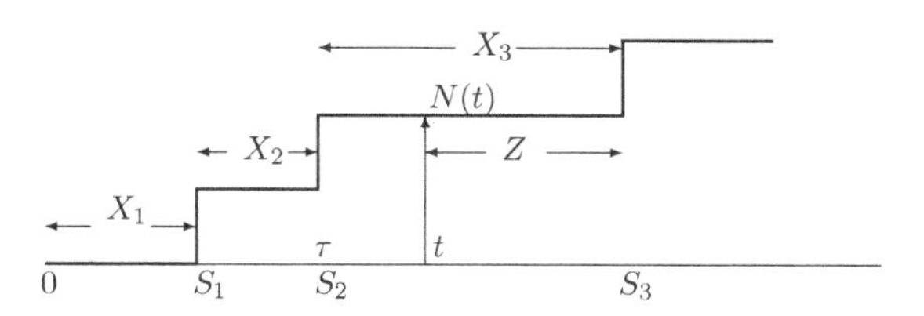

Figure 2.3 Given $N(t) = 2$, and $S_2 = \tau$, X_3 is equal to $Z + (t - \tau)$. Also, the event $\{N(t){=}2,\ S_2{=}\tau\}$ is the same as the event $\{S_2{=}\tau,\ X_3{>}t{-}\tau\}$.

where (2.9) follows because, given $S_n = \tau \le t$, we have $\{N(t) = n\} = \{X_{n+1} > t - \tau\}$ (see Figure 2.3). Equation (2.10) follows because X_{n+1} is independent of S_n. Equation (2.11) follows from the memoryless condition in (2.5) and the fact that X_{n+1} is exponential.

The same argument applies if, in (2.8), we condition not only on S_n but also on $S_1, \dots, S_{n-1}$. Since this is equivalent to conditioning on $N(\tau)$ for all τ in $(0, t]$, we have

$$\Pr\{Z > z \mid \{N(\tau), 0 < \tau \le t\}\} = \exp(-\lambda z). \tag{2.12}$$

□

Next consider the sequence of interarrival times starting from a given time t. The first of these is taken to be the rv Z (now renamed Z_1) of the time from t to the first arrival epoch after t. For $m \ge 2$, let Z_m be the time from the $(m-1)$th arrival after t to the mth arrival epoch after t. Thus, given $N(t) = n$ and $S_n = \tau$, we see that $Z_m = X_{m+n}$ for $m \ge 2$. It follows that, conditional on $N(t) = n$ and $S_n = \tau$, $Z_1, Z_2, \dots$ are IID exponentially distributed rv s (see Exercise 2.8). Since this is independent of $N(t)$ and S_n, it follows that $Z_1, Z_2, \dots$ are unconditionally IID and also independent of $N(t)$ and S_n. It should also be clear that $Z_1, Z_2, \dots$ are independent of $\{N(\tau); 0 < \tau \le t\}$.

The above argument shows that the portion of a Poisson process starting at an arbitrary time $t > 0$ is a probabilistic replica of the process starting at 0; i.e., the time until the first arrival after t is an exponentially distributed rv with parameter λ, and all subsequent interarrival times are independent of this first arrival epoch and of each other, and all have the same exponential distribution.

Definition 2.2.6 *A counting process $\{N(t);\ t > 0\}$ has the* ***stationary increment property*** *if $N(t') - N(t)$ has the same CDF as $N(t' - t)$ for every $t' > t > 0$.*

Let us define $\widetilde{N}(t, t') = N(t') - N(t)$ as the number of arrivals in the interval $(t, t']$ for any given $t' \ge t$. We have just shown that for a Poisson process, the rv $\widetilde{N}(t, t')$ has the same distribution as $N(t' - t)$, which means that a Poisson process has the stationary increment property. Thus, the distribution of the number of arrivals in an interval depends on the size of the interval but not on its starting point.

Definition 2.2.7 *A counting process $\{N(t);\ t > 0\}$ has the* ***independent increment property*** *if, for every integer $k > 0$ and every k-tuple of times $0 < t_1 < t_2 < \cdots < t_k$, the k-tuples of rv s $N(t_1)$, $\widetilde{N}(t_1, t_2), \dots, \widetilde{N}(t_{k-1}, t_k)$ are statistically independent.*

For the Poisson process, Theorem 2.2.5 says that for any t, the time Z_1 until the next arrival after t is independent of $N(\tau)$ for all $\tau \leq t$. Letting $t_1 < t_2 < \cdots < t_{k-1} < t$, this means that Z_1 is independent of $N(t_1), \widetilde{N}(t_1, t_2), \ldots, \widetilde{N}(t_{k-1}, t)$. We have also seen that the subsequent interarrival times after Z_1, and thus $\widetilde{N}(t, t')$ are independent of $N(t_1), \widetilde{N}(t_1, t_2), \ldots, \widetilde{N}(t_{k-1}, t)$. Renaming t as t_k and t' as t_{k+1}, we see that $\widetilde{N}(t_k, t_{k+1})$ is independent of $N(t_1), \widetilde{N}(t_1, t_2), \ldots, \widetilde{N}(t_{k-1}, t_k)$. Since this is true for all k, the Poisson process has the independent increment property. In summary, we have proved the following theorem.

Theorem 2.2.8 *Poisson processes have both the stationary increment and independent increment properties.*

Note that if we look only at integer times, then the Bernoulli process also has the stationary and independent increment properties.

2.2.2 Probability density of S_n and joint density of $S_1, \ldots, S_n$

Recall from (2.1) that, for a Poisson process, S_n is the sum of n IID rv s, each with the density function $f_X(x) = \lambda \exp(-\lambda x)$, $x \geq 0$. Also recall that the density of the sum of two independent rv s can be found by convolving their densities, and thus the density of S_2 can be found by convolving $f_X(x)$ with itself, that of S_3 by convolving the density of S_2 with $f_X(x)$, and so forth. The result, for $t \geq 0$, is called the *Erlang density*,[5]

$$f_{S_n}(t) = \frac{\lambda^n t^{n-1} \exp(-\lambda t)}{(n-1)!}. \tag{2.13}$$

We can understand this density (and other related matters) better by viewing the above mechanical derivation in a slightly different way involving the joint density of $S_1, \ldots, S_n$. For $n = 2$, the joint density of X_1 and S_2 (or equivalently, S_1 and S_2) is given by

$$f_{X_1 S_2}(x_1, s_2) = f_{X_1}(x_1) f_{S_2|X_1}(s_2|x_1) = \lambda e^{-\lambda x_1} \lambda e^{-\lambda(s_2 - x_1)} \qquad \text{for } 0 \leq x_1 \leq s_2,$$

where, since $S_2 = X_1 + X_2$, we have used the fact that the density of S_2 conditional on X_1 is just the exponential interarrival density evaluated at $S_2 - X_1$. Thus,

$$f_{X_1 S_2}(x_1, s_2) = \lambda^2 \exp(-\lambda s_2) \qquad \text{for } 0 \leq x_1 \leq s_2. \tag{2.14}$$

This says that the joint density does not contain x_1, except for the constraint $0 \leq x_1 \leq s_2$. Thus, for fixed s_2, the joint density, and thus the conditional density of X_1 given $S_2 = s_2$ is uniform over $0 \leq x_1 \leq s_2$. The integration over x_1 in the convolution equation is then simply multiplication by the interval size s_2, yielding the marginal PDF $f_{S_2}(s_2) = \lambda^2 s_2 \exp(-\lambda s_2)$, in agreement with (2.13) for $n = 2$.

The following theorem shows that this same curious behavior exhibits itself for the sum of an arbitrary number n of IID exponential rv s.

[5] Another (somewhat rarely used) name for the Erlang density is the *gamma density*.

Theorem 2.2.9 *Let $X_1, X_2, \ldots$ be IID rv s with the density $\mathsf{f}_X(x) = \lambda e^{-\lambda x}$ for $x \geq 0$. Let $S_n = X_1 + \cdots + X_n$ for each $n \geq 1$. Then for each $n \geq 2$*

$$\mathsf{f}_{S_1 \cdots S_n}(s_1, \ldots, s_n) = \lambda^n \exp(-\lambda s_n) \qquad \text{for } 0 \leq s_1 \leq s_2 \cdots \leq s_n. \tag{2.15}$$

Proof Replacing X_1 with S_1 in (2.14), we see the theorem holds for $n = 2$. This serves as the basis for the following inductive proof. Assume (2.15) holds for given n. Then

$$\begin{aligned}\mathsf{f}_{S_1 \cdots S_{n+1}}(s_1, \ldots, s_{n+1}) &= \mathsf{f}_{S_1 \cdots S_n}(s_1, \ldots, s_n)\mathsf{f}_{S_{n+1}|S_1 \cdots S_n}(s_{n+1}|s_1, \ldots, s_n) \\ &= \lambda^n \exp(-\lambda s_n)\, \mathsf{f}_{S_{n+1}|S_1 \cdots S_n}(s_{n+1}|s_1, \ldots, s_n),\end{aligned} \tag{2.16}$$

where we used (2.15) for the given n. Now $S_{n+1} = S_n + X_{n+1}$. Since X_{n+1} is independent of $S_1, \ldots, S_n$,

$$\mathsf{f}_{S_{n+1}|S_1 \cdots S_n}(s_{n+1}|s_1, \ldots, s_n) = \lambda \exp\big(-\lambda(s_{n+1} - s_n)\big).$$

Substituting this into (2.16) yields (2.15). □

The interpretation here is the same as with S_2. The joint density does not contain any arrival time other than s_n, except for the ordering constraint $0 \leq s_1 \leq s_2 \leq \cdots \leq s_n$, and thus this joint density is constant over all choices of arrival times satisfying the ordering constraint for a fixed s_n. Mechanically integrating this over s_1, then s_2, etc. we get the Erlang formula (2.13). The Erlang density then is the joint density in (2.15) times the volume $s_n^{n-1}/(n-1)!$ of the region of $s_1, \ldots, s_{n-1}$ satisfying $0 < s_1 < \cdots < s_n$. This will be discussed further later.

Note that (2.15), for all n, specifies the joint distribution for all arrival epochs, and thus fully specifies a Poisson process.

2.2.3 The probability mass function (PMF) for $N(t)$

The Poisson counting process, $\{N(t);\, t > 0\}$ consists of a non-negative integer rv $N(t)$ for each $t > 0$. In this section, we show that the PMF for this rv is the well-known Poisson PMF, as stated in the following theorem. We give two proofs for the theorem, each providing its own type of understanding and each showing the close relationship between $\{N(t) = n\}$ and $\{S_n = t\}$.

Theorem 2.2.10 *For a Poisson process of rate λ, and for any $t > 0$, the PMF for $N(t)$, i.e., the number of arrivals in $(0, t]$, is given by the Poisson PMF,*

$$\mathsf{p}_{N(t)}(n) = \frac{(\lambda t)^n \exp(-\lambda t)}{n!}. \tag{2.17}$$

Proof 1 This proof, for given n and t, is based on two ways of calculating $\Pr\{t < S_{n+1} \leq t + \delta\}$ for some vanishingly small δ. The first way is based on the already known density of S_{n+1} and gives

$$\Pr\{t < S_{n+1} \leq t + \delta\} \;=\; \int_t^{t+\delta} \mathsf{f}_{S_{n+1}}(\tau)\, d\tau \;=\; \mathsf{f}_{S_{n+1}}(t)\,(\delta + o(\delta)).$$

The term $o(\delta)$ is used to describe a function of δ that goes to 0 faster than δ as $\delta \to 0$. More precisely, a function $g(\delta)$ is said to be of order $o(\delta)$ if $\lim_{\delta \to 0}(g(\delta)/\delta) = 0$. Thus

$\Pr\{t < S_{n+1} \leq t+\delta\} = \mathsf{f}_{S_{n+1}}(t)(\delta + o(\delta))$ is simply a consequence of the fact that S_{n+1} has a continuous probability density in the interval $[t, t+\delta]$.

The second way to calculate $\Pr\{t < S_{n+1} \leq t+\delta\}$ is to first observe that the probability of more than 1 arrival in $(t, t+\delta]$ is $o(\delta)$. Ignoring this possibility, $\{t < S_{n+1} \leq t+\delta\}$ occurs if exactly n arrivals are in the interval $(0, t]$ and one arrival occurs in $(t, t+\delta]$. Because of the independent increment property, this is an event of probability $\mathsf{p}_{N(t)}(n)(\lambda\delta + o(\delta))$. Thus

$$\mathsf{p}_{N(t)}(n)(\lambda\delta + o(\delta)) + o(\delta) = \mathsf{f}_{S_{n+1}}(t)(\delta + o(\delta)).$$

Dividing by δ and taking the limit $\delta \to 0$, we get

$$\lambda \mathsf{p}_{N(t)}(n) = \mathsf{f}_{S_{n+1}}(t).$$

Using the density for f_{S_n} in (2.13), we get (2.17). □

Proof 2 The approach here is to use the fundamental relation that $\{N(t) \geq n\} = \{S_n \leq t\}$. Taking the probabilities of these events,

$$\sum_{i=n}^{\infty} \mathsf{p}_{N(t)}(i) = \int_0^t \mathsf{f}_{S_n}(\tau)\, d\tau \qquad \text{for all } n \geq 1 \text{ and } t > 0.$$

Since $\mathsf{p}_{N(t)}(0) = e^{-\lambda t}$, the sum on the left above, if specified for all $n \geq 1$, is equivalent to specifying $\mathsf{p}_{N(t)}(n)$ for all $n \geq 1$. Thus the theorem is equivalent to showing that for all $n \geq 0$,

$$\sum\nolimits_{i=n}^{\infty} \frac{(\lambda t)^i \exp(-\lambda t)}{i!} = \int_0^t \mathsf{f}_{S_n}(\tau)\, d\tau. \tag{2.18}$$

If we take the derivative with respect to t of each side of (2.18), we find that almost magically each term except the first on the left cancels out, leaving us with

$$\frac{\lambda^n t^{n-1} \exp(-\lambda t)}{(n-1)!} = \mathsf{f}_{S_n}(t).$$

Thus the derivative with respect to t of each side of (2.18) is equal to the derivative of the other for all $n \geq 1$ and $t > 0$. The two sides of (2.18) are also equal in the limit $t \to \infty$, so it follows that (2.18) is satisfied everywhere, completing the proof. □

2.2.4 Alternative definitions of Poisson processes

Definition 2 of a poisson process *A **Poisson counting process** $\{N(t); t > 0\}$ is a counting process that satisfies (2.17) (i.e., has the Poisson PMF) and has the independent and stationary increment properties.*

We have seen that the properties in Definition 2 are satisfied starting with Definition 1 (using IID exponential interarrival times), so Definition 1 implies Definition 2. Exercise 2.4 shows that IID exponential interarrival times are implied by Definition 2, so the two definitions are equivalent.

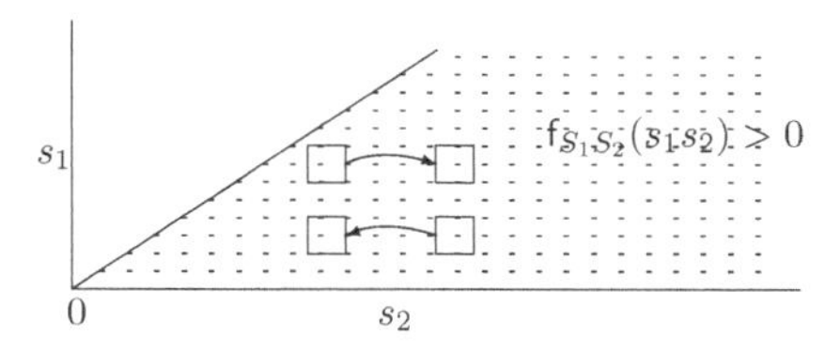

Figure 2.4 The joint density of S_1, S_2 is non-zero in the region shown. It can be changed, while holding the marginals constant, by reducing the joint density by ϵ in the upper left and lower right squares above and increasing it by ϵ in the upper right and lower left squares.

It may be somewhat surprising at first to realize that a counting process that has the Poisson PMF at each t is not necessarily a Poisson process, and that the independent and stationary increment properties are also necessary. One way to see this is to recall that the Poisson PMF for all t in a counting process is equivalent to the Erlang density for the successive arrival epochs. Specifying the probability density for $S_1, S_2, \ldots$ as Erlang specifies the *marginal* densities of $S_1, S_2, \ldots$ but need not specify the *joint* densities of these rv s. Figure 2.4 illustrates this in terms of the joint density of S_1, S_2, given as

$$\mathsf{f}_{S_1S_2}(s_1, s_2) = \lambda^2 \exp(-\lambda s_2) \qquad \text{for } 0 \leq s_1 \leq s_2$$

and 0 elsewhere. The figure illustrates how the joint density can be changed without changing the marginals.

There is a similar effect with the Bernoulli process in that a discrete counting process for which the number of arrivals from 0 to t, for each integer t, is a binomial rv, but the process is not Bernoulli. This is explored in Exercise 2.5.

The next definition of a Poisson process is based on its incremental properties. Consider the number of arrivals in some very small interval $(t, t+\delta]$. Since $\widetilde{N}(t, t+\delta)$ has the same distribution as $N(\delta)$, we can use (2.17) to get

$$\begin{aligned}
\Pr\{\widetilde{N}(t, t+\delta) = 0\} &= e^{-\lambda\delta} \approx 1 - \lambda\delta + o(\delta), \\
\Pr\{\widetilde{N}(t, t+\delta) = 1\} &= \lambda\delta e^{-\lambda\delta} \approx \lambda\delta + o(\delta), \\
\Pr\{\widetilde{N}(t, t+\delta) \geq 2\} &\approx o(\delta).
\end{aligned} \tag{2.19}$$

Definition 3 of a poisson process *A Poisson counting process is a counting process that satisfies (2.19) and has the stationary and independent increment properties.*

We have seen that Definition 1 implies Definition 3. The essence of the argument the other way is that for any interarrival interval X, $\mathsf{F}_X(x+\delta) - \mathsf{F}_X(x)$ is the probability of an arrival in an appropriate infinitesimal interval of width δ, which by (2.19) is $\lambda\delta + o(\delta)$. Turning this into a differential equation (see Exercise 2.7), we get the desired exponential interarrival intervals. Definition 3 has an intuitive appeal, since it is based on the idea of independent arrivals during arbitrary disjoint intervals. It has the disadvantage that one must do a considerable amount of work to be sure that these conditions are mutually consistent, and probably the easiest way is to start with Definition 1 and derive these properties. Showing that there is a unique process that satisfies the conditions of Definition 3 is even harder, but is not necessary at this point, since all we need is the use

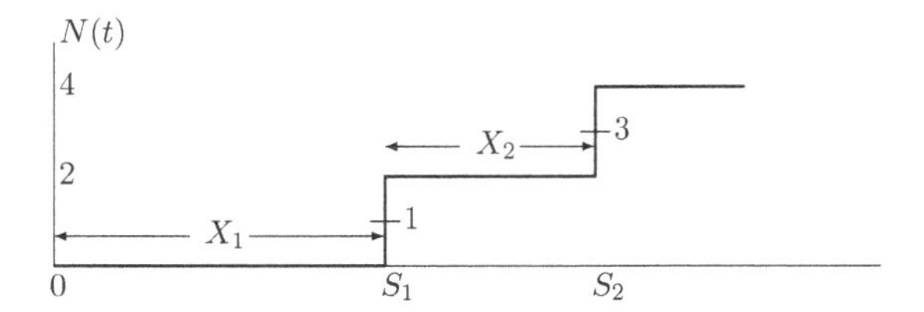

Figure 2.5 A counting process modeling bulk arrivals. X_1 is the time until the first pair of arrivals and X_2 is the interval between the first and second pair of arrivals.

of these properties. Section 2.2.5 will illustrate better how to use this definition (or more precisely, how to use (2.19)).

What (2.19) accomplishes in Definition 3, beyond the assumption of independent and stationary increments, is the prevention of bulk arrivals. For example, consider a counting process in which arrivals always occur in pairs, and the intervals between successive pairs are IID and exponentially distributed with parameter λ (see Figure 2.5). For this process, $\Pr\{\widetilde{N}(t, t+\delta)=1\} = 0$, and $\Pr\{\widetilde{N}(t, t+\delta)=2\} = \lambda\delta + o(\delta)$, thus violating (2.19). This process has stationary and independent increments, however, since the process formed by viewing a pair of arrivals as a single incident is a Poisson process.

2.2.5 The Poisson process as a limit of shrinking Bernoulli processes

The intuition of Definition 3 can be achieved in a less abstract way by starting with the Bernoulli process, which has the properties of Definition 3 in a discrete-time sense. We then go to an appropriate limit of a sequence of these processes, and find that this sequence of Bernoulli processes converges in some sense to the Poisson process.

Recall that a Bernoulli process is an IID sequence, $Y_1, Y_2, \ldots$ of binary rv s for which $\mathsf{p}_Y(1) = p$ and $\mathsf{p}_Y(0) = 1 - p$. We can visualize $Y_i = 1$ as an *arrival* at time i and $Y_i = 0$ as no arrival, but we can also 'shrink' the time scale of the process so that for some integer $j > 0$, Y_i is an arrival or no arrival at time $i2^{-j}$. We consider a sequence indexed by j of such shrinking Bernoulli processes, and in order to keep the arrival rate constant, we let $p = \lambda 2^{-j}$ for the jth process. Thus for each unit increase in j, the Bernoulli process shrinks by replacing each slot with two slots, each with half the previous arrival probability. The expected number of arrivals per unit time is then λ, matching the Poisson process that we are approximating.

If we look at this jth process relative to Definition 3 of a Poisson process, we see that for these regularly spaced increments of size $\delta = 2^{-j}$, the probability of one arrival in an increment is $\lambda\delta$ and that of no arrival is $1 - \lambda\delta$, and thus (2.19) is satisfied, and in fact the $o(\delta)$ terms are exactly zero. For arbitrary sized increments, it is clear that disjoint increments have essentially independent arrivals. The increments are not quite stationary, since, for example, an increment of size 2^{-j-1} might contain a time that is a multiple of 2^{-j} or might not, depending on its placement. However, for any fixed increment of size δ, the number of multiples of 2^{-j} (i.e., the number of possible arrival points) is either $\lfloor \delta 2^j \rfloor$ or $1 + \lfloor \delta 2^j \rfloor$. Thus in the limit $j \to \infty$, the increments are both stationary and independent.

For each j, the jth Bernoulli process has an associated Bernoulli counting process $N_j(t) = \sum_{i=1}^{\lfloor t2^j \rfloor} Y_i$. This is the number of arrivals up to time t and is a discrete rv with the binomial PMF. That is, $\mathsf{p}_{N_j(t)}(n) = \binom{\lfloor t2^j \rfloor}{n} p^n (1-p)^{\lfloor t2^j \rfloor - n}$ where $p = \lambda 2^{-j}$. We now show that this PMF approaches the Poisson PMF as j increases.[6]

Theorem 2.2.11 (Poisson's theorem) *Consider the sequence of shrinking Bernoulli processes with arrival probability $\lambda 2^{-j}$ and time-slot size 2^{-j}. Then for every fixed time $t > 0$ and fixed number of arrivals n, the counting PMF $\mathsf{p}_{N_j(t)}(n)$ approaches the Poisson PMF (of the same λ) with increasing j, i.e.,*

$$\lim_{j\to\infty} \mathsf{p}_{N_j(t)}(n) = \mathsf{p}_{N(t)}(n). \tag{2.20}$$

Proof We first rewrite the binomial PMF, for $\lfloor t2^j \rfloor$ variables with $p = \lambda 2^{-j}$ as

$$\begin{aligned}
\lim_{j\to\infty} \mathsf{p}_{N_j(t)}(n) &= \lim_{j\to\infty} \binom{\lfloor t2^j \rfloor}{n} \left(\frac{\lambda 2^{-j}}{1-\lambda 2^{-j}}\right)^n \exp[\lfloor t2^j \rfloor \ln(1-\lambda 2^{-j})] \\
&= \lim_{j\to\infty} \binom{\lfloor t2^j \rfloor}{n} \left(\frac{\lambda 2^{-j}}{1-\lambda 2^{-j}}\right)^n \exp(-\lambda t) && (2.21) \\
&= \lim_{j\to\infty} \frac{\lfloor t2^j \rfloor \cdot \lfloor t2^j - 1 \rfloor \cdots \lfloor t2^j - n + 1 \rfloor}{n!} \left(\frac{\lambda 2^{-j}}{1-\lambda 2^{-j}}\right)^n \exp(-\lambda t) && (2.22) \\
&= \frac{(\lambda t)^n \exp(-\lambda t)}{n!}. && (2.23)
\end{aligned}$$

We used $\ln(1-\lambda 2^{-j}) = -\lambda 2^{-j} + o(2^{-j})$ in (2.21) and expanded the combinatorial term in (2.22). In (2.23), we recognized that $\lim_{j\to\infty} \lfloor t2^j - i \rfloor \left(\frac{\lambda 2^{-j}}{1-\lambda 2^{-j}}\right) = \lambda t$ for $0 \le i \le n-1$. □

Since the binomial PMF (scaled as above) has the Poisson PMF as a limit for each n, the CDF of $N_j(t)$ also converges to the Poisson CDF for each t. In other words, for each $t > 0$, the counting rv s $N_j(t)$ of the Bernoulli processes converge in distribution to $N(t)$ of the Poisson process.

This does not say that the Bernoulli *counting processes* converge to the Poisson counting process in any meaningful sense, since the joint distributions are also of concern. The following corollary treats this.

Corollary 2.2.12 *For any finite integer $k > 0$, let $0 < t_1 < t_2 < \cdots < t_k$ be any set of time instants. Then the joint CDF of $N_j(t_1), N_j(t_2), \ldots N_j(t_k)$ approaches the joint CDF of $N(t_1), N(t_2), \ldots N(t_k)$ as $j \to \infty$.*

[6] This limiting result for the binomial distribution is very different from the asymptotic results in Chapter 1 for the binomial. Here the parameter p of the binomial is shrinking with increasing j, whereas there p is constant while the number of variables is increasing.

Proof We can rewrite the joint PMF for each Bernoulli process as

$$\mathsf{p}_{N_j(t_1),\ldots,N_j(t_k)}(n_1,\ldots,n_k) = \mathsf{p}_{N_j(t_1),\widetilde{N}_j(t_1,t_2),\ldots,\widetilde{N}_j(t_{k-1},t_k)}(n_1,n_2-n_1,\ldots,n_k-n_{k-1})$$
$$= \mathsf{p}_{N_j(t_1)}(n_1)\prod_{\ell=2}^{k}\mathsf{p}_{\widetilde{N}_j(t_\ell,t_{\ell-1})}(n_\ell - n_{\ell-1}), \tag{2.24}$$

where we have used the independent increment property for the Bernoulli process. For the Poisson process, we similarly have

$$\mathsf{p}_{N(t_1),\ldots,N(t_k)}(n_1,\ldots,n_k) = \mathsf{p}_{N(t_1)}(n_1)\prod_{\ell=2}^{k}\mathsf{p}_{\widetilde{N}(t_\ell,t_{\ell-1})}(n_\ell - n_{\ell-1}). \tag{2.25}$$

Taking the limit of (2.24) as $j \to \infty$, we recognize from Theorem 2.2.11 that each term of (2.24) goes to the corresponding term in (2.25). For the $\widetilde{N}$ rv s, this requires a trivial generalization in Theorem 2.2.11 to deal with the arbitrary starting time. □

We conclude from this that the sequence of Bernoulli processes above converges to the Poisson process in the sense of the corollary. Recall from Section 1.7.5 that there are a number of ways in which a sequence of rv s can converge. As one might imagine, there are many more ways in which a sequence of stochastic processes can converge, and the corollary simply establishes one of these. Note, however, that showing only that $p_{N_j(t)}(n)$ approaches $p_{N(t)}(n)$ for each t is too weak to be very helpful, because it shows nothing about the time evolution of the process. On the other hand, we do not even know how to define a joint distribution over an infinite number of epochs, let alone deal with limiting characteristics. Considering arbitrary finite sets of epochs forms a good middle ground.

Both the Poisson process and the Bernoulli process are so easy to analyze that the convergence of shrinking Bernoulli processes to Poisson is rarely the easiest way to establish properties about either. On the other hand, this convergence is a powerful aid to the intuition in understanding each process. In other words, the relation between Bernoulli and Poisson is very useful in suggesting new ways of looking at problems, but is usually not the best way to analyze those problems.

2.3 Combining and splitting Poisson processes

Suppose that $\{N_1(t);\, t > 0\}$ and $\{N_2(t);\, t > 0\}$ are independent Poisson counting processes[7] of rates λ_1 and λ_2 respectively. We want to look at the sum process where $N(t) = N_1(t)+N_2(t)$ for all $t \geq 0$. In other words, $\{N(t);\, t > 0\}$ is the process consisting of all arrivals to both process 1 and process 2. We shall show that $\{N(t);\, t > 0\}$ is a Poisson counting process of rate $\lambda = \lambda_1+\lambda_2$. We show this in three different ways, first using Definition 3 of a Poisson process (since that is most natural for this problem), then using

[7] Two processes $\{N_1(t);\, t > 0\}$ and $\{N_2(t);\, t > 0\}$ are said to be independent if for all positive integers k and all sets of times $0 < t_1 < t_2 < \cdots < t_k$, the rv s $N_1(t_1),\ldots,N_1(t_k)$ are independent of $N_2(t_1),\ldots,N_2(t_k)$. Here it is enough to extend the independent increment property to independence between increments over the two processes; equivalently, one can require the interarrival intervals for one process to be independent of the interarrivals for the other process.

Definition 2, and finally Definition 1. We then draw some conclusions about the way in which each approach is helpful. Since $\{N_1(t);\, t > 0\}$ and $\{N_2(t);\, t > 0\}$ are independent and both possess the stationary and independent increment properties, it follows from the definitions that $\{N(t);\, t > 0\}$ also possesses the stationary and independent increment properties. Using the approximations in (2.19) for the individual processes, we see that

$$\begin{aligned}\Pr\left\{\widetilde{N}(t, t+\delta) = 0\right\} &= \Pr\left\{\widetilde{N}_1(t, t+\delta) = 0\right\} \Pr\left\{\widetilde{N}_2(t, t+\delta) = 0\right\} \\ &= (1 - \lambda_1\delta)(1 - \lambda_2\delta) \;\approx\; 1 - \lambda\delta,\end{aligned}$$

where $\lambda_1\lambda_2\delta^2$ has been dropped. In the same way, $\Pr\{\widetilde{N}(t,\, t+\delta) = 1\}$ is approximated by $\lambda\delta$ and $\Pr\{\widetilde{N}(t, t+\delta) \geq 2\}$ is approximated by 0, both with errors proportional to δ^2. It follows that $\{N(t);\, t > 0\}$ is a Poisson process.

In the second approach, we have $N(t) = N_1(t) + N_2(t)$. Since $N(t)$, for any given t, is the sum of two independent Poisson rv s, it is also a Poisson rv with mean $\lambda t = \lambda_1 t + \lambda_2 t$. If the reader is not aware that the sum of two independent Poisson rv s is Poisson, it can be derived by discrete convolution of the two PMFs (see Exercise 1.22). More elegantly, one can observe that we have already implicitly shown this fact. That is, if we break an interval I into disjoint subintervals, I_1 and I_2, then the number of arrivals in I (which is Poisson) is the sum of the number of arrivals in I_1 and in I_2 (which are independent Poisson). Finally, since $N(t)$ is Poisson for each t, and since the stationary and independent increment properties are satisfied, $\{N(t);\, t > 0\}$ is a Poisson process.

In the third approach, X_1, the first interarrival interval for the sum process, is the minimum of X_{11}, the first interarrival interval for the first process, and X_{21}, the first interarrival interval for the second process. Thus $X_1 > t$ if and only if both X_{11} and X_{21} exceed t, so

$$\Pr\{X_1 > t\} = \Pr\{X_{11} > t\}\Pr\{X_{21} > t\} = \exp(-\lambda_1 t - \lambda_2 t) = \exp(-\lambda t).$$

Using the memoryless property, each subsequent interarrival interval can be analyzed in the same way.

The first approach above was the most intuitive for this problem, but it required constant care about the order of magnitude of the terms being neglected. The second approach was the simplest analytically (after recognizing that sums of independent Poisson rv s are Poisson), and required no approximations. The third approach was very simple in retrospect, but not very natural for this problem. If we add many independent Poisson processes together, it is clear, by adding them one at a time, that the sum process is again Poisson. What is more interesting is that when many independent counting processes (not necessarily Poisson) are added together, the sum process often tends to be approximately Poisson if the individual processes have small rates compared to the sum. To obtain some crude intuition about why this might be expected, note that the interarrival intervals for each process (assuming no bulk arrivals) will tend to be large relative to the mean interarrival interval for the sum process. Thus arrivals that are close together in time will typically come from different processes. The number of arrivals in an interval large relative to the combined mean interarrival interval, but small relative to the individual interarrival intervals, will be the sum of the number of arrivals from the different processes; each of these is 0 with large probability and 1 with small probability, so the sum will be approximately Poisson.

2.3.1 Subdividing a Poisson process

Next we look at how to break $\{N(t);\ t > 0\}$, a Poisson counting process of rate λ, into two processes, $\{N_1(t);\ t > 0\}$ and $\{N_2(t);\ t > 0\}$. Suppose that each arrival in $\{N(t);\ t > 0\}$ is sent to the first process with probability p and to the second process with probability $1 - p$ (see Figure 2.6). Each arrival is switched independently of each other arrival and independently of the arrival epochs. It may be helpful to visualize this as the combination of two independent processes. The first is the Poisson process of rate λ and the second is a Bernoulli process $\{X_n;\ n \geq 1\}$ where $\mathsf{p}_{X_n}(1) = p$ and $\mathsf{p}_{X_n}(2) = 1 - p$. The nth arrival of the Poisson process is, with probability p, labeled as a type 1 arrival, i.e., labeled as $X_n = 1$. With probability $1 - p$, it is labeled as type 2, i.e., labeled as $X_n = 2$.

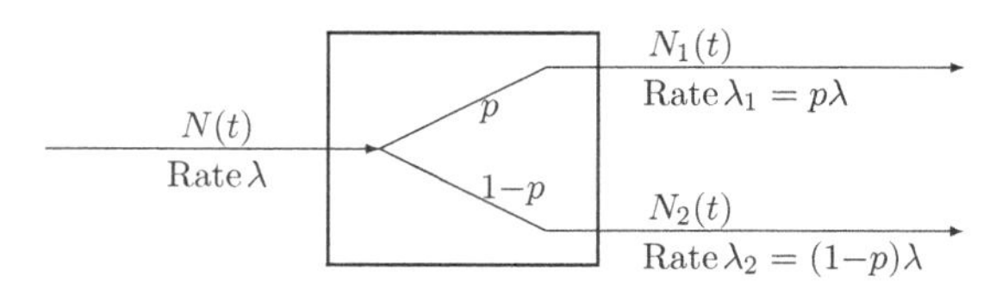

Figure 2.6 Each arrival is independently sent to process 1 with probability p and to process 2 otherwise.

We shall show that the resulting processes are each Poisson, with rates $\lambda_1 = \lambda p$ and $\lambda_2 = \lambda(1-p)$ respectively, and that furthermore the two processes are independent. Note that, conditional on the original process, the two new processes are not independent; in fact one completely determines the other. Thus this independence might be a little surprising.

First consider a small increment $(t, t + \delta]$. The original process has an arrival in this incremental interval with probability $\lambda\delta$ (ignoring δ^2 terms as usual), and thus the first process has an arrival with probability $\lambda\delta p$ and the second process with probability $\lambda\delta(1 - p)$. Because of the independent increment property of the original process and the independence of the division of each arrival between the two processes, the new processes each have the independent increment property, and from above have the stationary increment property. Thus each process is Poisson. Note now that we cannot verify that the two processes are independent from this small increment model. We would have to show that the numbers of arrivals for process 1 and process 2 are independent over $(t, t + \delta]$. Unfortunately, leaving out the terms of order δ^2, there is at most one arrival to the original process and no possibility of an arrival to each new process in $(t, t + \delta]$. *If* it is impossible for both processes to have an arrival in the same interval, they cannot be independent. It is possible, of course, for each process to have an arrival in the same interval, but this is a term of order δ^2. Thus, without paying attention to the terms of order δ^2, it is impossible to demonstrate that the processes are independent.

To demonstrate that processes 1 and 2 are independent, we first calculate the joint PMF for $N_1(t)$, $N_2(t)$ for arbitrary t. Conditioning on a given number of arrivals $N(t)$ for the original process, we have

$$\Pr\{N_1(t)=m, N_2(t)=k \mid N(t)=m+k\} = \frac{(m+k)!}{m!k!} p^m (1-p)^k. \tag{2.26}$$

Equation (2.26) is simply the binomial distribution, since, given $m+k$ arrivals to the original process, each independently goes to process 1 with probability p. Since the event $\{N_1(t) = m, N_2(t) = k\}$ is a subset of the conditioning event above,

$$\Pr\{N_1(t)=m, N_2(t)=k \mid N(t)=m+k\} = \frac{\Pr\{N_1(t)=m, N_2(t)=k\}}{\Pr\{N(t)=m+k\}}.$$

Combining this with (2.26), we have

$$\Pr\{N_1(t)=m, N_2(t)=k\} = \frac{(m+k)!}{m!k!} p^m (1-p)^k \frac{(\lambda t)^{m+k} e^{-\lambda t}}{(m+k)!}. \tag{2.27}$$

Rearranging terms, we get

$$\Pr\{N_1(t)=m, N_2(t)=k\} = \frac{(p\lambda t)^m e^{-\lambda p t}}{m!} \frac{[(1-p)\lambda t]^k e^{-\lambda(1-p)t}}{k!}. \tag{2.28}$$

This shows that $N_1(t)$ and $N_2(t)$ are independent. To show that the processes are independent, we must show that for any $k > 1$ and any set of times $0 \le t_1 \le t_2 \le \cdots \le t_k$, the sets $\{N_1(t_i);\ 1 \le i \le k\}$ and $\{N_2(t_j);\ 1 \le j \le k\}$ are independent of each other. It is equivalent to show that the sets $\{\widetilde{N}_1(t_{i-1}, t_i); 1 \le i \le k\}$ and $\{\widetilde{N}_2(t_{j-1}, t_j); 1 \le j \le k\}$ (where t_0 is 0) are independent. The argument above shows this independence for $i = j$, and for $i \ne j$ the independence follows from the independent increment property of $\{N(t);\ t > 0\}$.

2.3.2 Examples using independent Poisson processes

We have observed that if the arrivals of a Poisson process are split into two new arrival processes, with each arrival of the original process independently entering the first of the new processes with some fixed probability p, then each new process is Poisson and independent of the other. The most useful consequence of this is that any two independent Poisson processes can be viewed as being generated from a single process in this way. Thus, if one process has rate λ_1 and the other has rate λ_2, they can be viewed as coming from a process of rate $\lambda_1 + \lambda_2$. Each arrival to the combined process is then labeled as a first-process arrival with probability $p = \lambda_1/(\lambda_1 + \lambda_2)$ and as a second-process arrival with probability $1 - p$.

Example 2.3.1 The above point of view is very useful for finding probabilities such as $\Pr\left\{S_{1k} < S_{2j}\right\}$ where S_{1k} is the epoch of the kth arrival to the first process and S_{2j} is the epoch of the jth arrival to the second process. The problem can be rephrased in terms of a combined process to ask: out of the first $k + j - 1$ arrivals to the combined process, what is the probability that k or more of them are switched to the first process? (Note that if k or more of the first $k+j-1$ go to the first process, at most $j-1$ go to the second, so the kth arrival to the first precedes the jth arrival to the second; similarly if fewer than k of the first $k+j-1$ go to the first process, then the jth arrival to the second process precedes the kth arrival to the first). Since each of these first $k+j-1$ arrivals is switched independently with the same probability p, the answer is

$$\Pr\{S_{1k} < S_{2j}\} = \sum_{i=k}^{k+j-1} \frac{(k+j-1)!}{i!(k+j-1-i)!} p^i (1-p)^{k+j-1-i}. \tag{2.29}$$

Example 2.3.2 (The M/M/1 queue) Queueing theorists use a standard notation of characters separated by slashes to describe common types of queueing systems. The first character describes the arrival process to the queue. M stands for memoryless and means a Poisson arrival process; D stands for deterministic and means that the interarrival interval is fixed and non-random; G stands for general interarrival distribution. We assume that the interarrival intervals are IID (thus making the arrival process a renewal process), but many authors use GI to explicitly indicate IID interarrivals. The second character describes the service process. The same letters are used, with M indicating exponentially distributed service times. The third character gives the number of servers.[8] It is assumed, when this notation is used, that the service times are IID, independent of the arrival epochs, and independent of which server is used.

Consider an M/M/1 queue, i.e., a queueing system with a Poisson arrival system (say of rate λ) and a single server that serves arriving customers in order with a service time distribution $\mathsf{F}(y) = 1 - \exp[-\mu y]$. The service times are independent of each other and of the interarrival intervals. During any period when the server is busy, customers leave the system according to a Poisson process (process 2) of rate μ. We see that if j or more customers are waiting at a given time, then (2.29) gives the probability that the kth subsequent arrival comes before the jth departure.

Example 2.3.3 (The sum of geometrically many exponential rv s) Consider waiting for a taxi on a busy street during rush hour. We will model the epochs at which taxis pass our spot on the street as a Poisson process of rate λ. We assume that each taxi is independently empty (i.e., will stop to pick us up) with probability p. The empty taxis then form a Poisson process of rate $p\lambda$ and our waiting time is exponential with parameter λp. Looked at slightly differently, we will be picked by the first taxi with probability p, by the second with probability $(1-p)p$, and, in general, by the mth taxi with probability $(1-p)^{m-1}p$. In other words, the number of the taxi that picks us up (numbered in order of arrival) is a geometrically distributed rv, say M, and our waiting time is the sum of the first M interarrival times, i.e., S_M.

[8] Sometimes a fourth character is added which gives the number of customers that can be saved in the queue plus service facility. Thus, for example, an M/M/m/m queueing system would have Poisson arrivals, m independent servers, each with exponential service time. If an arrival occurs when all servers are busy, then that arrival is dropped.

As illustrated by taxis, waiting times modeled by a geometric sum of exponentials are relatively common. An important example that we shall analyze in Chapter 7 is that of the service time distribution of an M/M/1 queue in steady state (i.e., after the effect of the initially empty queue has disappeared). We will show there that the number M of customers in the system immediately after an arrival is geometric with the PMF

$$\Pr\{M = m\} = (\lambda/\mu)^{m-1}(1 - \lambda/\mu) \qquad \text{for } m \geq 1.$$

Now the time that the new arrival spends in the system is the sum of M service times, each of which is exponential with parameter μ. These service times are independent of each other and of M. Thus the system time (the time in queue and in service) of the new arrival is exponential with parameter $\mu(1 - \lambda/\mu) = \mu - \lambda$.

2.4 Non-homogeneous Poisson processes

The Poisson process, as we defined it, is characterized by a constant arrival rate λ. It is often useful to consider a more general type of process in which the arrival rate varies as a function of time. A *non-homogeneous Poisson process* with time-varying arrival rate $\lambda(t)$ is defined[9] as a counting process $\{N(t);\ t > 0\}$ which has the independent increment property and, for all $t \geq 0, \delta > 0$, also satisfies:

$$\begin{aligned}
\Pr\left\{\widetilde{N}(t, t+\delta) = 0\right\} &= 1 - \delta\lambda(t) + o(\delta), \\
\Pr\left\{\widetilde{N}(t, t+\delta) = 1\right\} &= \delta\lambda(t) + o(\delta), \\
\Pr\left\{\widetilde{N}(t, t+\delta) \geq 2\right\} &= o(\delta),
\end{aligned} \tag{2.30}$$

where $\widetilde{N}(t, t+\delta) = N(t+\delta) - N(t)$. The non-homogeneous Poisson process does not have the stationary increment property.

One common application occurs in optical communication where a non-homogeneous Poisson process is often used to model the stream of photons from an optical modulator; the modulation is accomplished by varying the photon intensity $\lambda(t)$. We shall see another application shortly in the next example. Sometimes a Poisson process, as we defined it earlier, is called a homogeneous Poisson process.

We can use a 'shrinking Bernoulli process' again to approximate a non-homogeneous Poisson process. To see how to do this, assume that $\lambda(t)$ is bounded away from zero. We partition the time axis into increments whose lengths δ vary inversely with $\lambda(t)$, thus holding the probability of an arrival in an increment at some fixed value $p = \delta\lambda(t)$. Thus, temporarily ignoring the variation of $\lambda(t)$ within an increment,

[9] We assume that $\lambda(t)$ is right continuous, i.e., that for each t, $\lambda(t)$ is the limit of $\lambda(t+\epsilon)$ as ϵ approaches 0 from above. This allows $\lambda(t)$ to contain discontinuities, as illustrated in Figure 2.7, but follows the convention that the value of the function at the discontinuity is the limiting value from the right. This convention is required in (2.30) to talk about the distribution of arrivals just to the right of time t.

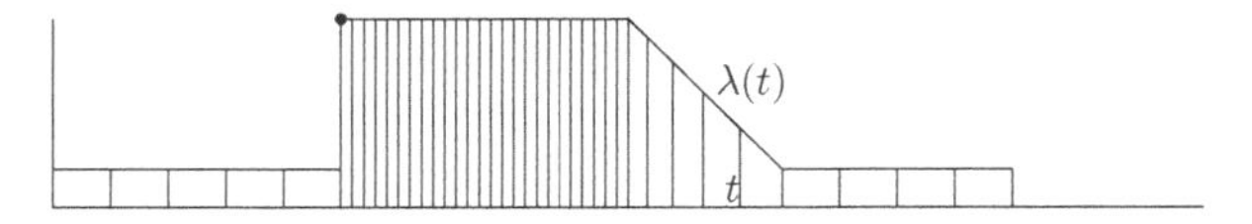

Figure 2.7 Partitioning the time axis into increments each with an expected number of arrivals equal to p. Each rectangle or trapezoid above has the same area, which ensures that the ith partition ends where $m(t) = i\,p$.

$$\begin{aligned}\Pr\left\{\widetilde{N}\left(t, t+\frac{p}{\lambda(t)}\right) = 0\right\} &= 1 - p + o(p),\\ \Pr\left\{\widetilde{N}\left(t, t+\frac{p}{\lambda(t)}\right) = 1\right\} &= p + o(p),\\ \Pr\left\{\widetilde{N}\left(t, t+\frac{p}{\lambda(t)}\right) \geq 2\right\} &= o(p).\end{aligned} \tag{2.31}$$

This partition is defined more precisely by defining $m(t)$ as

$$m(t) = \int_0^t \lambda(\tau)d\tau. \tag{2.32}$$

As shown in Figure 2.7, then the ith increment ends at that t for which $m(t) = i\,p$.

As before, let $\{Y_i;\ i \geq 1\}$ be a sequence of IID binary rv s with $\Pr\{Y_i = 1\} = p$ and $\Pr\{Y_i = 0\} = 1 - p$. Consider the counting process $\{N(t);\ t > 0\}$ in which Y_i, for each $i \geq 1$, denotes the number of arrivals in the interval $(t_{i-1}, t_i]$, where t_i satisfies $m(t_i) = i\,p$. Thus, $N(t_i) = Y_1 + Y_2 + \cdots + Y_i$. If p is decreased as 2^{-j}, each increment is successively split into a pair of increments. Thus by the same argument as in (2.23),

$$\Pr\{N(t) = n\} = \frac{[1 + o(p)][m(t)]^n \exp[-m(t)]}{n!}. \tag{2.33}$$

Similarly, for any interval $(t, \tau]$, taking $\widetilde{m}(t, \tau) = \int_t^\tau \lambda(u)du$, and taking $t = t_k$, $\tau = t_i$ for some k, i, we get

$$\Pr\left\{\widetilde{N}(t, \tau) = n\right\} = \frac{[1 + o(p)][\widetilde{m}(t, \tau)]^n \exp[-\widetilde{m}(t, \tau)]}{n!}. \tag{2.34}$$

Going to the limit $p \to 0$, the counting process $\{N(t);\ t > 0\}$ above approaches the non-homogeneous Poisson process under consideration, and we have the following theorem.

Theorem 2.4.1 *For a non-homogeneous Poisson process with right-continuous arrival rate $\lambda(t)$ bounded away from zero, the distribution of $\widetilde{N}(t, \tau)$, the number of arrivals in $(t, \tau]$, satisfies*

$$\Pr\left\{\widetilde{N}(t, \tau) = n\right\} = \frac{[\widetilde{m}(t, \tau)]^n \exp[-\widetilde{m}(t, \tau)]}{n!}, \qquad \text{where } \widetilde{m}(t, \tau) = \int_t^\tau \lambda(u)\,du. \tag{2.35}$$

Hence, one can view a non-homogeneous Poisson process as a (homogeneous) Poisson process over a non-linear time scale. That is, let $\{N^*(s); s \geq 0\}$ be a (homogeneous)

Poisson process with rate 1. The non-homogeneous Poisson process is then given by $N(t) = N^*(m(t))$ for each t.

Example 2.4.2 (The M/G/∞ queue) Using the queueing notation explained in Example 2.3.2, an M/G/∞ queue indicates a queue with Poisson arrivals, a general service distribution, and an infinite number of servers. Since the M/G/∞ queue has an infinite number of servers, no arriving customers are ever queued. Each arrival immediately starts to be served by some server, and the service time Y_i of customer i is IID over i with some given CDF $G(y)$; the service time is the interval from start to completion of service and is also independent of arrival epochs. We would like to find the CDF of the number of customers being served at a given epoch τ.

Let $\{N(t);\ t > 0\}$ be the Poisson counting process, at rate λ, of customer arrivals. Consider the arrival times of those customers that are still in service at some fixed time τ. In some arbitrarily small interval $(t, t+\delta]$, the probability of an arrival is $\delta\lambda + o(\delta)$ and the probability of two or more arrivals is negligible (i.e., $o(\delta)$). The probability that a customer arrives in $(t, t+\delta]$ and is still being served at time $\tau > t$ is then $\delta\lambda[1 - G(\tau - t)] + o(\delta)$. Consider a counting process $\{N_1(t);\ 0<t\leq\tau\}$, where $N_1(t)$ is the number of arrivals between 0 and t that are still in service at τ. This counting process has the independent increment property. To see this, note that the overall arrivals in $\{N(t);\ t > 0\}$ have the independent increment property; also the arrivals in $\{N(t);\ t > 0\}$ have independent service times, and thus are independently in or not in $\{N_1(t);\ 0 < t \leq \tau\}$. It follows that $\{N_1(t);\ 0 < t \leq \tau\}$ is a non-homogeneous Poisson process with rate $\lambda[1 - G(\tau - t)]$ at time $t \leq \tau$. The expected number of arrivals still in service at time τ is then

$$m(\tau) = \lambda \int_{t=0}^{\tau} [1 - G(\tau - t)]\, dt = \lambda \int_{t=0}^{\tau} [1 - G(t)]\, dt \tag{2.36}$$

and the PMF of the number in service at time τ is given by

$$\Pr\{N_1(\tau) = n\} = \frac{m(\tau)^n \exp(-m(\tau))}{n!}. \tag{2.37}$$

Note that as $\tau \to \infty$, the integral in (2.36) approaches the mean of the service time distribution (i.e., it is the integral of the complementary CDF, $1 - G(t)$, of the service time). This means that in steady state (as $\tau \to \infty$), the distribution of the number in service at τ depends on the service time distribution only through its mean. This example can be used to model situations such as the number of phone calls taking place at a given epoch. This requires arrivals of new calls to be modeled as a Poisson process and the holding time of each call to be modeled as a random variable independent of other holding times and of call arrival times. Finally, as shown in Figure 2.8, we can regard $\{N_1(t);\ 0<t \leq \tau\}$ as a splitting of the arrival process $\{N(t);\ t>0\}$. By the same type of argument as in Section 2.3, the number of customers who have completed service by time τ is independent of the number still in service.

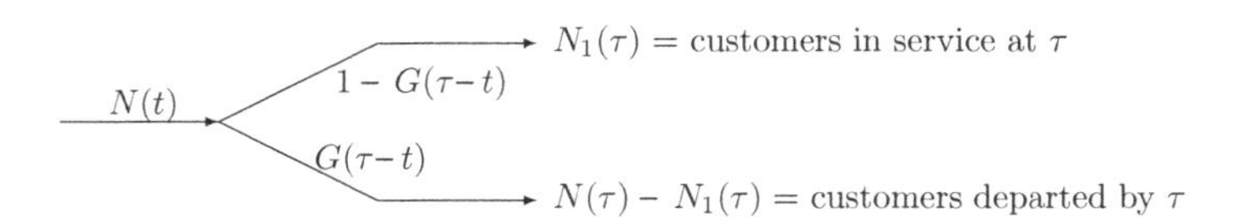

Figure 2.8 Poisson arrivals $\{N(t);\ t > 0\}$ can be considered to be split in a non-homogeneous way. An arrival at t is split with probability $1 - G(\tau - t)$ into a process of customers still in service at τ.

2.5 Conditional arrival densities and order statistics

A diverse range of problems involving Poisson processes are best tackled by conditioning on a given number n of arrivals in the interval $(0, t]$, i.e., on the event $N(t) = n$. Because of the incremental view of the Poisson process as independent and stationary arrivals in each incremental interval of the time axis, we would guess that the arrivals should have some sort of uniform distribution given $N(t) = n$. More precisely, the following theorem shows that the joint density of $\boldsymbol{S}^{(n)} = (S_1, S_2, \dots, S_n)$ given $N(t) = n$ is uniform over the region $0 < S_1 < S_2 < \cdots < S_n < t$.

Theorem 2.5.1 *Let $\mathsf{f}_{\boldsymbol{S}^{(n)}|N(t)}(\boldsymbol{s}^{(n)} \mid n)$ be the joint density of $\boldsymbol{S}^{(n)}$ conditional on $N(t) = n$. This density is constant over the region $0 < s_1 < \cdots < s_n < t$ and has the value*

$$\mathsf{f}_{\boldsymbol{S}^{(n)}|N(t)=n}(\boldsymbol{s}^{(n)} \mid n) = \frac{n!}{t^n}. \tag{2.38}$$

Two proofs are given, each illustrative of useful techniques.

Proof 1 Recall that the joint density of the first $n + 1$ arrivals $\boldsymbol{S}^{(n+1)} = (S_1, \dots, S_n, S_{n+1})$ with no conditioning is given in (2.15). We first use Bayes' law to calculate the joint density of $\boldsymbol{S}^{(n+1)}$ conditional on $N(t) = n$:

$$\mathsf{f}_{\boldsymbol{S}^{(n+1)}|N(t)}(\boldsymbol{s}^{(n+1)} \mid n)\, \mathsf{p}_{N(t)}(n) = \mathsf{p}_{N(t)|\boldsymbol{S}^{(n+1)}}(n|\boldsymbol{s}^{(n+1)})\mathsf{f}_{\boldsymbol{S}^{(n+1)}}(\boldsymbol{s}^{(n+1)}).$$

Note that $N(t) = n$ if and only if $S_n \le t$ and $S_{n+1} > t$. Thus $\mathsf{p}_{N(t)|\boldsymbol{S}^{(n+1)}}(n|\boldsymbol{s}^{(n+1)})$ is 1 if $S_n \le t$ and $S_{n+1} > t$ and is 0 otherwise. Restricting attention to the case $N(t) = n$, $S_n \le t$ and $S_{n+1} > t$,

$$\begin{aligned} \mathsf{f}_{\boldsymbol{S}^{(n+1)}|N(t)}(\boldsymbol{s}^{(n+1)} \mid n) &= \frac{\mathsf{f}_{\boldsymbol{S}^{(n+1)}}(\boldsymbol{s}^{(n+1)})}{\mathsf{p}_{N(t)}(n)} \\ &= \frac{\lambda^{n+1}\exp(-\lambda s_{n+1})}{(\lambda t)^n \exp(-\lambda t)\,/n!} \\ &= \frac{n!\lambda \exp[-\lambda(s_{n+1} - t)]}{t^n}. \end{aligned} \tag{2.39}$$

This is a useful expression, but we are interested in $\boldsymbol{S}^{(n)}$ rather than $\boldsymbol{S}^{(n+1)}$. Thus we break up the left-hand side of (2.39) as follows:

$$\mathsf{f}_{\boldsymbol{S}^{(n+1)}|N(t)}(\boldsymbol{s}^{(n+1)} \mid n) = \mathsf{f}_{\boldsymbol{S}^{(n)}|N(t)}(\boldsymbol{s}^{(n)} \mid n)\ \mathsf{f}_{S_{n+1}|\boldsymbol{S}^{(n)}N(t)}(s_{n+1}|\boldsymbol{s}^{(n)}, n).$$

Conditional on $N(t) = n$, S_{n+1} is the first arrival epoch after t, which by the memoryless property is conditionally independent of $\boldsymbol{S}^{(n)}$, again given $N(t) = n$. Thus that final term is simply $\lambda \exp(-\lambda(s_{n+1} - t))$ for $s_{n+1} > t$. Substituting this into (2.39), the result is (2.38). □

Proof 2 This alternative proof derives (2.38) by looking at arrivals in very small increments of size δ (see Figure 2.9). For a given t and a given set of n times, $0 < s_1 < \cdots < s_n < t$, we calculate the probability that there is a single arrival in each of the intervals $(s_i, s_i{+}\delta]$, $1 \le i \le n$ and no other arrivals in the interval $(0, t]$. Letting $A(\delta)$ be this event,

$$\Pr\{A(\delta)\} = \mathsf{p}_{N(s_1)}(0)\ \mathsf{p}_{\widetilde{N}(s_1,s_1+\delta)}(1)\ \mathsf{p}_{\widetilde{N}(s_1+\delta,s_2)}(0)\ \mathsf{p}_{\widetilde{N}(s_2,s_2+\delta)}(1)\cdots \mathsf{p}_{\widetilde{N}(s_n+\delta,t)}(0).$$

The sum of the lengths of the above intervals is t. Also $\mathsf{p}_{\widetilde{N}(s_i,s_i+\delta)}(1) = \lambda\delta \exp(-\lambda\delta)$ for each i, so

$$\Pr\{A(\delta)\} = (\lambda\delta)^n \exp(-\lambda t).$$

The event $A(\delta)$ can be characterized as the event that, first, $N(t) = n$ and, second, the n arrivals occur in $(s_i, s_i{+}\delta]$ for $1 \le i \le n$. Thus we conclude that

$$\mathsf{f}_{\boldsymbol{S}^{(n)}|N(t)}(\boldsymbol{s}^{(n)}|n) = \lim_{\delta\to 0} \frac{\Pr\{A(\delta)\}}{\delta^n \mathsf{p}_{N(t)}(n)},$$

which simplifies to (2.38). □

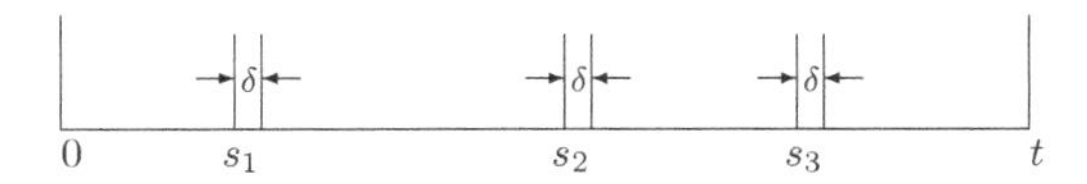

Figure 2.9 Intervals for the event $A(t)$ used to derive the joint arrival density.

The joint density of the interarrival intervals, $\boldsymbol{X}^{(n)} = (X_1, \ldots, X_n)$ given $N(t) = n$ can be found directly from Theorem 2.5.1 simply by making the linear transformation $X_1 = S_1$ and $X_i = S_i - S_{i-1}$ for $2 \le i \le n$. The density is unchanged, but the constraint region transforms into $\sum_{i=1}^n X_i < t$ with $X_i > 0$ for $1 \le i \le n$ (see Figure 2.10):

$$\mathsf{f}_{\boldsymbol{X}^{(n)}|N(t)}(\boldsymbol{x}^{(n)} \mid n) = \frac{n!}{t^n} \qquad \text{for } \boldsymbol{X}^{(n)} > 0,\ \sum_{i=1}^{n} X_i < t. \tag{2.40}$$

It is also instructive to compare the joint distribution of $\boldsymbol{S}^{(n)}$ conditional on $N(t) = n$ with the joint distribution of n IID uniformly distributed rv s, $\boldsymbol{U}^{(n)} = (U_1, \ldots, U_n)$ on $(0, t]$. For any point $\boldsymbol{U}^{(n)} = \boldsymbol{u}^{(n)}$, this joint density is

$$\mathsf{f}_{\boldsymbol{U}^{(n)}}(\boldsymbol{u}^{(n)}) = \frac{1}{t^n} \qquad \text{for } 0 < u_i \le t,\ 1 \le i \le n.$$

Both $\mathsf{f}_{\boldsymbol{S}^{(n)}}$ and $\mathsf{f}_{\boldsymbol{U}^{(n)}}$ are uniform over the volume of n-space where they are non-zero, but as illustrated in Figure 2.11 for $n = 2$, the volume for the latter is $n!$ times larger than the volume for the former. To explain this more fully, we can define a set of rv s $S_1, \ldots, S_n$, not as arrival epochs in a Poisson process, but rather as the order statistics function of the IID uniform variables $U_1, \ldots, U_n$, i.e.,

$$S_1 = \min(U_1, \ldots, U_n);\ S_2 = \text{2nd smallest } (U_1, \ldots, U_n);\ \text{etc.}$$

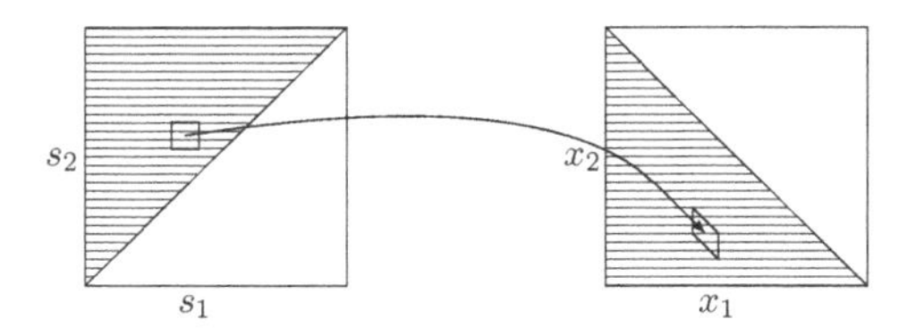

Figure 2.10 Mapping from arrival epochs to interarrival times. Note that incremental cubes in the arrival space map to parallelepipeds of the same volume in the interarrival space.

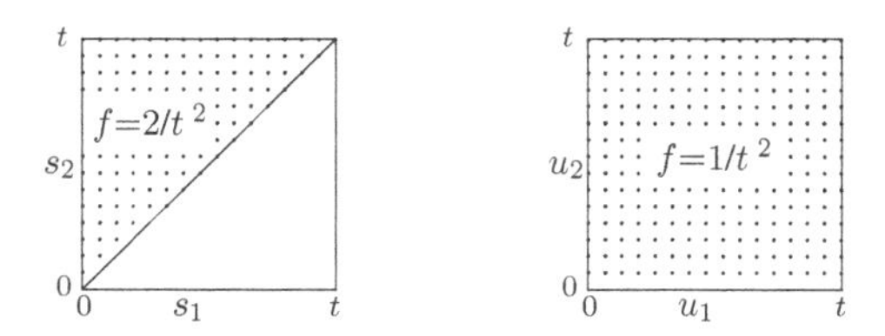

Figure 2.11 Density for the order statistics of an IID two-dimensional uniform distribution. Note that the square over which $f_{U^{(2)}}$ is non-zero contains one triangle where $u_2 > u_1$ and another of equal size where $u_1 > u_2$. Each of these maps, by a permutation mapping, to the single triangle where $s_2 > s_1$.

The n-cube is partitioned into $n!$ regions, one where $u_1 < u_2 < \cdots < u_n$. For each permutation $\pi(i)$ of the integers 1 to n, there is another region[10] where $u_{\pi(1)} < u_{\pi(2)} < \cdots < u_{\pi(n)}$. By symmetry, each of these regions has the same volume, which then must be $1/n!$ of the volume t^n of the n-cube.

All of these $n!$ regions map to the same region of ordered values. Thus these order statistics have the same joint probability density function (PDF) as the arrival epochs $S_1, \ldots, S_n$ conditional on $N(t) = n$. Anything we know (or can discover) about order statistics is valid for Poisson arrival epochs given $N(t) = n$ and vice versa.[11]

Next we want to find the marginal CDF's of the individual S_i, conditional on $N(t) = n$. Starting with S_1, and viewing it as the minimum of the IID uniformly distributed variables $U_1, \ldots, U_n$, we recognize that $S_1 > \tau$ if and only if $U_i > \tau$ for all i, $1 \leq i \leq n$. Thus,

$$\Pr\{S_1 > \tau \mid N(t)=n\} = \left[\frac{t-\tau}{t}\right]^n \qquad \text{for } 0 < \tau \leq t. \tag{2.41}$$

For S_2 to S_n, the density is slightly simpler in appearance than the CDF. To find $f_{S_i|N(t)}(s_i \mid n)$, look at n uniformly distributed rv s in $(0, t]$. The probability that one of these lies in the interval $(s_i, s_i + dt]$ is $(n\,dt)/t$. Out of the remaining $n-1$, the probability

[10] As usual, we are ignoring those points where $u_i = u_j$ for some i, j, since the set of such points has 0 probability.

[11] There is certainly also the intuitive notion, given n arrivals in $(0, t]$, and given the stationary and independent increment properties of the Poisson process, that those n arrivals can be viewed as uniformly distributed. One way to view this is to visualize the Poisson process as the sum of a very large number k of independent processes of rate λ/k each. Then, given $N(t) = n$, with $k >> n$, there is negligible probability of more than one arrival from any one process, and, for each of the n processes with arrivals, that arrival is uniformly distributed in $(0, t]$.

that $i-1$ lie in the interval $(0, s_i]$ is given by the binomial distribution with probability of success s_i/t. Thus the desired density is

$$\mathsf{f}_{S_i|N(t)}(s_i \mid n)\,dt = \frac{s_i^{i-1}(t-s_i)^{n-i}(n-1)!}{t^{n-1}(n-i)!(i-1)!} \quad \frac{n\,dt}{t}$$

$$\mathsf{f}_{S_i|N(t)}(s_i \mid n) = \frac{s_i^{i-1}(t-s_i)^{n-i}n!}{t^{n}(n-i)!(i-1)!}. \tag{2.42}$$

It is easy to find the expected value of S_1 conditional on $N(t) = n$ by integrating the complementary CDF in (2.41), getting

$$\mathsf{E}[S_1 \mid N(t)=n] = \frac{t}{n+1}. \tag{2.43}$$

We come back later to find $\mathsf{E}[S_i \mid N(t) = n]$ for $2 \le i \le n$. First, we look at the marginal distributions of the interarrival intervals. Recall from (2.40) that

$$\mathsf{f}_{\boldsymbol{X}^{(n)}|N(t)}(\boldsymbol{x}^{(n)} \mid n) = \frac{n!}{t^n} \qquad \text{for } \boldsymbol{X}^{(n)} > 0, \ \sum_{i=1}^{n} X_i < t. \tag{2.44}$$

The joint density is the same for all points in the constraint region, and the constraint does not distinguish between $X_1, \dots, X_n$. Thus $X_1, \dots, X_n$ must all have the same marginal distribution, and more generally the marginal distribution of any subset of the X_i can depend only on the size of the subset. We have found the distribution of S_1, which is the same as X_1, and thus

$$\Pr\{X_i > \tau \mid N(t)=n\} = \left[\frac{t-\tau}{t}\right]^n \qquad \text{for } 1 \le i \le n \text{ and } 0 < \tau \le t, \tag{2.45}$$

$$\mathsf{E}[X_i \mid N(t)=n] = \frac{t}{n+1} \qquad \text{for } 1 \le i \le n. \tag{2.46}$$

From this, we see immediately that for $1 \le i \le n$,

$$\mathsf{E}[S_i \mid N(t) = n] = \frac{it}{n+1}. \tag{2.47}$$

One could go on and derive joint distributions of all sorts at this point, but there is one additional type of interval that must be discussed. Define $X_{n+1}^* = t - S_n$ to be the interval from the largest arrival epoch before t to t itself. Rewriting (2.44),

$$\mathsf{f}_{\boldsymbol{X}^{(n)}|N(t)}(\boldsymbol{x}^{(n)} \mid n) = \frac{n!}{t^n} \qquad \text{for } \boldsymbol{X}^{(n)} > 0, \ X_{n+1}^* > 0, \ \sum_{i=1}^{n} X_i + X_{n+1}^* = t.$$

The constraints above are symmetric in $X_1, \dots, X_n, X_{n+1}^*$, and, within the constraint region, the joint density of $X_1, \dots, X_n$ (conditional on $N(t) = n$) is uniform. Note that there is no joint density over $X_1, \dots, X_n, X_{n+1}^*$ conditional on $N(t) = n$, since X_{n+1}^* is then a deterministic function of $X_1, \dots, X_n$. However, the density over $X_1, \dots, X_n$ can be replaced by a density over any other n rv s out of $X_1, \dots, X_n, X_{n+1}^*$ by a linear transformation with unit determinant. Thus X_{n+1}^* has the same marginal distribution as each of the X_i. This gives us a partial check on our work, since the interval $(0, t]$ is divided into $n+1$ intervals of sizes $X_1, X_2, \dots, X_n, X_{n+1}^*$, and each of these has a mean size $t/(n+1)$.

We also see that the joint CDF of any proper subset of $X_1, X_2, \ldots, X_n, X^*_{n+1}$ depends only on the size of the subset and not the particular rv s in the subset.

One important consequence of this is that we can look at the arrivals in a segment $(0, t)$ of a Poisson process either forward or backward[12] in time and they 'look the same.' Looked at backwards, the interarrival intervals are $X^*_{n+1}, X_n, \ldots, X_2$. These intervals are IID, and X_1 is then determined as $t - X^*_{n+1} - X_n - \cdots - X_2$. We will not make any particular use of this property here, but we will later explore this property of time-reversibility for other types of processes. For Poisson processes, this reversibility is intuitively obvious from the stationary and independent properties. It is less obvious how to express this condition by equations, but that is not really necessary at this point.

2.6 Summary

We started the chapter with three equivalent definitions of a Poisson process – first as a renewal process with exponentially distributed inter-renewal intervals, second as a stationary and independent increment counting process with a Poisson distributed number of arrivals in each interval, and third essentially as a limit of shrinking Bernoulli processes. We saw that each definition provided its own insights into the properties of the process. We emphasized the importance of the memoryless property of the exponential distribution, both as a useful tool in problem solving and as an underlying reason why the Poisson process is so simple.

We next showed that the sum of independent Poisson processes is again a Poisson process. We also showed that if the arrivals in a Poisson process are independently routed to different locations with some fixed probability assignment, then the arrivals at these locations form independent Poisson processes. This ability to view independent Poisson processes either independently or as a splitting of a combined process is a powerful technique for finding almost trivial solutions to many problems.

It was next shown that a non-homogeneous Poisson process could be viewed as a (homogeneous) Poisson process on a non-linear time scale. This allows all the properties of (homogeneous) Poisson processes to be applied directly to the non-homogeneous case. The simplest and most useful result from this is (2.35), showing that the number of arrivals in any interval has a Poisson PMF. This result was used to show that the number of customers in service at any given time τ in an M/G/∞ queue has a Poisson PMF with a mean approaching λ times the expected service time in the limit as $\tau \to \infty$.

Finally we looked at the distribution of arrival epochs conditional on n arrivals in the interval $(0, t]$. It was found that these arrival epochs had the same joint distribution as the order statistics of n uniform IID rv s in $(0, t]$. By using symmetry and going back and forth between the uniform variables and the Poisson process arrivals, we found the distribution of the interarrival times, the arrival epochs, and various conditional distributions.

[12] This must be interpreted carefully, since if we ask about the distribution to the next arrival from some $\tau \in (0, t)$, it is exponential (possibly occurring after t). But if we look for the distribution of the previous arrival before τ, we usually truncate the distribution at 0. Thus the proper interpretation is to look only at the interval $(0, t)$ under the condition of a given number of arrivals in that interval.

2.7 Exercises

Exercise 2.1 **(a)** Find the Erlang density $\mathsf{f}_{S_n}(t)$ by convolving $\mathsf{f}_X(x) = \lambda \exp(-\lambda x)$ with itself n times.

(b) Find the moment generating function (MGF) of X (or find the Laplace transform of $\mathsf{f}_X(x)$), and use this to find the moment generating function (or Laplace transform) of $S_n = X_1 + X_2 + \cdots + X_n$. Invert your result to find $\mathsf{f}_{S_n}(t)$.

(c) Find the Erlang density by starting with (2.15) and then calculating the marginal density for S_n.

Exercise 2.2 **(a)** Find the mean, variance, and MGF of $N(t)$, as given by (2.17).

(b) Show by discrete convolution that the sum of two independent Poisson rv s is again Poisson.

(c) Show by using the properties of the Poisson process that the sum of two independent Poisson rv s must be Poisson.

Exercise 2.3 The purpose of this exercise is to give an alternative derivation of the Poisson distribution for $N(t)$, the number of arrivals in a Poisson process up to time t. Let λ be the rate of the process.

(a) Find the conditional probability $\Pr\{N(t) = n \mid S_n = \tau\}$ for all $\tau \leq t$.

(b) Using the Erlang density for S_n, use (a) to find $\Pr\{N(t) = n\}$.

Exercise 2.4 Assume that a counting process $\{N(t); t{>}0\}$ has the independent and stationary increment properties and satisfies (2.17) (for all $t > 0$). Let X_1 be the epoch of the first arrival and X_n be the interarrival time between the $(n-1)$th and the nth arrival. Use only these assumptions in doing the following parts of this exercise.

(a) Show that $\Pr\{X_1 > x\} = e^{-\lambda x}$.

(b) Let S_{n-1} be the epoch of the $(n-1)$th arrival. Show that $\Pr\{X_n > x \mid S_{n-1} = \tau\} = e^{-\lambda x}$.

(c) For each $n > 1$, show that $\Pr\{X_n > x\} = e^{-\lambda x}$ and that X_n is independent of S_{n-1}.

(d) Argue that X_n is independent of $X_1, X_2, \ldots, X_{n-1}$.

Exercise 2.5 The point of this exercise is to show that the sequence of PMFs for a Bernoulli counting process does not specify the process. In other words, knowing that $N(t)$ satisfies the binomial distribution for all t does not mean that the process is Bernoulli. This helps us understand why the second definition of a Poisson process requires stationary and independent increments as well as the Poisson distribution for $N(t)$.

(a) Let $Y_1, Y_2, Y_3, \ldots$ be a sequence of binary rv s in which each rv is 0 or 1 with equal probability. Find a joint distribution for Y_1, Y_2, Y_3 that satisfies the binomial distribution, $\mathsf{p}_{N(t)}(k) = \binom{t}{k}2^{-t}$ for $t = 1, 2, 3$ and $0 \leq k \leq t$, but for which Y_1, Y_2, Y_3 are not independent.

One simple solution for this contains four 3-tuples with probability 1/8 each, two 3-tuples with probability 1/4 each, and two 3-tuples with probability 0. Note that by making the subsequent arrivals IID and equiprobable, you have an example where $N(t)$ is binomial for all t but the process is not Bernoulli. Hint: Use the binomial for $t = 3$ to find two 3-tuples that must have probability 1/8. Combine this with the binomial for

$t = 2$ to find two other 3-tuples that must have probability 1/8. Finally, look at the constraints imposed by the binomial distribution on the remaining four 3-tuples.

(b) Generalize (a) to the case where Y_1, Y_2, Y_3 satisfy $\Pr\{Y_i = 1\} = p$ and $\Pr\{Y_i = 0\} = 1 - p$. Assume $p < 1/2$ and find a joint distribution on Y_1, Y_2, Y_3 that satisfies the binomial distribution, but for which the 3-tuple (0, 1, 1) has zero probability.

(c) More generally yet, view a joint PMF on binary t-tuples as a non-negative vector in a 2^t dimensional vector space. Each binomial probability $\mathsf{p}_{N(\tau)}(k) = \binom{\tau}{k} p^k (1-p)^{\tau-k}$ constitutes a linear constraint on this vector. For each τ, show that one of these constraints may be replaced by the constraint that the components of the vector sum to 1.

(d) Using (c), show that at most $(t+1)t/2 + 1$ of the binomial constraints are linearly independent. Note that this means that the linear space of vectors satisfying these binomial constraints has dimension at least $2^t - (t+1)t/2 - 1$. This linear space has dimension 1 for $t = 3$, explaining the results in (a) and (b). It has a rapidly increasing dimension for $t > 3$, suggesting that the binomial constraints are relatively ineffectual for constraining the joint PMF of a joint distribution. More work is required for the case of $t > 3$ because of all the inequality constraints, but it turns out that this large dimensionality remains.

Exercise 2.6 Let $h(x)$ be a positive function of a real variable that satisfies $h(x+t) = h(x) + h(t)$ and let $h(1) = c$.

(a) Show that for integer $k > 0$, $h(k) = kc$.

(b) Show that for integer $j > 0$, $h(1/j) = c/j$.

(c) Show that for all integer k, j, $h(k/j) = ck/j$.

(d) The above parts show that $h(x)$ is linear in positive *rational* numbers. For very picky mathematicians, this does not guarantee that $h(x)$ is linear in positive *real* numbers. Show that if $h(x)$ is also monotonic in x, then $h(x)$ is linear in $x > 0$.

Exercise 2.7 Assume that a counting process $\{N(t);\, t{>}0\}$ has the independent and stationary increment properties and, for all $t > 0$, satisfies

$$\begin{aligned}
\Pr\left\{\widetilde{N}(t, t+\delta) = 0\right\} &= 1 - \lambda\delta + o(\delta),\\
\Pr\left\{\widetilde{N}(t, t+\delta) = 1\right\} &= \lambda\delta + o(\delta),\\
\Pr\left\{\widetilde{N}(t, t+\delta) > 1\right\} &= o(\delta).
\end{aligned}$$

(a) Let $\mathsf{F}_0(\tau) = \Pr\{N(\tau) = 0\}$ and show that $d\mathsf{F}_0(\tau)/d\tau = -\lambda \mathsf{F}_0(\tau)$.

(b) Show that X_1, the time of the first arrival, is exponential with parameter λ.

(c) Let $\mathsf{F}^c_n(\tau) = \Pr\{\widetilde{N}(t, t+\tau) = 0 \mid S_{n-1} = t\}$ and show that $d\mathsf{F}^c_n(\tau)/d\tau = -\lambda \mathsf{F}^c_n(\tau)$.

(d) Argue that X_n is exponential with parameter λ and independent of earlier arrival times.

Exercise 2.8 For a Poisson process, let $t > 0$ be an arbitrary time and let Z_1 be the duration of the interval from t until the next arrival after t. Let Z_m, for each $m > 1$, be the interarrival time from the epoch of the $(m-1)$th arrival after t until the mth arrival.

(a) Given that $N(t) = n$, explain why $Z_m = X_{m+n}$ for $m > 1$ and $Z_1 = X_{n+1} - t + S_n$.

(b) Conditional on $N(t) = n$ and $S_n = \tau$, show that $Z_1, Z_2, \ldots$ are IID.

(c) Show that $Z_1, Z_2, \ldots$ are IID.

Exercise 2.9 Consider a 'shrinking Bernoulli' approximation $N_\delta(m\delta) = Y_1 + \cdots + Y_m$ to a Poisson process as described in Section 2.2.5.

(a) Show that

$$\Pr\{N_\delta(m\delta) = n\} = \binom{m}{n}(\lambda\delta)^n(1 - \lambda\delta)^{m-n}.$$

(b) Let $t = m\delta$, and let t be fixed for the remainder of the exercise. Explain why

$$\lim_{\delta\to 0} \Pr\{N_\delta(t) = n\} = \lim_{m\to\infty} \binom{m}{n}\left(\frac{\lambda t}{m}\right)^n\left(1 - \frac{\lambda t}{m}\right)^{m-n},$$

where the limit on the left-hand side is taken over values of δ that divide t.

(c) Derive the following two equalities:

$$\lim_{m\to\infty} \binom{m}{n}\frac{1}{m^n} = \frac{1}{n!} \qquad \text{and} \qquad \lim_{m\to\infty}\left(1 - \frac{\lambda t}{m}\right)^{m-n} = e^{-\lambda t}.$$

(d) Conclude from this that for every t and every n, $\lim_{\delta\to 0}\Pr\{N_\delta(t)=n\} = \Pr\{N(t)=n\}$, where $\{N(t);\ t > 0\}$ is a Poisson process of rate λ.

Exercise 2.10 Let $\{N(t);\ t > 0\}$ be a Poisson process of rate λ.

(a) Find the joint PMF of $N(t)$, $N(t + s)$ for $s > 0$.

(b) Find $\mathsf{E}[N(t) \cdot N(t + s)]$ for $s > 0$.

(c) Find $\mathsf{E}[\widetilde{N}(t_1, t_3) \cdot \widetilde{N}(t_2, t_4)]$, where $\widetilde{N}(t, \tau)$ is the number of arrivals in $(t, \tau]$ and $t_1 < t_2 < t_3 < t_4$.

Exercise 2.11 An elementary experiment is independently performed N times, where N is a Poisson rv of mean λ. Let $\{a_1, a_2, \ldots, a_K\}$ be the set of sample points of the elementary experiment and let p_k, $1 \le k \le K$, denote the probability of a_k.

(a) Let N_k denote the number of elementary experiments performed for which the output is a_k. Find the PMF for N_k $(1 \le k \le K)$. Hint: No calculation is necessary.

(b) Find the PMF for $N_1 + N_2$.

(c) Find the conditional PMF for N_1 given that $N = n$.

(d) Find the conditional PMF for $N_1 + N_2$ given that $N = n$.

(e) Find the conditional PMF for N given that $N_1 = n_1$.

Exercise 2.12 Starting from time 0, northbound buses arrive at 77 Mass. Avenue according to a Poisson process of rate λ. Customers arrive according to an independent Poisson process of rate μ. When a bus arrives, all waiting customers instantly enter the bus and subsequent customers wait for the next bus.

(a) Find the PMF for the number of customers entering a bus (more specifically, for any given m, find the PMF for the number of customers entering the mth bus).

(b) Find the PMF for the number of customers entering the mth bus given that the interarrival interval between bus $m - 1$ and bus m is x.

(c) Given that a bus arrives at time 10:30 pm, find the PMF for the number of customers entering the next bus.

(d) Given that a bus arrives at 10:30 pm and no bus arrives between 10:30 and 11 pm, find the PMF for the number of customers on the next bus.

(e) Find the PMF for the number of customers waiting at some given time, say 2:30 pm (assume that the processes started infinitely far in the past). Hint: Think of what happens moving backward in time from 2:30 pm.

(f) Find the PMF for the number of customers getting on the next bus to arrive after 2:30 pm. Hint: This is different from (a); look carefully at part (e).

(g) Given that I arrive to wait for a bus at 2:30 pm, find the PMF for the number of customers getting on the next bus.

Exercise 2.13 **(a)** Show that the arrival epochs of a Poisson process satisfy

$$\mathsf{f}_{\boldsymbol{S}^{(n)}|S_{n+1}}(\boldsymbol{s}^{(n)}|s_{n+1}) = n!/s_{n+1}^n.$$

Hint: This is easy if you use only the results of Section 2.2.2.

(b) Contrast this with the result of Theorem 2.5.1.

Exercise 2.14 Equation (2.42) gives $\mathsf{f}_{S_i|N(t)}(S_i \mid n)$, which is the density of rv S_i conditional on $N(t) = n$ for $n \geq i$. Multiply this expression by $\Pr\{N(t) = n\}$ and sum over n to find $\mathsf{f}_{S_i}(s_i)$; verify that your answer is indeed the Erlang density.

Exercise 2.15 Consider generalizing the bulk arrival process in Figure 2.5. Assume that the epochs at which arrivals occur form a Poisson process $\{N(t);\, t > 0\}$ of rate λ. At each arrival epoch, S_n, the number of arrivals, Z_n, satisfies $\Pr\{Z_n{=}1\} = p$, $\Pr\{Z_n{=}2)\} = 1 - p$. The rv s Z_n are IID.

(a) Let $\{N_1(t); t > 0\}$ be the counting process of the epochs at which single arrivals occur. Find the PMF of $N_1(t)$ as a function of t. Similarly, let $\{N_2(t); t \geq 0\}$ be the counting process of the epochs at which double arrivals occur. Find the PMF of $N_2(t)$ as a function of t.

(b) Let $\{N_B(t); t \geq 0\}$ be the counting process of the total number of arrivals. Give an expression for the PMF of $N_B(t)$ as a function of t.

Exercise 2.16 **(a)** For a Poisson counting process of rate λ, find the joint probability density of $S_1, S_2, \ldots, S_{n-1}$ conditional on $S_n = t$.

(b) Find $\Pr\{X_1 > \tau \mid S_n{=}t\}$.

(c) Find $\Pr\{X_i > \tau \mid S_n{=}t\}$ for $1 \leq i \leq n$.

(d) Find the density $\mathsf{f}_{S_i|S_n}(s_i|t)$ for $1 \leq i \leq n-1$.

(e) Give an explanation for the striking similarity between the condition $N(t) = n - 1$ and the condition $S_n = t$.

Exercise 2.17 **(a)** For a Poisson process of rate λ, find $\Pr\{N(t){=}n \mid S_1{=}\tau\}$ for $t > \tau$ and $n \geq 1$.

(b) Using this, find $\mathsf{f}_{S_1}(\tau \mid N(t)=n)$.

(c) Check your answer against (2.41).

Exercise 2.18 Consider a counting process in which the rate is a rv Λ with probability density $\mathsf{f}_\Lambda(\lambda) = \alpha e^{-\alpha\lambda}$ for $\lambda > 0$. Conditional on a given sample value λ for the rate, the counting process is a Poisson process of rate λ (i.e., nature first chooses a sample value λ and then generates a sample path of a Poisson process of that rate λ).

(a) What is $\Pr\{N(t)=n \mid \Lambda=\lambda\}$, where $N(t)$ is the number of arrivals in the interval $(0, t]$ for some given $t > 0$?

(b) Show that $\Pr\{N(t)=n\}$, the unconditional PMF for $N(t)$, is given by

$$\Pr\{N(t)=n\} = \frac{\alpha t^n}{(t+\alpha)^{n+1}}.$$

(c) Find $\mathsf{f}_\Lambda(\lambda \mid N(t)=n)$, the density of λ conditional on $N(t)=n$.

(d) Find $\mathsf{E}[\Lambda \mid N(t)=n]$ and interpret your result for very small t with $n = 0$ and for very large t with n large.

(e) Find $\mathsf{E}[\Lambda \mid N(t)=n, S_1, S_2, \ldots, S_n]$. Hint: Consider the distribution of $S_1, \ldots, S_n$ conditional on $N(t)$ and Λ. Find $\mathsf{E}[\Lambda \mid N(t)=n, N(\tau)=m]$ for some $\tau < t$.

Exercise 2.19 **(a)** Use (2.42) to find $\mathsf{E}[S_i \mid N(t)=n]$. Hint: When you integrate $s_i \mathsf{f}_{S_i}(s_i \mid N(t)=n)$, compare this integral with $\mathsf{f}_{S_{i+1}}(s_i \mid N(t)=n+1)$ and use the fact that the latter expression is a probability density.

(b) Find the second moment and the variance of S_i conditional on $N(t)=n$. Hint: Extend the previous hint.

(c) Assume that n is odd, and consider $i = (n+1)/2$. What is the relationship between S_i, conditional on $N(t)=n$, and the sample median of n IID uniform rv s.

(d) Give a weak law of large numbers (WLLN) for the above median.

Exercise 2.20 Suppose cars enter a one-way infinite length, infinite lane highway at a Poisson rate λ. The ith car to enter chooses a velocity V_i and travels at this velocity. Assume that the V_i's are independent positive rv s having a common distribution F. Derive the distribution of the number of cars that are located in an interval $(0, a)$ at time t.

Exercise 2.21 Consider an M/G/∞ queue, i.e., a queue with Poisson arrivals of rate λ in which each arrival i, independent of other arrivals, remains in the system for a time X_i, where $\{X_i; i \geq 1\}$ is a set of IID rv s with some given CDF $\mathsf{F}(x)$.

You may assume that the number of arrivals in any interval $(t, t+\epsilon)$ that are still in the system at some later time $\tau \geq t+\epsilon$ is statistically independent of the number of arrivals in that same interval $(t, t+\epsilon)$ that have departed from the system by time τ.

(a) Let $N(\tau)$ be the number of customers in the system at time τ. Find the mean, $m(\tau)$, of $N(\tau)$ and find $\Pr\{N(\tau) = n\}$.

(b) Let $D(\tau)$ be the number of customers that have departed from the system by time τ. Find the mean, $\mathsf{E}[D(\tau)]$, and find $\Pr\{D(\tau) = d\}$.

(c) Find $\Pr\{N(\tau) = n, D(\tau) = d\}$.

(d) Let $A(\tau)$ be the total number of arrivals up to time τ. Find $\Pr\{N(\tau) = n \mid A(\tau) = a\}$.

(e) Find $\Pr\{D(\tau + \epsilon) - D(\tau) = d\}$.

Exercise 2.22 The voters in a given town arrive at the place of voting according to a Poisson process of rate $\lambda = 100$ voters per hour. The voters independently vote for candidate A and candidate B each with probability $1/2$. Assume that the voting starts at time 0 and continues indefinitely.

(a) Conditional on 1000 voters arriving during the first 10 hours of voting, find the probability that candidate A receives n of those votes.

(b) Again conditional on 1000 voters during the first 10 hours, find the probability that candidate A receives n votes in the first 4 hours of voting.

(c) Let T be the epoch of the arrival of the first voter voting for candidate A. Find the density of T.

(d) Find the PMF of the number of voters for candidate B who arrive before the first voter for A.

(e) Define the nth voter as a *reversal* if the nth voter votes for a different candidate than the $(n-1)$th. For example, in the sequence of votes $AABAABB$, the third, fourth, and sixth voters are reversals; the third and sixth are A to B reversals and the fourth is a B to A reversal. Let $N(t)$ be the number of reversals up to time t (t in hours). Is $\{N(t);\ t > 0\}$ a Poisson process? Explain.

(f) Find the expected time (in hours) between reversals.

(g) Find the probability density of the time between reversals.

(h) Find the density of the time from one A to B reversal to the next A to B reversal.

Exercise 2.23 Let $\{N_1(t);\ t > 0\}$ be a Poisson counting process of rate λ. Assume that the arrivals from this process are switched on and off by arrivals from a second independent Poisson process $\{N_2(t); t > 0\}$ of rate γ.

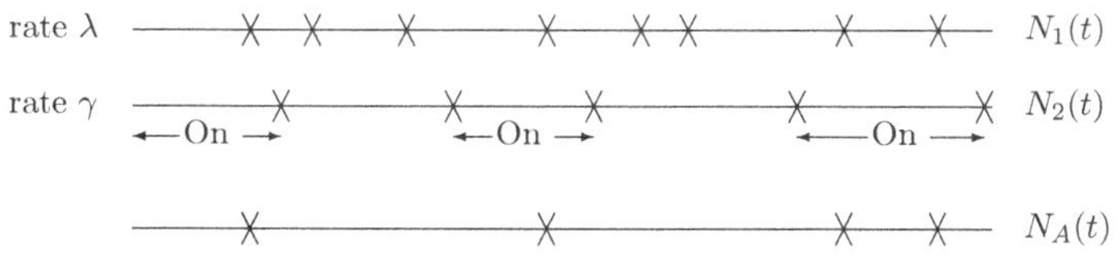

rate λ $N_1(t)$
rate γ $N_2(t)$
On On On
$N_A(t)$

Let $\{N_A(t); t{\geq}0\}$ be the switched process, i.e., $N_A(t)$ includes the arrivals from $\{N_1(t);\ t > 0\}$ during periods when $N_2(t)$ is even and excludes the arrivals from $\{N_1(t);\ t > 0\}$ while $N_2(t)$ is odd.

(a) Find the PMF for the number of arrivals of the first process, $\{N_1(t);\ t > 0\}$, during the nth period when the switch is on.

(b) Given that the first arrival for the second process occurs at epoch τ, find the conditional PMF for the number of arrivals of the first process up to τ.

(c) Given that the number of arrivals of the first process, up to the first arrival for the second process, is n, find the density for the epoch of the first arrival from the second process.

(d) Find the density of the interarrival time for $\{N_A(t); t \geq 0\}$. Note: This part is quite messy and is done most easily via Laplace transforms.

Exercise 2.24 Let us model the chess tournament between Fisher and Spassky as a stochastic process. Let X_i, for $i \geq 1$, be the duration of the ith game and assume that

$\{X_i;\ i{\geq}1\}$ is a set of IID exponentially distributed rv s each with density $\mathsf{f}_X(x) = \lambda e^{-\lambda x}$. Suppose that each game (independently of all other games, and independently of the length of the games) is won by Fisher with probability p, by Spassky with probability q, and is a draw with probability $1 - p - q$. The first player to win n games is defined to be the winner, but we consider the match up to the point of winning as being embedded in an unending sequence of games.

(a) Find the distribution of time, from the beginning of the match, until the completion of the first game that is won (i.e., that is not a draw). Characterize the process of the number $\{N(t);\ t > 0\}$ of games won up to and including time t. Characterize the process of the number $\{N_F(t); t \geq 0\}$ of games won by Fisher and the number $\{N_S(t); t \geq 0\}$ won by Spassky.

(b) For the remainder of the problem, assume that the probability of a draw is zero, i.e., that $p+q = 1$. How many of the first $2n - 1$ games must be won by Fisher in order to win the match?

(c) What is the probability that Fisher wins the match? Your answer should not involve any integrals. Hint: Consider the unending sequence of games and use (b).

(d) Let T be the epoch at which the match is completed (i.e., either Fisher or Spassky wins). Find the CDF of T.

(e) Find the probability that Fisher wins and that T lies in the interval $(t, t + \delta)$ for arbitrarily small δ.

Exercise 2.25 **(a)** Find the conditional density of S_{i+1}, conditional on $N(t) = n$ and $S_i = s_i$.

(b) Use (a) to find the joint density of $S_1, \ldots, S_n$ conditional on $N(t) = n$. Verify that your answer agrees with (2.38).

Exercise 2.26 A two-dimensional Poisson process is a process of randomly occurring special points in the plane such that (i) for any region of area A the number of special points in that region has a Poisson distribution with mean λA, and (ii) the number of special points in non-overlapping regions is independent. For such a process consider an arbitrary location in the plane and let X denote its distance from its nearest special point (where distance is measured in the usual Euclidean manner). Show that:

(a) $\Pr\{X > t\} = \exp(-\lambda\pi t^2)$;

(b) $\mathsf{E}[X] = 1/(2\sqrt{\lambda})$.

Exercise 2.27 This problem is intended to show that one can analyze the long-term behavior of queueing problems by using just notions of means and variances, but that such analysis is awkward, justifying understanding the strong law of large numbers (SLLN). Consider an M/G/1 queue. The arrival process is Poisson with $\lambda = 1$. The expected service time, $\mathsf{E}[Y]$, is $1/2$ and the variance of the service time is 1.

(a) Consider S_n, the time of the nth arrival, for $n = 10^{12}$. With high probability, S_n will lie within three standard derivations of its mean. Find and compare this mean and the 3σ range.

(b) Let V_n be the total amount of time during which the server is busy with these n arrivals (i.e., the sum of 10^{12} service times). Find the mean and 3σ range of V_n.

(c) Find the mean and 3σ range of I_n, the total amount of time the server is idle up until S_n (take I_n as $S_n - V_n$, thus ignoring any service time after S_n).

(d) An idle period starts when the server completes a service and there are no waiting arrivals; it ends on the next arrival. Find the mean and variance of an idle period. Are successive idle periods IID?

(e) Combine (c) and (d) to estimate the total number of idle periods up to time S_n. Use this to estimate the total number of busy periods.

(f) Combine (e) and (b) to estimate the expected length of a busy period.

Exercise 2.28 The purpose of this problem is to illustrate that for an arrival process with independent but not identically distributed interarrival intervals, $X_1, X_2, \ldots,$ the number of arrivals $N(t)$ in the interval $(0, t]$ can be a defective rv. In other words, the 'counting process' is not necessarily a stochastic process according to our definitions. This also suggests that it might be necessary to prove that the counting rv s for a renewal process are actually rv s.

(a) Let the CDF of the ith interarrival interval for an arrival process be $\mathsf{F}_{X_i}(x_i) = 1 - \exp(-\alpha^{-i} x_i)$ for some fixed $\alpha \in (0, 1)$. Let $S_n = X_1 + \cdots + X_n$ and show that

$$\mathsf{E}[S_n] = \frac{\alpha(1 - \alpha^n)}{1 - \alpha}.$$

(b) Sketch a 'reasonable' sample path for $N(t)$.

(c) Use the Markov inequality on $\Pr\{S_n \geq t\}$ to find an upper bound on $\Pr\{N(t) < n\}$ that is smaller than 1 for all n and for large enough t. Use this to show that $N(t)$ is defective for large enough t.

(d) (For those looking for a challenge) Show that $N(t)$ is defective for all $t > 0$. Hint: Use the Markov inequality to find an upper bound on $\Pr\{S_m - S_n \leq t/2\}$ for all $m > n$ for any fixed n. Show that, for any $t > 0$, this is bounded below 1 for large enough n. Then show that S_n has a density that is positive for all $t > 0$.

3 Gaussian random vectors and processes

3.1 Introduction

Poisson processes and Gaussian processes are similar in terms of their simplicity and beauty. When we first look at a new problem involving stochastic processes, we often start with insights from Poisson and/or Gaussian processes. Problems where queueing is a major factor tend to rely heavily on an understanding of Poisson processes, and those where noise is a major factor tend to rely heavily on Gaussian processes.

Poisson and Gaussian processes share the characteristic that the results arising from them are so simple, well known, and powerful that people often forget how much the results depend on assumptions that are rarely satisfied perfectly in practice. At the same time, these assumptions are often approximately satisfied, so the results, if used with insight and care, are often useful.

This chapter is aimed primarily at Gaussian processes, but starts with a study of Gaussian (normal[1]) random variables (rv s) and Gaussian random vectors (**rv** s), These initial topics are both important in their own right and also essential to an understanding of Gaussian processes. The material here is essentially independent of that on Poisson processes in Chapter 2.

3.2 Gaussian random variables

A rv W is defined to be a *normalized Gaussian rv* if it has the density

$$\mathsf{f}_W(w) = \frac{1}{\sqrt{2\pi}} \exp\left(\frac{-w^2}{2}\right) \qquad \text{for all } w \in \mathbb{R}. \tag{3.1}$$

Exercise 3.1 shows that $\mathsf{f}_W(w)$ integrates to 1 (i.e., it *is* a probability density), and that W has mean 0 and variance 1.

If we scale a normalized Gaussian rv W by a positive constant σ, i.e., if we consider the rv $Z = \sigma W$, then the distribution functions of Z and W are related by $\mathsf{F}_Z(\sigma w) =$

[1] Gaussian rv s are often called normal rv s. I prefer Gaussian, first, because the corresponding processes are usually called Gaussian, second, because Gaussian rv s (which have arbitrary means and variances) are often *normalized* to zero mean and unit variance and, third, because calling them normal gives the false impression that other rv s are abnormal.

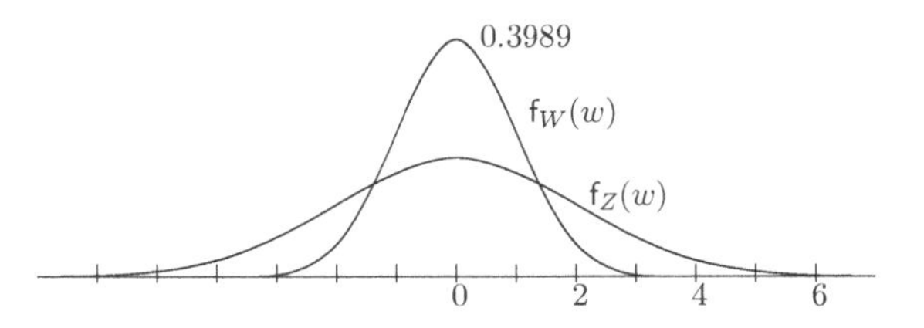

Figure 3.1 Graph of the PDF of a normalized Gaussian rv W (the taller curve) and of a zero-mean Gaussian rv Z with standard deviation 2 (the flatter curve).

$\mathsf{F}_W(w)$. This means that the probability densities are related by $\sigma\mathsf{f}_Z(\sigma w) = \mathsf{f}_W(w)$. Thus the probability density function (PDF) of Z is given by

$$\mathsf{f}_Z(z) = \frac{1}{\sigma}\mathsf{f}_W\left(\frac{z}{\sigma}\right) = \frac{1}{\sqrt{2\pi}\,\sigma}\exp\left(\frac{-z^2}{2\sigma^2}\right). \tag{3.2}$$

Thus the PDF for Z is scaled horizontally by the factor σ, and then scaled vertically by $1/\sigma$ (see Figure 3.1). This scaling leaves the integral of the density unchanged with value 1 and scales the variance by σ^2. If we let σ approach 0, this density approaches an impulse, i.e., Z becomes the atomic rv for which $\Pr\{Z{=}0\} = 1$. For convenience in what follows, we use (3.2) as the density for Z for all $\sigma \geq 0$, with the above understanding about the $\sigma = 0$ case. A rv with the density in (3.2), for any $\sigma \geq 0$, is defined to be a *zero-mean Gaussian rv*. The values $\Pr\{|Z| > \sigma\} = 0.318$, $\Pr\{|Z| > 3\sigma\} = 0.0027$, and $\Pr\{|Z| > 5\sigma\} = 2.2 \times 10^{-12}$ give us a sense of how small the tails of the Gaussian distribution are.

If we shift Z by an arbitrary $\mu \in \mathbb{R}$ to $U = Z + \mu$, then the density shifts so as to be centered at $\mathsf{E}[U] = \mu$, and the density satisfies $\mathsf{f}_U(u) = \mathsf{f}_Z(u - \mu)$. Thus

$$\mathsf{f}_U(u) = \frac{1}{\sqrt{2\pi}\,\sigma}\exp\left(\frac{-(u-\mu)^2}{2\sigma^2}\right). \tag{3.3}$$

A rv U with this density, for arbitrary μ and $\sigma \geq 0$, is defined to be a *Gaussian rv* and is denoted $U \sim \mathrm{N}(\mu, \sigma^2)$.

The added generality of a mean often obscures formulas; we usually assume zero-mean rv s and **rv** s and add means later if necessary. Recall that any rv U with a mean μ can be regarded as a constant μ plus the fluctuation, $U - \mu$, of U.

The moment generating function (MGF), $\mathsf{g}_Z(r)$, of a Gaussian rv $Z \sim \mathrm{N}(0, \sigma^2)$, can be calculated as follows:

$$\begin{aligned}\mathsf{g}_Z(r) = \mathsf{E}\left[\exp(rZ)\right] &= \frac{1}{\sqrt{2\pi}\,\sigma}\int_{-\infty}^{\infty}\exp(rz)\exp\left[\frac{-z^2}{2\sigma^2}\right]dz \\ &= \frac{1}{\sqrt{2\pi}\,\sigma}\int_{-\infty}^{\infty}\exp\left[\frac{-z^2+2\sigma^2 rz - r^2\sigma^4}{2\sigma^2} + \frac{r^2\sigma^2}{2}\right]dz \end{aligned} \tag{3.4}$$

$$= \exp\left[\frac{r^2\sigma^2}{2}\right]\left\{\frac{1}{\sqrt{2\pi}\,\sigma}\int_{-\infty}^{\infty}\exp\left[\frac{-(z-r\sigma)^2}{2\sigma^2}\right]dz\right\} \tag{3.5}$$

$$= \exp\left[\frac{r^2\sigma^2}{2}\right]. \tag{3.6}$$

We completed the square in the exponent in (3.4). We then recognized that the term in braces in (3.5) is the integral of a probability density and thus equal to 1.

Note that $\mathsf{g}_Z(r)$ exists for all real r, although it increases rapidly with $|r|$. As shown in Exercise 3.2, the moments for $Z \sim \mathrm{N}(0, \sigma^2)$, can be calculated from the MGF to be

$$\mathsf{E}\left[Z^{2k}\right] = \frac{(2k)!\,\sigma^{2k}}{k!\,2^k} = (2k-1)(2k-3)(2k-5)\cdots(3)(1)\sigma^{2k}. \tag{3.7}$$

Thus, $\mathsf{E}\left[Z^4\right] = 3\sigma^4$, $\mathsf{E}\left[Z^6\right] = 15\sigma^6$, etc. The odd moments of Z are all zero since z^{2k+1} is an odd function of z and the Gaussian density is even.

For an arbitrary Gaussian rv $U \sim \mathrm{N}(\mu, \sigma^2)$, let $Z = U - \mu$. Then $Z \sim \mathrm{N}(0, \sigma^2)$ and $\mathsf{g}_U(r)$ is given by

$$\mathsf{g}_U(r) = \mathsf{E}\left[\exp(r(\mu + Z))\right] = e^{r\mu}\mathsf{E}\left[e^{rZ}\right] = \exp(r\mu + r^2\sigma^2/2). \tag{3.8}$$

The characteristic function, $\mathsf{g}_Z(i\theta) = \mathsf{E}\left[e^{i\theta Z}\right]$ for $Z \sim \mathrm{N}(0, \sigma^2)$ and $i\theta$ imaginary can be shown to be (for example, see Chapter 2.12 in [27]):

$$\mathsf{g}_Z(i\theta) = \exp\left[\frac{-\theta^2\sigma^2}{2}\right]. \tag{3.9}$$

The argument in (3.4)–(3.6) does not show this since the term in braces in (3.5) is not a probability density for r imaginary. As explained in Section 1.5.5, the characteristic function is useful first because it exists for all rv s and second because an inversion formula (essentially the Fourier transform) exists to uniquely find the distribution of a rv from its characteristic function.

3.3 Gaussian random vectors

An $n \times \ell$ *matrix* $[A]$ is an array of $n\ell$ elements arranged in n rows and ℓ columns; A_{jk} denotes the kth element in the jth row. Unless specified to the contrary, the elements are real numbers. The *transpose* $[A^{\mathsf{T}}]$ of an $n \times \ell$ matrix $[A]$ is an $\ell \times n$ matrix $[B]$ with $B_{kj} = A_{jk}$ for all j, k. A matrix is *square* if $n = \ell$ and a square matrix $[A]$ is *symmetric* if $[A] = [A]^{\mathsf{T}}$. If $[A]$ and $[B]$ are each $n \times \ell$ matrices, $[A] + [B]$ is an $n \times \ell$ matrix $[C]$ with $C_{jk} = A_{jk} + B_{jk}$ for all j, k. If $[A]$ is $n \times \ell$ and $[B]$ is $\ell \times r$, the matrix $[A][B]$ is an $n \times r$ matrix $[C]$ with elements $C_{jk} = \sum_i A_{ji}B_{ik}$. A *vector* (or *column vector*) of dimension n is an $n \times 1$ matrix and a *row vector* of dimension n is a $1 \times n$ matrix. Since the transpose of a vector is a row vector, we denote a vector $\boldsymbol{a}$ as $(a_1, \ldots, a_n)^{\mathsf{T}}$. Note that if $\boldsymbol{a}$ is a (column) vector of dimension n, then $\boldsymbol{a}\boldsymbol{a}^{\mathsf{T}}$ is an $n \times n$ matrix whereas $\boldsymbol{a}^{\mathsf{T}}\boldsymbol{a}$ is a number. The reader is expected to be familiar with these vector and matrix manipulations.

The covariance matrix, $[K]$, (if it exists) of an arbitrary zero-mean n-rv $\mathbf{Z} = (Z_1, \ldots, Z_n)^{\mathsf{T}}$ is the matrix whose components are $K_{jk} = \mathsf{E}\left[Z_j Z_k\right]$. For a non-zero-mean n-rv $\boldsymbol{U}$, let $\boldsymbol{U} = \boldsymbol{m} + \boldsymbol{Z}$, where $\boldsymbol{m} = \mathsf{E}[\boldsymbol{U}]$ and $\boldsymbol{Z} = \boldsymbol{U} - \boldsymbol{m}$ is the fluctuation of $\boldsymbol{U}$. The covariance matrix $[K]$ of $\boldsymbol{U}$ is defined to be the same as the covariance matrix of the fluctuation $\boldsymbol{Z}$, i.e., $K_{jk} = \mathsf{E}\left[Z_j Z_k\right] = \mathsf{E}\left[(U_j - m_j)(U_k - m_k)\right]$. It can be seen that if an $n \times n$ covariance matrix $[K]$ exists, it must be symmetric, i.e., it must satisfy $K_{jk} = K_{kj}$ for $1 \leq j, k \leq n$.

3.3.1 Generating functions of Gaussian random vectors

The MGF of an n-rv $\boldsymbol{Z}$ is defined as $\mathsf{g}_{\boldsymbol{Z}}(\boldsymbol{r}) = \mathsf{E}\left[\exp(\boldsymbol{r}^{\mathsf{T}}\boldsymbol{Z})\right]$, where $\boldsymbol{r} = (r_1, \ldots, r_n)^{\mathsf{T}}$ is an n-dimensional real vector. The n-dimensional MGF might not exist for all $\boldsymbol{r}$ (just as the one-dimensional MGF discussed in Section 1.5.5 need not exist everywhere). As we will soon see, however, the MGF exists everywhere for Gaussian n-rv s.

The characteristic function, $\mathsf{g}_{\boldsymbol{Z}}(i\boldsymbol{\theta}) = \mathsf{E}[e^{i\boldsymbol{\theta}^{\mathsf{T}}\boldsymbol{Z}}]$, of an n-rv $\boldsymbol{Z}$, where $\boldsymbol{\theta} = (\theta_1, \ldots, \theta_n)^{\mathsf{T}}$ is a real n-vector, is equally important. As in the one-dimensional case, the characteristic function always exists for all real $\boldsymbol{\theta}$ and all n-rv $\boldsymbol{Z}$. In addition, there is a uniqueness theorem[2] stating that the characteristic function of an n-rv $\boldsymbol{Z}$ uniquely specifies the joint distribution of $\boldsymbol{Z}$.

If the components of an n-rv are independent and identically distributed (IID), we call the vector an IID n-rv.

3.3.2 IID normalized Gaussian random vectors

An example that will become familiar is that of an IID n-rv $\boldsymbol{W}$ where each component W_j, $1 \leq j \leq n$, is normalized Gaussian, $W_j \sim \mathrm{N}(0, 1)$. By taking the product of n densities as given in (3.1), the joint density of $\boldsymbol{W} = (W_1, W_2, \ldots, W_n)^{\mathsf{T}}$ is

$$\mathsf{f}_{\boldsymbol{W}}(\boldsymbol{w}) = \frac{1}{(2\pi)^{n/2}} \exp\left(\frac{-w_1^2 - w_2^2 - \cdots - w_n^2}{2}\right) = \frac{1}{(2\pi)^{n/2}} \exp\left(\frac{-\boldsymbol{w}^{\mathsf{T}}\boldsymbol{w}}{2}\right). \tag{3.10}$$

The joint density of $\boldsymbol{W}$ at a sample value $\boldsymbol{w}$ depends only on the squared distance $\boldsymbol{w}^{\mathsf{T}}\boldsymbol{w}$ of the sample value $\boldsymbol{w}$ from the origin. That is, $\mathsf{f}_{\boldsymbol{W}}(\boldsymbol{w})$ is spherically symmetric around the origin, and points of equal probability density lie on concentric spheres around the origin (see Figure 3.2).

[2] See Shiryaev [27] for a proof in the one-dimensional case and an exercise providing the extension to the n-dimensional case. It appears that the exercise is a relatively straightforward extension of the proof for one dimension, but the one-dimensional proof is measure theoretic and by no means trivial. The reader can get an engineering understanding of this uniqueness theorem by viewing the characteristic function and joint probability density essentially as n-dimensional Fourier transforms of each other.

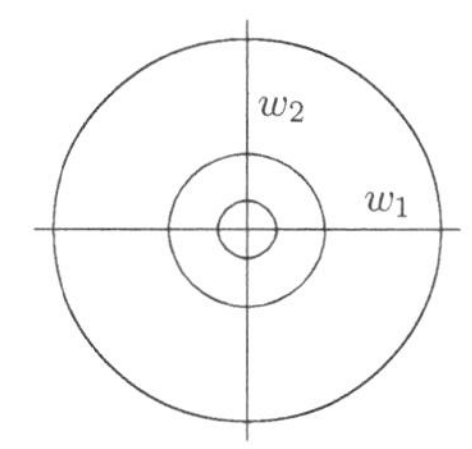

Figure 3.2 Equi-probability contours for an IID Gaussian 2-rv.

The MGF of $\boldsymbol{W}$ is easily calculated as follows:

$$\mathsf{g}_{\boldsymbol{W}}(\boldsymbol{r}) = \mathsf{E}\left[\exp \boldsymbol{r}^{\mathsf{T}}\boldsymbol{W})\right] = \mathsf{E}\left[\exp(r_1 W_1 + \cdots + r_n W_n\right] = \mathsf{E}\left[\prod_j \exp(r_j W_j)\right]$$

$$= \prod_j \mathsf{E}\left[\exp(r_j W_j)\right] = \prod_j \exp\left(\frac{r_j^2}{2}\right) = \exp\left[\frac{\boldsymbol{r}^{\mathsf{T}}\boldsymbol{r}}{2}\right]. \qquad (3.11)$$

The interchange of the expectation with the product above is justified because, first, the rv s W_j (and thus the rv s $\exp(r_j W_j)$) are independent, and, second, the expectation of a product of independent rv s is equal to the product of the expected values. The MGF of each W_j then follows from (3.6). The characteristic function of $\boldsymbol{W}$ is similarly calculated using (3.9),

$$\mathsf{g}_{\boldsymbol{W}}(i\boldsymbol{\theta}) = \exp\left[\frac{-\boldsymbol{\theta}^{\mathsf{T}}\boldsymbol{\theta}}{2}\right]. \qquad (3.12)$$

Next consider rv s that are linear combinations of $W_1, \ldots, W_n$, i.e., rv s of the form $Z = \boldsymbol{a}^{\mathsf{T}}\boldsymbol{W} = a_1 W_1 + \cdots + a_n W_n$. By convolving the densities of the components $a_j W_j$, it is shown in Exercise 3.4 that Z is Gaussian, $Z \sim \mathrm{N}(0, \sigma^2)$, where $\sigma^2 = \sum_{j=1}^n a_j^2$, i.e., $Z \sim \mathrm{N}(0, \sum_j a_j^2)$.

3.3.3 Jointly-Gaussian random vectors

We now go on to define the general class of zero-mean jointly-Gaussian n-**rv** s.

Definition 3.3.1 *$\{Z_1, Z_2, \ldots, Z_n\}$ is a set of* ***jointly-Gaussian zero-mean*** *rv s, and $\boldsymbol{Z} = (Z_1, \ldots, Z_n)^{\mathsf{T}}$ is a Gaussian* ***zero-mean*** *n-**rv**, if, for some finite set of IID $\mathrm{N}(0, 1)$ rv s, $W_1, \ldots, W_m$, each Z_j can be expressed as*

$$Z_j = \sum_{\ell=1}^{m} a_{j\ell} W_\ell, \qquad \text{i.e., } \boldsymbol{Z} = [A]\boldsymbol{W}, \qquad (3.13)$$

*where $\{a_{j\ell}, 1 \le j \le n, 1 \le \ell \le m,\}$ is a given array of real numbers. More generally, $\boldsymbol{U} = (U_1, \ldots, U_n)^{\mathsf{T}}$ is a Gaussian n-**rv** if $\boldsymbol{U} = \boldsymbol{Z} + \boldsymbol{\mu}$, where $\boldsymbol{Z}$ is a zero-mean Gaussian n-**rv** and $\boldsymbol{\mu}$ is a real n vector.*

As seen in Exercise 3.4, each linear combination of IID $\mathrm{N}(0,1)$ rv s is Gaussian. The definition here defines $Z_1, \dots, Z_n$ to be jointly Gaussian if all of them are linear combinations of a common set of IID normalized Gaussian rv s. This definition might not appear to restrict jointly-Gaussian rv s far beyond being individually Gaussian, but several examples later show that being jointly Gaussian in fact implies a great deal more than being individually Gaussian. We will also see that the remarkable properties of jointly Gaussian rv s depend very heavily on this linearity property.

Note from the definition that a Gaussian n-**rv** is a vector whose components are *jointly* Gaussian rather than only individually Gaussian. When we define Gaussian processes later, the requirement that the components be jointly Gaussian will again be present.

The intuition behind jointly-Gaussian rv s is that in many physical situations there are multiple rv s, each of which is a linear combination of a common large set of small essentially independent rv s. The central limit theorem (CLT) indicates that each such sum can be approximated by a Gaussian rv, and, more to the point here, linear combinations of those sums are also approximately Gaussian. For example, when a broadband noise waveform is passed through a narrowband linear filter, the output at any given time is usually well approximated as the sum of a large set of essentially independent rv s. The outputs at different times are different linear combinations of the same set of underlying small, essentially independent, rv s. Thus we would expect a set of outputs at different times to be jointly Gaussian according to the above definition.

The following simple theorem begins the process of specifying the properties of jointly-Gaussian rv s. These results are given for zero-mean rv s since the extension to non-zero mean is obvious.

Theorem 3.3.2 *Let $\boldsymbol{Z} = (Z_1, \dots, Z_n)^{\mathsf{T}}$ be a zero-mean Gaussian n-**rv**. Let $\boldsymbol{Y} = (Y_1, \dots, Y_k)^{\mathsf{T}}$ be a k-**rv** satisfying $\boldsymbol{Y} = [B]\boldsymbol{Z}$. Then $\boldsymbol{Y}$ is a zero-mean Gaussian k-**rv**.*

Proof Since $\boldsymbol{Z}$ is a zero-mean Gaussian n-**rv**, it can be represented as $\boldsymbol{Z} = [A]\boldsymbol{W}$, where the components of $\boldsymbol{W}$ are IID and $\mathrm{N}(0,1)$. Thus $\boldsymbol{Y} = [B][A]\boldsymbol{W}$. Since $[B][A]$ is a matrix, $\boldsymbol{Y}$ is a zero-mean Gaussian k-**rv**. □

For $k = 1$, this becomes the following trivial but important corollary.

Corollary 3.3.3 *Let $\boldsymbol{Z} = (Z_1, \dots, Z_n)^{\mathsf{T}}$ be a zero-mean Gaussian n-**rv**. Then for any real n-vector $\boldsymbol{a} = (a_1, \dots, a_n)^{\mathsf{T}}$, the linear combination $\boldsymbol{a}^{\mathsf{T}}\boldsymbol{Z}$ is a zero-mean Gaussian rv.*

We next give an example of two rv s, Z_1, Z_2 that are each zero-mean Gaussian but for which $Z_1 + Z_2$ is not Gaussian. From the theorem, then, Z_1 and Z_2 are not jointly Gaussian and the 2-**rv** $\boldsymbol{Z} = (Z_1, Z_2)^{\mathsf{T}}$ is not a Gaussian vector. This is the first of a number of examples of rv s that are marginally Gaussian but not jointly Gaussian.

Example 3.3.4 Let $Z_1 \sim \mathrm{N}(0,1)$, and let X be independent of Z_1 and take equiprobable values ± 1. Let $Z_2 = Z_1 X_1$. Then $Z_2 \sim \mathrm{N}(0,1)$ and $\mathsf{E}[Z_1 Z_2] = 0$. The joint probability density, $\mathsf{f}_{Z_1 Z_2}(z_1, z_2)$, however, is impulsive on the diagonals where $z_2 = \pm z_1$ and is zero elsewhere. Then $Z_1 + Z_2$ cannot be Gaussian, since it takes on the value 0 with probability one half.

This example also shows the falseness of the frequently heard statement that uncorrelated Gaussian rv s are independent. The correct statement, as we see later, is that uncorrelated *jointly*-Gaussian rv s are independent.

The next theorem specifies the MGF of an arbitrary zero-mean Gaussian n-rv $\mathbf{Z}$. The important feature is that the MGF depends only on the covariance function $[K]$. Essentially, as developed later, Z is characterized by a probability density that depends only on $[K]$.

Theorem 3.3.5 *Let $\mathbf{Z}$ be a zero-mean Gaussian n-rv with covariance matrix $[K]$. Then the MGF $\mathsf{g}_{\mathbf{Z}}(\boldsymbol{r}) = \mathsf{E}\left[\exp(\boldsymbol{r}^{\mathsf{T}}\mathbf{Z})\right]$ and the characteristic function $\mathsf{g}_{\mathbf{Z}}(i\boldsymbol{\theta}) = \mathsf{E}\left[\exp(i\boldsymbol{\theta}^{\mathsf{T}}\mathbf{Z})\right]$ are given by*

$$\mathsf{g}_{\mathbf{Z}}(\boldsymbol{r}) = \exp\left[\frac{\boldsymbol{r}^{\mathsf{T}}[K]\boldsymbol{r}}{2}\right], \qquad \mathsf{g}_{\mathbf{Z}}(i\boldsymbol{\theta}) = \exp\left[\frac{-\boldsymbol{\theta}^{\mathsf{T}}[K]\boldsymbol{\theta}}{2}\right]. \tag{3.14}$$

Proof For any given real n-vector $\boldsymbol{r} = (r_1, \ldots, r_n)^{\mathsf{T}}$, let $X = \boldsymbol{r}^{\mathsf{T}}\mathbf{Z}$. Then from Corollary 3.3.3, X is zero-mean Gaussian and from (3.6),

$$\mathsf{g}_X(s) = \mathsf{E}\left[\exp(sX)\right] = \exp(\sigma_X^2 s^2/2). \tag{3.15}$$

Thus for the given $\boldsymbol{r}$,

$$\mathsf{g}_{\mathbf{Z}}(r) = \mathsf{E}\left[\exp(\boldsymbol{r}^{\mathsf{T}}\mathbf{Z})\right] = \mathsf{E}\left[\exp(X)\right] = \exp(\sigma_X^2/2), \tag{3.16}$$

where the last step uses (3.15) with $s = 1$. Finally, since $X = \boldsymbol{r}^{\mathsf{T}}\mathbf{Z}$, we have

$$\sigma_X^2 = \mathsf{E}\left[|\boldsymbol{r}^{\mathsf{T}}\mathbf{Z}|^2\right] = \mathsf{E}\left[\boldsymbol{r}^{\mathsf{T}}\mathbf{Z}\mathbf{Z}^{\mathsf{T}}\boldsymbol{r}\right] = \boldsymbol{r}^{\mathsf{T}}\mathsf{E}\left[\mathbf{Z}\mathbf{Z}^{\mathsf{T}}\right]\boldsymbol{r} = \boldsymbol{r}^{\mathsf{T}}[K]\boldsymbol{r}. \tag{3.17}$$

Substituting (3.17) into (3.16), yields (3.14). The proof is the same for the characteristic function except (3.9) is used in place of (3.6). □

Since the characteristic function of an n-rv uniquely specifies the CDF, this theorem also shows that the joint CDF of a zero-mean Gaussian n-rv is completely determined by the covariance function. To make this story complete, we will show later that for any possible covariance function for any n-rv, there is a corresponding zero-mean Gaussian n-rv with that covariance.

As a slight generalization of (3.14), let $\boldsymbol{U}$ be a Gaussian n-rv with an arbitrary mean, i.e., $\boldsymbol{U} = \boldsymbol{m} + \mathbf{Z}$, where the n-vector $\boldsymbol{m}$ is the mean of $\boldsymbol{U}$ and the zero-mean Gaussian n-rv $\mathbf{Z}$ is the fluctuation of $\boldsymbol{U}$. Note that the covariance matrix $[K]$ of $\boldsymbol{U}$ is the same as that for $\mathbf{Z}$, yielding

$$\mathsf{g}_{\boldsymbol{U}}(\boldsymbol{r}) = \exp\left(\boldsymbol{r}^{\mathsf{T}}\boldsymbol{m} + \frac{\boldsymbol{r}^{\mathsf{T}}[K]\boldsymbol{r}}{2}\right), \qquad \mathsf{g}_{\boldsymbol{U}}(i\boldsymbol{\theta}) = \exp\left[i\boldsymbol{\theta}^{\mathsf{T}}\boldsymbol{m} - \frac{\boldsymbol{\theta}^{\mathsf{T}}[K]\boldsymbol{\theta}}{2}\right]. \tag{3.18}$$

We denote a Gaussian n-rv $\boldsymbol{U}$ of mean $\boldsymbol{m}$ and covariance $[K]$ as $\boldsymbol{U} \sim \mathrm{N}\,(\boldsymbol{m}, [K])$.

3.3.4 Joint probability density for Gaussian n-rv s (special case)

A zero-mean Gaussian n-rv, by definition, has the form $\boldsymbol{Z} = [A]\boldsymbol{W}$, where $\boldsymbol{W}$ is $\mathcal{N}(0, [I_n])$. In this section we look at the special case where $[A]$ is $n \times n$ and non-singular. The covariance matrix of $\boldsymbol{Z}$ is then

$$[K] = \mathsf{E}[\boldsymbol{Z}\boldsymbol{Z}^{\mathsf{T}}] = \mathsf{E}[[A]\boldsymbol{W}\boldsymbol{W}^{\mathsf{T}}[A]^{\mathsf{T}}]$$

$$= [A]\mathsf{E}[\boldsymbol{W}\boldsymbol{W}^{\mathsf{T}}][A]^{\mathsf{T}} = [A][A]^{\mathsf{T}} \tag{3.19}$$

since $\mathsf{E}[WW^{\mathsf{T}}]$ is the identity matrix, $[I_n]$.

To find $\mathsf{f}_{\boldsymbol{Z}}(\boldsymbol{z})$ in this case, we first consider the transformation of real-valued vectors, $\boldsymbol{z} = [A]\boldsymbol{w}$. Let $\boldsymbol{e}_j$ be the jth unit vector (i.e., the vector whose jth component is 1 and whose other components are 0). Then $[A]\boldsymbol{e}_j = \boldsymbol{a}_j$, where $\boldsymbol{a}_j$ is the jth column of $[A]$. Thus, $\boldsymbol{z} = [A]\boldsymbol{w}$ transforms each unit vector $\boldsymbol{e}_j$ into the column $\boldsymbol{a}_j$ of [A]. For $n=2$, Figure 3.3 shows how this transformation carries each vector $\boldsymbol{w}$ into the vector $\boldsymbol{z} = [A]\boldsymbol{w}$. Note that an incremental square of side δ is carried into a parallelogram with corners $\boldsymbol{0}, \boldsymbol{a}_1\delta, \boldsymbol{a}_2\delta$, and $(\boldsymbol{a}_1 + \boldsymbol{a}_2)\delta$.

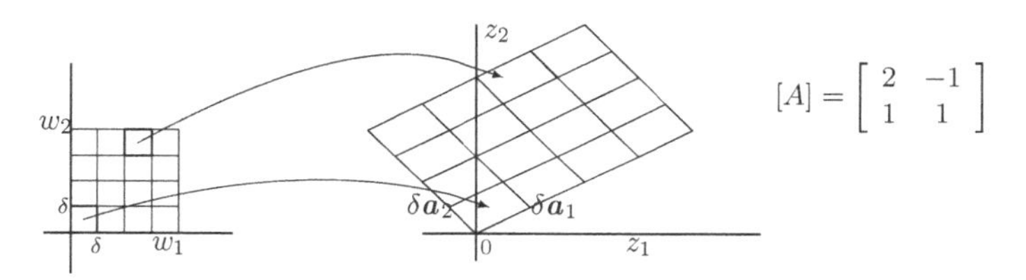

Figure 3.3 Example illustrating how $\boldsymbol{z} = [A]\boldsymbol{w}$ maps cubes into parallelepipeds. Let $z_1 = 2w_1 - w_2$ and $z_2 = w_1 + w_2$. Thus $\boldsymbol{w} = (1, 0)^{\mathsf{T}}$ transforms to $\boldsymbol{a}_1 = (2, 1)^{\mathsf{T}}$ and $\boldsymbol{w} = (0, 1)^{\mathsf{T}}$ transforms to $\boldsymbol{a}_2 = (-1, 1)^{\mathsf{T}}$. The lower left square in the first figure is the set $\{(w_1, w_2) : 0 \le w_1 \le \delta;\ 0 \le w_2 \le \delta\}$. This square is transformed into the parallelogram with sides $\delta\boldsymbol{a}_1$ and $\delta\boldsymbol{a}_2$. The figure also shows how the w_1, w_2 space can be quantized into adjoining squares, which map into corresponding adjoining parallelograms in the z_1, z_2 space.

For an arbitrary number of dimensions, the unit cube in the $\boldsymbol{w}$ space is the set of points $\{\boldsymbol{w} : 0 \le w_j \le 1; 1 \le j \le n\}$. There are 2^n corners of the unit cube, and each is some 0/1 combination of the unit vectors, i.e., each has the form $\boldsymbol{e}_{j_1} + \boldsymbol{e}_{j_2} + \cdots + \boldsymbol{e}_{j_k}$. The transformation $[A]\boldsymbol{w}$ carries the unit cube into a parallelepiped, where each corner of the cube, $\boldsymbol{e}_{j_1} + \boldsymbol{e}_{j_2} + \cdots + \boldsymbol{e}_{j_k}$, is carried into a corresponding corner $\boldsymbol{a}_{j_1} + \boldsymbol{a}_{j_2} + \cdots + \boldsymbol{a}_{j_n}$ of the parallelepiped. One of the most interesting and geometrically meaningful properties of the determinant, det$[A]$, of a square real matrix $[A]$ is that the *magnitude* of that determinant, $|\det[A]|$, is equal to the volume of that parallelepiped (see Strang [28]). If $\det[A] = 0$, i.e., if $[A]$ is singular, then the n-dimensional unit cube in the $\boldsymbol{w}$ space is transformed into a lower-dimensional parallelepiped whose volume (as a region of n-dimensional space) is 0. This case is considered in Section 3.4.4.

Now let $\boldsymbol{z}$ be a sample value of $\boldsymbol{Z}$, and let $\boldsymbol{w} = [A]^{-1}\boldsymbol{z}$ be the corresponding sample value of $\boldsymbol{W}$. The joint density at $\boldsymbol{z}$ must satisfy

$$\mathsf{f}_{\boldsymbol{Z}}(\boldsymbol{z})|d\boldsymbol{z}| = \mathsf{f}_{\boldsymbol{W}}(\boldsymbol{w})|d\boldsymbol{w}|, \tag{3.20}$$

where $|dw|$ is the volume of an incremental cube with dimension $\delta = dw_j$ on each side, and $|dz|$ is the volume of that incremental cube transformed by $[A]$. Thus $|dw| = \delta^n$ and $|dz| = \delta^n|\det[A]|$ so that $|dz|/|dw| = |\det[A]|$. Using this in (3.20), and using (3.10) for $\mathsf{f}_{\boldsymbol{W}}(\boldsymbol{w})$, we see that the density of a jointly-Gaussian **rv** $\boldsymbol{Z} = [A]\boldsymbol{W}$ is

$$\mathsf{f}_{\boldsymbol{Z}}(\boldsymbol{z}) = \frac{\exp\left(-\frac{1}{2}\boldsymbol{z}^{\mathsf{T}}[A^{-1}]^{\mathsf{T}}[A^{-1}]\boldsymbol{z}\right)}{(2\pi)^{n/2}|\det[A]|}. \tag{3.21}$$

From (3.19), we have $[K] = [AA^{\mathsf{T}}]$, so $[K^{-1}] = [A^{-1}]^{\mathsf{T}}[A^{-1}]$. Also, for arbitrary square real matrices $[A]$ and $[B]$, $\det[AB] = \det[A]\det[B]$ and $\det[A] = \det[A^{\mathsf{T}}]$. Thus $\det[K] = \det[A]\det[A^{\mathsf{T}}] = \big(\det[A]\big)^2 > 0$ and (3.21) becomes

$$\mathsf{f}_{\boldsymbol{Z}}(\boldsymbol{z}) = \frac{\exp\left(-\frac{1}{2}\boldsymbol{z}^{\mathsf{T}}[K^{-1}]\boldsymbol{z}\right)}{(2\pi)^{n/2}\sqrt{\det[K]}}. \tag{3.22}$$

Note that this density depends only on $[K]$, so the density depends on $[A]$ only through $[A][A^{\mathsf{T}}] = [K]$. This is not surprising, since we saw that the characteristic function of $\boldsymbol{Z}$ also depended only on the covariance matrix of $\boldsymbol{Z}$.

The expression in (3.22) is quite beautiful. It arises, first, because the density of $\boldsymbol{W}$ is spherically symmetric, and second, because $\boldsymbol{Z}$ is a linear transformation of $\boldsymbol{W}$. We show later that this density applies to any zero-mean Gaussian n-**rv** for which the covariance is a non-singular matrix $[K]$.

Example 3.3.6 Consider (3.22) for the two-dimensional case. Let $\mathsf{E}\left[Z_1^2\right] = \sigma_1^2$, $\mathsf{E}\left[Z_2^2\right] = \sigma_2^2$ and $\mathsf{E}[Z_1Z_2] = k_{12}$. Define the *normalized covariance*, ρ, as $k_{12}/(\sigma_1\sigma_2)$. Then $\det[K] = \sigma_1^2\sigma_2^2 - k_{12}^2 = \sigma_1^2\sigma_2^2(1-\rho^2)$. For $[A]$ to be non-singular, we need $\det[K] = \big(\det[A]\big)^2 > 0$, so we need $|\rho| < 1$. We then have

$$[K]^{-1} = \frac{1}{\sigma_1^2\sigma_2^2 - k_{12}^2}\begin{bmatrix} \sigma_2^2 & -k_{12} \\ -k_{12} & \sigma_1^2 \end{bmatrix} = \frac{1}{1-\rho^2}\begin{bmatrix} 1/\sigma_1^2 & -\rho/(\sigma_1\sigma_2) \\ -\rho/(\sigma_1\sigma_2) & 1/\sigma_2^2 \end{bmatrix}.$$

$$\mathsf{f}_{\boldsymbol{Z}}(\boldsymbol{z}) = \frac{1}{2\pi\sqrt{\sigma_1^2\sigma_2^2 - k_{12}^2}}\exp\left(\frac{-z_1^2\sigma_2^2 + 2z_1z_2k_{12} - z_2^2\sigma_1^2}{2(\sigma_1^2\sigma_2^2 - k_{12}^2)}\right)$$

$$= \frac{1}{2\pi\sigma_1\sigma_2\sqrt{1-\rho^2}}\exp\left(\frac{-\frac{z_1^2}{\sigma_1^2} + \frac{2\rho z_1z_2}{\sigma_1\sigma_2} - \frac{z_2^2}{\sigma_2^2}}{2(1-\rho^2)}\right). \tag{3.23}$$

The exponent in (3.23) is a quadratic in z_1, z_2 and from this it can be deduced that the equiprobability contours for $\boldsymbol{Z}$ are concentric ellipses. This will become clearer (both for $n = 2$ and $n > 2$) in Section 3.4.4.

Perhaps the more important lesson from (3.23), however, is that vector notation simplifies such equations considerably even for $n = 2$. We must learn to reason directly from the vector equations and use standard computer programs for required calculations.

For completeness, let $\boldsymbol{U} = \boldsymbol{\mu} + \boldsymbol{Z}$, where $\boldsymbol{\mu} = \mathsf{E}[\boldsymbol{U}]$ and $\boldsymbol{Z}$ is a zero-mean Gaussian n-**rv** with the density in (3.21). Then the density of $\boldsymbol{U}$ is given by

$$\mathsf{f}_{\boldsymbol{U}}(\boldsymbol{u}) = \frac{\exp\left(-\frac{1}{2}(\boldsymbol{u}-\boldsymbol{\mu})^{\mathsf{T}}[K^{-1}](\boldsymbol{u}-\boldsymbol{\mu})\right)}{(2\pi)^{n/2}\sqrt{\det[K]}}, \tag{3.24}$$

where $[K]$ is the covariance matrix of both $\boldsymbol{U}$ and $\boldsymbol{Z}$.

3.4 Properties of covariance matrices

In this section, we summarize some simple properties of covariance matrices that will be used frequently in what follows. We start with symmetric matrices before considering covariance matrices.

3.4.1 Symmetric matrices

A number λ is said to be an eigenvalue of an $n \times n$ matrix, $[B]$, if there is a non-zero n-vector $\boldsymbol{q}$ such that $[B]\boldsymbol{q} = \lambda\boldsymbol{q}$, i.e., such that $[B - \lambda I]\boldsymbol{q} = 0$. In other words, λ is an eigenvalue of $[B]$ if $[B - \lambda I]$ is singular. We are interested only in real matrices here, but the eigenvalues and eigenvectors might be complex. The values of λ that are eigenvalues of $[B]$ are the solutions to the characteristic equation, $\det[B - \lambda I] = 0$, i.e., they are the roots of $\det[B - \lambda I]$. As a function of λ, $\det[B - \lambda I]$ is a polynomial of degree n. From the fundamental theorem of algebra, it therefore has n roots (possibly complex and not necessarily distinct).

If $[B]$ is symmetric, then the eigenvalues are all real.[3] Also, the eigenvectors can all be chosen to be real. In addition, eigenvectors of distinct eigenvalues must be orthogonal, and if an eigenvalue λ has multiplicity ℓ (i.e., $\det[B - \lambda I]$ as a polynomial in λ has an ℓth order root at λ), then ℓ orthogonal eigenvectors can be chosen for that λ.

What this means is that we can list the eigenvalues as $\lambda_1, \lambda_2, \ldots, \lambda_n$ (where each distinct eigenvalue is repeated according to its multiplicity). To each eigenvalue λ_j, we can associate an eigenvector $\boldsymbol{q}_j$, where $\boldsymbol{q}_1, \ldots, \boldsymbol{q}_n$ are orthogonal. Finally, each eigenvector can be normalized so that $\boldsymbol{q}_j^{\mathsf{T}}\boldsymbol{q}_k = \delta_{jk}$, where $\delta_{jk} = 1$ for $j = k$ and $\delta_{jk} = 0$ otherwise; the set $\{\boldsymbol{q}_1, \ldots, \boldsymbol{q}_n\}$ is then called orthonormal.

If we take the resulting n equations, $[B]\boldsymbol{q}_j = \lambda_j\boldsymbol{q}_j$ and combine them into a matrix equation, we get

$$[BQ] = [Q\Lambda], \tag{3.25}$$

where $[Q]$ is the $n \times n$ matrix whose columns are the orthonormal vectors $\boldsymbol{q}_1, \ldots, \boldsymbol{q}_n$ and where $[\Lambda]$ is the $n \times n$ diagonal matrix whose diagonal elements are $\lambda_1, \ldots, \lambda_n$.

The matrix $[Q]$ is called an orthonormal or orthogonal matrix and, as we have seen, has the property that its columns are orthonormal. The matrix $[Q]^{\mathsf{T}}$ then has the rows $\boldsymbol{q}_j^{\mathsf{T}}$ for $1 \leq j \leq n$. If we multiply $[Q]^{\mathsf{T}}$ by $[Q]$, we see that the j, k element of the product

[3] See Strang [28] or other linear algebra texts for a derivation of these standard results.

is $q_j^{\mathsf{T}} q_k = \delta_{jk}$. Thus $[Q^{\mathsf{T}}Q] = [I]$ and $[Q^{\mathsf{T}}]$ is the inverse, $[Q^{-1}]$, of $[Q]$. Finally, since $[QQ^{-1}] = [I] = [QQ^{\mathsf{T}}]$, we see that the rows of Q are also orthonormal. This can be summarized in the following theorem.

Theorem 3.4.1 *Let $[B]$ be a real symmetric matrix and let $[\Lambda]$ be the diagonal matrix whose diagonal elements $\lambda_1, \ldots, \lambda_n$ are the eigenvalues of $[B]$, repeated according to multiplicity. Then a set of orthonormal eigenvectors $\boldsymbol{q}_1, \ldots, \boldsymbol{q}_n$ can be chosen so that $[B]\boldsymbol{q}_j = \lambda_j \boldsymbol{q}_j$ for $1 \le j \le n$. The matrix $[Q]$ with orthonormal columns $\boldsymbol{q}_1, \ldots, \boldsymbol{q}_n$ satisfies (3.25). Also $[Q^{\mathsf{T}}] = [Q^{-1}]$ and the rows of $[Q]$ are orthonormal. Finally $[B]$ and $[Q]$ satisfy*

$$[B] = [Q\Lambda Q^{-1}], \qquad [Q^{-1}] = [Q^{\mathsf{T}}]. \tag{3.26}$$

Proof The only new statement is the initial part of (3.26), which follows from (3.25) by post-multiplying both sides by $[Q^{-1}]$. □

3.4.2 Positive definite matrices and covariance matrices

Definition 3.4.2 *A real $n \times n$ matrix $[K]$ is **positive definite** if it is symmetric and if $\boldsymbol{b}^{\mathsf{T}}[K]\boldsymbol{b} > 0$ for all real n-vectors $\boldsymbol{b} \neq 0$. It is **positive semi-definite**[4] if $\boldsymbol{b}^{\mathsf{T}}[K]\boldsymbol{b} \ge 0$. It is a **covariance matrix** if there is a zero-mean n-rv $\mathbf{Z}$ such that $[K] = \mathsf{E}\left[\mathbf{Z}\mathbf{Z}^{\mathsf{T}}\right]$.*

We will see shortly that the class of positive semi-definite matrices is the same as the class of covariance matrices and that the class of positive definite matrices is the same as the class of non-singular covariance matrices. First we develop some useful properties of positive (semi-)definite matrices.

Theorem 3.4.3 *A symmetric matrix $[K]$ is positive definite*[5] *if and only if each eigenvalue of $[K]$ is positive. It is positive semi-definite if and only if each eigenvalue is non-negative.*

Proof Assume that $[K]$ is positive definite. It is symmetric by the definition of positive definiteness, so for each eigenvalue λ_j of $[K]$, we can select a real normalized eigenvector $\boldsymbol{q}_j$ as a vector $\boldsymbol{b}$ in Definition 3.4.2. Then

$$0 < \boldsymbol{q}_j^{\mathsf{T}}[K]\boldsymbol{q}_j = \lambda_j \boldsymbol{q}_j^{\mathsf{T}}\boldsymbol{q}_j = \lambda_j,$$

so each eigenvalue is positive. To go the other way, assume that each $\lambda_j > 0$ and use the expansion of (3.26) with $[Q^{-1}] = [Q^{\mathsf{T}}]$. Then for any real $\boldsymbol{b} \neq 0$,

$$\boldsymbol{b}^{\mathsf{T}}[K]\boldsymbol{b} = \boldsymbol{b}^{\mathsf{T}}[Q\Lambda Q^{\mathsf{T}}]\boldsymbol{b} = \boldsymbol{c}^{\mathsf{T}}[\Lambda]\boldsymbol{c}, \qquad \text{where } \boldsymbol{c} = [Q^{\mathsf{T}}]\boldsymbol{b}.$$

Now $[\Lambda]\boldsymbol{c}$ is a vector with components $\lambda_j c_j$. Thus $\boldsymbol{c}^{\mathsf{T}}[\Lambda]\boldsymbol{c} = \sum_j \lambda_j c_j^2$. Since each c_j is real, $c_j^2 \ge 0$ and thus $c_j^2 \lambda_j \ge 0$. Since $\boldsymbol{c} \neq 0$, $c_j \neq 0$ for at least one j and thus $\lambda_j c_j^2 > 0$

[4] Positive semi-definite is sometimes referred to as non-negative definite, which is more transparent but less common.

[5] Do not confuse the positive definite and positive semi-definite matrices here with the positive and non-negative matrices we soon study as the stochastic matrices of Markov chains. The terms positive definite and semi-definite relate to the eigenvalues of symmetric matrices, whereas the terms positive and non-negative matrices relate to the elements of typically non-symmetric matrices.

for at least one j, so $\boldsymbol{c}^{\mathsf{T}}[\Lambda]\boldsymbol{c} > 0$. The proof for the positive semi-definite case follows by replacing the strict inequalitites above with non-strict inequalities. □

Theorem 3.4.4 *If $[K] = [AA^{\mathsf{T}}]$ for some real $n \times n$ matrix $[A]$, then $[K]$ is positive semi-definite. If $[A]$ is also non-singular, then $[K]$ is positive definite.*

Proof For the hypothesized $[A]$ and any real n-vector $\boldsymbol{b}$,

$$\boldsymbol{b}^{\mathsf{T}}[K]\boldsymbol{b} = \boldsymbol{b}^{\mathsf{T}}[AA^{\mathsf{T}}]\boldsymbol{b} = \boldsymbol{c}^{\mathsf{T}}\boldsymbol{c} \geq 0, \qquad \text{where } \boldsymbol{c} = [A^{\mathsf{T}}]\boldsymbol{b}.$$

Thus $[K]$ is positive semi-definite. If $[A]$ is non-singular, then $\boldsymbol{c} \neq 0$ if $\boldsymbol{b} \neq 0$. Thus $\boldsymbol{c}^{\mathsf{T}}\boldsymbol{c} > 0$ for $\boldsymbol{b} \neq 0$ and $[K]$ is positive definite. □

A converse can be established showing that if $[K]$ is positive (semi-)definite, then an $[A]$ exists such that $[K] = [A][A^{\mathsf{T}}]$. It seems more productive, however, to actually specify a matrix with this property.

From (3.26) and Theorem 3.4.3, we have

$$[K] = [Q\Lambda Q^{-1}],$$

where, for $[K]$ positive semi-definite, each element λ_j on the diagonal matrix $[\Lambda]$ is non-negative. Now define $[\Lambda^{1/2}]$ as the diagonal matrix with the elements $\sqrt{\lambda_j}$. We then have

$$[K] = [Q\Lambda^{1/2}\Lambda^{1/2}Q^{-1}] = [Q\Lambda^{1/2}Q^{-1}][Q\Lambda^{1/2}Q^{-1}]. \tag{3.27}$$

Define the square-root matrix $[R]$ for $[K]$ as

$$[R] = [Q\Lambda^{1/2}Q^{-1}]. \tag{3.28}$$

Comparing (3.27) with (3.28), we see that $[K] = [RR]$. However, since $[Q^{-1}] = [Q^{\mathsf{T}}]$, we see that $[R]$ is symmetric and consequently $[R] = [R^{\mathsf{T}}]$. Thus

$$[K] = [RR^{\mathsf{T}}], \tag{3.29}$$

and $[R]$ is one choice for the desired matrix $[A]$. If $[K]$ is positive definite, then each $\lambda_j > 0$ so each $\sqrt{\lambda_j} > 0$ and $[R]$ is non-singular. This then provides a converse to Theorem 3.4.4, using the square-root matrix for $[A]$. We can also use the square-root matrix in the following simple theorem.

Theorem 3.4.5 *Let $[K]$ be an $n \times n$ semi-definite matrix and let $[R]$ be its square-root matrix. Then $[K]$ is the covariance matrix of the Gaussian zero-mean n-rv $\boldsymbol{Y} = [R]\boldsymbol{W}$ where $\boldsymbol{W} \sim \mathrm{N}(0, [I_n])$.*

Proof

$$\mathsf{E}\left[\boldsymbol{Y}\boldsymbol{Y}^{\mathsf{T}}\right] = [R]\mathsf{E}\left[\boldsymbol{W}\boldsymbol{W}^{\mathsf{T}}\right][R^{\mathsf{T}}] = [RR^{\mathsf{T}}] = [K].$$ □

We can now finally relate covariance matrices to positive (semi-)definite matrices.

Theorem 3.4.6 *An $n \times n$ real matrix $[K]$ is a covariance matrix if and only if it is positive semi-definite. It is a non-singular covariance matrix if and only if it is positive definite.*

Proof First assume $[K]$ is a covariance matrix, i.e., assume there is a zero-mean n-**rv** $\boldsymbol{Z}$ such that $[K] = \mathsf{E}\left[\boldsymbol{Z}\boldsymbol{Z}^{\mathsf{T}}\right]$. For any given real n-vector $\boldsymbol{b}$, let the zero-mean rv X satisfy $X = \boldsymbol{b}^{\mathsf{T}}\boldsymbol{Z}$. Then

$$0 \leq \mathsf{E}[X^2] = \mathsf{E}\left[\boldsymbol{b}^{\mathsf{T}}\boldsymbol{Z}\boldsymbol{Z}^{\mathsf{T}}\boldsymbol{b}\right] = \boldsymbol{b}^{\mathsf{T}}\mathsf{E}\left[\boldsymbol{Z}\boldsymbol{Z}^{\mathsf{T}}\right]\boldsymbol{b} = \boldsymbol{b}^{\mathsf{T}}[K]\boldsymbol{b}.$$

Since $\boldsymbol{b}$ is arbitrary, this shows that $[K]$ is positive semi-definite. If in addition, $[K]$ is non-singular, then its eigenvalues are all non-zero and thus positive. Consequently $[K]$ is positive definite.

Conversely, if $[K]$ is positive semi-definite, Theorem 3.4.5 shows that $[K]$ is a covariance matrix. If, in addition, $[K]$ is positive definite, then $[K]$ is non-singular and $[K]$ is then a non-singular covariance matrix. □

3.4.3 Joint probability density for Gaussian n-rv s (general case)

Recall that the joint probability density for a Gaussian n-**rv** $\boldsymbol{Z}$ was derived in Section 3.3.4 only for the special case where $\boldsymbol{Z} = [A]\boldsymbol{W}$ where the $n \times n$ matrix $[A]$ is non-singular and $\boldsymbol{W} \sim \mathrm{N}(0, [I_n])$. The above theorem lets us generalize this as follows:

Theorem 3.4.7 *Let a Gaussian zero-mean n-**rv** $\boldsymbol{Z}$ have a non-singular covariance matrix $[K]$. Then the probability density of $\boldsymbol{Z}$ is given by (3.22).*

Proof Let $[R]$ be the square-root matrix of $[K]$ as given in (3.28). From Theorem 3.4.5, the Gaussian vector $\boldsymbol{Y} = [R]\boldsymbol{W}$ has covariance $[K]$. Also $[K]$ is positive definite, so from Theorem 3.4.4 $[R]$ is non-singular. Thus $\boldsymbol{Y}$ satisfies the conditions under which (3.22) was derived, so $\boldsymbol{Y}$ has the probability density in (3.22). Since $\boldsymbol{Y}$ and $\boldsymbol{Z}$ have the same covariance and are both Gaussian zero-mean n-**rv** s, they have the same characteristic function, and thus the same distribution. □

The question still remains about the distribution of a zero-mean Gaussian n-**rv** $\boldsymbol{Z}$ with a singular covariance matrix $[K]$. In this case $[K^{-1}]$ does not exist and thus the density in (3.22) has no meaning. From Theorem 3.4.5, $\boldsymbol{Y} = [R]\boldsymbol{W}$ has covariance $[K]$ but $[R]$ is singular. This means that the individual sample vectors $\boldsymbol{w}$ are mapped into a proper linear subspace of $\mathbb{R}^n$. The n-**rv** $\boldsymbol{Z}$ has zero probability outside of that subspace and, viewed as an n-dimensional density, is impulsive within that subspace.

In this case $[R]$ has one or more linearly dependent combinations of rows. As a result, one or more components Z_j of $\boldsymbol{Z}$ can be expressed as a linear combination of the other components. Very messy notation can then be avoided by viewing a maximal linearly independent set of components of $\boldsymbol{Z}$ as a vector $\boldsymbol{Z}'$. All other components of $\boldsymbol{Z}$ are linear combinations of $\boldsymbol{Z}'$. Thus $\boldsymbol{Z}'$ has a non-singular covariance matrix and its probability density is given by (3.22).

Jointly-Gaussian rv s are often defined as rv s all of whose linear combinations are Gaussian. The next theorem shows that this definition is equivalent to the one we have given.

Theorem 3.4.8 *Let $Z_1, \dots, Z_n$ be zero-mean rv s. These rv s are jointly Gaussian if and only if $\sum_{j=1}^{n} a_j Z_j$ is zero-mean Gaussian for all real $a_1, \dots, a_n$.*

Proof First, assume that $Z_1,\dots,Z_n$ are zero-mean jointly Gaussian, i.e., $\mathbf{Z} = (Z_1,\dots,Z_n)^{\mathsf{T}}$ is a zero-mean Gaussian n-**rv**. Corollary 3.3.3 then says that $\boldsymbol{a}^{\mathsf{T}}\mathbf{Z}$ is zero-mean Gaussian for all real $\boldsymbol{a} = (a_1,\dots,a_n)^{\mathsf{T}}$.

Second, assume that for all real vectors $\boldsymbol{\theta} = (\theta_1,\dots,\theta_n)^{\mathsf{T}}$, $\boldsymbol{\theta}^{\mathsf{T}}\mathbf{Z}$ is zero-mean Gaussian. For any given $\boldsymbol{\theta}$, let $X = \boldsymbol{\theta}^{\mathsf{T}}\mathbf{Z}$, from which it follows that $\sigma_X^2 = \boldsymbol{\theta}^{\mathsf{T}}[K]\boldsymbol{\theta}$, where $[K]$ is the covariance matrix of $\mathbf{Z}$. By assumption, X is zero-mean Gaussian, so from (3.9) the characteristic function, $\mathsf{g}_X(i\phi) = \mathsf{E}\left[\exp(i\phi X\right]$, of X is

$$\mathsf{g}_X(i\phi) = \exp\left(\frac{-\phi^2\sigma_X^2}{2}\right) = \exp\left(\frac{-\phi^2\boldsymbol{\theta}^{\mathsf{T}}[K]\boldsymbol{\theta}}{2}\right). \tag{3.30}$$

Setting $\phi = 1$, we see that

$$\mathsf{g}_X(i) = \mathsf{E}\left[\exp(iX)\right] = \mathsf{E}\left[\exp(i\boldsymbol{\theta}^{\mathsf{T}}\mathbf{Z})\right].$$

In other words, the characteristic function of $X = \boldsymbol{\theta}^{\mathsf{T}}\mathbf{Z}$, evaluated at $\phi = 1$, is the characteristic function of $\mathbf{Z}$ evaluated at the given $\boldsymbol{\theta}$. Since this applies for all choices of $\boldsymbol{\theta}$,

$$\mathsf{g}_{\mathbf{Z}}(i\boldsymbol{\theta}) = \exp\left(\frac{-\boldsymbol{\theta}^{\mathsf{T}}[K]\boldsymbol{\theta}}{2}\right). \tag{3.31}$$

From (3.14), this is the characteristic function of an arbitrary $\mathbf{Z} \sim \mathrm{N}(0,[K])$. Since the characteristic function uniquely specifies the distribution of $\mathbf{Z}$, we have shown that $\mathbf{Z}$ is a zero-mean Gaussian n-rv. □

The following theorem summarizes the conditions under which a set of zero-mean rv s are jointly Gaussian

Theorem 3.4.9 *The following four sets of conditions are each necessary and sufficient for a zero-mean n-**rv** Z to be a zero-mean Gaussian n-**rv**, i.e., for the components $Z_1,\dots,Z_n$ of **Z** to be jointly Gaussian:*

- *$\mathbf{Z}$ can be expressed as $\mathbf{Z} = [A]\mathbf{W}$, where $[A]$ is real and $\mathbf{W}$ is $\mathrm{N}(0,[I])$.*
- *For all real n-vectors $\mathbf{a}$, the rv $\mathbf{a}^{\mathsf{T}}\mathbf{Z}$ is zero-mean Gaussian.*
- *The linearly independent components of **Z** have the probability density in (3.22).*
- *The characteristic function of **Z** is given by (3.9).*

We emphasize once more that the distribution of a zero-mean Gaussian n-**rv** depends only on the covariance, and for every covariance matrix, zero-mean Gaussian n-**rv** s exist with that covariance. If that covariance matrix is diagonal (i.e., the components of the Gaussian n-rv are uncorrelated), then the components are also independent. As we have seen from several examples, this depends on the definition of a Gaussian n-**rv** as having jointly-Gaussian components.

3.4.4 Geometry and principal axes for Gaussian densities

The purpose of this section is to explain the geometry of the probability density contours of a zero-mean Gaussian n-**rv** with a non-singular covariance matrix $[K]$. From (3.22), the density is constant over the region of vectors z for which $z^{\mathsf{T}}[K^{-1}]z = c$ for any given $c > 0$. We shall see that this region is an ellipsoid centered on 0 and that the ellipsoids for different c are concentric and expanding with increasing c.

First consider a simple special case where $Z_1, \dots, Z_n$ are independent with different variances, i.e., $Z_j \sim \mathrm{N}(0, \lambda_j)$, where $\lambda_j = \mathsf{E}\left[Z_j^2\right]$. Then $[K]$ is diagonal with elements $\lambda_1, \dots, \lambda_n$ and $[K^{-1}]$ is diagonal with elements $\lambda_1^{-1}, \dots, \lambda_n^{-1}$. Then the contour for a given c is

$$z^{\mathsf{T}}[K^{-1}]z = \sum_{j=1}^{n} z_j^2 \lambda_j^{-1} = c. \tag{3.32}$$

This is the equation of an ellipsoid which is centered at the origin and has axes lined up with the coordinate axes. We can view this ellipsoid as a deformed n-dimensional sphere where the original sphere has been expanded or contracted along each coordinate axis j by a linear factor of $\sqrt{\lambda_j}$. An example is given in Figure 3.4.

For the general case with $\boldsymbol{Z} \sim \mathrm{N}(0, [K])$, the equi-probability contours are similar, except that the axes of the ellipsoid become the eigenvectors of $[K]$. To see this, we represent $[K]$ as $[Q\Lambda Q^{\mathsf{T}}]$, where the orthonormal columns of $[Q]$ are the eigenvectors of $[K]$ and $[\Lambda]$ is the diagonal matrix of eigenvalues, all of which are positive. Thus we want to find the set of vectors z for which

$$z^{\mathsf{T}}[K^{-1}]z = z^{\mathsf{T}}[Q\Lambda^{-1}Q^{\mathsf{T}}]z = c. \tag{3.33}$$

Since the eigenvectors $\boldsymbol{q}_1, \dots, \boldsymbol{q}_n$ are orthonormal, they span $\mathbb{R}^n$ and any vector $z \in \mathbb{R}^n$ can be represented as a linear combination, say $\sum_j v_j \boldsymbol{q}_j$, of $\boldsymbol{q}_1, \dots, \boldsymbol{q}_n$. In vector terms this is $z = [Q]\boldsymbol{v}$. Thus $\boldsymbol{v}$ represents z in the coordinate basis in which the axes are the eigenvectors $\boldsymbol{q}_1, \dots, \boldsymbol{q}_n$. Substituting $z = [Q]\boldsymbol{v}$ in (3.33),

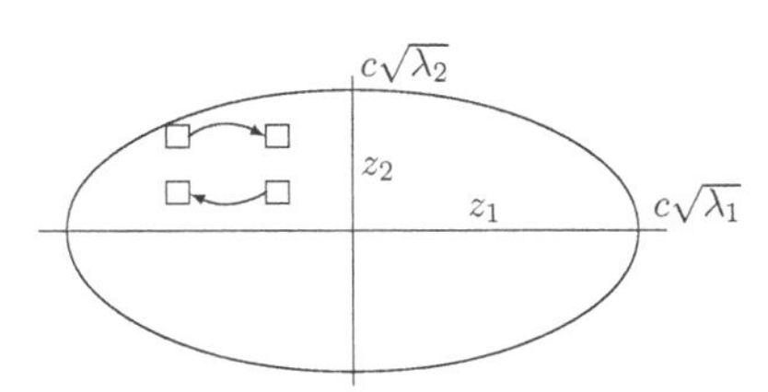

Figure 3.4 A contour of equal probability density for two dimensions with diagonal $[K]$. The figure assumes that $\lambda_1 = 4\lambda_2$. The figure also shows how the joint probability density can be changed without changing the Gaussian marginal probability densities. For any rectangle aligned with the coordinate axes, incremental squares can be placed at the vertices of the rectangle and ϵ probability can be transferred from left to right on top and right to left on bottom with no change in the marginals. This transfer can be done simultaneously for any number of rectangles, and by reversing the direction of the transfers appropriately, zero covariance can be maintained. Thus the elliptical contour property depends critically on the variables being jointly Gaussian rather than merely individually Gaussian.

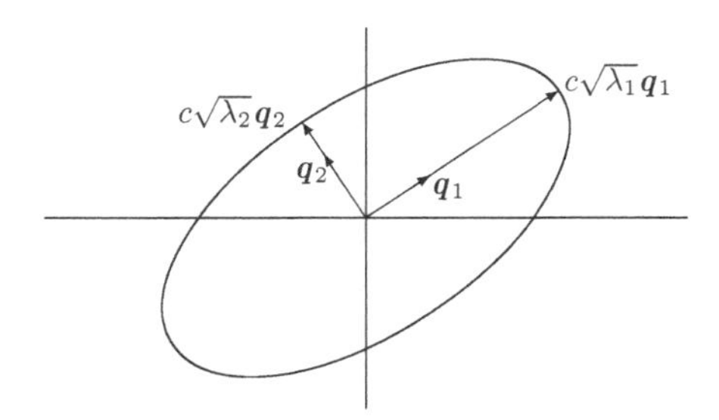

Figure 3.5 Contours of equal probability density. Points $\boldsymbol{z}$ on the $\boldsymbol{q}_j$ axis are points for which $v_k = 0$ for all $k \neq j$. Points on the illustrated ellipse satisfy $\boldsymbol{z}^{\mathsf{T}}[K^{-1}]\boldsymbol{z} = c$.

$$\boldsymbol{z}^{\mathsf{T}}[K^{-1}]\boldsymbol{z} \;=\; \boldsymbol{v}^{\mathsf{T}}[\Lambda^{-1}]\boldsymbol{v} \;=\; \sum_{j=1}^{n} v_j^2 \lambda_j^{-1} \;=\; c. \tag{3.34}$$

This is the same as (3.32) except that here the ellipsoid is defined in terms of the representation $v_j = \boldsymbol{q}_j^{\mathsf{T}}\boldsymbol{z}$ for $1 \le j \le n$. Thus the equi-probability contours are ellipsoids whose axes are the eigenfunctions of $[K]$ (see Figure 3.5). We can also substitute this into (3.22) to obtain what is often a more convenient expression for the probability density of $\boldsymbol{Z}$.

$$\mathsf{f}_{\boldsymbol{Z}}(\boldsymbol{z}) = \frac{\exp\left(-\frac{1}{2}\sum_{j=1}^{n} v_j^2\lambda_j^{-1}\right)}{(2\pi)^{n/2}\sqrt{\det[K]}} \tag{3.35}$$

$$= \prod_{j=1}^{n} \frac{\exp(-v_j^2/(2\lambda_j))}{\sqrt{2\pi\lambda_j}}, \tag{3.36}$$

where $v_j = \boldsymbol{q}_j^{\mathsf{T}}\boldsymbol{z}$ and we have used the fact that $\det[K] = \prod_j \lambda_j$.

3.5 Conditional PDFs for Gaussian random vectors

Next consider the conditional probability $\mathsf{f}_{X|Y}(x|y)$ for two zero-mean jointly-Gaussian rv s X and Y with a non-singular covariance matrix. From (3.23),

$$\mathsf{f}_{X,Y}(x,y) = \frac{1}{2\pi\sigma_X\sigma_Y\sqrt{1-\rho^2}} \exp\left[\frac{-(x/\sigma_X)^2 + 2\rho(x/\sigma_X)(y/\sigma_Y) - (y/\sigma_Y)^2}{2(1-\rho^2)}\right],$$

where $\rho = \mathsf{E}[XY]/(\sigma_X\sigma_Y)$. Since $\mathsf{f}_Y(y) = (2\pi\sigma_Y^2)^{-1/2}\exp(-y^2/2\sigma_Y^2)$, we have

$$\mathsf{f}_{X|Y}(x|y) = \frac{1}{\sigma_X\sqrt{2\pi(1-\rho^2)}} \exp\left[\frac{-(x/\sigma_X)^2 + 2\rho(x/\sigma_X)(y/\sigma_Y) - \rho^2(y/\sigma_Y)^2}{2(1-\rho^2)}\right].$$

The numerator of the exponent is the negative of the square $\left(x/\sigma_x - \rho y/\sigma_y\right)^2$. Thus

$$\mathsf{f}_{X|Y}(x|y) = \frac{1}{\sigma_X\sqrt{2\pi(1-\rho^2)}}\exp\left[\frac{-\left[x-\rho(\sigma_X/\sigma_Y)y\right]^2}{2\sigma_X^2(1-\rho^2)}\right]. \qquad (3.37)$$

This says that, given any particular sample value y for the rv Y, the conditional density of X is Gaussian with variance $\sigma_X^2(1-\rho^2)$ and mean $\rho(\sigma_X/\sigma_Y)y$. Given $Y=y$, we can view X as a rv in the restricted sample space where $Y=y$. In that restricted sample space, X is $\mathcal{N}\left(\rho(\sigma_X/\sigma_Y)y,\ \sigma_X^2(1-\rho^2)\right)$.

We see that the variance of X, given $Y=y$, has been reduced by a factor of $1-\rho^2$ from the variance before the observation. It is not surprising that this reduction is large when $|\rho|$ is close to 1 and negligible when ρ is close to 0. It is surprising that this conditional variance is the same for all values of y. It is also surprising that the conditional mean of X is linear in y and that the conditional distribution is Gaussian with a variance constant in y.

Another way to interpret this conditional distribution of X, conditional on Y, is to use the above observation that the conditional fluctuation of X, conditional on $Y=y$, does not depend on y. This fluctuation can then be denoted as a rv V that is independent of Y. Thus we can represent X (conditional on Y) as $X = \rho(\sigma_X/\sigma_Y)Y + V$, where $V \sim \mathcal{N}(0, (1-\rho^2)\sigma_X^2)$ and V is independent of Y.

As will be seen in Chapter 10, this simple form for the conditional distribution leads to important simplifications in estimating X from Y. We now go on to show that this same kind of simplification occurs when we study the conditional density of one Gaussian **rv** conditional on another Gaussian **rv**, assuming that all the variables are jointly Gaussian.

Let $\boldsymbol{X} = (X_1,\dots,X_n)^{\mathsf{T}}$ and $\boldsymbol{Y} = (Y_1,\dots,Y_m)^{\mathsf{T}}$ be zero-mean jointly-Gaussian **rv** s of length n and m (i.e., $X_1,\dots,X_n,Y_1,\dots,Y_m$ are jointly Gaussian). Let their covariance matrices be $[K_{\boldsymbol{X}}]$ and $[K_{\boldsymbol{Y}}]$ respectively. Let $[K]$ be the covariance matrix of the $(n+m)$-**rv** $(X_1,\dots,X_n,Y_1,\dots,Y_m)^{\mathsf{T}}$.

The $(n+m)\times(n+m)$ covariance matrix $[K]$ can be partitioned into n rows on top and m rows on bottom, and then further partitioned into n and m columns, yielding

$$[K] = \begin{bmatrix} [K_{\boldsymbol{X}}] & [K_{\boldsymbol{X}\cdot\boldsymbol{Y}}] \\ [K_{\boldsymbol{X}\cdot\boldsymbol{Y}}^{\mathsf{T}}] & [K_{\boldsymbol{Y}}] \end{bmatrix}. \qquad (3.38)$$

Here $[K_{\boldsymbol{X}}] = \mathsf{E}[\boldsymbol{X}\boldsymbol{X}^{\mathsf{T}}]$, $[K_{\boldsymbol{X}\cdot\boldsymbol{Y}}] = \mathsf{E}[\boldsymbol{X}\boldsymbol{Y}^{\mathsf{T}}]$, and $[K_{\boldsymbol{Y}}] = \mathsf{E}[\boldsymbol{Y}\boldsymbol{Y}^{\mathsf{T}}]$. Note that if $\boldsymbol{X}$ and $\boldsymbol{Y}$ have means, then $[K_{\boldsymbol{X}}] = \mathsf{E}\left[(\boldsymbol{X}-\overline{\boldsymbol{X}})(\boldsymbol{X}-\overline{\boldsymbol{X}})^{\mathsf{T}}\right]$, $[K_{\boldsymbol{X}\cdot\boldsymbol{Y}}] = \mathsf{E}\left[(\boldsymbol{X}-\overline{\boldsymbol{X}})(\boldsymbol{Y}-\overline{\boldsymbol{Y}})^{\mathsf{T}}\right]$, etc.

In what follows, assume that $[K]$ is non-singular. We then say that $\boldsymbol{X}$ and $\boldsymbol{Y}$ are *jointly non-singular*, which implies that none of the rv s $X_1,\dots,X_n,Y_1,\dots,Y_m$ can be expressed as a linear combination of the others. The inverse of $[K]$ then exists and can be denoted in block form as

$$[K^{-1}] = \begin{bmatrix} [B] & [C] \\ [C^{\mathsf{T}}] & [D] \end{bmatrix}. \qquad (3.39)$$

The blocks $[B], [C], [D]$ can be calculated directly from $[KK^{-1}] = [I]$ (see Exercise 3.16), but for now we simply use them to find $\mathsf{f}_{X|Y}(x|y)$.

We shall find that for any given y, $\mathsf{f}_{X|Y}(x|y)$ is a jointly-Gaussian density with a conditional covariance matrix equal to $[B^{-1}]$ (Exercise 3.11 shows that $[B]$ is non-singular). As in (3.37), where X and Y are one-dimensional, this covariance does not depend on y. Also, the conditional mean of X, given $Y = y$, will turn out to be $-[B^{-1}C]y$. More precisely, we have the following theorem:

Theorem 3.5.1 *Let X and Y be zero-mean, jointly-Gaussian, jointly-non-singular* **rv** *s. Then X, conditional on* $Y = y$, *is* $\mathrm{N}\left(-[B^{-1}C]y,\, [B^{-1}]\right)$, *i.e.,*

$$\mathsf{f}_{X|Y}(x|y) = \frac{\exp\left\{-\frac{1}{2}\left(x + [B^{-1}C]y\right)^{\mathsf{T}} [B] \left(x + [B^{-1}C]y\right)\right\}}{(2\pi)^{n/2}\sqrt{\det[B^{-1}]}}. \tag{3.40}$$

Proof Express $\mathsf{f}_{X|Y}(x|y)$ as $\mathsf{f}_{XY}(x,y)/\mathsf{f}_Y(y)$. From (3.22),

$$\mathsf{f}_{XY}(x,y) = \frac{\exp\left\{-\frac{1}{2}(x^{\mathsf{T}},y^{\mathsf{T}})[K^{-1}](x^{\mathsf{T}},y^{\mathsf{T}})^{\mathsf{T}}\right\}}{(2\pi)^{(n+m)/2}\sqrt{\det[K^{-1}]}}$$

$$= \frac{\exp\left\{-\frac{1}{2}\left(x^{\mathsf{T}}[B]x + x^{\mathsf{T}}[C]y + y^{\mathsf{T}}[C^{\mathsf{T}}]x + y^{\mathsf{T}}[D]y\right)\right\}}{(2\pi)^{(n+m)/2}\sqrt{\det[K^{-1}]}}.$$

Note that x appears only in the first three terms of the exponent above, and that x does not appear at all in $\mathsf{f}_Y(y)$. Thus we can express the dependence on x in $\mathsf{f}_{X|Y}(x|y)$ by

$$\mathsf{f}_{X|Y}(x \mid y) = \phi(y)\exp\left\{-\frac{1}{2}\left[x^{\mathsf{T}}[B]x + x^{\mathsf{T}}[C]y + y^{\mathsf{T}}[C^{\mathsf{T}}]x\right]\right\}, \tag{3.41}$$

where $\phi(y)$ is some function of y. We now complete the square around $[B]$ in the exponent above, getting

$$\mathsf{f}_{X|Y}(x \mid y) = \phi(y)\exp\left\{-\frac{1}{2}\left[(x + [B^{-1}C]y)^{\mathsf{T}}[B](x + [B^{-1}C]y) + y^{\mathsf{T}}[C^{\mathsf{T}}B^{-1}C]y\right]\right\}.$$

Since the last term in the exponent does not depend on x, we can absorb it into $\phi(y)$. The remaining expression has the form of the density of a Gaussian n-**rv** with non-zero mean as given in (3.24). Comparison with (3.24) also shows that $\phi(y)$ must be $(2\pi)^{-n/2}(\det[B^{-1}])^{-1/2}$. With this substituted for $\phi(y)$, we have (3.40). □

To interpret (3.40), note that for any sample value y for Y, the conditional distribution of X has a mean given by $-[B^{-1}C]y$ and a Gaussian fluctuation around the mean of variance $[B^{-1}]$. This fluctuation has the same distribution for all y and thus can be represented as a **rv** V that is independent of Y. Thus we can represent X as

$$X = [G]Y + V, \quad Y, V \text{ independent}, \tag{3.42}$$

where

$$[G] = -[B^{-1}C] \quad \text{and} \quad V \sim \mathrm{N}(0, [B^{-1}]). \tag{3.43}$$

We often call $\boldsymbol{V}$ an *innovation*, because it is the part of $\boldsymbol{X}$ that is independent of $\boldsymbol{Y}$. It is also called a *noise term* for the same reason. We will call $[K_{\boldsymbol{V}}] = [B^{-1}]$ the *conditional covariance* of $\boldsymbol{X}$ given a sample value $\boldsymbol{y}$ for $\boldsymbol{Y}$. In summary, the unconditional covariance, $[K_{\boldsymbol{X}}]$, of $\boldsymbol{X}$ is given by the upper left block of $[K]$ in (3.38), while the conditional covariance $[K_{\boldsymbol{V}}]$ is the inverse of the upper left block, $[B]$, of the inverse of $[K]$.

The following theorem expresses (3.42) and (3.43) directly in terms of the covariances of $\boldsymbol{X}$ and $\boldsymbol{Y}$.

Theorem 3.5.2 *Let $\boldsymbol{X}$ and $\boldsymbol{Y}$ be zero-mean, jointly Gaussian, and jointly non-singular. Then $\boldsymbol{X}$ can be expressed as $\boldsymbol{X} = [G]\boldsymbol{Y} + \boldsymbol{V}$, where $\boldsymbol{V}$ is statistically independent of $\boldsymbol{Y}$ and*

$$G = [K_{\boldsymbol{X}\cdot\boldsymbol{Y}} K_{\boldsymbol{Y}}^{-1}], \tag{3.44}$$

$$[K_{\boldsymbol{V}}] = [K_{\boldsymbol{X}}] - [K_{\boldsymbol{X}\cdot\boldsymbol{Y}} K_{\boldsymbol{Y}}^{-1} K_{\boldsymbol{X}\cdot\boldsymbol{Y}}^{\mathsf{T}}]. \tag{3.45}$$

Proof From (3.42), we know that $\boldsymbol{X}$ can be represented as $[G]\boldsymbol{Y} + \boldsymbol{V}$ with $\boldsymbol{Y}$ and $\boldsymbol{V}$ independent, so we simply have to evaluate $[G]$ and $[K_{\boldsymbol{V}}]$. Using (3.42), the covariance of $\boldsymbol{X}$ and $\boldsymbol{Y}$ is given by

$$[K_{\boldsymbol{X}\cdot\boldsymbol{Y}}] \;=\; \mathsf{E}[\boldsymbol{X}\boldsymbol{Y}^{\mathsf{T}}] \;=\; \mathsf{E}[[G]\boldsymbol{Y}\boldsymbol{Y}^{\mathsf{T}} + \boldsymbol{V}\boldsymbol{Y}^{\mathsf{T}}] \;=\; [GK_{\boldsymbol{Y}}],$$

where we used the fact that $\boldsymbol{V}$ and $\boldsymbol{Y}$ are independent. Post-multiplying both sides by $[K_{\boldsymbol{Y}}^{-1}]$ yields (3.44). To verify (3.45), we use (3.42) to express $[K_{\boldsymbol{X}}]$ as

$$\begin{aligned}[K_{\boldsymbol{X}}] = \mathsf{E}[\boldsymbol{X}\boldsymbol{X}^{\mathsf{T}}] \;&=\; \mathsf{E}[([G]\boldsymbol{Y} + \boldsymbol{V})([G]\boldsymbol{Y} + \boldsymbol{V})^{\mathsf{T}}] \\ &= [GK_{\boldsymbol{Y}}G^{\mathsf{T}}] + [K_{\boldsymbol{V}}],\end{aligned}$$

so

$$[K_{\boldsymbol{V}}] = [K_{\boldsymbol{X}}] - [GK_{\boldsymbol{Y}}G^{\mathsf{T}}].$$

This yields (3.45) when (3.44) is used for $[G]$. □

We have seen that $[K_{\boldsymbol{V}}]$ is the covariance of $\boldsymbol{X}$ conditional on $\boldsymbol{Y} = \boldsymbol{y}$ for each sample value $\boldsymbol{y}$. The expression in (3.45) provides some insight into how this covariance is reduced from $[K_{\boldsymbol{X}}]$. More particularly, for any n-vector $\boldsymbol{b}$,

$$\boldsymbol{b}^{\mathsf{T}}[K_{\boldsymbol{X}}]\boldsymbol{b} \geq \boldsymbol{b}^{\mathsf{T}}[K_{\boldsymbol{V}}]\boldsymbol{b},$$

i.e., the unconditional variance of $\boldsymbol{b}^{\mathsf{T}}\boldsymbol{X}$ is always greater than or equal to the variance of $\boldsymbol{b}^{\mathsf{T}}\boldsymbol{X}$ conditional on $\boldsymbol{Y} = \boldsymbol{y}$.

In the process of deriving these results, we have also implicity evaluated the matrices $[C]$ and $[B]$ in the inverse of $[K]$ in (3.39). Combining the second part of (3.43) with (3.45),

$$[B] = \Big([K_{\boldsymbol{X}}] - [K_{\boldsymbol{X}\cdot\boldsymbol{Y}} K_{\boldsymbol{Y}}^{-1} K_{\boldsymbol{X}\cdot\boldsymbol{Y}}^{\mathsf{T}}]\Big)^{-1}. \tag{3.46}$$

Combining the first part of (3.43) with (3.44), we get

$$[C] = -[BK_{\boldsymbol{X}\cdot\boldsymbol{Y}} K_{\boldsymbol{Y}}^{-1}]. \tag{3.47}$$

Finally, reversing the roles of $\boldsymbol{X}$ and $\boldsymbol{Y}$, we can express D as

$$[D] = \left([K_Y] - [K_{Y \cdot X} K_X^{-1} K_{Y \cdot X}^{\mathsf{T}}]\right)^{-1}. \tag{3.48}$$

Reversing the roles of $\boldsymbol{X}$ and $\boldsymbol{Y}$ is even more important in another way, since Theorem 3.5.2 then also says that $\boldsymbol{X}$ and $\boldsymbol{Y}$ are related by

$$\boldsymbol{Y} = [H]\boldsymbol{X} + \boldsymbol{Z}, \tag{3.49}$$

where $\boldsymbol{X}$ and $\boldsymbol{Z}$ are independent and

$$[H] = [K_{Y \cdot X} K_X^{-1}], \tag{3.50}$$

$$[K_Z] = [K_Y] - [K_{Y \cdot X} K_X^{-1} K_{Y \cdot X}^{\mathsf{T}}]. \tag{3.51}$$

This gives us three ways of representing any pair $\boldsymbol{X}, \boldsymbol{Y}$ of zero-mean jointly-Gaussian **rv** s whose combined covariance is non-singular. First, they can be represented simply as an overall **rv**, $(X_1, \ldots, X_n Y_1, \ldots, Y_m)^{\mathsf{T}}$, second, as $\boldsymbol{X} = [G]\boldsymbol{Y} + \boldsymbol{V}$, where $\boldsymbol{Y}$ and $\boldsymbol{V}$ are independent and, third, as $\boldsymbol{Y} = [H]\boldsymbol{X} + \boldsymbol{Z}$, where $\boldsymbol{X}$ and $\boldsymbol{Z}$ are independent.

Each of these formulations essentially implies the existence of the other two. If we start with formulation 3, for example, Exercise 3.17 shows simply that if $\boldsymbol{X}$ and $\boldsymbol{Z}$ are each zero-mean Gaussian **rv** s, the independence between them assures that they are jointly Gaussian, and thus that $\boldsymbol{X}$ and $\boldsymbol{Y}$ are also jointly Gaussian. Similarly, if $[K_X]$ and $[K_Z]$ are non-singular, the overall $[K]$ for $(X_1, \ldots, X_n, Y_1, \ldots, Y_m)^{\mathsf{T}}$ must be non-singular. In Chapter 10, we will find that this provides a very simple and elegant solution to jointly-Gaussian estimation problems.

3.6 Gaussian processes

Recall that a stochastic process (or random process) $\{X(t); t \in \mathrm{T}\}$ is a collection of rv s, one for each value of the parameter t in some parameter set T. The parameter t usually denotes time, so there is one rv for each instant of time. For discrete-time processes, T is usually limited to the set of integers, $\mathbb{Z}$, and for continuous-time processes, T is usually limited to $\mathbb{R}$. In each case, t is sometimes additionally restricted to $t \geq 0$; this is denoted $\mathbb{Z}^+$ and $\mathbb{R}^+$ respectively. We use the word *epoch* to denote a value of t within T.

Definition 3.6.1 *A* ***Gaussian process*** $\{X(t); t \in \mathrm{T}\}$ *is a stochastic process such that for all positive integers k and all choices of epochs* $t_1, \ldots, t_k \in \mathrm{T}$, *the set of rv s* $X(t_1), \ldots, X(t_k)$ *is a jointly-Gaussian set of rv s.*

The previous sections of this chapter should motivate both the simplicity and usefulness associated with this jointly-Gaussian requirement. In particular, the joint probability density of any k-**rv** $(X(t_1), \ldots, X(t_k))^{\mathsf{T}}$ is essentially specified by (3.24), using only the covariance matrix and the mean for each rv. If the rv s are individually Gaussian but not jointly Gaussian, none of this holds.

Definition 3.6.2 *The* **covariance function**, $K_X(t, \tau)$, *of a stochastic process* $\{X(t);\ t\in\mathrm{T}\}$ *is defined for all* $t, \tau \in \mathrm{T}$ *by*

$$K_X(t,\tau) = \mathsf{E}\left[(X(t) - \overline{X}(t))(X(\tau) - \overline{X}(\tau))\right]. \tag{3.52}$$

Note that for each k-**rv** $(X(t_1), \ldots, X(t_k))^{\mathsf{T}}$, the (j, ℓ) element of the covariance matrix is simply $K_X(t_j, t_\ell)$. Thus the covariance function and the mean of a process specify the covariance matrix and mean of each k-**rv**. This establishes the following simple but important result.

Theorem 3.6.3 *For a Gaussian process* $\{X(t); t \in \mathrm{T}\}$, *the covariance function* $K_X(t, \tau)$ *and the mean* $\mathsf{E}[X(t)]$ *for each* $t, \tau \in \mathrm{T}$ *specify the joint probability density for all* k-**rv** *s* $(X(t_1), \ldots, X(t_k))^{\mathsf{T}}$ *for all* $k > 1$.

We now give several examples of discrete-time Gaussian processes and their covariance functions. As usual, we look at the zero-mean case, since a mean can always be simply added later. Continuous-time Gaussian processes are considerably more complicated and are considered in Section 3.6.3.

Example 3.6.4 (Discrete-time IID Gaussian process) Consider the stochastic process $\{W(n);\ n \in \mathbb{Z}\}$, where $\ldots, W(-1), W(0), W(1), \ldots$ is a sequence of IID Gaussian rv s, $W(n) \sim \mathrm{N}(0, \sigma^2)$. The mean is zero for all n and the covariance function is $K_W(n, k) = \sigma^2 \delta_{nk}$. For any k epochs, $n_1, n_2, \ldots, n_k$, the joint density is

$$\mathsf{f}_{W(n_1),\ldots,W(n_k)}(w_1, \ldots, w_k) = \frac{1}{(2\pi\sigma^2)^{k/2}} \exp\left(-\sum_{i=1}^{k} \frac{w_i^2}{2\sigma^2}\right). \tag{3.53}$$

Note that this process is very much like the IID Gaussian vectors we have studied. The only difference is that we now have an infinite number of dimensions (i.e., an infinite number of IID rv s) for which all finite subsets are jointly Gaussian.

Example 3.6.5 (Discrete-time Gaussian sum process) Consider the stochastic process $\{S(n); n \geq 1\}$ which is defined from the discrete-time IID Gaussian process by $S(n) = W(1) + W(2) + \cdots + W(n)$. Viewing $(S_1, \ldots, S_n)^{\mathsf{T}}$ as a linear transformation of $(W_1, \ldots, W_n)^{\mathsf{T}}$, we see that $S_1, \ldots, S_n$ is a zero-mean jointly-Gaussian set of rv s. Since this is true for all $n \geq 1$, $\{S(n); n \geq 1\}$ is a zero-mean Gaussian process. For $n \leq k$, the covariance function is

$$K_S(n, k) = \mathsf{E}\left[\sum_{j=1}^{n} W_j \sum_{\ell=1}^{k} W_\ell\right] = \sum_{j=1}^{n} \mathsf{E}\left[W_j^2\right] = n\sigma^2.$$

Using a similar argument for $n > k$, the general result is

$$K_S(n, k) = \min(n, k)\sigma^2.$$

Example 3.6.6 (Discrete-time Gauss–Markov process) Let α be a real number, $|\alpha| < 1$ and consider a stochastic process $\{X(n); n \in \mathbb{Z}^+\}$ which is defined in terms of the previous example of an IID Gaussian process $\{W_n; n \in \mathbb{Z}\}$ by

$$X(n+1) = \alpha X(n) + W(n) \quad \text{for } n \in \mathbb{Z}^+, \quad X(0) = 0. \tag{3.54}$$

The process in (3.54) can be visualized as being implemented by Figure 3.6. By applying (3.54) recursively, or by using Figure 3.6,

$$X(n) = W(n-1) + \alpha W(n-2) + \alpha^2 W(n-3) + \cdots + \alpha^{n-1} W(0). \tag{3.55}$$

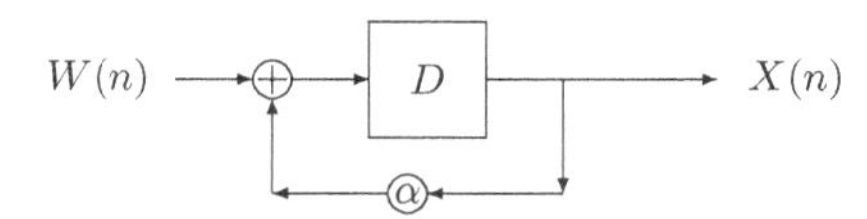

Figure 3.6 Schematic of the generation of $\{X(n); n \geq 1\}$ from $X(0) = 0$ and $\{W(n); n \geq 0\}$. The element D is a unit delay. It can be seen from the figure that X_{n+1} depends probabilistically on the past history $X_1, \dots, X_n$ only through X_n. This is called a Gauss–Markov process, and the sample value x_n of X_n is called the *state* of the process at time n. This process differs from the Markov processes developed in Chapters 4, 6, and 7 in the sense that the state is an arbitrary real number rather than a discrete value.

This is another example in which the new process $\{X(n); n \geq 1\}$ is a linear transformation of another process $\{W(n); n \geq 0\}$. Since $\{W(n); n \geq 0\}$ is a zero-mean Gaussian process, $\{X_n; n \geq 0\}$ is also. Thus $\{X(n); n \geq 0\}$ is specified by its covariance function, calculated in Exercise 3.22 to be

$$\mathsf{E}[X(n)X(n+k)] = \frac{\sigma^2(1-\alpha^{2n})\alpha^k}{1-\alpha^2}. \tag{3.56}$$

Since $|\alpha| < 1$, the coefficients α^k in (3.55) are geometrically decreasing in k, and therefore, for large n it makes little difference whether the sum stops with the term $\alpha^{n-1}W(0)$ or whether terms $\alpha^n W(-1)$, $\alpha^{n+1} W_{-2}, \dots,$ are added.[6] Similarly, from (3.56), we see that $\lim_{n\to\infty} \mathsf{E}[X(n)X(n+k)] = \sigma^2\alpha^k/(1-\alpha^2)$. This suggests that the starting time of this process is irrelevant if it is far enough into the past, and thus suggests that we could define essentially the same process over all integer times n by

$$X(n+1) = \alpha X(n) + W(n); \quad \text{for all } n \in \mathbb{Z}. \tag{3.57}$$

By applying (3.57) recursively, $X(n) = \sum_{j=1}^{\infty} \alpha^{j-1} W(n-j)$.

3.6.1 Stationarity and related concepts

Many of the most useful stochastic processes have the property that the location of the time origin is irrelevant, i.e., that the process 'behaves' the same way at one time as

[6] One might ask whether the limit $\sum_{j=1}^{\infty} \alpha^{j-1} W(n-j)$ exists as a rv. As intuition almost demands, the answer is yes. We will show this in Section 9.9.2 as a consequence of the martingale convergence theorem.

at any other time. This property is called *stationarity* and such a process is called a *stationary process*. A precise definition will be given shortly.

An obvious requirement for stationarity is that $X(t)$ must be identically distributed for all $t \in \mathtt{T}$. A more subtle requirement is that for every $k > 1$ and set of epochs $t_1, \ldots, t_k \in \mathtt{T}$, the joint distribution over these epochs should be the same as that over a shift in time of these epochs to, say, $t_1+\tau, \ldots, t_k+\tau \in \mathtt{T}$.

This shift requirement for stationarity becomes quite obscure and meaningless unless $\mathtt{T}$ is chosen so that a shift of a set of epochs in $\mathtt{T}$ is also in $\mathtt{T}$. This explains why the definition of $\mathtt{T}$ is restricted in the following definition.

Definition 3.6.7 *Let a stochastic process $\{X(t);\ t \in \mathtt{T}\}$ be defined over a set of epochs $\mathtt{T}$ where $\mathtt{T}$ is either $\mathbb{Z}$, $\mathbb{R}$, $\mathbb{Z}^+$, or $\mathbb{R}^+$. The process is* ***stationary*** *if, for all positive integers k and all $\tau, t_1, \ldots, t_k$ in $\mathtt{T}$,*

$$\mathsf{F}_{X(t_1),\ldots,X(t_k)}(x_1 \ldots, x_k) = \mathsf{F}_{X(t_1+\tau),\ldots,X(t_k+\tau)}(x_1 \ldots, x_k). \tag{3.58}$$

Note that the restriction on $\mathtt{T}$ in the definition guarantees that if $X(t_1), \ldots, X(t_k) \in \mathtt{T}$, then $X(t_1+\tau), \ldots, X(t_k+\tau) \in \mathtt{T}$ also. In this chapter, $\mathtt{T}$ is usually $\mathbb{Z}$ or $\mathbb{R}$, whereas in Chapters 4, 6, and 7, $\mathtt{T}$ is usually restricted to $\mathbb{Z}^+$ or $\mathbb{R}^+$.

The discrete-time IID Gaussian process in Example 3.6.4 is stationary since all joint distributions of a given number of distinct variables from $\{W(n);\ n \in \mathbb{Z}\}$ are the same. More generally, for any Gaussian process, the joint distribution of $X(t_1), \ldots, X(t_k)$ depends only on the mean and covariance of those variables. In order for this distribution to be the same as that of $X(t_1+\tau), \ldots, X(t_k+\tau)$, it is necessary that $\mathsf{E}[X(t)] = \mathsf{E}[X(0)]$ for all epochs t and also that $K_X(t_1, t_2) = K_X(t_1+\tau,\ t_2+\tau)$ for all epochs t_1, t_2, and τ. This latter condition can be simplified to the statement that $K_X(t,\ t+u)$ is a function only of u and not of t. It can be seen that these conditions are also sufficient for a Gaussian process $\{X(t)\}$ to be stationary. We summarize this in the following theorem.

Theorem 3.6.8 *A Gaussian process $\{X(t);\ t \in \mathtt{T}\}$ (where $\mathtt{T}$ is $\mathbb{Z}$, $\mathbb{R}$, $\mathbb{Z}^+$, or $\mathbb{R}^+$) is stationary if and only if $\mathsf{E}[X(t)] = \mathsf{E}[X(0)]$ and $K_X(t,\ t+u) = K_X(0, u)$ for all $t, u \in \mathtt{T}$.*

With this theorem, we see that the Gauss–Markov process of Example 3.6.6, extended to the set of all integers, is a discrete-time stationary process. The Gaussian sum process of Example 3.6.5, however, is non-stationary.

For non-Gaussian processes, it is frequently difficult to calculate joint distributions in order to determine if the process is stationary. There are a number of results that depend only on the mean and the covariance function, and these make it convenient to have the following more relaxed definition:

Definition 3.6.9 *A stochastic process* $\{X(t);\ t \in \mathrm{T}\}$ *(where* T *is* $\mathbb{Z}$, $\mathbb{R}$, $\mathbb{Z}^+$, *or* $\mathbb{R}^+$*) is* ***wide sense stationary***[7] (WSS) *if* $\mathsf{E}[X(t)] = \mathsf{E}[X(0)]$ *and* $K_X(t, t+u) = K_X(0, u)$ *for all* $t, u \in \mathrm{T}$.

Since the covariance function $K_X(t, t+u)$ of a stationary or WSS process is a function of only one variable u, we will often write the covariance function of a WSS process as a function of one variable, namely $K_X(u)$ in place of $K_X(t, t+u)$. The single variable in the single-argument form represents the difference between the two arguments in the two-argument form. Thus, the covariance function $K_X(t, \tau)$ of a WSS process must be a function only of $t - \tau$ and is expressed in single-argument form as $K_X(t - \tau)$. Note also that since $K_X(t, \tau) = K_X(\tau, t)$, the covariance function of a WSS process must be symmetric, i.e., $K_X(u) = K_X(-u)$.

The reader should not conclude from the frequent use of the term WSS in the literature that there are many important processes that are WSS but not stationary. Rather, the use of WSS in a result is used primarily to indicate that the result depends only on the mean and covariance.

3.6.2 Orthonormal expansions

The previous Gaussian process examples were discrete-time processes. The simplest way to generate a broad class of continuous-time Gaussian processes is to start with a discrete-time process (i.e., a sequence of jointly-Gaussian rv s) and use these rv s as the coefficients in an orthonormal expansion. We describe some of the properties of orthonormal expansions in this section and then describe how to use these expansions to generate continuous-time Gaussian processes in Section 3.6.3.

A set of functions $\{\phi_n(t);\ n \geq 1\}$ is defined to be orthonormal if

$$\int_{-\infty}^{\infty} \phi_n(t)\phi_k^*(t)\,dt = \delta_{nk} \qquad \text{for all integers } n, k. \tag{3.59}$$

These functions can be either complex or real functions of the real variable t; the complex case (using the reals as a special case) is most convenient.

The most familiar orthonormal set is that used in the Fourier series. Letting $i = \sqrt{-1}$,

$$\phi_n(t) = \begin{cases} (1/\sqrt{T})\exp[i2\pi nt/T] & \text{for } |t| \leq T/2, \\ 0 & \text{for } |t| > T/2. \end{cases} \tag{3.60}$$

We can then take any square-integrable real or complex function $x(t)$ over $(-T/2, T/2)$ and essentially[8] represent it by

$$x(t) = \sum_n x_n\phi_n(t), \quad \text{where } x_n = \int_{-T/2}^{T/2} x(t)\phi_n^*(t)dt. \tag{3.61}$$

[7] This is also called weakly stationary, covariance stationary, and second-order stationary.

[8] More precisely, the difference between $x(t)$ and its Fourier series $\sum_n x_n\phi_n(t)$ has zero energy, i.e., $\int \left|x(t) - \sum_n x_n\phi_n(t)\right|^2 dt = 0$. This allows $x(t)$ and $\sum_n x_n\phi_n(t)$ to differ at isolated values of t such as points of discontinuity in $x(t)$. Engineers view this as essential equality and mathematicians define it carefully and call it L_2 equivalence.

The complex exponential form of the Fourier series could be replaced by the sine/cosine form when expanding real functions (as here). This has the conceptual advantage of keeping everything real, but does not warrant the added analytical complexity.

Many features of the Fourier transform are due not to the special nature of sinusoids, but rather to the fact that the function is being represented as a series of orthonormal functions. To see this, let $\{\phi_n(t);\ n \in \mathbb{Z}\}$ be any set of orthonormal functions, and assume that a function $x(t)$ can be represented as

$$x(t) = \sum_n x_n \phi_n(t). \tag{3.62}$$

Multiplying both sides of (3.62) by $\phi_m^*(t)$ and integrating,

$$\int x(t)\phi_m^*(t)dt = \int \sum_n x_n \phi_n(t)\phi_m^*(t)dt.$$

Using (3.59) to see that only one term on the right-hand side is non-zero, we get

$$\int x(t)\phi_m^*(t)dt = x_m. \tag{3.63}$$

We do not have the mathematical tools to easily justify this interchange and it would take us too far afield to acquire those tools. Thus for the remainder of this section, we will concentrate on the results and ignore a number of mathematical fine points.

If a function can be represented by orthonormal functions as in (3.62), then the coefficients $\{x_n\}$ must be determined as in (3.63), which is the same pair of relations as in (3.61). We can also represent the energy in $x(t)$ in terms of the coefficients $\{x_n;\ n \in \mathbb{Z}\}$. Since $|x^2(t)| = (\sum_n x_n\phi_n(t))(\sum_m x_m^*\phi_m^*(t))$, we get

$$\int |x^2(t)|dt = \int \sum_n \sum_m x_n x_m^* \phi_n(t)\phi_m^*(t)dt = \sum_n |x_n|^2. \tag{3.64}$$

Next suppose $x(t)$ is any square-integrable function and $\{\phi_n(t);\ n \in \mathbb{Z}\}$ is an orthonormal set. Let $x_n = \int x(t)\phi_n^*(t)dt$. Let $\epsilon_k(t) = x(t) - \sum_{n=1}^{k} x_n\phi_n(t)$ be the error when $x(t)$ is represented by the first k of these orthonormal functions. First we show that $\epsilon_k(t)$ is orthogonal to $\phi_m(t)$ for $1 \le m \le k$.

$$\int \epsilon_k(t)\phi_m^*(t)dt = \int x(t)\phi_m^*(t)dt - \int \sum_{n=1}^{k} x_n\phi_n(t)\phi_m^*(t)dt = x_m - x_m = 0. \tag{3.65}$$

Viewing functions as vectors, $\epsilon_k(t)$ is the difference between $x(t)$ and its projection on the linear subspace spanned by $\{\phi_n(t);\ 1 \le n \le k\}$. The integral of the magnitude squared error is given by

$$\int |x^2(t)|dt = \int \left| \epsilon_k(t) + \sum_{n=1}^{k} x_n\phi_n(t) \right|^2 dt \tag{3.66}$$

$$= \int |\epsilon_k^2(t)|dt + \int \sum_{n=1}^{k} \sum_{n=1}^{k} x_n x_m^* \phi_n(t)\phi_m^*(t)dt \tag{3.67}$$

$$= \int |\epsilon_k^2(t)|dt + \sum_{n=1}^{k} |x_n^2|. \tag{3.68}$$

Since $|\epsilon_k^2(t)|dt \geq 0$, this implies the following inequality, known as Bessel's inequality, follows:

$$\sum_{n=1}^{k} |x_n^2| \leq \int |x^2(t)|dt. \tag{3.69}$$

We see from (3.68) that $\int |\epsilon_k^2(t)|^2 dt$ is non-increasing with k. Thus, in the limit $k \to \infty$, either the energy in $\epsilon_k(t)$ approaches 0 or it approaches some positive constant. A set of orthonormal functions is said to *span* a class C of functions if this error energy approaches 0 for all $x(t) \in$ C. For example, the Fourier series set of functions in (3.60) spans the set of functions that are square integrable and zero outside of $[-T/2, T/2]$. There are many other countable sets of functions that span this class of functions and many others that span the class of square-integrable functions over $(-\infty, \infty)$.

In the next subsection, we use a sequence of independent Gaussian rv s as coefficients in these orthonormal expansions to generate a broad class of continuous-time Gaussian processes.

3.6.3 Continuous-time Gaussian processes

Given an orthonormal set of real-valued functions, $\{\phi_n(t); n \in \mathbb{Z}\}$ and given a sequence $\{X_n; n \in \mathbb{Z}\}$ of independent rv s[9] with $X_n \sim \mathrm{N}(0, \sigma_n^2)$, consider the following expression:

$$X(t) = \lim_{\ell \to \infty} \sum_{n=-\ell}^{\ell} X_n \phi_n(t). \tag{3.70}$$

Note that for any given t and ℓ, the sum above is a Gaussian rv of variance $\sum_{n=-\ell}^{\ell} \sigma_n^2 \phi_n^2(t)$. If this variance increases without bound as $\ell \to \infty$, then it is not hard to convince oneself that there cannot be a limiting distribution, so there is no limiting rv. The more important case of bounded variance is covered in the following theorem. Note that the theorem does not require the functions $\phi_n(t)$ to be orthonormal.

Theorem 3.6.10 *Let $\{X_n; n \in \mathbb{Z}\}$ be a sequence of independent rv s, $X_n \sim \mathrm{N}(0, \sigma_n^2)$, and let $\{\phi_n(t); n \in \mathbb{Z}\}$ be a sequence of real-valued functions. Assume that $\sum_{n=-\ell}^{\ell} \sigma_n^2 \phi_n^2(t)$ converges to a finite value as $\ell \to \infty$ for each t. Then $\{X(t); t \in \mathbb{R}\}$ as given in (3.70) is a Gaussian process.*

Proof The difficult part of the proof is showing that $X(t)$ is a rv for any given t under the conditions of the theorem. This means that, for a given t, the sequence of rv s $\{\sum_{n=-\ell}^{\ell} X_n \phi_n(t); \ell \geq 1\}$ must converge WP1 to a rv as $\ell \to \infty$. This is proven in Section 9.9.2 as a special case of the martingale convergence theorem, so we simply accept that result for now. Since this sequence converges WP1, it also converges in distribution, so, since each term in the sequence is Gaussian, the limit is also Gaussian. Thus $X(t)$ exists and is Gaussian for each t.

[9] Previous sections have considered possibly complex orthonormal functions, but we restrict them here to be real. Using rv s (which are real by definition) with complex orthonormal functions is an almost trivial extension, but using complex rv s and complex functions is less trivial and is treated in Section 3.7.8.

Next, we must show that for any k, any $t_1, \dots, t_k$, and any $a_1, \dots, a_k$, the sum $a_1X(t_1) + \cdots + a_kX(t_k)$ is Gaussian. This sum, however, is just the limit

$$\lim_{\ell\to\infty} \sum_{n=-\ell}^{\ell} [a_1X_n\phi_n(t_1) + \cdots + a_kX_n\phi_n(t_k)].$$

This limit exists and is Gaussian by the same argument as used above for $k = 1$. Thus the process is Gaussian. □

Example 3.6.11 First consider an almost trivial example. Let $\{\phi_n(t); n \in \mathbb{Z}\}$ be a sequence of unit pulses each of unit duration, i.e., $\phi_n(t) = 1$ for $n \le t < n+1$ and $\phi_n(t) = 0$ elsewhere. Then $X(t) = X_{\lfloor t \rfloor}$. In other words, we have converted the discrete-time process $\{X_n; n \in \mathbb{Z}\}$ into a continuous-time process simply by maintaining the value of X_n as a constant over each unit interval.

Note that $\{X_n; n \in \mathbb{Z}\}$ is stationary as a discrete-time process, but the resulting continuous-time process is non-stationary because the covariance of two points within a unit interval differs from that between the two points shifted so that an integer lies between them.

Example 3.6.12 (The Fourier series expansion) Consider the real-valued orthonormal functions in the sine/cosine form of the Fourier series over an interval $[-T/2, T/2)$, i.e.,

$$\phi_n(t) = \begin{cases} \sqrt{2/T}\cos(2\pi nt/T) & \text{for } n > 0,\ |t| \le T/2, \\ \sqrt{1/T} & \text{for } n = 0,\ |t| \le T/2, \\ \sqrt{2/T}\sin(-2\pi nt/T) & \text{for } n < 0,\ |t| \le T/2, \\ 0 & \text{for } |t| > T/2. \end{cases}$$

If we represent a real-valued function $x(t)$ over $(-T/2, T/2)$ as $x(t) = \sum_n x_n\phi_n(t)$, then the coefficients x_n and x_{-n} essentially represent how much of the frequency n/T is contained in $x(t)$. If an orchestra plays a particular chord during $(-T/2, T/2)$, then the corresponding coefficients of X_n will tend to be larger in magnitude than the coefficients of frequencies not in the chord. If there is considerable randomness in what the orchestra is playing then these coefficients might be modeled as rv s.

When we represent a zero-mean Gaussian process, $X(t) = \sum_n X_n\phi_n(t)$, by these orthonormal functions, then the variances σ_n^2 signify, in some sense that will be refined later, how the process is distributed between different frequencies. We assume for this example that the variances σ_n^2 of the X_n satisfy $\sum_n \sigma_n^2 < \infty$, since this is required to ensure that $\mathsf{E}\left[X^2(t)\right]$ is finite for each t. The only intention of this example is to show, first, that a Gaussian process can be defined in this way, second, that joint probability densities over any finite set of epochs, $-T/2 < t_1 < t_2 < \cdots < t_n < T/2$ are in principle determined by $\{\sigma_n^2;\ n\in\mathbb{Z}\}$ and, third, that these variances have some sort of relation to the frequency content of the Gaussian process.

The above example is very appropriate if we want to model noise over some finite time interval. As suggested in Section 3.6.1, however, we often want to model noise as being stationary over $(-\infty, \infty)$. Neither the interval $(-T/2, T/2)$ nor its limit as $T \to \infty$ turn out to be very productive in this case. The next example, based on the sampling theorem of linear systems, turns out to work much better.

3.6.4 Gaussian sinc processes

The sinc function is defined to be $\text{sinc}(t) = \sin(\pi t)/\pi t$ and is sketched in Figure 3.7.

The Fourier transform of $\text{sinc}(t)$ is a square pulse that is 1 for $|f| \leq 1/2$ and 0 elsewhere. This can be verified easily by taking the inverse transform of the square pulse. The most remarkable (and useful) feature of the sinc function is that it and its translates over integer intervals form an orthonormal set, i.e.,

$$\int \text{sinc}(t-n)\text{sinc}(t-k)\,dt = \delta_{nk} \qquad \text{for } n, k \in \mathbb{Z}. \tag{3.71}$$

This can be verified (with effort) by direct integration, but the following approach is more insightful: the Fourier transform of $\text{sinc}(t-n)$ is $e^{-i2\pi nf}$ for $|f| \leq 1/2$ and is 0 elsewhere. Thus the Fourier transform of $\text{sinc}(t-n)$ is easily seen to be orthonormal to that of $\text{sinc}(t-k)$ for $n \neq k$. By Parseval's theorem, then, $\text{sinc}(t-n)$ and $\text{sinc}(t-k)$ are themselves orthonormal for $n \neq k$.

If we now think of representing any square-integrable function of *frequency*, say $v(f)$ over the frequency interval $(-1/2, 1/2)$ by a Fourier *series*, we see that $v(f) = \sum_n v_n e^{i2\pi nf}$ over $f \in (-1/2, 1/2)$, where $v_n = \int_{-1/2}^{1/2} v(f)e^{-i2\pi nf}\,df$. Taking the inverse Fourier *transform* we see that any function of time that is frequency limited to $(-1/2, 1/2)$ can be represented by the set $\{\text{sinc}(t-n);\ n \in \mathbb{Z}\}$. In other words, if $x(t)$ is a square-integrable continuous[10] function whose Fourier transform is limited to $f \in [-1/2, 1/2]$, then

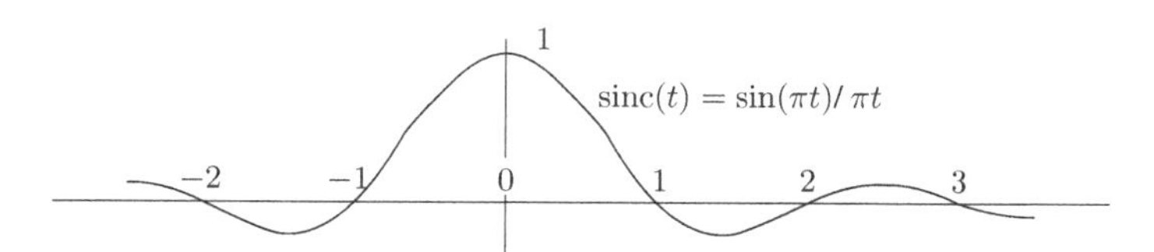

Figure 3.7 The function $\text{sinc}(t)$ is 1 at $t = 0$ and 0 at every other integer t. The amplitude of its oscillations goes to 0 with increasing $|t|$ as $1/|t|$.

[10] The reason for requiring continuity here is that a function can be altered at a finite (or even countable) number of points without changing its Fourier transform. The inverse transform of the Fourier transform of a band limited function, however, is continuous and is the function referred to. It is the same as the original function except at those originally altered points. The reader who wants a more complete development here is referred to [10].

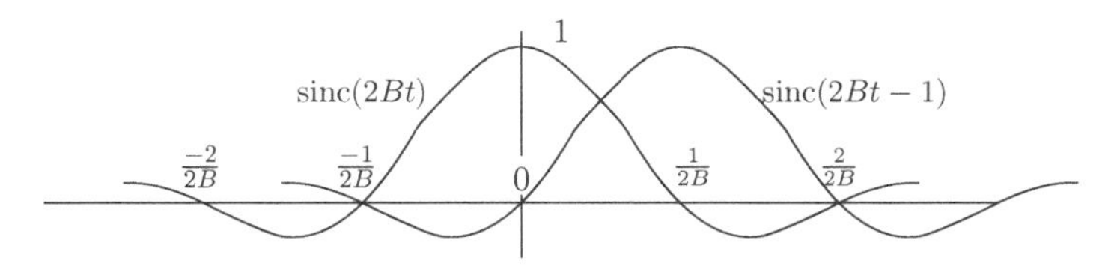

Figure 3.8 The function sinc($2Bt$) is 1 at $t = 0$ and 0 at every other integer multiple of $(2B)^{-1}$. The function sinc($2Bt - 1$) is 1 at $t = (2B)^{-1}$ and 0 at every other integer multiple of $(2B)^{-1}$.

$$x(t) = \sum_n x_n \mathrm{sinc}(t - n), \qquad \text{where } x_n = \int x(t)\mathrm{sinc}(t - n)\, dt. \tag{3.72}$$

There is one further simplification that occurs here: for any integer value of t, say $t = k$, $\mathrm{sinc}(t - n) = \delta_{kn}$, so $x(n) = x_n$. Thus for any square-integrable continuous function, limited in frequency to $[-1/2, 1/2]$,

$$x(t) = \sum_n x(n)\mathrm{sinc}(t - n). \tag{3.73}$$

This sinc function expansion (better known as the sampling theorem expansion) is much more useful when it is linearly scaled in time, replacing the functions $\mathrm{sinc}(t - n)$ with $\mathrm{sinc}(2Bt - n)$ for some given bandwidth $B > 0$ (see Figure 3.8). The set of functions $\{\mathrm{sinc}(2Bt - n); n \in \mathbb{Z}\}$ is still an orthogonal set, but the scaling in time by a factor of $(2B)^{-1}$ causes the squared integral to become $(2B)^{-1}$. Since the scaling by $(2B)^{-1}$ in time causes a scaling of $2B$ in frequency, these orthogonal function are now limited in frequency to $[-B, B]$. The argument above, applied to this scaled orthogonal set, leads to the well-known sampling theorem.

Theorem 3.6.13 *Let $x(t)$ be a continuous square-integrable real or complex function of $t \in \mathbb{R}$ which is limited in frequency to $[-B, B]$ for any given $B > 0$. Then*

$$x(t) = \sum_n x\left(\frac{n}{2B}\right) \mathrm{sinc}(2Bt - n). \tag{3.74}$$

This theorem adds precision to the notion that any well-behaved function of a real variable can be approximated by its samples, saying that if the function is frequency limited, then sufficiently close samples represent the function perfectly when the points between the samples are filled in by this sinc expansion.

Now suppose that $\{X_n; n \in \mathbb{Z}\}$ is a sequence of IID Gaussian rv s and consider the following *Gaussian sinc process*,

$$X(t) = \sum_{-\infty}^{\infty} X_n \mathrm{sinc}(2Bt - n), \qquad \text{where } X_n \sim \mathrm{N}(0, \sigma^2). \tag{3.75}$$

The following theorem shows that the Gaussian sinc process of (3.75) is indeed a Gaussian process, calculates its covariance function, and shows that the process is stationary.

Theorem 3.6.14 *The Gaussian sinc process $\{X(t);\ t \in \mathbb{R}\}$ in (3.75) is a stationary Gaussian process with*

$$K_X(t) = \sigma^2 \text{sinc}(2Bt). \tag{3.76}$$

Proof From (3.75), we have

$$K_X(t,\tau) = \mathsf{E}\left[\left(\sum_n X_n \text{sinc}(2Bt - n)\right)\left(\sum_k X_k \text{sinc}(2B\tau - k)\right)\right]$$

$$= \mathsf{E}\left[\sum_n X_n^2 \,\text{sinc}(2Bt - n)\text{sinc}(2B\tau - n)\right] \tag{3.77}$$

$$= \sigma^2 \sum_n \text{sinc}(2Bt - n)\text{sinc}(2B\tau - n) \tag{3.78}$$

$$= \sigma^2 \text{sinc}(2B(t - \tau)), \tag{3.79}$$

where (3.77) results from $\mathsf{E}[X_n X_k] = 0$ for $k \neq n$ and (3.78) results from $\mathsf{E}\left[X_n^2\right] = \sigma^2$ for all n. To establish the identity between (3.78) and (3.79), let $y(t) = \text{sinc}(2B(t - \tau))$ for any given τ. The Fourier transform of $y(t)$ is $Y(f) = \sqrt{(2B)^{-1}} \exp(-i2\pi B\tau f)$ for $-B \le f \le B$ and 0 elsewhere. Thus $y(t)$ is frequency limited to $[-B, B]$ and therefore satisfies (3.74), which is the desired identity.

Now note that $K_X(t,t) = \sigma^2 = \sigma^2 \sum_n \text{sinc}^2(2Bt - n)$. Thus this series converges, and from Theorem 3.6.10, $\{X(t);\ t \in \mathbb{R}\}$ is a Gaussian process. Finally, since the covariance depends only on $t - \tau$, the process is stationary and the covariance in single-variable form is $K_X(t) = \sigma^2 \text{sinc}(2Bt)$. □

3.6.5 Filtered Gaussian sinc processes

Many important applications of stochastic processes involve linear filters where the filter input is one stochastic process and the output is another. The filter might be some physical phenomenon, or it might be a filter being used to detect or estimate something from the input stochastic process. It might also be used simply to demonstrate the existence of a stochastic process with certain properties. In this section, we restrict our attention to the case where the input stochastic process is the Gaussian sinc process described in the previous section. We then show that the output is a stationary Gaussian process and find its covariance function. Figure 3.9 illustrates the situation.

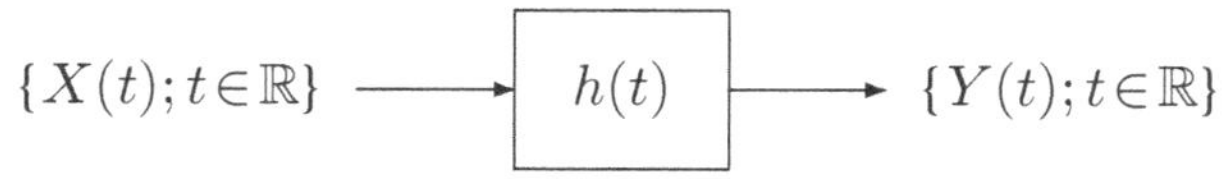

$$\{X(t); t \in \mathbb{R}\} \longrightarrow \boxed{h(t)} \longrightarrow \{Y(t); t \in \mathbb{R}\}$$

Figure 3.9 A stochastic process $\{X(t);\ t \in \mathbb{R}\}$ is the input to a linear time-invariant filter, and the output is another stochastic process. A WSS input leads to a WSS output and a Gaussian input leads to a Gaussian output.

A linear time-invariant filter with impulse response $h(t)$ creates a linear transformation from an input function $x(t)$ to an output function $y(t)$ defined by $y(t) = \int_{-\infty}^{\infty} x(\tau)h(t - \tau)\,d\tau$. In other words, the output at time t is a linear combination of the inputs over all time. The time invariance refers to the property that if the input function is translated by a given d, then the output function is translated by the same d.

In many situations, $h(t)$ is restricted to be *realizable*, meaning that $h(t) = 0$ for $t < 0$. This indicates that the output at a given t is a function only of the inputs up to and including t. In other situations, the filtering is done 'off-line,' meaning that the entire function $x(t)$ is available before performing the filtering. In some cases, the time reference at the filter output might have a delay of d relative to that at the input. This often occurs when a communication channel is subject to both filtering and propagation delay, and in these cases, $h(t)$ might be non-zero for all $t \geq -d$; this can still be regarded as a realizable filter, since only the time reference at the output has been altered.

In this section, we assume that $h(t)$ has a Fourier transform that is 0 for $|f| > B$, where B is the bandwidth of the input Gaussian sinc process. We shall find later that this implies that the filter is non-realizable. This is of no concern here since our purpose is simply to characterize a family of Gaussian processes at the filter output.

Suppose a stochastic process $\{X(t);\ t \in \mathbb{R}\}$ is the input to a linear time-invariant (LTI) filter. Let Ω be the underlying sample space and let ω be a sample point of Ω. The corresponding sample function of the process $\{X(t);\ t \in \mathbb{R}\}$ is then $X(t, \omega)$. The output of the LTI filter with impulse response $h(t)$ and input $X(t, \omega)$ is given by

$$Y(t, \omega) = \int_{-\infty}^{\infty} X(\tau, \omega)h(t - \tau)d\tau.$$

If the integrals exist for each ω, this (in principle) defines a rv for each t and thus (in principle) defines a stochastic process $\{Y(t);\ t \in \mathbb{R}\}$. Developing a theory of integration for a continuum of rv s is quite difficult[11] and would take us too far afield. This is why we are concentrating on stochastic processes that can be represented as orthonormal expansions using a sequence of rv s as coefficients. The next section generalizes this to other input processes that can also be represented as orthogonal expansions.

If we express the input sinc process $X(t)$ as in (3.75), then the output process is given by

$$Y(t) = \int_{-\infty}^{\infty} \sum_n X_n \mathrm{sinc}(2B\tau - n)h(t - \tau)\,d\tau, \quad \text{where } X_n \sim \mathrm{N}\,(0, \sigma^2). \tag{3.80}$$

Assuming that the integration and summation can be interchanged, we see that

$$\begin{aligned} \int_{-\infty}^{\infty} \mathrm{sinc}(2B\tau - n)\,h(t - \tau)\,d\tau &= \int_{-\infty}^{\infty} \mathrm{sinc}(2B\tau)\,h\left(t - \frac{n}{2B} - \tau\right)\,d\tau \\ &= \frac{1}{2B}\,h\left(t - \frac{n}{2B}\right), \end{aligned} \tag{3.81}$$

[11] Readers who believe that stochastic processes are sort of like ordinary functions and can be integrated and understood in the same way should look at Example 3.6.16.

where we have viewed the convolution as a product in the frequency domain and used the fact that the transform of the sinc function is constant over $[-B, B]$ and that $H(f)$ is zero outside that range. Thus, substituting (3.81) into (3.80) we have

$$Y(t) = \sum_n \frac{X_n}{2B}\, h\left(t - \frac{n}{2B}\right). \tag{3.82}$$

From Theorem 3.6.10, if $\sum_n h^2(t - n/2B)$ is finite for each t, then $\{Y(t);\ t \in \mathbb{R}\}$ is a Gaussian process (and the previous interchange of integration and summation is justified). Exercise 3.22 shows that $\sum_n h^2(t - n/2B) = \int_{-\infty}^{\infty} h^2(\tau)\, d\tau$ for each t. This shows that if $h(t)$ is square integrable, then $Y(t)$ is a Gaussian process.

In the next subsection, we show that when a WSS stochastic process is filtered by an LTI filter, the output is also WSS. Thus our final result is that if the X_n are IID and $\mathrm{N}(0, \sigma^2)$ and if $h(t)$ is square integrable and bandlimited to $[-B, B]$, then $Y(t)$ in (3.82) is a stationary Gaussian process. We discuss this further in the next subsection.

3.6.6 Filtered continuous-time stochastic processes

As discussed in the previous subsection, if a sample function $X(t, \omega)$ is the input to an LTI filter with impulse response $h(t)$, then the output sample function (assuming that the integral converges) is $Y(t, \omega) = \int X(\tau, \omega) h(t - \tau)\, d\tau$. If $\{X_n;\ n \in \mathbb{Z}\}$ is a sequence of IID rv s, $\{\phi_n(t);\ n \in \mathbb{Z}\}$ is a sequence of orthonormal functions, and $X(t) = \sum_n X_n \phi_n(t)$, then we can visualize $Y(t)$ as an output stochastic process described as

$$Y(t) = \int_{-\infty}^{\infty} X(\tau) h(t - \tau) d\tau = \int_{-\infty}^{\infty} \sum_n X_n \phi_n(\tau) h(t - \tau) d\tau. \tag{3.83}$$

There are some mathematical issues about whether the infinite summation in the final expression converges and whether the integral converges, but we saw how to treat these questions for the case in which $X(t)$ is a Gaussian sinc process and $h(t)$ is band limited and square integrable. More broadly, if the X_n are IID Gaussian, we saw how to use Theorem 3.6.10 to show that the output process is also Gaussian if $Y(t)$ has a finite variance for all t.

We will ignore these convergence questions in more general cases for the time being, and also use the middle expression in (3.83), viewing it as shorthand for the final expression.

In what follows, we will find the covariance function $K_Y(t, \tau)$ of the output process in terms of the covariance function K_X of the input and the impulse response of the filter. We also introduce and interpret the spectral density if the process is WSS. We will simply assume that any needed limits exist, although a few examples will be given later where more care is needed.

Assume throughout that the input stochastic process, $\{X(t);\ t \in \mathbb{R}\}$, is real and zero mean and that $h(t)$ is real. It follows then that $Y(t)$ is real and zero mean. If we rewrite $\int X(\tau) h(t - \tau)\, d\tau$ as $\int X(t - \tau) h(\tau)\, d\tau$, then the covariance function of $Y(t)$ can be expressed as

$$K_Y(t,u) = \mathsf{E}\left[\int_{-\infty}^{\infty} X(t-\tau)h(\tau)d\tau \int_{-\infty}^{\infty} X(u-s)h(s)\,ds\right]. \tag{3.84}$$

Interchanging expectation and integration,

$$K_Y(t,u) = \int_{-\infty}^{\infty}\int_{-\infty}^{\infty} K_X(t-\tau,\ u-s)h(\tau)h(s)\,d\tau ds. \tag{3.85}$$

This equation is valid whether or not X is WSS. Assuming that X is WSS, we can rewrite $K_X(t-\tau,\ u-s)$ in the single argument form as $K_X(t-u-\tau+s)$,

$$K_Y(t,u) = \int_{-\infty}^{\infty}\int_{-\infty}^{\infty} K_X(t-u-\tau+s)h(\tau)h(s)d\tau\, ds. \tag{3.86}$$

This is a function only of $t-u$, showing that Y is WSS. Thus $K_Y(t,u)$ can be written in the single argument form $K_Y(t-u)$. Replacing $t-u$ by v, we have

$$K_Y(v) = \int_{-\infty}^{\infty}\int_{-\infty}^{\infty} K_X(v-\tau+s)h(\tau)h(s)d\tau\, ds. \tag{3.87}$$

We now interpret the right-hand side of (3.87) as the convolution of three functions. To do this, we first rewrite (3.87) as

$$K_Y(v) = \int_s h(s)\left[\int_\tau K_X(v+s-\tau)h(\tau)\,d\tau\right]ds. \tag{3.88}$$

The term in square brackets is the convolution of h and K_X evaluated at $v+s$, which we denote as $[h * K_X](v+s)$. Now define $h_b(s) = h(-s)$. That is, h_b is h reversed in time. Replacing s with $-s$, (3.88) becomes

$$K_Y(v) = \int_s h_b(s)[h * K_X](v-s)\,ds = [h_b * h * K_X](v). \tag{3.89}$$

One of the simplest and best-known results of linear systems is that convolution in the time domain corresponds to multiplication in the Fourier transform domain. This leads us to define spectral density.

Definition 3.6.15 *The* ***spectral density*** $S_Y(f)$ *of a WSS stochastic process* $\{Y(t);\ t \in \mathbb{R}\}$ *is the Fourier transform of its covariance function* $K_Y(t)$*, i.e.,*

$$S_Y(f) = \int_{-\infty}^{\infty} K_Y(t)e^{-i2\pi ft}\,dt. \tag{3.90}$$

We now express (3.89) in terms of spectral densities. Let $H(f)$ be the Fourier transform of the impulse response $h(t)$,

$$H(f) = \int_{-\infty}^{\infty} h(t)e^{-i2\pi ft}dt\,. \tag{3.91}$$

The Fourier transform of the backward impulse response, $h_b(t)$, is then

$$H_b(f) = \int_{-\infty}^{\infty} h_b(t)e^{-i2\pi ft}\,dt = \int_{-\infty}^{\infty} h(\tau)e^{i2\pi f\tau}\,d\tau = H^*(f). \tag{3.92}$$

The transform of (3.89) is then

$$S_Y(f) = H^*(f)H(f)S_X(f) = |H(f)|^2 S_X(f). \tag{3.93}$$

Thus the covariance function of $Y(t)$ is best expressed as the inverse Fourier transform, $\mathrm{F}^{-1}[|H(f)|^2 S_X(f)]$.

3.6.7 Interpretation of spectral density and covariance

First note that the covariance function of a real WSS process must be real and symmetric around 0. Thus $S_X(f)$ and $S_Y(f)$ are real and symmetric around 0. Also, since $h(t)$ is real, $|H(f)|^2$ must be real and symmetric around 0, which is also implied by (3.93).

Now consider a very narrow band filter around some given frequency f_0. In particular, assume a filter with frequency response

$$H(f) = \begin{cases} 1 & \text{for } f_0 - \epsilon/2 \le |f| \le f_0 + \epsilon/2, \\ 0 & \text{elsewhere.} \end{cases}$$

If we pass a zero-mean WSS stochastic process $\{X(t)\}$ through this filter, then from (3.93),

$$S_Y(f) = \begin{cases} S_X(f) & \text{for } f_0 - \epsilon/2 \le |f| \le f_0 + \epsilon/2, \\ 0 & \text{elsewhere.} \end{cases} \tag{3.94}$$

The expected power out of this filter, i.e., $\mathsf{E}\left[Y^2(t)\right] = K_Y(0)$, is independent of t because Y is WSS. Since $K_Y(t)$ is the inverse Fourier transform of $S_Y(f)$, $K_Y(0) = \mathsf{E}\left[Y^2(t)\right]$ is given by

$$\mathsf{E}[Y^2(t)] = \int S_Y(f)\, df \approx 2\epsilon\, S_X(f_0), \tag{3.95}$$

where we assume that $S_X(f)$ is continuous and ϵ is so small that $S_X(f) = S_X(-f_0)$ does not vary appreciably from $f_0 - \epsilon/2$ to $f_0 + \epsilon/2$. This means that the expected output power from the filter is proportional to $S_X(f_0)$. This output power can then be interpreted as the input power over the range of frequencies $\pm(f_0 - \epsilon/2, f_0 + \epsilon/2)$. Since this is proportional to 2ϵ (the aggregate range of positive and negative frequencies in the filter passband), we interpret spectral density as the power per unit frequency in the WSS process. This also says (with a little care about points of discontinuity in $S_X(f)$) that $S_X(f) \ge 0$ for all f.

Now consider the class of filtered Gaussian sinc processes again. For a Gaussian sinc process band limited to B, we have $S_X(f) = \sigma^2/2B$ for $|f| \le B$ and $S_X(f) = 0$ for $|f| > B$. If this is filtered with frequency response $H(f)$, band limited to B, then from (3.93)), we have

$$S_Y(f) = \begin{cases} \dfrac{\sigma^2 |H(f)|^2}{2B} & \text{for } f \le B, \\ 0 & \text{for } f > B. \end{cases}$$

There are several remarkable features about this. First, the covariance function $K_Y(t)$ is determined by this spectral density, and since $Y(t)$ is Gaussian, this determines all the joint distributions of $Y(t)$, i.e., it determines the stochastic process, aside from possible interactions with other stochastic processes.

Second, the spectral density, and thus all joint cumulative distribution functions (CDFs) depend on $H(f)$ only through $|H(f)|^2$. A given choice for $|H(f)|^2$ can leave significant freedom in choosing $H(f)$, but that freedom does not change the joint probability distributions of the process.

Third, since there are essentially no constraints on $|H(f)|$ other than being non-negative and limited to B, any desired stationary Gaussian process band limited to B can be chosen in this way. Since B is arbitrary, this limitation does not at first appear significant, although this will be discussed shortly.

Fourth, we have seen that $S_Y(f) \geq 0$ for all f for all WSS stochastic processes. If we restrict our attention to WSS processes for which $K_Y(0) < \infty$ (and we really cannot make much sense out of other processes), then $\int S_Y(f)\,df < \infty$, so it follows that we can approximate $S_Y(f)$ by a band limited choice with large enough B. Since any band limited choice for $S_Y(f) \geq 0$ is the spectral density of a filtered Gaussian sinc process, we see that the only constraint on $K_Y(t)$ to be a covariance function is that $K_Y(0) < \infty$ and $S_Y(f) \geq 0$ for all f. Furthermore, any such spectral density can be approximated (in some sense) by the spectral density of a filtered Gaussian sinc process.

The trouble with filtered Gaussian sinc processes is that realizable filters cannot be band limited. In fact, the Paley–Wiener theorem (see [24]) says that a necessary and sufficient condition on the Fourier transform for a non-zero square-integrable function $h(t)$ to be 0 for all $t < 0$ is that $\int \frac{|\ln |H(f)||}{1+f^2}\,df < \infty$. This is more than an obscure mathematical issue, since it turns out that band limited stationary Gaussian processes have some peculiar properties even though their spectral densities closely approximate those of processes that are not band limited. We will not resolve these issues here, and readers are urged to exercise care when approximating non-band-limited processes by band limited processes.

The purpose of this subsection has not been to imply that filtered Gaussian sinc processes provide a universally acceptable way to deal with continuous-time stationary Gaussian processes. Rather these band limited processes provide a great deal of insight into more general stationary Gaussian processes and, with care, can be used to solve a large number of engineering problems concerning these processes.

3.6.8 White Gaussian noise

Physical noise processes are often well modeled as stationary Gaussian processes, as we have pointed out earlier. Often they also have the characteristic that their spectral density is quite flat over the bandwidths of interest in a given situation. In this latter situation, we can simplify and idealize the model by assuming that the spectral density is constant over all frequencies. This idealization is called *white Gaussian noise*. Unfortunately, this simplification comes at a steep price – the power in such a process $\{W(t);\, t \in \mathbb{R}\}$ is

$$\mathsf{E}[|W(t)|^2] = K_W(0) = \frac{1}{2\pi}\int_{-\infty}^{\infty} S_W(f)df \;=\; \infty. \tag{3.96}$$

Thus $W(t)$ at any given t cannot even be approximated as a Gaussian rv. On the other hand, if a stationary Gaussian process has spectral density $S_X(f)$ and is filtered with frequency function $H(f)$, then, from (3.93), the output process $Y(t)$ has spectral density $S_Y(f) = |H(f)|^2 S_X(f)$. If $S_X(f)$ is flat over the range of f where $H(f)$ is non-zero, then the output Gaussian process $\{Y(t);\ t \in \mathbb{R}\}$ has the same joint distributions no matter what $S_X(f)$ is outside the bandwidth of $H(f)$.

If a Gaussian noise process is looked at only through the outputs of various linear filters, and if its spectral density is constant over the frequency response of each filter, then we might as well assume that the process is white and not concern ourselves with the spectral density outside of the range of interest. Since measurement devices generally involve filtering to some extent (even if flat over such a broad bandwidth that it can usually be ignored), this view of white Gaussian noise as an idealization is usually the only view that is physically meaningful.

In summary then, white Gaussian noise is an idealization of a Gaussian process with spectral density[12] $S_W(f) = N_0/2$ over all f such that $|f| \leq B$, where B is larger than all frequencies of interest. In the limit $B \to \infty$, the covariance function of white noise can be taken to be $(N_0/2)\delta(t)$, where $\delta(t)$ is the Dirac unit impulse. This is a generalized function, roughly defined by the property that for any well-behaved function $a(t)$, we have $\int a(t)\delta(t)\,dt = a(0)$. We can visualize $\delta(t)$ as an idealization of a narrow pulse of unit area, narrow relative to the smallest interval over which any $a(t)$ of interest can change. With a little thought, it can be seen that this is just another way of saying, as before, that white Gaussian noise is an idealization of a stationary Gaussian noise process whose spectral density is constant over the frequency range of interest.

An important feature of white Gaussian noise is that we can view any zero-mean stationary Gaussian process as a filtered version of white Gaussian noise. That is, a zero-mean stationary Gaussian process $\{Y(t);\ t \in R\}$ with spectral density $S_Y(f)$ can be viewed as white noise of unit spectral density passed through a filter with frequency response $H(f)$ such that $|H(f)|^2 = S_X(f)$. Recall that this view was quite valuable in studying Gaussian vectors, and it is equally valuable here.

It almost appears that white Gaussian noise can be viewed as the limit of a sequence of Gaussian sinc processes where process ℓ has bandwidth B_ℓ and power $\mathsf{E}\left[X^2(t)\right] = B_\ell N_0$. Thus the spectral density for process ℓ is $N_0/2$ for $|f| \leq B_\ell$. For any realizable filter with frequency response $H(f)$, we have observed from the Paley–Wiener theorem that $H(f)$ can only approach 0 at a limited rate as $f \to \infty$. Thus there is no B large enough that white noise filtered by $H(f)$ is quite the same as a filtered sinc Gaussian process, although it could certainly be approximated that way.

The following two examples show that we have slightly oversimplified matters in viewing zero-mean stationary Gaussian processes as being characterized by their spectral densities.

[12] $N_0/2$ is the standard term among engineers to denote the spectral density of white noise. Spectral density is the power per unit frequency in a process, counting positive and negative frequencies separately. Thus if we look at the power in a bandwidth B, i.e., a frequency interval of width B in positive frequencies and another B in negative frequencies, the noise power in that band is $N_0 B$.

Example 3.6.16 (Pathological barely visible Gaussian noise) Consider a stationary Gaussian process $\{X(t)\,;\, t \in \mathbb{R}\}$ for which $X(t) \sim \mathrm{N}(0,1)$ for each $t \in \mathbb{R}$. Assume that $X(t)$ and $X(\tau)$ are independent for all t, τ with $\tau \neq t$. Thus $K_X(t,\tau)$ is 1 for $t = \tau$ and 0 otherwise. This process is Gaussian and stationary, and its single-variable covariance function $K_X(t)$ is 1 for $t = 0$ and 0 elsewhere. It follows that $S_X(f) = 0$ for all f. Also, if we express $X(t)$ in terms of any set of orthonormal functions, we see that $\int X(t)\phi_n(t)\, dt = 0$ WP1 for all n. In the same way, if $X(t)$ is passed through any square-integrable linear filter, the output process is 0 WP1 for all t. Thus, in a very real sense, this Gaussian process is effectively 0. From a physical point of view, one could never observe such a process, because any physical measurement requires some type of averaging over a very small but non-zero interval of time. The sample-average measurement over any such interval would then be 0 WP1.

We can compare this pathological process to a sequence of Gaussian sinc processes with bandwidths $B_1 \leq B_2, \ldots \to \infty$ as before. Here, however, we take the power in each process to be 1. Thus the spectral density of the ℓth process is $(2B_\ell)^{-1}$ for $|f| \leq B_\ell$, so the spectral density at each f approaches 0 with increasing ℓ. As before, however, there is not any decent kind of limit for the process.

There are a number of broadband communication systems where the transmitted channel waveform can be roughly modeled as a Gaussian sinc process with large bandwidth and negligible spectral density. Such a process appears almost non-existent to other communication systems, but as will be seen in Chapter 8 on detection, the signals can still be detected, in contrast to the strange process here.

The strangeness of the pathological process in this example arises largely from the fact that the covariance function is not continuous. Exercise 3.25 shows that if a WSS process has a covariance function $K_X(t)$ that is continuous at $t = 0$, then it is continuous everywhere. A large part of the theory for constructing orthonormal expansions for continuous random processes depends on a continuous covariance function. From a more application oriented viewpoint, the properties arising from discontinuities in the covariance cannot be observed (as in the example here). Thus a continuous covariance function is almost always assumed.

Example 3.6.17 (Pathological invisible Gaussian noise) Let $X \sim \mathrm{N}(0,1)$ and let Y be a uniform rv over $(0, 1]$. Let $Z(t)$ be a stochastic process where $Z(Y) = X$ and $Z(t) = 0$ for all $t \neq Y$. Now for any given t, the probability that $Y = t$ is 0, so $Z(t) = 0$ WP1. Thus $Z(t)$ can be viewed as Gaussian with variance 0 and, of course, $K_Z(t) = 0$ for all t.

This is more pathological than the previous example, but is important in showing that specifying the covariance function of a Gaussian process does not fully specify the process. Here every sample function of this process is discontinuous, whereas the conventional zero function is continuous. For any finite set of epochs $t_1, \ldots, t_k$, we see that $X(t_1), \ldots, X(t_k)$ are all 0 WP1, so these joint distributions do not distingish this process from the all-zero process. The difference between this process and the all-zero process could never be measured, so essentially, the covariance function, and thus the set of finite joint distributions, specifies a process. However, as shown here, more intricate tools (common sense for engineers and measure theory for mathematicians) are needed to make sense of continuity for these sample functions.

3.6.9 The Wiener process/Brownian motion

Recall that one of the major properties of the Poisson counting process (see Chapter 2) is that it has stationary and independent increments. These properties can be be defined for arbitrary stochastic processes as well as for counting processes. They are fundamental properties of the Wiener process, which is also known as Brownian motion.[13]

Definition 3.6.18 *Let a stochastic process $\{X(t);\ t \in \mathrm{T}\}$ be defined over a set of epochs T, where T is either the non-negative reals or non-negative integers. Then $\{X(t);\ t \in \mathrm{T}\}$ has **stationary increments** if for any epochs $t_1 < t_2$, the increment $X(t_2) - X(t_1)$ has the same distribution as $X(t_2 - t_1) - X(0)$.*

Definition 3.6.19 *Let a stochastic process $\{X(t);\ t \in \mathrm{T}\}$ be defined over a set of epochs T, where T is either the non-negative reals or non-negative integers. Then $\{X(t);\ t \geq 0\}$ has **independent increments** if for any sequence of epochs $t_1 < t_2 < t_3 < \cdots < t_k$, the rv s*

$$[X(t_2) - X(t_1)],\ \ [X(t_3) - X(t_2)],\ \ \ldots, [X(t_k) - X(t_{k-1})]$$

are statistically independent.

Now consider an arbitrary process $\{X(t);\ t \geq 0\}$ with independent and stationary increments and with $X(0) = 0$. Let Δ be an arbitrary increment size and, for an arbitrary positive integer n, write $X(n\Delta)$ as

$$X(n\Delta) = [X(n\Delta) - X((n-1)\Delta)] + [X((n-1)\Delta) - X((n-2)\Delta)] + \cdots + [X(\Delta) - X(0)].$$

Because of this, we see that $\mathsf{E}[X(n\Delta)] = n\mathsf{E}[X(\Delta)]$ and $\mathsf{VAR}[X(n\Delta)] = n\,\mathsf{VAR}[X(\Delta)]$. Thus the mean and variance of $X(t)$ must each be linear in t. Because of the independent increments, we can also see that $K_X(t, \tau)$, for any $\tau \geq t$, is equal to $\mathsf{VAR}[X(t)]$. We summarize this in the following theorem.

Theorem 3.6.20 *Let $\{X(t);\ 0 \leq t\}$ have independent and stationary increments and let $X(0) = 0$. Then for any epochs t and $\tau > t$,*

$$\mathsf{E}[X(t)] = t\,\mathsf{E}[X(1)]]; \qquad K_X(t, \tau) = t\,\mathsf{VAR}[X(1)]. \tag{3.97}$$

One interesting consequence of this is that (except in the uninteresting case of zero variance) processes with independent and stationary increments cannot be stationary. That is, $\{X(t);\ t \geq 0\}$ has stationary increments if the *changes* $X(t) - X(t - \Delta)$ do not depend

[13] Brownian motion is a standard model for the motion of small particles in a gas. Norbert Wiener substantially developed its mathematical analysis. We will discuss only the one-dimensional version of the process.

probabilistically on t, whereas (essentially) the process is stationary if the *process values themselves*, $X(t)$, do not depend probabilistically on t. Another consequence is that these processes are not meaningful over the entire time interval from $-\infty$ to $+\infty$. This is because the variance is growing linearly with t and must remain non-negative for all epochs t.

The restriction that $X(0) = 0$ in the theorem is inessential, and the extension to the case where $X(0)$ is an arbitrary rv is contained in Exercise 3.20.

Definition 3.6.21 *A* ***Wiener process (Brownian motion process)*** *is a zero-mean Gaussian process* $\{X(t);\ t \geq 0\}$ *which has stationary and independent increments, satisfies* $X(0) = 0$*, and has continuous sample functions WP1.*

The continuity restriction rules out the addition of 'invisible' processes such as that in Example 3.6.17 to a continuous process with stationary and independent increments. See Feller [9] for a proof that Wiener processes exist. Given this existence, we see that $\mathsf{E}[X(t)] = 0$ for all $t \geq 0$ and also, from Theorem 3.6.20, $K_X(t, \tau) = \min(t, \tau)\sigma^2$ where $\sigma^2 = \mathsf{E}\left[X^2(1)\right]$. Since a zero-mean Gaussian process is essentially specified by its covariance function, we see that the Wiener process is essentially specified by the single parameter σ^2. Also, since the covariance function has been derived using only the stationary and independent increments property, we see that the added assumption about continuity is not required for specifying all the joint CDFs of the process.

A type of continuity also follows directly from the independent and stationary increment property without the added requirement of continuity in the definition of a Wiener process. Consider the increment $X(t + \Delta) - X(t)$ for very small Δ. The increment has the variance $\Delta\sigma^2$, and by the Chebyshev inequality,

$$\Pr\{X(t + \Delta) - X(t) > \epsilon\} \leq \frac{\Delta\sigma^2}{\epsilon^2}.$$

This means that as $\Delta \to 0$, the probability that $X(t)$ changes by more than ϵ goes to zero. To understand the type of continuity implied by this, consider the sequence $X(t + \Delta), X(t + \Delta/2), X(t + \Delta/3), \ldots$ for some given t, Δ. This sequence approaches $X(t)$ in probability (see Section 1.7.6). This is weaker, of course, than the continuity of sample functions WP1 required in the definition of a Wiener process.

Strangely enough, although the sample functions of a Wiener process have these continuity properties, they are essentially not differentiable. To see this, note that $[X(t + \Delta) - X(t)]/\Delta$ has variance σ^2/Δ. This goes to ∞ as $\Delta \to 0$. Despite these strange properties, the Wiener process is widely used by engineers, and often provides sound insights into real issues.

In a sense, the Poisson counting process and Wiener process both model an 'integral' of independent objects. In the Poisson case, we are interested in random point arrivals. If we view a sample function of these arrivals as a sequence of unit impulses, then the corresponding sample function of the counting process is the integral of that impulse chain. The Wiener process models an accumulation or integral of individually small but very dense independent disturbances (noise). One can envision the process being integrated as white Gaussian noise, although, as we have seen, the derivative of the Wiener process does not exist and white Gaussian noise does not exist except as a generalized form of stochastic process.

We now show that the Wiener process can be viewed as a limit of a sum of IID rv s if the limit uses the appropriate kind of scaling. Let $\{Y_n;\ n \geq 1\}$ be a sequence of zero-mean IID rv s each with finite variance σ^2. Consider a sequence of processes $\{X_\ell(t);\ t \geq 0\}$, where the ℓth process is defined in terms of $\{Y_n;\ n \geq 1\}$ by

$$X_\ell(t) = \sum_{k=1}^{\lfloor 2^\ell t \rfloor} 2^{-\ell/2} Y_k.$$

Then $\mathsf{E}[X_\ell(t)] = 0$ and $\mathsf{E}\left[X_\ell^2(t)\right] = \sigma^2 t$, where we are ignoring the difference between $\lfloor 2^\ell t \rfloor$ and $2^\ell t$.

Note that each unit increase in ℓ doubles the number of IID rv s added together in each unit of time. Note also that the magnitudes of the IID rv s are scaled down by the square root of this rate doubling. Thus the variance of the scaled sum for a given t remains constant as ℓ increases. By the CLT, the distribution of this scaled sum approaches the Gaussian. It is easy to see that the covariance of $X_\ell(t)$ approaches that of the Wiener process (in fact, it is only the integer approximation $\lfloor 2^\ell t \rfloor \approx 2^\ell t$ that is involved in the covariance).

We do not want to address the issue of how a limit of a sequence of stochastic processes approaches another process. The important thing is that a sum of (finite variance zero-mean) rv s can be modeled as a Wiener process with the appropriate scaling; this helps explain why the Wiener process appears in so many applications.

This completes our discussion of (real) Gaussian processes. The next section discusses the complex case.

3.7 Circularly-symmetric complex random vectors

Many of the (real-valued) waveforms used for communication and other purposes have the property that their Fourier transforms are 0 except in two relatively narrow bands of frequencies, one around a positive carrier frequency f_0, and the other around $-f_0$. Such waveforms are often represented as

$$x(t) = z_{\text{re}}(t)\cos(2\pi f_0 t) + z_{\text{im}}(t)\sin(2\pi f_0 t) = \Re\left[z(t)e^{-2\pi i f_0 t}\right].$$

Representing $x(t)$ in terms of a complex 'baseband waveform' $z(t) = z_{\text{re}}(t) + i z_{\text{im}}(t)$ or in terms of two real baseband waveforms, $z_{\text{re}}(t)$ and $z_{\text{im}}(t)$ is often convenient analytically, since if the bandwidth is small and f_0 is large, then $z(t)$ changes slowly relative to $x(t)$, while still specifying the waveform exactly for a given f_0.

The same relationship, $X(t) = \Re[Z(t)\exp[-2\pi i f_0 t]$, is equally convenient for a stochastic process rather than an individual waveform in a limited bandwidth. Note, however, that $\sin(2\pi f_0 t)$ is the same as $\cos(2\pi f_0 t)$ except for a small delay, $1/(4f_0)$. Normally, we would not expect the statistics of the noise to be sensitive to this small delay; in more graphic terms, we would not expect the noise to 'know' where our time reference $t = 0$ is. Thus we often model bandpass noise so that $Z_{\text{re}}(t)$ and $Z_{\text{im}}(t)$ are identically distributed. By extending this slightly, we often model bandpass noise so that $Z(t)$ and $Z(t)e^{-i\theta}$ are identically distributed for all phases θ. More specifically, we often model bandpass noise so that for each $t_1, t_2, \ldots, t_n$, the joint distribution of the complex random vector $(Z(t_1), \ldots, Z(t_n))^{\mathsf{T}}$ is the same as that of $(Z(t_1)e^{i\theta}, \ldots, Z(t_n)e^{i\theta})^{\mathsf{T}}$ for each real θ.

The purpose of the above argument is not to convince the reader that this joint distribution property is 'necessary' for band-pass noise, but simply to motivate why this kind of phase invariance, which is called circular symmetry, might be useful to understand. The results here are widely used in many branches of engineering, mathematics, and physics, but not widely accessible in a systematic form.

3.7.1 Circular symmetry and complex Gaussian random variables

Definition 3.7.1 *A complex rv $Z = Z_{\text{re}} + iZ_{\text{im}}$ is **Gaussian** if Z_{re} and Z_{im} are jointly Gaussian; Z is **circularly symmetric** if Z and $Ze^{i\theta}$ have the same distribution for all real θ.*

Note that if Z has a PDF and is circularly symmetric, then the PDF is constant on any circle centered on the origin. If Z is Gaussian, then its equal probability contours are ellipses; these are circular and centered on the origin if and only if Z_{re} and Z_{im} are IID zero-mean Gaussian. The amplitude $|Z|$ of a circularly-symmetric Gaussian rv is Rayleigh-distributed and the phase is uniformly distributed.

If we multiply a circularly-symmetric rv Z by a complex constant c, then the amplitude of cZ is the product of the amplitudes of Z and c; the phase is the sum of the individual phases. It is intuitively clear (from the original uniform phase of Z) that such an addition of phases maintains the circular symmetry.

A circularly-symmetric Gaussian rv Z is fully described by its variance, $\sigma^2 = \mathsf{E}[ZZ^*] = \mathsf{E}[|Z|^2]$. The complex conjugate is necessary in the definition of variance, since $\mathsf{E}[ZZ^*] = \mathsf{E}[Z_{\text{re}}^2] + \mathsf{E}[Z_{\text{im}}^2]$, whereas $\mathsf{E}[Z^2] = \mathsf{E}[Z_{\text{re}}^2] - \mathsf{E}[Z_{\text{im}}^2]$.

Just as a Gaussian rv X of mean a and variance σ^2 is described as $X \sim \mathrm{N}(a, \sigma^2)$, a circularly-symmetric Gaussian rv Z of variance σ^2 is described as $Z \sim \mathrm{CN}(0, \sigma^2)$. Note that the real and imaginary parts of Z are then IID with variance $\sigma^2/2$ each. The terminology allows for a complex rv with a mean a and a fluctuation that is circularly-symmetric Gaussian to be referred to as $\mathrm{CN}(a, \sigma^2)$.

Definition 3.7.2 *An n-dimensional complex **rv** (complex n-**rv**) $\mathbf{Z} = (Z_1, \ldots, Z_n)^{\mathsf{T}}$ is **Gaussian** if the $2n$ real and imaginary components of $\mathbf{Z}$ are jointly Gaussian. It is **circularly symmetric** if the distribution of $\mathbf{Z}$ (i.e., the joint distribution of the real and imaginary parts) is the same as that of $e^{i\theta}\mathbf{Z}$ for all phase angles θ. It is **circularly-symmetric Gaussian** if it is Gaussian and circularly symmetric.*

Example 3.7.3 An important example of a circularly-symmetric Gaussian rv is $\boldsymbol{W} = (W_1, \ldots, W_n)^{\mathsf{T}}$, where the components W_k, $1 \le k \le n$ are statistically independent and each is $\mathcal{CN}(0,1)$. Since each W_k, is $\mathcal{CN}(0,1)$, it can be seen that $e^{i\theta}W_k$ has the same distribution as W_k. Using the independence, $e^{i\theta}\boldsymbol{W}$ then has the same distribution as $\boldsymbol{W}$. The $2n$ real and imaginary components of $\boldsymbol{W}$ are IID and $\mathcal{N}(0,1/2)$ so that the probability density (being careful about the factors of 1/2) is

$$f_{\boldsymbol{W}}(\boldsymbol{w}) = \frac{1}{\pi^n}\exp\left[\sum_{k=1}^{n} -|w_k|^2\right], \tag{3.98}$$

where we have used the fact that $|w_k|^2 = [\Re(w_k)]^2 + [\Im(w_k)]^2$ for each k to replace a sum over $2n$ terms with a sum over n terms.

3.7.2 Covariance and pseudo-covariance of complex n-dimensional random vectors

We saw in Section 3.3.4 that the distribution of a real zero-mean Gaussian n-rv (i.e., a vector with jointly-Gaussian components) is completely determined by its covariance matrix. Here we will find that the distribution of a *circularly-symmetric* Gaussian n-rv is also determined by its covariance matrix. *Without circular symmetry, the covariance matrix is not sufficient to determine the distribution.* In order to understand this, we first define both the covariance matrix and the pseudo-covariance matrix of a complex n-rv.

Definition 3.7.4 *The* ***covariance matrix*** $[K_{\boldsymbol{Z}}]$ *and the* ***pseudo-covariance matrix*** $[M_{\boldsymbol{Z}}]$ *of a zero-mean complex n-rv* $\boldsymbol{Z} = (Z_1,\dots,Z_n)^{\mathsf{T}}$ *are the* $n\times n$ *matrices of complex components given respectively by*

$$[K_{\boldsymbol{Z}}] = \mathsf{E}\left[\boldsymbol{Z}\boldsymbol{Z}^{\dagger}\right], \qquad [M_{\boldsymbol{Z}}] = \mathsf{E}\left[\boldsymbol{Z}\boldsymbol{Z}^{\mathsf{T}}\right], \tag{3.99}$$

where $\boldsymbol{Z}^{\dagger}$ *is the the complex conjugate of the transpose, i.e.,* $\boldsymbol{Z}^{\dagger} = \boldsymbol{Z}^{\mathsf{T}*}$.

As shown below, $[K_{\boldsymbol{Z}}]$ and $[M_{\boldsymbol{Z}}]$ determine the covariance matrix of the real $2n$-rv $\begin{bmatrix}\boldsymbol{Z}_{\rm re}\\ \boldsymbol{Z}_{\rm im}\end{bmatrix}$:

$$\begin{aligned}
\mathsf{E}\left[\Re(Z_k)\Re(Z_j)\right] &= \frac{1}{2}\Re\left([K_{\boldsymbol{Z}}]_{kj} + [M_{\boldsymbol{Z}}]_{kj}\right),\\
\mathsf{E}\left[\Im(Z_k)\Im(Z_j)\right] &= \frac{1}{2}\Re\left([K_{\boldsymbol{Z}}]_{kj} - [M_{\boldsymbol{Z}}]_{kj}\right),\\
\mathsf{E}\left[\Re(Z_k)\Im(Z_j)\right] &= \frac{1}{2}\Im\left(-[K_{\boldsymbol{Z}}]_{kj} + [M_{\boldsymbol{Z}}]_{kj}\right),\\
\mathsf{E}\left[\Im(Z_k)\Re(Z_j)\right] &= \frac{1}{2}\Im\left([K_{\boldsymbol{Z}}]_{kj} + [M_{\boldsymbol{Z}}]_{kj}\right).
\end{aligned} \tag{3.100}$$

If $\boldsymbol{Z}$ is also Gaussian, this shows that $[K_{\boldsymbol{Z}}]$ and $[M_{\boldsymbol{Z}}]$ together specify not only the covariance but also the distribution of $\begin{bmatrix}\boldsymbol{Z}_{\rm re}\\ \boldsymbol{Z}_{\rm im}\end{bmatrix}$, and thus specify the distribution of $\boldsymbol{Z}$. We next start to connect the notion of circular symmetry with the pseudo-covariance matrix.

Lemma 3.7.5 *Let* $\boldsymbol{Z}$ *be a circularly-symmetric complex n-rv. Then the pseudo-covariance matrix satisfies* $[M_{\boldsymbol{Z}}] = 0$.

Proof Since $\mathbf{Z}$ and $e^{i\theta}\mathbf{Z}$ have the same joint distribution for any given θ, they have the same pseudo-covariance matrix, i.e., $[M_{e^{i\theta}\mathbf{Z}}] = [M_{\mathbf{Z}}]$. Denote the j, ℓ component of $[M_{e^{i\theta}\mathbf{Z}}]$ as $[M_{e^{i\theta}\mathbf{Z}}]_{j,\ell}$. Then

$$[M_{e^{i\theta}\mathbf{Z}}]_{j,\ell} = \mathsf{E}\left[e^{i\theta}Z_j \cdot e^{i\theta}Z_\ell\right] = e^{i2\theta}[M_{\mathbf{Z}}]_{j\ell}.$$

For $\theta = \pi/2$ then, $[M_{\mathbf{Z}}]_{j,\ell} = -[M_{\mathbf{Z}}]_{j,\ell}$. Thus $[M_{\mathbf{Z}}]_{j,\ell} = 0$ for all j, ℓ. □

In general, $[M_{\mathbf{Z}}] = 0$ is not enough to ensure that $\mathbf{Z}$ is circularly symmetric. For example, in the one-dimensional case, if Z_{re} and Z_{im} are IID, binary equiprobable $(1, -1)$, then $[M_Z] = 0$ but Z is obviously not circularly symmetric. The next theorem, however, shows that $[M_{\mathbf{Z}}] = 0$ is enough in the Gaussian case.

Theorem 3.7.6 *Let $\mathbf{Z}$ be a zero-mean complex Gaussian n–rv. Then $[M_{\mathbf{Z}}] = 0$ if and only if $\mathbf{Z}$ is circularly-symmetric Gaussian.*

Proof The lemma shows that $[M_{\mathbf{Z}}] = 0$ if $\mathbf{Z}$ is circularly symmetric. For the only-if side, assume $[M_{\mathbf{Z}}] = 0$. Then $[M_{e^{i\theta}\mathbf{Z}}] = 0$ also, so $[M_{\mathbf{Z}}] = [M_{e^{i\theta}\mathbf{Z}}]$.

We must next consider $[K_{e^{i\theta}\mathbf{Z}}]$. The j, ℓ component of this matrix for any j, ℓ is

$$\mathsf{E}\left[e^{i\theta}Z_k \cdot e^{-i\theta}Z_\ell^*\right] = \mathsf{E}\left[Z_k \cdot Z_\ell^*\right] = [K_{\mathbf{Z}}]_{j\ell}.$$

Thus, $[K_{e^{i\theta}\mathbf{Z}}] = [K_{\mathbf{Z}}]$, so $e^{i\theta}\mathbf{Z}$ has the same covariance and pseudo-covariance as $\mathbf{Z}$.

Since $e^{i\theta}\mathbf{Z}$ and $\mathbf{Z}$ are each zero-mean complex Gaussian, each distribution is specified by its covariance and pseudo-covariance. Since these are the same, $e^{i\theta}\mathbf{Z}$ and $\mathbf{Z}$ must have the same distribution. This holds for all real θ, so $\mathbf{Z}$ is circularly symmetric Gaussian. □

Since $[M_{\mathbf{Z}}]$ is zero for any circularly-symmetric Gaussian n-**rv** $\mathbf{Z}$, the distribution of $\mathbf{Z}$ is determined solely by $[K_{\mathbf{Z}}]$ and is denoted as $\mathbf{Z} \sim \mathrm{CN}\ (0, [K_{\mathbf{Z}}])$, where C denotes that $\mathbf{Z}$ is both complex and circularly symmetric. The complex normalized IID **rv** of Example 3.7.3 is thus denoted as $\boldsymbol{W} \sim \mathrm{CN}\ (0, [I_n])$.

The following two examples illustrate some subtleties in Theorem 3.7.6.

Example 3.7.7 Let $\mathbf{Z} = (Z_1, Z_2)^\mathsf{T}$, where $Z_1 \sim \mathrm{CN}\ (0, 1)$ and $Z_2 = XZ_1$ where X is statistically independent of Z_1 and has possible values ± 1 with probability 1/2 each. It is easy to see that $Z_2 \sim \mathrm{CN}\ (0, 1)$, but the real and imaginary parts of Z_1 and Z_2 together are not jointly Gaussian. In fact, the joint distribution of $\Re(Z_1)$ and $\Re(Z_2)$ is concentrated on the two diagonal axes and the distribution of $\Im(Z_1)$ and $\Im(Z_2)$ is similarly concentrated. Thus, $\mathbf{Z}$ is not Gaussian. Even though Z_1 and Z_2 are individually circularly symmetric Gaussian, $\mathbf{Z}$ is not circularly symmetric Gaussian according to the definition. In this example, it turns out that $\mathbf{Z}$ *is* circularly symmetric and $[M_{\mathbf{Z}}] = \begin{bmatrix} 0 & 0 \\ 0 & 0 \end{bmatrix}$. The example can be changed slightly, changing the definition of Z_2 to $\Re(Z_2) = X\Re(Z_1)$ and $\Im(Z_2) \sim \mathrm{N}\ (0, 1/2)$, where $\Im(Z_2)$ is statistically independent of all the other variables. Then $[M_{\mathbf{Z}}]$ is still 0, but $\mathbf{Z}$ is not circularly symmetric. Thus, without the jointly-Gaussian property, the relation between circular symmetry and $[M_{\mathbf{Z}}] = 0$ is not an if-and-only-if relation.

Example 3.7.8 Consider a vector $\boldsymbol{Z} = (Z_1, Z_2)^{\mathsf{T}}$, where $Z_1 \sim \mathcal{CN}(0, 1)$ and $Z_2 = Z_1^*$. Since $\Re(Z_2) = \Re(Z_1)$ and $\Im(Z_2) = -\Im(Z_1)$, we see that the four real and imaginary components of $\boldsymbol{Z}$ are jointly Gaussian, so $\boldsymbol{Z}$ is complex Gaussian and the theorem applies. We see that $[M_{\boldsymbol{Z}}] = \begin{bmatrix} 0 & 1 \\ 1 & 0 \end{bmatrix}$, and thus $\boldsymbol{Z}$ is Gaussian but not circularly symmetric. This makes sense, since when Z_1 is real (or approximately real), $Z_2 = Z_1$ (or $Z_2 \approx Z_1$) and when Z_1 is pure imaginary (or close to pure imaginary), Z_2 is the negative of Z_1 (or $Z_2 \approx -Z_1$). Thus the relationship of Z_2 to Z_1 is certainly not phase invariant.

What makes this example interesting is that both $Z_1 \sim \mathcal{CN}(0, 1)$ and $Z_2 \sim \mathcal{CN}(0, 1)$. Thus, as in Example 3.7.7, it is the relationship between Z_1 and Z_2 that breaks up the circularly-symmetric Gaussian property. Here it is the circular symmetry that causes the problem, whereas in Example 3.7.7 it was the lack of a jointly Gaussian distribution.

3.7.3 Covariance matrices of complex *n*-dimensional random vectors

The covariance matrix of a complex *n*-**rv** $\boldsymbol{Z}$ is $[K_{\boldsymbol{Z}}] = \mathsf{E}\left[\boldsymbol{Z}\boldsymbol{Z}^{\dagger}\right]$. The properties of these covariance matrices are quite similar to those for a real *n*-**rv** except that $[K_{\boldsymbol{Z}}]$ is no longer symmetric ($K_{kj} = K_{jk}$), but rather is *Hermitian*, defined as a square matrix $[K]$ for which $K_{kj} = K_{jk}^*$ for all j, k. These matrices are analyzed in virtually the same way as the symmetric matrices considered in Section 3.4.1, so we simply summarize the results we need here.

If $[K]$ is Hermitian, then the eigenvalues are all real and the eigenvectors $\boldsymbol{q}_j$ and $\boldsymbol{q}_k$ of distinct eigenvalues are orthogonal in the sense that $\boldsymbol{q}_j^{\dagger}\boldsymbol{q}_k = 0$. Also if an eigenvalue has multiplicity ℓ, then ℓ orthogonal eigenvectors can be chosen for that eigenvalue.

The eigenvalues $\lambda_1, \ldots, \lambda_n$, repeating each distinct eigenvalue according to its multiplicity, can be used as the elements of a diagonal matrix $[\Lambda]$. To each λ_j, we can associate an eigenvector $\boldsymbol{q}_j$, where the eigenvectors are chosen to be orthonormal ($\boldsymbol{q}_j^{\dagger}\boldsymbol{q}_k = \delta_{jk}$). Letting $[Q]$ be the matrix with orthonormal columns[14] $\boldsymbol{q}_1, \ldots, \boldsymbol{q}_n$, we have the relationship

$$[K] = [Q\Lambda Q^{-1}] \qquad [Q^{\dagger}] = [Q^{-1}] \qquad \text{for } [K] \text{ Hermitian.} \tag{3.101}$$

An $n \times n$ Hermitian matrix $[K]$ is positive semi-definite if, for all complex n-vectors $\boldsymbol{b}$, the equation $\boldsymbol{b}^{\dagger}[K]\boldsymbol{b} \geq 0$ holds. It is positive definite if $\boldsymbol{b}^{\dagger}[K]\boldsymbol{b} > 0$ for all $\boldsymbol{b} \neq 0$. By the same arguments as in the real case, we have the following lemma.

Lemma 3.7.9 *If $\boldsymbol{Z}$ is a complex n-**rv** with covariance matrix $[K]$, then $[K]$ satisfies (3.101) and is positive semi-definite. It is positive definite if $[K]$ is non-singular. Also, for any complex $n \times n$ matrix $[A]$, the matrix $[AA^{\dagger}]$ is positive semi-definite and is positive definite if $[A]$ is non-singular. For any positive semi-definite $[K]$, there is a square-root matrix $[R] = [Q\sqrt{\Lambda}Q^{-1}]$ as given in (3.101) such that $\boldsymbol{Z} = [R]\boldsymbol{W}$ (where $\boldsymbol{W} \sim \mathcal{CN}(0, I)$) is circularly symmetric Gaussian with $[K_{\boldsymbol{Z}}] = [K]$.*

14 A square complex matrix with orthonormal columns is said to be *unitary*. Viewed as a transformation, Qz has the same length as z where the length of z is $\sqrt{z^{\dagger}z}$.

We have seen that the major change in going from real n-**rv** s to complex n-**rv** s is a judicious conversion of transposes into complex-conjugate transposes.

3.7.4 Linear transformations of $\boldsymbol{W} \sim \mathcal{CN}(0, [I_\ell])$

One of the best ways to understand real Gaussian n-**rv** s is to view them as linear transformations of an ℓ-**rv** (for given ℓ) with IID components, each $\mathtt{N}\ (0,1)$. The same approach turns out to work equally well for circularly-symmetric Gaussian vectors. Thus let $[A]$ be an arbitrary complex $n \times \ell$ matrix and let the complex n-**rv** $\boldsymbol{Z} = (Z_1, \ldots, Z_n)^{\mathsf{T}}$ be defined by

$$\boldsymbol{Z} = [A]\boldsymbol{W}, \qquad \text{where } \boldsymbol{W} \sim \mathtt{CN}\ (0, [I_\ell]). \tag{3.102}$$

The complex n-**rv** defined by this complex linear transformation has jointly-Gaussian real and imaginary parts. To see this, represent the complex n-dimensional transformation in (3.102) by the following $2n$-dimensional real linear transformation:

$$\begin{bmatrix} \boldsymbol{Z}_{\text{re}} \\ \boldsymbol{Z}_{\text{im}} \end{bmatrix} = \begin{bmatrix} [A_{\text{re}}] & -[A_{\text{im}}] \\ [A_{\text{im}}] & [A_{\text{re}}] \end{bmatrix} \begin{bmatrix} \boldsymbol{W}_{\text{re}} \\ \boldsymbol{W}_{\text{im}} \end{bmatrix}, \tag{3.103}$$

where $\boldsymbol{Z}_{\text{re}} = \Re(\boldsymbol{Z})$, $\boldsymbol{Z}_{\text{im}} = \Im(\boldsymbol{Z})$, $[A_{\text{re}}] = \Re([A])$, and $[A]_{\text{im}} = \Im([A])$. By definition, real linear transformations on real IID Gaussian **rv** s have jointly-Gaussian components. Thus $\boldsymbol{Z}_{\text{re}}$ and $\boldsymbol{Z}_{\text{im}}$ are jointly Gaussian and $\boldsymbol{Z}$ is a complex Gaussian n-**rv**.

The **rv** $\boldsymbol{Z}$ is also circularly symmetric.[15] To see this, note that

$$[K_{\boldsymbol{Z}}] = \mathsf{E}\Big[[A]\boldsymbol{W}\boldsymbol{W}^{\dagger}[A^{\dagger}]\Big] = [AA^{\dagger}], \qquad [M_{\boldsymbol{Z}}] = \mathsf{E}[[A]\boldsymbol{W}\boldsymbol{W}^{\mathsf{T}}[A^{\mathsf{T}}]] = 0. \tag{3.104}$$

Thus, from Theorem 3.7.6, $\boldsymbol{Z}$ is circularly symmetric Gaussian and $\boldsymbol{Z} \sim \mathtt{CN}\ (0, [AA^{\dagger}])$.

This proves the *if* part of the following theorem.

Theorem 3.7.10 *A complex* **rv** $\boldsymbol{Z}$ *is circularly symmetric Gaussian if and only if it can be expressed as* $\boldsymbol{Z} = [A]\boldsymbol{W}$ *for a complex matrix* $[A]$ *and an IID circularly-symmetric Gaussian* **rv** $\boldsymbol{W} \sim \mathtt{CN}\ (0, [I])$.

Proof For the *only if* part, choose $[A]$ to be the square-root matrix $[R]$ of Lemma 3.7.9. Then $\boldsymbol{Z} = [R]\boldsymbol{W}$ is circularly symmetric Gaussian with $[K_{\boldsymbol{Z}}] = [RR^{\dagger}]$. □

We now have three equivalent characterizations for circularly-symmetric Gaussian n-**rv** s: first, phase invariance, second, zero pseudo-covariance and, third, linear transformations of IID circularly-symmetric Gaussian vectors. One advantage of the third characterization is that the jointly-Gaussian requirement is automatically met, whereas the other two depend on that as a separate requirement. Another advantage of the third characterization is that the usual motivation for modeling **rv** s as circularly symmetric Gaussian is that they are linear transformations of essentially IID circularly-symmetric Gaussian **rv** s.

[15] Conversely, as shown later, all circularly-symmetric Gaussian **rv** s can be defined this way.

3.7.5 Linear transformations of $\boldsymbol{Z} \sim \mathcal{CN}(0, [K])$

Let $\boldsymbol{Z} \sim \text{CN}(0, [K])$. If some other **rv** $\boldsymbol{Y}$ can be expressed as $\boldsymbol{Y} = [B]\boldsymbol{Z}$, then $\boldsymbol{Y}$ is also a circularly-symmetric Gaussian **rv**. To see this, represent $\boldsymbol{Z}$ as $\boldsymbol{Z} = [A]\boldsymbol{W}$, where $\boldsymbol{W} \sim \text{CN}(0, [I])$. Then $\boldsymbol{Y} = [BA]\boldsymbol{W}$, so $\boldsymbol{Y} \sim \text{CN}(0, [BKB^{\dagger}])$. This helps show why circular symmetry is important – it is invariant to linear transformations.

If $[B]$ is 1 by n (i.e., if it is a row vector $\boldsymbol{b}^{\mathsf{T}}$) then $Y = \boldsymbol{b}^{\mathsf{T}}\boldsymbol{Z}$ is a complex rv. Thus all linear combinations of a circularly-symmetric Gaussian **rv** are circularly-symmetric Gaussian rv s.

Conversely, we now want to show that if all linear combinations of a complex **rv** $\boldsymbol{Z}$ are circularly symmetric Gaussian, then $\boldsymbol{Z}$ must also be circularly-symmetric Gaussian. The question of being Gaussian can be separated from that of circular symmetry. Thus assume that for all complex n-vectors $\boldsymbol{b}$, the complex rv $\boldsymbol{b}^{\mathsf{T}}\boldsymbol{Z}$ is complex Gaussian. It follows that $\Re(\boldsymbol{b}^{\mathsf{T}}\boldsymbol{Z}) = \boldsymbol{b}_{\text{re}}^{\mathsf{T}}\boldsymbol{Z}_{\text{re}} - \boldsymbol{b}_{\text{im}}^{\mathsf{T}}\boldsymbol{Z}_{\text{im}}$ is a real Gaussian rv for all choices of $\boldsymbol{b}_{\text{re}}$ and $\boldsymbol{b}_{\text{im}}$. Thus from Theorem 3.4.8, the real $2n$-rv $\begin{bmatrix} \boldsymbol{Z}_{\text{re}} \\ \boldsymbol{Z}_{\text{im}} \end{bmatrix}$ is a Gaussian $2n$-**rv**. By defininition, then, $\boldsymbol{Z}$ is complex Gaussian.

We could now show that $\boldsymbol{Z}$ is also circularly-symmetric Gaussian if $\boldsymbol{b}^{\mathsf{T}}\boldsymbol{Z}$ is circularly symmetric for all $\boldsymbol{b}$, but it is just as easy, and yields a slightly stronger result, to show that if $\boldsymbol{Z}$ is Gaussian and the pairwise linear combinations $Z_j + Z_k$ are circularly symmetric for all j, k, then $\boldsymbol{Z} \sim \text{CN}(0, [K_{\boldsymbol{Z}}])$. If $Z_j + Z_j$ is circularly symmetric for all j, then $\mathsf{E}\left[Z_j^2\right] = 0$, so that the main diagonal of $[M_{\boldsymbol{Z}}]$ is zero. If, in addition, $Z_j + Z_k$ is circularly symmetric, then $\mathsf{E}\left[(Z_j + Z_k)^2\right] = 0$. But since $\mathsf{E}\left[Z_j^2\right] = \mathsf{E}\left[Z_k^2\right] = 0$, we must have $2\mathsf{E}\left[Z_j Z_k\right] = 0$. Thus the j, k element of $[M_{\boldsymbol{Z}}] = 0$. Thus if $Z_j + Z_k$ is circularly symmetric for all j, k, it follows that $[M_{\boldsymbol{Z}}] = 0$ and $\boldsymbol{Z}$ is circularly symmetric.[16] Summarizing,

Theorem 3.7.11 *A complex **rv** $\boldsymbol{Z} = (Z_1, \ldots, Z_n)^{\mathsf{T}}$ is circularly-symmetric Gaussian if and only if all linear combinations of $\boldsymbol{Z}$ are complex Gaussian and $Z_j + Z_k$ is circularly symmetric for all j, k.*

3.7.6 The PDF of circularly-symmetric Gaussian n-dimensional random vectors

Since the probability density of a complex rv or **rv** is defined in terms of the real and imaginary parts of that rv or **rv**, we now pause to discuss these relationships. The major reason for using complex vector spaces and complex **rv** s is to avoid all the detail of the real and imaginary parts, but our intuition comes from $\mathbb{R}^2$ and $\mathbb{R}^3$, and the major source of confusion in treating complex **rv** s comes from assuming that $\mathbb{C}^n$ is roughly the same as $\mathbb{R}^n$. This assumption causes additional confusion when dealing with circular symmetry.

Assume that $\boldsymbol{Z} \sim \text{CN}(0, [K_{\boldsymbol{Z}}])$, and let $\boldsymbol{U} = \begin{bmatrix} \boldsymbol{Z}_{\text{re}} \\ \boldsymbol{Z}_{\text{im}} \end{bmatrix}$ be the corresponding real $2n$-**rv**. Let $[K_{\boldsymbol{U}}]$ be the covariance of the real $2n$-**rv** $\boldsymbol{U}$. From (3.100), with $[M_{\boldsymbol{Z}}] = 0$, we can express $[K_{\boldsymbol{U}}]$ as

[16] Example 3.7.8 showed that if $\boldsymbol{Z}$ is Gaussian with individually circularly-symmetric components, then $\boldsymbol{Z}$ is not necessarily circularly symmetric Gaussian. This shows that the only additional requirement is for $Z_k + Z_j$ to be circularly symmetric for all k, j.

$$[K_U] = \begin{bmatrix} \frac{1}{2}[K_{\text{re}}] & -\frac{1}{2}[K_{\text{im}}] \\ \\ \frac{1}{2}[K_{\text{im}}] & \frac{1}{2}[K_{\text{re}}] \end{bmatrix}, \tag{3.105}$$

where $[K_{\text{re}}]$ is the $n \times n$ matrix whose components are the real parts of the components of $[K_Z]$ and correspondingly $[K_{\text{im}}]$ is the matrix of imaginary parts.

Now suppose that $(\lambda, \boldsymbol{q})$ is an eigenvalue, eigenvector pair for $[K_Z]$. Separating $[K_Z]\boldsymbol{q} = \lambda\boldsymbol{q}$ into real and imaginary parts,

$$[K_{\text{re}}]\boldsymbol{q}_{\text{re}} - [K_{\text{im}}]\boldsymbol{q}_{\text{im}} = \lambda\boldsymbol{q}_{\text{re}}\ ; \qquad [K_{\text{im}}]\boldsymbol{q}_{\text{re}} + [K_{\text{re}}]\boldsymbol{q}_{\text{im}} = \lambda\boldsymbol{q}_{\text{im}}.$$

Comparing this with $[K_U]\begin{bmatrix} q_{\text{re}} \\ q_{\text{im}} \end{bmatrix}$, where $[K_U]$ is given in (3.105), we see that $\lambda/2$ is an eigenvalue of $[K_U]$ with eigenvector $\begin{bmatrix} q_{\text{re}} \\ q_{\text{im}} \end{bmatrix}$. Furthermore, assuming that $\boldsymbol{q}$ is normalized over complex n-space, $\begin{bmatrix} q_{\text{re}} \\ q_{\text{im}} \end{bmatrix}$ is normalized over real $2n$-space. As a complex n-vector, $i\boldsymbol{q}$ is a complex scalar times $\boldsymbol{q}$. It is an eigenvector of $[K_Z]$ but not independent of $\boldsymbol{q}$. The corresponding real $2n$-vector $\begin{bmatrix} -q_{im} \\ q_{\text{re}} \end{bmatrix}$, is orthonormal to $\begin{bmatrix} q_{\text{re}} \\ q_{\text{im}} \end{bmatrix}$ and is also an eigenvector of $[K_U]$. In addition, for any orthonormal complex n-vectors, the corresponding real $2n$-vectors are orthonormal. This establishes the following lemma.

Lemma 3.7.12 *Let $(\lambda_1, \boldsymbol{q}_1), \dots, (\lambda_n, \boldsymbol{q}_n)$ denote the n pairs of eigenvalues and orthonormal eigenvectors of the covariance matrix $[K_Z]$ of a circularly-symmetric n-rv $\boldsymbol{Z}$. Then the real $2n$-rv $\boldsymbol{U} = \begin{bmatrix} Z_{\text{re}} \\ Z_{\text{im}} \end{bmatrix}$ has a covariance matrix $[K_U]$ with the $2n$ eigenvalue, orthonormal eigenvector pairs*

$$\left(\frac{\lambda_1}{2}, \begin{bmatrix} q_{1,\text{re}} \\ q_{1,\text{im}} \end{bmatrix}\right), \dots, \left(\frac{\lambda_n}{2}, \begin{bmatrix} q_{n,\text{re}} \\ q_{n,\text{im}} \end{bmatrix}\right) \left(\frac{\lambda_1}{2}, \begin{bmatrix} -q_{1,\text{im}} \\ q_{1,\text{re}} \end{bmatrix}\right), \dots, \left(\frac{\lambda_n}{2}, \begin{bmatrix} -q_{n,\text{im}} \\ q_{n,\text{re}} \end{bmatrix}\right). \tag{3.106}$$

Since the determinant of a matrix is the product of the eigenvalues, we see that

$$\det[K_U] \;=\; \prod_{j=1}^{n} \left(\frac{\lambda_j}{2}\right)^2 \;=\; 2^{-2n}(\det[K_Z])^2. \tag{3.107}$$

Recall that the probability density of $\boldsymbol{Z}$ (if it exists) is the same as the probability density of $\boldsymbol{U} = \begin{bmatrix} Z_{\text{re}} \\ Z_{\text{im}} \end{bmatrix}$, i.e., it is the probability density taken over the real and imaginary components of $\boldsymbol{Z}$. This plus (3.107) makes it easy to find the probability density for $\boldsymbol{Z}$ assuming that $\boldsymbol{Z} \sim \mathrm{CN}\ (0, [K_Z])$.

Theorem 3.7.13 *Assume that $\boldsymbol{Z} \sim \mathrm{CN}\ (0, [K_Z])$ and assume that $[K_Z]$ is non-singular. Then the probability density of $\boldsymbol{Z}$ exists everywhere and is given by*

$$\mathsf{f}_Z(z) = \frac{1}{\pi^n \det[K_Z]} \exp(-z^{\dagger}[K_Z^{-1}]z). \tag{3.108}$$

Proof Since $[K_Z]$ is non-singular, its eigenvalues are all positive, so the eigenvalues of $[K_U]$ are also positive and $[K_U]$ is non-singular. Since $\boldsymbol{Z}$ is circularly symmetric

Gaussian, $\boldsymbol{U} = \begin{bmatrix} \boldsymbol{Z}_{\rm re} \\ \boldsymbol{Z}_{\rm im} \end{bmatrix}$ must be zero-mean Gaussian. Since $\boldsymbol{U}$ is a zero-mean Gaussian $2n$-**rv**, its PDF is given from (3.22) as

$$\mathsf{f}_{\boldsymbol{U}}(\boldsymbol{u}) = \prod_{j=1}^{2n} \frac{1}{\sqrt{2\pi\mu_j}} \exp(-v_j^2/(2\mu_j), \tag{3.109}$$

where μ_j is the jth eigenvalue of $[K_{\boldsymbol{U}}]$ and $v_j = \boldsymbol{a}_j^{\mathsf{T}}\boldsymbol{u}_j$, where $\boldsymbol{a}_j$ is the jth orthonormal eigenvector of $[K_{\boldsymbol{U}}]$. We have seen that the eigenvalues λ_j of $[K_{\boldsymbol{Z}}]$ are related to those of $[K_{\boldsymbol{U}}]$ by $\mu_j = \lambda_j/2$ and $\mu_{j+n} = \lambda_j/2$ for $1 \le j \le n$. Similarly the eigenvectors can be related by $\boldsymbol{a}_j^{\mathsf{T}} = (\boldsymbol{q}_{{\rm re},j}^{\mathsf{T}}, \boldsymbol{q}_{{\rm im},j}^{\mathsf{T}})$ and $\boldsymbol{a}_{j+n}^{\mathsf{T}} = (-\boldsymbol{q}_{{\rm im},j}^{\mathsf{T}}, \boldsymbol{q}_{{\rm re},j}^{\mathsf{T}})$. With a little calculation, we get

$$\begin{aligned} v_j^2 + v_{j+n}^2 &= (\boldsymbol{q}_{{\rm re},j}^{\mathsf{T}} z_{\rm re} + \boldsymbol{q}_{{\rm im},j}^{\mathsf{T}} z_{\rm im})^2 + (-\boldsymbol{q}_{{\rm im},j}^{\mathsf{T}} z_{\rm re} + \boldsymbol{q}_{{\rm re},j}^{\mathsf{T}} z_{\rm im})^2 \\ &= [\Re(\boldsymbol{q}_j^{\dagger} z)]^2 + [\Im(\boldsymbol{q}_j^{\dagger} z)]^2 \; = \; |\boldsymbol{q}_j^{\dagger} z|^2. \end{aligned}$$

Substituting this into (3.109) and recognizing that the density is now given directly in terms of $\boldsymbol{Z}$,

$$\mathsf{f}_{\boldsymbol{Z}}(z) = \prod_{j=1}^{n} \frac{1}{\pi\lambda_j} \exp(-|\boldsymbol{q}_j^{\dagger} z|^2/(\lambda_j).$$

$$= \frac{1}{\pi^n \det[K_{\boldsymbol{Z}}]} \exp\left(\sum_{j=1}^{n} -|\boldsymbol{q}_j^{\dagger} z|^2/(\lambda_j)\right). \tag{3.110}$$

Finally, recalling that $\boldsymbol{q}_j$ is the jth column of $[Q]$,

$$\sum_{j=1}^{n} |\boldsymbol{q}_j^{\dagger} z|^2/\lambda_j = z^{\dagger}[Q\Lambda^{-1}Q^{-1}]z = z^{\dagger}K_{\boldsymbol{Z}}^{-1}z.$$

Substituting this into (3.110) completes the proof. □

Note that (3.110) is also a useful expression for the density of circularly-symmetric Gaussian n-**rv** s. The geometric picture is not as easy to interpret as for real zero-mean Gaussian n-**rv** s, but the regions of equal density are still ellipsoids. In this case, however, $e^{i\theta}z$ is on the same ellipsoid for all phases θ.

The following theorem summarizes circularly-symmetric Gaussian n-**rv** s.

Theorem 3.7.14 *A complex n-**rv** $\boldsymbol{Z}$ is circularly symmetric Gaussian if and only if any one of the following conditions is satisfied.*

- *$\boldsymbol{Z}$ is a Gaussian n-**rv** and has the same distribution as $e^{i\theta}\boldsymbol{Z}$ for all real ϕ.*
- *$\boldsymbol{Z}$ is a zero-mean Gaussian n-**rv** and the pseudo-covariance matrix $[M_{\boldsymbol{Z}}]$ is zero.*
- *$\boldsymbol{Z}$ can be expressed as $\boldsymbol{Z} = [A]\boldsymbol{W}$, where $\boldsymbol{W} \sim \mathrm{CN}\ (0, [I_n])$.*
- *For non-singular $[K_{\boldsymbol{Z}}]$, the probability density of $\boldsymbol{Z}$ is given in (3.108). For singular $[K_{\boldsymbol{Z}}]$, (3.108) gives the density of $\boldsymbol{Z}$ after removal of the deterministically dependent components.*
- *All linear combinations of $\boldsymbol{Z}$ are complex Gaussian and $Z_j + Z_k$ is circularly symmetric for all j, k.*

Note that either all or none of these conditions are satisfied. The significance of the theorem is that any one of the conditions may be used to either establish the circularly-symmetric Gaussian property or to show that it does not hold. We have also seen (in Lemma 3.7.9) that if K is the covariance matrix for any complex n-**rv**, then it is also the covariance matrix of some circularly-symmetric Gaussian n-**rv**.

3.7.7 Conditional PDFs for circularly-symmetric Gaussian random vectors

It turns out that conditional PDFs for circularly-symmetric Gaussian **rv** s are virtually the same as those for real-valued **rv** s. Operationally, the only difference is that transposes must be replaced with Hermitian conjugates and the basic form for the unconditional real Gaussian PDF must be replaced with the basic form of the unconditional complex circularly-symmetric Gaussian PDF. This is not obvious without going through all the calculations used to find conditional PDFs for real **rv** s, but the calculations are virtually the same, so we will not repeat them here. We simply repeat and discuss Theorem 3.5.1, modified as above.

Theorem 3.7.15 *Let $(X_1, \ldots, X_n, Y_1, \ldots, Y_m)^{\mathsf{T}}$ be circularly symmetric and jointly Gaussian with the non-singular covariance matrix $[K]$ partitioned into $n + m$ columns and rows as*

$$[K] = \begin{bmatrix} [K_X] & [K_{X\cdot Y}] \\ [K^{\dagger}_{X\cdot Y}] & [K_Y] \end{bmatrix}, \qquad [K^{-1}] = \begin{bmatrix} [B] & [C] \\ [C^{\dagger}] & [D] \end{bmatrix}.$$

Then the joint PDF of $\boldsymbol{X} = (X_1, \ldots, X_n)^{\mathsf{T}}$ conditional on $\boldsymbol{Y} = (Y_1, \ldots, Y_m)^{\mathsf{T}}$ is given by

$$\mathsf{f}_{\boldsymbol{X}|\boldsymbol{Y}}(\boldsymbol{x}|\boldsymbol{y}) = \frac{\exp\left\{-\left(\boldsymbol{x} + [B^{-1}C]\boldsymbol{y}\right)^{\dagger}[B]\left(\boldsymbol{x} + [B^{-1}C]\boldsymbol{y}\right)\right\}}{(\pi)^n \det[B^{-1}]}. \tag{3.111}$$

For a given $\boldsymbol{Y} = \boldsymbol{y}$, the conditional density of $\boldsymbol{X}$ can be denoted as $\mathrm{CN}\ (-[B^{-1}C]\boldsymbol{y}, [B^{-1}])$. The notation $\mathrm{CN}\ (\boldsymbol{a}, [K])$ here means that the **rv** has a mean $\boldsymbol{a}$ and a fluctuation around $\boldsymbol{a}$ which is circularly-symmetric Gaussian with covariance $[K]$. Thus, in this case, $\boldsymbol{X}$, conditional on $\boldsymbol{Y} = \boldsymbol{y}$, is circularly symmetric with covariance $[B]^{-1}$ around the mean $[B^{-1}C]\boldsymbol{y}$.

The theorem can be interpreted as saying that $\boldsymbol{X}$ can be represented as $\boldsymbol{X} = [G]\boldsymbol{Y} + \boldsymbol{V}$, where $\boldsymbol{Y}$ and $\boldsymbol{V}$ are independent, circularly symmetric, and Gaussian.

The matrix $[G] = -[B^{-1}C]$ and the covariance matrix of $\boldsymbol{V}$ is $[B^{-1}]$. As in Theorem 3.5.2, the matrices $[G]$ and $[K_V]$ can be expressed directly in terms of the joint covariances of $\boldsymbol{X}$ and $\boldsymbol{Y}$ as

$$G = [K_{X\cdot Y}K_Y^{-1}], \tag{3.112}$$

$$[K_V] = [K_X] - [K_{X\cdot Y}K_Y^{-1}K^{\dagger}_{X\cdot Y}]. \tag{3.113}$$

Conversely, if $\boldsymbol{X}$ can be expressed as $\boldsymbol{X} = G\boldsymbol{Y} + \boldsymbol{V}$, where $\boldsymbol{Y}$ and $\boldsymbol{V}$ are independent and each circularly symmetric Gaussian, then it is easy to see that $(\boldsymbol{X}^{\mathsf{T}}, \boldsymbol{Y}^{\mathsf{T}})^{\mathsf{T}}$ must be circularly symmetric Gaussian. Using the resultant symmetry between $\boldsymbol{X}$ and $\boldsymbol{Y}$, we see that there must be a matrix $[H]$ and a rv $\boldsymbol{Z}$ so that $\boldsymbol{Y} = \boldsymbol{HX} + \boldsymbol{Z}$, where $\boldsymbol{X}$ and $\boldsymbol{Z}$ are independent. We will see how this is used for estimation in Section 10.5

3.7.8 Circularly-symmetric Gaussian processes

In this section, we modify Section 3.6 on continuous-time Gaussian processes to briefly outline the properties of circularly-symmetric Gaussian processes.

Definition 3.7.16 *A **circularly-symmetric Gaussian** process $\{X(t); t \in \mathbb{R}\}$ is a complex stochastic process such that for all positive integers k and all choices of epochs $t_1, \ldots, t_k \in \mathbb{R}$, the complex n-rv with components $X(t_1), \ldots, X(t_k)$ is a circularly-symmetric Gaussian n-rv.*

Now assume that $\{X(t);\ t \in \mathbb{R}\}$ is a circularly-symmetric Gaussian process. Since each n-rv $(X(t_1), \ldots, X(t_k))^{\mathsf{T}}$ is circularly symmetric, the corresponding pseudo-covariance matrix is 0 and the covariance matrix specifies the distribution of $(X(t_1), \ldots, X(t_k))^{\mathsf{T}}$. It follows then that the pseudo-covariance function, $M_X(t, \tau) = \mathsf{E}[X(t)X(\tau)] = 0$ for all t, τ and the covariance function $K_X(t, \tau) = \mathsf{E}\left[X(t)X^*(\tau)\right]$ for all t, τ specifies all finite joint distributions.

A convenient way of generating a circularly-symmetric Gaussian process is to start with a sequence of (complex) orthonormal functions $\{\phi_n(t);\ n \in \mathbb{Z}\}$ and a sequence of independent circularly-symmetric Gaussian rv s $\{X_n \sim \mathcal{CN}(0, \sigma_n^2);\ n \in \mathbb{Z}\}$. Then if $\sum_n \sigma_n^2 \phi_n^2(t) < \infty$ for all t, it follows, as in Theorem 3.6.10 for ordinary Gaussian processes, that $X(t) = \sum_n X_n \phi_n(t)$ is a circularly-symmetric Gaussian process.

One convenient such orthonormal expansion is the set of functions $\phi_n(t) = e^{i2\pi nt/T}$ for $t \in (-T/2,\ T/2)$ used in the Fourier series over that time interval. The interpretation here is very much like that in Example 3.6.12, but here the functions are complex, the rv s are circularly symmetric Gaussian, and the arithmetic is over $\mathbb{C}$.

Another particularly convenient such expansion is the sinc-function expansion of Section 3.6.4. The sinc functions are real, but the expansion is now over the complex field using circularly symmetric rv s. It is intuitively plausible in this case that the process is circularly symmetric, since the real and imaginary parts of the process are identically distributed.

A complex stochastic process $\{X(t);\ t \in \mathbb{R}\}$ can be filtered by a complex filter with impulse response $h(t)$. The output is then the complex convolution $Y(\tau) = \int X(t)h(\tau - t)\,dt$. If $X(t)$ is a circularly-symmetric Gaussian process expressed as an orthonormal expansion, then by looking at $Y(\tau)$ over say $\tau_1, \ldots, \tau_k$, we can argue as before that $\{Y(\tau); \tau \in \mathbb{R}\}$ is a circularly-symmetric process if its power is finite at all τ. When circularly-symmetric sinc processes are passed through filters, we have a broad class of circularly-symmetric processes.

The definition of stationarity is the same for complex stochastic processes as for (real) stochastic processes, but the CDF over say $X(t_1), \ldots, X(t_k)$ is now over both the real and imaginary parts of those complex rv s. If $X(t_1), \ldots, X(t_k)$ are circularly symmetric Gaussian, however, then these distributions are determined by the covariance matrices. Thus, circularly-symmetric Gaussian processes are stationary if and only if the covariance function satisfies $K_X(t, t+u)) = K_X(0, u)$.

For a stationary circularly-symmetric Gaussian process $\{X(t);\ t \in \mathbb{R}\}$, the covariance function can be expressed as a function of a single variable, $K_X(u)$. This function must be Hermitian (i.e., it must satisfy $K_X(t) = K_X^*(-t)$). The Fourier transform of a Hermitian function must be real, and by repeating the argument in Section 3.6.7, we see that this Fourier transform must be non-negative. This Fourier transform is called the spectral density of $\{X(t);\ t \in \mathbb{R}\}$.

The spectral density of a stationary circularly-symmetric Gaussian process has the same interpretation as the spectral density of a (real) stationary Gaussian process. White Gaussian noise is defined and interpreted the same way as in the real case, and can be approximated in the same way by Gaussian sinc processes.

It is important to understand that these very close analogies between real and complex Gaussian processes are actually between real and *circularly-symmetric* Gaussian processes. A complex Gaussian process that is not circularly symmetric has rather messy properties and is perhaps better thought of as a pair of processes, one real and one pure imaginary.

3.8 Summary

The sum of sufficiently many rv s that are not too dependent tends toward the Gaussian distribution, and multiple such sums tend toward a jointly-Gaussian distribution. The requirements for a set of n rv s to be jointly Gaussian are far more stringent than the requirement that each be Gaussian alone, but fortunately, as above, the conditions that lead individual rv s to be Gaussian often also lead multiple rv s to be jointly Gaussian. Theorem 3.4.9 collects four sets of necessary and sufficient conditions for zero-mean rv s to be jointly Gaussian. Non-zero-mean rv s are jointly Gaussian if their fluctuations are jointly Gaussian. Finally, a **rv** is defined to be Gaussian if its components are jointly Gaussian.

The distribution of a Gaussian vector $\boldsymbol{Z}$ is completely specified by its mean $\overline{\boldsymbol{Z}}$ and covariance matrix $[K_{\boldsymbol{Z}}]$. The distribution is denoted as $\mathrm{N}\,(\overline{\boldsymbol{Z}}, [K_{\boldsymbol{Z}}])$.

If $X_1, X_2, \ldots, X_n, Y_1, \ldots, Y_m$ are zero mean and jointly Gaussian with a non-singular covariance matrix, then the conditional density $\mathsf{f}_{\boldsymbol{X}|\boldsymbol{Y}}(\boldsymbol{x} \mid \boldsymbol{y})$ is jointly Gaussian for each $\boldsymbol{y}$. The covariance of this conditional distribution is $[K_{\boldsymbol{X}}] - [K_{\boldsymbol{X}\cdot\boldsymbol{Y}} K_{\boldsymbol{Y}}^{-1} K_{\boldsymbol{X}\cdot\boldsymbol{Y}}^{\mathsf{T}}]$, which does not depend on the particular sample value $\boldsymbol{y}$. The conditional mean, $[K_{\boldsymbol{X}\cdot\boldsymbol{Y}} K_{\boldsymbol{Y}}^{-1}]\boldsymbol{y}$, depends linearly on $\boldsymbol{y}$. This situation can be equivalently formulated as $\boldsymbol{X} = [G]\boldsymbol{Y} + \boldsymbol{V}$, where $\boldsymbol{V}$ is a zero-mean Gaussian n-**rv** independent of $\boldsymbol{Y}$. Using the symmetry between the roles of $\boldsymbol{X}$ and $\boldsymbol{Y}$, we have $\boldsymbol{Y} = [H]\boldsymbol{X} + \boldsymbol{Z}$, where $\boldsymbol{X}$ and $\boldsymbol{Z}$ are independent.

A stochastic process $\{X(t);\ t \in \mathrm{T}\}$ is a Gaussian process if, for all finite sets $t_1, \ldots, t_k$ of epochs, the rv s $X(t_1), \ldots, X(t_k)$ are jointly Gaussian. If T is $\mathbb{R}$ or $\mathbb{R}^+$, then Gaussian processes can be easily generated as orthonormal expansions. When a Gaussian orthonormal expansion is used as the input to a linear filter, the output is essentially also a Gaussian process.

A stochastic process is stationary if all finite joint distributions are invariant to time shifts. It is WSS if the covariance function and mean are invariant to time shifts. A zero-mean Gaussian process is stationary if the covariance function is invariant to time shifts, i.e., if $K_X(t, t+u) = K_X(0, u)$. Thus a stationary zero-mean Gaussian process is determined by its single-variable covariance function $K_X(u)$. A stationary zero-mean Gaussian process is also essentially determined by its spectral density, which is the Fourier transform of $K_X(u)$.

The spectral density $S_X(f)$ is interpreted as the process's power per unit frequency at frequency f. If a stationary zero-mean Gaussian process has a positive spectral density equal to a constant value, say $N_0/2$, over all frequencies of interest, it is called white Gaussian noise. Approximating $S_X(f)$ as constant over all f is often convenient, but implies infinite aggregate power, which is both mathematically and physically absurd. The Gaussian sinc process also models white Gaussian noise over an arbitrarily broad but finite band of frequencies.

Circularly-symmetric rv s are complex rv s for which the distribution over the real and imaginary plane is circularly symmetric. A **rv** $\boldsymbol{Z}$ is circularly symmetric if $\boldsymbol{Z}$ and $e^{i\theta}\boldsymbol{Z}$ have the same distribution for all phases ϕ. Theorem 3.7.14 collects five sets of necessary and sufficient conditions for complex **rv** s to be circularly symmetric Gaussian.

Vectors and processes of circularly-symmetric Gaussian rv s have many analogies with ordinary Gaussian **rv** s and processes, and many of the equations governing the real case can be converted to the circularly-symmetric case simply by replacing transposes by Hermitian transposes. This is true for conditional distributions also. Unfortunately, this extreme simplicity relating the equations sometimes hides more fundamental differences. Complex Gaussian **rv** s and processes that are not circularly symmetric are usually best modeled as separate real and imaginary parts, since almost all of the insights that we might try to transfer from the real to the complex case fail except when circular symmetry is present.

3.9 Exercises

Exercise 3.1 **(a)** Let X, Y be IID rv s, each with density $\mathsf{f}_X(x) = \alpha \exp(-x^2/2)$. In (b), we show that α must be $1/\sqrt{2\pi}$ in order for $\mathsf{f}_X(x)$ to integrate to 1, but in this part, we leave α undetermined. Let $S = X^2 + Y^2$. Find the probability density of S in terms of α.

(b) Prove from (a) that α must be $1/\sqrt{2\pi}$ in order for S, and thus X and Y, to be rv s. Show that $\mathsf{E}[X] = 0$ and that $\mathsf{E}\left[X^2\right] = 1$.

(c) Find the probability density of $R = \sqrt{S}$. R is called a *Rayleigh* rv.

Exercise 3.2 **(a)** By expanding in a power series in $(1/2)r^2\sigma^2$, show that

$$\exp\left(\frac{r^2\sigma^2}{2}\right) = 1 + \frac{r^2\sigma^2}{2} + \frac{r^4\sigma^4}{2(2^2)} + \cdots + \frac{r^{2k}\sigma^{2k}}{k!2^k} + \cdots .$$

(b) By expanding e^{rZ} in a power series in rZ, show that

$$\mathsf{g}_Z(r) = \mathsf{E}\left[e^{rZ}\right] = 1 + r\mathsf{E}[Z] + \frac{r^2\mathsf{E}\left[Z^2\right]}{2} + \cdots + \frac{r^k\mathsf{E}\left[Z^k\right]}{(k)!} + \cdots.$$

(c) By matching powers of r between (a) and (b), show that for all integer $k \geq 1$,

$$\mathsf{E}\left[Z^{2k}\right] = \frac{(2k)!\sigma^{2k}}{k!2^k} = (2k-1)(2k-3)\cdots(3)(1)\sigma^{2k} \quad ; \qquad \mathsf{E}\left[Z^{2k+1}\right] = 0.$$

Exercise 3.3 Let X and Z be IID normalized Gaussian rv s. Let $Y = |Z|\,\mathrm{Sgn}(X)$, where $\mathrm{Sgn}(X)$ is 1 if $X \geq 0$ and -1 otherwise. Show that X and Y are each Gaussian, but are not jointly Gaussian. Sketch the contours of equal joint probability density.

Exercise 3.4 **(a)** Let $X_1 \sim \mathrm{N}(0, \sigma_1^2)$ and let $X_2 \sim \mathrm{N}(0, \sigma_2^2)$ be independent of X_1. Convolve the density of X_1 with that of X_2 to show that $X_1 + X_2$ is Gaussian, $\mathrm{N}(0, \sigma_1^2 + \sigma_2^2)$.

(b) Let W_1, W_2 be IID normalized Gaussian rv s. Show that $a_1W_1 + a_2W_2$ is Gaussian, $\mathrm{N}(0, a_1^2 + a_2^2)$. Hint: You could repeat all the equations of (a), but the insightful approach is to let $X_i = a_iW_i$ for $i = 1, 2$ and then use (a) directly.

(c) Combine (b) with induction to show that all linear combinations of IID normalized Gaussian rv s are Gaussian.

Exercise 3.5 **(a)** Let $\boldsymbol{U}$ be an n-**rv** with mean $\boldsymbol{m}$ and covariance $[K]$ whose MGF is given by (3.18). Let $X = \boldsymbol{r}^{\mathsf{T}}\boldsymbol{U}$ for an arbitrary real n-vector $\boldsymbol{r}$. Show that the MGF of X is given by $\mathsf{g}_X(r) = \exp\left[r\mathsf{E}[X] + r^2\sigma_X^2/2\right]$ and relate $\mathsf{E}[X]$ and σ_X^2 to $\boldsymbol{m}$ and $[K]$.

(b) Show that $\boldsymbol{U}$ is a Gaussian **rv**.

Exercise 3.6 **(a)** Let $\boldsymbol{Z} \sim \mathrm{N}(0, [K])$ be n-dimensional. By expanding in a power series in $(1/2)\boldsymbol{r}^{\mathsf{T}}[K]\,\boldsymbol{r}$, show that

$$\mathsf{g}_{\boldsymbol{Z}}(\boldsymbol{r}) = \exp\left[\frac{\boldsymbol{r}^{\mathsf{T}}[K]\boldsymbol{r}}{2}\right] = 1 + \frac{\sum_{j,k} r_j r_k K_{j,k}}{2} + \cdots + \frac{\left(\sum_{j,k} r_j r_k K_{j,k}\right)^m}{2^m m!} + \cdots.$$

(b) By expanding $e^{r_jZ_j}$ in a power series in r_jZ_j for each j, show that

$$\mathsf{g}_{\boldsymbol{Z}}(\boldsymbol{r}) = \mathsf{E}\left[\exp\left(\sum_j r_jZ_j\right)\right] = \sum_{l_1=0}^{\infty}\cdots\sum_{l_n=0}^{\infty}\frac{r_1^{l_1}}{(l_1)!}\cdots\frac{r_n^{l_n}}{(l_n)!}\mathsf{E}\left[Z_1^{l_1}\ldots Z_n^{l_n}\right].$$

(c) Let $D = \{i_1, i_2, \ldots, i_{2m}\}$ be a set of $2m$ distinct integers each between 1 and n and let $j_1, \ldots, j_{2m}$ be a permutation of D. Consider the term $r_{j_1}r_{j_2}\cdots r_{j_{2m}}\mathsf{E}\left[Z_{j_1}Z_{j_2}\cdots Z_{j_{2m}}\right]$ in (b) and let $(j_1, \ldots j_n)$ be a permutation of D. By comparing with the set of terms in (a) containing the same product $r_{j_1}r_{j_2}\cdots r_{j_{2m}}$, show that

$$\mathsf{E}\left[Z_{j_1}Z_{j_2}\cdots Z_{j_{2m}}\right] = \frac{\sum_{j_1j_2\ldots j_{2m}} K_{j_1j_2}K_{j_3j_4}\cdots K_{j_{2m-1}j_{2m}}}{2^m m!},$$

where the sum is over all permutations $(j_1, j_2, \ldots, j_{2m})$ of the set D.

(d) Find the number of permutations of D that contain the same set of unordered pairs $(\{j_1, j_2\}, \ldots, \{j_{2m-1}, j_{2m}\})$. For example, $(\{1, 2\}, \{3, 4\})$ is the same set of unordered pairs as $(\{3, 4\}, \{2, 1\})$. Show that

$$\mathsf{E}\left[Z_{j_1} Z_{j_2} \cdots Z_{j_{2m}}\right] = \sum_{j_1, j_2, \ldots, j_{2m}} K_{j_1 j_2} K_{j_3 j_4} \cdots K_{j_{2m-1} j_{2m}}, \tag{3.114}$$

where the sum is over distinct sets of unordered pairs of the set D. Note: Another way to say the same thing is that the sum is over the set of all permutations of D for which $j_{2k-1} < j_{2k}$ for $1 \le k \le m$ and $j_{2k-1} < j_{2k+1}$ for $1 \le k \le m-1$.

(e) To find $\mathsf{E}\left[Z_1^{j_1} \cdots Z_n^{j_n}\right]$, where $j_1 + j_2 + \cdots + j_n = 2m$, construct the rv s $U_1, \ldots, U_{2m}$, where $U_1, \ldots, U_{j_1}$ are all identically equal to Z_1, where $U_{j_1+1}, \ldots, U_{j_1+j_2}$ are identically equal to Z_2, etc., and use (3.114) to find $\mathsf{E}[U_1 U_2 \cdots U_{2m}]$. Use this formula to find $\mathsf{E}\left[Z_1^2 Z_2 Z_3\right]$, $\mathsf{E}\left[Z_1^2 Z_2^2\right]$, and $\mathsf{E}\left[Z_1\right]^4$.

Exercise 3.7 Let $[Q]$ be an orthonormal matrix. Show that the squared distance between any two vectors $\boldsymbol{z}$ and $\boldsymbol{y}$ is equal to the squared distance between $[Q]\boldsymbol{z}$ and $[Q]\boldsymbol{y}$.

Exercise 3.8 **(a)** Let $[K] = \begin{bmatrix} 0.75 & 0.25 \\ 0.25 & 0.75 \end{bmatrix}$. Show that 1 and 1/2 are eigenvalues of $[K]$ and find the normalized eigenvectors. Express $[K]$ as $[Q \Lambda Q^{-1}]$, where $[\Lambda]$ is diagonal and $[Q]$ is orthonormal.

(b) Let $[K'] = \alpha[K]$ for real $\alpha \neq 0$. Find the eigenvalues and eigenvectors of $[K']$. Do not use brute force – think!

(c) Find the eigenvalues and eigenvectors of $[K^m]$, where $[K^m]$ is the mth power of $[K]$.

Exercise 3.9 Let X and Y be jointly Gaussian with means m_X, m_Y, variances σ_X^2, σ_Y^2, and normalized covariance ρ. Find the conditional density $\mathsf{f}_{X|Y}(x \mid y)$.

Exercise 3.10 **(a)** Let X and Y be zero-mean jointly Gaussian with variances σ_X^2, σ_Y^2, and normalized covariance ρ. Let $V = Y^3$. Find the conditional density $\mathsf{f}_{X|V}(x \mid v)$. Hint: This requires no computation.

(b) Let $U = Y^2$ and find the conditional density of $\mathsf{f}_{X|U}(x \mid u)$. Hint: First understand why this is harder than (a).

Exercise 3.11 **(a)** Let $(\boldsymbol{X}^{\mathsf{T}}, \boldsymbol{Y}^{\mathsf{T}})$ have a non-singular covariance matrix $[K]$. Show that $[K_{\boldsymbol{X}}]$ and $[K_{\boldsymbol{Y}}]$ are positive definite, and thus non-singular.

(b) Show that the matrices $[B]$ and $[D]$ in (3.39) are also positive definite and thus non-singular.

Exercise 3.12 Let $\boldsymbol{X}$ and $\boldsymbol{Y}$ be jointly-Gaussian **rv** s with means $\boldsymbol{m}_{\boldsymbol{X}}$ and $\boldsymbol{m}_{\boldsymbol{Y}}$, covariance matrices $[K_{\boldsymbol{X}}]$ and $[K_{\boldsymbol{Y}}]$ and cross covariance matrix $[K_{\boldsymbol{X}\cdot\boldsymbol{Y}}]$. Find the conditional probability density $\mathsf{f}_{\boldsymbol{X}|\boldsymbol{Y}}(\boldsymbol{x} \mid \boldsymbol{y})$. Assume that the covariance of $(\boldsymbol{X}^{\mathsf{T}}, \boldsymbol{Y}^{\mathsf{T}})$ is non-singular. Hint: Think of the fluctuations of $\boldsymbol{X}$ and $\boldsymbol{Y}$.

Exercise 3.13 **(a)** Let $\boldsymbol{W}$ be a normalized IID Gaussian n-**rv** and let $\boldsymbol{Y}$ be a Gaussian m-**rv**. Suppose we would like the joint covariance $\mathsf{E}\left[\boldsymbol{W}\boldsymbol{Y}^{\mathsf{T}}\right]$ to be some arbitrary real-valued $n \times m$ matrix $[K]$. Find the matrix $[A]$ such that $\boldsymbol{Y} = [A]\boldsymbol{W}$ achieves the desired joint covariance. Note: This shows that any real-valued $n \times m$ matrix is the joint covariance matrix for some choice of random **rv** s.

(b) Let $\boldsymbol{Z}$ be a zero-mean Gaussian n-**rv** with non-singular covariance $[K_{\boldsymbol{Z}}]$, and let $\boldsymbol{Y}$ be a Gaussian m-**rv**. Suppose we would like the joint covariance $\mathsf{E}\left[\boldsymbol{Z}\boldsymbol{Y}^{\mathsf{T}}\right]$ to be some arbitrary $n \times m$ matrix $[K']$. Find the matrix $[B]$ such that $\boldsymbol{Y} = [B]\boldsymbol{Z}$ achieves the desired joint covariance. Note: This shows that any real-valued $n \times m$ matrix is the joint covariance matrix for some choice of **rv** s $\boldsymbol{Z}$ and $\boldsymbol{Y}$ where $[K_{\boldsymbol{Z}}]$ is given (and non-singular).

(c) Now assume that $\boldsymbol{Z}$ has a singular covariance matrix in (b). Explain the constraints this places on possible choices for the joint covariance $\mathsf{E}\left[\boldsymbol{Z}\boldsymbol{Y}^{\mathsf{T}}\right]$. Hint: Your solution should involve the eigenvectors of $[K_{\boldsymbol{Z}}]$.

Exercise 3.14 **(a)** Let $\boldsymbol{W} = (W_1, W_2, \ldots, W_{2n})^{\mathsf{T}}$ be a $2n$-dimensional IID normalized Gaussian **rv**. Let $S_{2n} = W_1^2 + W_2^2 + \cdots + W_{2n}^2$. Show that S_{2n} is an nth-order Erlang rv with parameter $1/2$, i.e., that $\mathsf{f}_{S_{2n}}(s) = 2^{-n}s^{n-1}e^{-s/2}/(n-1)!$. Hint: Look at S_2 from Exercise 3.1.

(b) Let $R_{2n} = \sqrt{S_{2n}}$. Find the probability density of R_{2n}.

(c) Let $v_{2n}(r)$ be the volume of a $2n$-dimensional sphere of radius r and let $b_{2n}(r)$ be the surface area of that sphere, i.e., $b_{2n}(r) = dv_{2n}(r)/dr$. The point of this exercise is to show how to calculate these quantities. By considering an infinitesimally thin spherical shell of thickness δ at radius r, show that

$$\mathsf{f}_{R_{2n}}(r) = b_{2n}(r)\mathsf{f}_{\boldsymbol{W}}(\boldsymbol{w})\,|_{\boldsymbol{W}:\boldsymbol{W}^{\mathsf{T}}\boldsymbol{W}=r^2}\,.$$

(d) Calculate $b_{2n}(r)$ and $v_{2n}(r)$. Note that for any fixed $\delta \ll r$, the limiting ratio as $n \to \infty$ of that of a sphere of radius r to that of a spherical shell of width δ and outer radius r is 1.

Exercise 3.15 **(a)** Solve directly for $[B]$, $[C]$, and $[D]$ in (3.39) for the one-dimensional case where $n = m = 1$. Show that (3.40) agrees with (3.37).

Exercise 3.16 **(a)** Express $[B]$, $[C]$, and $[D]$, as defined in (3.39), in terms of $[K_{\boldsymbol{X}}]$, $[K_{\boldsymbol{Y}}]$ and $[K_{\boldsymbol{X}\cdot\boldsymbol{Y}}]$ by multiplying the block expression for $[K]$ by that for $[K]^{-1}$. You can check your solutions against those in (3.46)–(3.48). Hint: You can solve for $[B]$ and $[C]$ by looking at only two of the four block equations in $[KK^{-1}]$. You can use the symmetry between $\boldsymbol{X}$ and $\boldsymbol{Y}$ to solve for $[D]$.

(b) Use your result in (a) for $[C]$ plus the symmetry between $\boldsymbol{X}$ and $\boldsymbol{Y}$ to show that

$$[BK_{\boldsymbol{X}\cdot\boldsymbol{Y}}K_{\boldsymbol{Y}}^{-1}] = [K_{\boldsymbol{X}}^{-1}K_{\boldsymbol{X}\cdot\boldsymbol{Y}}D].$$

(c) For the formulations $\boldsymbol{X} = [G]\boldsymbol{Y} + \boldsymbol{V}$ and $\boldsymbol{Y} = [H]\boldsymbol{X} + \boldsymbol{Z}$, where $\boldsymbol{X}$ and $\boldsymbol{Y}$ are zero-mean, jointly Gaussian and have a non-singular combined covariance matrix, show that

$$[K_{\boldsymbol{V}}^{-1}G] = [H^{\mathsf{T}}K_{\boldsymbol{Z}}^{-1}]. \tag{3.115}$$

Hint: This is almost trivial from (b), (3.43), (3.44), and the symmetry.

Exercise 3.17 Let $\boldsymbol{X}$ and $\boldsymbol{Z}$ be statistically independent Gaussian **rv** s of arbitrary dimension n and m respectively. Let $\boldsymbol{Y} = [H]\boldsymbol{X} + \boldsymbol{Z}$, where $[H]$ is an arbitrary real $n \times m$ matrix.

(a) Explain why $X_1, \ldots, X_n, Z_1, \ldots, Z_m$ must be jointly Gaussian rv s. Then explain why $X_1, \ldots, X_n, Y_1, \ldots, Y_m$ must be jointly Gaussian.

(b) Show that if $[K_{\boldsymbol{X}}]$ and $[K_{\boldsymbol{Z}}]$ are non-singular, then the combined covariance matrix $[K]$ for $(X_1, \ldots, X_n, Y_1, \ldots, Y_m)^{\mathsf{T}}$ must be non-singular.

Exercise 3.18 **(a)** Verify (3.56) for $k = 0$ by the use of induction on (3.54).

(b) Verify (3.56) for $k = 0$ directly from (3.55).

(c) Verify (3.56) for $k > 0$ by using induction on k.

(d) Verify that

$$\lim_{n\to\infty} \mathsf{E}[X(n)X(n+k)] = \frac{\sigma^2\alpha^k}{1-\alpha^2}.$$

Exercise 3.19 Let $\{X(t);\ t \in \Re\}$ be defined by $X(t) = tA$ for all $t \in \Re$ where $A \sim \mathrm{N}(0,1)$. Show that this is a Gaussian process. Find its mean for each t and find its covariance function. Note: The purpose of this exercise is to show that Gaussian processes can be very degenerate and trivial.

Exercise 3.20 Let $\{X(t);\ t \geq 0\}$ be a stochastic process with independent and stationary increments and let $X(0)$ be an arbitrary rv. Show that $\mathsf{E}[X(t)] = \mathsf{E}[X(0)] + t\,\mathsf{E}[X(1) - X(0)]$ and that

$$K_X(t,\tau) = \mathsf{VAR}[X(0)] + t\,[\mathsf{VAR}[X(1)] - \mathsf{VAR}[X(0)]].$$

Exercise 3.21 **(a)** Let $X(t) = R\cos(2\pi ft + \theta)$ where R is a Rayleigh rv and the rv θ is independent of R and uniformly distributed over the interval 0 to 2π. Show that $\mathsf{E}[X(t)] = 0$.

(b) Show that $\mathsf{E}[X(t)X(t+\tau)] = \frac{1}{2}\mathsf{E}\left[R^2\right]\cos(2\pi f\tau)$.

(c) Show that $X(t);\, t\in\Re$ is a Gaussian process.

Exercise 3.22 Let $h(t)$ be a real square-integrable function whose Fourier transform is 0 for $|f| > B$ for some $B > 0$. Show that $\sum_n h^2(t - n/2B) = (1/2B)\int h^2(\tau)\,d\tau$ for all $t \in \Re$. Hint: Find the sampling theorem expansion for a time shifted sinc function.

Exercise 3.23 **(a)** Let $\mathbf{Z} = (Z_1, \ldots, Z_n)^{\mathsf{T}}$ be a circularly-symmetric n-**rv**. Show that Z_k is circularly symmetric for each k, $1 \leq k \leq n$. Hint: Use the definition directly (you cannot assume that $\mathbf{Z}$ is also Gaussian).

(b) Show that $Z_1 + Z_2$ is circularly symmetric. For any complex n-vector $\boldsymbol{c}$, show that $\boldsymbol{c}^{\mathsf{T}}\mathbf{Z}$ is a circularly symmetric rv.

Exercise 3.24 Let A be a complex Gaussian rv, i.e., $A = A_1 + iA_2$ where A_1 and A_2 are zero-mean jointly-Gaussian real rv s with variances σ_1^2 and σ_2^2 respectively.

(a) Show that $\mathsf{E}\left[AA^*\right] = \sigma_1^2 + \sigma_2^2$.

(b) Show that

$$\mathsf{E}\left[(AA^*)^2\right] = 3\sigma_1^4 + 3\sigma_2^4 + 2\sigma_1^2\sigma_2^2 + 4\,(\mathsf{E}[A_1A_2])^2.$$

(c) Show that $\mathsf{E}\left[(AA^*)^2\right] \geq 2\left(\mathsf{E}\left[AA^*\right]\right)^2$ with equality if and only if A_1 and A_2 are IID. Hint: Lower bound $(\mathsf{E}[A_1A_2])^2$ by 0.

(d) Show that $\mathsf{VAR}\left[AA^*\right] \geq \left(\mathsf{E}\left[AA^*\right]\right)^2$.

Exercise 3.25 Let $K_X(t)$ be the covariance function of a WSS process $\{X(t);\ t \in \Re\}$. Show that if $K_X(t)$ is continuous at $t = 0$, then it is continuous everywhere. Hint: You must show that $\lim_{\delta\to 0}\mathsf{E}[X(0)(X(t+\delta) - X(t))] = 0$ for all t. Use the Schwarz inequality.

4 Finite-state Markov chains

4.1 Introduction

The counting processes $\{N(t);\ t > 0\}$ described in Section 2.1.1 have the property that $N(t)$ *changes* at discrete instants of time, but is *defined* for all real $t > 0$. The Markov chains to be discussed in this chapter are stochastic processes *defined* only at integer values of time, $n = 0, 1, \ldots$. At each integer time $n \geq 0$, there is an integer-valued random variable (rv) X_n, called the *state* at time n, and the process is the family of rv s $\{X_n;\ n \geq 0\}$. We refer to these processes as *integer-time processes*. An integer-time process $\{X_n;\ n \geq 0\}$ can also be viewed as a process $\{X(t);\ t \geq 0\}$ defined for all real t by taking $X(t) = X_n$ for $n \leq t < n+1$, but since changes occur only at integer times, it is usually simpler to view the process only at those integer times.

In general, for Markov chains, the set of possible values for each rv X_n is a countable set S. If S is countably infinite, it is usually taken to be $\mathsf{S} = \{0, 1, 2, \ldots\}$, whereas if S is finite, it is usually taken to be $\mathsf{S} = \{1, \ldots, \mathsf{M}\}$. In this chapter (except for Theorems 4.2.8 and 4.2.9), we restrict attention to the case in which S is finite, i.e., processes whose sample functions are sequences of integers, each between 1 and M. There is no special significance to using integer labels for states, and no compelling reason to include 0 for the countably infinite case and not for the finite case. For the countably infinite case, the most common applications come from queueing theory, where the state often represents the number of waiting customers, which might be zero. For the finite case, we often use vectors and matrices, where positive integer labels simplify the notation. In some examples, it will be more convenient to use more illustrative labels for states.

Definition 4.1.1 *A* ***Markov chain*** *is an integer-time process,* $\{X_n, n \geq 0\}$, *for which the sample values for each rv* X_n, $n \geq 1$, *lie in a countable set* S *and depend on the past only through the most recent rv* X_{n-1}. *More specifically, for all positive integers n, and for all choices of* $i, j, k, \ldots, \ell$ *in* S,

$$\Pr\{X_n{=}j \mid X_{n-1}{=}i, X_{n-2}{=}k, \ldots, X_0{=}\ell\} = \Pr\{X_n{=}j \mid X_{n-1}{=}i\} \tag{4.1}$$

for all conditioning events $X_{n-1} = i, X_{n-2} = k, \ldots, X_0 = \ell$ *of positive probability. Furthermore,* $\Pr\{X_n{=}j \mid X_{n-1}{=}i\}$ *depends only on i and j (not n) and is denoted by*

$$\Pr\{X_n{=}j \mid X_{n-1}{=}i\} = P_{ij}. \tag{4.2}$$

The initial state X_0 has an arbitrary probability distribution. A finite-state Markov chain is a Markov chain in which S is finite.

Equations such as (4.1) are often easier to read if they are abbreviated as

$$\Pr\{X_n \mid X_{n-1}, X_{n-2}, \dots, X_0\} = \Pr\{X_n \mid X_{n-1}\}.$$

This abbreviation means that equality holds for all sample values of each of the rv s. i.e., it means the same thing as (4.1).

The rv X_n is called the state of the chain at time n. The possible values for the state at time n, namely $\{1, \dots, \mathsf{M}\}$ or $\{0, 1, \dots\}$ are also generally called states, usually without too much confusion. Thus P_{ij} is the probability of going to state j given that the previous state is i; the new state, given the previous state, is independent of all earlier states. The use of the word *state* here conforms to the usual idea of the state of a system – the state at a given time summarizes everything about the past that is relevant to the future.

Definition 4.1.1 is used by some people as the definition of a *homogeneous Markov chain*. For them, Markov chains include more general cases where the transition probabilities can vary with n. Thus they replace (4.1) and (4.2) by

$$\Pr\{X_n{=}j \mid X_{n-1}{=}i, X_{n-2}{=}k, \dots, X_0{=}\ell\} = \Pr\{X_n{=}j \mid X_{n-1}{=}i\} = P_{ij}(n). \tag{4.3}$$

We will call a process that obeys (4.3), with a dependence on n, a *non-homogeneous Markov chain*. We will discuss only the homogeneous case, with no dependence on n, and thus restrict the definition to that case. Not much of general interest can be said about non-homogeneous chains.[1]

An initial probability distribution for X_0 combined with the transition probabilities $\{P_{ij}\}$ (or $\{P_{ij}(n)\}$ for the non-homogeneous case) define the probabilities for all events in the Markov chain.

Markov chains can be used to model an enormous variety of physical phenomena and can be used to approximate many other kinds of stochastic processes such as the following example.

Example 4.1.2 Consider an integer-time process $\{Z_n;\ n \geq 0\}$, where the Z_n are finite integer-valued rv s as in a Markov chain, but each Z_n depends probabilistically on the previous m rv s, $Z_{n-1}, Z_{n-2}, \dots, Z_{n-m}$. In other words, using abbreviated notation,

$$\Pr\{Z_n \mid Z_{n-1}, Z_{n-2}, \dots, Z_0\} = \Pr\{Z_n \mid Z_{n-1}, \dots, Z_{n-m}\}. \tag{4.4}$$

We now show how to view the condition on the right-hand side of (4.4), i.e., $(Z_{n-1}, Z_{n-2}, \dots, Z_{n-m})$ as the state of the process at time $n-1$. We can rewrite (4.4) as

$$\Pr\{Z_n, Z_{n-1}, \dots, Z_{n-m+1} \mid Z_{n-1}, \dots, Z_0\} = \Pr\{Z_n, \dots, Z_{n-m+1} \mid Z_{n-1}, \dots, Z_{n-m}\},$$

[1] On the other hand, we frequently find situations where a small set of rv s, say W, X, Y, Z, satisfy the *Markov property* that $\Pr\{Z \mid Y, X, W\} = \Pr\{Z \mid Y\}$ and $\Pr\{Y \mid X, W\} = \Pr\{Y \mid X\}$ but where the conditional probability $\Pr\{Z{=}j \mid Y{=}i\}$ is not necessarily the same as $\Pr\{Y{=}j \mid X{=}i\}$. In other words, *Markov chains* imply homogeneity here, whereas the *Markov property* does not.

since, for each side of the equation, any given set of values for $Z_{n-1}, \dots, Z_{n-m+1}$ on the right-hand side of the conditioning sign specifies those values on the left-hand side. Thus if we define $X_{n-1} = (Z_{n-1}, \dots, Z_{n-m})$ for each n, this simplifies to

$$\Pr\{X_n \mid X_{n-1}, \dots, X_{m-1}\} = \Pr\{X_n \mid X_{n-1}\}.$$

We see that by expanding the state space to include m-tuples of the rv s Z_n, we have converted the m dependence over time to a unit dependence over time, i.e., a Markov chain is defined using the expanded state space.

Note that in this new Markov chain, the initial state is $X_{m-1} = (Z_{m-1}, \dots, Z_0)$, so one might want to shift the time axis to start with X_0.

Markov chains are often described by a directed graph (see Figure 4.1(a)). In this graphical representation, there is one node for each state and a directed arc for each non-zero transition probability. If $P_{ij} = 0$, then the arc from node i to node j is omitted, so the difference between zero and non-zero transition probabilities stands out clearly in the graph. The classification of states, as discussed in Section 4.2, is determined by the set of transitions with non-zero probabilities, and thus the graphical representation is ideal for that topic.

A finite-state Markov chain is also often described by a matrix $[P]$ (see Figure 4.1(b)). If the chain has M states, then $[P]$ is an $\mathsf{M} \times \mathsf{M}$ matrix with elements P_{ij}. The matrix representation is ideally suited for studying algebraic and computational issues.

4.2 Classification of states

This section, except where indicated otherwise, applies to Markov chains with both finite and countable state spaces. We start with several definitions.

Definition 4.2.1 *An (n-step)* ***walk*** *is an ordered string of nodes,* $(i_0, i_1, \dots, i_n)$, $n \geq 1$, *in which there is a directed arc from* i_{m-1} *to* i_m *for each* m, $1 \leq m \leq n$. *A* ***path*** *is a walk in which no nodes are repeated. A* ***cycle*** *is a walk in which the first and last nodes are the same and no other node is repeated.*

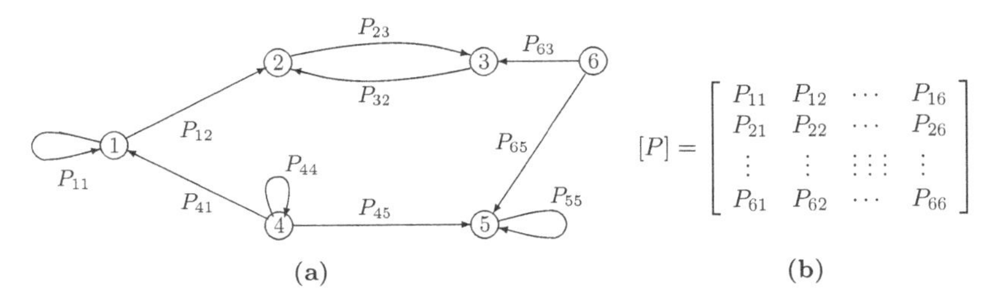

Figure 4.1 (a) Graphical and (b) matrix representations of a six-state Markov chain; a directed arc from i to j is included in the graph if and only if $P_{ij} > 0$.

Note that a walk can start and end on the same node, whereas a path cannot. Also the number of steps in a walk can be arbitrarily large, whereas a path can have at most $\mathsf{M}-1$ steps and a cycle at most M steps for a finite-state Markov chain with $|\mathsf{S}| = \mathsf{M}$.

Definition 4.2.2 *A state j is **accessible** from i (abbreviated as $i \to j$) if there is a walk in the graph from i to j.*

For example, in Figure 4.1(a), there is a walk from node 1 to node 3 (passing through node 2), so state 3 is accessible from state 1. There is no walk from node 5 to node 3, so state 3 is not accessible from state 5. State 2 is accessible from itself, but state 6 is not accessible from itself. To see the probabilistic meaning of accessibility, suppose that a walk $i_0, i_1, \ldots, i_n$ exists from node i_0 to node i_n. Then, conditional on $X_0 = i_0$, there is a positive probability, $P_{i_0 i_1}$, that $X_1 = i_1$, and consequently (since $P_{i_1 i_2} > 0$), there is a positive probability that $X_2 = i_2$. Continuing this argument, there is a positive probability that $X_n = i_n$, so that $\Pr\{X_n{=}i_n \mid X_0{=}i_0\} > 0$. Similarly, if $\Pr\{X_n{=}i_n \mid X_0{=}i_0\} > 0$, then an n-step walk from i_0 to i_n must exist. Summarizing, $i \to j$ if and only if $\Pr\{X_n{=}j \mid X_0{=}i\} > 0$ for some $n \geq 1$. We denote $\Pr\{X_n{=}j \mid X_0{=}i\}$ by P_{ij}^n. Thus, for $n \geq 1$, $P_{ij}^n > 0$ if and only if the graph has an n step walk from i to j (perhaps visiting the same node more than once). For the example in Figure 4.1(a), $P_{13}^2 = P_{12}P_{23} > 0$. On the other hand, $P_{53}^n = 0$ for all $n \geq 1$. An important relation that we use often in what follows is that if there is an n-step walk from state i to j and an m-step walk from state j to k, then there is a walk of $m+n$ steps from i to k. Thus

$$P_{ij}^n > 0 \text{ and } P_{jk}^m > 0 \quad \text{imply} \quad P_{ik}^{n+m} > 0. \tag{4.5}$$

This also shows that

$$i \to j \text{ and } j \to k \quad \text{imply} \quad i \to k. \tag{4.6}$$

Definition 4.2.3 *Two distinct states i and j **communicate** (abbreviated $i \leftrightarrow j$) if i is accessible from j and j is accessible from i.*

An important fact about communicating states is that if $i \leftrightarrow j$ and $m \leftrightarrow j$ then $i \leftrightarrow m$. To see this, note that $i \leftrightarrow j$ and $m \leftrightarrow j$ imply that $i \to j$ and $j \to m$, so that $i \to m$. Similarly, $m \to i$, so $i \leftrightarrow m$.

Definition 4.2.4 *A **class** C of states is a non-empty set of states such that for each $i \in \mathsf{C}$, each state $j \neq i$ satisfies $j \in \mathsf{C}$ if $i \leftrightarrow j$ and $j \notin \mathsf{C}$ if $j \not\leftrightarrow i$.*

For the example of Figure 4.1(a), $\{2, 3\}$ is one class of states, $\{1\}$, $\{4\}$, $\{5\}$, and $\{6\}$ are the other classes. Note that states 1, 4, 5, and 6 do not communicate with any other states so that the corresponding classes are singleton classes. State 6 does not even communicate with itself, but $\{6\}$ is still a class according to the definition. The entire set of states in a given Markov chain is partitioned into one or more disjoint classes in this way.

Definition 4.2.5 *For finite-state Markov chains, a **recurrent state** is a state i that is accessible from all states that are accessible from i (i is recurrent if $i \to j$ implies that $j \to i$). A **transient state** is a state that is not recurrent.*

Recurrence is more complicated for Markov chains with countably-infinite state spaces, and will be defined and discussed in Chapter 6.

According to the definition, a state i in a finite-state Markov chain is recurrent if there is no possibility of going to a state j from which there can be no return. As we shall see later, if a Markov chain ever enters a recurrent state, it returns to that state eventually with probability 1, and thus keeps returning infinitely often (in fact, this essentially serves as the definition of recurrence for Markov chains with countably-infinite state spaces). A state i is transient if there is some j that is accessible from i but from which there is no possible return. Each time the system returns to i, there is a possibility of going to j; eventually this possibility will occur and there can be no further returns to i.

Theorem 4.2.6 *For finite-state Markov chains, either all states in a class are transient or all are recurrent.*[2]

Proof Assume that state i is transient (i.e., for some j, $i \to j$ but $j \not\to i$) and suppose that i and m are in the same class (i.e., $i \leftrightarrow m$). Then $m \to i$ and $i \to j$, so $m \to j$. Now if $j \to m$, then the walk from j to m could be extended to i; this is a contradiction, and therefore there is no walk from j to m, and m is transient. Since we have just shown that all nodes in a class are transient if any are, it follows that the states in a class are either all recurrent or all transient. □

For the example of Figure 4.1(a), $\{2, 3\}$ and $\{5\}$ are recurrent classes and the other classes are transient. In terms of the graph of a Markov chain, a class is transient if there are any directed arcs going from a node in the class to a node outside the class. Every finite-state Markov chain must have at least one recurrent class of states (see Exercise 4.2), and can have arbitrarily many additional classes of recurrent states and transient states.

States can also be classified according to their periods (see Figure 4.2). For $X_0 = 2$ in Figure 4.2(a), X_n must be 2 or 4 for n even and 1 or 3 for n odd. On the other hand, if X_0 is 1 or 3, then X_n is 2 or 4 for n odd and 1 or 3 for n even. Thus the effect of the starting state never dies out. Figure 4.2(b) illustrates another example in which the memory of the starting state never dies out. The states in both of these Markov chains are said to be periodic with period 2. Other examples of periodic states are states 2 and 3 in Figure 4.1(a).

Definition 4.2.7 *The* ***period of a state*** i*, denoted* $d(i)$*, is the greatest common divisor (gcd) of those values of* n *for which* $P^n_{ii} > 0$*. If the period is* 1*, the state is* ***aperiodic****, and if the period is* 2 *or more, the state is periodic.*

For example, in Figure 4.2(a), $P^n_{11} > 0$ for $n = 2, 4, 6, \ldots$. Thus $d(1)$, the period of state 1, is 2. Similarly, $d(i) = 2$ for the other states in Figure 4.2(a). For Figure 4.2(b), we have $P^n_{11} > 0$ for $n = 4, 8, 10, 12, \ldots$; thus $d(1) = 2$, and it can be seen that $d(i) = 2$ for all the states. These examples suggest the following theorem.

[2] As shown in Chapter 6, this theorem is also true for Markov chains with a countably-infinite state space, but the proof given here is inadequate. Also recurrent classes with a countably-infinite state space are further classified into either *positive recurrent* or *null recurrent*, a distinction that does not appear in the finite-state case.

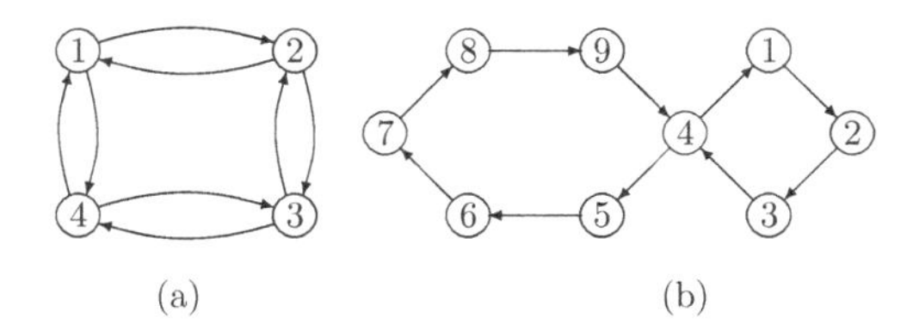

Figure 4.2 The four-state chain in (a) and the nine-state chain in (b) are examples of periodic Markov chains, each of period 2.

Theorem 4.2.8 *For any Markov chain (with either a finite or countably-infinite number of states), all states in the same class have the same period.*

Proof Let i and j be any distinct pair of states in a class C. Then $i \leftrightarrow j$ and there is some n such that $P_{ij}^n > 0$ and some m such that $P_{ji}^m > 0$. Since there is a walk of length $n+m$ going from i to j and back to i, $n+m$ must be divisible by $d(i)$. Let t be any integer such that $P_{jj}^t > 0$. Since there is a walk of length $n+t+m$ from i to j, then back to j, and then to i, $n+t+m$ is divisible by $d(i)$, and thus t is divisible by $d(i)$. Since this is true for every t such that $P_{jj}^t > 0$, $d(j)$ is divisible by $d(i)$. Reversing the roles of i and j, $d(i)$ is divisible by $d(j)$, so $d(i) = d(j)$. □

Since the states in a class C all have the same period and are either all recurrent or all transient, we refer to C itself as having the period of its states and as being recurrent or transient. Similarly if a Markov chain has a single class of states, we refer to the chain in the same way as the class.

Theorem 4.2.9 *If a class C in a finite-state Markov chain has period d, then the states in C can be partitioned into d subsets, $\mathsf{S}_0, \mathsf{S}_1, \dots, \mathsf{S}_{d-1}$, in such a way that all transitions from S_ℓ go to $\mathsf{S}_{\ell+1}$ for $\ell < d-1$, and go to S_0 for $\ell = d-1$.*

Proof See Figure 4.3 for an illustration of the theorem. For any given state in C, say state 1, define the subsets S_0 to S_{d-1} by

$$\mathsf{S}_\ell = \{j \in \mathsf{C} : P_{1j}^n > 0 \quad \text{for some } n \text{ such that } n \bmod d = \ell\},$$

where $n \bmod d$ is the remainder when n is divided by d. We first show that the subsets $\mathsf{S}_0, \dots, \mathsf{S}_{d-1}$ are disjoint. For any given $j \in \mathsf{C}$, let n be the length of any particular walk from 1 to j, i.e., n satisfies $P_{ij}^n > 0$. There must be a walk from j back to 1, and we denote

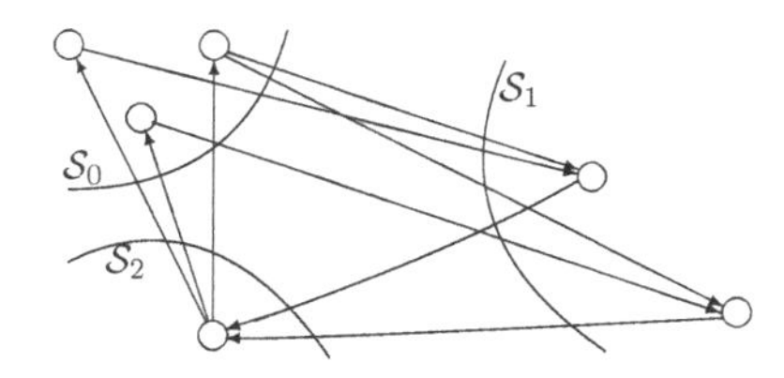

Figure 4.3 Structure of a periodic Markov chain with $d = 3$. Note that transitions only go from one subset $\mathcal{S}_\ell$ to the next subset $\mathcal{S}_{\ell+1}$ (or from $\mathcal{S}_{d-1}$ to $\mathcal{S}_0$).

its length by m. Then $n+m$ is the length of a walk from 1 back to 1, so $n+m$ is divisible by d. Finally, if n' is the length of any other walk from 1 to j, then $n'+m$ is also divisible by d. Thus $n-n'$ is divisible by d and n mod $d = n'$ mod d. Thus $j \in \mathsf{S}_\ell$ only for $\ell = n$ mod d. Since j is arbitrary, the sets $\mathsf{S}_0, \dots, \mathsf{S}_{d-1}$ are disjoint.

Finally, suppose $j \in \mathsf{S}_\ell$ and $P_{jk} > 0$ for some $k \in \mathsf{C}$. Then $P^n_{1j} > 0$ for some n such that n mod $d = \ell$. It follows that $P^{n+1}_{1k} \geq P^n_{1j}P_{jk} > 0$. Thus $k \in \mathsf{S}_{(n+1) \bmod d}$. This means that $k \in \mathsf{S}_{\ell+1}$ for $\ell < d-1$ and $k \in \mathsf{S}_0$ for $\ell = d-1$. □

We have seen that each class of states (for a finite-state chain) can be classified both in terms of its period and in terms of whether or not it is recurrent. The most important case is that in which a class is both recurrent and aperiodic.

Definition 4.2.10 *For a finite-state Markov chain, an* ***ergodic class of states*** *is a class that is both recurrent and aperiodic. A Markov chain consisting entirely of one ergodic class is called an* ***ergodic chain****.*

We shall see later that these chains have the desirable property that P^n_{ij} becomes independent of the starting state i as $n \to \infty$. The next theorem establishes the first part of this by showing that $P^n_{ij} > 0$ for all i and j when n is sufficiently large. A guided proof is given in Exercise 4.5.

Theorem 4.2.11 *For an ergodic* M *state Markov chain,* $P^m_{ij} > 0$ *for all* i, j*, and all* $m \geq (\mathsf{M}-1)^2 + 1$.

Figure 4.4 illustrates a situation where the bound $(\mathsf{M}-1)^2+1$ is met with equality. Note that there is one cycle of length $\mathsf{M}-1$ and the single node not on this cycle, node 1, is the unique starting node at which the bound is met with equality.

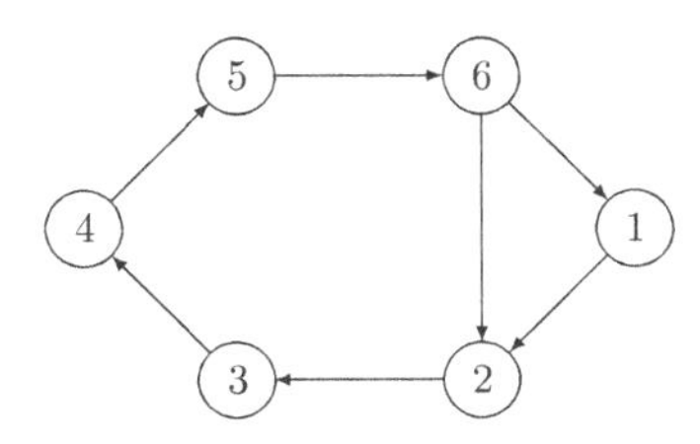

Figure 4.4 An ergodic chain with $\mathsf{M} = 6$ states in which $P^m_{ij} > 0$ for all $m > (\mathsf{M}-1)^2$ and all i, j but $P^{(\mathsf{M}-1)^2}_{11} = 0$. The figure also illustrates that an M state ergodic Markov chain with $\mathsf{M} \geq 2$ must have a cycle with $\mathsf{M}-1$ or fewer nodes. To see this, note that an ergodic chain must have cycles, since each node must have a walk to itself, and subcycles of repeated nodes can be omitted from that walk, converting it into a cycle. Such a cycle might have M nodes, but a chain with only an M-node cycle would be periodic. Thus some nodes must be on smaller cycles, such as the cycle of length 5 in the figure.

4.3 The matrix representation

The matrix $[P]$ of transition probabilities of a Markov chain is called a stochastic matrix, i.e., a *stochastic matrix* is a square matrix of non-negative terms in which the elements in each row sum to 1. We first consider the n-step transition probabilities P^n_{ij} in terms of [P]. The probability, starting in state i, of going to state j in two steps is the sum over k of the probability of going first to k and then to j. Using the Markov property in (4.1),

$$P^2_{ij} = \sum_{k=1}^{\mathsf{M}} P_{ik}P_{kj}.$$

It can be seen that this is just the ij term of the product of the matrix $[P]$ with itself; denoting $[P][P]$ as $[P^2]$, this means that P^2_{ij} is the (i,j) element of the matrix $[P^2]$. Similarly, P^n_{ij} is the ij element of the nth power of the matrix $[P]$. Since $[P^{m+n}] = [P^m][P^n]$, this means that

$$P^{m+n}_{ij} = \sum_{k=1}^{\mathsf{M}} P^m_{ik}\, P^n_{kj}. \tag{4.7}$$

This is known as the *Chapman–Kolmogorov* equation. An efficient approach to compute $[P^n]$ (and thus P^n_{ij}) for large n is to multiply $[P^2]$ by $[P^2]$, then $[P^4]$ by $[P^4]$ and so forth. Then $[P], [P^2], [P^4], \dots$ can be multiplied as needed to get $[P^n]$.

4.3.1 Steady state and $[P^n]$ for large n

The matrix $[P^n]$ (i.e., the matrix of transition probabilities raised to the nth power) is important for a number of reasons. The i,j element of this matrix is $P^n_{ij} = \Pr\{X_n{=}j \mid X_0{=}i\}$. If memory of the past dies out with increasing n, then we would expect the dependence of P^n_{ij} on both n and i to disappear as $n \to \infty$. This means, first, that $[P^n]$ should converge to a limit as $n \to \infty$ and, second, that for each column j, the elements in that column, $P^n_{1j}, P^n_{2j}, \dots, P^n_{\mathsf{M}j}$ should all tend toward the same value, say π_j, as $n \to \infty$. If this type of convergence occurs (and we later determine the circumstances under which it occurs), then $P^n_{ij} \to \pi_j$ and each row of the limiting matrix will be $(\pi_1, \dots, \pi_{\mathsf{M}})$, i.e., each row is the same as each other row.

If we now look at the equation $P^{n+1}_{ij} = \sum_k P^n_{ik}P_{kj}$, and assume the above type of convergence as $n \to \infty$, then the limiting equation becomes $\pi_j = \sum_k \pi_k P_{kj}$. In vector form, this equation is $\boldsymbol{\pi} = \boldsymbol{\pi}[P]$. We will do this more carefully later, but what it says is that if P^n_{ij} approaches a limit denoted π_j as $n \to \infty$, then $\boldsymbol{\pi} = (\pi_1, \dots, \pi_{\mathsf{M}})$ satisfies $\boldsymbol{\pi} = \boldsymbol{\pi}[P]$. If nothing else, it is easier to solve the linear equations $\boldsymbol{\pi} = \boldsymbol{\pi}[P]$ than to multiply $[P]$ by itself an infinite number of times.

Definition 4.3.1 *A **steady-state vector** (or a **steady-state distribution**) for an* M*-state Markov chain with transition matrix $[P]$ is a row vector $\boldsymbol{\pi}$ that satisfies*

$$\boldsymbol{\pi} = \boldsymbol{\pi}[P]\,, \quad \textit{where} \sum_i \pi_i = 1 \textit{ and } \pi_i \ge 0\,,\ 1 \le i \le \mathsf{M}. \tag{4.8}$$

The last half of (4.8) says that $\boldsymbol{\pi}$ must be a probability vector, i.e., a vector whose components are non-negative and sum to 1. If $\boldsymbol{\pi}$ is taken as the initial probability mass function (PMF) of the chain at time 0, then that PMF is maintained forever. To see this, assume that $\mathsf{p}_{X_0}(i) = \pi_i$ for each state i. Then

$$\mathsf{p}_{X_1}(j) = \sum_i \mathsf{p}_{X_0}(i) P_{ij} = \sum_i \pi_i P_{ij} = \pi_j.$$

Repeating this iteratively for each time n, $n = 1, 2, \ldots$,

$$\mathsf{p}_{X_{n+1}}(j) = \sum_i \mathsf{p}_{X_n}(i) P_{ij} = \sum_i \pi_i P_{ij} = \pi_j.$$

Another way of seeing this is to post-multiply both sides of (4.8) by $[P]$, getting $\boldsymbol{\pi}[P] = \boldsymbol{\pi}[P^2]$, and iterating this, $\boldsymbol{\pi} = \boldsymbol{\pi}[P^2] = \boldsymbol{\pi}[P^3] = \cdots$.

It is important to recognize that we have shown that *if* $[P^n]$ converges to a matrix all of whose rows are equal to $\boldsymbol{\pi}$, *then* $\boldsymbol{\pi}$ is a steady-state vector, i.e., $\boldsymbol{\pi}$ satisfies (4.8). However, *if* $\boldsymbol{\pi}$ satisfies (4.8), it *does not* necessarily follow that $[P^n]$ converges to $\boldsymbol{\pi}$ as $n \to \infty$. For the example of Figure 4.1, it can be seen that if we choose $\pi_2 = \pi_3 = 1/2$ with $\pi_i = 0$ otherwise, then $\boldsymbol{\pi}$ is a steady-state vector. Reasoning more physically, we see that if the chain is in either state 2 or state 3, it simply oscillates between those states for all time. If it starts at time 0 being in state 2 or state 3 with equal probability, then it persists forever being in state 2 or state 3 with equal probability. Although this choice of $\boldsymbol{\pi}$ satisfies the definition in (4.8) and also is a steady-state distribution in the sense of not changing over time, it is not a very satisfying form of steady state, and almost seems to be concealing the fact that we are dealing with a simple oscillation between states.

This example raises one of a number of questions that should be answered concerning steady-state distributions and the convergence of $[P^n]$:

1 Under what conditions does $\boldsymbol{\pi} = \boldsymbol{\pi}[P]$ have a probability vector solution?
2 Under what conditions does $\boldsymbol{\pi} = \boldsymbol{\pi}[P]$ have a unique probability vector solution?
3 Under what conditions does each row of $[P^n]$ converge to a probability vector $\boldsymbol{\pi}$ satisfying $\boldsymbol{\pi} = \boldsymbol{\pi}[P]$?

We first give the answers to these questions for finite-state Markov chains and then derive them. First, $\boldsymbol{\pi} = \boldsymbol{\pi}[P]$ *always* has a probability vector solution (although this is not necessarily true for infinite-state chains). The answers to the second and third questions are simplified if we use the following definition.

Definition 4.3.2 *A* ***unichain*** *is a finite-state Markov chain that contains a single recurrent class plus, perhaps, some transient states. An* ***ergodic unichain*** *is a unichain for which the recurrent class is ergodic.*

A unichain, as we shall see, is the natural generalization of a recurrent chain to allow for some initial transient behavior without disturbing the long-term aymptotic behavior of the underlying recurrent chain.

The answer to the second question above is that $\boldsymbol{\pi} = \boldsymbol{\pi}[P]$ has a unique probability vector solution if and only if [P] is the transition matrix of a *unichain*. If there are c recurrent classes, then (4.8) has c linearly independent solutions, each non-zero only over the elements of the corresponding recurrent class.

For the third question, each row of $[P^n]$ converges to the unique solution of $\boldsymbol{\pi} = \boldsymbol{\pi}[P]$ if and only if [P] is the transition matrix of an *ergodic unichain*. If there are multiple recurrent classes, and each one is aperiodic, then $[P^n]$ still converges, but to a matrix with non-identical rows. If the Markov chain has one or more periodic recurrent classes, then $[P^n]$ does not converge.

We first look at these answers from the standpoint of the transition matrices of finite-state Markov chains, and then proceed in Chapter 6 to look at the more general problem of Markov chains with a countably-infinite number of states. There we use renewal theory to answer these same questions (and to discover the differences that occur for infinite-state Markov chains).

The matrix approach is useful computationally and also has the advantage of telling us something about rates of convergence. The approach using renewal theory is very simple (given an understanding of renewal processes), but is more abstract.

In answering the above questions (plus a few more) for finite-state Markov chains, it is simplest to first consider the third question,[3] i.e., the convergence of each row of $[P^n]$ to the solution of (4.8). The simplest approach to this, for each column j of $[P^n]$, is to study the difference between the largest and the smallest element of that column and how this difference changes with n. The following almost trivial lemma starts this study, and is valid for all finite-state Markov chains.

Lemma 4.3.3 *Let $[P]$ be the transition matrix of a finite-state Markov chain and let $[P^n]$ be the nth power of $[P]$, i.e., the matrix of nth-order transition probabilities, P^n_{ij}. Then for each state j and each integer $n \geq 1$*

$$\max_i P^{n+1}_{ij} \leq \max_\ell P^n_{\ell j}, \qquad \min_i P^{n+1}_{ij} \geq \min_\ell P^n_{\ell j}. \tag{4.9}$$

Discussion The lemma says that for each column j, the maximum over the elements of that column is non-increasing with n and the minimum is non-decreasing with n. The position i in column j at which a maximum occurs can vary with n, and P^n_{ij} might consitute the maximum over i for one value of n and the minimum for another. However, the range over i from the smallest to largest P^n_{ij} must shrink or stay the same as n increases.

Proof For each i, j, n, we use the Chapman–Kolmogorov equation, (4.7), followed by the fact that $P^n_{kj} \leq \max_\ell P^n_{\ell j}$, to see that

$$P^{n+1}_{ij} = \sum_k P_{ik} P^n_{kj} \leq \sum_k P_{ik} \max_\ell P^n_{\ell j} = \max_\ell P^n_{\ell j}. \tag{4.10}$$

[3] One might naively try to show that a steady-state vector exists by first noting that each row of P sums to 1. The column vector $\boldsymbol{e} = (1, 1, \ldots, 1)^\mathsf{T}$ then satisfies the eigenvector equation $\boldsymbol{e} = [P]\boldsymbol{e}$. Thus there must also be a left eigenvector satisfying $\boldsymbol{\pi}[P] = \boldsymbol{\pi}$. The problem here is showing that $\boldsymbol{\pi}$ is real and non-negative.

Since this holds for all states i, and thus for the maximizing i, the first half of (4.9) follows. The second half of (4.9) is the same, with minima replacing maxima, i.e.,

$$P_{ij}^{n+1} = \sum_k P_{ik}P_{kj}^n \geq \sum_k P_{ik} \min_\ell P_{\ell j}^n = \min_\ell P_{\ell j}^n. \qquad \square$$

For some Markov chains, the maximizing value for each column decreases with n and reaches the same limit as that of the minimizing value, i.e., for each column, all elements in that column are the same. This means that each row converges to the same limiting row vector. For other Markov chains, the maximizing value for a column might converge with n to a limit strictly above the limit of the minimizing elements. Then $[P^n]$ does not converge to a matrix where all rows are the same, and might not converge at all.

The following three subsections establish the above kind of convergence (and a number of subsidiary results) for three cases of increasing complexity. The first assumes that $P_{ij} > 0$ for all i, j. This is denoted as $[P] > 0$ and provides a needed step for the other cases. Ergodic Markov chains constitute the second case, and ergodic unichains the third.

4.3.2 Steady state assuming $[P] > 0$

Lemma 4.3.4 *Let the transition matrix of a finite-state Markov chain satisfy* $[P] > 0$ *(i.e.,* $P_{ij} > 0$ *for all* i, j*), and let* $\alpha = \min_{i,j} P_{ij}$*. Then for all states* j *and all* $n \geq 1$*:*

$$\max_i P_{ij}^{n+1} - \min_i P_{ij}^{n+1} \leq \left(\max_\ell P_{\ell j}^n - \min_\ell P_{\ell j}^n\right)(1 - 2\alpha). \tag{4.11}$$

$$\left(\max_\ell P_{\ell j}^n - \min_\ell P_{\ell j}^n\right) \leq (1 - 2\alpha)^n. \tag{4.12}$$

$$\lim_{n\to\infty} \max_\ell P_{\ell j}^n = \lim_{n\to\infty} \min_\ell P_{\ell j}^n \geq \alpha > 0. \tag{4.13}$$

Discussion Since $P_{ij} > 0$ for all i, j, we must have $\alpha > 0$. Thus the theorem says that for each j, the elements P_{ij}^n in column j of $[P^n]$ approach equality over both i and n as $n \to \infty$, i.e., the state at time n becomes independent of the state at time 0 as $n \to \infty$. The approach is geometric in n.

Proof We first tighten the inequality in (4.10) slightly . For a given j and n, let $\ell_{\min}$ be a value of ℓ that minimizes $P_{\ell j}^n$. Then

$$\begin{aligned} P_{ij}^{n+1} &= \sum_k P_{ik}P_{kj}^n \\ &\leq \sum_{k \neq \ell_{\min}} P_{ik} \max_\ell P_{\ell j}^n + P_{i\ell_{\min}} \min_\ell P_{\ell j}^n \\ &= \max_\ell P_{\ell j}^n - P_{i\ell_{\min}}\left(\max_\ell P_{\ell j}^n - \min_\ell P_{\ell j}^n\right) &(4.14) \\ &\leq \max_\ell P_{\ell j}^n - \alpha\left(\max_\ell P_{\ell j}^n - \min_\ell P_{\ell j}^n\right), &(4.15) \end{aligned}$$

where in (4.14) we added and subtracted $P_{i\ell_{\min}} \max_\ell P^n_{\ell j}$ on the right-hand side, and in (4.15), we used $\alpha \le P_{i\ell_{\min}}$ in conjuction with the fact that the term in parentheses must be non-negative.

Repeating the same argument with the roles of max and min reversed,

$$P^{n+1}_{ij} \ge \min_\ell P^n_{\ell j} + \alpha\left(\max_\ell P^n_{\ell j} - \min_\ell P^n_{\ell j}\right). \tag{4.16}$$

Applying the upper bound, (4.15), to $\max_i P^{n+1}_{ij}$ and the lower bound, (4.16), to $\min_i P^{n+1}_{ij}$, we can subtract the lower bound from the upper bound to get (4.11).

Next, note that

$$\min_\ell P_{\ell j} \ge \alpha > 0, \qquad \max_\ell P_{\ell j} \le 1 - \alpha.$$

Thus $\max_\ell P_{\ell j} - \min_\ell P_{\ell j} \le 1 - 2\alpha$. Using this as the base for iterating (4.11) over n, we get (4.12). This, in conjuction with the monotonicity property of (4.9), shows not only that the limits in (4.13) exist and are positive and equal, but that the limits are approached geometrically in n. Finally, since $\min_\ell P^n_{\ell j}$ is monotonic non-decreasing in n and $\min_\ell P_{\ell j} \ge \alpha$, we see that the limits in (4.13) are lower bounded by α. □

4.3.3 Ergodic Markov chains

Lemma 4.3.4 extends quite easily to arbitrary ergodic finite-state Markov chains. The key to this comes from Theorem 4.2.11, which shows that if $[P]$ is the matrix for an M-state ergodic Markov chain, then the matrix $[P^h]$ is positive for any $h \ge (\mathsf{M}-1)^2 + 1$. Thus, choosing $h = (\mathsf{M}-1)^2 + 1$, we can apply Lemma 4.3.4 to $[P^h] > 0$. For each integer $m \ge 1$,

$$\max_i P^{h(m+1)}_{ij} - \min_i P^{h(m+1)}_{ij} \le \left(\max_\ell P^{hm}_{\ell j} - \min_\ell P^{hm}_{\ell j}\right)(1-2\beta)$$

$$\left(\max_m P^{hm}_{\ell j} - \min_\ell P^{hm}_{\ell j}\right) \le (1-2\beta)^m \tag{4.17}$$

$$\lim_{m\to\infty} \max_\ell P^{hm}_{\ell j} = \lim_{m\to\infty} \min_\ell P^{hm}_{\ell j} \ \ge\ \beta\ >\ 0, \tag{4.18}$$

where $\beta = \min_{i,j} P^h_{ij}$. Lemma 4.3.3 states that $\max_i P^n_{ij}$ is non-increasing and $\min_i P^n_{ij}$ non-decreasing in n. Thus (4.17) and (4.18) can be replaced with

$$\left(\max_\ell P^n_{\ell j} - \min_m P^n_{\ell j}\right) \le (1-2\beta)^{\lfloor n/h \rfloor} \tag{4.19}$$

$$\lim_{n\to\infty} \max_\ell P^n_{\ell j} = \lim_{n\to\infty} \min_\ell P^n_{\ell j}\ >\ 0. \tag{4.20}$$

Now define $\boldsymbol{\pi} > 0$ by

$$\pi_j = \lim_{n\to\infty} \max_\ell P^n_{\ell j} = \lim_{n\to\infty} \min_\ell P^n_{\ell j}\ >\ 0. \tag{4.21}$$

Since π_j lies between the minimum and maximum over i of P^n_{ij} for each n,

$$\left|P^n_{ij} - \pi_j\right| \le (1-2\beta)^{\lfloor n/h \rfloor}. \tag{4.22}$$

In the limit, then,

$$\lim_{n\to\infty} P_{ij}^n = \pi_j \quad \text{for each } i,j. \tag{4.23}$$

This says that the matrix $[P^n]$ has a limit as $n \to \infty$ and the i,j term of that matrix is π_j for all i,j. In other words, each row of $[P^n]$ converges to this same vector $\boldsymbol{\pi}$. This is represented most compactly by

$$\lim_{n\to\infty} [P^n] = \boldsymbol{e}\boldsymbol{\pi}, \qquad \text{where } \boldsymbol{e} = (1, 1, \dots, 1)^{\mathsf{T}}. \tag{4.24}$$

The following theorem[4] summarizes these results and adds one small additional result.

Theorem 4.3.5 *Let $[P]$ be the matrix of an ergodic finite-state Markov chain. Then there is a unique steady-state vector $\boldsymbol{\pi}$, that vector is positive and satisfies (4.23) and (4.24). The convergence in n is geometric, satisfying (4.19).*

Proof We need to show that $\boldsymbol{\pi}$ as defined in (4.21) is the unique steady-state vector. Let $\boldsymbol{\mu}$ be any steady-state vector, i.e., any probability vector solution to $\boldsymbol{\mu}[P] = \boldsymbol{\mu}$. Then $\boldsymbol{\mu}$ must satisfy $\boldsymbol{\mu} = \boldsymbol{\mu}[P^n]$ for all $n > 1$. Going to the limit,

$$\boldsymbol{\mu} = \boldsymbol{\mu} \lim_{n\to\infty} [P^n] = \boldsymbol{\mu}\boldsymbol{e}\boldsymbol{\pi} = \boldsymbol{\pi}.$$

We have used the fact that $\boldsymbol{\mu}$ is a probability vector and thus its elements sum to 1, i.e., $\boldsymbol{\mu}\boldsymbol{e} = 1$. Thus $\boldsymbol{\pi}$ is a steady-state vector and is unique. □

4.3.4 Ergodic unichains

Understanding how P_{ij}^n approaches a limit as $n \to \infty$ for ergodic unichains is a straightforward extension of the results in Section 4.3.3, but the details require some care. Let $\mathtt{T}$ denote the set of transient states (which might contain several transient classes), and assume the states of $\mathtt{T}$ are numbered $1, 2, \dots, t$. Let $\mathtt{R}$ denote the recurrent class, assumed to be numbered $t+1, \dots, t+r$ (see Figure 4.5).

If i and j are both recurrent states, then there is no possibility of leaving the recurrent class in going from i to j. Assuming this class to be ergodic, the transition matrix $[P_{\mathcal{R}}]$ as shown in Figure 4.5 has been analyzed in Section 4.3.3.

If the initial state is a transient state, then eventually the recurrent class is entered, and after that, the distribution gradually approaches steady state within the recurrent class. This suggests (and we next show) that there is a steady-state vector $\boldsymbol{\pi}$ for $[P]$ itself such that $\pi_j = 0$ for $j \in \mathtt{T}$ and π_j is as given in Section 4.3.3 for each $j \in \mathtt{R}$.

Initially we will show that P_{ij}^n converges to 0 for $i,j \in \mathtt{T}$. The exact nature of how and when the recurrent class is entered starting in a transient state is an interesting problem

[4] This is essentially the Frobenius theorem for non-negative irreducible matrices, specialized to Markov chains. A non-negative matrix $[P]$ is *irreducible* if its graph (containing an edge from node i to node j if $P_{ij} > 0$) is the graph of a recurrent Markov chain. There is no constraint that each row of $[P]$ sums to 1. The proof of the Frobenius theorem requires some fairly intricate analysis and seems to be far more complex than the simple proof here for Markov chains. A proof of the Frobenius theorem can be found in [13].

$$[P] = \left[\begin{array}{c|c} [P_{\mathcal{T}}] & [P_{\mathcal{TR}}] \\ \hline [0] & [P_{\mathcal{R}}] \end{array}\right], \qquad \text{where} \qquad [P_{\mathcal{T}}] = \begin{bmatrix} P_{11} & \cdots & P_{1t} \\ \cdots & \cdots & \cdots \\ P_{t1} & \ldots & P_{tt} \end{bmatrix}.$$

$$[P_{\mathcal{TR}}] = \begin{bmatrix} P_{1,t+1} & \cdots & P_{1,t+r} \\ \cdots & \cdots & \cdots \\ P_{t,t+1} & \ldots & P_{t,t+r} \end{bmatrix} \qquad [P_{\mathcal{R}}] = \begin{bmatrix} P_{t+1,t+1} & \cdots & P_{t+1,t+r} \\ \cdots & \cdots & \cdots \\ P_{t+r,t+1} & \ldots & P_{t+r,t+r} \end{bmatrix}$$

Figure 4.5 The transition matrix of a unichain. The block of zeroes in the lower left corresponds to the absence of transitions from recurrent to transient states.

in its own right, and is discussed in both Sections 4.4 and 4.5. For now, a crude bound will suffice.

For each transient state, there must be a walk to some recurrent state, and since there are only t transient states, there must be some such walk of length at most t. Each such walk has positive probability, and thus for each $i \in \mathtt{T}$, $\sum_{j\in\mathcal{R}} P_{ij}^t > 0$. It follows that for each $i \in \mathtt{T}$, $\sum_{j\in\mathcal{T}} P_{ij}^t < 1$. Let $\gamma < 1$ be the maximum of these probabilities over $i \in \mathtt{T}$, i.e.,

$$\gamma = \max_{i\in\mathcal{T}} \sum_{j\in\mathcal{T}} P_{ij}^t < 1.$$

Lemma 4.3.6 *Let $[P]$ be a unichain, let $\mathtt{T}$ be the set of transient states, and let $t = |\mathtt{T}|$.*

$$\max_{\ell\in\mathcal{T}} \sum_{j\in\mathcal{T}} P_{\ell j}^n \le \gamma^{\lfloor n/t \rfloor}. \tag{4.25}$$

Proof For each integer multiple νt of t and each $i \in \mathtt{T}$,

$$\sum_{j\in\mathcal{T}} P_{ij}^{(\nu+1)t} \;=\; \sum_{k\in\mathcal{T}} P_{ik}^t \sum_{j\in\mathcal{T}} P_{kj}^{\nu t} \;\le\; \sum_{k\in\mathcal{T}} P_{ik}^t \max_{\ell\in\mathcal{T}} \sum_{j\in\mathcal{T}} P_{\ell j}^{\nu t} \;\le\; \gamma \max_{\ell\in\mathcal{T}} \sum_{j\in\mathcal{T}} P_{\ell j}^{\nu t}.$$

Recognizing that this applies to all $i \in \mathtt{T}$, and thus to the maximum over i, we can iterate this equation, getting

$$\max_{\ell\in\mathcal{T}} \sum_{j\in\mathcal{T}} P_{\ell j}^{\nu t} \le \gamma^{\nu}.$$

This establishes (4.25) for all n that are multiples of t. Since $\sum_{j\in\mathcal{T}} P_{lj}^n$ is non-increasing in n, (4.25) follows in general. □

We now proceed to the case where the initial state is $i \in \mathtt{T}$ and the final state is $j \in \mathtt{R}$. For any integer $n \ge 1$, any $i \in \mathtt{T}$ and any $j \in \mathtt{R}$, the Chapman–Kolmogorov equation says that

$$P_{ij}^{2n} = \sum_{k\in\mathcal{T}} P_{ik}^n P_{kj}^n + \sum_{k\in\mathcal{R}} P_{ik}^n P_{kj}^n.$$

Upper bounding and lower bounding each of these terms,

$$P_{ij}^{2n} \le \sum_{k\in\mathcal{T}} P_{ik}^n + \sum_{k\in\mathcal{R}} P_{ik}^n \max_{\ell\in\mathcal{R}} P_{\ell j}^n \;\le\; \sum_{k\in\mathcal{T}} P_{ik}^n + \max_{\ell\in\mathcal{R}} P_{\ell j}^n.$$

$$P_{ij}^{2n} \ge \sum_{k\in\mathcal{R}} P_{ik}^n \min_{\ell\in\mathcal{R}} P_{\ell j}^n \;=\; \left(1 - \sum_{k\in\mathcal{T}} P_{ik}^n\right) \min_{\ell\in\mathcal{R}} P_{\ell j}^n$$

$$\ge \min_{\ell\in\mathcal{R}} P_{\ell j}^n - \sum_{k\in\mathcal{T}} P_{ik}^n.$$

Let π_j be the steady-state probability of state $j \in \mathrm{R}$ in the recurrent Markov chain with states R. Then π_j lies between the maximum and minimum, over $\ell \in \mathrm{R}$, of $P_{\ell j}^n$. Using (4.20), we then have

$$\left|P_{ij}^{2n} - \pi_j\right| \le (1-2\beta)^{\lfloor n/h \rfloor} + \gamma^{\lfloor n/t \rfloor},$$

where $h = (r-1)^2 + 1$ and $\beta = \min_{i,j\in\mathcal{R}} P_{ij}^h > 0$. Since $\max_i P_{ij}^n$ is non-increasing in n and $\min_i P_{ij}^n$ is non-decreasing, This can be restated for all n as

$$\left|P_{ij}^{n} - \pi_j\right| \le (1-2\beta)^{\lfloor n/2h \rfloor} + \gamma^{\lfloor n/2t \rfloor}. \tag{4.26}$$

This is summarized in the following theorem.

Theorem 4.3.7 *Let $[P]$ be the transition matrix of a finite-state ergodic unichain. Then $\lim_{n\to\infty}[P^n] = \boldsymbol{e\pi}$, where $\boldsymbol{e} = (1,1,\ldots,1)^{\mathsf{T}}$ and $\boldsymbol{\pi}$ is the steady-state vector of the recurrent class of states, expanded by zeros for each transient state of the unichain. The convergence is geometric in n for all i,j.*

4.3.5 Arbitrary finite-state Markov chains

The asymptotic behavior of $[P^n]$ as $n \to \infty$ for arbitrary finite-state Markov chains can mostly be deduced from the ergodic unichain case by simple extensions and common sense.

First, consider the case of $m > 1$ aperiodic classes plus a set of transient states. If the initial state is in the κth of the recurrent classes, say R^{κ} then the chain remains in R^{κ} and there is a unique finite-state vector $\boldsymbol{\pi}^{\kappa}$ that is non-zero only in R^{κ} that can be found by viewing class κ in isolation.

If the initial state i is transient, then, for each R^{κ}, there is a certain probability that R^{κ} is eventually reached, and once it is reached there is no exit, so the steady state over that recurrent class is approached. The question of finding the probability of entering each recurrent class from a given transient class will be discussed in the next section.

Next consider a recurrent Markov chain that is periodic with period d. The dth-order transition probability matrix, $[P^d]$, is then constrained by the fact that $P_{ij}^d = 0$ for all j not in the same periodic subset as i. In other words, $[P^d]$ is the matrix of a chain with d recurrent classes. We will obtain greater facility in working with this in the next section when eigenvalues and eigenvectors are discussed.

4.4 The eigenvalues and eigenvectors of stochastic matrices

For ergodic unichains, the previous section showed that the dependence of a state on the distant past disappears with increasing n, i.e., $P^n_{ij} \to \pi_j$. In this section we look more carefully at the eigenvalues and eigenvectors of $[P]$ to sharpen our understanding of how fast $[P^n]$ converges for ergodic unichains and what happens for other finite-state Markov chains.

Definition 4.4.1 *A row vector $\boldsymbol{\pi}$ is a left*[5] ***eigenvector*** *of $[P]$ of* ***eigenvalue*** *λ if $\boldsymbol{\pi} \neq \boldsymbol{0}$ and $\boldsymbol{\pi}[P] = \lambda\boldsymbol{\pi}$, i.e., $\sum_i \pi_i P_{ij} = \lambda\pi_j$ for all j. A column vector $\boldsymbol{\nu}$ is a* ***right eigenvector of eigenvalue*** *λ if $\boldsymbol{\nu} \neq \boldsymbol{0}$ and $[P]\boldsymbol{\nu} = \lambda\boldsymbol{\nu}$, i.e., $\sum_j P_{ij}\nu_j = \lambda\nu_i$ for all i.*

We showed that for an ergodic unichain, there is a unique steady-state vector $\boldsymbol{\pi}$ that is a left eigenvector with $\lambda = 1$ and (within a scale factor) a unique right eigenvector $\boldsymbol{e} = (1, \dots, 1)^{\mathsf{T}}$. In this section we look at the other eigenvalues and eigenvectors and also look at Markov chains other than ergodic unichains. We start by limiting the number of states to $\mathsf{M} = 2$. This provides insight without requiring much linear algebra. After that, the general case with arbitrary $\mathsf{M} < \infty$ is analyzed.

4.4.1 Eigenvalues and eigenvectors for $\mathsf{M} = 2$ states

The eigenvalues and eigenvectors can be found by elementary (but slightly tedious) algebra. The left and right eigenvector equations can be written out as

$$\begin{aligned} \pi_1 P_{11} + \pi_2 P_{21} &= \lambda\pi_1 \\ \pi_1 P_{12} + \pi_2 P_{22} &= \lambda\pi_2 \end{aligned} \quad \text{(left)}, \qquad \begin{aligned} P_{11}\nu_1 + P_{12}\nu_2 &= \lambda\nu_1 \\ P_{21}\nu_1 + P_{22}\nu_2 &= \lambda\nu_2 \end{aligned} \quad \text{(right)}. \tag{4.27}$$

Each set of equations has a non-zero solution if and only if the matrix $[P - \lambda I]$, where $[I]$ is the identity matrix, is singular (i.e., there must be a non-zero $\boldsymbol{\nu}$ for which $[P - \lambda I]\boldsymbol{\nu} = \boldsymbol{0}$). Thus λ must be such that the determinant of $[P - \lambda I]$, namely $(P_{11} - \lambda)(P_{22} - \lambda) - P_{12}P_{21}$, is equal to 0. Solving this quadratic equation in λ, we find that λ has two solutions:

$$\lambda_1 = 1, \qquad \lambda_2 = 1 - P_{12} - P_{21}.$$

Assuming initially that P_{12} and P_{21} are not both 0, the solutions for the left and right eigenvectors, $\boldsymbol{\pi}^{(1)}$ and $\boldsymbol{\nu}^{(1)}$, of λ_1 and $\boldsymbol{\pi}^{(2)}$ and $\boldsymbol{\nu}^{(2)}$ of λ_2, are given by

$$\begin{array}{llll} \pi_1^{(1)} = \dfrac{P_{21}}{P_{12}+P_{21}} & \pi_2^{(1)} = \dfrac{P_{12}}{P_{12}+P_{21}}, & \nu_1^{(1)} = 1 & \nu_2^{(1)} = 1 \\ \pi_1^{(2)} = 1 & \pi_2^{(2)} = -1 & \nu_1^{(2)} = \dfrac{P_{12}}{P_{12}+P_{21}} & \nu_2^{(2)} = \dfrac{-P_{21}}{P_{12}+P_{21}} \end{array}.$$

These solutions contain arbitrarily chosen normalization factors, where $\boldsymbol{\pi}^{(1)} = (\pi_1^{(1)}, \pi_2^{(1)})$ has been chosen so that $\boldsymbol{\pi}^{(1)}$ is a steady-state vector (i.e., the components

[5] Right eigenvectors are much more common in linear algebra than left eigenvectors, which are often ignored in abstract linear algebra. Here a less abstract view is desirable because of the direct connection of $[P^n]$ with transition probabilities. When Markov chains with rewards are considered in Section 4.5, both left and right eigenvectors will play important roles.

sum to 1). The solutions have also been normalized so that $\boldsymbol{\pi}_i \boldsymbol{\nu}_i = 1$ for $i = 1, 2$. Now define

$$[\Lambda] = \begin{bmatrix} \lambda_1 & 0 \\ 0 & \lambda_2 \end{bmatrix} \quad \text{and} \quad [U] = \begin{bmatrix} \nu_1^{(1)} & \nu_1^{(2)} \\ \nu_2^{(1)} & \nu_2^{(2)} \end{bmatrix},$$

i.e., $[U]$ is a matrix whose columns are the eigenvectors $\boldsymbol{\nu}^{(1)}$ and $\boldsymbol{\nu}^{(2)}$. Then the two right eigenvector equations in (4.27) can be combined compactly as $[P][U] = [U][\Lambda]$. It turns out (for the given normalization of the eigenvectors) that the inverse of $[U]$ is just the matrix whose rows are the left eigenvectors of $[P]$ (this can be verified by noting that $\boldsymbol{\pi}_1 \boldsymbol{\nu}_2 = \boldsymbol{\pi}_2 \boldsymbol{\nu}_1 = 0$. We then see that $[P] = [U][\Lambda][U^{-1}]$ and consequently $[P^n] = [U][\Lambda]^n[U^{-1}]$. Multiplying this out, we get

$$[P^n] = \begin{bmatrix} \pi_1 + \pi_2\lambda_2^n & \pi_2 - \pi_2\lambda_2^n \\ \pi_1 - \pi_1\lambda_2^n & \pi_2 + \pi_1\lambda_2^n \end{bmatrix}, \tag{4.28}$$

where $\boldsymbol{\pi} = (\pi_1, \pi_2)$ is the steady-state vector $\boldsymbol{\pi}^{(1)}$. Recalling that $\lambda_2 = 1 - P_{12} - P_{21}$, we see that $|\lambda_2| \leq 1$. There are two trivial cases where $|\lambda_2| = 1$. In the first, $P_{12} = P_{21} = 0$, so that $[P]$ is just the identity matrix. The Markov chain then has two recurrent classes and stays forever where it starts. In the other trivial case, $P_{12} = P_{21} = 1$. Then $\lambda_2 = -1$ so that $[P^n]$ alternates between the identity matrix for n even and $[P]$ for n odd. In all other cases, $|\lambda_2| < 1$ and $[P^n]$ approaches the steady-state matrix $\lim_{n\to\infty}[P^n] = \boldsymbol{e}\boldsymbol{\pi}$.

What we have learned from this is the exact way in which $[P^n]$ approaches $\boldsymbol{e}\boldsymbol{\pi}$. Each term in $[P^n]$ approaches the steady-state value geometrically in n as λ_2^n. Thus, in place of the upper bound in (4.22), we have an exact expression, which in this case is simpler than the bound. As we see shortly, this result is representative of the general case, but the simplicity is lost.

4.4.2 Eigenvalues and eigenvectors for M > 2 states

For the general case of a stochastic matrix, we start with the fact that the set of eigenvalues is given by the set of (possibly complex) values of λ that satisfy the determinant equation $\det[P - \lambda I] = 0$. Since $\det[P - \lambda I]$ can be expressed as a polynomial of degree M in λ, this equation has M roots (i.e., M eigenvalues), not all of which need be distinct.[6]

Case with M *distinct eigenvalues:* We start with the simplest case in which the M eigenvalues, say $\lambda_1, \dots, \lambda_{\mathsf{M}}$, are all distinct. The matrix $[P - \lambda_i I]$ is singular for each i, so there must be a right eigenvector $\boldsymbol{\nu}^{(i)}$ and a left eigenvector $\boldsymbol{\pi}^{(i)}$ for each eigenvalue λ_i. The right eigenvectors span M-dimensional space and thus the matrix U with columns $(\boldsymbol{\nu}^{(1)}, \dots, \boldsymbol{\nu}^{(\mathsf{M})})$ is non-singular. The left eigenvectors, if normalized to satisfy $\boldsymbol{\pi}^{(i)}\boldsymbol{\nu}^{(i)} = 1$ for each i, then turn out to be the rows of $[U^{-1}]$ (see Exercise 4.11). As in the two-state case, we can then express $[P^n]$ as

$$[P^n] = [U][\Lambda^n][U^{-1}], \tag{4.29}$$

where Λ is the diagonal matrix with entries $\lambda_1, \dots, \lambda_{\mathsf{M}}$.

[6] Readers with little exposure to linear algebra can either accept the linear algebra results in this section (without a great deal of lost insight) or can find them in Strang [28] or many other linear algebra texts.

If Λ is broken into the sum of M diagonal matrices,[7] each with only a single non-zero element, then (see Exercise 4.11) $[P^n]$ can be expressed as

$$[P^n] = \sum_{i=1}^{\mathsf{M}} \lambda_i^n \boldsymbol{\nu}^{(i)} \boldsymbol{\pi}^{(i)}. \tag{4.30}$$

Note that this is the same form as (4.28), where in (4.28), the eigenvalue $\lambda_1 = 1$ simply appears as the value 1. Since each row of $[P]$ sums to 1, the vector $\boldsymbol{e} = (1, 1, \ldots, 1)^{\mathsf{T}}$ is a right eigenvector of eigenvalue 1, so there must also be a left eigenvector $\boldsymbol{\pi}$ of eigenvalue 1. The other eigenvalues and eigenvectors can be complex, but it is almost self evident from the fact that $[P^n]$ is a stochastic matrix that $|\lambda_i| \leq 1$ for each i. A simple guided proof of this is given in Exercise 4.12.

We have seen that $\lim_{n\to\infty}[P^n] = \boldsymbol{e\pi}$ for ergodic unichains. This implies that all terms except $i = 1$ in (4.30) die out with n, which further implies that $|\lambda_i| < 1$ for all eigenvalues except $\lambda = 1$. In this case, we see that the rate at which $[P^n]$ approaches steady state is given by the second largest eigenvalue in magnitude, i.e., $\max_{i:|\lambda_i|<1} |\lambda_i|$.

If a recurrent chain is periodic with period d, it turns out that there are d eigenvalues of magnitude 1, and these are uniformly spaced around the unit circle in the complex plane. Exercise 4.19 contains a guided proof of this.

Case with repeated eigenvalues and M *linearly independent eigenvectors:* If some of the M eigenvalues of $[P]$ are not distinct, the question arises as to how many linearly independent left (or right) eigenvectors exist for an eigenvalue λ_i of a given multiplicity k_i, i.e., a λ_i that is a k_ith-order root of $\det[P - \lambda I]$. Perhaps the ugliest part of linear algebra is the fact that an eigenvalue of multiplicity k need not have k linearly independent eigenvectors. An example of a very simple Markov chain with $\mathsf{M} = 3$ but only two linearly independent eigenvectors is given in Exercise 4.14. These eigenvectors do not span M-space, and thus the expansion in (4.30) cannot be used.

Before looking at this ugly case, we look at the case where the right eigenvectors, say, span the space, i.e., where the number of linearly-independent eigenvectors of each distinct eigenvalue is equal to the multiplicity of that eigenvalue. We can again form a matrix $[U]$ whose columns are the M linearly independent right eigenvectors, and again $[U^{-1}]$ is a matrix whose rows are the corresponding left eigenvectors of $[P]$. We then get (4.30) again. Thus, so long as the eigenvectors span the space, the asymptotic expression for the limiting transition probabilities can be found in the same way.

The most important situation where these repeated eigenvalues make a major difference is for Markov chains with $\kappa > 1$ recurrent classes. In this case, κ is the multiplicity of the eigenvalue 1. It is easy to see that there are κ different steady-state vectors. The steady-state vector for recurrent class ℓ, $1 \leq \ell \leq \kappa$, is strictly positive for each state of the ℓth recurrent class and is zero for all other states.

The set of eigenvalues for $[P]$ in this case is the aggregation of the set of eigenvalues for each separately viewed recurrent class. If class j contains r_j states, then r_j of the eigenvalues (counting repetitions) of $[P]$ are the eigenvalues of the $r_j \times r_j$ matrix for the

[7] If 0 is one of the M eigenvalues, then only $\mathsf{M} - 1$ such matrices are required.

states in that recurrent class. Thus the rate of convergence of $[P^n]$ within that submatrix is determined by the second largest eigenvalue (in magnitude) in that class.

What this means is that this general theory using eigenvalues says exactly what common sense says: if there are κ recurrent classes, look at each one separately, since they have nothing to do with each other. This also lets us see that for any recurrent class that is aperiodic, all the other eigenvalues for that class are strictly less than 1 in magnitude.

The situation is less obvious if there are κ recurrent classes plus a set of t transient states. All but t of the eigenvalues (counting repetitions) are associated with the recurrent classes, and the remaining t eigenvalues are the eigenvalues of the $t \times t$ matrix, say $[P_t]$, between the transient states. Each of these t eigenvalues is strictly less than 1 (as seen in Section 4.3.4) and neither these eigenvalues nor their eigenvectors depend on the transition probabilities from the transient to recurrent states. The left eigenvectors for the recurrent classes also do not depend on these transient to recurrent states. The right eigenvector for $\lambda = 1$ for each recurrent class R_ℓ is very interesting, however. Its value is 1 for each state in R_ℓ, is 0 for each state in the other recurrent classes, and is equal to $\lim_{n\to\infty} \Pr\{X_n \in \mathsf{R}_\ell \mid X_0 = i\}$ for each transient state i (see Exercise 4.13).

The Jordan form case: As mentioned before, there are cases in which one or more eigenvalues of $[P]$ are repeated (as roots of $\det[P - \lambda I]$) but the number of linearly independent right eigenvectors for a given repeated eigenvalue is less than the multiplicity of that eigenvalue. In this case, there are not enough eigenvectors to span the space, so there is no $\mathsf{M} \times \mathsf{M}$ matrix whose columns are linearly independent eigenvectors. Thus $[P]$ cannot be expressed as $[U][\Lambda][U^{-1}]$, where Λ is the diagonal matrix of the eigenvalues, repeated according to their multiplicity.

The Jordan form is the cure for this unfortunate situation. The Jordan form for a given $[P]$ is the following modification of the diagonal matrix of eigenvalues: we start with the diagonal matrix of eigenvalues, with the repeated eigenvalues as neighboring elements. Then for each missing eigenvector for a given eigenvalue, a 1 is placed immediately to the right and above a neighboring pair of appearances of that eigenvalue, as seen by example[8] below:

$$[J] = \begin{bmatrix} \lambda_1 & 1 & 0 & 0 & 0 \\ 0 & \lambda_1 & 0 & 0 & 0 \\ 0 & 0 & \lambda_2 & 1 & 0 \\ 0 & 0 & 0 & \lambda_2 & 1 \\ 0 & 0 & 0 & 0 & \lambda_2 \end{bmatrix}.$$

There is a theorem in linear algebra that says that an invertible matrix $[U]$ exists and a Jordan form exists such that $[P] = [U][J][U^{-1}]$. The major value to us of this result is that it makes it relatively easy to calculate $[J^n]$ for large n (see Exercise 4.15). This exercise also shows that for all stochastic matrices, each eigenvalue of magnitude 1 has precisely one associated eigenvector. This is usually expressed by the statement that all the eigenvalues of magnitude 1 are *simple*, meaning that their multiplicity equals

[8] See Strang [28], for example, for a more complete description of how to construct a Jordan form

their number of linearly independent eigenvectors. Finally the exercise shows that the components of $[P^n] - \boldsymbol{e\pi}$ for an ergodic chain converge to 0 at least as $n^k\lambda_s^n$, where λ_s is the eigenvalue of largest magnitude less than 1 and k is the size of the largest Jordan block of the eigenvalue λ_s.

The most important results of this section on eigenvalues and eigenvectors can be summarized in the following theorem.

Theorem 4.4.2 *The transition matrix of a finite-state unichain has a single eigenvalue* $\lambda = 1$ *with an accompanying left eigenvector* $\boldsymbol{\pi}$ *satisfying (4.8) and a right eigenvector* $\boldsymbol{e} = (1, 1, \dots, 1)^\mathsf{T}$. *The other eigenvalues* λ_i *all satisfy* $|\lambda_i| \le 1$. *The inequality is strict unless the unichain is periodic, say with period d, and then there are d eigenvalues of magnitude 1 spaced equally around the unit circle. If the unichain is ergodic, then* $[P^n]$ *converges to steady state* $\boldsymbol{e\pi}$ *with an error in each component that goes to 0 at least as* $n^k\lambda_s^n$, *where* λ_s *is the eigenvalue of largest magnitude less than 1 and k is the size of the largest Jordan block for the eigenvalue* λ_s.

Arbitrary Markov chains can be split into their recurrent classes, and this theorem can be applied separately to each class.

4.5 Markov chains with rewards

Suppose that each state i in a Markov chain is associated with a reward, r_i. As the Markov chain proceeds from state to state, there is an associated sequence of rewards that are not independent, but are related by the statistics of the Markov chain. The concept of a reward in each state[9] is quite graphic for modeling corporate profits or portfolio performance, and is also useful for studying queueing delay, the time until some given state is entered, and many other phenomena. The reward r_i associated with a state could equally well be veiwed as a cost or any given real-valued function of the state.

In Section 4.6, we study dynamic programming and Markov decision theory. These topics include a 'decision maker,' 'policy maker,' or 'control' that modifies both the transition probabilities and the rewards at each trial of the 'Markov chain.' The decision maker attempts to maximize the expected reward, but is typically faced with compromising between immediate reward and the longer-term reward arising from choosing transition probabilities that lead to 'high-reward' states. This is a more challenging problem than the rewards discussed in this section, but a thorough understanding of rewards provides the machinery to understand Markov decision theory.

Some problems involving rewards on Markov chains involve only transient behavior, looking at the reward until some particular event occurs and others involve rewards that

[9] Occasionally it is more natural to associate rewards with transitions rather than states. If r_{ij} denotes a reward associated with a transition from i to j and P_{ij} denotes the corresponding transition probability, then defining $r_i = \sum_j P_{ij}r_{ij}$ essentially simplifies these transition rewards to rewards over the initial state for the transition. These transition rewards are ignored here, since the details add complexity to a topic that is complex enough for a first treatment.

continue over all time, so that a significant parameter is the steady-state reward per unit time. We look first at some purely transient problems.

4.5.1 Expected first-passage times

In this subsection, we study the expected number of trials required to first reach some given state of a Markov chain, say state 1, starting from some other given state i. These problems arise in many contexts, such as finding the average time until a queue is empty or finding the average time until a queue overflows.

The number of trials until the first entry into state 1 is independent of the transitions after the first entry to state 1. Thus we can modify the chain to convert the final state, say state 1, into a trapping state (state 1 is a *trapping state* if it has no exit, i.e., if $P_{11} = 1$). That is, we modify P_{11} to 1 and P_{1j} to 0 for all $j \neq 1$. We leave P_{ij} unchanged for all $i \neq 1$ and all j (see Figure 4.6). This modification of the chain will not change the probability of any walk that includes state 1 only at the final transition.

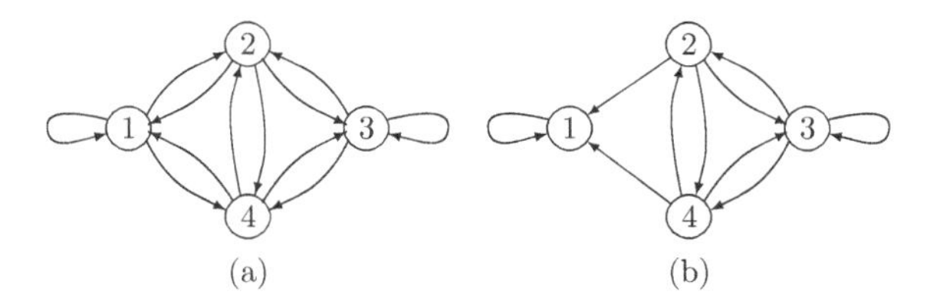

Figure 4.6 The conversion of the four-state recurrent Markov chain in (a) into the chain in (b) for which state 1 is a trapping state, i.e., the outgoing arcs from node 1 have been removed.

Let v_i be the expected number of steps to first reach state 1 starting in state $i \neq 1$. This number of steps includes the first step plus the expected number of remaining steps to reach state 1 starting from whatever state is entered next (if state 1 is the next state entered, this remaining number is 0). Thus, for the chain in Figure 4.6(b), we have the equations

$$\begin{aligned} v_2 &= 1 + P_{23}v_3 + P_{24}v_4, \\ v_3 &= 1 + P_{32}v_2 + P_{33}v_3 + P_{34}v_4, \\ v_4 &= 1 + P_{42}v_2 + P_{43}v_3. \end{aligned}$$

For an arbitrary chain of M states where 1 is a trapping state and all other states are transient, this set of equations becomes

$$v_i = 1 + \sum_{j \neq 1} P_{ij}v_j, \quad i \neq 1. \tag{4.31}$$

If we define $r_i = 1$ for $i \neq 1$ and $r_i = 0$ for $i = 1$, then r_i is a unit reward for not yet entering the trapping state, and v_i is the expected aggregate reward before entering the trapping state. Thus by taking $r_1 = 0$, the reward ceases upon entering the trapping state, and v_i is the expected transient reward, i.e., the expected first-passage time from state i to state 1. Note that, in this example, rewards occur only in transient states. Since transient states have zero steady-state probabilities, the steady-state gain per unit time is 0.

If we define $v_1 = 0$, then (4.31), in vector form (using the modified transition probabilities with $P_{11} = 1$) is given by

$$\boldsymbol{v} = \boldsymbol{r} + [P]\boldsymbol{v}, \quad v_1 = 0. \tag{4.32}$$

For a Markov chain with M states, (4.31) is a set of M $-$ 1 linear equations in the M $-$ 1 variables v_2 to v_{M}. The equation $\boldsymbol{v} = \boldsymbol{r} + [P]\boldsymbol{v}$ is a set of M linear equations, of which the first is the vacuous equation $v_1 = 0 + v_1$, and, with $v_1 = 0$, the last M $-$ 1 correspond to (4.31). It is not hard to show that (4.32) has a unique solution for $\boldsymbol{v}$ under the condition that states $2, \ldots, \mathsf{M}$ are all transient and 1 is a trapping state, but we prove this later, in Theorem 4.5.4, under more general circumstances.

Example 4.5.1 (Finding a string in a binary sequence) A problem that arises frequently in many fields is to find the expected number of trials until the first appearance of a given binary string in a sequence of binary digits. For simplicity, assume the binary sequence is IID with $\Pr\{X_i = 1\} = p_1$ and $\Pr\{X_i = 0\} = p_0 = 1 - p_1$. A binary string $\boldsymbol{a} = (a_1, a_2, \ldots, a_k)$ is said to occur at time n if $(X_{n-k+1}, \ldots, X_n) = (a_1, \ldots, a_k)$. We want to find the expected time n until this first occurrence.

First consider a trivial case where the string is a single binary digit, say 1. The string then occurs at time 1 with probability p_1, at time 2 with probability p_0p_1, and at time n with probability $p_0^{n-1}p_1$. The expected value is then $1/p_1$. This can also be posed as an expected-first-passage-time problem as below, and the result is again $1/p_1$.

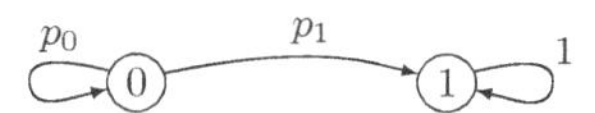

A slightly less trivial case occurs with the string (1,0). Here the expected time until the first 1 appears is $1/p_0$ as before. There is then a subsequent delay of zero or more subsequent 1's until the next 0 appears. At this point, the string (1, 0) has just occurred and the additional expected delay is $1/p_1$. Thus the overall delay is $1/p_0 + 1/p_1 = 1/p_0p_1$. This can also be posed as an expected first-passage time as below, and the expected first-passage time can be easily calculated to be $1/p_0p_1$.

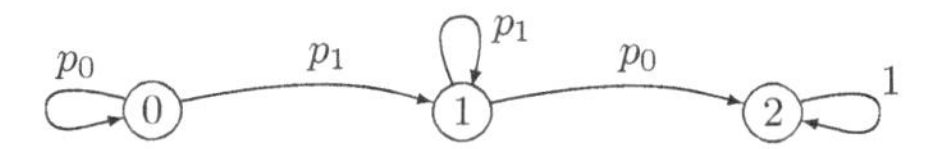

For an arbitrary string, the problem can again be formulated as a first-passage time problem, but the construction of the Markov chain is a little more complicated. For any binary string $\boldsymbol{a} = (a_1, a_2, \ldots, a_k)$, where the length $k \geq 1$ is arbitrary, we now show how to construct a corresponding Markov chain with $k+1$ states, labeled $0, \ldots, k$. Along with explaining the general case, we consider a particular example with $k = 5$ and $\boldsymbol{a} = (1, 0, 1, 0, 0)$. The Markov chain for the example will be shown to be that of Figure 4.7.

Both for the example and the general case, we define the Markov chain (if it exists) to be in state $i < k$ at time n if $(a_1, \ldots, a_k)$ has not yet occurred and if i is the largest

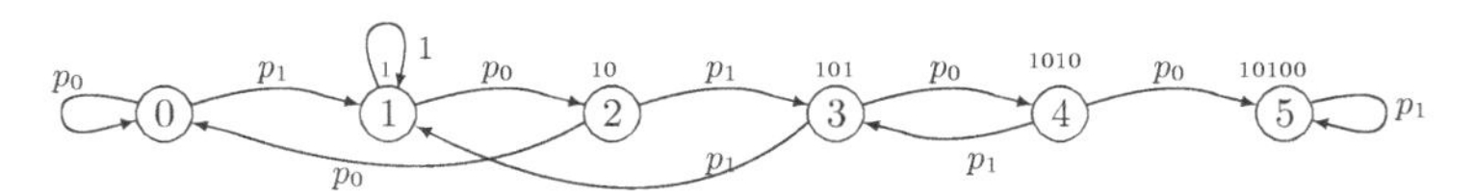

Figure 4.7 The Markov chain for finding the first occurrence of (10100). State 4, for example, indicates that the string 10100 will occur if the next bit is 0. If the next bit is 1, then the most recent fivebits will be 10101, and the final three of these are the first three bits of 10100, leading to state 3.

integer (less than k) such that $(X_{n-i+1}, \dots, X_n) = (a_1, \dots, a_i)$; i.e., $(X_{n-1-i}, \dots, X_n)$ is the longest suffix of $X_1, \dots, X_n$ that forms a proper prefix of $(a_1, \dots, a_k)$. In the special case $i = 0$, $(X_1, \dots, X_n)$ has no suffix of any positive length equal to a proper prefix of $(a_1, \dots, a_k)$.

We raise the question of existence because it is not obvious, given that the chain is in state i at time n, whether the new state at time $n+1$ can be determined by X_{n+1} and the state i. In other words, is the suffix of length i sufficient to determine the longest suffix at time $n+1$? The answer is yes, and we use induction on n to demonstrate this.

To establish the basis for the induction, we consider X_1. The initial state at time 0, according to the definition, is 0. If $X_1 = a_1$, then the state at time 1 is 1, since X_1 is the only suffix of the binary sequence up to the first bit, and it is equal to a_1. If $X_1 = \overline{a}_i$ (where $\overline{a}_1$ is the complement of a_1), then no suffix of (X_1) is equal to any prefix of $\boldsymbol{a}$. This determines the transitions out of state 0 if indeed a Markov chain exists. This also shows that at $n = 1$, the possible states, namely 0 and 1, have the properties of the definition.

Now, using induction, assume the states have the properties of the definition at time n. If the state at time n is $i < k$, then $(a_1, \dots, a_i) = (X_{n-i+1}, \dots, X_n)$ for the given i and no larger i. If $X_{n+1} = a_{i+1}$, then $(a_1, \dots, a_i, a_{i+1}) = (X_{n-i+1}, \dots, X_n, X_{n+1})$ for the given i and no larger, so the state at time $N+1$ is $i+1$ and the conditions of the definition are satisfied.

Alternatively, if $X_{n+1} = \overline{a}_{n+1}$, then $(a_1, \dots, a_i, a_{i+1}) \neq (X_{n-i+1}, \dots, X_n, X_{n+1})$, and inequality must hold for all larger i by the inductive assumption about the state at time n. This means, however, that the largest j that satisfies the definition is less than or equal to i. It then follows that the longest suffix of $X_1, \dots, X_{n+1}$ that satisfies the definition is determined by $X_{n-1+1}, \dots, X_{n+1}$, which is determined by state i and X_{n+1}. This shows that the state at time $n+1$ is determined according to the definition by X_{n+1} and the state at time n.

This completes the inductive argument, showing that a Markov chain of $k+1$ states exists for any string $\boldsymbol{a}$ of arbitrary length $k \geq 1$. The expected time to the first occurrence of $\boldsymbol{a}$ is then the expected first-passage time from state 0 to state $k+1$ in the Markov chain.

To construct the Markov chain, we use Figure 4.7 to illustrate the procedure. From each state $i < k$, there are two outgoing transitions, one of probability $p_{a_{i+1}}$ corresponding to $X_{n+1} = a_{i+1}$ and one of probability $p_{\overline{a}_{i+1}}$ corresponding to $X_{n+1} = \overline{a}_{i+1}$. The first goes to state $i+1$, as seen in Figure 4.7. The second goes to the largest j for which $(a_{i-j+2}, \dots, a_i, \overline{a}_{i+1}) = (a_1, \dots, a_j)$. In terms of prefixes and suffixes, this is the largest prefix of $(a_1, \dots, a_k)$ that is equal to a suffix of $(a_1, \dots, a_i, \overline{a}_{i+1})$.

This might look obscure as a general rule, but for any given example, it is quite simple. For example, from state 4 in Figure 4.7, $(a_1, \dots, a_i, \overline{a}_{i+1})$ is (1,0,1,0,1), and the longest suffix of this that is a prefix of $\boldsymbol{a} = (1, 0, 1, 0, 0)$ is (1,0,1), so there is a transition from state 4 to state 3.

Exercise 4.28 provides some simple strings for which the expected time until the string occurs has some special properties and is easily calculated. Renewal theory is then used in Exercise 5.35 to develop an actual formula for the expected time until the string occurs. Finally, a martingale is used in Exercise 9.26 to derive the same formula in an analytically simpler but less motivated way.

Example 4.5.2 Assume that a Markov chain has M states, $\{0, 1, \dots, \mathsf{M} - 1\}$, and that the state represents the number of customers in an integer-time queueing system. Suppose we wish to find the expected sum of the customer waiting times, starting with i customers in the system at some given integer time t and ending at the first integer time at which the system becomes idle. That is, for each of the i customers in the system at time t, the waiting time is counted from t until the integer time at which that customer exits the system. For each new customer entering before the system next becomes idle, the waiting time is counted from entry to exit. When we discuss Little's theorem in Section 5.5.4, it will be seen that this sum of waiting times is equal to the sum over τ of the state X_τ at time τ, taken from $\tau = t$ to the first subsequent time the system is empty.

As in the previous example, we modify the Markov chain to make state 0 a trapping state and assume the other states are then all transient. We take $r_i = i$ as the 'reward' in state i, and v_i as the expected aggregate reward until the trapping state is entered. Using the same reasoning as in the previous example, v_i is equal to the immediate reward $r_i = i$ plus the expected aggregate reward from whatever state is entered next. Thus $v_i = r_i + \sum_{j \geq 1} P_{ij} v_j$. This is $\boldsymbol{v} = \boldsymbol{r} + [P]\boldsymbol{v}$, with the additional constraint $v_0 = 0$. This has a unique solution for $\boldsymbol{v}$, as will be shown later in Theorem 4.5.4. This same analysis is valid for any choice of reward r_i for each transient state i; the reward in the trapping state must be 0 so as to keep the expected aggregate reward finite.

In the above examples, the final state of the first-passage problem is converted into a trapping state with zero gain, and thus the expected reward is a transient phenomenon with no reward after entering the trapping state. We now look at the more general case of a unichain. In this more general case, there can be some gain per unit time, along with some transient expected reward depending on the initial state. We first look at the aggregate gain over a finite number of time units, thus providing a clean way of going to the limit.

Example 4.5.3 The example in Figure 4.8 provides some intuitive appreciation for the general problem. Note that the chain tends to persist in whatever state it is in. Thus if the chain starts in state 2, not only is an immediate reward of 1 achieved, but there is a high probability of additional unit rewards on many successive transitions. Thus the

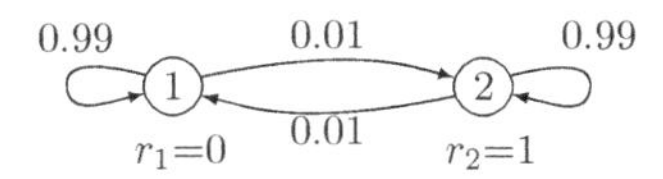

Figure 4.8 Markov chain with rewards and non-zero steady-state gain.

aggregate value of starting in state 2 is considerably more than the immediate reward of 1. On the other hand, we see from symmetry that the gain per unit time, over a long time period, must be one half.

4.5.2 The expected aggregate reward over multiple transitions

Returning to the general case, let X_m be the state at time m and let $R_m = R(X_m)$ be the reward at time m, i.e., if the sample value of X_m is i, then r_i is the sample value of R_m. Conditional on $X_m = i$, the aggregate expected reward $v_i(n)$ over n trials from X_m to X_{m+n-1} is

$$\begin{aligned} v_i(n) &= \mathsf{E}\left[R(X_m) + R(X_{m+1}) + \cdots + R(X_{m+n-1}) \mid X_m = i\right] \\ &= r_i + \sum_j P_{ij} r_j + \cdots + \sum_j P_{ij}^{n-1} r_j. \end{aligned}$$

This expression does not depend on the starting time m because of the homogeneity of the Markov chain. Since it gives the expected reward for each initial state i, it can be combined into the following vector expression $\boldsymbol{v}(n) = (v_1(n), v_2(n), \ldots, v_{\mathsf{M}}(n))^{\mathsf{T}}$:

$$\boldsymbol{v}(n) \;=\; \boldsymbol{r} + [P]\boldsymbol{r} + \cdots + [P^{n-1}]\boldsymbol{r} \;=\; \sum_{h=0}^{n-1} [P^h]\boldsymbol{r}, \tag{4.33}$$

where $\boldsymbol{r} = (r_1, \ldots, r_{\mathsf{M}})^{\mathsf{T}}$ and $[P^0]$ is the identity matrix. Now assume that the Markov chain is an ergodic unichain. Then $\lim_{n\to\infty}[P^n] = \boldsymbol{e\pi}$ and $\lim_{n\to\infty}[P^n]\boldsymbol{r} = \boldsymbol{e\pi r} = g\boldsymbol{e}$, where $g = \boldsymbol{\pi r}$ is the steady-state reward per unit time. If $g \neq 0$, then $\boldsymbol{v}(n)$ changes by approximately $g\boldsymbol{e}$ for each unit increase in n, so $\boldsymbol{v}(n)$ does not have a limit as $n \to \infty$. As shown below, however, $\boldsymbol{v}(n) - ng\boldsymbol{e}$ does have a limit, given by

$$\lim_{n\to\infty} (\boldsymbol{v}(n) - ng\boldsymbol{e}) = \lim_{n\to\infty} \sum_{h=0}^{n-1} [P^h - \boldsymbol{e\pi}]\boldsymbol{r}. \tag{4.34}$$

To see that this limit exists, note from (4.26) that $\epsilon > 0$ can be chosen small enough that $P_{ij}^n - \pi_j = o(\exp(-n\epsilon))$ for all states i,j and all $n \geq 1$. Thus $\sum_{h=n}^{\infty}(P_{ij}^h - \pi_j) = o(\exp(-n\epsilon))$ also. This shows that the limits on each side of (4.34) must exist for an ergodic unichain.

The limit in (4.34) is a vector over the states of the Markov chain. This vector gives the asymptotic relative expected advantage of starting the chain in one state relative to another. This is an important quantity in both the next section and the remainder of this one. It is called the relative-gain vector and denoted by $\boldsymbol{w}$:

$$\boldsymbol{w} = \lim_{n\to\infty} \sum_{h=0}^{n-1} [P^h - \boldsymbol{e\pi}]\boldsymbol{r} \tag{4.35}$$

$$= \lim_{n\to\infty} (\boldsymbol{v}(n) - ng\boldsymbol{e}). \tag{4.36}$$

Note from (4.36) that if $g > 0$, then $ng\boldsymbol{e}$ increases linearly with n and $\boldsymbol{v}(n)$ must asymptotically increase linearly with n. Thus the relative-gain vector $\boldsymbol{w}$ becomes small relative to both $ng\boldsymbol{e}$ and $\boldsymbol{v}(n)$ for large n. As we will see, $\boldsymbol{w}$ is still important, particularly in the next section on Markov decisions.

We can get some feel for $\boldsymbol{w}$ and how $v_i(n) - n\pi_i$ converges to w_i from Example 4.5.3 (as described in Figure 4.8). Since this chain has only two states, $[P^n]$ and $v_i(n)$ can be calculated easily from (4.28). The result is tabulated in Figure 4.9, and it is seen numerically that $\boldsymbol{w} = (-25, +25)^{\mathsf{T}}$. The rather significant advantage of starting in state 2 rather than state 1, however, requires hundreds of transitions before the gain is fully apparent.

n	$\boldsymbol{\pi v}(n)$	$v_1(n)$	$v_2(n)$
1	0.5	0	1
2	1	0.01	1.99
4	2	0.0592	3.9408
10	5	0.4268	9.5732
40	20	6.1425	33.8575
100	50	28.3155	71.6845
400	200	175.007	224.9923

Figure 4.9 The expected aggregate reward, as a function of starting state and stage, for the example of Figure 4.8. Note that $\boldsymbol{w} = (-25, +25)^{\mathsf{T}}$, but the convergence is quite slow.

This example also shows that it is somewhat inconvenient to calculate $\boldsymbol{w}$ from (4.35), and this inconvenience grows rapidly with the number of states. Fortunately, as shown in the following theorem, $\boldsymbol{w}$ can also be calculated simply by solving a set of linear equations.

Theorem 4.5.4 *Let $[P]$ be the transition matrix for an ergodic unichain. Then the relative-gain vector $\boldsymbol{w}$ given in (4.35) satisfies the following linear vector equation.*

$$\boldsymbol{w} + g\boldsymbol{e} = [P]\boldsymbol{w} + \boldsymbol{r} \qquad \textit{and} \qquad \boldsymbol{\pi w} = 0. \tag{4.37}$$

Furthermore (4.37) has a unique solution if $[P]$ is the transition matrix for a unichain (either ergodic or periodic).

Discussion For an ergodic unichain, the interpretation of $\boldsymbol{w}$ as an asymptotic relative gain comes from (4.35) and (4.36). For a periodic unichain, (4.37) still has a unique solution, but (4.35) no longer converges, so the solution to (4.37) no longer has a clean interpretation as an asymptotic limit of relative gain. This solution is still called a relative-gain vector, and can be interpreted as an asymptotic relative gain over a period, but the important result is that this equation has a unique solution for arbitrary unichains.

Definition 4.5.5 *The* ***relative-gain vector*** $\boldsymbol{w}$ *of a unichain is the unique vector that satisfies (4.37).*

Proof of Theorem 4.5.4 Premultiplying both sides of (4.35) by $[P]$,

$$
\begin{aligned}
[P]\boldsymbol{w} &= \lim_{n\to\infty} \sum_{h=0}^{n-1} \left[P^{h+1} - \boldsymbol{e\pi}\right]\boldsymbol{r} = \lim_{n\to\infty} \sum_{h=1}^{n} \left[P^{h} - \boldsymbol{e\pi}\right]\boldsymbol{r} \\
&= \lim_{n\to\infty} \left(\sum_{h=0}^{n} \left[P^{h} - \boldsymbol{e\pi}\right]\boldsymbol{r} \right) - \left[P^{0} - \boldsymbol{e\pi}\right]\boldsymbol{r} \\
&= \boldsymbol{w} - \left[P^{0} - \boldsymbol{e\pi}\right]\boldsymbol{r} \\
&= \boldsymbol{w} - \boldsymbol{r} + g\boldsymbol{e},
\end{aligned}
$$

where the next to last equality used (4.35) and the last equality used $[P^0] = I$ and $\boldsymbol{\pi r} = g$. Rearranging terms, we get (4.37). For a unichain, the eigenvalue 1 of $[P]$ has multiplicity 1, and the existence and uniqueness of the solution to (4.37) is then a simple result in linear algebra (see Exercise 4.23). □

Consider applying Theorem 4.5.4 to the special case of the first-passage-time example. In that case, $\pi_1 = 1$ and $r_1 = 0$ so that the steady-state gain $\boldsymbol{\pi r} = 0$. Thus (4.37) becomes $\boldsymbol{w} = [P]\boldsymbol{w} + \boldsymbol{r}$ with $\boldsymbol{\pi w} = 0$. The constraint $\boldsymbol{\pi w} = 0$ simply says that $w_1 = 0$. Thus the relative-gain vector $\boldsymbol{w}$ for this example is simply the expected aggregate reward $\boldsymbol{v}$ before entering the trapping state as expressed in (4.32).

The reason that the derivation of aggregate reward was so simple for first-passage time is that there was no steady-state gain in that example, and thus no need to separate the gain per transition g from the relative gain $\boldsymbol{w}$ between starting states.

One way to apply the intuition of the $g = 0$ case to the general case is as follows: given a reward vector $\boldsymbol{r}$, find the steady-state gain $g = \boldsymbol{\pi r}$, and then define a modified reward vector $\boldsymbol{r}' = \boldsymbol{r} - g\boldsymbol{e}$. Changing the reward vector from $\boldsymbol{r}$ to $\boldsymbol{r}'$ in this way does not change $\boldsymbol{w}$, but the modified limiting aggregate gain, say $\boldsymbol{v}'(n)$, then has a limit, which is in fact $\boldsymbol{w}$. The intuitive derivation used in (4.32) again gives us $\boldsymbol{w} = [P]\boldsymbol{w} + \boldsymbol{r}'$. This is equivalent to (4.37) since $\boldsymbol{r}' = \boldsymbol{r} - g\boldsymbol{e}$.

There are many generalizations of the first-passage-time example in which the reward in each recurrent state of a unichain is 0. Thus reward is accumulated only until a recurrent state is entered. The following corollary provides a monotonicity result about the relative-gain vector for these circumstances that might seem obvious.[10] Thus we simply state it and give a guided proof in Exercise 4.25.

Corollary 4.5.6 *Let $[P]$ be the transition matrix of a unichain with the recurrent class $\mathcal{R}$. Let $\boldsymbol{r} \geq 0$ be a reward vector for $[P]$ with $r_i = 0$ for $i \in \mathcal{R}$. Then the relative-gain vector $\boldsymbol{w}$ satisfies $\boldsymbol{w} \geq 0$ with $w_i = 0$ for $i \in \mathcal{R}$ and $w_i > 0$ for $r_i > 0$. Furthermore, if $\boldsymbol{r}'$ and $\boldsymbol{r}''$ are different reward vectors for $[P]$ and $\boldsymbol{r}' \geq \boldsymbol{r}''$ with $r'_i = r''_i$ for $i \in \mathcal{R}$, then $\boldsymbol{w}' \geq \boldsymbol{w}''$ with $w'_i = w''_i$ for $i \in \mathcal{R}$ and $w'_i > w''_i$ for $r'_i > r''_i$.*

[10] An obvious counterexample if we omit the condition $r_i = 0$ for $i \in \mathcal{R}$ is given by Figure 4.8 where $\boldsymbol{r} = (0, 1)^{\mathsf{T}}$ and $\boldsymbol{w} = (-25, 25)^{\mathsf{T}}$.

4.5.3 The expected aggregate reward with an additional final reward

Frequently when a reward is aggregated over n transitions of a Markov chain, it is appropriate to assign some added reward, say u_i, as a function of the final state i. For example, it might be particularly advantageous to end in some particular state. Also, if we wish to view the aggregate reward over $n+\ell$ transitions as the reward over the first n transitions plus that over the following ℓ transitions, we can model the expected reward over the final ℓ transitions as a final reward at the end of the first n transitions. Note that this final expected reward depends only on the state at the end of the first n transitions.

As before, let $R(X_{m+h})$ be the reward at time $m+h$ for $0 \le h \le n-1$ and $U(X_{m+n})$ be the final reward at time $m+n$, where $U(X) = u_i$ for $X = i$. Let $v_i(n, \boldsymbol{u})$ be the expected reward from time m to $m+n$, using the reward $\boldsymbol{r}$ from time m to $m+n-1$ and using the final reward $\boldsymbol{u}$ at time $m+n$. The expected reward is then the following simple modification of (4.33):

$$\boldsymbol{v}(n, \boldsymbol{u}) = \boldsymbol{r} + [P]\boldsymbol{r} + \cdots + [P^{n-1}]\boldsymbol{r} + [P^n]\boldsymbol{u} = \sum_{h=0}^{n-1}[P^h]\boldsymbol{r} + [P^n]\boldsymbol{u}. \tag{4.38}$$

This simplifies considerably if $\boldsymbol{u}$ is taken to be the relative-gain vector $\boldsymbol{w}$.

Theorem 4.5.7 *Let $[P]$ be the transition matrix of a unichain and let $\boldsymbol{w}$ be the corresponding relative-gain vector. Then for each $n \ge 1$,*

$$\boldsymbol{v}(n, \boldsymbol{w}) = ng\boldsymbol{e} + \boldsymbol{w}. \tag{4.39}$$

Also, for an arbitrary final reward vector $\boldsymbol{u}$,

$$\boldsymbol{v}(n, \boldsymbol{u}) = ng\boldsymbol{e} + \boldsymbol{w} + [P^n](\boldsymbol{u} - \boldsymbol{w}). \tag{4.40}$$

Discussion An important special case of (4.40) arises from setting the final reward $\boldsymbol{u}$ to 0, thus yielding the following expression for $\boldsymbol{v}(n)$:

$$\boldsymbol{v}(n) = ng\boldsymbol{e} + \boldsymbol{w} - [P^n]\boldsymbol{w}. \tag{4.41}$$

For an ergodic unichain, $\lim_{n\to\infty}[P^n] = \boldsymbol{e}\boldsymbol{\pi}$. Since $\boldsymbol{\pi}\boldsymbol{w} = 0$ by definition of $\boldsymbol{w}$, the limit of (4.41) as $n \to \infty$ is

$$\lim_{n\to\infty}(\boldsymbol{v}(n) - ng\boldsymbol{e}) = \boldsymbol{w},$$

which agrees with (4.36). The advantage of (4.41) over (4.36) is that it provides an explicit expression for $\boldsymbol{v}(n)$ for each n and also that it continues to hold for a periodic unichain.

Proof For $n = 1$, we see from (4.38) that

$$\boldsymbol{v}(1, \boldsymbol{w}) = \boldsymbol{r} + [P]\boldsymbol{w} = g\boldsymbol{e} + \boldsymbol{w},$$

so the theorem is satisfied for $n = 1$. For $n > 1$,

$$\boldsymbol{v}(n, \boldsymbol{w}) = \sum_{h=0}^{n-1}[P^h]\boldsymbol{r} + [P^n]\boldsymbol{w}$$

$$= \sum_{h=0}^{n-2} [P^h]\boldsymbol{r} + [P^{n-1}]\,(\boldsymbol{r} + [P]\boldsymbol{w})$$
$$= \sum_{h=0}^{n-2} [P^h]\boldsymbol{r} + [P^{n-1}]\,(g\boldsymbol{e} + \boldsymbol{w})$$
$$= \boldsymbol{v}(n-1, \boldsymbol{w}) + g\boldsymbol{e},$$

where the last step uses (4.38) plus the fact that $\boldsymbol{e}$ is a right eigenvector of any stochastic matrix. Using induction, this implies (4.39).

To establish (4.40), note from (4.38) that

$$\boldsymbol{v}(n, \boldsymbol{u}) - \boldsymbol{v}(n, \boldsymbol{w}) \;=\; [P^n](\boldsymbol{u} - \boldsymbol{w}).$$

Then (4.40) follows by using (4.39) for the value of $\boldsymbol{v}(n, \boldsymbol{w})$. □

4.6 Markov decision theory and dynamic programming

In the previous section, we analyzed the behavior of a Markov chain with rewards. In this section, we consider a much more elaborate structure in which a decision maker can choose among various possible rewards and transition probabilities. In place of the reward r_i and the transition probabilities $\{P_{ij}; 1 \leq j \leq \mathsf{M}\}$ associated with a given state i, there is a choice between some number K_i of different rewards, say $r_i^{(1)}, r_i^{(2)}, \ldots, r_i^{(K_i)}$ and a corresponding choice between K_i different sets of transition probabilities, say $\{P_{ij}^{(1)}; 1 \leq j \leq \mathsf{M}\}, \{P_{ij}^{(2)}, 1 \leq j \leq \mathsf{M}\}, \ldots, \{P_{ij}^{(K_i)}; 1 \leq j \leq \mathsf{M}\}$. At each time m, a decision maker, given $X_m = i$, selects one of the K_i possible choices for state i. Note that if decision k is chosen in state i, then the reward is $r_i^{(k)}$ and the transition probabilities from i are $\{P_{ij}^{(k)}; 1 \leq j \leq \mathsf{M}\}$; it is not permissable to choose $r_i^{(k)}$ for one k and $\{P_{ij}^{(k)}; 1 \leq j \leq \mathsf{M}\}$ for another k. We also assume that if decision k is selected at time m, the probability of entering state j at time $m+1$ is $P_{ij}^{(k)}$, independent of earlier states and decisions.

Figure 4.10 shows an example of this situation in which the decision maker can choose between two possible decisions in state 2 ($K_2 = 2$), and has no freedom of choice in state 1 ($K_1 = 1$). This figure illustrates the familiar tradeoff between instant gratification (alternative 2) and long-term gratification (alternative 1).

The set of rules used by the decision maker in selecting an alternative at each time is called a *policy*. We want to consider the expected aggregate reward over n steps of the 'Markov chain' as a function of the policy used by the decision maker. If for each

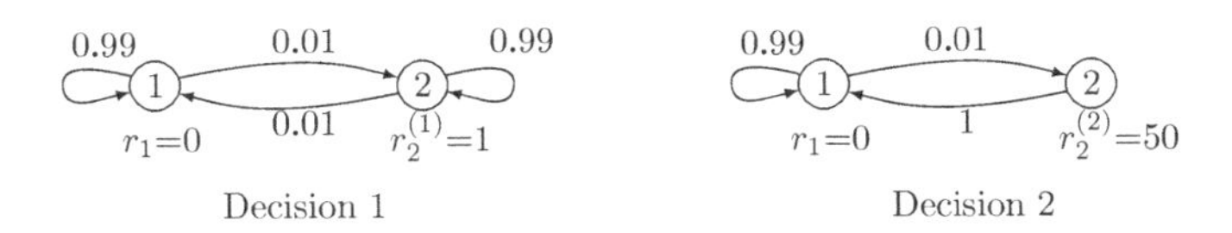

Figure 4.10 A Markov decision problem with two alternatives in state 2.

state i, the policy uses the same decision, say k_i, at each occurrence of i, then that policy corresponds to a homogeneous Markov chain with transition probabilities $P_{ij}^{(k_i)}$. We denote the matrix of these transition probabilities as $[P^{\boldsymbol{k}}]$, where $\boldsymbol{k} = (k_1, \dots, k_{\mathsf{M}})$. Such a policy, i.e., mapping each state i into a fixed decision k_i, independent of time, of past decisions, and of past transitions, is called a stationary policy. The aggregate gain for any such stationary policy was found in the previous section. Since both rewards and transition probabilities depend only on the state and the corresponding decision, and not on time, one feels intuitively that stationary policies make a certain amount of sense over a long period of time. On the other hand, if we look at the example of Figure 4.10, it is clear that decision 2 is the best choice in state 2 at the nth of n trials, but it is less obvious what to do at earlier trials.

In what follows, we first derive the optimal policy for maximizing expected aggregate reward over an arbitrary number n of trials, say at times m to $m+n-1$. We shall see that the decision at time $m+h$, $0 \leq h < n$, for the optimal policy can in fact depend on h and n (but not m). It turns out to simplify matters considerably if we include a final reward $\{u_i; 1 \leq i \leq \mathsf{M}\}$ at time $m+n$. This final reward $\boldsymbol{u}$ is considered as a fixed vector, to be chosen as appropriate, rather than as part of the choice of policy.

This optimized strategy, as a function of the number of steps n and the final reward $\boldsymbol{u}$, is called an *optimal dynamic policy* for that $\boldsymbol{u}$. This policy is found from the dynamic programming algorithm, which, as we shall see, is conceptually very simple. We then go on to find the relationship between optimal dynamic policies and optimal stationary policies. We shall find that, under fairly general conditions, each has the same long-term gain per trial.

4.6.1 Dynamic programming algorithm

As in our development of Markov chains with rewards, we consider the expected aggregate reward over n time periods, say m to $m+n-1$, with a final reward at time $m+n$. First, consider the optimal decision with $n=1$. Given $X_m = i$, a decision k is made with immediate reward $r_i^{(k)}$. With probability $P_{ij}^{(k)}$ the next state X_{m+1} is state j and the final reward is then u_j. The expected aggregate reward over times m and $m+1$, maximized over the decision k, is then

$$v_i^*(1, \boldsymbol{u}) = \max_k \left\{ r_i^{(k)} + \sum_j P_{ij}^{(k)} u_j \right\}. \tag{4.42}$$

Being explicit about the maximizing decision k', (4.42) becomes

$$v_i^*(1, \boldsymbol{u}) = r_i^{(k')} + \textstyle\sum_j P_{ij}^{(k')} u_j \qquad \text{for } k' \text{ such that}$$
$$r_i^{(k')} + \textstyle\sum_j P_{ij}^{(k')} u_j = \max_k \left\{ r_i^{(k)} + \sum_j P_{ij}^{(k)} u_j \right\}. \tag{4.43}$$

Note that a decision is made only at time m, but that there are two rewards, one at time m and the other, the final reward, at time $m+1$. We use the notation $v_i^*(n, \boldsymbol{u})$ to represent the maximum expected aggregate reward from times m to $m+n$ starting at $X_m = i$.

Decisions (with the reward vector $\boldsymbol{r}$) are made at the n times m to $m+n-1$, and this is followed by a final reward vector $\boldsymbol{u}$ (without any decision) at time $m+n$. It often simplifies notation to define the vector of maximal expected aggregate rewards

$$\boldsymbol{v}^*(n,\boldsymbol{u}) = (v_1^*(n,\boldsymbol{u}), v_2^*(n,\boldsymbol{u}), \ldots, v_{\mathsf{M}}^*(n,\boldsymbol{u}))^{\mathsf{T}}.$$

With this notation, (4.42) and (4.43) become

$$\boldsymbol{v}^*(1,\boldsymbol{u}) = \max_{\boldsymbol{k}}\{\boldsymbol{r}^{\boldsymbol{k}} + [P^{\boldsymbol{k}}]\boldsymbol{u}\}, \quad \text{where } \boldsymbol{k} = (k_1,\ldots,k_{\mathsf{M}})^{\mathsf{T}},\ \boldsymbol{r}^{\boldsymbol{k}} = (r_1^{k_1},\ldots,r_{\mathsf{M}}^{k_{\mathsf{M}}})^{\mathsf{T}}, \tag{4.44}$$

$$\boldsymbol{v}^*(1,\boldsymbol{u}) = \boldsymbol{r}^{\boldsymbol{k}'} + [P^{\boldsymbol{k}'}]\boldsymbol{u}, \qquad \text{where } \boldsymbol{r}^{\boldsymbol{k}'} + [P^{\boldsymbol{k}'}]\boldsymbol{u} = \max_{\boldsymbol{k}} \boldsymbol{r}^{\boldsymbol{k}} + [P^{\boldsymbol{k}}]\boldsymbol{u}. \tag{4.45}$$

Now consider $v_i^*(2,\boldsymbol{u})$, i.e., the maximal expected aggregate reward starting at $X_m = i$ with decisions made at times m and $m+1$ and a final reward at time $m+2$. The key to dynamic programming is that an optimal decision at time $m+1$ can be selected based only on the state j at time $m+1$; this decision (given $X_{m+1} = j$) is optimal independent of the decision at time m. That is, whatever decision is made at time m, the maximal expected reward at times $m+1$ and $m+2$, given $X_{m+1} = j$, is $\max_k(r_j^{(k)} + \sum_\ell P_{j\ell}^{(k)} u_\ell)$. Note that this maximum is $v_j^*(1,\boldsymbol{u})$, as found in (4.42).

Using this optimized decision at time $m+1$, it is seen that if $X_m = i$ and decision k is made at time m, then the sum of expected rewards at times $m+1$ and $m+2$ is $\sum_j P_{ij}^{(k)} v_j^*(1,\boldsymbol{u})$. Adding the expected reward at time m and maximizing over decisions at time m,

$$v_i^*(2,\boldsymbol{u}) = \max_k \left(r_i^{(k)} + \sum\nolimits_j P_{ij}^{(k)} v_j^*(1,\boldsymbol{u}) \right). \tag{4.46}$$

In other words, the maximum aggregate gain over times m to $m+2$ (using the final reward $\boldsymbol{u}$ at $m+2$) is the maximum over choices at time m of the sum of the reward at m plus the maximum aggregate expected reward for $m+1$ and $m+2$. The simple expression of (4.46) results from the fact that the maximization over the choice at time $m+1$ depends on the state at $m+1$ but, given that state, is independent of the policy chosen at time m.

This same argument can be used for all larger numbers of trials. To find the maximum expected aggregate reward from time m to $m+n$, we first find the maximum expected aggregate reward from $m+1$ to $m+n$, conditional on $X_{m+1} = j$ for each state j. This is the same as the maximum expected aggregate reward from time m to $m+n-1$, which is $v_j^*(n-1,\boldsymbol{u})$. This gives us the general expression for $n \geq 2$:

$$v_i^*(n,\boldsymbol{u}) = \max_k \left(r_i^{(k)} + \sum\nolimits_j P_{ij}^{(k)} v_j^*(n-1,\boldsymbol{u}) \right). \tag{4.47}$$

We can also write this in vector form as

$$\boldsymbol{v}^*(n,\boldsymbol{u}) = \max_{\boldsymbol{k}} \left(\boldsymbol{r}^{\boldsymbol{k}} + [P^{\boldsymbol{k}}]\boldsymbol{v}^*(n-1,\boldsymbol{u}) \right). \tag{4.48}$$

Here $\boldsymbol{k}$ is a set (or vector) of decisions, $\boldsymbol{k} = (k_1, k_2, \ldots, k_{\mathsf{M}})^{\mathsf{T}}$, where k_i is the decision for state i; $[P^{\boldsymbol{k}}]$ denotes a matrix whose (i,j) element is $P_{ij}^{(k_i)}$; and $\boldsymbol{r}^{\boldsymbol{k}}$ denotes a vector whose ith element is $r_i^{(k_i)}$. The maximization over $\boldsymbol{k}$ in (4.48) is really M separate and independent maximizations, one for each state, i.e., (4.48) is simply a vector form of (4.47). Another frequently useful way to rewrite (4.48) is as follows:

$$\boldsymbol{v}^*(n,\boldsymbol{u}) = \boldsymbol{r}^{\boldsymbol{k}'} + [P^{\boldsymbol{k}'}]\boldsymbol{v}^*(n-1,\boldsymbol{u}) \quad \text{for } \boldsymbol{k}' \text{ such that}$$
$$\boldsymbol{r}^{\boldsymbol{k}'} + [P^{\boldsymbol{k}'}]\boldsymbol{v}^*(n-1,\boldsymbol{u}) = \max_{\boldsymbol{k}}\left(\boldsymbol{r}^{\boldsymbol{k}} + [P^{\boldsymbol{k}}]\boldsymbol{v}^*(n-1,\boldsymbol{u})\right). \tag{4.49}$$

If $\boldsymbol{k}'$ satisfies (4.49), then $\boldsymbol{k}'$ is an optimal decision at an arbitrary time m given, first, that the objective is to maximize the aggregate gain from time m to $m+n$, second, that optimal decisions for this objective are to be made at times $m+1$ to $m+n-1$, and, third, that $\boldsymbol{u}$ is the final reward vector at $m+n$. In the same way, $\boldsymbol{v}^*(n,\boldsymbol{u})$ is the maximum expected reward over this finite sequence of n decisions from m to $m+n-1$ with the final reward $\boldsymbol{u}$ at $m+n$.

Note that (4.47), (4.48), and (4.49) are valid with no restrictions (such as recurrent or aperiodic states) on the possible transition probabilities $[P^{\boldsymbol{k}}]$. These equations are also valid in principle if the size of the state space is infinite. However, the optimization for each n can then depend on an infinite number of optimizations at $n-1$, which is often infeasible.

The *dynamic programming algorithm* is just the calculation of (4.47), (4.48), or (4.49), performed iteratively for $n = 1, 2, 3, \ldots$. The development of this algorithm, as a systematic tool for solving this class of problems, is due to Bellman [1]. Note that the algorithm is independent of the starting time m; the parameter n, usually referred to as stage n, is the number of decisions over which the aggregate gain is being optimized. This algorithm yields the optimal dynamic policy for any fixed final reward vector $\boldsymbol{u}$ and any given number of trials. Along with the calculation of $\boldsymbol{v}^*(n,\boldsymbol{u})$ for each n, the algorithm also yields the optimal decision at each stage (under the assumption that the optimal policy is to be used for each lower numbered stage, i.e., for each later trial of the process).

The surprising simplicity of the algorithm is due to the Markov property. That is, $v_i^*(n,\boldsymbol{u})$ is the aggregate present and future reward conditional on the present state. Since it is conditioned on the present state, it is independent of the past (i.e., how the process arrived at state i from previous transitions and choices).

Although dynamic programming is computationally straightforward and convenient,[11] the asymptotic behavior of $\boldsymbol{v}^*(n,\boldsymbol{u})$ as $n \to \infty$ is not evident from the algorithm. After working out some simple examples, we look at the general question of asymptotic behavior.

Example 4.6.1 Consider Figure 4.10, repeated below, with the final rewards $u_2 = u_1 = 0$.

[11] Unfortunately, many dynamic programming problems of interest have enormous numbers of states and possible choices of decision (the so-called curse of dimensionality), and thus, even though the equations are simple, the computational requirements might be beyond the range of practical feasibility.

Since $r_1 = 0$ and $u_1 = u_2 = 0$, the aggregate gain in state 1 at stage 1 is

$$v_1^*(1, \boldsymbol{u}) = r_1 + \sum_j P_{1j} u_j = 0.$$

Similarly, since policy 1 has an immediate reward $r_2^{(1)} = 1$ in state 2, and policy 2 has an immediate reward $r_2^{(2)} = 50$,

$$v_2^*(1, \boldsymbol{u}) = \max\left\{\left[r_2^{(1)} + \sum_j P_{2j}^{(1)} u_j\right],\ \left[r_2^{(2)} + \sum_j P_{2j}^{(2)} u_j\right]\right\} = \max\{1,\ 50\} = 50.$$

We can now go on to stage 2, using the results above for $v_j^*(1, \boldsymbol{u})$. From (4.46),

$$v_1^*(2, \boldsymbol{u}) = r_1 + P_{11} v_1^*(1, \boldsymbol{u}) + P_{12} v_2^*(1, \boldsymbol{u}) = P_{12} v_2^*(1, \boldsymbol{u}) = 0.5,$$

$$v_2^*(2, \boldsymbol{u}) = \max\left\{\left[r_2^{(1)} + \sum_j P_{2j}^{(1)} v_j^*(1, \boldsymbol{u})\right],\ \left[r_2^{(2)} + P_{21}^{(2)} v_1^*(1, \boldsymbol{u})\right]\right\}$$

$$= \max\left\{[1 + P_{22}^{(1)} v_2^*(1, \boldsymbol{u})],\ 50\right\} = \max\{50.5,\ 50\} = 50.5.$$

Thus for two trials, decision 1 is optimal in state 2 for the first trial (stage 2), and decision 2 is optimal in state 2 for the second trial (stage 1). What is happening is that the choice of decision 2 at stage 1 has made it very profitable to be in state 2 at stage 1. Thus if the chain is in state 2 at stage 2, it is preferable to choose decision 1 (i.e., the small unit gain) at stage 2 with the corresponding high probability of remaining in state 2 at stage 1. Continuing this computation for larger n, one finds that $v_1^*(n, \boldsymbol{u}) = n/2$ and $v_2^*(n, \boldsymbol{u}) = 50 + n/2$. The optimum dynamic policy (for $\boldsymbol{u} = 0$) is decision 2 for stage 1 (i.e., for the last decision to be made) and decision 1 for all stages $n > 1$ (i.e., for all decisions before the last).

This example also illustrates that the maximization of expected gain is not necessarily what is most desirable in all applications. For example, risk-averse people might well prefer decision 2 at the next to final decision (stage 2). This guarantees a reward of 50, rather than taking a small chance of losing that reward.

Example 4.6.2 (Shortest path problems) The problem of finding the shortest paths between nodes in a directed graph arises in many situations, from routing in communication networks to calculating the time to complete complex tasks. The problem is quite similar to the expected first-passage time in Section 4.5.1. In that problem, arcs in a directed graph were selected according to a probability distribution, whereas here decisions must be made about which arcs to take. Although this is not a probabilistic problem, the decisions can be posed as choosing a given arc with probability 1, thus viewing the problem as a special case of dynamic programming.

Consider finding the shortest path from each node in a directed graph to some particular node, say node 1 (see Figure 4.11). Each arc (except the special arc (1, 1)) has a positive *link length* associated with it that might reflect physical distance or an arbitrary type of cost. The special arc (1, 1) has 0 link length. The length of a path is the sum

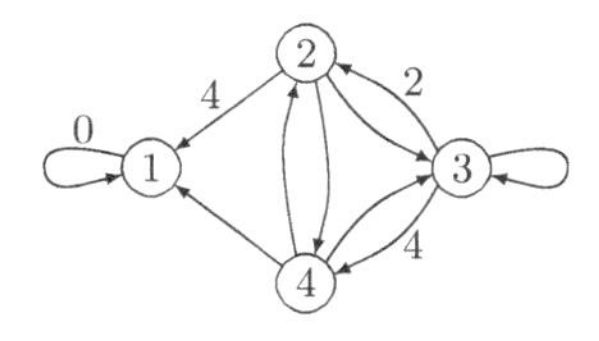

Figure 4.11 A shortest path problem. The arcs are marked with their lengths. Unmarked arcs have unit length.

of the lengths of the arcs on that path. In terms of dynamic programming, a policy is a choice of arc out of each node (state). Here we want to minimize cost (i.e., path length) rather than maximize reward, so we simply replace the maximum in the dynamic programming algorithm with a minimum (or, if one wishes, all costs can be replaced with negative rewards).

We start the dynamic programming algorithm with a final cost vector that is 0 for node 1 and infinite for all other nodes. In stage 1, the minimal cost decision for node (state) 2 is arc (2, 1) with a cost equal to 4. The minimal cost decision for node 4 is (4, 1) with unit cost. The cost from node 3 (at stage 1) is infinite whichever decision is made. The stage 1 costs are then

$$v_1^*(1, \boldsymbol{u}) = 0, \quad v_2^*(1, \boldsymbol{u}) = 4, \quad v_3^*(1, \boldsymbol{u}) = \infty, \quad v_4^*(1, \boldsymbol{u}) = 1.$$

In stage 2, the cost $v_3^*(2, \boldsymbol{u})$, for example, is

$$v_3^*(2, \boldsymbol{u}) = \min\left[2 + v_2^*(1, \boldsymbol{u}), \quad 4 + v_4^*(1, \boldsymbol{u})\right] = 5.$$

The set of costs at stage 2 is

$$v_1^*(2, \boldsymbol{u}) = 0, \quad v_2^*(2, \boldsymbol{u}) = 2, \quad v_3^*(2, \boldsymbol{u}) = 5, \quad v_4^*(2, \boldsymbol{u}) = 1.$$

The decision at stage 2 is for node 2 to go to 4, node 3 to 4, and 4 to 1. At stage 3, node 3 switches to node 2, reducing its path length to 4, and nodes 2 and 4 are unchanged. Further iterations yield no change, and the resulting policy is also the optimal stationary policy.

The above results at each stage n can be interpreted as the shortest paths constrained to at most n hops. As n is increased, this constraint is successively relaxed, reaching the true shortest paths in less than M stages.

It can be seen without too much difficulty that these final aggregate costs (path lengths) also result no matter what final cost vector $\boldsymbol{u}$ (with $u_1 = 0$) is used. This is a useful feature for many types of networks where link lengths change very slowly with time and a shortest path algorithm is desired that can track the corresponding changes in the shortest paths.

4.6.2 Optimal stationary policies

In Example 4.6.1, we saw that there was a final transient (at stage 1) in which decision 1 was taken, and in all other stages, decision 2 was taken. Thus, the optimal dynamic policy consisted of a long-term stationary policy, followed by a transient period (for a single stage in this case) over which a different policy was used. It turns out that this final transient can be avoided by choosing an appropriate final reward vector $\boldsymbol{u}$ for the dynamic programming algorithm. Good intuition might lead to guessing that the final reward vector $\boldsymbol{u}$ should be the relative-gain vector $\boldsymbol{w}$ of the above long-term stationary policy.

It seems reasonable to expect this same type of behavior for typical but more complex Markov decision problems. In order to understand this, we start by considering an arbitrary stationary policy $\boldsymbol{k}' = (k'_1, \dots, k'_{\mathsf{M}})$ and denote the transition matrix of the associated Markov chain as $[P^{\boldsymbol{k}'}]$. We assume that the associated Markov chain is a unichain, or, abbreviating terminology, that $\boldsymbol{k}'$ is a unichain. Let $\boldsymbol{w}'$ be the unique relative-gain vector for $\boldsymbol{k}'$. We then find some necessary conditions for $\boldsymbol{k}'$ to be the optimal dynamic policy at each stage using $\boldsymbol{w}'$ as the final reward vector.

From (4.45) $\boldsymbol{k}'$ is an optimal dynamic decision (with the final reward vector $\boldsymbol{w}'$ for $[P^{\boldsymbol{k}'}]$) at stage 1 if

$$\boldsymbol{r}^{\boldsymbol{k}'} + [P^{\boldsymbol{k}'}]\boldsymbol{w}' = \max_{\boldsymbol{k}}\{\boldsymbol{r}^{\boldsymbol{k}} + [P^{\boldsymbol{k}}]\boldsymbol{w}'\}. \tag{4.50}$$

Note that this is more than a simple statement that $\boldsymbol{k}'$ can be found by maximizing $\boldsymbol{r}^{\boldsymbol{k}} + [P^{\boldsymbol{k}}]\boldsymbol{w}'$ over $\boldsymbol{k}$. It also involves the fact that $\boldsymbol{w}'$ is the relative-gain vector for $\boldsymbol{k}'$, so there is no immediately obvious way to find a $\boldsymbol{k}'$ that satisfies (4.50), and no a-priori assurance that this equation even has a solution. The following theorem, however, says that this is the only condition required to ensure that $\boldsymbol{k}'$ is the optimal dynamic policy at every stage (again using $\boldsymbol{w}'$ as the final reward vector).

Theorem 4.6.3 *Assume that (4.50) is satisfied for some policy $\boldsymbol{k}'$ where the Markov chain for $\boldsymbol{k}'$ is a unichain and $\boldsymbol{w}'$ is the relative-gain vector of $\boldsymbol{k}'$. Then the optimal dynamic policy, using $\boldsymbol{w}'$ as the final reward vector, is the stationary policy $\boldsymbol{k}'$. Furthermore the optimal gain at each stage n is given by*

$$\boldsymbol{v}^*(n, \boldsymbol{w}') = \boldsymbol{w}' + ng'\boldsymbol{e}, \tag{4.51}$$

where $g' = \boldsymbol{\pi}'\boldsymbol{r}^{\boldsymbol{k}'}$ and $\boldsymbol{\pi}'$ is the steady-state vector for $\boldsymbol{k}'$.

Proof We have seen from (4.45) that $\boldsymbol{k}'$ is an optimal dynamic decision at stage 1. Also, since $\boldsymbol{w}'$ is the relative-gain vector for $\boldsymbol{k}'$, Theorem 4.5.7 asserts that if decision $\boldsymbol{k}'$ is used at each stage, then the aggregate gain satisfies $\boldsymbol{v}(n, \boldsymbol{w}') = ng'\boldsymbol{e} + \boldsymbol{w}'$. Since $\boldsymbol{k}'$ is optimal at stage 1, it follows that (4.51) is satisfied for $n = 1$.

We now use induction on n, with $n = 1$ as a basis, to verify (4.51) and the optimality of this same $\boldsymbol{k}'$ at each stage n. Thus, assume that (4.51) is satisfied for n. Then, from (4.48),

$$\boldsymbol{v}^*(n+1, \boldsymbol{w}') = \max_{\boldsymbol{k}}\{\boldsymbol{r}^{\boldsymbol{k}} + [P^{\boldsymbol{k}}]\boldsymbol{v}^*(n, \boldsymbol{w}')\} \tag{4.52}$$

$$= \max_{\boldsymbol{k}}\left\{\boldsymbol{r}^{\boldsymbol{k}} + [P^{\boldsymbol{k}}]\{\boldsymbol{w}' + ng'\boldsymbol{e}\}\right\} \tag{4.53}$$

$$= ng'\boldsymbol{e} + \max_{\boldsymbol{k}}\{\boldsymbol{r}^{\boldsymbol{k}} + [P^{\boldsymbol{k}}]\boldsymbol{w}'\} \tag{4.54}$$

$$= ng'\boldsymbol{e} + \boldsymbol{r}^{\boldsymbol{k}'} + [P^{\boldsymbol{k}'}]\boldsymbol{w}' \tag{4.55}$$
$$= (n+1)g'\boldsymbol{e} + \boldsymbol{w}'. \tag{4.56}$$

Equation (4.53) follows from the inductive hypothesis of (4.51), (4.54) follows because $[P^{\boldsymbol{k}}]\boldsymbol{e} = \boldsymbol{e}$ for all $\boldsymbol{k}$, (4.55) follows from (4.50), and (4.56) follows from the definition of $\boldsymbol{w}'$ as the relative-gain vector for $\boldsymbol{k}'$. This verifies (4.51) for $n + 1$. Also, since $\boldsymbol{k}'$ maximizes (4.54), it also maximizes (4.52), showing that $\boldsymbol{k}'$ is the optimal dynamic decision at stage $n + 1$. This completes the inductive step. □

Since our major interest in stationary policies is to help understand the relationship between the optimal dynamic policy and stationary policies, we define an optimal stationary policy as follows.

Definition 4.6.4 *A unichain stationary policy* $\mathbf{k}'$ *is* ***optimal*** *if the optimal dynamic policy with* $\mathbf{w}'$ *as the final reward uses* $\mathbf{k}'$ *at each stage.*

This definition side-steps several important issues. We might be interested in dynamic programming for some other final reward vector. Is it possible that dynamic programming performs much better in some sense with a different final reward vector? Is it possible that there is another stationary policy, especially one with a larger gain per stage? We answer these questions later and find that stationary policies that are optimal according to the definition do have maximal gain per stage compared with dynamic policies with arbitrary final reward vectors.

From Theorem 4.6.3, we see that if there is a policy $\boldsymbol{k}'$ which is a unichain with relative-gain vector $\boldsymbol{w}'$, and if that $\boldsymbol{k}'$ is a solution to (4.50), then $\boldsymbol{k}'$ is an optimal stationary policy.

It is easy to imagine Markov decision models for which each policy corresponds to a Markov chain with multiple recurrent classes. There are many special cases of such situations, and their detailed study is inappropriate in an introductory treatment. The essential problem with such models is that it is possible to get into various sets of states from which there is no exit, no matter what decisions are used. These sets might have different gains, so that there is no meaningful overall gain per stage. We avoid these situations by a modeling assumption called *inherent reachability*, which assumes, for each pair (i, j) of states, that there is some decision vector $\boldsymbol{k}$ containing a path from i to j.

The concept of inherent reachability is a little tricky, since it does not say the same $\boldsymbol{k}$ can be used for all pairs of states (i.e., that there is some $\boldsymbol{k}$ for which the Markov chain is recurrent). As shown in Exercise 4.31, however, inherent reachability does imply that for any state j, there is a $\boldsymbol{k}$ for which j is accessible from all other states. As we have seen a number of times, this implies that the Markov chain for $\boldsymbol{k}$ is a unichain in which j is a recurrent state.

Any desired model can be modified to satisfy inherent reachability by creating some new decisions with very large negative rewards; these allow for such paths but very much discourage them. This will allow us to construct optimal unichain policies, but also to use the appearance of these large negative rewards to signal that there was something questionable in the original model.

4.6.3 Policy improvement and the search for optimal stationary policies

The general idea of policy improvement is to start with an arbitrary unichain stationary policy $\boldsymbol{k}'$ with a relative-gain vector $\boldsymbol{w}'$ (as given by (4.37)). We assume inherent reachability throughout this section, so such unichains must exist. We then check whether (4.50), is satisfied, and if so, we know from Theorem 4.6.3 that $\boldsymbol{k}'$ is an optimal stationary policy. If not, we find another stationary policy $\boldsymbol{k}$ that is 'better' than $\boldsymbol{k}'$ in a sense to be described later. Unfortunately, the 'better' policy that we find might not be a unichain, so it will also be necessary to convert this new policy into an equally 'good' unichain policy. This is where the assumption of inherent reachability is needed. The algorithm then iteratively finds better and better unichain stationary policies, until eventually one of them satisfies (4.50) and is thus optimal.

We now state the policy-improvement algorithm for inherently reachable Markov decision problems. This algorithm is a generalization of Howard's policy-improvement algorithm, [16].

Policy-improvement algorithm

1. Choose an arbitrary unichain policy $\boldsymbol{k}'$.
2. For policy $\boldsymbol{k}'$, calculate $\boldsymbol{w}'$ and g' from $\boldsymbol{w}' + g'\boldsymbol{e} = \boldsymbol{r}^{\boldsymbol{k}'} + [P^{\boldsymbol{k}'}]\boldsymbol{w}'$ and $\boldsymbol{\pi}'\boldsymbol{w}' = 0$.
3. If $\boldsymbol{r}^{\boldsymbol{k}'} + [P^{\boldsymbol{k}'}]\boldsymbol{w}' = \max_{\boldsymbol{k}}\{\boldsymbol{r}^{\boldsymbol{k}} + [P^{\boldsymbol{k}}]\boldsymbol{w}'\}$, then stop; $\boldsymbol{k}'$ is optimal.
4. Otherwise, choose ℓ and k_ℓ so that $r_\ell^{(k'_\ell)} + \sum_j P_{\ell j}^{(k'_\ell)} w'_j < r_\ell^{(k_\ell)} + \sum_j P_{\ell j}^{(k_\ell)} w'_j$. For $i \neq \ell$, let $k_i = k'_i$.
5. If $\boldsymbol{k} = (k_1, \dots k_{\mathsf{M}})$ is not a unichain, then let R be the recurrent class in $\boldsymbol{k}$ that contains state ℓ, and let $\widetilde{\boldsymbol{k}}$ be a unichain policy for which $\widetilde{k}_i = k_i$ for each $i \in \mathsf{R}$. Alternatively, if $\boldsymbol{k}$ is already a unichain, let $\widetilde{\boldsymbol{k}} = \boldsymbol{k}$.
6. Update $\boldsymbol{k}'$ to the value of $\widetilde{\boldsymbol{k}}$ and return to step 2.

If the stopping test in step 3 fails, there must be an ℓ and k_ℓ for which $r_\ell^{(k'_\ell)} + \sum_j P_{\ell j}^{(k'_\ell)} w'_j < r_\ell^{(k_\ell)} + \sum_j P_{\ell j}^{(k_\ell)} w'_j$. Thus step 4 can always be executed if the algorithm does not stop in step 3, and since the decision is changed only for the single state ℓ, the resulting policy $\boldsymbol{k}$ satisfies

$$\boldsymbol{r}^{\boldsymbol{k}'} + [P^{\boldsymbol{k}'}]\boldsymbol{w}' \le \boldsymbol{r}^{\boldsymbol{k}} + [P^{\boldsymbol{k}}]\boldsymbol{w}' \qquad \text{with strict inequality for component } \ell. \tag{4.57}$$

The next three lemmas consider the different cases for the state ℓ whose decision is changed in step 4 of the algorithm. Taken together, they show that each iteration of the algorithm either increases the gain per stage or keeps the gain per stage constant while increasing the relative-gain vector. After proving these lemmas, we return to show that the algorithm must converge and explain the sense in which the resulting stationary algorithm is optimal.

For each of the lemmas, let $\boldsymbol{k}'$ be the decision vector in step 1 of a given iteration of the policy improvement algorithm and assume that the Markov chain for $\boldsymbol{k}'$ is a unichain. Let $g', \boldsymbol{w}'$, and R' respectively be the gain per stage, the relative gain vector, and the recurrent set of states for $\boldsymbol{k}'$. Assume that the stopping condition in step 3 is not satisfied and that ℓ denotes the state whose decision is changed. Let k_ℓ be the new decision in step 4 and let $\boldsymbol{k}$ be the new decision vector.

Lemma 4.6.5 *Assume that $\ell \in \mathrm{R}'$. Then the Markov chain for $\boldsymbol{k}$ is a unichain and ℓ is recurrent in $\boldsymbol{k}$. The gain per stage g for $\boldsymbol{k}$ satisfies $g > g'$.*

Proof The Markov chain for $\boldsymbol{k}$ is the same as that for $\boldsymbol{k}'$ except for the transitions out of state ℓ. Thus every path into ℓ in $\boldsymbol{k}'$ is still a path into ℓ in $\boldsymbol{k}$. Since ℓ is recurrent in the unichain $\boldsymbol{k}'$, it is accessible from all states in $\boldsymbol{k}'$ and thus in $\boldsymbol{k}$. It follows (see Exercise 4.3) that ℓ is recurrent in $\boldsymbol{k}$ and $\boldsymbol{k}$ is a unichain. Since $\boldsymbol{r}^{\boldsymbol{k}'} + [P^{\boldsymbol{k}'}]\boldsymbol{w}' = \boldsymbol{w}' + g'\boldsymbol{e}$ (see (4.37)), we can rewrite (4.57) as

$$\boldsymbol{w}' + g'\boldsymbol{e} \leq \boldsymbol{r}^{\boldsymbol{k}} + [P^{\boldsymbol{k}}]\boldsymbol{w}' \qquad \text{with strict inequality for component } \ell. \tag{4.58}$$

Premultiplying both sides of (4.58) by the steady-state vector $\boldsymbol{\pi}$ of the Markov chain $\boldsymbol{k}$ and using the fact that ℓ is recurrent and thus $\pi_\ell > 0$,

$$\boldsymbol{\pi}\boldsymbol{w}' + g' < \boldsymbol{\pi}\boldsymbol{r}^{\boldsymbol{k}} + \boldsymbol{\pi}[P^{\boldsymbol{k}}]\boldsymbol{w}'.$$

Since $\boldsymbol{\pi}[P^{\boldsymbol{k}}] = \boldsymbol{\pi}$, this simplifies to

$$g' < \boldsymbol{\pi}\boldsymbol{r}^{\boldsymbol{k}}. \tag{4.59}$$

The gain per stage g for $\boldsymbol{k}$ is $\boldsymbol{\pi}\boldsymbol{r}^{\boldsymbol{k}}$, so we have $g' < g$. □

Lemma 4.6.6 *Assume that $\ell \notin \mathrm{R}'$ (i.e., ℓ is transient in $\boldsymbol{k}'$) and that the states of R' are not accessible from ℓ in $\boldsymbol{k}$. Then $\boldsymbol{k}$ is not a unichain and ℓ is recurrent in $\boldsymbol{k}$. A decision vector $\widetilde{\boldsymbol{k}}$ exists that is a unichain for which $\widetilde{k}_i = k_i$ for $i \in \mathrm{R}$, and its gain per stage $\widetilde{g}$ satisfies $\widetilde{g} > g$.*

Proof Since $\ell \notin \mathrm{R}'$, the transition probabilities from the states of R' are unchanged in going from $\boldsymbol{k}'$ to $\boldsymbol{k}$. Thus the set of states accessible from R' remains unchanged, and R' is a recurrent set of $\boldsymbol{k}$. Since R' is not accessible from ℓ, there must be another recurrent set, R, in $\boldsymbol{k}$, and thus $\boldsymbol{k}$ is not a unichain. The states accessible from R no longer include R', and since ℓ is the only state whose transition probabilities have changed, all states in R have paths to ℓ in $\boldsymbol{k}$. It follows that $\ell \in \mathrm{R}$.

Now let $\boldsymbol{\pi}$ be the steady-state vector for R in the Markov chain for $\boldsymbol{k}$. Since $\pi_\ell > 0$, (4.58) and (4.59) are still valid for this situation. Let $\widetilde{\boldsymbol{k}}$ be a decision vector for which $\widetilde{k}_i = k_i$ for each $i \in \mathrm{R}$. Using inherent reachability, we can also choose $\widetilde{k}_i$ for each $i \notin \mathrm{R}$ so that ℓ is reachable from i (see Exercise 4.31). Thus $\widetilde{\boldsymbol{k}}$ is a unichain with the recurrent class R. Since $\widetilde{\boldsymbol{k}}$ has the same transition probabilities and rewards in R as $\boldsymbol{k}$, we see that $\widetilde{g} = \pi \boldsymbol{r}^{\boldsymbol{k}}$ and thus $\widetilde{g} > g'$. □

The final lemma now includes all cases not in Lemmas 4.6.5 and 4.6.6

Lemma 4.6.7 *Assume that $\ell \notin \mathrm{R}'$ and that R' is accessible from ℓ in $\boldsymbol{k}$. Then $\boldsymbol{k}$ is a unichain with the same recurrent set R' as $\boldsymbol{k}'$. The gain per stage g is equal to g' and the relative-gain vector $\boldsymbol{w}$ of $\boldsymbol{k}$ satisfies*

$$\boldsymbol{w}' \leq \boldsymbol{w} \qquad \text{with } w'_\ell < w_\ell \text{ and } w'_i = w_i \text{ for } i \in \mathrm{R}'. \tag{4.60}$$

Proof Since $\boldsymbol{k}'$ is a unichain, $\boldsymbol{k}'$ contains a path from each state to R'. If such a path does not go through state ℓ, then $\boldsymbol{k}$ also contains that path. If such a path does go through ℓ, then that path can be replaced in $\boldsymbol{k}$ by the same path to ℓ followed by a path in $\boldsymbol{k}$ from ℓ to R'. Thus R' is accessible from all states in $\boldsymbol{k}$. Since the states accessible from R' are unchanged from $\boldsymbol{k}'$ to $\boldsymbol{k}$, $\boldsymbol{k}$ is still a unichain with the recurrent set R' and state ℓ is still transient.

If we write out the defining equation (4.37) for $\boldsymbol{w}'$ component by component, we get

$$w'_i + g' = r_i^{k'_i} + \sum_j P_{ij}^{k'_i} w'_j. \tag{4.61}$$

Consider the set of these equations for which $i \in \mathsf{R}'$. Since $P_{ij}^{k'_i} = 0$ for all transient j in $\boldsymbol{k}'$, these are the same relative-gain equations as for the Markov chain restricted to R'. Therefore $\boldsymbol{w}'$ is uniquely defined for $i \in \mathsf{R}'_i$ by this restricted set of equations. These equations are not changed in going from $\boldsymbol{k}'$ to $\boldsymbol{k}$, so it follows that $w_i = w'_i$ for $i \in \mathsf{R}'$. We have also seen that the steady-state vector $\boldsymbol{\pi}'$ is determined solely by the transition probabilities in the recurrent class, so $\boldsymbol{\pi}'$ is unchanged from $\boldsymbol{k}'$ to $\boldsymbol{k}$, and $g = g'$.

Finally, consider the difference between the relative-gain equations for $\boldsymbol{k}'$ in (4.61) and those for $\boldsymbol{k}$. Since $g' = g$,

$$w_i - w'_i = r_i^{k_i} - r_i^{k'_i} + \sum_j \left(P_{ij}^{k_i} w_j - P_{ij}^{k'_i} w'_j \right). \tag{4.62}$$

For all $i \neq \ell$, this simplifies to

$$w_i - w'_i = \sum_j P_{ij}^{k_i} (w_j - w'_j). \tag{4.63}$$

For $i = \ell$, (4.62) can be rewritten as

$$w_\ell - w'_\ell = \sum_j P_{\ell j}^{k_\ell} (w_j - w'_j) + \left[r_\ell^{k_\ell} - r_\ell^{k'_\ell} + \sum_j \left(P_{\ell j}^{k_\ell} w'_j - P_{\ell j}^{k'_\ell} w'_j \right) \right]. \tag{4.64}$$

The quantity in square brackets must be positive because of step 4 of the algorithm, and we denote it as $\hat{r}_\ell - \hat{r}'_\ell$. If we also define $\hat{r}_i = \hat{r}'_i$ for $i \neq \ell$, then we can apply the last part of Corollary 4.5.6 (using $\hat{\boldsymbol{r}}$ and $\hat{\boldsymbol{r}}'$ as reward vectors) to conclude that $\boldsymbol{w} \geq \boldsymbol{w}'$ with $w_\ell > w'_\ell$. □

We now see that each iteration of the algorithm either increases the gain per stage or holds the gain per stage the same and increases the relative-gain vector $\boldsymbol{w}$. Thus the sequence of policies found by the algorithm can never repeat. Since there is a finite number of stationary policies, the algorithm must eventually terminate at step 3. This means that the optimal dynamic policy using the final reward vector $\boldsymbol{w}'$ for the terminating decision vector $\boldsymbol{k}'$ must in fact be the stationary policy $\boldsymbol{k}'$.

The question now arises whether the optimal dynamic policy using some other final reward vector can be substantially better than that using $\boldsymbol{w}'$. The answer is quite simple and is developed in Exercise 4.30. It is shown there that if $\boldsymbol{u}$ and $\boldsymbol{u}'$ are arbitrary final reward vectors used on the dynamic programming algorithm, then $v^*(n, \boldsymbol{u})$ and $v^*(n, \boldsymbol{u}')$ are related by

$$v^*(n, \boldsymbol{u}) \leq v^*(n, \boldsymbol{u}') + \alpha \boldsymbol{e},$$

where $\alpha = \max_i(u_i - u'_i)$. Using $\boldsymbol{w}'$ for $\boldsymbol{u}'$, it is seen that the gain per stage of dynamic programming, with any final reward vector, is at most the gain g' of the stationary policy at the termination of the policy-improvement algorithm.

The above results are summarized in the following theorem.

Theorem 4.6.8 *For any inherently reachable finite-state Markov decision problem, the policy-improvement algorithm terminates with a stationary policy $\boldsymbol{k}'$ that is the same as the solution to the dynamic programming algorithm using $\boldsymbol{w}'$ as the final reward vector. The gain per stage g' of this stationary policy maximizes the gain per stage over all stationary policies and over all final-reward vectors for the dynamic programming algorithm.*

One remaining issue is the question whether the relative-gain vector found by the policy-improvement algorithm is in any sense optimal. The example in Figure 4.12 illustrates two different solutions terminating the policy-improvement algorithm. They each have the same gain (as guaranteed by Theorem 4.6.8) but their relative-gain vectors are not ordered.

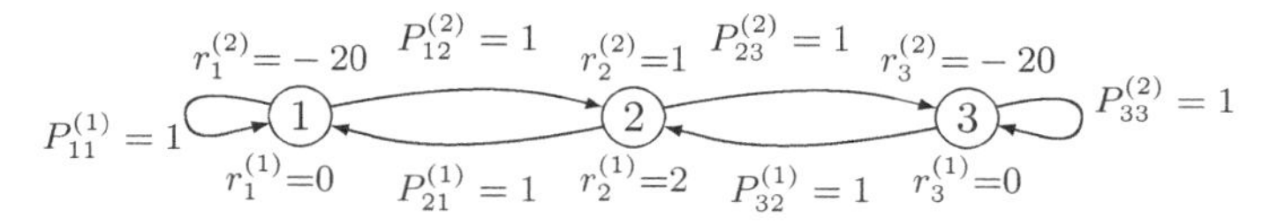

Figure 4.12 A Markov decision problem in which there are two unichain decision vectors (one left-going, and the other right-going). For each, (4.50) is satisfied and the gain per stage is 0. The dynamic programming algorithm (with no final reward) is stationary but has two recurrent classes, one of which is {3}, using decision 2 and the other of which is {1}, using decision 1 in each state.

In many applications such as variations on the shortest path problem, the interesting issue is what happens before the recurrent class is entered, and there is often only one recurrent class and one set of decisions within that class of interest. The following corollary shows that in this case, the relative-gain vector for the stationary policy that terminates the algorithm is maximal not only among the policies visited by the algorithm but among all policies with the same recurrent class and the same decisions within that class. The proof is almost the same as that of Lemma 4.6.7 and is carried out in Exercise 4.33.

Corollary 4.6.9 *Assume the policy-improvement algorithm terminates with the recurrent class R', the decision vector $\boldsymbol{k}'$, and the relative-gain vector $\boldsymbol{w}'$. Then for any stationary policy that has the recurrent class R' and a decision vector $\boldsymbol{k}$ satisfying $k_i = k'_i$ for all $i \in \mathsf{R}'$, the relative-gain vector $\boldsymbol{w}$ satisfies $\boldsymbol{w} \leq \boldsymbol{w}'$.*

4.7 Summary

This chapter has developed the basic results about finite-state Markov chains. It was shown that the states of any finite-state chain can be partitioned into classes, where each class is either transient or recurrent, and each class is periodic or aperiodic. If a recurrent class is periodic of period d, then the states in that class can be partitioned into d cyclically-ordered subsets where each subset has transitions only into the next subset.

The transition probabilities in the Markov chain can be represented as a matrix $[P]$, and the n-step transition probabilities are given by the matrix product $[P^n]$. If the chain is ergodic, i.e., one aperiodic recurrent class, then the limit of the n-step transition probabilities is independent of the initial state, i.e., $\lim_{n\to\infty} P_{ij}^n = \pi_j$, where $\boldsymbol{\pi} = (\pi_1, \dots, \pi_{\mathsf{M}})$ is called the steady-state probability. Thus the limiting value of $[P^n]$ is an $\mathsf{M} \times \mathsf{M}$ matrix whose rows are all the same, i.e., the limiting matrix is the product $\boldsymbol{e}\boldsymbol{\pi}$. The steady-state probabilities are uniquely specified by $\sum_i \pi_i P_{ij} = \pi_j$ and $\sum_i \pi_i = 1$. That unique solution must satisfy $\pi_i > 0$ for all i. The same result holds (see Theorem 4.3.7) for aperiodic unichains with the exception that $\pi_i = 0$ for all transient states.

The eigenvalues and eigenvectors of $[P]$ are useful in many ways, but in particular provide precise results about how P_{ij}^n approaches π_j with increasing n. An eigenvalue equal to 1 always exists, and its multiplicity is equal to the number of recurrent classes. For each recurrent class, there is a left eigenvector $\boldsymbol{\pi}$ of eigenvalue 1. It is the steady-state vector for the given recurrent class. If a recurrent class is periodic with period d, then there are d corresponding eigenvalues of magnitude 1 uniformly spaced around the unit circle. The left eigenvector corresponding to each is non-zero only on that periodic recurrent class.

All other eigenvalues of $[P]$ are less than 1 in magnitude. If the eigenvectors of the entire set of eigenvalues span M-dimensional space, then $[P^n]$ can be represented by (4.30) which shows explicitly how steady state is approached for aperiodic recurrent classes of states. If the eigenvectors do not span M-space, then (4.30) can be replaced by a Jordan form.

For an arbitrary finite-state Markov chain, if the initial state is transient, then the Markov chain will eventually enter a recurrent state, and the probability that this requires n or more steps approaches zero geometrically in n. Exercise 4.18 shows how to find the probability that each recurrent class is entered. Given an entry into a particular recurrent class, the results about recurrent chains can be used to analyze the behavior within that class.

The results about Markov chains were extended to Markov chains with rewards. The use of reward functions (or cost functions) provides a systematic way to approach a large class of problems ranging from first-passage times to dynamic programming. For unichains, the key result here is Theorem 4.5.7, which provides both an exact expression and an asymptotic expression for the expected aggregate reward over n stages. Markov chains with rewards and multiple recurrent classes are best handled by considering the individual recurrent classes separately.

Finally, the results on Markov chains with rewards were used to understand Markov decision theory. The Bellman dynamic programming algorithm was developed, and the policy-improvement algorithm was discussed and analyzed. Theorem 4.6.8 demonstrated the relationship between the optimal dynamic policy and the optimal stationary policy. This section provided only an introduction to dynamic programming and omitted all discussion of discounting (in which future gain is considered worth less than present gain because of interest rates). The development was also restricted to finite state-spaces.

For a review of vectors, matrices, and linear algebra, see any introductory text on linear algebra such as Strang [28]. For further reading on Markov decision theory and dynamic programming, see Bertsekas [3]. Bellman [1] is of historic interest and quite readable.

4.8 Exercises

Exercise 4.1 Let $[P]$ be the transition matrix for a finite state Markov chain and let state i be recurrent. Prove that i is aperiodic if $P_{ii} > 0$.

Exercise 4.2 Show that every Markov chain with $\mathsf{M} < \infty$ states contains at least one recurrent set of states. Explaining each of the following statements is sufficient.

(a) If state i_1 is transient, then there is some other state i_2 such that $i_1 \to i_2$ and $i_2 \not\to i_1$.

(b) If the i_2 of (a) is also transient, there is a third state i_3 such that $i_2 \to i_3$, $i_3 \not\to i_2$; that state must satisfy $i_3 \neq i_2$, $i_3 \neq i_1$.

(c) Continue iteratively to repeat (b) for successive states, $i_1, i_2, \ldots$. That is, if $i_1, \ldots, i_k$ are generated as above and are all transient, generate i_{k+1} such that $i_k \to i_{k+1}$ and $i_{k+1} \not\to i_k$. Then $i_{k+1} \neq i_j$ for $1 \leq j \leq k$.

(d) Show that for some $k \leq \mathsf{M}$, k is not transient, i.e., it is recurrent, so a recurrent class exists.

Exercise 4.3 Consider a finite-state Markov chain in which some given state, say state 1, is accessible from every other state. Show that the chain has exactly one recurrent class $\mathtt{R}$ of states and state $1 \in \mathtt{R}$. (Note that the chain is then a unichain.)

Exercise 4.4 Show how to generalize the graph in Figure 4.4 to an arbitrary number of states $\mathsf{M} \geq 3$ with one cycle of M nodes and one of $\mathsf{M} - 1$ nodes. For $\mathsf{M} = 4$, let node 1 be the node not in the cycle of $\mathsf{M} - 1$ nodes. List the set of states accessible from node 1 in n steps for each $n \leq 12$ and show that the bound in Theorem 4.2.11 is met with equality. Explain why the same result holds for all larger M.

Exercise 4.5 (Proof of Theorem 4.2.11) **(a)** Show that an ergodic Markov chain with M states must contain a cycle with $\tau < \mathsf{M}$ states. Hint: Use ergodicity to show that the smallest cycle cannot contain M states.

(b) Let ℓ be a fixed state on this cycle of length τ. Let $\mathtt{T}(m)$ be the set of states accessible from ℓ in m steps. Show that for each $m \geq 1$, $\mathtt{T}(m) \subseteq \mathtt{T}(m + \tau)$. Hint: For any given state $j \in \mathtt{T}(m)$, show how to construct a walk of $m + \tau$ steps from ℓ to j from the assumed walk of m steps.

(c) Define $\mathrm{T}(0)$ to be the singleton set $\{\ell\}$ and show that

$$\mathrm{T}(0) \subseteq \mathrm{T}(\tau) \subseteq \mathrm{T}(2\tau) \subseteq \cdots \subseteq \mathrm{T}(n\tau) \subseteq \cdots.$$

(d) Show that if one of the inclusions above is satisfied with equality, then all subsequent inclusions are satisfied with equality. Show from this that at most the first $M-1$ inclusions can be satisfied with strict inequality and that $\mathrm{T}(n\tau) = \mathrm{T}((M-1)\tau)$ for all $n \geq M-1$.

(e) Show that all states are included in $\mathrm{T}((M-1)\tau)$.

(f) Show that $P_{ij}^{(\mathsf{M}-1)^2+1} > 0$ for all i, j.

Exercise 4.6 Consider a Markov chain with one ergodic class of r states, say $\{1, 2, \ldots, r\}$ and $\mathsf{M} - r$ other states that are all transient. Show that $P_{ij}^n > 0$ for all $j \leq r$ and $n \geq (r-1)^2 + 1 + \mathsf{M} - r$.

Exercise 4.7 **(a)** Let τ be the number of states in the smallest cycle of an arbitrary ergodic Markov chain of $\mathsf{M} \geq 3$ states. Show that $P_{ij}^n > 0$ for all $n \geq (\mathsf{M}-2)\tau + \mathsf{M}$. Hint: Look at the proof of Theorem 4.2.11 in Exercise 4.5.

(b) For $\tau = 1$, draw the graph of an ergodic Markov chain (generalized for arbitrary $\mathsf{M} \geq 3$) for which there is an i, j for which $P_{ij}^n = 0$ for $n = 2\mathsf{M} - 3$. Hint: Look at Figure 4.4.

(c) For arbitrary $\tau < \mathsf{M} - 1$, draw the graph of an ergodic Markov chain (generalized for arbitrary M) for which there is an i, j for which $P_{ij}^n = 0$ for $n = (\mathsf{M}-2)\tau + \mathsf{M} - 1$.

Exercise 4.8 A transition probability matrix $[P]$ is said to be doubly stochastic if

$$\sum_j P_{ij} = 1 \quad \text{for all } i, \qquad \sum_i P_{ij} = 1 \quad \text{for all } j.$$

That is, the row sum and the column sum each equal 1. If a doubly stochastic chain has M states and is ergodic (i.e., has a single class of states and is aperiodic), calculate its steady-state probabilities.

Exercise 4.9 **(a)** Find the steady-state probabilities $\pi_0, \ldots, \pi_{k-1}$ for the Markov chain below. Express your answer in terms of the ratio $\rho = p/q$, where $q = 1 - p$. Pay particular attention to the special case $\rho = 1$.

(b) Sketch $\pi_0, \ldots, \pi_{k-1}$. Give one sketch for $\rho = 1/2$, one for $\rho = 1$, and one for $\rho = 2$.

(c) Find the limit of π_0 as k approaches ∞; give separate answers for $\rho < 1$, $\rho = 1$, and $\rho > 1$. Find limiting values of π_{k-1} for the same cases.

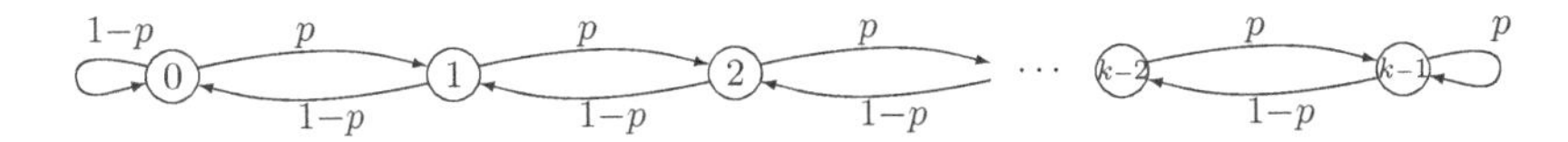

Exercise 4.10 **(a)** Find the steady-state probabilities for each of the Markov chains in Figure 4.2. Assume that all clockwise probabilities in the first graph are the same, say p, and assume that $P_{4,5} = P_{4,1}$ in the second graph.

(b) Find the matrices $[P^2]$ for the same chains. Draw the graphs for the Markov chains represented by $[P^2]$, i.e., the graph of two step transitions for the original chains. Find the steady-state probabilities for these two-step chains. Explain why your steady-state probabilities are not unique.

(c) Find $\lim_{n\to\infty}[P^{2n}]$ for each of the chains.

Exercise 4.11 **(a)** Assume that $\boldsymbol{\nu}^{(i)}$ is a right eigenvector and $\boldsymbol{\pi}^{(j)}$ is a left eigenvector of an $\mathsf{M} \times \mathsf{M}$ stochastic matrix $[P]$ where $\lambda_i \neq \lambda_j$. Show that $\boldsymbol{\pi}^{(j)}\boldsymbol{\nu}^{(i)} = 0$. Hint: Consider two ways of finding $\boldsymbol{\pi}^{(j)}[P]\boldsymbol{\nu}^{(i)}$.

(b) Assume that $[P]$ has M distinct eigenvalues. The right eigenvectors of $[P]$ then span M space (see Section 5.2 of Strang [28]), so the matrix $[U]$ with those eigenvectors as columns is non-singular. Show that U^{-1} is a matix whose rows are the M left eigenvectors of $[P]$. Hint: Use (a).

(c) For any given integer $i \in \{1, \mathsf{M}\}$, let $[A]$ be a diagonal matrix with a single non-zero element, $A_{ii} = a \neq 0$. Using the assumptions and results of (b), show that

$$[UAU^{-1}] = a\boldsymbol{\nu}^{(i)}\boldsymbol{\pi}^{(i)}.$$

Hint: Visualize straightforward vector/matrix multiplication.

(d) Verify (4.30).

Exercise 4.12 **(a)** Let λ_k be an eigenvalue of a stochastic matrix $[P]$ and let $\boldsymbol{\pi}^{(k)}$ be a left eigenvector for λ_k. Show that for each component $\pi_j^{(k)}$ of $\boldsymbol{\pi}^{(k)}$ and each n that

$$\lambda_k^n \pi_j^{(k)} = \sum_i \pi_i^{(k)} P_{ij}^n.$$

(b) By taking magnitudes of each side and looking at the appropriate j, show that

$$\left|\lambda_k\right|^n \leq \mathsf{M}.$$

(c) Show that $\left|\lambda_k\right| \leq 1$.

Exercise 4.13 Consider a finite-state Markov chain with matrix $[P]$ which has κ aperiodic recurrent classes, $\mathtt{R}_1, \ldots, \mathtt{R}_\kappa$ and a set $\mathtt{T}$ of transient states. For any given recurrent class ℓ, consider a vector $\boldsymbol{\nu}$ such that $\nu_i = 1$ for each $i \in \mathtt{R}_\ell$, $\nu_i = \lim_{n\to\infty} \Pr\{X_n \in \mathtt{R}_\ell | X_0 = i\}$ for each $i \in \mathtt{T}$, and $\nu_i = 0$ otherwise. Show that $\boldsymbol{\nu}$ is a right eigenvector of $[P]$ with eigenvalue 1. Hint: Redraw Figure 4.5 for multiple recurrent classes and first show that $\boldsymbol{\nu}$ is an eigenvector of $[P^n]$ in the limit.

Exercise 4.14 Answer the following questions for the following stochastic matrix $[P]$:

$$[P] = \begin{bmatrix} 1/2 & 1/2 & 0 \\ 0 & 1/2 & 1/2 \\ 0 & 0 & 1 \end{bmatrix}.$$

(a) Find $[P^n]$ in closed form for arbitrary $n > 1$.

(b) Find all distinct eigenvalues and the multiplicity of each distinct eigenvalue for $[P]$.

(c) Find a right eigenvector for each distinct eigenvalue, and show that the eigenvalue of multiplicity 2 does not have two linearly independent eigenvectors.

(d) Use (c) to show that there is no diagonal matrix $[\Lambda]$ and no invertible matrix $[U]$ for which $[P][U] = [U][\Lambda]$.

(e) Rederive the result of (d) using the result of (a) rather than (c).

Exercise 4.15 **(a)** Let $[J_i]$ be a 3×3 block of a Jordan form, i.e.,

$$[J_i] = \begin{bmatrix} \lambda_i & 1 & 0 \\ 0 & \lambda_i & 1 \\ 0 & 0 & \lambda_i \end{bmatrix}.$$

Show that the nth power of $[J_i]$ is given by

$$[J_i^n] = \begin{bmatrix} \lambda_i^n & n\lambda_i^{n-1} & \binom{n}{2}\lambda_i^{n-2} \\ 0 & \lambda_i^n & n\lambda_i^{n-1} \\ 0 & 0 & \lambda_i^n \end{bmatrix}.$$

Hint: Perhaps the easiest way is to calculate $[J_i^2]$ and $[J_i^3]$ and then use iteration.

(b) Generalize (a) to a $k \times k$ block of a Jordan form. Note that the nth power of an entire Jordan form is composed of these blocks along the diagonal of the matrix.

(c) Let $[P]$ be a stochastic matrix represented by a Jordan form $[J]$ as $[P] = [U][J][U^{-1}]$ and consider $[U^{-1}][P][U] = [J]$. Show that any repeated eigenvalue of $[P]$ (and, in particular, any eigenvalue represented by a Jordan block of 2×2 or more) must be strictly less than 1. Hint: Upper bound the elements of $[U^{-1}][P^n][U]$ by taking the magnitude of the elements of $[U]$ and $[U^{-1}]$ and upper bounding each element of a stochastic matrix by 1.

(d) Let λ_s be the eigenvalue of largest magnitude less than 1. Assume that the Jordan blocks for λ_s are at most of size k. Show that each ergodic class of $[P]$ converges at least as fast as $n^k\lambda_s^n$.

Exercise 4.16 **(a)** Let λ be an eigenvalue of a matrix $[A]$, and let $\boldsymbol{\nu}$ and $\boldsymbol{\pi}$ be right and left eigenvectors respectively of λ, normalized so that $\boldsymbol{\pi\nu} = 1$. Show that

$$[[A] - \lambda\boldsymbol{\nu\pi}]^2 = [A^2] - \lambda^2\boldsymbol{\nu\pi}.$$

(b) Show that $[[A^n] - \lambda^n\boldsymbol{\nu\pi}][[A] - \lambda\boldsymbol{\nu\pi}] = [A^{n+1}] - \lambda^{n+1}\boldsymbol{\nu\pi}$.

(c) Use induction to show that $[[A] - \lambda\boldsymbol{\nu\pi}]^n = [A^n] - \lambda^n\boldsymbol{\nu\pi}$.

Exercise 4.17 Let $[P]$ be the transition matrix for an aperiodic Markov unichain with the states numbered as in Figure 4.5.

(a) Show that $[P^n]$ can be partitioned as

$$[P^n] = \begin{bmatrix} [P_{\mathcal{T}}{}^n] & [P_x^n] \\ 0 & [P_{\mathcal{R}}^n] \end{bmatrix}.$$

That is, the blocks on the diagonal are simply products of the corresponding blocks of $[P]$, and the upper right block is whatever it turns out to be.

(b) Let q_i be the probability that the chain will be in a recurrent state after t transitions, starting from state i, i.e., $q_i = \sum_{t<j\leq t+r} P_{ij}^t$. Show that $q_i > 0$ for all transient i.

(c) Let q be the minimum q_i over all transient i and show that $P_{ij}^{nt} \leq (1-q)^n$ for all transient i, j (i.e., show that $[P_{\mathcal{T}}^n]$ approaches the all zero matrix $[0]$ with increasing n).

(d) Let $\boldsymbol{\pi} = (\boldsymbol{\pi}_{\mathcal{T}}, \boldsymbol{\pi}_{\mathcal{R}})$ be a left eigenvector of $[P]$ of eigenvalue 1. Show that $\boldsymbol{\pi}_{\mathcal{T}} = \mathbf{0}$ and show that $\boldsymbol{\pi}_{\mathcal{R}}$ must be positive and be a left eigenvector of $[P_{\mathcal{R}}]$. Thus show that $\boldsymbol{\pi}$ exists and is unique (within a scale factor).

(e) Show that $\boldsymbol{e}$ is the unique right eigenvector of $[P]$ of eigenvalue 1 (within a scale factor).

Exercise 4.18 Generalize Exercise 4.17 to the case of a Markov chain $[P]$ with m recurrent classes and one or more transient classes. In particular, show that:

(a) $[P]$ has exactly κ linearly independent left eigenvectors, $\boldsymbol{\pi}^{(1)}, \boldsymbol{\pi}^{(2)}, \dots, \boldsymbol{\pi}^{(\kappa)}$ of eigenvalue 1, and that the mth can be taken as a probability vector that is positive on the mth recurrent class and zero elsewhere.

(b) $[P]$ has exactly κ linearly independent right eigenvectors, $\boldsymbol{\nu}^{(1)}, \boldsymbol{\nu}^{(2)}, \dots, \boldsymbol{\nu}^{(\kappa)}$ of eigenvalue 1, and the mth can be taken as a vector with $\nu_i^{(m)}$ equal to the probability that recurrent class m will ever be entered starting from state i.

(c) Show that

$$\lim_{n\to\infty} [P^n] = \sum_m \boldsymbol{\nu}^{(m)} \boldsymbol{\pi}^{(m)}.$$

Exercise 4.19 Suppose a recurrent Markov chain has period d and let S_m, $0 \le m \le d-1$, be the mth subset in the sense of Theorem 4.2.9. Assume the states are numbered so that the first s_0 states are the states of S_0, the next s_1 are those of S_1, and so forth. Thus the matrix $[P]$ for the chain has the block form given by

$$[P] = \begin{bmatrix} 0 & [P_0] & \ddots & \ddots & 0 \\ 0 & 0 & [P_1] & \ddots & \ddots \\ \ddots & \ddots & \ddots & \ddots & \ddots \\ 0 & 0 & \ddots & \ddots & [P_{d-2}] \\ [P_{d-1}] & 0 & \ddots & \ddots & 0 \end{bmatrix},$$

where $[P_m]$ has dimension $s_m \times s_{m+1}$ for $0 \le m \le d-1$, with $(d-1)+1$ interpreted as 0 throughout. In what follows it is often more convenient to express $[P_m]$ as an $\mathsf{M} \times \mathsf{M}$ matrix $[P'_m]$ whose entries are 0 except for the rows of S_m and the columns of S_{m+1}, where the entries are equal to those of $[P_m]$. In this view, $[P] = \sum_{m=0}^{d-1} [P'_m]$.

(a) Show that $[P^d]$ has the form

$$[P^d] = \begin{bmatrix} [Q_0] & 0 & \ddots & 0 \\ 0 & [Q_1] & \ddots & \ddots \\ 0 & 0 & \ddots & [Q_{d-1}] \end{bmatrix},$$

where $[Q_m] = [P_m][P_{m+1}] \cdots [P_{d-1}][P_0] \cdots [P_{m-1}]$. Expressing $[Q_m]$ as an $\mathsf{M} \times \mathsf{M}$ matrix $[Q'_m]$ whose entries are 0 except for the rows and columns of S_m where the entries are equal to those of $[Q_m]$, this becomes $[P^d] = \sum_{m=0}^{d-1} [Q'_m]$.

(b) Show that $[Q_m]$ is the matrix of an ergodic Markov chain and let its eigenvector of eigenvalue 1 (scaled to be a probability vector) be $\hat{\boldsymbol{\pi}}_m$ and let $\hat{\boldsymbol{\nu}}_m$ be the corresponding right eigenvector scaled so that $\hat{\boldsymbol{\pi}}_m\hat{\boldsymbol{\nu}}_m = 1$. Let $\hat{\boldsymbol{\pi}}'_m$ and $\hat{\boldsymbol{\nu}}'_m$ be the corresponding M-vectors. Show that $\lim_{n\to\infty}[P^{nd}] = \sum_{m=0}^{d-1}\hat{\boldsymbol{\nu}}'_m\hat{\boldsymbol{\pi}}'_m$.

(c) Show that $\hat{\boldsymbol{\pi}}'_m[P'_m] = \hat{\boldsymbol{\pi}}'_{m+1}$. Note that $\hat{\boldsymbol{\pi}}'_m$ is an M-tuple that is non-zero only on the components of S_m.

(d) Let $\phi = 2\pi\sqrt{-1}/d$ and let $\boldsymbol{\pi}^{(k)} = \sum_{m=0}^{d-1}\hat{\boldsymbol{\pi}}'_m e^{mk\phi}$. Show that $\boldsymbol{\pi}^{(k)}$ is a left eigenvector of $[P]$ of eigenvalue $e^{-\phi k}$.

Exercise 4.20 **(a)** Show that, with the eigenvectors defined in Exercise 4.19,

$$\lim_{n\to\infty}[P^{nd}][P] = \sum_{i=0}^{d-1}\boldsymbol{\nu}^{(i)}\boldsymbol{\pi}^{(i+1)},$$

where, as before, $(d-1)+1$ is taken to be 0.

(b) Show that, for $1 \le j < d$,

$$\lim_{n\to\infty}[P^{nd}][P^j] = \sum_{i=1}^{d}\boldsymbol{\nu}^{(i)}\boldsymbol{\pi}^{(i+j)}.$$

(c) Show that

$$\lim_{n\to\infty}[P^{nd}]\left\{I + [P] + \cdots + [P^{d-1}]\right\} = \left(\sum_{i=0}^{d-1}\boldsymbol{\nu}^{(i)}\right)\left(\sum_{i=0}^{d-1}\boldsymbol{\pi}^{(i+j)}\right).$$

(d) Show that

$$\lim_{n\to\infty}\frac{1}{d}\left([P^n] + [P^{n+1}] + \cdots + [P^{n+d-1}]\right) = \boldsymbol{e}\boldsymbol{\pi},$$

where $\boldsymbol{\pi}$ is the steady-state probability vector for $[P]$. Hint: Show that $\boldsymbol{e} = \sum_m \boldsymbol{\nu}^{(m)}$ and $\boldsymbol{\pi} = (1/d)\sum_m \boldsymbol{\pi}^{(m)}$.

(e) Show that the above result is also valid for periodic unichains.

Exercise 4.21 Suppose A and B are each ergodic Markov chains with transition probabilities $\{P_{A_i,A_j}\}$ and $\{P_{B_i,B_j}\}$ respectively. Denote the steady-state probabilities of A and B by $\{\pi_{A_i}\}$ and $\{\pi_{B_i}\}$ respectively. The chains are now connected and modified as shown below. In particular, states A_1 and B_1 are connected and the new transition probabilities P' for the combined chain are given by

$$P'_{A_1,B_1} = \varepsilon, \quad P'_{A_1,A_j} = (1-\varepsilon)P_{A_1,A_j} \qquad \text{for all } A_j;$$
$$P'_{B_1,A_1} = \delta, \quad P'_{B_1,B_j} = (1-\delta)P_{B_1,B_j} \qquad \text{for all } B_j.$$

All other transition probabilities remain the same. Think intuitively of ε and δ as being small, but do not make any approximations in what follows. Give your answers to the following questions as functions of ε, δ, $\{\pi_{A_i}\}$ and $\{\pi_{B_i}\}$.

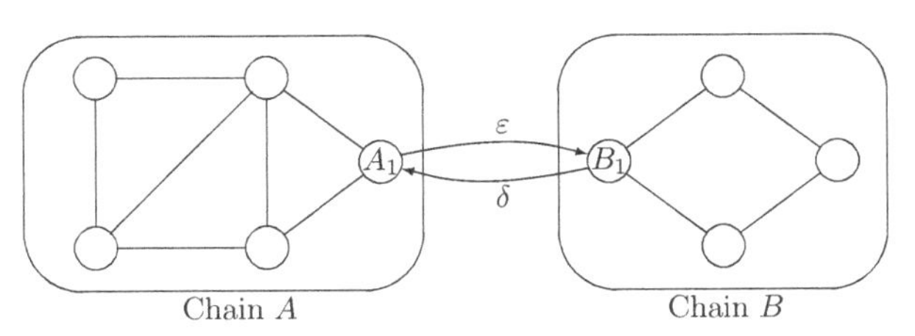

(a) Assume that $\epsilon > 0$, $\delta = 0$ (i.e., that A is a set of transient states in the combined chain). Starting in state A_1, find the conditional expected time to return to A_1 given that the first transition is to some state in chain A.

(b) Assume that $\epsilon > 0$, $\delta = 0$. Find $T_{A,B}$, the expected time to first reach state B_1 starting from state A_1. Your answer should be a function of ϵ and the original steady-state probabilities $\{\pi_{A_i}\}$ in chain A.

(c) Assume $\varepsilon > 0$, $\delta > 0$. Find $T_{B,A}$, the expected time to first reach state A_1, starting in state B_1. Your answer should depend only on δ and $\{\pi_{B_i}\}$.

(d) Assume $\varepsilon > 0$ and $\delta > 0$. Find $P'(A)$, the steady-state probability that the combined chain is in one of the states $\{A_j\}$ of the original chain A.

(e) Assume $\varepsilon > 0$, $\delta = 0$. For each state $A_j \neq A_1$ in A, find v_{A_j}, the expected number of visits to state A_j, starting in state A_1, before reaching state B_1. Your answer should depend only on ε and $\{\pi_{A_i}\}$.

(f) Assume $\varepsilon > 0$, $\delta > 0$. For each state A_j in A, find π'_{A_j}, the steady-state probability of being in state A_j in the combined chain. Hint: Be careful in your treatment of state A_1.

Exercise 4.22 Section 4.5.1 showed how to find the expected first-passage times to a fixed state, say 1, from all other states. It is often desirable to include the expected first recurrence time from state 1 to return to state 1. This can be done by splitting state 1 into two states, first an initial state with no transitions coming into it but the original transitions going out, and, second, a final trapping state with the original transitions coming in.

(a) For the chain on the left side of Figure 4.6, draw the graph for the modified chain with five states where state 1 has been split into two states.

(b) Suppose one has found the expected first-passage times v_j for states $j = 2, \ldots, 4$ (or in general from 2 to M). Find an expression for v_1, the expected first recurrence time for state 1 in terms of $v_2, v_3, \ldots, v_\mathsf{M}$ and $P_{12}, \ldots, P_{1\mathsf{M}}$.

Exercise 4.23 **(a)** Assume throughout that $[P]$ is the transition matrix of a unichain (and thus the eigenvalue 1 has multiplicity 1). Show that a solution to the equation $[P]\boldsymbol{w} - \boldsymbol{w} = \boldsymbol{r} - g\boldsymbol{e}$ exists if and only if $\boldsymbol{r} - g\boldsymbol{e}$ lies in the column space of $[P - I]$, where $[I]$ is the identity matrix.

(b) Show that this column space is the set of vectors $\boldsymbol{x}$ for which $\boldsymbol{\pi x} = 0$. Then show that $\boldsymbol{r} - g\boldsymbol{e}$ lies in this column space.

(c) Show that, with the extra constraint that $\boldsymbol{\pi w} = 0$, the equation $[P]\boldsymbol{w} - \boldsymbol{w} = \boldsymbol{r} - g\boldsymbol{e}$ has a unique solution.

Exercise 4.24 For the Markov chain with rewards in Figure 4.8:

(a) Find the solution to (4.37) and find the gain g.

(b) Modify Figure 4.8 by letting P_{12} be an arbitrary probability. Find g and $\boldsymbol{w}$ again and give an intuitive explanation of why P_{12} affects w_2.

Exercise 4.25 (Proof of Corollary 4.5.6) **(a)** Show that the gain per stage g is 0. Hint: Show that $\boldsymbol{r}$ is zero where the steady-state vector $\boldsymbol{\pi}$ is non-zero.

(b) Let $[P_{\mathcal{R}}]$ be the transition matrix for the recurrent states and let $\boldsymbol{r}_{\mathcal{R}} = 0$ be the reward vector and $\boldsymbol{w}_{\mathcal{R}}$ the relative-gain vector for $[P_{\mathcal{R}}]$. Show that $\boldsymbol{w}_{\mathcal{R}} = 0$. Hint: Use Theorem 4.5.4.

(c) Show that $w_i = 0$ for all $i \in \mathrm{R}$. Hint: Compare the relative-gain equations for $[P]$ to those for $[P_{\mathcal{R}}]$.

(d) Show that for each $n \geq 0$, $[P^n]\boldsymbol{w} = [P^{n+1}]\boldsymbol{w} + [P^n]\boldsymbol{r}$. Hint: Start with the relative-gain equation for $[P]$.

(e) Show that $\boldsymbol{w} = [P^{n+1}]\boldsymbol{w} + \sum_{m=0}^{n}[P^m]\boldsymbol{r}$. Hint: Sum the result in (b).

(f) Show that $\lim_{n\to\infty}[P^{n+1}]\boldsymbol{w} = 0$ and that $\lim_{n\to\infty}\sum_{m=0}^{n}[P^m]\boldsymbol{r}$ is finite, non-negative, and has positive components for $r_i > 0$. Hint: Use Lemma 4.3.6.

(g) Demonstrate the final result of the corollary by using the previous results on $\boldsymbol{r} = \boldsymbol{r}' - \boldsymbol{r}''$.

Exercise 4.26 Consider the Markov chain below:

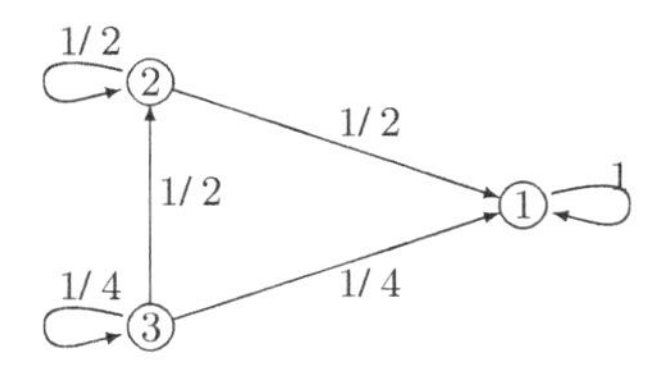

(a) Suppose the chain is started in state i and goes through n transitions; let $v_i(n)$ be the expected number of transitions (out of the total of n) until the chain enters the trapping state, state 1. Find an expression for $\boldsymbol{v}(n) = (v_1(n), v_2(n), v_3(n))^{\mathsf{T}}$ in terms of $\boldsymbol{v}(n-1)$ (take $v_1(n) = 0$ for all n). Hint: View the system as a Markov reward system; what is the value of $\boldsymbol{r}$?

(b) Find the numerical value of $\lim_{n\to\infty}\boldsymbol{v}(n)$. Interpret the meaning of the elements v_i in the solution of (4.32).

(c) Give a direct argument why (4.32) provides the solution directly to the expected time from each state to enter the trapping state.

Exercise 4.27 **(a)** Show that (4.48) can be rewritten in the more compact form

$$\boldsymbol{v}^*(n, \boldsymbol{u}) = \boldsymbol{v}^*(1, \boldsymbol{v}^*(n{-}1, \boldsymbol{u})).$$

(b) Explain why it is also true that

$$\boldsymbol{v}^*(2n, \boldsymbol{u}) = \boldsymbol{v}^*(n, \boldsymbol{v}^*(n, \boldsymbol{u})). \tag{4.65}$$

(c) One might guess that (4.65) could be used iteratively, finding $\boldsymbol{v}^*(2^{n+1}, \boldsymbol{u})$ from $\boldsymbol{v}^*(2^n, \boldsymbol{u})$. Explain why this is not possible in any straighttforward way. Hint: Think through explicitly how one might calculate $\boldsymbol{v}^*(n, \boldsymbol{v}^*(n, \boldsymbol{u}))$ from $\boldsymbol{v}^*(n, \boldsymbol{u})$.

Exercise 4.28 Consider finding the expected time until a given string appears in a IID binary sequence with $\Pr\{X_n = 1\} = p_1$, $\Pr\{X_n = 0\} = p_0 = 1 - p_1$.

(a) Following the procedure in Example 4.5.1, draw the three-state Markov chain for the string (0,1). Find the expected number of trials until the first occurrence of the string.

(b) For (b) and (c), let $(a_1, a_2, a_3, \ldots, a_k) = (0, 1, 1, \ldots, 1)$, i.e., zero followed by $k-1$ ones. Draw the corresponding Markov chain for $k = 4$.

(c) Let v_i, $1 \leq i \leq k$ be the expected first-passage time from state i to state k. Note that $v_k = 0$. For each i, $1 \leq i < k$, show that $v_i = \alpha_i + v_{i+1}$ and $v_0 = \beta_i + v_{i+1}$, where

α_i and β_i are each expressed as a product of powers of p_0 and p_1. Hint: Use induction on i taking $i = 1$ as the base. For the inductive step, first find β_{i+1} as a function of β_i starting with $i = 1$ and using the equation $v_0 = 1/p_0 + v_1$.

(d) Let $\boldsymbol{a} = (0, 1, 0)$. Draw the corresponding Markov chain for this string. Evaluate v_0, the expected time for $(0, 1, 0)$ to occur.

Exercise 4.29 **(a)** Find $\lim_{n\to\infty}[P^n]$ for the Markov chain below. Hint: Think in terms of the long-term transition probabilities. Recall that the edges in the graph for a Markov chain correspond to the positive transition probabilities.

(b) Let $\boldsymbol{\pi}^{(1)}$ and $\boldsymbol{\pi}^{(2)}$ denote the first two rows of $\lim_{n\to\infty}[P^n]$ and let $\boldsymbol{\nu}^{(1)}$ and $\boldsymbol{\nu}^{(2)}$ denote the first two columns of $\lim_{n\to\infty}[P^n]$. Show that $\boldsymbol{\pi}^{(1)}$ and $\boldsymbol{\pi}^{(2)}$ are independent left eigenvectors of $[P]$, and that $\boldsymbol{\nu}^{(1)}$ and $\boldsymbol{\nu}^{(2)}$ are independent right eigenvectors of $[P]$. Find the eigenvalue for each eigenvector.

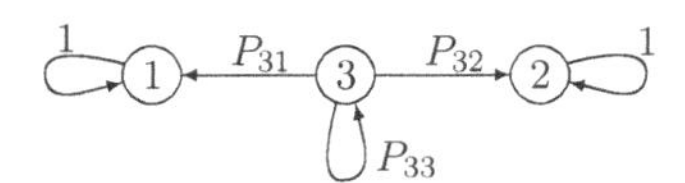

(c) Let $\boldsymbol{r}$ be an arbitrary reward vector and consider the equation

$$\boldsymbol{w} + g^{(1)}\boldsymbol{\nu}^{(1)} + g^{(2)}\boldsymbol{\nu}^{(2)} = \boldsymbol{r} + [P]\boldsymbol{w}. \tag{4.66}$$

Determine what values $g^{(1)}$ and $g^{(2)}$ must have in order for (4.66) to have a solution. Argue that with the additional constraints $w_1 = w_2 = 0$, (4.66) has a unique solution for $\boldsymbol{w}$ and find that $\boldsymbol{w}$.

Exercise 4.30 Let $\boldsymbol{u}$ and $\boldsymbol{u}'$ be arbitrary final reward vectors with $\boldsymbol{u} \le \boldsymbol{u}'$.

(a) Let $\boldsymbol{k}$ be an arbitrary stationary policy and prove that $\boldsymbol{v}^{\boldsymbol{k}}(n, \boldsymbol{u}) \le \boldsymbol{v}^{\boldsymbol{k}}(n, \boldsymbol{u}')$ for each $n \ge 1$.

(b) For the optimal dynamic policy, prove that $\boldsymbol{v}^*(n, \boldsymbol{u}) \le \boldsymbol{v}^*(n, \boldsymbol{u}')$ for each $n \ge 1$. This is known as the monotonicity theorem.

(c) Now let $\boldsymbol{u}$ and $\boldsymbol{u}'$ be arbitrary. Let $\alpha = \max_i(u_i - u_i')$. Show that

$$\boldsymbol{v}^*(n, \boldsymbol{u}) \le \boldsymbol{v}^*(n, \boldsymbol{u}') + \alpha\boldsymbol{e}.$$

Exercise 4.31 Consider a Markov decision problem with M states in which some state, say state 1, is inherently reachable from each other state.

(a) Show that there must be some other state, say state 2, and some decision, k_2, such that $P_{21}^{(k_2)} > 0$.

(b) Show that there must be some other state, say state 3, and some decision, k_3, such that either $P_{31}^{(k_3)} > 0$ or $P_{32}^{(k_3)} > 0$.

(c) Assume, for some i, and some set of decisions $k_2, \dots, k_i$ that, for each j, $2 \le j \le i$, $P_{jl}^{(k_j)} > 0$ for some $l < j$ (i.e., that each state from 2 to j has a non-zero transition to a lower numbered state). Show that there is some state (other than $1, \dots, i$), say $i + 1$ and some decision k_{i+1} such that $P_{i+1,l}^{(k_{i+1})} > 0$ for some $l \le i$.

(d) Use (a), (b), and (c) to observe that there is a stationary policy $\boldsymbol{k} = k_1, \dots, k_{\mathsf{M}}$ for which state 1 is accessible from each other state.

Exercise 4.32 George drives his car to the theater, which is at the end of a one-way street. There are parking places along the side of the street and a parking garage that costs \$5 at the theater. Each parking place is independently occupied or unoccupied with probability $1/2$. If George parks n parking places away from the theater, it costs him n cents (in time and shoe leather) to walk the rest of the way. George is myopic and can only see the parking place he is currently passing. If George has not already parked by the time he reaches the nth place, he first decides whether or not he will park if the place is unoccupied, and then observes the place and acts according to his decision. George can never go back and must park in the parking garage if he has not parked before.

(a) Model the above problem as a two-state dynamic programming problem. In the 'driving' state, state 2, there are two possible decisions: park if the current place is unoccupied or drive on whether or not the current place is unoccupied.

(b) Find $v_i^*(n, \boldsymbol{u})$, the *minimum* expected aggregate cost for n stages (i.e., immediately before observation of the nth parking place) starting in state $i = 1$ or 2; it is sufficient to express $v_i^*(n, \boldsymbol{u})$ in terms of $v_i^*(n-1)$. The final costs, in cents, at stage 0 should be $v_2(0) = 500$, $v_1(0) = 0$.

(c) For what values of n is the optimal decision the decision to drive on?

(d) What is the probability that George will park in the garage, assuming that he follows the optimal policy?

Exercise 4.33 (Proof of Corollary 4.6.9) **(a)** Show that if two stationary policies $\boldsymbol{k}'$ and $\boldsymbol{k}$ have the same recurrent class R' and if $k_i' = k_i$ for all $i \in \mathsf{R}'$, then $w_i' = w_i$ for all $i \in \mathsf{R}'$. Hint: See the first part of the proof of Lemma 4.6.7.

(b) Assume that $\boldsymbol{k}'$ satisfies (4.50) (i.e., that it satisfies the termination condition of the policy improvement algorithm) and that $\boldsymbol{k}$ satisfies the conditions of (a). Show that (4.64) is satisfied for all states ℓ.

(c) Show that $\boldsymbol{w} \leq \boldsymbol{w}'$. Hint: Follow the reasoning at the end of the proof of Lemma 4.6.7.

Exercise 4.34 Consider the dynamic programming problem below with two states and two possible policies, denoted $\boldsymbol{k}$ and $\boldsymbol{k}'$. The policies differ only in state 2.

(a) Find the steady-state gain per stage, g and g', for stationary policies $\boldsymbol{k}$ and $\boldsymbol{k}'$. Show that $g = g'$.

(b) Find the relative-gain vectors, $\boldsymbol{w}$ and $\boldsymbol{w}'$, for stationary policies $\boldsymbol{k}$ and $\boldsymbol{k}'$.

(c) Suppose the final reward, at stage 0, is $u_1 = 0$, $u_2 = u$. For what range of u does the dynamic programming algorithm use decision $\boldsymbol{k}$ in state 2 at stage 1?

(d) For what range of u does the dynamic programming algorithm use decision $\boldsymbol{k}$ in state 2 at stage 2? At stage n? You should find that (for this example) the dynamic programming algorithm uses the same decision at each stage n as it uses in stage 1.

(e) Find the optimal gain $v_2^*(n, \boldsymbol{u})$ and $v_1^*(n, \boldsymbol{u})$ as a function of stage n assuming $u = 10$.

(f) Find $\lim_{n\to\infty}[\boldsymbol{v}^*(n, \boldsymbol{u} - n g\boldsymbol{e})]$ and show how it depends on $\boldsymbol{u}$.

Exercise 4.35 Consider a Markov decision problem in which the stationary policies $\boldsymbol{k}$ and $\boldsymbol{k}'$ each satisfy (4.50) and each correspond to ergodic Markov chains.

(a) Show that if $\boldsymbol{r}^{\boldsymbol{k}'} + [P^{\boldsymbol{k}'}]\boldsymbol{w}' \geq \boldsymbol{r}^{\boldsymbol{k}} + [P^{\boldsymbol{k}}]\boldsymbol{w}'$ is not satisfied with equality, then $g' > g$.

(b) Show that $\boldsymbol{r}^{\boldsymbol{k}'} + [P^{\boldsymbol{k}'}]\boldsymbol{w}' = \boldsymbol{r}^{\boldsymbol{k}} + [P^{\boldsymbol{k}}]\boldsymbol{w}'$ Hint: Use (a).

(c) Find the relationship between the relative gain vector $\boldsymbol{w}^{\boldsymbol{k}}$ for policy $\boldsymbol{k}$ and the relative-gain vector $\boldsymbol{w}'$ for policy $\boldsymbol{k}'$. Hint: Show that $\boldsymbol{r}^{\boldsymbol{k}} + [P^{\boldsymbol{k}}]\boldsymbol{w}' = g\boldsymbol{e} + \boldsymbol{w}'$; what does this say about $\boldsymbol{w}$ and $\boldsymbol{w}'$?

(d) Suppose that policy $\boldsymbol{k}$ uses decision 1 in state 1 and policy $\boldsymbol{k}'$ uses decision 2 in state 1 (i.e., $k_1 = 1$ for policy $\boldsymbol{k}$ and $k_1 = 2$ for policy $\boldsymbol{k}'$). What is the relationship between $r_1^{(k)}, P_{11}^{(k)}, P_{12}^{(k)}, \ldots, P_{1J}^{(k)}$ for k equal to 1 and 2?

(e) Now suppose that policy $\boldsymbol{k}$ uses decision 1 in each state and policy $\boldsymbol{k}'$ uses decision 2 in each state. Is it possible that $r_i^{(1)} > r_i^{(2)}$ for all i? Explain carefully.

(f) Now assume that $r_i^{(1)}$ is the same for all i. Does this change your answer to (e)? Explain.

Exercise 4.36 Consider a Markov decision problem with three states. Assume that each stationary policy corresponds to an ergodic Markov chain. It is known that a particular policy $\boldsymbol{k}' = (k_1, k_2, k_3) = (2, 4, 1)$ is the unique optimal stationary policy (i.e., the gain per stage in steady state is maximized by always using decision 2 in state 1, decision 4 in state 2, and decision 1 in state 3). As usual, $r_i^{(k)}$ denotes the reward in state i under decision k, and $P_{ij}^{(k)}$ denotes the probability of a transition to state j given state i and given the use of decision k in state i. Consider the effect of changing the Markov decision problem in each of the following ways (the changes in each part are to be considered in the absence of the changes in the other parts):

(a) $r_1^{(1)}$ is replaced by $r_1^{(1)} - 1$.

(b) $r_1^{(2)}$ is replaced by $r_1^{(2)} + 1$.

(c) $r_1^{(k)}$ is replaced by $r_1^{(k)} + 1$ for all state 1 decisions k.

(d) For all i, $r_i^{(k_i)}$ is replaced by $r_i^{(k_i)} + 1$ for the decision k_i of policy $\boldsymbol{k}'$.

For each of the above changes, answer the following questions; *give explanations*:

(1) Is the gain per stage, g', increased, decreased, or unchanged by the given change?

(2) Is it possible that another policy, $\boldsymbol{k} \neq \boldsymbol{k}'$, is optimal after the given change?

Exercise 4.37 (The Odoni bound) Let $\boldsymbol{k}'$ be the optimal stationary policy for a Markov decision problem and let g' and $\boldsymbol{\pi}'$ be the corresponding gain and steady-state probability respectively. Let $v_i^*(n, \boldsymbol{u})$ be the optimal dynamic expected reward for starting in state i at stage n with final reward vector $\boldsymbol{u}$.

(a) Show that $\min_i[v_i^*(n, \boldsymbol{u}) - v_i^*(n-1, \boldsymbol{u})] \leq g' \leq \max_i[v_i^*(n, \boldsymbol{u}) - v_i^*(n-1, \boldsymbol{u})]$; $n \geq 1$. Hint: Consider premultiplying $\boldsymbol{v}^*(n, \boldsymbol{u}) - \boldsymbol{v}^*(n-1, \boldsymbol{u})$ by $\boldsymbol{\pi}'$ or $\boldsymbol{\pi}$ where $\boldsymbol{k}$ is the optimal dynamic policy at stage n.

(b) Show that the lower bound is non-decreasing in n and the upper bound is non-increasing in n.

Exercise 4.38 Consider an integer-time queueing system with a finite buffer of size 2. At the beginning of the nth time interval, the queue contains at most two customers. There is a cost of one unit for each customer in queue (i.e., the cost of delaying that customer). If there is one customer in queue, that customer is served. If there are two customers, an extra server is hired at a cost of 3 units and both customers are served. Thus the total immediate cost for two customers in queue is 5, the cost for one customer is 1, and the cost for 0 customers is 0. At the end of the nth time interval, either 0, 1, or 2 new customers arrive (each with probability 1/3).

(a) Assume that the system starts with $0 \leq i \leq 2$ customers in queue at time -1 (i.e., in stage 1) and terminates at time 0 (stage 0) with a final cost $\boldsymbol{u}$ of 5 units for each customer in queue (at the beginning of interval 0). Find the expected aggregate cost $v_i(1, \boldsymbol{u})$ for $0 \leq i \leq 2$.

(b) Assume now that the system starts with i customers in queue at time -2 with the same final cost at time 0. Find the expected aggregate cost $v_i(2, \boldsymbol{u})$ for $0 \leq i \leq 2$.

(c) For an arbitrary starting time $-n$, find the expected aggregate cost $v_i(n, \boldsymbol{u})$ for $0 \leq i \leq 2$.

(d) Find the cost per stage and find the relative cost (gain) vector.

(e) Now assume that there is a decision maker who can choose whether or not to hire the extra server when there are two customers in queue. If the extra server is not hired, the three-unit fee is saved, but only one of the customers is served. If there are two arrivals in this case, assume that one is turned away at a cost of 5 units. Find the minimum dynamic aggregate expected cost $v_i^*(1)$, $0 \leq i \leq 2$, for stage 1 with the same final cost as before.

(f) Find the minimum dynamic aggregate expected cost $v_i^*(n, \boldsymbol{u})$ for stage n, $0 \leq i \leq 2$.

(g) Now assume a final cost $\boldsymbol{u}$ of 1 unit per customer rather than 5, and find the new minimum dynamic aggregate expected cost $v_i^*(n, \boldsymbol{u})$, $0 \leq i \leq 2$.

5 Renewal processes

5.1 Introduction

Recall that a renewal process is an arrival process in which the interarrival intervals are positive,[1] independent, and identically distributed (IID) random variables (rv s). Renewal processes (since they are arrival processes) can be specified in three standard ways: first, by the joint distributions of the arrival epochs $S_1, S_2, \ldots$, second, by the joint distributions of the interarrival times $X_1, X_2, \ldots$, and, third, by the joint distributions of the counting rv s, $N(t)$ for $t > 0$. Recall that $N(t)$ represents the number of arrivals to the system in the interval $(0, t]$. Figure 5.1 reviews the connection between the sample values of these rv s.

The simplest characterization is through the interarrival times X_i, since they are IID. Each arrival epoch S_n is simply the sum $X_1 + X_2 + \cdots + X_n$ of n IID rv s. The characterization of greatest interest in this chapter is the renewal counting process, $\{N(t);\ t > 0\}$. Recall from (2.2) and (2.3) that the arrival epochs and the counting rv s are related in each of the following equivalent ways:

$$\{S_n \leq t\} = \{N(t) \geq n\}; \qquad \{S_n > t\} = \{N(t) < n\}. \tag{5.1}$$

The reason for calling these processes *renewal processes* is that the process probabilistically starts over at each arrival epoch, S_n. That is, if the nth arrival occurs at $S_n = \tau$, then, counting from $S_n = \tau$, the jth subsequent arrival epoch is at $S_{n+j} - S_n = X_{n+1} + \cdots + X_{n+j}$. Thus, given $S_n = \tau$, $\{N(\tau + t) - N(\tau); t \geq 0\}$ is a renewal counting process with IID interarrival intervals of the same distribution as the original renewal process. This interpretation of arrivals as renewals will be discussed in more detail later.

The major reason for studying renewal processes is that many complicated processes have randomly occurring instants at which the system returns to a state probabilistically equivalent to the starting state. These embedded renewal epochs allow us to separate the

[1] Renewal processes are often defined in a slightly more general way, allowing the interarrival intervals X_i to include the possibility $1 > \Pr\{X_i = 0\} > 0$. All of the theorems in this chapter are valid under this more general assumption, as can be verified by complicating the proofs somewhat. Allowing $\Pr\{X_i = 0\} > 0$ allows multiple arrivals at the same instant, which makes it necessary to allow $N(0)$ to take on positive values, and appears to inhibit intuition about renewals. Exercise 5.3 shows how to view these more general renewal processes under the definition here, obviating the need for this generality.

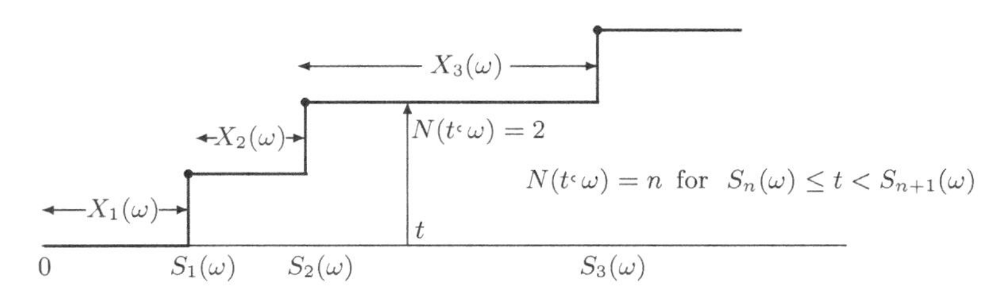

Figure 5.1 A sample function of an arrival process for a given sample point ω with its arrival epochs $\{S_1(\omega), S_2(\omega), \dots\}$, its interarrival times $\{X_1(\omega), X_2(\omega), \dots\}$, and its counting process $\{N(t, \omega);\ t > 0\}$. The sample function of the counting process is the step function illustrated with a unit step at each arrival epoch.

long-term behavior of the process (which can be studied through renewal theory) from the behavior within each renewal period.

Example 5.1.1 (Visits to a given state for a Markov chain) Assume that a recurrent finite-state Markov chain with transition matrix $[P]$ starts in state i at time 0. Then on the first return to state i, say at time n, the Markov chain, from time n on, is a probabilistic replica of the chain starting at time 0. That is, the state at time 1 is j with probability P_{ij}, and, given a return to i at time n, the probability of state j at time $n+1$ is P_{ij}. In the same way, for any $m > 0$,

$$\Pr\{X_1 = j, \dots, X_m = k \mid X_0 = i\} = \Pr\{X_{n+1} = j, \dots, X_{n+m} = k \mid X_n = i\}. \tag{5.2}$$

Each subsequent return to state i at a given time n starts a new probabilistic replica of the Markov chain starting in state i at time 0. Thus the sequence of entry times to state i can be viewed as the arrival epochs of a renewal process.

This example is important, and will form the key to the analysis of Markov chains with a countably-infinite set of states in Chapter 6. At the same time, (5.2) does not quite justify viewing successive returns to state i as a renewal process. The problem is that the time of the first entry to state i after time 0 is a rv rather than the given time n indicated in (5.2). This will not be a major problem to sort out, but the resolution will be more insightful after developing some basic properties of renewal processes.

Example 5.1.2 (The G/G/m queue) The customer arrivals to a G/G/m queue form a renewal counting process, $\{N(t);\ t > 0\}$. Each arriving customer enters one of the m servers if the server is not busy and otherwise waits in a common queue with first-come first-served (FCFS) service until a server becomes free. The service time required by each customer is a rv, IID over customers, and independent of arrival times and servers. The system is assumed to be empty for $t < 0$, and an arrival, viewed as customer number 0, is assumed at time 0. The subsequent interarrival intervals $X_1, X_2, \dots,$ are IID. Note that $N(t)$ for each $t > 0$ is the number of arrivals in $(0, t]$, so arrival number 0 at $t = 0$ is not counted[2] in $N(t)$.

We define a new counting process, $\{N^r(t);\ t > 0\}$, for which the renewal epochs are those arrival epochs in the original arrival process $\{N(t);\ t > 0\}$ at which an arriving customer sees an empty system (i.e., no customer in queue and none in service).[3] We will show in Section 5.5.3 that $\{N^r(t);\ t > 0\}$ is actually a renewal process, but give an intuitive explanation here. Note that customer 0 arrives at time 0 to an empty system, and given a first subsequent arrival to an empty system, at say epoch $S_1^r > 0$, the subsequent customer interarrival intervals are independent of the arrivals in $(0, S_1^r)$ and are identically distributed to those earlier arrivals. The service times after S_1^r are also IID from those earlier. Finally, the number of customers in the queue is the same function of inter-arrival intervals and service completions for the system starting at time 0 as for any subsequent arrival to an empty system.

In most situations, we use the words *arrivals* and *renewals* interchangeably, but for this type of example, the word *arrival* is used for the counting process $\{N(t);\ t > 0\}$ of the actual arrivals to the queueing system and the word *renewal* is used for $\{N^r(t);\ t > 0\}$. The reason for being interested in $\{N^r(t);\ t > 0\}$ is that it allows us to analyze very complicated queues such as this in two stages. First, $\{N(t);\ t > 0\}$ lets us analyze the distribution of the inter-renewal intervals X_n^r of $\{N^r(t);\ t > 0\}$. Second, the general renewal results developed in this chapter can be applied to the distribution on X_n^r to understand the overall behavior of the queueing system.

Throughout our study of renewal processes, we use $\overline{X}$ and $\mathsf{E}[X]$ interchangeably to denote the mean inter-renewal interval, and use σ_X^2 or simply σ^2 to denote the variance of the inter-renewal interval. We will usually assume that $\overline{X}$ is finite, but, except where explicitly stated, we need not assume that σ^2 is finite. This means, first, that σ^2 need not be calculated (which is often difficult if renewals are embedded into a more complex process), and, second, since modeling errors on the far tails of the inter-renewal distribution typically affect σ^2 more than $\overline{X}$, the results are relatively robust to these kinds of modeling errors.

Much of this chapter will be devoted to understanding the behavior of $N(t)$ and $N(t)/t$ as t becomes large. As might appear to be intuitively obvious, and as is proven in Exercise 5.1, $N(t)$ is a rv (i.e., not defective) for each $t > 0$. Also, as proven in Exercise 5.2, $\mathsf{E}[N(t)]$ is finite for all $t > 0$. It is then also clear that $N(t)/t$, which is interpreted as the time-average renewal rate over $(0,t]$, is also a rv with finite expectation.

One of the major results about renewal theory, which we establish shortly, concerns the behavior of the rv s $N(t)/t$ as $t \to \infty$. For each sample point $\omega \in \Omega$ of the overall process, $N(t, \omega)/t$ is a non-negative number for each t, so that, for given ω, we can

[2] There is always a certain amount of awkwardness in 'starting' a renewal process, and the assumption of an arrival at time 0 which is not counted in $N(t)$ seems strange, but simplifies the notation. The process is defined in terms of the IID inter-renewal intervals $X_1, X_2, \ldots$. The first renewal epoch is at $S_1 = X_1$, and this is the point at which $N(t)$ changes from 0 to 1.

[3] Readers who accept without question that $\{N^r(t)\ t > 0\}$ is a renewal process should be proud of their probabilistic intuition, but should also question exactly how such a conclusion can be proven.

view $\{N(t, \omega)/t;\ t > 0\}$ as a function of time. Thus $\lim_{t\to\infty} N(t, \omega)/t$, if it exists, is the time-average renewal rate over $(0, \infty)$ for the sample point ω.

The *strong law for renewal processes* states that this limiting time-average renewal rate exists and has value $1/\overline{X}$ for a set of ω that has probability 1. We shall often refer to this result by the less precise statement that the time-average renewal rate is $1/\overline{X}$. This result is a direct consequence of the strong law of large numbers (SLLN) for IID rv s. In the next section, we first state and prove the SLLN for IID rv s and then establish the strong law for renewal processes.

There is a certain intuitive appeal to the idea of taking a limit over a sample path as opposed to showing, for example, that the rv $N(t)/t$ has a variance that approaches 0 as $t \to \infty$, but the reader will need some patience to understand both how to deal with limits over individual sample paths and the reason for wanting to deal with such quantities.

Another important theoretical result in this chapter is the elementary renewal theorem, which states that $\mathsf{E}\left[N(t)/t\right]$ also approaches $1/\overline{X}$ as $t \to \infty$. Surprisingly, this is more than a trival consequence of the strong law for renewal processes, and we shall develop several widely useful results such as Wald's equality, in establishing this theorem.

The final major theoretical result of the chapter is Blackwell's theorem, which shows that, for appropriate values of δ, the expected number of renewals in an interval $(t, t+\delta]$ approaches $\delta/\overline{X}$ as $t \to \infty$. We shall thus interpret $1/\overline{X}$ as an ensemble-average renewal rate. This rate is the same as the above time-average renewal rate. We shall see the benefits of being able to work with both time averages and ensemble averages.

There is a wide range of other results, from standard queueing results to results that are needed in all subsequent chapters.

5.2 The strong law of large numbers and convergence with probability 1

The concept of a sequence of rv s converging with probability 1 (WP1) was introduced briefly in Section 1.7.6. We discuss this type of convergence more fully here and establish some conditions under which it holds. Next the *SLLN* is stated for IID rv s (this is essentially the result that the partial sample averages of IID rv s converge to the mean WP1). A proof is given under the added condition that the rv s have a finite fourth moment. Finally, in the following section, we state the strong law for renewal processes and use the SLLN for IID rv s to prove it.

5.2.1 Convergence with probability 1 (WP1)

Recall that a sequence $\{Z_n;\ n \geq 1\}$ of rv s on a sample space Ω is defined to converge WP1 to a rv Z on Ω if

$$\Pr\left\{\omega \in \Omega : \lim_{n\to\infty} Z_n(\omega) = Z(\omega)\right\} = 1,$$

i.e., if the set of sample sequences $\{Z_n(\omega);\ n \geq 1\}$ that converge to $Z(\omega)$ is an event and has probability 1. This becomes slightly easier to understand if we define $Y_n = Z_n - Z$ for each n. The sequence $\{Y_n;\ n \geq 1\}$ then converges to 0 WP1 if and only if the sequence $\{Z_n;\ n \geq 1\}$ converges to Z WP1. Dealing only with convergence to 0 rather than to an arbitrary rv does not cut any steps from the following proofs, but it simplifies the notation and the concepts at no cost to generality.

We start with a simple lemma that provides a useful condition under which convergence to 0 WP1 occurs. We shall see later how to use this lemma in an indirect way to prove the SLLN.

Lemma 5.2.1 *Let $\{Y_n;\ n \geq 1\}$ be a sequence of rv s, each with finite expectation. If $\sum_{n=1}^{\infty} \mathsf{E}[|Y_n|] < \infty$, then* $\Pr\{\omega : \lim_{n\to\infty} Y_n(\omega) = 0\} = 1$.

Discussion Note that the Y_n are not IID and usually not independent. If one wants a simple example of such rv s, think of $Y_n = (X_1 + \cdots + X_n)/n$, where the X_n are IID. Note also that the condition $\sum_{n=1}^{\infty} \mathsf{E}[|Y_n|] < \infty$ implies that $\lim_{n\to\infty} \mathsf{E}[|Y_n|] = 0$, but also implies that the convergence of $\mathsf{E}[|Y_n|]$ is fast enough to make the sum converge.

Proof For any $\alpha > 0$ and any integer $m \geq 1$, the Markov inequality says that

$$\Pr\left\{\sum_{n=1}^{m} |Y_n| > \alpha\right\} \leq \frac{\mathsf{E}\left[\sum_{n=1}^{m} |Y_n|\right]}{\alpha} = \frac{\sum_{n=1}^{m} \mathsf{E}[|Y_n|]}{\alpha}. \tag{5.3}$$

Since $|Y_n|$ is non-negative, $\sum_{n=1}^{m} |Y_n| > \alpha$ implies that $\sum_{n=1}^{m+1} |Y_n| > \alpha$. Thus the left-hand side of (5.3) is non-decreasing in m and is upper bounded by the limit of the right-hand side. Thus both sides have a limit, and

$$\lim_{m\to\infty} \Pr\left\{\sum_{n=1}^{m} |Y_n| > \alpha\right\} \leq \frac{\sum_{n=1}^{\infty} \mathsf{E}[|Y_n|]}{\alpha}. \tag{5.4}$$

We next show that the limit on the left-hand side of (5.4) can be brought inside the probability. Let $A_m = \{\omega : \sum_{n=1}^{m} |Y_n(\omega)| > \alpha\}$. As seen above, the sequence $\{A_m;\ m \geq 1\}$ is nested, $A_1 \subseteq A_2 \cdots$, so from property (1.9) of the axioms of probability,

$$\begin{aligned}\lim_{m\to\infty} \Pr\left\{\sum_{n=1}^{m} |Y_n| > \alpha\right\} &= \Pr\left\{\bigcup\nolimits_{m=1}^{\infty} A_m\right\} \\ &= \Pr\left\{\omega : \sum\nolimits_{n=1}^{\infty} |Y_n(\omega)| > \alpha\right\},\end{aligned} \tag{5.5}$$

where the second equality uses the fact that for any given ω, $\sum_{n=1}^{\infty} |Y_n(\omega)| > \alpha$ if and only if $\sum_{n=1}^{m} |Y_n(\omega)| > \alpha$ for some $m \geq 1$. Combining (5.4) with (5.5),

$$\Pr\left\{\omega : \sum_{n=1}^{\infty} |Y_n(\omega)| > \alpha\right\} \leq \frac{\sum_{n=1}^{\infty} \mathsf{E}[|Y_n|]}{\alpha}.$$

Looking at the complementary set and assuming $\alpha > \sum_{n=1}^{\infty} \mathsf{E}[|Y_n|]$,

$$\Pr\left\{\omega : \sum_{n=1}^{\infty} |Y_n(\omega)| \leq \alpha\right\} \geq 1 - \frac{\sum_{n=1}^{\infty} \mathsf{E}[|Y_n|]}{\alpha}. \tag{5.6}$$

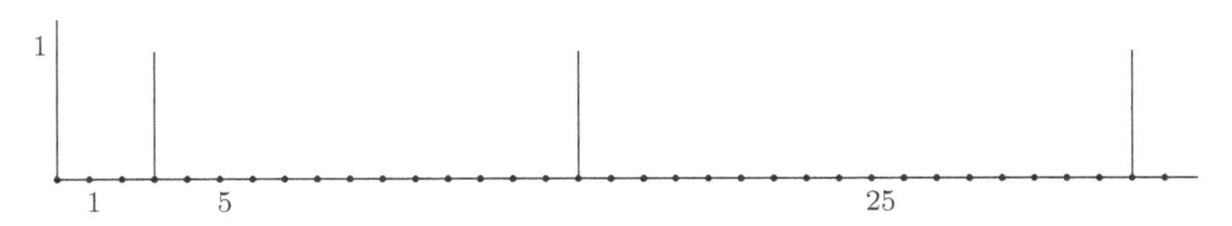

Figure 5.2 Illustration of a sample path of a sequence of rv s $\{Y_n; n \geq 0\}$, where, for each $j \geq 0$, $Y_n = 1$ for an equiprobable choice of $n \in [5^j, 5^{j+1})$ and $Y_n = 0$ otherwise.

For any ω such that $\sum_{n=1}^{\infty} |Y_n(\omega)| \leq \alpha$, we see that $\{|Y_n(\omega)|; n \geq 1\}$ is simply a sequence of non-negative numbers with a finite sum. Thus the individual numbers in that sequence must approach 0, i.e., $\lim_{n\to\infty} |Y_n(\omega)| = 0$ for each such ω. It follows then that

$$\Pr\left\{\omega : \lim_{n\to\infty} |Y_n(\omega)| = 0\right\} \geq \Pr\left\{\omega : \sum_{n=1}^{\infty} |Y_n(\omega)| \leq \alpha\right\}.$$

Combining this with (5.6),

$$\Pr\left\{\omega : \lim_{n\to\infty} |Y_n(\omega)| = 0\right\} \geq 1 - \frac{\sum_{n=1}^{\infty} \mathsf{E}[|Y_n|]}{\alpha}.$$

This is true for all α, so $\Pr\{\omega : \lim_{n\to\infty} |Y_n(\omega)| = 0\} = 1$. The same limit holds for $Y_n(\omega)$ in place of $|Y_n(\omega)|$, completing the proof. □

Note that the interchange of limits in (5.5) also shows that $\{\omega : \sum_{n=1}^{\infty} |Y_n(\omega)| > \alpha\}$ is an event, which is by no means obvious for a countably-infinite sum of rv s.

It is instructive to recall Example 1.7.9, illustrated in Figure 5.2, where $\{Y_n;\ n \geq 1\}$ converges in probability but does not converge with probability 1. Note that $\mathsf{E}[Y_n] = 1/(5^{j+1} - 5^j)$ for $n \in [5^j,\ 5^{j+1})$. Thus $\lim_{n\to\infty} \mathsf{E}[Y_n] = 0$, but $\sum_{n=1}^{\infty} \mathsf{E}[Y_n] = \infty$. Thus this sequence does not satisfy the conditions of the lemma. This helps explain how the conditions in the lemma exclude such sequences.

Before proceeding to the SLLN, we want to show that convergence WP1 implies convergence in probability. We give an incomplete argument here with precise versions in both Exercise 5.5 and Exercise 5.7. Exercise 5.7 has the added merit of expressing the set $\{\omega : \lim_n Y_n(\omega) = 0\}$ explicitly in terms of countable unions and intersections of simple events involving finite sets of the Y_n. This representation is valid whether or not the conditions of the lemma are satisfied and shows that this set is indeed an event.

Assume that $\{Y_n;\ n \geq 1\}$ is a sequence of rv s such that $\lim_{n\to\infty} Y_n = 0$ WP1. Then for any $\epsilon > 0$, each sample sequence $\{Y_n(\omega);\ n \geq 1\}$ that converges to 0 satisfies $|Y_n| \leq \epsilon$ for all sufficiently large n. This means (see Exercise 5.5) that $\lim_{n\to\infty} \Pr\{|Y_n| \leq \epsilon\} = 1$. Since this is true for all $\epsilon > 0$, $\{Y_n;\ n \geq 0\}$ converges in probability to 0.

5.2.2 Strong law of large numbers

We next develop the SLLN. We do not have the mathematical tools to prove the theorem in its full generality, but will give a fairly insightful proof under the additional assumption that the rv under discussion has a finite fourth moment. The theorem has a remarkably simple and elementary form, considering that it is certainly one of the most important theorems in probability theory. Most of the hard work in understanding the theorem comes from understanding what convergence WP1 means, and that has already been discussed. Given this understanding, the theorem is relatively easy to understand and surprisingly easy to prove (assuming a fourth moment).

Theorem 5.2.2 (SLLN) *For each integer $n \geq 1$, let $S_n = X_1 + \cdots + X_n$, where $X_1, X_2, \ldots$ are IID rv s satisfying $\mathsf{E}[|X|] < \infty$. Then*

$$\Pr\left\{\omega : \lim_{n\to\infty} \frac{S_n(\omega)}{n} = \overline{X}\right\} = 1. \tag{5.7}$$

Proof (for the case where $\overline{X} = 0$ and $\mathsf{E}\left[X^4\right] < \infty$) Assume that $\overline{X} = 0$ and $\mathsf{E}\left[X^4\right] < \infty$. Denote $\mathsf{E}\left[X^4\right]$ by γ. Let x be a real number. If $|x| \leq 1$, then $x^2 \leq 1$, and if $|x| > 1$, then $x^2 < x^4$. Thus $x^2 \leq 1 + x^4$ for all x. It follows that $\sigma^2 = \mathsf{E}\left[X^2\right] \leq 1 + \mathsf{E}\left[X^4\right]$. Thus σ^2 is finite if $\mathsf{E}\left[X^4\right]$ is.

Now let $S_n = X_1 + \cdots + X_n$, where $X_1, \ldots, X_n$ are IID with the distribution of X.

$$\begin{aligned}
\mathsf{E}\left[S_n^4\right] &= \mathsf{E}\left[(X_1 + \cdots + X_n)(X_1 + \cdots + X_n)(X_1 + \cdots + X_n)(X_1 + \cdots + X_n)\right] \\
&= \mathsf{E}\left[\left(\sum_{i=1}^n X_i\right)\left(\sum_{j=1}^n X_j\right)\left(\sum_{k=1}^n X_k\right)\left(\sum_{\ell=1}^n X_\ell\right)\right] \\
&= \sum_{i=1}^n \sum_{j=1}^n \sum_{k=1}^n \sum_{\ell=1}^n \mathsf{E}\left[X_i X_j X_k X_\ell\right],
\end{aligned}$$

where we have multiplied out the product of sums to get a sum of n^4 terms.

For each i, $1 \leq i \leq n$, there is a term in this sum with $i = j = k = \ell$. For each such term, $\mathsf{E}\left[X_i X_j X_k X_\ell\right] = \mathsf{E}\left[X^4\right] = \gamma$. There are n such terms (one for each choice of i, $1 \leq i \leq n$) and they collectively contribute $n\gamma$ to the sum $\mathsf{E}\left[S_n^4\right]$. Also, for each $i, k \neq i$, there is a term with $j = i$ and $\ell = k$. For each of these $n(n-1)$ terms, $\mathsf{E}\left[X_i X_i X_k X_k\right] = \sigma^4$. There are another $n(n-1)$ terms with $j \neq i$ and $k = i$, $\ell = j$. Each such term contributes σ^4 to the sum. Finally, for each $i \neq j$, there is a term with $\ell = i$ and $k = j$. Collectively all of these terms contribute $3n(n-1)\sigma^4$ to the sum. Each of the remaining terms is 0 since at least one of i, j, k, ℓ is different from all the others, Thus we have

$$\mathsf{E}\left[S_n^4\right] = n\gamma + 3n(n-1)\sigma^4.$$

Now consider the sequence of rv s $\{S_n^4/n^4;\ n \geq 1\}$.

$$\sum_{n=1}^{\infty} \mathsf{E}\left[\left|\frac{S_n^4}{n^4}\right|\right] = \sum_{n=1}^{\infty} \frac{n\gamma + 3n(n-1)\sigma^4}{n^4} < \infty,$$

where we have used the facts that the series $\sum_{n\geq 1} 1/n^2$ and the series $\sum_{n\geq 1} 1/n^3$ converge.

Using Lemma 5.2.1 applied to $\{S_n^4/n^4;\ n \geq 1\}$, we see that $\lim_{n\to\infty} S_n^4/n^4 = 0$ WP1. For each ω such that $\lim_{n\to\infty} S_n^4(\omega)/n^4 = 0$, the non-negative fourth root of that sequence of non-negative numbers also approaches 0. Thus $\lim_{n\to\infty} |S_n/n| = 0$ WP1. □

The above proof assumed that $\mathsf{E}[X] = 0$. It can be extended trivially to the case of an arbitrary finite $\overline{X}$ by replacing X in the proof with $X - \overline{X}$. A proof using the weaker condition that $\sigma_X^2 < \infty$ will be given in Section 9.9.1.

The technique that was used at the end of this proof provides a clue about why the concept of convergence WP1 is so powerful. The technique showed that if one sequence of rv s ($\{S_n^4/n^4;\ n \geq 1\}$) converges to 0 WP1, then another sequence ($|S_n/n|;\ n \geq 1\}$) also converges WP1. We will formalize and generalize this technique in Lemma 5.3.3 as a major step toward establishing the strong law for renewal processes.

The SLLN appears to be fairly intuitive. We start with a sequence of IID rv s $\{X_i;\ i \geq 1\}$, next form a sequence of sample averages, $\{Y_n;\ n \geq 1\}$, where $Y_n = (X_1 + \cdots X_n)/n$, and finally look at a sample sequence $\{y_n;\ n \geq 1\}$ of the sample averages. Each sample sequence $\{y_n;\ n \geq 1\}$ is a sequence of real numbers and we would expect that sequence to converge to $\overline{X}$. The SLLN says that this convergence takes place WP1.

The intuitive nature of the SLLN fades slightly if we look at it in the following way: consider a Bernoulli process $\{X_i;\ i \geq 1\}$ with $p = \mathsf{p}_X(1)$. Then $\{\omega : \lim_n Y_n(\omega) = p\}$ is an event in the sample space with probability 1. Consider a different Bernoulli process $\{X_n';\ n \geq 1\}$, with probability of success $p' \neq p$. The set of sample points and the set of events for this process is the same as before. The set $\{\omega : \lim_n Y_n'(\omega) = p\}$ is still the same event as before, but it has probability 0 under this new probability measure.

The parameter p of a Bernoulli process can take on an uncountably infinite number of values in the interval (0, 1). Thus there are uncountably many events $\{\omega : Y_n(\omega) = p\}$, one for each value of p. Each such event has probability 1 for the Bernoulli process with that value of p and probability 0 for each other value of p. There is no mathematical contradiction here, but we should realize that intuition about these events requires more subtlety than might be imagined initially.

5.3 Strong law for renewal processes

To get an intuitive idea why $N(t)/t$ should approach $1/\overline{X}$ for large t, consider Figure 5.3. For any given sample function of $\{N(t);\ t > 0\}$, note that, for any given t, $N(t)/t$ is the slope of a straight line from the origin to the point $(t, N(t))$. As t increases, this slope decreases in the interval between each adjacent pair of arrival epochs and then jumps up at the next arrival epoch. In order to express this as an equation for arbitrary $t > 0$, let the rv $S_{N(t)}$ denote the arrival epoch of the $N(t)$th arrival and let $S_{N(t)+1}$ denote the arrival epoch of the first arrival following t (see Figure 5.3). If $N(t) = 0$, we take $S_{N(t)} = 0$ by convention. Thus we have $S_{N(t)} \leq t < S_{N(t)+1}$. For all $t > 0$, then

$$\frac{N(t)}{S_{N(t)}} \geq \frac{N(t)}{t} > \frac{N(t)}{S_{N(t)+1}}. \tag{5.8}$$

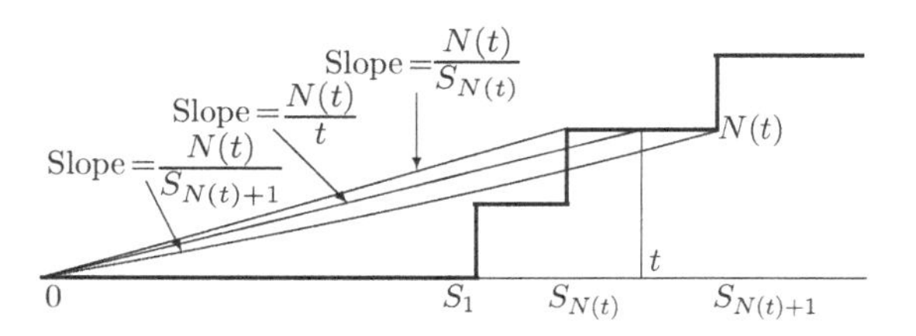

Figure 5.3 Comparison of a sample function of $N(t)/t$ with $N(t)/S_{N(t)}$ and $N(t)/S_{N(t)+1}$ for the same sample point. Note that for the given sample point, $N(t)$ is the number of arrivals up to and including t, and thus $S_{N(t)}$ is the epoch of the last arrival before or at time t. Similarly, $S_{N(t)+1}$ is the epoch of the first arrival strictly after time t.

We want to show intuitively why the slope $N(t)/t$ in the figure approaches $1/\overline{X}$ as $t \to \infty$. As t increases, we would guess that $N(t)$ increases without bound, i.e., that for each arrival, another arrival occurs eventually. Assuming this, the left-hand side of (5.8) increases with increasing t as $1/S_1, 2/S_2, \ldots, n/S_n, \ldots$, where $n = N(t)$. Since S_n/n converges to $\overline{X}$ WP1 from the SLLN, we might be brave enough or insightful enough to guess that n/S_n converges to $1/\overline{X}$.

We are now ready to state the strong law for renewal processes as a theorem. Before proving the theorem, we formulate the above two guesses as lemmas and prove their validity.

Theorem 5.3.1 (Strong law for renewal processes) *For a renewal process with mean inter-renewal interval $\overline{X} < \infty$, $\lim_{t\to\infty} N(t)/t = 1/\overline{X}$ WP1.*

Lemma 5.3.2 *Let $\{N(t);\ t > 0\}$ be a renewal counting process with inter-renewal rv s $\{X_n;\ n \geq 1\}$. Then (whether or not $\overline{X} < \infty$), $\lim_{t\to\infty} N(t) = \infty$ WP1 and $\lim_{t\to\infty} \mathsf{E}[N(t)] = \infty$.*

Proof Note that for each sample point ω, $N(t, \omega)$ is a non-decreasing real-valued function of t and thus either has a finite limit or an infinite limit. The first part of the lemma, i.e., $\lim_{t\to\infty} N(t) = \infty$ WP1, will be proven by showing that the set of ω for which this limit is finite has probability 0.

$$\Pr\left\{\omega : \lim_{t\to\infty} N(t,\omega) < \infty\right\} = \Pr\left\{\omega : \bigcup\nolimits_{n\geq 1} \lim_{t\to\infty} N(t,\omega) < n\right\}$$
$$\leq \sum_{n\geq 1} \Pr\left\{\omega : \lim_{t\to\infty} N(t,\omega) < n\right\}. \tag{5.9}$$

It is sufficient to show that each term in this sum is 0. Writing $\Pr\{\omega : \lim_{t\to\infty} N(t,\omega) < n\}$ as $\Pr\{\lim_{t\to\infty} N(t) < n\}$, we now show that $\Pr\{\lim_{t\to\infty} N(t) < n\} = 0$ for each $n \geq 1$. Using (5.1),

$$\lim_{t\to\infty} \Pr\{N(t) < n\} = \lim_{t\to\infty} \Pr\{S_n > t\} = 1 - \lim_{t\to\infty} \Pr\{S_n \leq t\}.$$

Since the X_i are rv s, the sums S_n are also rv s (i.e., non-defective) for each n (see Section 1.5.1), and thus $\lim_{t\to\infty} \Pr\{S_n \leq t\} = 1$ for each n. Thus $\lim_{t\to\infty} \Pr\{N(t) < n\} = 0$ for each n, and from (5.9), $\lim_{t\to\infty} N(t) = \infty$ WP1.

Next, $\mathsf{E}[N(t)]$ is non-decreasing in t, and thus has either a finite or infinite limit as $t \to \infty$. For each n, $\Pr\{N(t) \geq n\} \geq 1/2$ for large enough t, and therefore $\mathsf{E}[N(t)] \geq n/2$ for such t. Thus $\mathsf{E}[N(t)]$ can have no finite limit as $t \to \infty$, and $\lim_{t\to\infty} \mathsf{E}[N(t)] = \infty$. □

The following lemma is quite a bit more general than the second guess above, but it will be useful elsewhere. This is the formalization of the technique used at the end of the proof of the SLLN.

Lemma 5.3.3 *Let $\{Z_n;\, n \geq 1\}$ be a sequence of rv s such that $\lim_{n\to\infty} Z_n = \alpha$ WP1. Let f be a real valued function of a real variable that is continuous at α. Then*

$$\lim_{n\to\infty} f(Z_n) = f(\alpha) \qquad \text{WP1.} \tag{5.10}$$

Proof First let $z_1, z_2, \ldots$ be a sequence of real numbers such that $\lim_{n\to\infty} z_n = \alpha$. Continuity of f at α means that for every $\epsilon > 0$, there is a $\delta > 0$ such that $|f(z) - f(\alpha)| < \epsilon$ for all z such that $|z - \alpha| < \delta$. Also, since $\lim_{n\to\infty} z_n = \alpha$, we know that for every $\delta > 0$, there is an m such that $|z_n - \alpha| \leq \delta$ for all $n \geq m$. Putting these two statements together, we know that for every $\epsilon > 0$, there is an m such that $|f(z_n) - f(\alpha)| < \epsilon$ for all $n \geq m$. Thus $\lim_{n\to\infty} f(z_n) = f(\alpha)$.

If ω is any sample point such that $\lim_{n\to\infty} Z_n(\omega) = \alpha$, then $\lim_{n\to\infty} f(Z_n(\omega)) = f(\alpha)$. Since this set of sample points has probability 1, (5.10) follows. □

Proof of Theorem 5.3.1 (Strong law for renewal processes) Since $\Pr\{X > 0\} = 1$ for a renewal process, we see that $\overline{X} > 0$. Choosing $f(x) = 1/x$, we see that $f(x)$ is continuous at $x = \overline{X}$. It follows from Lemma 5.3.3 that

$$\lim_{n\to\infty} \frac{n}{S_n} = \frac{1}{\overline{X}} \qquad \text{WP1.}$$

From Lemma 5.3.2, we know that $\lim_{t\to\infty} N(t) = \infty$ with probability 1, so, with probability 1, $N(t)$ increases through all the non-negative integers as t increases from 0 to ∞. Thus

$$\lim_{t\to\infty} \frac{N(t)}{S_{N(t)}} = \lim_{n\to\infty} \frac{n}{S_n} = \frac{1}{\overline{X}} \qquad \text{WP1.}$$

Recall that $N(t)/t$ is sandwiched between $N(t)/S_{N(t)}$ and $N(t)/S_{N(t)+1}$, so we can complete the proof by showing that $\lim_{t\to\infty} N(t)/S_{N(t)+1} = 1/\overline{X}$. To show this,

$$\lim_{t\to\infty} \frac{N(t)}{S_{N(t)+1}} = \lim_{n\to\infty} \frac{n}{S_{n+1}} = \lim_{n\to\infty} \frac{n+1}{S_{n+1}} \frac{n}{n+1} = \frac{1}{\overline{X}} \qquad \text{WP1.}$$

□

We have gone through the proof of this theorem in great detail, since a number of the techniques are probably unfamiliar to many readers. If one reads the proof again, after becoming familiar with the details, the simplicity of the result will be quite striking. The theorem is also true if the mean inter-renewal interval is infinite; this can be seen by a truncation argument (see Exercise 5.9).

As explained in Section 5.2.1, Theorem 5.3.1 also implies the corresponding weak law of large numbers (WLLN) for $N(t)$, i.e., for any $\epsilon > 0$, $\lim_{t\to\infty} \Pr\{|N(t)/t - 1/\overline{X}| \geq$

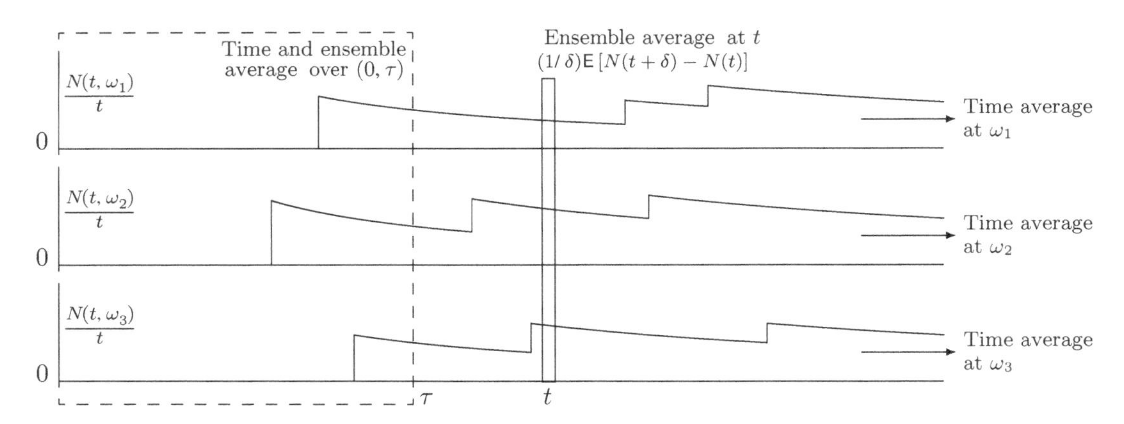

Figure 5.4 The time average at a sample point ω, the time and ensemble average from 0 to a given τ, and the ensemble average in an interval $(t, t+\delta]$.

$\epsilon\} = 0$. This weak law could also be derived from the WLLN for S_n (Theorem 1.7.4). We do not pursue that here, since the derivation is tedious and uninstructive. As we will see, it is the SLLN that is most useful for renewal processes.

Figure 5.4 helps give some appreciation of what the strong law for $N(t)$ does and does not say. The strong law deals with time averages, $\lim_{t\to\infty} N(t,\omega)/t$, for individual sample points ω; these are indicated in the figure as horizontal averages, one for each ω. It is also of interest to look at time and ensemble averages, $\mathsf{E}\left[N(t)/t\right]$, shown in the figure as vertical averages. Note that $N(t,\omega)/t$ is the time-average number of renewals from 0 to t for a given ω, whereas $\mathsf{E}\left[N(t)/t\right]$ averages also over the ensemble. Finally, to focus on arrivals in the vicinity of a particular time t, it is of interest to look at the ensemble average $\mathsf{E}\left[N(t+\delta) - N(t)\right]/\delta$.

Given the strong law for $N(t)$, one would hypothesize that $\mathsf{E}\left[N(t)/t\right]$ approaches $1/\overline{X}$ as $t \to \infty$. One might also hypothesize that $\lim_{t\to\infty} \mathsf{E}\left[N(t+\delta) - N(t)\right]/\delta = 1/\overline{X}$, subject to some minor restrictions on δ. These hypotheses are correct and are discussed in detail in what follows. This equality of time averages and limiting ensemble averages for renewal processes carries over to a large number of stochastic processes, and forms the basis of *ergodic theory*. These results are important for both theoretical and practical purposes. It is sometimes easy to find time averages (just like it was easy to find the time-average $N(t,\omega)/t$ from the SLLN), and it is sometimes easy to find limiting ensemble averages. Being able to equate the two then allows us to alternate at will between time and ensemble averages.

Note that in order to equate time averages and limiting ensemble averages, quite a few conditions are required. First, the time average must exist in the limit $t \to \infty$ with probability 1; that time average must also be the same for all sample points in a set of probability 1. Second, the ensemble average must approach a limit as $t \to \infty$. Third, the time average and ensemble average must be the same. The following example, for a stochastic process very different from a renewal process, shows that equality between time and ensemble averages is not always satisfied for arbitrary processes.

Example 5.3.4 Let $\{X_i; i \geq 1\}$ be a sequence of binary IID rv s, each taking the value 0 with probability 1/2 and 2 with probability 1/2. Let $\{M_n; n \geq 1\}$ be the product process in which $M_n = X_1 X_2 \cdots X_n$. Since $M_n = 2^n$ if $X_1, \ldots, X_n$ each take the value 2 (an event of probability 2^{-n}) and $M_n = 0$ otherwise, we see that $\lim_{n\to\infty} M_n = 0$ with probability 1. Also $\mathsf{E}[M_n] = 1$ for all $n \geq 1$. Thus the time average exists and equals 0 WP1 and the ensemble average exists and equals 1 for all n, but the two are different. The problem is that as n increases, the atypical event in which $M_n = 2^n$ has a probability approaching 0, but still has a significant effect on the ensemble average.

Further discussion of ensemble averages is postponed to Section 5.6. Before that, we briefly state and discuss the central limit theorem (CLT) for counting renewal processes and then introduce the notion of rewards associated with renewal processes.

Theorem 5.3.5 (CLT for $N(t)$) *Assume that the inter-renewal intervals for a renewal counting process $\{N(t); t > 0\}$ have finite standard deviation $\sigma > 0$. Then*

$$\lim_{t\to\infty} \Pr\left\{ \frac{N(t) - t/\overline{X}}{\sigma \overline{X}^{-3/2} \sqrt{t}} < \alpha \right\} = \Phi(\alpha), \tag{5.11}$$

where

$$\Phi(y) = \int_{-\infty}^{y} \frac{1}{\sqrt{2\pi}} \exp(-x^2/2)dx.$$

This says that the cumulative distribution function (CDF) of $N(t)$ tends to the Gaussian distribution with mean $t/\overline{X}$ and standard deviation $\sigma\overline{X}^{-3/2}\sqrt{t}$.

The theorem can be proved by applying Theorem 1.7.2 (the CLT for a sum of IID rv s) to S_n and then using the identity $\{S_n \leq t\} = \{N(t) \geq n\}$. The general idea is illustrated in Figure 5.5, but the details are somewhat tedious, and can be found, for example, in [22]. We simply outline the argument here. For any real α, the CLT states that

$$\Pr\left\{S_n \leq n\overline{X} + \alpha\sqrt{n}\sigma\right\} \approx \Phi(\alpha),$$

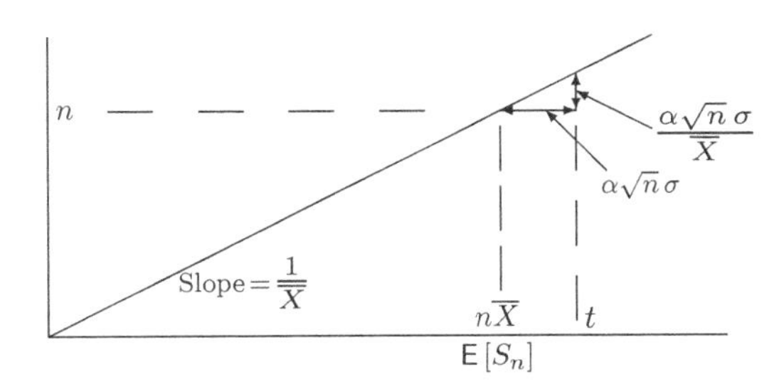

Figure 5.5 Illustration of the CLT for renewal processes. A given integer n is shown on the vertical axis, and the corresponding mean, $\mathsf{E}[S_n] = n\overline{X}$, is shown on the horizontal axis. The horizontal line with arrows at height n indicates α standard deviations of S_n from $\mathsf{E}[S_n]$, and the vertical line with arrows indicates the distance below $(t/\overline{X})$.

where $\Phi(\alpha) = \int_{-\infty}^{\alpha} (2\pi)^{-1/2} \exp(-x^2/2)\, dx$ and where the approximation becomes exact in the limit $n \to \infty$. Letting

$$t = n\overline{X} + \alpha\sqrt{n}\sigma,$$

and using $\{S_n \le t\} = \{N(t) \ge n\}$,

$$\Pr\{N(t) \ge n\} \approx \Phi(\alpha). \tag{5.12}$$

Since t is monotonic in n for fixed α, we can express n in terms of t, getting

$$n = \frac{t}{\overline{X}} - \frac{\alpha\sigma\sqrt{n}}{\overline{X}} \approx \frac{t}{\overline{X}} - \alpha\sigma t^{1/2}(\overline{X})^{-3/2}.$$

Substituting this into (5.12) establishes the theorem for $-\alpha$, which establishes the theorem since α is arbitrary. The omitted details involve handling the approximations carefully.

5.4 Renewal–reward processes; time averages

There are many situations in which, along with a renewal counting process $\{N(t);\ t > 0\}$, there is another randomly varying function of time $\{R(t); t > 0\}$, called a *reward function*, that models a rate at which the process is accumulating a reward. We shall illustrate many examples of such processes and see that a 'reward' could also be a cost or any randomly varying quantity of interest. The important restriction on these *reward functions* is that $R(t)$ at a given t depends only on the location of t within the inter-renewal interval containing t and perhaps other rv s local to that interval. Before defining this precisely, we start with several examples. These examples also illustrate that reward functions can be defined to study fundamental questions concerning the renewal process itself. The treatment here considers time averages for these questions, whereas Section 5.7 views the corresponding ensemble averages.

Example 5.4.1 (Time-average residual life) For a renewal counting process $\{N(t), t > 0\}$, let $Y(t)$ be the residual life at time t. The *residual life* is defined as the interval from t until the next renewal epoch, i.e., as $S_{N(t)+1} - t$. For example, if we arrive at a bus stop at time t and buses arrive according to a renewal process, $Y(t)$ is the time we have to wait for a bus to arrive (see Figure 5.6). We interpret $\{Y(t);\ t \ge 0\}$ as a reward function. The time average of $Y(t)$, over the interval $(0, t]$, is given by[4] $(1/t)\int_0^t Y(\tau)d\tau$. We are interested in the limit of this average as $t \to \infty$ (assuming that it exists in some sense). Figure 5.6 illustrates a sample function of a renewal counting process $\{N(t);\ t > 0\}$ and shows the residual life $Y(t)$ for that sample function. Note that, for a given sample

[4] $\int_0^t Y(\tau)d\tau$ is a rv just like any other function of a set of rv s. It has a sample value for each sample function of $\{N(t);\ t > 0\}$, and its CDF could be calculated in a straightforward but tedious way. For arbitrary stochastic processes, integration and differentiation can require great mathematical sophistication, but none of those subtleties occurs here.

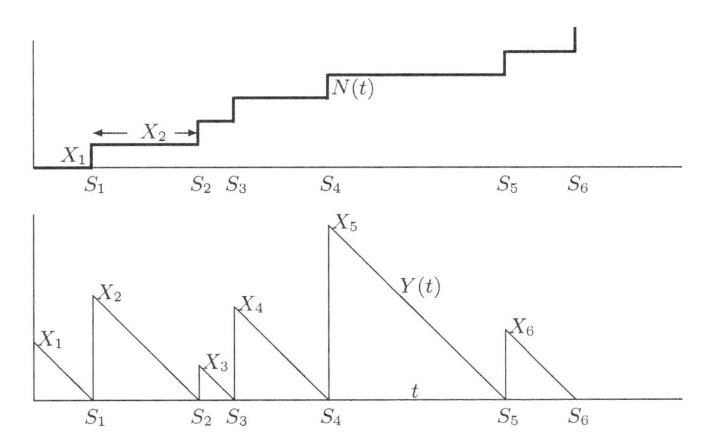

Figure 5.6 Residual life at time t. For any given sample function of the renewal process, the sample function of residual life decreases linearly with a slope of -1 from the beginning to the end of each inter-renewal interval.

function $\{Y(t) = y(t)\}$, the integral $\int_0^t y(\tau)d\tau$ is simply a sum of isosceles right triangles, with part of a final triangle at the end. Thus it can be expressed as

$$\int_0^t y(\tau)d\tau = \frac{1}{2}\sum_{i=1}^{n(t)} x_i^2 + \int_{\tau=s_{n(t)}}^t y(\tau)d\tau,$$

where $\{x_i;\ 0 < i < \infty\}$ is the set of sample values for the inter-renewal intervals.

Since this relationship holds for every sample point, we see that the rv $\int_0^t Y(\tau)d\tau$ can be expressed in terms of the inter-renewal random variables X_n as

$$\int_{\tau=0}^t Y(\tau)d\tau = \frac{1}{2}\sum_{n=1}^{N(t)} X_n^2 + \int_{\tau=S_{N(t)}}^t Y(\tau)d\tau.$$

Although the final term above can be easily evaluated for a given $S_{N(t)}(t)$, it is more convenient to use the following bound:

$$\frac{1}{2t}\sum_{n=1}^{N(t)} X_n^2 \le \frac{1}{t}\int_{\tau=0}^t Y(\tau)d\tau \le \frac{1}{2t}\sum_{n=1}^{N(t)+1} X_n^2. \tag{5.13}$$

The term on the left can now be evaluated in the limit $t \to \infty$ (for all sample functions except a set of probability zero) as follows:

$$\lim_{t\to\infty} \frac{\sum_{n=1}^{N(t)} X_n^2}{2t} = \lim_{t\to\infty} \frac{\sum_{n=1}^{N(t)} X_n^2}{N(t)} \frac{N(t)}{2t}. \tag{5.14}$$

Consider each term on the right-hand side of (5.14) separately. For the first term, recall that $\lim_{t\to 0} N(t) = \infty$ WP1. Thus as $t \to \infty$, $\sum_{n=1}^{N(t)} X_n^2/N(t)$ goes through the same set of values as $\sum_{n=1}^{k} X_n^2/k$ as $k \to \infty$. Thus, assuming that $\mathsf{E}\left[X^2\right] < \infty$, we can apply the SLLN to $\{X_n^2;\, n \geq 1\}$,

$$\lim_{t\to\infty} \frac{\sum_{n=1}^{N(t)} X_n^2}{N(t)} = \lim_{k\to\infty} \frac{\sum_{n=1}^{k} X_n^2}{k} = \mathsf{E}\left[X^2\right] \qquad \text{WP1.}$$

The second term on the right-hand side of (5.14) is simply $N(t)/2t$. Since $\overline{X} < \infty$, the strong law for renewal processes says that $\lim_{t\to\infty} N(t)/2t = 1/(2\mathsf{E}[X])$ WP1. Thus both limits exist WP1 and

$$\lim_{t\to\infty} \frac{\sum_{n=1}^{N(t)} X_n^2}{2t} = \frac{\mathsf{E}\left[X^2\right]}{2\mathsf{E}[X]} \qquad \text{WP1.} \tag{5.15}$$

The right-hand term of (5.13) is handled almost the same way:

$$\lim_{t\to\infty} \frac{\sum_{n=1}^{N(t)+1} X_n^2}{2t} = \lim_{t\to\infty} \frac{\sum_{n=1}^{N(t)+1} X_n^2}{N(t)+1}\, \frac{N(t)+1}{N(t)}\, \frac{N(t)}{2t} = \frac{\mathsf{E}\left[X^2\right]}{2\mathsf{E}[X]}. \tag{5.16}$$

Combining these two results, we see that, WP1, the time-average residual life is given by

$$\lim_{t\to\infty} \frac{\int_{\tau=0}^{t} Y(\tau)\, d\tau}{t} = \frac{\mathsf{E}\left[X^2\right]}{2\mathsf{E}[X]}. \tag{5.17}$$

Note that this time average depends on the second moment of X; this is $\overline{X}^2 + \sigma^2 \geq \overline{X}^2$, so the time-average residual life is at least half the expected inter-renewal interval (which is not surprising). On the other hand, the second moment of X can be arbitrarily large (even infinite) for any given value of $\mathsf{E}[X]$, so that the time-average residual life can be arbitrarily large relative to $\mathsf{E}[X]$. This can be explained intuitively by observing that large inter-renewal intervals are weighted more heavily in this time average than small inter-renewal intervals.

Example 5.4.2 As an example of the effect of improbable but large inter-renewal intervals, let X take on the value ϵ with probability $1-\epsilon$ and value $1/\epsilon$ with probability ϵ. Then, for small ϵ, $\mathsf{E}[X] \sim 1, \mathsf{E}\left[X^2\right] \sim 1/\epsilon$, and the time average residual life is approximately $1/(2\epsilon)$ (see Figure 5.7).

Example 5.4.3 (Time-average age) Let $Z(t)$ be the age of a renewal process at time t where *age* is defined as the interval from the most recent arrival before (or at) t until t, i.e., $Z(t) = t - S_{N(t)}$. By convention, if no arrivals have occurred by time t, we take the age to be t (i.e., in this case, $N(t) = 0$ and we take S_0 to be 0).

As seen in Figure 5.8, the age process, for a given sample function of the renewal process, is almost the same as the residual life process – the isosceles right triangles are

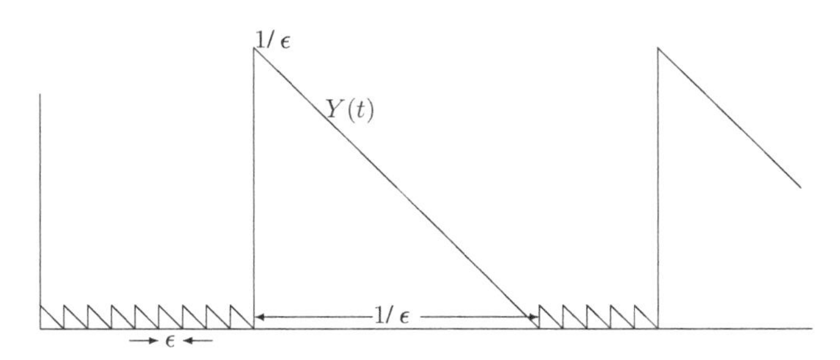

Figure 5.7 Average residual life is dominated by large interarrival intervals. Each large interval has duration $1/\epsilon$, and the expected aggregate duration between successive large intervals is $1 - \epsilon$.

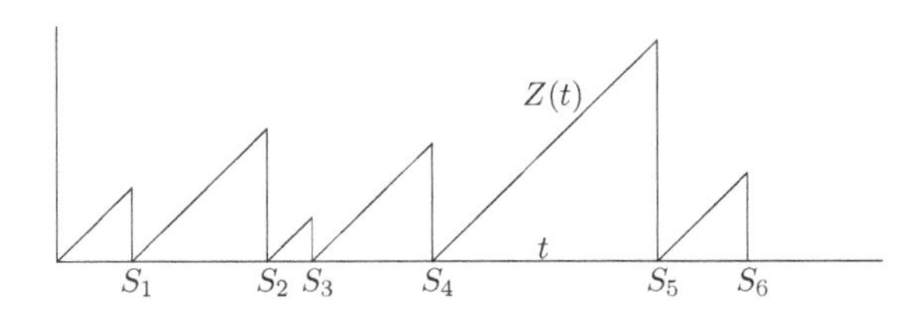

Figure 5.8 Age at time t: for any given sample function of the renewal process, the sample function of age increases linearly with a slope of 1 from the beginning to the end of each inter-renewal interval.

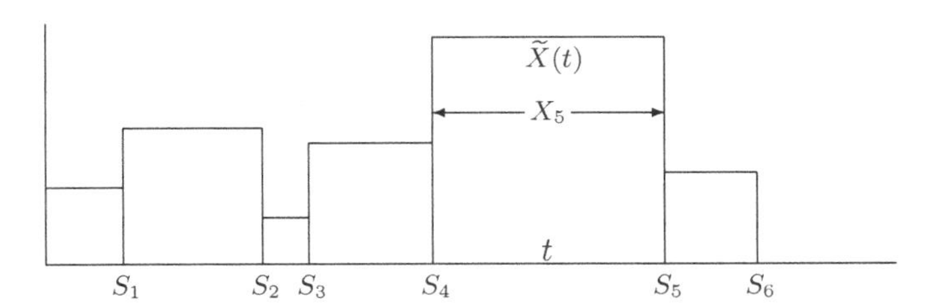

Figure 5.9 Duration $\widetilde{X}(t) = X_{N(t)}$ of the inter-renewal interval containing t.

simply turned around. Thus the same analysis as before can be used to show that the time average of $Z(t)$ is the same as the time average of the residual life,

$$\lim_{t\to\infty} \frac{\int_{\tau=0}^{t} Z(\tau)\,d\tau}{t} = \frac{\mathsf{E}[X^2]}{2\mathsf{E}[X]} \qquad \text{WP1}. \tag{5.18}$$

Example 5.4.4 (Time-average duration) Let $\widetilde{X}(t)$ be the duration of the inter-renewal interval containing time t, i.e., $\widetilde{X}(t) = X_{N(t)+1} = S_{N(t)+1} - S_{N(t)}$ (see Figure 5.9). It is clear that $\widetilde{X}(t) = Z(t) + Y(t)$, and thus the time average of the duration is given by

$$\lim_{t\to\infty} \frac{\int_{\tau=0}^{t} \widetilde{X}(\tau)\,d\tau}{t} = \frac{\mathsf{E}[X^2]}{\mathsf{E}[X]} \qquad \text{WP1}. \tag{5.19}$$

Again, long intervals are heavily weighted in this average, so that the time-average duration is at least as large as the mean inter-renewal interval and often much larger.

5.4.1 General renewal–reward processes

In each of these examples, and in many other situations, we have a random function of time (i.e., $Y(t)$, $Z(t)$, or $\widetilde{X}(t)$) whose value at time t depends only on where t is in the current inter-renewal interval (i.e., on the age $Z(t)$ and the duration $\widetilde{X}(t)$ of the current inter-renewal interval). We now investigate the general class of reward functions for which the reward at time t depends at most on the age and the duration at t, i.e., the reward $R(t)$ at time t is given explicitly as a function $\mathrm{R}\,(Z(t), \widetilde{X}(t))$ of the age and duration at t. For the three examples above, the function R is trivial. That is, the residual life, $Y(t)$, is given by $\mathrm{R}\,(Z(t), \widetilde{X}(t)) = \widetilde{X}(t) - Z(t)$. Similarly, the age is given directly as $\mathrm{R}\,(Z(t), \widetilde{X}(t)) = Z(t)$. This illustrates that the general function $\mathrm{R}\,(Z(t), \widetilde{X}(t))$ need not involve both $Z(t)$ and $\widetilde{X}(t)$; it simply cannot involve anything else.

We now find the time-average value of $R(t)$, namely, $\lim_{t\to\infty} [\int_0^t R(\tau)\,d\tau]/t$. As in Examples 5.4.1–5.4.4 above, we first want to look at the accumulated reward over each inter-renewal period separately. Define R_n as the accumulated reward in the nth renewal interval,

$$\mathrm{R}_n = \int_{S_{n-1}}^{S_n} R(\tau)\,d(\tau) = \int_{S_{n-1}}^{S_n} \mathrm{R}\,[Z(\tau), \widetilde{X}(\tau)]\,d\tau. \tag{5.20}$$

For residual life (see Example 5.4.1), R_n is the area of the nth isosceles right triangle in Figure 5.6. In general, for τ in $(S_{n-1}, S_n]$, we have $Z(\tau) = \tau - S_{n-1}$ and $\widetilde{X}(\tau) = S_n - S_{n-1} = X_n$. It follows that R_n is given by

$$\mathrm{R}_n = \int_{S_{n-1}}^{S_n} \mathrm{R}\,(\tau - S_{n-1},\, X_n)\,d\tau = \int_{z=0}^{X_n} \mathrm{R}\,(z,\, X_n)\,dz. \tag{5.21}$$

Note that z is integrated out in this expression, so R_n is a function only of X_n although, of course, the form of that function is determined by the function $\mathrm{R}\,(z, x)$. Assuming that this integral exists for all sample values of X_n and that the resulting R_n is a rv, it is clear that $\{\mathrm{R}_n; n \geq 1\}$ is a sequence of IID rv s. For residual life, $\mathrm{R}\,(z, X_n) = X_n - z$, so the integral in (5.21) is $X_n^2/2$, as calculated by inspection before. In general, from (5.21), the expected value of R_n is given by

$$\mathsf{E}\,[\mathrm{R}_n] = \int_{x=0}^{\infty} \left(\int_{z=0}^{x} \mathrm{R}\,(z, x)\,dz \right) d\mathsf{F}_X(x). \tag{5.22}$$

Breaking $\int_0^t R(\tau)\,d\tau$ into the reward over successive renewal periods, we get

$$\begin{aligned}\int_0^t R(\tau)\,d\tau &= \int_0^{S_1} R(\tau)\,d\tau + \int_{S_1}^{S_2} R(\tau)\,d\tau + \cdots + \int_{S_{N(t)-1}}^{S_{N(t)}} R(\tau)\,d\tau + \int_{S_{N(t)}}^{t} R(\tau)d\tau \\ &= \sum_{n=1}^{N(t)} \mathrm{R}_n + \int_{S_{N(t)}}^{t} R(\tau)\,d\tau. \end{aligned} \tag{5.23}$$

The following theorem now generalizes the results of Examples 5.4.1, 5.4.3, and 5.4.4 to general renewal–reward functions.

Theorem 5.4.5 *Let $\{R(t); t > 0\}$ be a non-negative renewal–reward function for a renewal process with expected inter-renewal time $\mathsf{E}[X] = \overline{X} < \infty$. If each R_n is a rv with $\mathsf{E}[\mathrm{R}_n] < \infty$, then with probability 1,*

$$\lim_{t\to\infty} \frac{1}{t} \int_{\tau=0}^{t} R(\tau)\, d\tau \;=\; \frac{\mathsf{E}[\mathrm{R}_n]}{\overline{X}}. \tag{5.24}$$

Proof Using (5.23), the accumulated reward up to time t can be bounded between the accumulated reward up to the renewal before t and that to the next renewal after t,

$$\frac{\sum_{n=1}^{N(t)} \mathrm{R}_n}{t} \;\le\; \frac{\int_{\tau=0}^{t} R(\tau)\, d\tau}{t} \;\le\; \frac{\sum_{n=1}^{N(t)+1} \mathrm{R}_n}{t}. \tag{5.25}$$

The left-hand side of (5.25) can be separated into

$$\frac{\sum_{n=1}^{N(t)} \mathrm{R}_n}{t} \;=\; \frac{\sum_{n=1}^{N(t)} \mathrm{R}_n}{N(t)} \frac{N(t)}{t}. \tag{5.26}$$

Each R_n is a given function of X_n, so the R_n are IID. As $t \to \infty$, $N(t) \to \infty$, and, thus, as we have seen before, the SLLN can be used on the first term on the right-hand side of (5.26), getting $\mathsf{E}[\mathrm{R}_n]$ WP1. Also the second term approaches $1/\overline{X}$ by the strong law for renewal processes. Since $0 < \overline{X} < \infty$ and $\mathsf{E}[\mathrm{R}_n]$ is finite, the product of the two terms approaches the limit $\mathsf{E}[\mathrm{R}_n]/\overline{X}$. The right-hand inequality of (5.25) is handled in almost the same way,

$$\frac{\sum_{n=1}^{N(t)+1} \mathrm{R}_n}{t} \;=\; \frac{\sum_{n=1}^{N(t)+1} \mathrm{R}_n}{N(t)+1} \frac{N(t)+1}{N(t)} \frac{N(t)}{t}. \tag{5.27}$$

It is seen that the terms on the right-hand side of (5.27) approach limits as before and thus the term on the left approaches $\mathsf{E}[\mathrm{R}_n]/\overline{X}$ WP1. Since the upper and lower bound in (5.25) approach the same limit, $(1/t)\int_0^t R(\tau)\, d\tau$ approaches the same limit and the theorem is proved. □

The restriction to non-negative renewal–reward functions in Theorem 5.4.5 is slightly artificial. The same result holds for non-positive reward functions simply by changing the directions of the inequalities in (5.25). Assuming that $\mathsf{E}[\mathrm{R}_n]$ exists (i.e., that both its positive and negative parts are finite), the same result applies in general by splitting an arbitrary reward function into a positive and negative part. This gives us the following corollary.

Corollary 5.4.6 *Let $\{R(t); t > 0\}$ be a renewal–reward function for a renewal process with expected inter-renewal time $\mathsf{E}[X] = \overline{X} < \infty$. If each R_n is a rv with $\mathsf{E}[|\mathrm{R}_n|] < \infty$, then*

$$\lim_{t\to\infty} \frac{1}{t} \int_{\tau=0}^{t} R(\tau)\, d\tau = \frac{\mathsf{E}[\mathrm{R}_n]}{\overline{X}} \qquad \textit{WP1}. \tag{5.28}$$

Example 5.4.7 (Distribution of residual life) Example 5.4.1 treated the time-average value of the residual life $Y(t)$. Suppose, however, that we would like to find the time-average CDF of $Y(t)$, i.e., the fraction of time that $Y(t) \le y$ for any given y. The

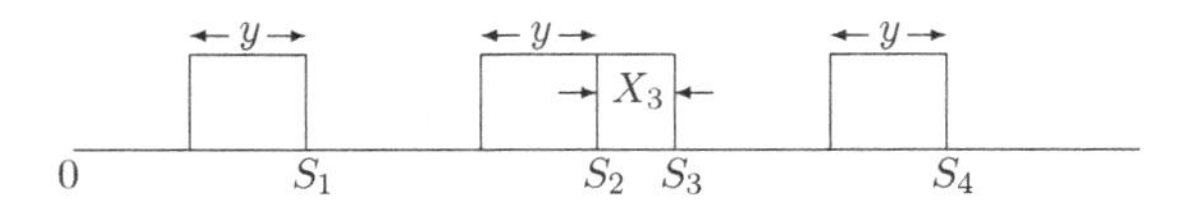

Figure 5.10 Reward function to find the time-average fraction of time that $\{Y(t) \leq y\}$. For the sample function in the figure, $X_1 > y$, $X_2 > y$, and $X_4 > y$, but $X_3 < y$.

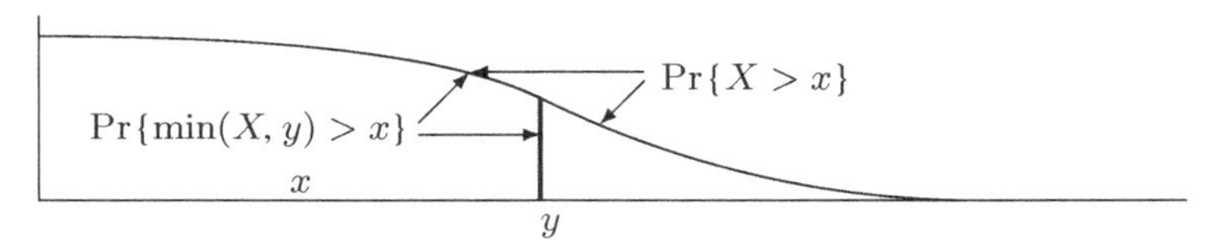

Figure 5.11 R_n for distribution of residual life.

approach, which applies to a wide variety of applications, is to use an indicator function (for a given value of y) as a reward function. That is, define $R(t)$ to have the value 1 for all t such that $Y(t) \leq y$ and to have the value 0 otherwise. Figure 5.10 illustrates this function for a given sample path. Expressing this reward function in terms of $Z(t)$ and $\widetilde{X}(t)$, we have

$$R(t) = \mathrm{R}\,(Z(t), \widetilde{X}(t)) = \begin{cases} 1, & \widetilde{X}(t) - Z(t) \leq y, \\ 0, & \text{otherwise.} \end{cases}$$

Note that if an inter-renewal interval is smaller than y (such as the third interval in Figure 5.10), then $R(t)$ has the value 1 over the entire interval, whereas if the interval is greater than y, then $R(t)$ has the value 1 only over the final y units of the interval. Thus $\mathrm{R}_n = \min[y, X_n]$. Note that the rv $\min[y, X_n]$ is equal to X_n for $X_n \leq y$, and thus has the same CDF as X_n in the range 0–y. Figure 5.11 illustrates this in terms of the complementary CDF. From the figure, we see that

$$\mathsf{E}[\mathrm{R}_n] = \mathsf{E}\big[\min(X, y)\big] = \int_{x=0}^{\infty} \Pr\{\min(X, y) > x\}\, dx = \int_{x=0}^{y} \Pr\{X > x\}\, dx. \quad (5.29)$$

Let $\mathsf{F}_Y(y) = \lim_{t\to\infty}(1/t)\int_0^t R(\tau)\,d\tau$ denote the time-average fraction of time that the residual life is less than or equal to y. From Theorem 5.4.5 and (5.29), we then have

$$\mathsf{F}_Y(y) = \frac{\mathsf{E}[\mathrm{R}_n]}{\overline{X}} = \frac{1}{\overline{X}} \int_{x=0}^{y} \Pr\{X > x\}\, dx \qquad \text{WP1.} \quad (5.30)$$

As a check, note that this integral is increasing in y and approaches 1 as $y \to \infty$. Note also that the expected value of Y, calculated from (5.30), is given by $\mathsf{E}\left[X^2\right]/2\overline{X}$, in agreement with (5.17).

The same argument can be applied to the time-average distribution of age (see Exercise 5.14). The time-average fraction of time, $\mathsf{F}_Z(z)$, that the age is at most z is given by

$$\mathsf{F}_Z(z) = \frac{1}{\overline{X}} \int_{x=0}^{z} \Pr\{X > x\}\, dx \qquad \text{WP1.} \quad (5.31)$$

In the development so far, the reward function $R(t)$ has been a function solely of the age and duration intervals, and the aggregate reward over the nth inter-renewal interval is a function only of X_n. In more general situations, where the renewal process is embedded in some more complex process, it is often desirable to define $R(t)$ to depend on other aspects of the process as well. The important thing here is for $\{R_n;\ n \geq 1\}$ to be an IID sequence, and Theorem 5.4.5 clearly remains valid if $\{R_n;\ n \geq 1\}$ is IID. This more general type of renewal–reward function will be required and further discussed in Sections 5.5.3–5.5.5 where we discuss Little's theorem and the M/G/1 expected queueing delay, both of which use this more general structure.

Limiting time averages are sometimes visualized by the following type of experiment. For some given large time t, let T be a uniformly distributed rv over $(0, t]$; T is independent of the renewal–reward process under consideration. Then $(1/t)\int_0^t R(\tau)\,d\tau$ is the expected value (over T) of $R(T)$ for a given sample path of $\{R(\tau);\ \tau>0\}$. Theorem 5.4.5 states that in the limit $t \to \infty$, all sample paths (except a set of probability 0) yield the same expected value over T. This approach of viewing a time average as a random choice of time is referred to as *random incidence*. Random incidence is awkward mathematically, since the rv T changes with the overall time t and has no reasonable limit. We will not use it in what follows, since, while somewhat intuitive, it is often confusing.

5.5 Stopping times for repeated experiments

Visualize performing an experiment repeatedly, observing successive sample outputs from a sequence of rv s. Depending on the sample values, $x_1, x_2, \ldots$ the observations are stopped after some trial n, where n is chosen based on the sample values $x_1, \ldots, x_n$ already observed.

This type of situation occurs frequently in applications. For example, we might be required to choose between several hypotheses, and might repeat an experiment until the hypotheses are sufficiently discriminated. We will find that if the number of trials is allowed to depend on the outcome, then the mean number of trials required to achieve a given error probability is typically a small fraction of the number of trials required when the number is chosen in advance. Another example occurs in tree searches where a path is explored until further extensions of the path appear to be unprofitable.

The first careful study of experimental situations where the number of trials depends on the data was made by the statistician Abraham Wald and led to the field of sequential analysis (see [29]). We study these situations now since one of the major results, Wald's equality, will be useful in studying $\mathsf{E}[N(t)]$ in the next section. Stopping times are frequently useful in the study of random processes, and in particular will be used in Section 5.7 for the analysis of queues, and again in Chapter 9 as central topics in the study of random walks and martingales.

An important part of experiments that stop after a random number of trials is the rule for stopping. Such a rule must specify, for each sample path, the trial at which the experiment stops, i.e., the final trial after which no more trials are performed. Thus the rule for stopping should specify a positive integer-valued, rv J, called the *stopping time*, or *stopping trial*, mapping sample paths to this final trial at which the experiment stops.

We view the sample space as including all possible outcomes for the never-ending sequence of rv s $X_1, X_2, \ldots$. That is, even if the experiment is stopped at the end of the second trial, we still visualize the third, fourth, ... rv s as having sample values as part of the sample path. In other words, we visualize that the experiment continues forever, but the observer stops watching after the occurrence of the stopping point. From an application standpoint, the experiment might or might not continue after the observer stops watching. From a mathematical standpoint, however, it is far preferable to view the experiment as continuing. This avoids confusion and ambiguity about the meaning of rv s when the very existence of later rv s depends on the sample values of earlier rv s.

The intuitive notion of stopping a sequential experiment should involve stopping based on the data (i.e., the sample values) gathered up to and including the stopping point. For example, if $X_1, X_2, \ldots$ represent the successive changes in our fortune when gambling, we might want to stop when our cumulative gain exceeds some fixed value. The stopping trial n then depends on the sample values of $X_1, X_2, \ldots, X_n$.

The whole notion of stopping on the basis of past sample values would be violated by rules that allow the experimenter to peek at subsequent values before making the decision to stop or not. For example, poker players do not take kindly to a player who attempts to withdraw his ante if he does not like his cards. Similarly, a statistician gathering data on product failures should not respond to a failure by choosing an earlier trial as a stopping time, thus not recording the failure.

It is not immediately obvious how to specify a rv J that is defined as a function of a sequence $X_1, X_2, \ldots$, but has the property that the event $J = n$ depends only on the sample values of $X_1, \ldots, X_n$. A sensible approach, embodied in the following definition, defines J in terms of the successive indicator functions $\mathbb{I}_{\{J=n\}}$ of J.

Definition 5.5.1 *A **stopping trial** (or stopping time[5]) J for a sequence of rv s $X_1, X_2, \ldots$ is a positive integer-valued rv such that for each $n \geq 1$, the indicator rv $\mathbb{I}_{\{J=n\}}$ is a function of $\{X_1, X_2, \ldots, X_n\}$.*

The last clause of the definition means that any given sample values $x_1, \ldots, x_n$ for $X_1, \ldots, X_n$ uniquely determine whether the corresponding sample value of J is n or not. Note that since the stopping trial J is defined to be a positive integer-valued rv, the events $\{J = n\}$ and $\{J = m\}$ for $m < n$ are disjoint events, so stopping at trial m makes it impossible to also stop at n for a given sample path. Also the union of the events $\{J = n\}$ over $n \geq 1$ has probability 1. Aside from this final restriction, the definition does not depend on the probability measure and depends solely on the set of events $\{J = n\}$ for each n. In many situations, it is useful to relax the definition further to allow

[5] Stopping trials are more often called stopping times or optional stopping times in the literature. In our first major application of a stopping trial, however, the stopping trial is the first trial n at which a renewal epoch S_n exceeds a given time t. Viewing this *trial* as a *time* generates considerable confusion.

J to be a possibly-defective rv. In this case the question of whether stopping occurs with probability 1 can be postponed until after specifying the disjoint events $\{J = n\}$ over $n \geq 1$.

Example 5.5.2 Consider a Bernoulli process $\{X_n;\ n \geq 1\}$. A very simple stopping trial for this process is to stop at the first occurrence of the string $(1, 0)$. Figure 5.12 illustrates this stopping trial by viewing it as a truncation of the tree of possible binary sequences.

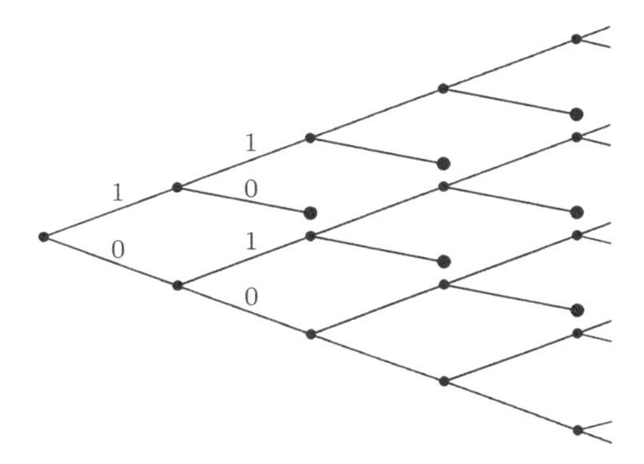

Figure 5.12 A tree representing the set of binary sequences, with a stopping rule viewed as a pruning of the tree. The particular stopping rule here is to stop on the first occurrence of the string $(1, 0)$. The leaves of the tree (i.e., the nodes at which stopping occurs) are marked with large dots and the intermediate nodes (the other nodes) with small dots. Note that each leaf in the pruned tree has a one-to-one correspondence with an initial segment of the tree, so the stopping nodes can be unambiguously viewed either as leaves of the tree or initial segments of the sample sequences.

The event $\{J = 2\}$, i.e., the event that stopping occurs at trial 2, is the event $\{X_1{=}1,\ X_2{=}0\}$. Similarly, the event $\{J = 3\}$ is $\{X_1{=}1,\ X_2{=}1,\ X_3{=}0\} \bigcup \{X_1{=}0,\ X_2{=}1,\ X_3{=}0\}$. The disjointness of $\{J = n\}$ and $\{J = m\}$ for $n \neq m$ is represented in the figure by terminating the tree at each stopping node. It can be seen that the tree never dies out completely, and in fact, for each trial n, the number of stopping nodes is $n - 1$. However, the probability that stopping has not occurred by trial n goes to zero geometrically with n, which ensures that J is a random variable.

This example is generalized in Exercise 5.35 to stopping rules that stop on the first occurrence of some arbitrary string of binary digits. This also uses renewal theory to provide a simple way to find the expected number of trials until stopping.

Representing a stopping rule by a pruned tree can be used for any discrete random sequence, although the tree becomes quite unwieldy in all but trivial cases. Visualizing a stopping rule in terms of a pruned tree is useful conceptually, but stopping rules are usually stated in other terms. For example, we shortly consider a stopping trial for the interarrival intervals of a renewal process as the first n for which the arrival epoch S_n satisfies $S_n > t$ for some given $t > 0$.

5.5.1 Wald's equality

An important question that arises with stopping trials is to evaluate the sum S_J of the rv s up to the stopping trial, i.e., $S_J = \sum_{n=1}^{J} X_n$. Many gambling strategies and investing strategies involve some sort of rule for when to stop, and it is important to understand the rv S_J (which can model the overall gain or loss up to and including that trial). Wald's equality relates the expected value of S_J to $\mathsf{E}[J]$, and thus makes it easy to find $\mathsf{E}[S_J]$ if $\mathsf{E}[J]$ can be found. We will find a surprising number of cases also where $\mathsf{E}[S_J]$ can be directly found (or estimated), thus making it easy to find (or estimate) $\mathsf{E}[J]$.

Theorem 5.5.3 (Wald's equality) *Let $\{X_n; n \geq 1\}$ be a sequence of IID rv s, each of mean $\overline{X}$. If J is a stopping trial for $\{X_n; n \geq 1\}$ and if $\mathsf{E}[J] < \infty$, then the sum $S_J = X_1 + X_2 + \cdots + X_J$ at the stopping trial J satisfies*

$$\mathsf{E}[S_J] = \overline{X}\mathsf{E}[J]. \tag{5.32}$$

Proof Note that X_n is included in $S_J = \sum_{n=1}^{J} X_n$ whenever $n \leq J$, i.e., whenever the indicator function $\mathbb{I}_{\{J \geq n\}} = 1$. Thus

$$S_J = \sum_{n=1}^{\infty} X_n \mathbb{I}_{\{J \geq n\}} . \tag{5.33}$$

This includes X_n as part of the sum if stopping has not occurred before trial n. The event $\{J \geq n\}$ is the complement of $\{J < n\} = \{J = 1\} \bigcup \cdots \bigcup \{J = n-1\}$. All of these latter events are determined by $X_1, \ldots, X_{n-1}$ and are thus independent of X_n. It follows that $\{J < n\}$ is independent of X_n. Since $\{J \geq n\} = \{J < n\}^{\mathsf{c}}$, it follows[6] that $\{J \geq n\}$ is independent of X_n. Thus

$$\mathsf{E}\left[X_n \mathbb{I}_{\{J \geq n\}}\right] = \overline{X}\mathsf{E}\left[\mathbb{I}_{\{J \geq n\}}\right].$$

We then have

$$\begin{aligned}
\mathsf{E}[S_J] &= \mathsf{E}\left[\sum_{n=1}^{\infty} X_n \mathbb{I}_{\{J \geq n\}}\right] \\
&= \sum_{n=1}^{\infty} \mathsf{E}\left[X_n \mathbb{I}_{\{J \geq n\}}\right] && (5.34) \\
&= \sum_{n=1}^{\infty} \overline{X}\mathsf{E}\left[\mathbb{I}_{\{J \geq n\}}\right] \\
&= \overline{X}\mathsf{E}[J]. && (5.35)
\end{aligned}$$

The interchange of expectation and infinite sum in (5.34) is obviously valid for a finite sum, and is shown in Exercise 5.20 to be valid for an infinite sum if $\mathsf{E}[J] < \infty$. The example below shows that Wald's equality can be invalid when $\mathsf{E}[J] = \infty$. The final step above comes from the observation that $\mathsf{E}\left[\mathbb{I}_{\{J \geq n\}}\right] = \Pr\{J \geq n\}$. Since J is a positive integer rv, $\mathsf{E}[J] = \sum_{n=1}^{\infty} \Pr\{J \geq n\}$. One can also obtain the last step by using $J = \sum_{n=1}^{\infty} \mathbb{I}_{\{J \geq n\}}$ (see Exercise 5.15). □

[6] This can be quite confusing initially, since (as seen in the example of Figure 5.12) the event $\{J = n\}$ is not necessarily independent of X_n (and is usually highly influenced by X_n). Similarly $\{J = n + i\}$ for $i \geq 1$ is not necessarily independent of X_n. The resolution of this paradox is that *given that* stopping has not occurred before trial n, then X_n can have a great deal to do with *the trial* at which stopping occurs. However, as shown above, X_n has nothing to do with *whether* $\{J < n\}$ or $\{J \geq n\}$.

What this result essentially says in terms of gambling is that strategies for stopping under various conditions are not effective as far as the mean is concerned, i.e., the mean gain is simply the mean gain per play times the average number of plays. This sometimes appears obvious and sometimes appears very surprising, depending on the application.

Example 5.5.4 (Stop when you are ahead in coin tossing) We can model a (perhaps biased) coin-tossing game as a sequence of IID rv s $X_1, X_2, \ldots$, where each X_i is 1 with probability p and -1 with probability $1 - p$. Consider the possibly-defective stopping trial J, where J is the first n for which $S_n = X_1 + \cdots + X_n = 1$, i.e., the first trial at which the gambler is ahead.

We first want to see if J is a rv, i.e., if the probability of eventual stopping, say $\theta = \Pr\{J < \infty\}$, is 1. We solve this by a frequently useful trick, but will use other more systematic approaches in Chapters 6 and 9 when we look at this same example as a birth–death Markov chain and then as a simple random walk. Note that stopping occurs at trial 1 (i.e., $J = 1$) if $S_1 = 1$. Thus $\Pr\{J = 1\} = \Pr\{S_1 = 1\} = p$. With probability $1-p$, $S_1 = -1$. If $S_1 = -1$, the only way to eventually become 1 ahead is to first return to $S_n = 0$ for some $n > 1$, and, after the first such return, go on to $S_m = 1$ at some later trial m. The probability of eventually going from -1 to 0 is the same as that of going from 0 to 1, i.e., θ. Also, given a first return to 0 from -1, the probability of reaching 1 from 0 is θ. Thus,

$$\theta = p + (1 - p)\theta^2.$$

This is a quadratic equation in θ with two solutions, $\theta = 1$ and $\theta = p/(1 - p)$. For $p > 1/2$, the second solution is impossible since θ is a probability. Thus we conclude that J is a rv. For $p = 1/2$ (and this is the most interesting case), both solutions are the same, $\theta = 1$, and again J is a rv. For $p < 1/2$, we will see in Section 7.5 on birth–death Markov processes that the correct solution is $\theta = p/(1-p)$. Thus $\theta < 1$ in this case and J is a defective rv.

For the cases where $p \geq 1/2$, i.e., where J is a rv, we can use the same trick to evaluate $\mathsf{E}[J]$:

$$\mathsf{E}[J] = p + (1 - p)(1 + 2\mathsf{E}[J]).$$

The solution to this is

$$\mathsf{E}[J] = \frac{1}{2p - 1}.$$

We see that $\mathsf{E}[J]$ is finite for $p > 1/2$ and infinite for $p = 1/2$.

For $p > 1/2$, we can check that these results agree with Wald's equality. In particular, since S_J is 1 with probability 1, we also have $\mathsf{E}[S_J] = 1$. Since $\overline{X} = 2p - 1$ and $\mathsf{E}[J] = 1/(2p - 1)$, Wald's equality is satisfied (which of course it has to be).

For $p = 1/2$, we still have $S_J = 1$ with probability 1 and thus $\mathsf{E}[S_J] = 1$. However, $\overline{X} = 0$ so $\overline{X}\mathsf{E}[J]$ has no meaning and Wald's equality breaks down. Thus we see that the restriction $\mathsf{E}[J] < \infty$ in Wald's equality is indeed needed. These results are summarized below.

$$\Pr(J < \infty) = \begin{cases} \frac{p}{1-p} & \text{for } p < 0.5, \\ 1- & \text{for } p = 0.5, \\ 0 & \text{for } p > 0.5, \end{cases} \qquad \mathsf{E}[J] = \begin{cases} \infty & \text{for } p < 0.5, \\ \infty & \text{for } p = 0.5, \\ \frac{1}{2p-1} & \text{for } p > 0.5. \end{cases}$$

It is surprising that with $p = 1/2$, the gambler can eventually become 1 ahead with probability 1. This has little practical value, first because the required expected number of trials is infinite, and second (as will be seen later) because the gambler must risk a potentially infinite capital.

5.5.2 Applying Wald's equality to $\mathsf{E}[N(t)]$

Let $\{N(t);\ t > 0\}$ be a renewal counting process with arrival epochs $\{S_n;\ n \geq 1\}$ and interarrival intervals $\{X_n;\ n \geq 1\}$. For any given $t > 0$, let J be the trial n for which S_n first exceeds t. Note that $J = n$ is specified by the sample values of $\{X_1, \ldots, X_n\}$ and thus J is a (possibly-defective) stopping trial for $\{X_n;\ n \geq 1\}$.

Since J is the first trial n for which $S_n > t$, we see that $S_{J-1} \leq t$ and $S_J > t$. Thus $N(t) = J - 1$, so $J = N(t) + 1$. Since $N(t)$ is non-defective and has a finite expectation for all $t > 0$ (see Exercise 5.2), J is also, so J is a stopping trial for $\{X_n;\ n \geq 1\}$ and $\mathsf{E}[J] < \infty$.

We can then employ Wald's equality to obtain

$$\mathsf{E}\left[S_{N(t)+1}\right] = \overline{X}\mathsf{E}[N(t) + 1] \tag{5.36}$$

$$\mathsf{E}[N(t)] = \frac{\mathsf{E}\left[S_{N(t)+1}\right]}{\overline{X}} - 1. \tag{5.37}$$

As is often the case with Wald's equality, this provides a relationship between two quantities, $\mathsf{E}[N(t)]$ and $\mathsf{E}\left[S_{N(t)+1}\right]$, that are both unknown. This will be used in proving the elementary renewal theorem (i.e., the statement that $\lim_{t\to\infty} \mathsf{E}[N(t)]/t = 1/\overline{X}$) by upper and lower bounding $\mathsf{E}\left[S_{N(t)+1}\right]$. The lower bound is easy, since $\mathsf{E}\left[S_{N(t)+1}\right] > t$, and thus $\mathsf{E}[N(t)] > t/\overline{X} - 1$. It follows that

$$\frac{\mathsf{E}[N(t)]}{t} > \frac{1}{\overline{X}} - \frac{1}{t}. \tag{5.38}$$

We derive an upper bound on $\mathsf{E}\left[S_{N(t)+1}\right]$ in the next section. First, however, as a sanity check, consider Figure 5.13 which illustrates (5.36) for the case where each X_n is a deterministic rv where $X_n = \overline{X}$ with probability 1.

It might be puzzling why we used $N(t) + 1$ rather than $N(t)$ as a stopping trial for the epochs $\{S_i;\ i \geq 1\}$ in this application of Wald's equality. To understand this, assume, for example, that $N(t) = n$. On observing only the sample values of $S_1, \ldots, S_n$, with $S_n < t$, there is typically no assurance that other arrivals will not occur in the interval $(S_n, t]$. In other words, $N(t) = n$ implies that $S_n \leq t$, but $S_n < t$ does not imply that $N(t) = n$.

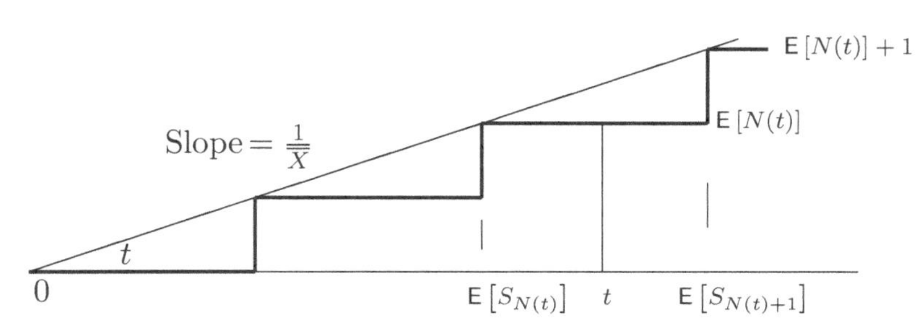

Figure 5.13 Illustration of (5.36) for the special case where X is deterministic. Note that $\mathsf{E}[N(t)]$, as a function of t, is then the illustrated staircase function. On each increment of t by $\overline{X}$, $\mathsf{E}[N(t)]$ increases by 1. Then $\mathsf{E}[N(t)+1]$ and $\mathsf{E}\left[S_{N(t)+1}\right]$ are two sides of a right triangle of slope $1/\overline{X}$, yielding (5.36).

5.5.3 Generalized stopping trials, embedded renewals, and G/G/1 queues

The above definition of a stopping trial is quite restrictive in that it refers only to a single sequence of rv s. Many queueing situations involve both a sequence of interarrival times and a sequence of service times, so it is useful to consider stopping rules that depend on the sample values of both of these sequences. In what follows, we consider G/G/1 queues, both to understand how these queueing systems work and to understand why a more general stopping rule is needed for them.

Example 5.5.5 (G/G/1 queues) Consider a G/G/1 queue (the single server case of the G/G/m queue described in Example 5.1.2). Let $\{X_i;\ i \geq 1\}$ denote the sequence of IID interarrival intervals, where X_i is the interval from the arrival of customer $i-1$ to i. Let $\{V_i;\ i \geq 0\}$ denote the IID required service times for these customers; these service times are independent of the interarrival times. Customer 0 arrives at time 0 and immediately enters the server. Subsequent arrivals are stored in a queue if the server is busy. Each enqueued arrival enters the server, in first-come-first-served (FCFS) order,[7] when the server completes the previous service. Figure 5.14 illustrates a sample path for these arrivals and departures.

The figure illustrates a sample path for which $X_1 < V_0$, so the first arrival waits in queue for $W_1^q = V_0 - X_1$. If $X_1 \geq V_0$, on the other hand, then customer 1 enters service immediately, i.e., customer 1 'sees an empty system.' In general, $W_1^q = \max(V_0 - X_1,\ 0)$. In the same way, as illustrated in the figure, if $W_1^q > 0$, then customer 2 waits for $W_1^q + V_1 - X_2$ if positive and does not wait otherwise. This same formula works if $W_1^q = 0$, so $W_2^q = \max(W_1^q + V_1 - X_2,\ 0)$. In general, it can be seen that

$$W_i^q = \max(W_{i-1}^q + V_{i-1} - X_i,\ 0). \tag{5.39}$$

This equation will be analyzed further in Section 9.2 where we are interested in queueing delay and system delay. Here our objectives are simpler, since we only want to show that the subsequence of customer arrivals i for which $W_i^q = 0$ forms the renewal epochs of a renewal process. To do this, note from (5.39) that the first arrival n that does not enter the queue, i.e., for which $W_n^q = 0$, is the smallest n for which $\sum_{i=1}^{n}(X_i - V_{i-1}) \geq 0$.

[7] For single server queues, this is sometimes referred to as first-in-first-out (FIFO) service.

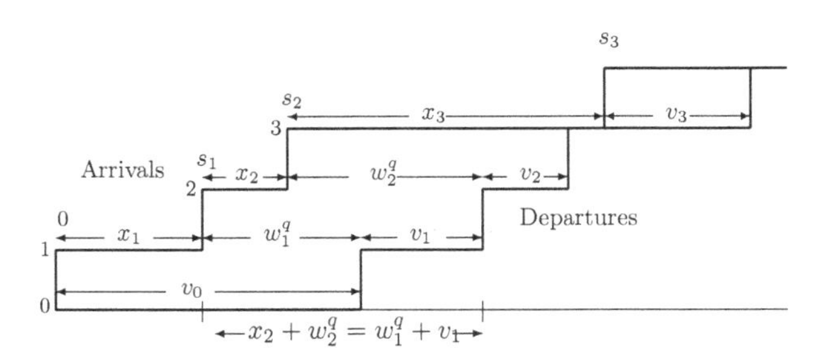

Figure 5.14 Sample path of arrivals and departures from a G/G/1 queue. Customer 0 arrives at time 0 and enters service immediately. Customer 1 arrives at time $s_1 = x_1$. For the case shown above, customer 0 has not yet departed, i.e., $x_1 < v_0$, so customer 1 is queued for the interval $w_1^q = v_0 - x_1$ before entering service. As illustrated, customer 1's system waiting time (queueing time plus service time) is $w_1 = w_1^q + v_1$.
Customer 2 arrives at $s_2 = x_1 + x_2$. For the case shown above, this is before customer 1 departs at $v_0 + v_1$. Thus, customer 2's wait in queue is $w_2^q = v_0 + v_1 - x_1 - x_2$. As illustrated above, $x_2 + w_2^q$ is also equal to customer 1's system time, so $w_2^q = w_1^q + v_1 - x_2$. Customer 3 arrives when the system is empty for this example and so enters service immediately with no wait in queue, i.e., $w_3^q = 0$.

The nth arrival then enters the server immediately and the situation appears to be the same as at time 0, i.e., it appears that this nth arrival constitutes a renewal of the queueing system.

To clarify that this is actually a renewal, we interrupt the discussion of G/G/1 queues with the following generalization of the definition of a stopping trial.

Definition 5.5.6 *A (possibly-defective)* ***generalized stopping trial*** *J for a sequence of pairs of rv s $(X_1, Y_1), (X_2, Y_2), \ldots$ is a positive integer-valued (possibly-degenerate) rv such that, for each $n \geq 1$, the indicator rv $\mathbb{I}_{\{J=n\}}$ is a function of $X_1, Y_1, X_2, Y_2, \ldots, X_n, Y_n$.*

Theorem 5.5.7 (Generalized Wald's equality) *Let $\{(X_n, Y_n);\ n \geq 1\}$ be a sequence of pairs of rv s, where each pair is IID to all other pairs. Assume that each X_i has finite mean $\overline{X}$. If J is a stopping trial for $\{(X_n, Y_n);\ n \geq 1\}$ and if $\mathsf{E}[J] < \infty$, then the sum $S_J = X_1 + X_2 + \cdots + X_J$ satisfies*

$$\mathsf{E}[S_J] = \overline{X}\mathsf{E}[J]. \tag{5.40}$$

The proof of this will be omitted, since it is the same as the proof of Theorem 5.5.3. In fact, the definition of stopping trials could be further generalized by replacing the rv s Y_i by vector rv s or by a random number of rv s, and Wald's equality still holds.[8]

[8] Some people define J to be a stopping rule if $\mathbb{I}_{\{J \geq n\}}$ is independent of $X_n, X_{n+1}, \ldots$ for each n. This includes the generalized definition here and makes it easy to prove Wald's equality, but ignores the intuitive notion of stopping and makes it difficult to see when the definition holds.

Returning to the G/G/1 queue, consider the sequence of pairs $\{(X_1, V_0), (X_2, V_1), \ldots\}$ and let J be the smallest n for which $\sum_{i=1}^{n}(X_i - V_{i-1}) \geq 0$. Then J is the first arrival to see an empty queue, and is also a possibly-degenerate generalized stopping rule for $\{(X_n, V_{n-1});\ n \geq 1\}$. Note also that $\{(X_n, V_{n-1});\ n \geq 1\}$ satisfies the IID conditions of Theorem 5.5.7. Finally, if $\mathsf{E}[X] > \mathsf{E}[V]$, we can apply the WLLN to the sequence $\{(X_n - V_{n-1});\ n \geq 1\}$. This shows that

$$\lim_{n\to\infty} \Pr\left\{\sum_{i=1}^{n}(X_i - V_{i-1}) \geq 0\right\} = 1 \qquad \text{for } \mathsf{E}\left[X_i - V_{i-1}\right] > 0.$$

Thus J is a (non-degenerate) stopping rule. If $\mathsf{E}[J] < \infty$, then the generalized Wald equality holds and $\mathsf{E}[S_J] < \infty$.

It is important here, as in many applications, to avoid the confusion created by viewing J as a stopping *time*. We have seen that J is the *number* of the first customer to see an empty system, and S_J is the *time* until that customer arrives. Thus $\mathsf{E}[S_J]$ is the expected time until an arrival $n \geq 1$ first sees an empty system.

There is a further possible timing confusion about whether a customer's service time is determined when the customer arrives or completes service. This makes no difference, since the ordered sequence of pairs is well defined and satisfies the appropriate IID condition for using the Wald equality.

It is also interesting to see that, although successive pairs (X_i, Y_i) are assumed independent, it is not necessary for X_i and Y_i to be independent. This lack of independence does not occur for the G/G/1 (or G/G/m) queue, but can be useful in situations such as packet networks where the interarrival time between two packets at a given node can depend on the service time (the length) of the first packet if both packets are coming from the same node.

The first arrival to see an empty system is perhaps better described as the beginning of a busy period for the G/G/1 queue. The most important aspect of characterizing this as a stopping trial is the ability to show that successive beginnings of busy periods form a renewal process. To do this, let $X_{2,1}, X_{2,2}, \ldots$ be the interarrival times following J, the first arrival to see an empty queue. Conditioning on $J = j$, we have $X_{2,1}{=}X_{j+1},\ X_{2,2}{=}X_{j+2}, \ldots,$. Thus $\{X_{2,k};\ k \geq 1\}$ is an IID sequence with the original interarrival distribution. Similarly $\{(X_{2,k}, V_{2,k-1});\ k \geq 1\}$ is a sequence of IID pairs with the original distribution. This is valid for all sample values j of the stopping trial J. Thus $\{(X_{2,k}, V_{2,k-1});\ k \geq 1\}$ is statistically independent of J and $(X_i, V_{i-1});\ 1 \leq i \leq J$.

The argument above can be repeated for subsequent arrivals to an empty system, so we have shown that successive arrivals to an empty system, i.e., beginnings of busy periods, actually form a renewal process.

One can define many different stopping rules for queues, such as the first trial at which a given number of customers are in the queue. Wald's equality can be applied to any such stopping rule, but much more is required for the stopping trial to also form a renewal point. At the first time when n customers are in the system, the subsequent departure times depend partly on the old service times and partly on the new arrival and service times, so the required independence for a renewal point does not exist. Stopping rules are helpful in understanding embedded renewal points, but are by no means equivalent to embedded renewal points.

Finally, nothing in the argument above for the G/G/1 queue made any use of the FCFS service discipline. One can use any service discipline for which the choice of which customer to serve at a given time t is based solely on the arrival and service times of customers in the system by time t. In fact, if the server is never idle when customers are in the system, the renewal epochs will not depend on the service descipline. It is also possible to extend these arguments to the G/G/m queue, although the service discipline can affect the renewal points in this case.

We can summarize the results of this section, and add a little, with the following theorem.

Theorem 5.5.8 *Consider a G/G/1 queue with finite expected interarrival time* $\mathsf{E}[X]$ *and finite expected service time* $\mathsf{E}[V] < \mathsf{E}[X]$. *Then the subsequence of arrivals that see an empty system forms a renewal process. The expected number of arrivals* $\mathsf{E}[J]$ *between arrivals to an empty system is finite and the expected time between successive busy periods (i.e., the expected inter-renewal time) is equal to* $\mathsf{E}[J]\,\mathsf{E}[X]$.

The new part here is that $\mathsf{E}[J] < \infty$ and we simply outline the proof, leaving the details to Exercise 5.21. Since $\mathsf{E}[X] > \mathsf{E}[V]$, the server is not busy all the time, and with probability 1, the fraction of time not busy is $(\mathsf{E}[X] - \mathsf{E}[V])/\mathsf{E}[X]$. Consider the special case where X is bounded, with $\Pr\{X > b\} = 0$. Then the idle period at the end of each busy period has duration at most b, so the number of busy periods (and thus inter-renewal intervals) per arrival grows at least as $(\mathsf{E}[X] - \mathsf{E}[V])/b$. This is bounded away from 0 so $\mathsf{E}[J]$ is bounded below infinity. The case where X is not bounded is handled by a truncation argument.

5.5.4 Little's theorem

Little's theorem is an important queueing result stating that the expected number of customers in a queueing system is equal to the product of the arrival rate and the expected time each customer waits in the system. This result is true under very general conditions; we use the G/G/1 queue with FCFS service as a specific example, but the reason for the greater generality will be clear as we proceed. Note that the theorem does not tell us how to find either the expected number or expected wait; it only says that if one can be found, the other can also be found.

Figure 5.15 illustrates a sample path for a G/G/1 queue with FCFS service. It illustrates a sample path $a(t)$ for the arrival process $A(t) = N(t) + 1$, i.e., the number of customer arrivals in $[0, t]$, specifically including customer number 0 arriving at $t = 0$. Similarly, it illustrates the departure process $D(t)$, which is the number of departures up to time t, again including customer 0. The difference, $L(t) = A(t) - D(t)$, is then the number in the system at time t.

Recall from Section 5.5.3 that the subsequence of customer arrivals for $t > 0$ that see an empty system forms a renewal process. Actually, we showed a little more than that. Not only are the inter-renewal intervals, $X_i^r = S_i^r - S_{i-1}^r$ IID, but the numbers

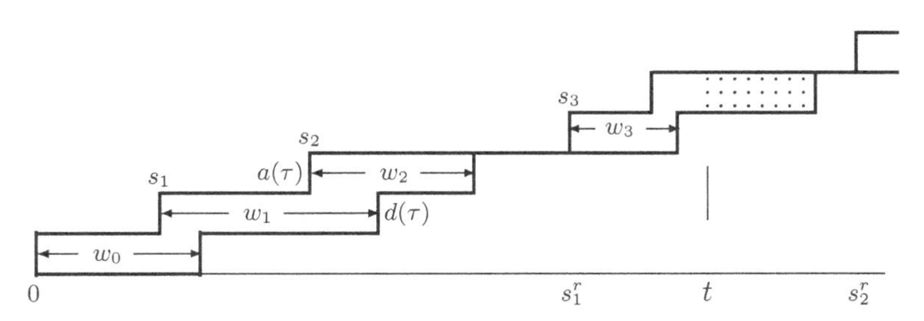

Figure 5.15 Sample path of arrivals, departures, and system waiting times for a G/G/1 queue with FCFS service. The upper step function is the number of customer arrivals, including the customer at time 0 and is denoted $a(\tau)$. Thus $a(\tau)$ is a sample path of $A(\tau) = N(\tau) + 1$, i.e., the arrival counting process incremented by 1 for the initial arrival at $\tau = 0$. The lower step function, $d(\tau)$ is a sample path for $D(\tau)$, which is the number of departures (including customer 0) up to time τ. For each $i \geq 0$, w_i is the sample value of the system waiting time W_i for customer i. Note that $W_i = W_i^q + V_i$.

The figure also shows the sample values s_1^r and s_2^r of the first two arrivals that see an empty system (recall from Section 5.5.3 that the subsequence of arrivals that see an empty system forms a renewal process).

of customer arrivals in the subsequent inter-renewal intervals are IID, and the inter-arrival intervals and service times between subsequent inter-renewal intervals are IID. The sample values, s_1^r and s_2^r of the first two renewal epochs are shown in the figure.

The essence of Little's theorem can be seen by observing that $\int_0^{S_1^r} L(\tau)\,d\tau$ in the figure is the area between the upper and lower step functions, integrated out to the first time that the two step functions become equal (i.e., the system becomes empty). For the sample value in the figure, this integral is equal to $w_0 + w_1 + w_2$. In terms of the rv s,

$$\int_0^{S_1^r} L(\tau)\,d\tau = \sum_{i=0}^{N(S_1^r)-1} W_i. \tag{5.41}$$

The same relationship exists in each inter-renewal interval, and in particular we can define L_n for each $n \geq 1$ as

$$L_n = \int_{S_{n-1}^r}^{S_n^r} L(\tau)\,d\tau \;=\; \sum_{i=N(S_{n-1}^r)}^{N(S_n^r)-1} W_i. \tag{5.42}$$

The interpretation of this is far simpler than the notation. The arrival step function and the departure step function in Figure 5.15 are separated whenever there are customers in the system (the system is busy) and are equal whenever the system is empty. Renewals occur when the system goes from empty to busy, so the nth renewal is at the beginning of the nth busy period. Then L_n is the area of the region between the two step functions over the nth busy period. By simple geometry, this area is also the sum of the customer waiting times over that busy period. Finally, since the interarrival intervals and service times in each busy period are IID with respect to those in each other busy period, the sequence $L_1, L_2, \ldots$ is a sequence of IID rv s.

The function $L(\tau)$ has the same behavior as a renewal reward function, but it is slightly more general, being a function of more than the age and duration of the renewal counting process $\{N^r(t);\ t > 0\}$ at $t = \tau$. However, the fact that $\{L_n;\ n \geq 1\}$ is an IID sequence lets us use the same methodology to treat $L(\tau)$ as was used earlier to treat renewal–reward functions. We now state and prove Little's theorem. The proof is almost the same as that of Theorem 5.4.5, so we will not dwell on it.

Theorem 5.5.9 (Little) *For a FCFS G/G/1 queue in which the expected inter-renewal interval is finite, the limiting time-average number of customers in the system is equal, WP1, to a constant denoted as $\overline{L}$. The sample-path-average waiting time per customer is also equal, WP1, to a constant denoted as $\overline{W}$. Finally $\overline{L} = \lambda\overline{W}$ where λ is the customer arrival rate, i.e., the reciprocal of the expected interarrival time.*

Proof Note that for any $t > 0$, $\int_0^t L(\tau)\,d\tau$ can be expressed as the sum over the busy periods completed before t plus a residual term involving the busy period including t. The residual term can be upper bounded by the integral over that complete busy period. Using this with (5.42), we have

$$\sum_{n=1}^{N^r(t)} \mathrm{L}_n \leq \int_{\tau=0}^{t} L(\tau)\,d\tau \leq \sum_{i=0}^{N(t)} W_i \leq \sum_{n=1}^{N^r(t)+1} \mathrm{L}_n. \tag{5.43}$$

Assuming that the expected inter-renewal interval, $\mathsf{E}[X^r]$, is finite, we can divide both sides of (5.43) by t and go to the limit $t \to \infty$. From the same argument as in Theorem 5.4.5,

$$\lim_{t\to\infty} \frac{\sum_{i=0}^{N(t)} W_i}{t} = \lim_{t\to\infty} \frac{\int_{\tau=0}^{t} L(\tau)\,d\tau}{t} = \frac{\mathsf{E}[\mathrm{L}_n]}{\mathsf{E}[X^r]} \qquad \text{WP1}. \tag{5.44}$$

The equality on the right shows that the limiting time average of $L(\tau)$ exists WP1 and is equal to $\overline{L} = \mathsf{E}[L_n]/\mathsf{E}[X^r]$. The quantity on the left of (5.44) can now be broken up as waiting time per customer multiplied by number of customers per unit time, i.e.,

$$\lim_{t\to\infty} \frac{\sum_{i=0}^{N(t)} W_i}{t} = \lim_{t\to\infty} \frac{\sum_{i=0}^{N(t)} W_i}{N(t)} \lim_{t\to\infty} \frac{N(t)}{t}. \tag{5.45}$$

From (5.44), the limit on the left-hand side of (5.45) exists (and equals $\overline{L}$) WP1. The second limit on the right also exists WP1 by the strong law for renewal processes, applied to $\{N(t); t > 0\}$. This limit is called the *arrival rate* λ, and is equal to the reciprocal of the mean interarrival interval for $\{N(t)\}$. Since these two limits exist WP1, the first limit on the right, which is the sample-path-average waiting time per customer, denoted $\overline{W}$, also exists WP1. □

Reviewing this proof and the development of the G/G/1 queue before the theorem, we see that there was a simple idea, expressed by (5.41), combined with a lot of notational complexity due to the fact that we were dealing with both an arrival counting process $\{N(t);\ t > 0\}$ and an embedded renewal counting process $\{N^r(t);\ t > 0\}$. The difficult thing, mathematically, was showing that $\{N^r(t);\ t > 0\}$ is actually a renewal process and showing that the L_n are IID, and this was where we needed to understand stopping rules.

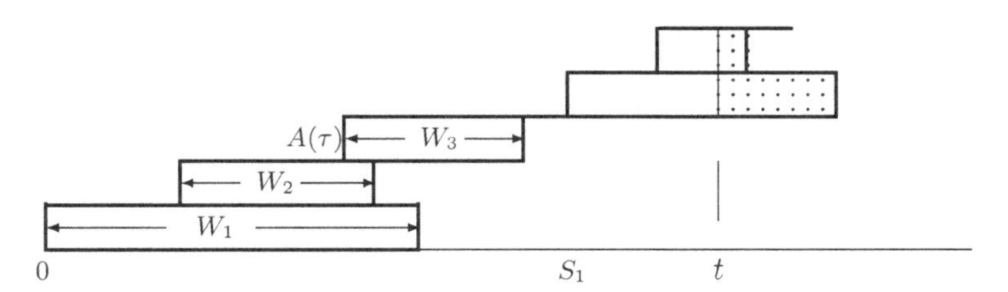

Figure 5.16 Arrivals and departures in a non-FCFS systems. The server, for example, could work simultaneously (at a reduced rate) on all customers in the system, and thus complete service for customers with small service needs before completing earlier arrivals with greater service needs. Note that the jagged right-hand edge of the diagram does not represent number of deparatures, but this is not essential for the argument.

Recall that we assumed earlier that customers departed from the queue in the same order in which they arrived. From Figure 5.16, however, it is clear that FCFS order is not required for the argument. Thus the theorem generalizes to systems with multiple servers and arbitrary service disciplines in which customers do not follow FCFS order. In fact, all that the argument requires is that the system has renewals (which are IID by definition of a renewal) and that the inter-renewal interval is finite WP1.

For example, if higher priority is given to customers with small service times, then it is not hard to see that the average number of customers in the system and the average waiting time per customer will be decreased. However, if the server is always busy when there is work to be done, it can be seen that the renewal times are unaffected. Service disciplines will be discussed further in Section 6.8.

The same argument as in Little's theorem can be used to relate the average number of customers in a single server queue (not counting service) to the average wait in the queue (not counting service). Renewals still occur on arrivals to an empty system, and the integral of customers in queue over a busy period is still equal to the sum of the queue waiting times. Let $L^q(t)$ be the number in the queue at time t and let $\overline{L}^q = \lim_{t\to\infty}(1/t)\int_0^t L^q(\tau)d\tau$ be the time-average queue wait. Letting $\overline{W}^q$ be the sample-path-average waiting time in queue,

$$\overline{L}^q = \lambda\overline{W}^q. \tag{5.46}$$

The same argument can also be applied to the service facility of a single server queue. The time average of the number of customers in the server is just the fraction of time that the server is busy. Denoting this fraction by ρ and the expected service time by $\overline{V}$, we get

$$\rho = \lambda\overline{V}. \tag{5.47}$$

5.5.5 M/G/1 queues

The above results for G/G/1 queues provide many relationships between quantities but do not allow interesting quantites such as expected delay (either time or ensemble average) or expected duration of a busy period to be actually found. For the M/G/1 queue, we can go further. First, we find the expected renewal period between successive customers entering an empty system, and then we find a simple equation (the Pollaczek–Khinchin formula) for the average queueing delay. The arrival sequence $\{X_i;\ i \geq 1\}$ is Poisson with rate $\lambda = 1/\overline{X}$. Let $\{V_i;\ i \geq 0\}$ be the sequence of service requirements; these are IID and also independent of the arrivals. Assume that $\overline{X} > \overline{V} > 0$.

As we saw for the more general G/G/1 queue, the first arrival $J > 0$ to see an empty system is the smallest $n > 0$ for which $\sum_{i=1}^{n}(X_i - V_{i-1}) \geq 0$. We saw in Theorem 5.5.8 that J is a generalized stopping rule and that $\mathsf{E}[J] < \infty$. We saw that $\mathsf{E}[J]$ is the expected number of trials between renewals at which an arrival sees an empty system. We also saw, by Wald's equality, that the expected duration of a renewal period is $\overline{X}\mathsf{E}[J]$. The special feature of the M/G/1 queue that allows us to find a simple formula for $\overline{J}$, and the expected duration of the renewal interval, is the memoryless feature of the Poisson distribution.

In particular, the duration of the first renewal period is $\sum_{i=1}^{J} X_i$. The time required by the server to complete service for customers 1 to $J-1$ is $\sum_{i=0}^{J-1} V_i$. Since none of these customers sees an empty system, the server is busy until some time between the arrivals of customers $J-1$ and J. By the memorylessness of the Poisson process, the time from the completion of service of customer $J-1$ to the arrival of customer J is exponential with mean $\overline{X}$. What this means is that

$$\mathsf{E}\left[\sum_{i=1}^{J} X_i - \sum_{i=0}^{J-1} V_i\right] = \overline{X}.$$

Applying Wald's equality to $\{X_i - V_{i-1}; i \geq 1\}$, we see that $(\overline{X} - \overline{V})\mathsf{E}[J] = \overline{X}$. Thus

$$\mathsf{E}[J] = \frac{\overline{X}}{\overline{X} - \overline{V}}, \qquad \overline{X}\mathsf{E}[J] = \frac{\overline{X}^2}{\overline{X} - \overline{V}}. \tag{5.48}$$

Summarizing:

Theorem 5.5.10 *For an M/G/1 queue with $\overline{V} < \overline{X} < \infty$, the expected number of arrivals $\mathsf{E}[J]$ and the expected time $\overline{X}\mathsf{E}[J]$ between successive initiations of busy periods is given by (5.48).*

We next proceed to the expected waiting time in queue. The added assumption that the service time V has a finite second moment is now needed. Let $N(t)$ denote the number of arrivals in $(0, t]$. There is also an arrival at $t = 0$ which is not counted in $\{N(t);\ t > 0\}$.

At any given time t, let $L^q(t)$ be the number of customers in the queue (not counting the customer in service, if any) and let $R(t)$ be the residual life of the customer in service. If no customer is in service, $R(t) = 0$, and otherwise $R(t)$ is the remaining time until the current service is completed. Let $U(t)$ be the waiting time in queue that would be experienced by a customer arriving at time t, assuming FCFS service. This is often called the unfinished work in the queueing literature and represents the delay until all the customers currently in the system complete service. Thus the rv $U(t)$ is equal to $R(t)$, the residual life of the customer in service, plus the service times of each of the $L^q(t)$ customers currently waiting in the queue:

$$U(t) = \sum_{i=0}^{L^q(t)-1} V_{N(t)-i} + R(t). \tag{5.49}$$

Note that if the queue contains $L^q(t) \geq 1$ customers at time t, then customer number $N(t)$ is at the back of the queue and customer number $N(t) - L^q(t) - 1$ is at the front. Now $L^q(t)$ is the number of arrivals in $[0, t]$ (counting the arrival at time 0) less the number of services completed in $(0, t]$ and less 1 for the customer in service. The service times and inter-arrival times of these completed services are independent of the service times of the customers in the queue, and thus, given $L^q(t) = \ell$,

$$\mathsf{E}\left[\sum_{i=0}^{L^q(t)-1} V_{N(t)-i} \,|\, L^q(t) = \ell\right] = \ell\overline{V}.$$

Taking the expectation over $L^q(t)$ and substituting into (5.49),

$$\mathsf{E}[U(t)] = \mathsf{E}\left[L^q(t)\right]\overline{V} + \mathsf{E}[R(t)]. \tag{5.50}$$

Figure 5.17 illustrates how to find the time average of $R(t)$. Viewing $R(t)$ as a reward function, we can find the accumulated reward up to time t as the sum of triangular areas.

First, consider $\int R(\tau)d\tau$ from 0 to $S^r_{N^r(t)}$, i.e., the accumulated reward up to the final renewal epoch in $(0, t]$. Note that $S^r_{N^r(t)}$ is not only a renewal epoch for the renewal process, but also an arrival epoch for the arrival process; in particular, it is the $N(S^r_{N^r(t)})$th arrival epoch, and the $N(S^r_{N^r(t)}) - 1$ earlier arrivals are the customers that have received service up to time $S^r_{N^r(t)}$. Thus,

$$\int_0^{S^r_{N^r(t)}} R(\tau)\,d\tau \;=\; \sum_{i=0}^{N(S^r_{N^r(t)})-1} \frac{V_i^2}{2} \;\leq\; \sum_{i=0}^{N(t)-1} \frac{V_i^2}{2}.$$

We can similarly upper bound the term on the right above by $\int_0^{S^r_{N^r(t)+1}} R(\tau)\,d\tau$. We also know (from going through virtually the same argument several times) that

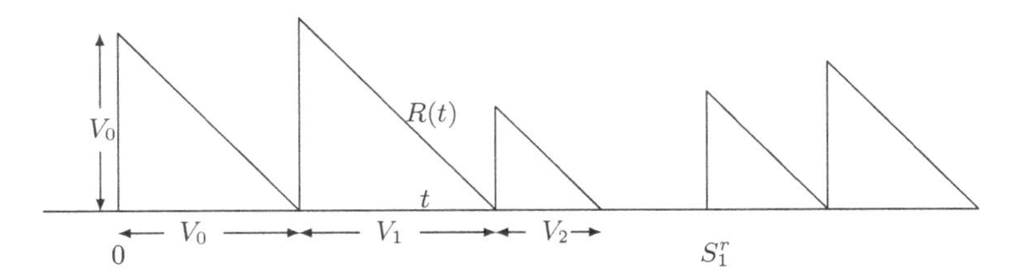

Figure 5.17 Sample value of the residual life function of customers in service.

$(1/t)\int_{\tau=0}^{t} R(\tau)d\tau$ will approach a limit with probability 1 as $t \to \infty$, and that the limit will be unchanged if t is replaced with $S^r_{N^r(t)}$ or $S^r_{N^r(t)+1}$. Thus, taking λ as the arrival rate,

$$\lim_{t\to\infty} \frac{\int_0^t R(\tau)\,d\tau}{t} = \lim_{t\to\infty} \frac{\sum_{i=0}^{N(t)-1} V_i^2}{2N(t)} \frac{N(t)}{t} = \frac{\lambda \mathsf{E}\left[V^2\right]}{2} \qquad \text{WP1}.$$

As will be seen in Section 5.7.5, the time average above can be replaced with a limiting ensemble average so that

$$\lim_{t\to\infty} \mathsf{E}\left[R(t)\right] = \frac{\lambda \mathsf{E}\left[V^2\right]}{2}. \tag{5.51}$$

Similarly, there is a limiting ensemble-average form of Little's theorem which shows that $\lim_{t\to\infty} \mathsf{E}\left[L^q(t)\right] = \lambda \overline{W}^q$. Substituting this plus (5.51) into (5.50), we get

$$\lim_{t\to\infty} \mathsf{E}\left[U(t)\right] = \lambda \mathsf{E}\left[V\right] \overline{W}^q + \frac{\lambda \mathsf{E}\left[V^2\right]}{2}. \tag{5.52}$$

Thus $\mathsf{E}\left[U(t)\right]$ is asymptotically independent of t. It is now important to distinguish between $\mathsf{E}\left[U(t)\right]$ and $\overline{W}^q$. The first is the expected unfinished work at time t; this is the queueing delay that a customer would incur by arriving at t; the second is the sample-path-average expected queueing delay. For Poisson arrivals, the probability of an arrival in $(t, t+\delta]$ is independent of all earlier arrivals and service times,[9] so it is independent of $U(t)$. Thus, in the limit $t \to \infty$, each arrival faces an expected waiting time in the queue equal to $\lim_{t\to\infty} \mathsf{E}\left[U(t)\right]$. It follows that $\lim_{t\to\infty} \mathsf{E}\left[U(t)\right]$ must be equal to $\overline{W}^q$. Substituting this into (5.52), we obtain the celebrated *Pollaczek–Khinchin* formula,

$$\overline{W}^q = \frac{\lambda \mathsf{E}\left[V^2\right]}{2(1 - \lambda \mathsf{E}\left[V\right])}. \tag{5.53}$$

This queueing delay has some of the peculiar features of residual life, and in particular, if $\mathsf{E}\left[V^2\right] = \infty$, the limiting expected queueing delay is infinite even when the expected service time is less than the expected interarrival interval.

In trying to visualize why the queueing delay is so large when $\mathsf{E}\left[V^2\right]$ is large, note that while a particularly long service is taking place, numerous arrivals are coming into the system, and all are being delayed by this single long service. In other words, the number of new customers held up by a long service is proportional to the length of the service, and the amount each of them is held up is also proportional to the length of the service. This visualization is rather crude, but does serve to explain the second moment of V in (5.53). This phenomenon is sometimes called the 'slow truck effect' because of the pile up of cars behind a slow truck on a single-lane road.

For a G/G/1 queue, (5.52) is still valid, but arrival times are no longer independent of $U(t)$, so that typically $\mathsf{E}\left[U(t)\right] \neq \overline{W}^q$. As an example, suppose that the service time

[9] This is often called the *PASTA* property, standing for Poisson arrivals see time averages. This name confuses rv s, sample values of rv s, and expectations of rv s, but still suggests an approach to problem solving – find some quantity like unfinished work at time t that depends only on the past, and use the memorylessness of the Poisson process to associate the expectation of that quantity with a new arrival at t.

is uniformly distributed between $1-\epsilon$ and $1+\epsilon$ and that the interarrival interval is uniformly distributed between $2-\epsilon$ and $2+\epsilon$. Assuming that $\epsilon < 1/2$, the system has no queueing and $\overline{W}^q = 0$. On the other hand, for small ϵ, $\lim_{t\to\infty} \mathsf{E}[U(t)] \sim 1/4$ (i.e., the server is busy half the time with unfinished work ranging from 0 to 1).

5.6 Expected number of renewals; ensemble averages

The purpose of this section is to evaluate $\mathsf{E}[N(t)]$, denoted $m(t)$, as a function of $t > 0$ for arbitrary renewal processes. We first find an exact expression, in the form of an integral equation, for $m(t)$. This can be easily solved by Laplace transform methods in special cases. For the general case, however, $m(t)$ becomes increasingly messy for large t, so we then find the asymptotic behavior of $m(t)$. Since $N(t)/t$ approaches $1/\overline{X}$ with probability 1, we might expect $m(t)$ to grow with a derivative $m'(t)$ that asymptotically approaches $1/\overline{X}$. This is not true in general. Two somewhat weaker results, however, are true. The first, called the elementary renewal theorem (Theorem 5.6.2), states that $\lim_{t\to\infty} m(t)/t = 1/\overline{X}$. The second result, called Blackwell's theorem (Theorem 5.6.3), states that, subject to some limitations on $\delta > 0$, $\lim_{t\to\infty}[m(t+\delta) - m(t)] = \delta/\overline{X}$. This says essentially that the expected renewal rate approaches steady state as $t \to \infty$. We will find a number of applications of Blackwell's theorem throughout the remainder of the text.

The exact calculation of $m(t)$ makes use of the fact that the expectation of a non-negative rv is defined as the integral of its complementary CDF,

$$m(t) = \mathsf{E}[N(t)] = \sum_{n=1}^{\infty} \Pr\{N(t) \geq n\}.$$

Since the event $\{N(t) \geq n\}$ is the same as $\{S_n \leq t\}$, $m(t)$ is expressed in terms of the CDFs of S_n, $n \geq 1$, as follows:

$$m(t) = \sum_{n=1}^{\infty} \Pr\{S_n \leq t\}. \tag{5.54}$$

Although this expression looks fairly simple, it becomes increasingly complex with increasing t. As t increases, there is an increasing set of values of n for which $\Pr\{S_n \leq t\}$ is significant, and $\Pr\{S_n \leq t\}$ itself is not that easy to calculate if the interarrival distribution $\mathsf{F}_X(x)$ is complicated. The main utility of (5.54) comes from the fact that it leads to an integral equation for $m(t)$. Since $S_n = S_{n-1} + X_n$ for each $n \geq 1$ (interpreting S_0 as 0), and since X_n and S_{n-1} are independent, we can use the convolution equation (1.11) to get

$$\Pr\{S_n \leq t\} = \int_{x=0}^{t} \Pr\{S_{n-1} \leq t - x\}\, d\mathsf{F}_X(x) \quad \text{for } n \geq 2.$$

Substituting this in (5.54) for $n \geq 2$ and using the fact that $\Pr\{S_1 \leq t\} = \mathsf{F}_X(t)$, we can interchange the order of integration and summation to get

$$m(t) = \mathsf{F}_X(t) + \int_{x=0}^{t} \sum_{n=2}^{\infty} \Pr\{S_{n-1} \le t - x\}\, d\mathsf{F}_X(x)$$

$$= \mathsf{F}_X(t) + \int_{x=0}^{t} \sum_{n=1}^{\infty} \Pr\{S_n \le t - x\}\, d\mathsf{F}_X(x)$$

$$= \mathsf{F}_X(t) + \int_{x=0}^{t} m(t-x) d\mathsf{F}_X(x), \qquad t \ge 0. \tag{5.55}$$

An alternative derivation is given in Exercise 5.25. This integral equation is called the *renewal equation*. The following alternative form is achieved by integration by parts:[10]

$$m(t) = \mathsf{F}_X(t) + \int_{\tau=0}^{t} F_X(t-\tau)\, dm(\tau), \qquad t \ge 0. \tag{5.56}$$

5.6.1 Laplace transform approach

If we assume that $X \ge 0$ has a density $\mathsf{f}_X(x)$, and that this density has a Laplace transform[11] $L_X(s) = \int_0^\infty \mathsf{f}_X(x) e^{-sx} dx$, then we can take the Laplace transform of both sides of (5.55). Note that the final term in (5.55) is the convolution of m with f_X, so that the Laplace transform of $m(t)$ satisfies

$$L_m(s) = \frac{L_X(s)}{s} + L_m(s) L_X(s).$$

Solving for $L_m(s)$,

$$L_m(s) = \frac{L_X(s)}{s[1 - L_X(s)]}. \tag{5.57}$$

Example 5.6.1 As a simple example of how this can be used to calculate $m(t)$, suppose $\mathsf{f}_X(x) = (1/2)e^{-x} + e^{-2x}$ for $x \ge 0$. The Laplace transform is given by

$$L_X(s) = \frac{1}{2(s+1)} + \frac{1}{s+2} = \frac{(3/2)s + 2}{(s+1)(s+2)}.$$

Substituting this into (5.57) yields

$$L_m(s) = \frac{(3/2)s + 2}{s^2(s + 3/2)} = \frac{4}{3s^2} + \frac{1}{9s} - \frac{1}{9(s+3/2)}.$$

We can solve for $m(t), t \ge 0$, by taking the inverse Laplace transform:

$$m(t) = \frac{4t}{3} + \frac{1 - \exp[-(3/2)t]}{9}.$$

The procedure in this example can be used for any inter-renewal density $\mathsf{f}_X(x)$ for which the Laplace transform is a rational function, i.e., a ratio of polynomials. In such cases,

[10] A mathematical subtlety with the Stieltjes integrals (5.55) and (5.56) will be discussed in Section 5.7.3.

[11] Note that $L_X(s) = \mathsf{E}\left[e^{-sX}\right] = \mathsf{g}_X(-s)$, where g is the moment generating function (MGF) of X. Thus the argument here could be carried out using the MGF. We use the Laplace transform since the mechanics here are so familiar to most engineering students.

$L_m(s)$ will also be a rational function. The Heaviside inversion formula (i.e., factoring the denominator and expressing $L_m(s)$ as a sum of individual poles as done above) can then be used to calculate $m(t)$. In the example above, there was a second-order pole at $s = 0$ leading to the linear term $4t/3$ in $m(t)$, there was a first-order pole at $s = 0$ leading to the constant $1/9$, and there was a pole at $s = -3/2$ leading to the exponentially decaying term.

We now show that a second-order pole at $s = 0$ always occurs when $L_X(s)$ is a rational function. To see this, note that $L_X(0)$ is just the integral of $\mathsf{f}_X(x)$, which is 1; thus $1-L_X(s)$ has a zero at $s = 0$ and, from (5.57), $L_m(s)$ has a second-order pole at $s = 0$. To evaluate the residue for this second-order pole, we recall that the first and second derivatives of $L_X(s)$ at $s = 0$ are $-\mathsf{E}[X]$ and $\mathsf{E}[X^2]$ respectively. Expanding $L_X(s)$ in a power series around $s = 0$ then yields $L_X(s) = 1 - s\mathsf{E}[X] + (s^2/2)\mathsf{E}[X^2]$ plus terms of order s^3 or higher. This gives us

$$L_m(s) = \frac{1 - s\overline{X} + (s^2/2)\mathsf{E}[X^2] + \cdots}{s^2\left[\overline{X} - (s/2)\mathsf{E}[X^2] + \cdots\right]} = \frac{1}{s^2\overline{X}} + \frac{1}{s}\left(\frac{\mathsf{E}[X^2]}{2\overline{X}^2} - 1\right) + \cdots. \tag{5.58}$$

The remaining terms are the other poles of $L_m(s)$ with their residues. For values of s with $\Re(s) \geq 0$, we have $|L_X(s)| = |\int \mathsf{f}_X(x)e^{-sx}dx| \leq \int \mathsf{f}_X(x)|e^{-sx}|dx \leq \int \mathsf{f}_X(x)dx = 1$ with strict inequality except for $s = 0$. Thus $L_X(s)$ cannot have any poles on the imaginary axis or the right half plane, and $1 - L_X(s)$ cannot have any zeros there other than the one at $s = 0$. It follows that all the remaining poles of $L_m(s)$ are strictly in the left half plane. This means that the inverse transforms for all these remaining poles die out as $t \to \infty$. Thus the inverse Laplace transform of $L_m(s)$ is

$$\begin{aligned} m(t) &= \frac{t}{\overline{X}} + \frac{\mathsf{E}[X^2]}{2\overline{X}^2} - 1 + \epsilon(t) \\ &= \frac{t}{\overline{X}} + \frac{\sigma^2}{2\overline{X}^2} - \frac{1}{2} + \epsilon(t) \qquad \text{for } t \geq 0, \end{aligned} \tag{5.59}$$

where $\lim_{t\to\infty} \epsilon(t) = 0$.

We have derived (5.59) only for the special case in which $\mathsf{f}_X(x)$ has a rational Laplace transform. For this case, (5.59) implies both the elementary renewal theorem ($\lim_{t\to\infty} m(t)/t = 1/\overline{X}$) and also Blackwell's theorem ($\lim_{t\to\infty}[m(t+\delta)-m(t)] = \delta/\overline{X}$). We will interpret the meaning of the constant term $\sigma^2/(2\overline{X}^2) - 1/2$ in Section 5.8.

5.6.2 The elementary renewal theorem

Theorem 5.6.2 (The elementary renewal theorem) *Let $\{N(t);\, t > 0\}$ be a renewal counting process with mean inter-renewal interval $\overline{X}$ where $\overline{X}$ can be finite or infinite. Then $\lim_{t\to\infty} \mathsf{E}[N(t)]/t = 1/\overline{X}$.*

Discussion We have already seen that $m(t) = \mathsf{E}[N(t)]$ is finite for all $t > 0$ (see Exercise 5.2). The theorem is proven by establishing lower and upper bounds to $m(t)/t$

and showing that each approaches $1/\mathsf{E}[X]$ as $t \to \infty$. The key element for each bound is (5.37), repeated below, which comes from the Wald equality.

$$m(t) = \frac{\mathsf{E}\left[S_{N(t)+1}\right]}{\overline{X}} - 1. \tag{5.60}$$

Proof The lower bound to $m(t)/t$ comes by recognizing that $S_{N(t)+1}$ is the epoch of the first arrival after t. Thus $\mathsf{E}\left[S_{N(t)+1}\right] > t$. Substituting this into (5.60),

$$\frac{m(t)}{t} > \frac{1}{\mathsf{E}[X]} - \frac{1}{t}.$$

Clearly this lower bound approaches $1/\mathsf{E}[X]$ as $t \to \infty$. The upper bound, which is more difficult[12] and might be omitted on a first reading, is next established by first truncating $X(t)$ and then applying (5.60) to the truncated process.

For an arbitrary constant $b > 0$, let $\breve{X}_i = \min(b, X_i)$. Since these truncated rv s are IID, they form a related renewal counting process $\{\breve{N}(t); t > 0\}$ with $\breve{m}(t) = \mathsf{E}\left[\breve{N}(t)\right]$ and $\breve{S}_n = \breve{X}_1 + \cdots + \breve{X}_n$. Since $\breve{X}_i \leq X_i$ for all i, we see that $\breve{S}_n \leq S_n$ for all n. Since $\{S_n \leq t\} = \{N(t) \geq n\}$, it follows that $\breve{N}(t) \geq N(t)$ and thus $\breve{m}(t) \geq m(t)$. Finally, in the truncated process, $\breve{S}_{\breve{N}(t)+1} \leq t + b$ and thus $\mathsf{E}\left[\breve{S}_{\breve{N}(t)+1}\right] \leq t + b$. Thus, applying (5.60) to the truncated process,

$$\frac{m(t)}{t} \leq \frac{\breve{m}(t)}{t} = \frac{\mathsf{E}\left[S_{\breve{N}(t)+1}\right]}{t\mathsf{E}\left[\breve{X}\right]} - \frac{1}{t} \leq \frac{t+b}{t\mathsf{E}\left[\breve{X}\right]}.$$

Next, choose $b = \sqrt{t}$. Then

$$\frac{m(t)}{t} \leq \frac{1}{\mathsf{E}\left[\breve{X}\right]} + \frac{1}{\sqrt{t}\,\mathsf{E}\left[\breve{X}\right]}.$$

Note finally that $\mathsf{E}[\breve{X}] = \int_0^b [1-\mathsf{F}_X(x)]\,dx$. Since $b = \sqrt{t}$, we have $\lim_{t\to\infty} \mathsf{E}[\breve{X}] = \mathsf{E}[X]$, Note that the above steps apply both for $\overline{X} < \infty$ and $\overline{X} = \infty$. □

Recall that $N[t, \omega]/t$ is the time-average renewal rate from 0 to t for a sample function ω, and $m(t)/t$ is the average of this over ω. Combining this with Theorem 5.3.1, we see that the limiting time and ensemble average equals the time-average renewal rate for each sample function except for a set of probability 0.

Another interesting question is to determine the expected renewal rate in the limit of large t without averaging from 0 to t. That is, are there some values of t at which renewals are more likely than others for large t? If the inter-renewal intervals have an integer CDF (i.e., each inter-renewal interval must last for an integer number of time units), then each renewal epoch S_n must also be an integer. This means that $N(t)$ can increase only at integer times and the expected rate of renewals is zero at all non-integer times.

[12] The difficulty here, and the reason for using a truncation argument, comes from the fact that the residual life, $S_{N(t)+1} - t$, at t might be arbitrarily large. We saw in Section 5.4 that the time-average residual life is infinite if $\mathsf{E}\left[X^2\right]$ is infinite. Figure 5.7 also illustrates why residual life can be so large.

An obvious generalization of integer-valued inter-renewal intervals is that of inter-renewals that occur only at integer multiples of some real number $\lambda > 0$. Such a distribution is called an *arithmetic distribution*. The *span* of an arithmetic distribution is the largest number λ such that this property holds. Thus, for example, if X takes on only the values 0, 2, and 6, its distribution is arithmetic with span $\lambda = 2$. Similarly, if X takes on only the values $1/3$ and $1/5$, then the span is $\lambda = 1/15$. The remarkable thing, for our purposes, is that any inter-renewal distribution that is not an arithmetic distribution leads to an essentially uniform expected rate of renewals in the limit of large t. This result is contained in Blackwell's renewal theorem, which we state without proof.[13] Recall, however, that for the special case of an inter-renewal density with a rational Laplace transform, Blackwell's renewal theorem is a simple consequence of (5.59).

Theorem 5.6.3 (Blackwell) *If a renewal process has an inter-renewal distribution that is non-arithmetic, then for each $\delta > 0$,*

$$\lim_{t\to\infty} [m(t+\delta) - m(t)] = \frac{\delta}{\mathsf{E}[X]}. \tag{5.61}$$

If the inter-renewal distribution is arithmetic with span λ, then

$$\lim_{t\to\infty} [m(t+\lambda) - m(t)] = \frac{\lambda}{\mathsf{E}[X]}. \tag{5.62}$$

Equation (5.61) says that for non-arithmetic distributions, the expected number of arrivals in the interval $(t, t+\delta]$ is equal to $\delta/\mathsf{E}[X]$ in the limit $t \to \infty$. Since the theorem is true for arbitrarily small δ, the theorem almost seems to be saying that $m(t)$ has a derivative for large t, but this is not true. One can see the reason by looking at an example where X can take on only the values 1 and π. Then no matter how large t is, $N(t)$ can only increase at discrete points of time of the form $k + j\pi$, where k and j are non-negative integers. Thus $dm(t)/dt$ is either 0 or ∞ for all t. As t gets larger, the jumps in $m(t)$ become both smaller in magnitude and more closely spaced from one to the next. Thus $\big[m(t+\delta) - m(t)\big]/\delta$ can approach $1/\mathsf{E}[X]$ as $t \to \infty$ for any fixed δ (as the theorem says), but as δ gets smaller, the convergence in t gets slower. For the above example (and for all discrete non-arithmetic distributions), $\big[m(t+\delta) - m(t)\big]/\delta$ does not approach[14] $1/\mathsf{E}[X]$ for any t as $\delta \to 0$.

For an arithmetic renewal process with span λ, the asymptotic behavior of $m(t)$ as $t \to \infty$ is much simpler. Renewals can only occur at multiples of λ, and since simultaneous renewals are not allowed, either 0 or 1 renewal occurs at each time $k\lambda$. Thus for any k. we have

$$\Pr\{\text{renewal at } \lambda k\} = m(\lambda k) - m(\lambda(k-1)), \tag{5.63}$$

where, by convention, we take $m(0) = 0$. Thus (5.62) can be restated as

$$\lim_{k\to\infty} \Pr\{\text{renewal at } k\lambda\} = \frac{\lambda}{\overline{X}}. \tag{5.64}$$

The limiting behavior of $m(t)$ is discussed further in the next section.

[13] See Theorem 1 of Section 11.1, of [9] for a proof.

[14] This might seem like mathematical nitpicking. However, $m(t)$ is the expected number of renewals in $(0, t]$, and how $m(t)$ varies with t is central to this chapter and keeps reappearing.

5.7 Renewal–reward processes; ensemble averages

Theorem 5.4.5 showed that if a renewal–reward process has an expected inter-renewal interval $\overline{X}$ and an expected inter-renewal reward $\mathsf{E}[R_n]$, then the time-average reward is $\mathsf{E}[R_n]/\overline{X}$ WP1. In this section, we explore the ensemble average, $\mathsf{E}[R(t)]$, as a function of time t. It is easy to see that $\mathsf{E}[R(t)]$ typically changes with t, especially for small t, but a question of major interest here is whether $\mathsf{E}[R(t)]$ approaches a constant as $t \to \infty$.

In more concrete terms, if the arrival times of buses at a bus station form a renewal process, then the waiting time for the next bus, i.e., the residual life, starting at time t, can be represented as a reward function $R(t)$. We would like to know if the expected waiting time depends critically on t, where t is the time since the renewal process started, i.e., the time since a hypothetical bus number 0 arrived. If $\mathsf{E}[R(t)]$ varies significantly with t, even as $t \to \infty$, it means that the choice of $t = 0$ as the beginning of the initial interarrival interval never dies out as $t \to \infty$.

Blackwell's renewal theorem (and common sense) tells us that there is a large difference between arithmetic inter-renewal times and non-arithmetic inter-renewal times. For the arithmetic case, all renewals occur at multiples of the span λ. Thus, for example, the expected waiting time (i.e., the expected residual life) decreases at rate 1 from each multiple of λ to the next, and it increases with a jump equal to the probability of an arrival at each multiple of λ. For this reason, we usually consider reward functions for arithmetic processes only at multiples of λ. We would guess, then, that $\mathsf{E}[R(n\lambda)]$ approaches a constant as $n \to \infty$.

For the non-arithmetic case, on the other hand, the expected number of renewals in any small interval of length δ becomes independent of t as $t \to \infty$, so we might guess that $\mathsf{E}[R(t)]$ approaches a limit as $t \to \infty$. We would also guess that these asymptotic ensemble averages are equal to the appropriate time averages from Section 5.4.

The bottom line for this section is that under very broad conditions, the above guesses are essentially correct. Thus the limit as $t \to \infty$ of a given ensemble-average reward can usually be computed simply by finding the time average and vice-versa. Sometimes time averages are simpler, and sometimes ensemble averages are. The advantages of the ensemble-average approach are both the ability to find $\mathsf{E}[R(t)]$ for finite values of t and the ability to understand the rate of convergence to the asymptotic result.

The following subsection is restricted to the arithmetic case. We will derive the joint CDF of age and duration for any given time t, and then look at the limit as $t \to \infty$. This leads us to arbitrary reward functions (such as residual life) for the arithmetic case. We will not look specifically at generalized reward functions that depend on other processes, but this generalization is quite similar to that for time averages.

The non-arithmetic case is analyzed in the remainder of the subsections of this section. The basic ideas are the same as the arithmetic case, but a number of subtle mathematical limiting issues arise. The reader is advised to understand the arithmetic case first, since the limiting issues in the non-arithmetic case can then be viewed within the intuitive context of the arithmetic case.

5.7.1 Age and duration for arithmetic processes

Let $\{N(t); t > 0\}$ be an arithmetic renewal counting process with inter-renewal intervals $X_1, X_2, \ldots$ and arrival epochs $S_1, S_2, \ldots$, where $S_n = X_1 + \cdots + X_n$. To keep the notation as uncluttered as possible, we take the span to be 1 and then scale to an arbitrary λ later. Thus each X_i is a positive integer-valued rv.

Recall that the age $Z(t)$ at any given $t > 0$ is $Z(t) = t - S_{N(t)}$ (where by convention $S_0 = 0$) and the duration $\widetilde{X}(t)$ is $\widetilde{X}(t) = S_{N(t)+1}(t) - S_{N(t)}$. Since arrivals occur only at integer times, we initially consider age and duration only at integer times also. If an arrival occurs at integer time t, then $S_{N(t)} = t$ and $Z(t) = 0$. Also, if $S_1 > t$, then $N(t) = 0$ and $Z(t) = t$ (i.e., the age is taken to be t if no arrivals occur up to and including time t). Thus, for integer t, $Z(t)$ is an integer-valued rv taking values from $[0, t]$. Since $S_{N(t)+1} > t$, it follows that $\widetilde{X}(t)$ is an integer-valued rv satisfying $\widetilde{X}(t) > Z(t)$. Since both are integer-valued, $\widetilde{X}(t)$ must exceed $Z(t)$ by at least 1 (or by λ in the more general case of span λ).

In order to satisfy $Z(t) = i$ and $\widetilde{X}(t) = k$ for given $i < t$, it is necessary and sufficient to have an arrival epoch at $t - i$ followed by an interarrival interval of length k, where $k \geq i + 1$. For $Z(t) = t$ and $\widetilde{X}(t) = k$, it is necessary and sufficient that $k > t$, i.e., that the first inter-renewal epoch occurs at $k > t$.

Theorem 5.7.1 *Let $\{X_n; n \geq 1\}$ be the interarrival intervals of an arithmetic renewal process with unit span. Then the joint probability mass function (PMF) of the age and duration at integer time $t \geq 1$ is given by*

$$\mathsf{p}_{Z(t),\widetilde{X}(t)}(i,k) = \begin{cases} \mathsf{p}_X(k) & \text{for } i = t,\ k > t, \\ q_{t-i}\,\mathsf{p}_X(k) & \text{for } 0 \leq i < t,\ k > i, \\ 0 & \text{otherwise.} \end{cases} \tag{5.65}$$

where $q_j = \Pr\{$arrival at time $j\}$. The limit as $t \to \infty$ for any given $0 \leq i < k$ is given by

$$\lim_{\text{integer } t \to \infty} \mathsf{p}_{Z(t),\widetilde{X}(t)}(i,k) = \frac{\mathsf{p}_X(k)}{\overline{X}}. \tag{5.66}$$

Proof The idea is quite simple. For the top part of (5.65), note that the age is t if and only *if* there are no arrivals in $(0, t]$, which corresponds to $X_1 = k$ for some $k > t$. For the middle part, the age is i for a given $i < t$ if and only if there is an arrival at $t - i$ and the next arrival epoch is after t, which means that the corresponding interarrival interval k exceeds i. The probability of an arrival at $t - i$, i.e., q_{t-i}, depends only on the arrival epochs up to and including time $t - i$, which should be independent of the subsequent interarrival time, leading to the product in the middle term of (5.65). To be more precise about this independence, note that for $i < t$,

$$q_{t-i} = \Pr\{\text{arrival at } t - i\} = \sum_{n \geq 1} \mathsf{p}_{S_n}(t - i). \tag{5.67}$$

Given that $S_n = t - i$, the probability that $X_{n+1} = k$ is $\mathsf{p}_X(k)$. This is the same for all n, establishing (5.65).

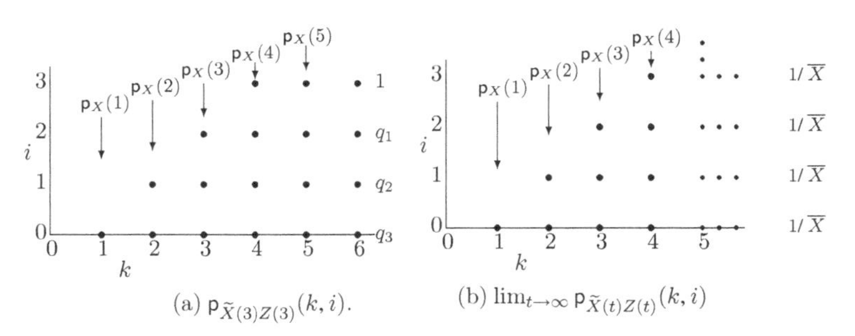

(a) $p_{\widetilde{X}(3)Z(3)}(k,i)$. (b) $\lim_{t\to\infty} p_{\widetilde{X}(t)Z(t)}(k,i)$

Figure 5.18 Joint PMF, $p_{\widetilde{X}(t)Z(t)}(k,i)$, of $\widetilde{X}(t)$ and $Z(t)$ in an arithmetic renewal process with span 1. In (a), $t=3$ and the PMF at each sample point is the product of two terms, $q_{t-i} = \Pr\{\text{arrival at } t-i\}$ and $p_X(k)$. Part (b) is the asymptotic case where $t \to \infty$. Here the arrival probabilities q_{t-i} become uniform.

For any fixed i,k with $i < k$, note that only the middle term in (5.65) is relevant as $t \to \infty$. Using Blackwell's theorem (5.64) to take the limit as $t \to \infty$, we get (5.66). □

The probabilities in the theorem, both for finite t and asymptotically as $t \to \infty$, are illustrated in Figure 5.18. The product form of the probabilities in (5.65) (as illustrated in the figure) might lead one to think that $Z(t)$ and $\widetilde{X}(t)$ are independent, but this is incorrect because of the constraint that $\widetilde{X}(t) > Z(t)$. It is curious that in the asymptotic case, (5.66) shows that, for a given duration $\widetilde{X}(t) = k$, the age is equally likely to have any integer value from 0 to $k-1$, i.e., for a given duration, the interarrival interval containing t is uniformly distributed around t.

The marginal PMF for $Z(t)$ is calculated below using (5.65):

$$p_{Z(t)}(i) = \begin{cases} F^c_X(i) & \text{for } i = t, \\ q_{t-i}\, F^c_X(i) & \text{for } 0 \le i < t, \\ 0 & \text{otherwise.} \end{cases} \tag{5.68}$$

where $F^c_X(i) = p_X(i+1) + p_X(i+2) + \cdots$. The marginal PMF for $\widetilde{X}(t)$ can be calculated directly from (5.65), but it is simplified somewhat by recognizing that

$$q_j = m(j) - m(j-1). \tag{5.69}$$

Substituting this into (5.65) and summing over age,

$$p_{\widetilde{X}(t)}(k) = \begin{cases} p_X(k)\big[m(t) - m(t-k)\big] & \text{for } k < t, \\ p_X(k)\, m(t) & \text{for } k = t, \\ p_X(k)\big[m(t) + 1\big] & \text{for } k > t. \end{cases} \tag{5.70}$$

The term $+1$ in the expression for $k > t$ corresponds to the uppermost point for the given k in Figure 5.18(a). This accounts for the possibility of no arrivals up to time t. It is not immediately evident from (5.70) that $\sum_k p_{\widetilde{X}(t)}(k) = 1$, but this can be verified from the renewal equation, (5.55).

Blackwell's theorem shows that the arrival probabilities tend to $1/\overline{X}$ as $t \to \infty$, so the limiting marginal probabilities for age and duration become

$$\lim_{\text{integer } t\to\infty} \mathsf{p}_{Z(t)}(i) = \frac{\mathsf{F}^{\mathsf{c}}_X(i)}{\overline{X}}, \tag{5.71}$$

$$\lim_{\text{integer } t\to\infty} \mathsf{p}_{\widetilde{X}(t)}(k) = \frac{k\,\mathsf{p}_X(k)}{\overline{X}}. \tag{5.72}$$

The expected value of $Z(t)$ and $\widetilde{X}(t)$ can also be found for all t from (5.65) and (5.66) respectively, but they do not have a particularly interesting form. The asymptotic values as $t \to \infty$ are more simple and interesting. The asymptotic expected value for age is derived below by using (5.71) and taking the limit[15] for integer $t \to \infty$:

$$\begin{aligned}
\lim_{t\to\infty} \mathsf{E}\left[Z(t)\right] &= \sum_i i \lim_{t\to\infty} \mathsf{p}_{Z(t)}(i) \\
&= \frac{1}{\overline{X}} \sum_{i=1}^{\infty} \sum_{j=i+1}^{\infty} i\,\mathsf{p}_X(j) = \frac{1}{\overline{X}} \sum_{j=2}^{\infty} \sum_{i=1}^{j-1} i\,\mathsf{p}_X(j) \\
&= \frac{1}{\overline{X}} \sum_{j=2}^{\infty} \frac{j(j-1)}{2}\,\mathsf{p}_X(j) \\
&= \frac{\mathsf{E}\left[X^2\right]}{2\overline{X}} - \frac{1}{2}.
\end{aligned} \tag{5.73}$$

This limiting ensemble-average age has the same dependence on $\mathsf{E}\left[X^2\right]$ as the time average in (5.18), but, perhaps surprisingly, it is reduced from that amount by 1/2. To understand this, note that we have only calculated the expected age at integer values of t. Since arrivals occur only at integer values, the age for each sample function must increase with unit slope as t is increased from one integer to the next. The expected age thus also increases, and then at the next integer value, it drops discontinuously due to the probability of an arrival at that next integer. Thus the limiting value of $\mathsf{E}\left[Z(t)\right]$ has a saw tooth shape and the value at each discontinuity is the lower side of that discontinuity. Averaging this asymptotic expected age over a unit of time, the average is $\mathsf{E}\left[X^2\right]/2\overline{X}$, in agreement with (5.18).

As with the time average, the limiting expected age is infinite if $\mathsf{E}\left[X^2\right] = \infty$. However, for each t, $Z(t) \le t$, so $\mathsf{E}\left[Z(t)\right] < \infty$ for all t, increasing without bound as $t \to \infty$

The asymptotic expected duration is derived in a similar way, starting from (5.72) and taking the limit over integer $t \to \infty$:

[15] Finding the limiting expectations from the limiting PMFs requires interchanging a limit with an expectation. This can be justified (in both (5.73) and (5.74)) by assuming that X has a finite second moment and noting that all the terms involved are positive, that $\Pr\{\text{arrival at } j\} \le 1$ for all j, and that $\mathsf{p}_X(k) \le 1$ for all k.

$$\lim_{t\to\infty} \mathsf{E}\left[\widetilde{X}(t)\right] = \sum_k \lim_{t\to\infty} k\, \mathsf{p}_{\widetilde{X}(t)}(k)$$
$$= \sum_k \frac{k^2 \mathsf{p}_X(k)}{\overline{X}} = \frac{\mathsf{E}\left[X^2\right]}{\overline{X}}. \tag{5.74}$$

This agrees with the time average in (5.19). The reduction by 1/2 seen in (5.73) is not present here, since as t is increased in the interval $[t, t+1)$, $\widetilde{X}(t)$ remains constant.

Since the asymptotic ensemble-average age differs from the time-average age in only a trivial way, and the asymptotic ensemble-average duration is the same as the time-average duration, it might appear that we have gained little by this exploration of ensemble averages. What we have gained, however, is a set of results that apply to all t. Thus they show how (in principle) these results converge as $t \to \infty$. Perhaps more important, they provide a different way to understand the somewhat paradoxical dependence of these quantities on $\mathsf{E}\left[X^2\right]$.

5.7.2 Joint age and duration: non-arithmetic case

Non-arithmetic renewal processes are mathematically more complicated than arithmetic renewal processes, but the concepts are the same. We start by looking at the joint probability of the age and duration, each over an incremental interval (see Figure 5.19(a)).

Theorem 5.7.2 *Consider an arbitrary renewal process with age $Z(t)$ and duration $\widetilde{X}(t)$ at any given time $t > 0$. Let A be the event*

$$A = \{z \le Z(t) < z+\delta\} \bigcap \{x-\delta < \widetilde{X}(t) \le x\}, \tag{5.75}$$

where $0 \le z < z+\delta \le t$ and $z+2\delta \le x$. Then

$$\Pr\{A\} = \left[m(t-z) - m(t-z-\delta)\right]\left[\mathsf{F}_X(x) - \mathsf{F}_X(x-\delta)\right]. \tag{5.76}$$

If in addition the renewal process is non-arithmetic,

$$\lim_{t\to\infty} \Pr\{A\} = \frac{\delta\left[\mathsf{F}_X(x) - \mathsf{F}_X(x-\delta)\right]}{\overline{X}}. \tag{5.77}$$

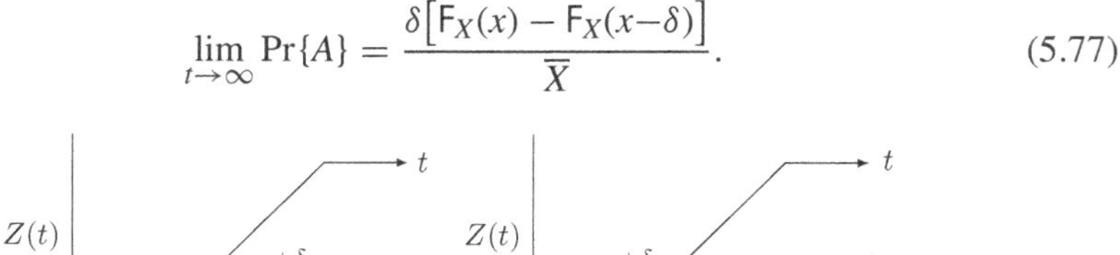
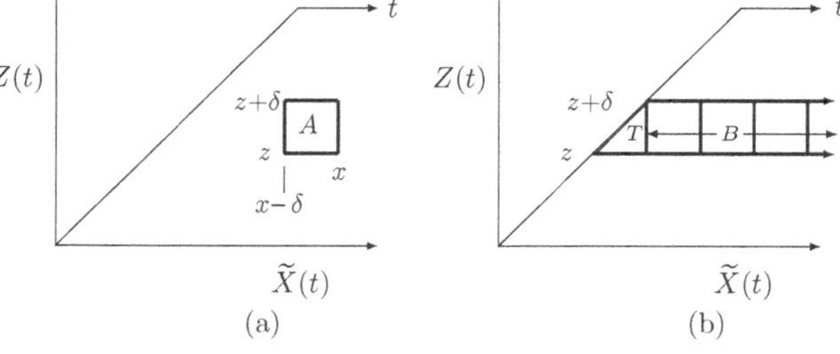

Figure 5.19 (a) The incremental square region A of sample values for $Z(t)$ and $\widetilde{X}(t)$ whose joint probability is specified in (5.76). The square is assumed to be inside the indicated semi-infinite trapezoidal region, i.e., to satisfy $0 \le z < z+\delta \le t$ and $z+2\delta \le x$.

(b) Summing over discrete sample regions of $\widetilde{X}(t)$ to find the marginal probability $\{z \le Z(t) < z+\delta\} = \Pr\{T\} + \Pr\{B\}$, where T is the triangular area and B the set of indicated squares.

Proof Note that A is the box illustrated in Figure 5.19(a) and that under the given conditions, $\widetilde{X}(t) > Z(t)$ for all sample points in A. Recall that $Z(t) = t - S_{N(t)}$ and $\widetilde{X}(t) = S_{N(t)+1} - S_{N(t)} = X_{N(t)+1}$, so A can also be expressed as

$$A = \{t{-}z{-}\delta < S_{N(t)} \le t - z\} \bigcap \{x - \delta < X_{N(t)+1} \le x\}. \tag{5.78}$$

We now argue that A can also be rewritten as

$$A = \bigcup_{n=1}^{\infty} \Big\{ \{t{-}z{-}\delta < S_n \le t - z\} \bigcap \{x - \delta < X_{n+1} \le x\} \Big\}. \tag{5.79}$$

To see this, first assume that the event in (5.78) occurs. Then $N(t)$ must have some positive sample value n, so (5.79) occurs. Next assume that the event in (5.79) occurs, which means that one of the events, say the nth, in the union occurs. Since $S_{n+1} = S_n + X_{n+1} > t$, we see that n is the sample value of $S_{N(t)}$ and (5.78) must occur.

To complete the proof, we must find $\Pr\{A\}$. First note that A is a union of disjoint events. That is, although more than one arrival epoch might occur in $(t{-}z{-}\delta,\ t{-}z]$, the following arrival epoch can exceed t for only one of them. Thus

$$\Pr\{A\} = \sum_{n=1}^{\infty} \Pr\Big\{ \{t{-}z{-}\delta < S_n \le t - z\} \bigcap \{x - \delta < X_{n+1} \le x\} \Big\}. \tag{5.80}$$

For each n, X_{n+1} is independent of S_n, so

$$\Pr\Big\{ \{t{-}z{-}\delta < S_n \le t - z\} \bigcap \{x - \delta < X_{n+1} \le x\} \Big\}$$
$$= \Pr\big\{t{-}z{-}\delta < S_n \le t - z\big\}\big[\mathsf{F}_X(x) - \mathsf{F}_X(x{-}\delta\big].$$

Substituting this into (5.80) and using (5.54) to sum the series, we get

$$\Pr\{A\} = \big[m(t{-}z) - m(t{-}z{-}\delta)\big]\big[\mathsf{F}_X(x) - \mathsf{F}_X(x{-}\delta)\big].$$

This establishes (5.76). Blackwell's theorem then establishes (5.77). □

It is curious that $\Pr\{A\}$ has such a simple product expression, where one term depends only on the function $m(t)$ and the other only on the CDF F_X. Although the theorem is most useful as $\delta \to 0$, the expression is exact for all δ such that the square region A satisfies the given constraints (i.e., A lies in the indicated semi-infinite trapezoidal region).

5.7.3 Age $Z(t)$ for finite t: non-arithmetic case

In this section, we first use Theorem 5.7.2 to find bounds on the marginal incremental probability of $Z(t)$. We then find the CDF, $\mathsf{F}_{Z(t)}(z)$, and the expected value, $\mathsf{E}[Z(t)]$ of $Z(t)$.

Corollary 5.7.3 *For $0 \le z < z + \delta \le t$, the following bounds hold on $\Pr\{z \le Z(t) < z{+}\delta\}$:*

$$\Pr\{z \le Z(t) < z{+}\delta\} \ge \big[m(t{-}z) - m(t{-}z{-}\delta)\big]\mathsf{F}^{\mathsf{c}}_X(z{+}\delta); \tag{5.81}$$
$$\Pr\{z \le Z(t) < z{+}\delta\} \le \big[m(t{-}z) - m(t{-}z{-}\delta)\big]\mathsf{F}^{\mathsf{c}}_X(z). \tag{5.82}$$

Proof As indicated in Figure 5.19(b), $\Pr\{z \le Z(t) < z+\delta\} = \Pr\{T\} + \Pr\{B\}$, where T is the triangular region

$$T = \{z \le Z(t) < z+\delta\} \bigcap \{Z(t) < \widetilde{X}(t) \le z+\delta\},$$

and B is the rectangular region

$$B = \{z \le Z(t) < z+\delta\} \bigcap \{\widetilde{X}(t) > z+\delta\}.$$

It is easy to find $\Pr\{B\}$ by summing the probabilities in (5.76) for the squares indicated in Figure 5.19(b). The result is

$$\Pr\{B\} = \big[m(t-z) - m(t-z-\delta)\big]\mathsf{F}^{\mathsf{c}}_X(z+\delta). \tag{5.83}$$

Since $\Pr\{T\} \ge 0$, this establishes the lower bound in (5.81). We next need an upper bound to $\Pr\{T\}$ and start by finding an event that includes T:

$$\begin{aligned} T &= \{z \le Z(t) < z+\delta\} \bigcap \{Z(t) < \widetilde{X}(t) \le z+\delta\} \\ &= \bigcup_{n\ge 1} \Big[\{t-z-\delta < S_n \le t-z\} \bigcap \{t - S_n < X_{n+1} \le z+\delta\}\Big] \\ &\subseteq \bigcup_{n\ge 1} \Big[\{t-z-\delta < S_n \le t-z\} \bigcap \{z < X_{n+1} \le z+\delta\}\Big], \end{aligned}$$

where the inclusion follows from the fact that $S_n \le t - z$ for the event on the left to occur, and this implies that $z \le t - S_n$.

Using the union bound, we then have

$$\begin{aligned} \Pr\{T\} &\le \Big[\sum_{n\ge 1} \Pr\{\{t-z-\delta < S_n \le t-z\}\Big]\big[\mathsf{F}^{\mathsf{c}}_X(z) - \mathsf{F}^{\mathsf{c}}_X(z+\delta)\big] \\ &= \big[m(t-z) - m(t-z-\delta)\big]\big[\mathsf{F}^{\mathsf{c}}_X(z) - \mathsf{F}^{\mathsf{c}}_X(z+\delta)\big]. \end{aligned}$$

Combining this with (5.83), we have (5.82). □

The following corollary uses Corollary 5.7.3 to determine the CDF of $Z(t)$. The rather strange dependence on the existence of a Stieltjes integral will be explained after the proof.

Corollary 5.7.4 *If the Stieltjes integral* $\int_{t-z}^{t} \mathsf{F}^{\mathsf{c}}_X(t-\tau)\, dm(\tau)$ *exists for given* $t > 0$ *and* $0 < z < t$, *then*

$$\Pr\{Z(t) \le z\} = \int_{t-z}^{t} \mathsf{F}^{\mathsf{c}}_X(t-\tau)\, dm(\tau). \tag{5.84}$$

Proof First, partition the interval $[0, z)$ into a given number ℓ of increments, each of size $\delta = z/\ell$. Then

$$\Pr\{Z(t) < z\} = \sum_{k=0}^{\ell-1} \Pr\{k\delta \le Z(t) < k\delta + \delta\}.$$

Applying the bounds in (5.81) and (5.82) to the terms in this sum,

$$\Pr\{Z(t) < z\} \geq \sum_{k=0}^{\ell-1} \Big[m\big(t - k\delta\big) - m\big(t - k\delta - \delta\big)\Big] \mathsf{F}^{\mathsf{c}}_X(k\delta + \delta), \tag{5.85}$$

$$\Pr\{Z(t) < z\} \leq \sum_{k=0}^{\ell-1} \Big[m\big(t - k\delta\big) - m\big(t - k\delta - \delta\big)\Big] \mathsf{F}^{\mathsf{c}}_X\big(k\delta\big). \tag{5.86}$$

These are, respectively, lower and upper Riemann sums for the Stieltjes integral $\int_0^z \mathsf{F}^{\mathsf{c}}_X(t-\tau)\,dm(\tau)$. Thus, if this Stieltjes integral exists, then, letting $\delta = z/\ell \to 0$,

$$\Pr\{Z(t) < z\} = \int_{t-z}^{t} \mathsf{F}^{\mathsf{c}}_X(t-\tau)\,dm(\tau).$$

This is a convolution and thus the Stieltjes integral exists unless $m(\tau)$ and $\mathsf{F}^{\mathsf{c}}_X(t-\tau)$ both have a discontinuity at some $\tau \in [0, z]$ (see Exercise 1.15). If no such discontinuity exists, then $\Pr\{Z(t) < z\}$ cannot have a discontinuity at z. Thus, if the Stieltjes integral exists, $\Pr\{Z(t) < z\} = \Pr\{Z(t) \leq z\}$, and, for $z < t$,

$$\mathsf{F}_{Z(t)}(z) = \int_{t-z}^{t} \mathsf{F}^{\mathsf{c}}_X(t-\tau)\,dm(\tau).$$

□

The above argument showed us that the values of z at which the Stieltjes integral in (5.84) fails to exist are those at which $\mathsf{F}_{Z(t)}(z)$ has a step discontinuity. At these values we know that $\mathsf{F}_{Z(t)}(z)$ (as a CDF) should have the value at the top of the step (thus including the discrete probability that $\Pr\{Z(t) = z\}$). In other words, at any point z of discontinuity where the Stieltjes integral does not exist, $\mathsf{F}_{Z(t)}(z)$ is the limit[16] of $\mathsf{F}_{Z(t)}(z+\epsilon)$ as $\epsilon > 0$ approaches 0. Another way of expressing this is that for $0 \leq z < t$, $\mathsf{F}_{Z(t)}(z)$ is the limit of the upper Riemann sum on the right-hand side of (5.86).

The next corollary uses an almost identical argument to find $\mathsf{E}[Z(t)]$. As we will see, the Stieltjes integral fails to exist at those values of t at which there is a discrete positive probability of arrival. The expected value at these points is the lower Riemann sum for the Stieltjes integral.

Corollary 5.7.5 *If the Stieltjes integral $\int_0^t \mathsf{F}^{\mathsf{c}}_X(t-\tau)\,dm(\tau)$ exists for given $t > 0$, then*

$$\mathsf{E}[Z(t)] = \mathsf{F}^{\mathsf{c}}_X(t) + \int_0^t (t-\tau)\mathsf{F}^{\mathsf{c}}_X(t-\tau)\,dm(\tau). \tag{5.87}$$

Proof Note that $Z(t) = t$ if and only if $X_1 > t$, which has probability $\mathsf{F}^{\mathsf{c}}_X(t)$. For the other possible values of $Z(t)$, we divide $[0, t)$ into ℓ equal intervals of length $\delta = t/\ell$ each. Then $\mathsf{E}[Z(t)]$ can be lower bounded by

[16] This seems to be rather abstract mathematics, but, as engineers, we often evaluate functions with step discontinuities by ignoring the values at the discontinuities or evaluating these points by ad-hoc means.

$$\begin{aligned}\mathsf{E}[Z(t)] &\geq \mathsf{F}^{\mathsf{c}}_X(t) + \sum_{k=0}^{\ell-1} k\delta \Pr\{k\delta \leq Z(t) < k\delta + \delta\} \\ &\geq \mathsf{F}^{\mathsf{c}}_X(t) + \sum_{k=0}^{\ell-1} k\delta \left[m(t-k\delta) - m(t-k\delta-\delta)\right]\mathsf{F}^{c}_X(k\delta + \delta),\end{aligned}$$

where we used (5.81) for the second step. Similarly, $\mathsf{E}[Z(t)]$ can be upper bounded by

$$\begin{aligned}\mathsf{E}[Z(t)] &\leq \mathsf{F}^{\mathsf{c}}_X(t) + \sum_{k=0}^{\ell-1} (k\delta + \delta)\Pr\{k\delta \leq Z(t) < k\delta + \delta\} \\ &\leq \mathsf{F}^{\mathsf{c}}_X(t) + \sum_{k=0}^{\ell-1} (k\delta + \delta)\left[m(t-k\delta) - m(t-k\delta-\delta)\right]\mathsf{F}^{c}_X(k\delta),\end{aligned}$$

where we used (5.82) for the second step. These provide lower and upper Riemann sums to the Stieltjes integral in (5.84), completing the proof in the same way as for the previous corollary. □

5.7.4 Age $Z(t)$ as $t \to \infty$: non-arithmetic case

Next, for non-arithmetic renewal processes, we want to find the limiting values, as $t \to \infty$, for $\mathsf{F}_{Z(t)}(z)$ and $\mathsf{E}[Z(T)]$. Temporarily ignoring any subtleties about the limit, we first view $dm(t)$ as going to $(\overline{X})^{-1}\, dt$ as $t \to \infty$. Thus from (5.84),

$$\lim_{t\to\infty} \Pr\{Z(t) \leq z\} = \frac{1}{\overline{X}} \int_0^z \mathsf{F}^{\mathsf{c}}_X(\tau)d\tau. \tag{5.88}$$

Note that this agrees with the time-average result in (5.31). If X has a probability density function (PDF), (5.88) simplifies further to

$$\lim_{t\to\infty} \mathsf{f}_{Z(t)}(z) = \frac{1}{\overline{X}}\mathsf{F}^{\mathsf{c}}_X(z). \tag{5.89}$$

Taking these limits carefully requires more mathematics than seems justified, especially since the result uses Blackwell's theorem, which we did not prove. Thus we state (without proof) another theorem, equivalent to Blackwell's theorem, called the key renewal theorem, that simplifies taking this type of limit. Essentially Blackwell's theorem is easier to interpret, but the key renewal theorem is often easier to use.

Theorem 5.7.6 (Key renewal theorem) *Let $r(x) \geq 0$ be a directly Riemann integrable function, and let $m(t) = \mathsf{E}[N(t)]$ for a non-arithmetic renewal process with $\overline{X} < \infty$. Then*

$$\lim_{t\to\infty} \int_{\tau=0}^{t} r(t-\tau)dm(\tau) = \frac{1}{\overline{X}} \int_0^\infty r(x)dx. \tag{5.90}$$

We first explain what directly Riemann integrable means. If $r(x)$ is non-zero only over finite limits, say $[0, b]$, then direct Riemann integration is the same as ordinary Riemann integration (as learned in elementary calculus). However, if $r(x)$ is non-zero over $[0, \infty)$, then the ordinary Riemann integral (if it exists) is the result of integrating from 0 to b

and then taking the limit as $b \to \infty$. The direct Riemann integral (if it exists) is the result of taking a Riemann sum over the entire half line, $[0, \infty)$, and then taking the limit as the grid becomes finer. Exercise 5.28 gives an example of a simple but bizarre function that is Riemann integrable but not directly Riemann integrable. If $r(x) \geq 0$ can be upper bounded by a decreasing Riemann integrable function, however, then, as shown in Exercise 5.28, $r(x)$ must be directly Riemann integrable. The bottom line is that restricting $r(x)$ to be directly Riemann integrable is not a major restriction.

Next we interpret the theorem. If $m(t)$ has a derivative, then Blackwell's theorem would suggest that $dm(t)/dt \to (1/\overline{X})\,dt$, which leads to (5.90) (leaving out the mathematical details). On the other hand, if X is discrete but non-arithmetic, then $dm(t)/dt$ can be intuitively viewed as a sequence of impulses that become smaller and more closely spaced as $t \to \infty$. Then $r(t)$ acts like a smoothing filter which, as $t \to \infty$, smoothes these small impulses. The theorem says that the required smoothing occurs whenever $r(t)$ is directly Riemann integrable. The theorem does not assert that the Stieltjes integral exists for all t, but only that the limit exists. For most applications to discrete inter-renewal intervals, the Stieltjes integral does not exist everywhere. Using the key renewal theorem, we can finally determine the CDF and expected value of $Z(t)$ as $t \to \infty$. These limiting ensemble averages are, of course, equal to the time averages found earlier.

Theorem 5.7.7 *For any non-arithmetic renewal process, the limiting CDF of the age $Z(t)$ is given by*

$$\lim_{t\to\infty} \mathsf{F}_{Z(t)}(z) = \frac{1}{\overline{X}} \int_0^z \mathsf{F}^{\mathsf{c}}_X(x)\,dx. \tag{5.91}$$

Furthermore, if $\mathsf{E}\left[X^2\right] < \infty$, then the limiting expected value is

$$\lim_{t\to z} \mathsf{E}\left[Z(t)\right] = \frac{\mathsf{E}\left[X^2\right]}{2\overline{X}}. \tag{5.92}$$

Proof For any given $z > 0$, let $r(x) = \mathsf{F}^{\mathsf{c}}_X(x)$ for $0 \leq x \leq z$ and $r(x) = 0$ elsewhere. Then (5.84) becomes

$$\Pr\{Z(t) \leq z\} = \int_0^t r(t-\tau)\,dm(\tau).$$

Taking the limit as $t \to \infty$,

$$\begin{aligned}\lim_{t\to\infty} \Pr\{Z(t) \leq z\} &= \lim_{t\to\infty} \int_0^t r(t-\tau)\,dm(\tau) \\ &= \frac{1}{\overline{X}} \int_0^\infty r(x)\,dx \;=\; \frac{1}{\overline{X}} \int_0^z \mathsf{F}^{\mathsf{c}}_X(x)\,dx,\end{aligned} \tag{5.93}$$

where in (5.93) we used the fact that $\mathsf{F}^{\mathsf{c}}_X(x)$ is decreasing to justify using (5.90). This establishes (5.91).

To establish (5.92), we again use the key renewal theorem, but here we let $r(x) = x\,\mathsf{F}^{\mathsf{c}}_X(x)$. Exercise 5.28 shows that $x\,\mathsf{F}^{\mathsf{c}}_X(x)$ is directly Riemann integrable if $\mathsf{E}\left[X^2\right] < \infty$. Then, taking the limit of (5.87) and finally using (5.90), we have

$$\lim_{t\to\infty} \mathsf{E}[Z(t)] = \lim_{t\to\infty} \mathsf{F}^{\mathsf{c}}_X(t) + \int_0^t r(t-\tau)\, dm(\tau)$$
$$= \frac{1}{\overline{X}} \int_0^\infty r(x)\, dx = \frac{1}{\overline{X}} \int_0^\infty x\mathsf{F}^{\mathsf{c}}_X(x)\, dx.$$

Integrating this by parts, we get (5.92). □

5.7.5 Arbitrary renewal–reward functions: non-arithmetic case

If we omit all the mathematical precision from the previous three subsections, we get a very simple picture. We started with (5.75), which gave the probability of an incremental square region A in the $(Z(t), \widetilde{X}(t))$ plane for given t. We then converted various sums over an increasingly fine grid of such regions into Stieltjes integrals. These integrals evaluated the distribution and expected value of age at arbitrary values of t. Finally, the key renewal theorem let us take the limit of these values as $t \to \infty$.

In this subsection, we will go through the same procedure for an arbitrary reward function, say $R(t) = \mathrm{R}(Z(t), \widetilde{X}(t))$, and show how to find $\mathsf{E}[R(T)]$. Note that $\Pr\{Z(t) \le z\} = \mathsf{E}\left[\mathbb{I}_{Z(t)\le z}\right]$ is a special case of $\mathsf{E}[R(T)]$ where $R(t)$ is chosen to be $\mathbb{I}_{Z(t)\le z}$. Similarly, finding the CDF at a given argument for any rv can be converted to the expectation of an indicator function. Thus, having a methodology for finding the expectation of an arbitrary reward function also covers CDFs and many other quantities of interest.

We will leave out all the limiting arguments here about converting finite incremental sums of areas into integrals, since we have seen how to do that in treating $Z(t)$. In order to make this general case more transparent, we use the following shorthand for A when it is incrementally small:

$$\Pr\{A\} = m'(t-z) f_X(x)\, dx\, dz, \tag{5.94}$$

where, if the derivatives exist, $m'(\tau) = dm(\tau)/d\tau$ and $f_X(x) = d\mathsf{F}_X(x)/dx$. If the derivatives do not exist, we can view $m'(\tau)$ and $f_X(x)$ as generalized functions including impulses, or, more appropriately, view them simply as shorthand. After using the shorthand as a guide, we can put the results in the form of Stieltjes integrals and verify the mathematical details at whatever level seems appropriate.

We do this first for the example of the CDF of duration, $\Pr\{\widetilde{X}(t) \le x_0\}$, where we first assume that $x_0 \le t$. As illustrated in Figure 5.20, the corresponding reward function $R(t)$ is 1 in the triangular region where $\widetilde{X}(t) \le x_0$ and $Z(t) < \widetilde{X}(t)$. It is 0 elsewhere.

$$\Pr\left\{\widetilde{X}(t) \le x_0\right\} = \int_{z=0}^{x_0} \int_{x=z}^{x_0} m'(t-z) \mathsf{f}_X(x)\, dx\, dz$$
$$= \int_{z=0}^{x_0} m'(t-z)\big[\mathsf{F}_X(x_0) - \mathsf{F}_X(z)\big]\, dz$$
$$= \mathsf{F}_X(x_0)\big[m(t) - m(t-x_0)\big] - \int_{t-x_0}^{t} \mathsf{F}_X(t-\tau)\, dm(\tau). \tag{5.95}$$

For the opposite case, where $x_0 > t$, it is easier to find $\Pr\{\widetilde{X}(t) > x_0\}$. As shown in the figure, this is the region where $0 \le Z(t) \le t$ and $\widetilde{X}(t) > x_0$. There is a subtlety here in

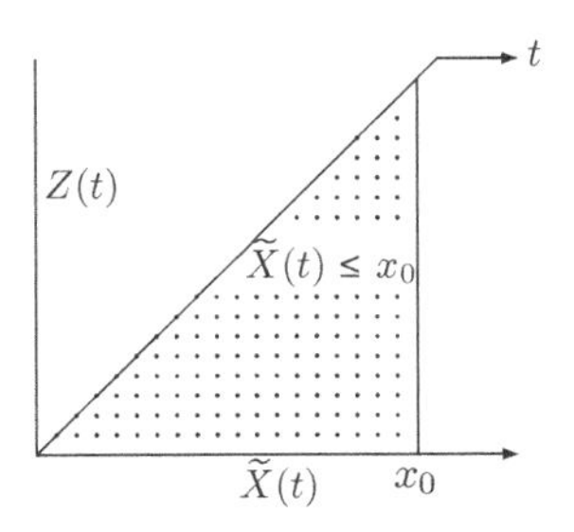

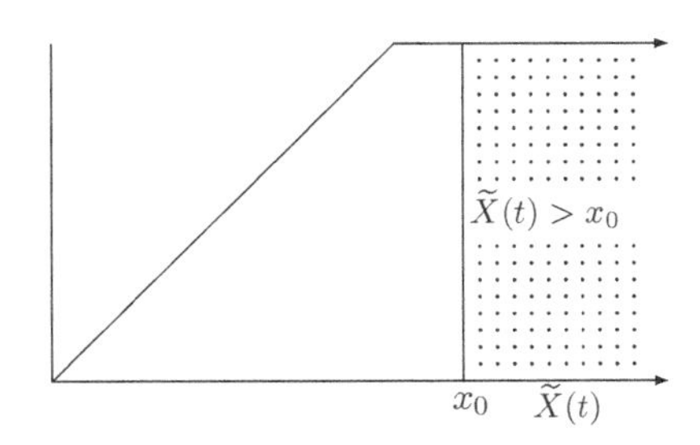

Figure 5.20 Finding $\mathsf{F}_{\widetilde{X}(t)}(x_0)$ for $x_0 \le t$ and for $x_0 > t$.

that the incremental areas we are using are only valid for $Z(t) < t$. If the age is equal to t, then no renewals have occured in $(0, t]$, so that $\Pr\{\widetilde{X}(t) > x_0; Z(t) = t\} = \mathsf{F}^{\mathsf{c}}_X(x_0)$. Thus

$$\Pr\{\widetilde{X}(t) > x_0\} = \mathsf{F}^{\mathsf{c}}_X(x_0) + \int_{z=0}^{t^-}\int_{x=x_0}^{\infty} m'(t-z)\mathsf{f}_X(x)\,dx\,dz$$

$$= \mathsf{F}^{\mathsf{c}}_X(x_0) + m(t)\mathsf{F}^{\mathsf{c}}_X(x_0). \tag{5.96}$$

As a sanity test, the renewal equation, (5.56), can be used to show that the sum of (5.95) and (5.96) at $x_0 = t$ is equal to 1 (as it must be if the equations are correct).

We can now take the limit, $\lim_{t\to\infty} \Pr\{\widetilde{X}(t) \le x_0\}$. For any given x_0, (5.95) holds for sufficiently large t, and the key renewal theorem can be used since the integral has a finite range. Thus,

$$\lim_{t\to\infty} \Pr\{\widetilde{X}(t) \le x_0\} = \frac{1}{\overline{X}}\Big[x_0\mathsf{F}_X(x_0) - \int_0^{x_0} \mathsf{F}_X(x)\,dx\Big]$$

$$= \frac{1}{\overline{X}}\int_0^{x_0}\Big[\mathsf{F}_X(x_0) - \mathsf{F}_X(x)\Big]\,dx$$

$$= \frac{1}{\overline{X}}\int_0^{x_0}\Big[\mathsf{F}^{\mathsf{c}}_X(x) - dF^{\mathsf{c}}_X(x_0)\Big]\,dx. \tag{5.97}$$

It is easy to see that the right-hand side of (5.97) is increasing from 0 to 1 with increasing x_0, i.e., it is a CDF.

After this example, it is now straightforward to write down the expected value of an arbitrary renewal–reward function $R(t)$ whose sample value at $Z(t) = z$ and $X(t) = x$ is denoted by $\mathrm{R}(z, x)$. We have

$$\mathsf{E}[R(t)] = \int_{x=t}^{\infty} \mathrm{R}(t,x)\,d\mathsf{F}_X(x) + \int_{z=0}^{t}\int_{x=z}^{\infty} \mathrm{R}(z,x)\,d\mathsf{F}_X(x)\,dm(t-z). \tag{5.98}$$

The first term above arises from the subtlety discussed for the case where $Z(t) = t$. The second term is simply the integral over the semi-infinite trapezoidal area in Figure 5.20.

The analysis up to this point applies to both the arithmetic and non-arithmetic cases, but we now must assume again that the renewal process is non-arithmetic. If the inner integral, i.e., $\int_{x=z}^{\infty} \mathrm{R}(z,x)\,d\mathsf{F}_X(x)$, as a function of z, is directly Riemann integrable, then not only can the key renewal theorem be applied to this second term, but also the first term must approach 0 as $t \to \infty$. Thus the limit of (5.98) as $t \to \infty$ is

$$\lim_{t\to\infty} \mathsf{E}[R(t)] = \frac{1}{\overline{X}} \int_{z=0}^{\infty} \int_{x=z}^{\infty} \mathrm{R}\,(z,x)\, d\mathsf{F}_X(x)\, dz. \tag{5.99}$$

This is the same expression as found for the time-average renewal reward in Theorem 5.4.5. Thus, as indicated earlier, we can now equate any time-average result for the non-arithmetic case with the corresponding limiting ensemble average, and the same equations have been derived in both cases.

As a simple example of (5.99), let $\mathrm{R}\,(z,x) = x$. Then $\mathsf{E}[R(t)] = \mathsf{E}[\widetilde{X}(t)]$ and

$$\begin{aligned}\lim_{t\to\infty} \mathsf{E}\left[\widetilde{X}(t)\right] &= \frac{1}{\overline{X}} \int_{z=0}^{\infty} \int_{x=z}^{\infty} x\, d\mathsf{F}_X(x)\, dz \;=\; \int_{x=0}^{\infty} \int_{z=0}^{x} x\, dz\, d\mathsf{F}_X(x) \\ &= \frac{1}{\overline{X}} \int_{x=0}^{\infty} x^2\, d\mathsf{F}_X(x) \;=\; \frac{\mathsf{E}\left[X^2\right]}{\overline{X}}.\end{aligned} \tag{5.100}$$

After calculating the integral above by interchanging the order of integration, we can go back and assert that the key renewal theorem applies if $\mathsf{E}\left[X^2\right]$ is finite. If it is infinite, then it is not hard to see that $\lim_{t\to\infty} \mathsf{E}[\widetilde{X}(t)]$ is infinite also.

It has been important, and theoretically reassuring, to be able to find ensemble averages for non-arithmetic renewal–reward functions in the limit of large t and to show (not surprisingly) that they are the same as the time-average results. The ensemble-average results are quite subtle, though, and it is wise to check results achieved that way with the corresponding time-average results.

5.8 Delayed renewal processes

We have seen a certain awkwardness in our discussion of Little's theorem and the M/G/1 delay result because an arrival was assumed, but not counted, at time 0; this was necessary for the first inter-renewal interval to be statistically identical to the others. In this section, we correct that defect by allowing the epoch at which the first renewal occurs to be arbitrarily distributed. The resulting type of process is a generalization of the class of renewal processes known as *delayed renewal processes*. The word *delayed* does not necessarily imply that the first renewal epoch is in any sense larger than the other inter-renewal intervals. Rather, it means that the usual renewal process, with IID inter-renewal times, is delayed until after the epoch of the first renewal. What we shall discover is intuitively satisfying – both the time-average behavior and, in essence, the limiting ensemble behavior are not affected by the distribution of the first renewal epoch. It might be somewhat surprising, however, to find that this irrelevance of the distribution of the first renewal epoch holds even when the mean of the first renewal epoch is infinite.

To be more precise, we let $\{X_i;\ i{\geq}1\}$ be a set of independent non-negative rv s. X_1 has a given CDF $\mathsf{G}(x)$, whereas $\{X_i;\ i \geq 2\}$ are identically distributed with a given CDF $\mathsf{F}(x)$. Thus a renewal process is a special case of a delayed renewal process for which

$\mathsf{G}(x) = \mathsf{F}(x)$. Let $S_n = \sum_{i=1}^{n} X_i$ be the nth renewal epoch. We first show that the SLLN still holds despite the deviant behavior of X_1.

Lemma 5.8.1 *Let $\{X_i;\ i \geq 2\}$ be IID with a mean $\overline{X}$ (i.e., $\mathsf{E}[|X|] < \infty$) and let X_1 be a rv, independent of $\{X_i;\ i \geq 2\}$. Let $S_n = \sum_{i=1}^{n} X_i$. Then $\lim_{n\to\infty} S_n/n = \overline{X}$ WP1.*

Proof Note that

$$\frac{S_n}{n} = \frac{X_1}{n} + \frac{\sum_{i=2}^{n} X_i}{n}.$$

Since X_1 is finite WP1, the first term above goes to 0 WP1 as $n \to \infty$. The second term goes to $\overline{X}$, proving the lemma (which is thus a trivial variation of the SLLN). □

Now, for the given delayed renewal process, let $N(t)$ be the number of renewal epochs up to and including time t. This is still determined by the fact that $\{N(t) \geq n\}$ if and only if $\{S_n \leq t\}$). $\{N(t);\ t > 0\}$ we call a delayed renewal counting process. The following simple lemma follows from Lemma 5.3.2.

Lemma 5.8.2 *Let $\{N(t);\ t > 0\}$ be a delayed renewal counting process. Then $\lim_{t\to\infty} N(t) = \infty$ with probability 1 and $\lim_{t\to\infty} \mathsf{E}[N(t)] = \infty$.*

Proof Conditioning on $X_1 = x$, we can write $N(t) = 1 + N'(t - x)$, where $N'\{t; t \geq 0\}$ is the ordinary renewal counting process with inter-renewal intervals $X_2, X_3, \ldots$. From Lemma 5.3.2, $\lim_{t\to\infty} N'(t - x) = \infty$ with probability 1, and $\lim_{t\to\infty} \mathsf{E}\left[N'(t - x)\right] = \infty$. Since this is true for every finite $x > 0$, and X_1 is finite with probability 1, the lemma is proven. □

Theorem 5.8.3 (Strong law for delayed renewal processes) *Let $N(t);\ t > 0$ be the renewal counting process for a delayed renewal process where the CDF of the inter-renewal intervals $X_2, X_3, \ldots$ is F and the mean $\overline{X} = \int_0^\infty [1 - \mathsf{F}(x)]\, dx$ is finite. Then*

$$\lim_{t\to\infty} \frac{N(t)}{t} = \frac{1}{\overline{X}} \qquad \textit{WP1}. \tag{5.101}$$

Proof Using Lemma 5.8.1, the conditions for Theorem 5.3.5 are fulfilled, so the proof follows exactly as the proof of Theorem 5.3.1. □

Next we look at the elementary renewal theorem and Blackwell's theorem for delayed renewal processes. To do this, we view a delayed renewal counting process $\{N(t);\ t > 0\}$ as an ordinary renewal counting process that starts at a random non-negative epoch X_1 with some CDF $\mathsf{G}(t)$. Define $N_o(t - X_1)$ as the number of renewals that occur in the interval $(X_1, t]$. Conditional on any given sample value x for X_1, $\{N_o(t - x);\ t - x > 0\}$ is an ordinary renewal counting process and thus, given $X_1 = x$, $\lim_{t\to\infty} \mathsf{E}[N_o(t - x)]/(t - x) = 1/\overline{X}$. Since $N(t) = 1 + N_o(t - X_1)$ for $t > X_1$, we see that, conditional on $X_1 = x$,

$$\lim_{t\to\infty} \frac{\mathsf{E}[N(t) \mid X_1 {=} x]}{t} = \lim_{t\to\infty} \frac{\mathsf{E}[N_o(t - x)]}{t - x} \, \frac{t - x}{t} = \frac{1}{\overline{X}}. \tag{5.102}$$

Since this is true for every finite sample value x for X_1, we have established the following theorem.

Theorem 5.8.4 (Elementary delayed renewal theorem) *For a delayed renewal process with* $\mathsf{E}[X_i] = \overline{X}$ *for* $i \geq 2$,

$$\lim_{t\to\infty} \frac{\mathsf{E}[N(t)]}{t} = \frac{1}{\overline{X}}. \tag{5.103}$$

The same approach gives us Blackwell's theorem. Specifically, if $\{X_i;\ i\geq 2\}$ is a sequence of IID non-arithmetic rv s, then, for any $\delta > 0$, Blackwell's theorem for ordinary renewal processes implies that

$$\lim_{t\to\infty} \frac{\mathsf{E}[N_o(t-x+\delta) - N_o(t-x)]}{\delta} = \frac{1}{\overline{X}}. \tag{5.104}$$

Thus, conditional on any sample value $X_1 = x$, $\lim_{t\to\infty} \mathsf{E}[N(t+\delta) - N(t) \mid X_1 = x] = \delta/\overline{X}$. Taking the expected value over X_1 gives us $\lim_{t\to\infty} \mathsf{E}[N(t+\delta) - N(t)] = \delta/\overline{X}$. The case in which $\{X_i;\ i\geq 2\}$ are arithmetic with span λ is somewhat more complicated. If X_1 is arithmetic with span λ (or a multiple of λ), then the first renewal epoch must be at some multiple of λ and $\lambda/\overline{X}$ gives the expected number of arrivals at time $i\lambda$ in the limit as $i \to \infty$. If X_1 is non-arithmetic or arithmetic with a span other than a multiple of λ, then the effect of the first renewal epoch never dies out, since all subsequent renewals occur at multiples of λ from this first epoch. We ignore this rather ugly case and state the following theorem for the straightforward situations.

Theorem 5.8.5 (Blackwell for delayed renewal) *If* $\{X_i;\ i\geq 2\}$ *are non-arithmetic, then, for all* $\delta > 0$,

$$\lim_{t\to\infty} \frac{\mathsf{E}[N(t+\delta) - N(t)]}{\delta} = \frac{1}{\overline{X}}. \tag{5.105}$$

If $\{X_i; i \geq 2\}$ *are arithmetic with span* λ *and mean* $\overline{X}$ *and* X_1 *is arithmetic with span* $m\lambda$ *for some positive integer* m, *then*

$$\lim_{i\to\infty} \Pr\{\text{renewal at } t = i\lambda\} = \frac{\lambda}{\overline{X}}. \tag{5.106}$$

5.8.1 Delayed renewal–reward processes

We have seen that the distribution of the first renewal epoch has no effect on the time or ensemble-average behavior of a renewal process (other than the ensemble dependence on time for an arithmetic process). This carries over to reward functions with almost no change. In particular, the generalized version of Theorem 5.4.5 is as follows.

Theorem 5.8.6 *Let* $\{N(t);\ t > 0\}$ *be a delayed renewal counting process where* F *is the CDF of the inter-renewal intervals* $X_2, X_3, \ldots$. *Let* $Z(t) = t - S_{N(t)}$, *let* $\widetilde{X}(t) = S_{N(t)+1} - S_{N(t)}$, *and let* $R(t) = \mathrm{R}(Z(t), \widetilde{X}(t))$ *be a reward function. Assume that*

$$\mathsf{E}[\mathrm{R}_n] = \int_{x=0}^{\infty} \int_{z=0}^{x} \mathrm{R}(z,x)\, dz\, d\mathsf{F}(x) < \infty\,.$$

Then, WP1,

$$\lim_{t\to\infty} \frac{1}{t} \int_{\tau=0}^{t} R(\tau)d\tau = \frac{\mathsf{E}[\mathrm{R}_n]}{\overline{X}_2} \quad \textit{for } n \geq 2. \tag{5.107}$$

We omit the proof of this since it is a minor variation of that of Theorem 5.4.5. Finally, the equivalence of time and limiting ensemble averages holds as before, yielding

$$\lim_{t\to\infty} \mathsf{E}[R(t)] = \frac{\mathsf{E}[\mathsf{R}_n]}{\overline{X}_2}. \tag{5.108}$$

5.8.2 Transient behavior of delayed renewal processes

Let $m(t) = \mathsf{E}[N(t)]$ for a delayed renewal process. As in (5.54), we have

$$m(t) = \sum_{n=1}^{\infty} \Pr\{N(t) \geq n\} = \sum_{n=1}^{\infty} \Pr\{S_n \leq t\}. \tag{5.109}$$

For $n \geq 2$, $S_n = S_{n-1} + X_n$, where X_n and S_{n-1} are independent. From the convolution equation (1.12),

$$\Pr\{S_n \leq t\} = \int_{x=0}^{t} \Pr\{S_{n-1} \leq t - x\}\, d\mathsf{F}(x) \quad \text{for } n \geq 2. \tag{5.110}$$

For $n = 1$, $\Pr\{S_n \leq t\} = \mathsf{G}(t)$. Substituting this in (5.109) and interchanging the order of integration and summation,

$$\begin{aligned} m(t) &= \mathsf{G}(t) + \int_{x=0}^{t} \sum_{n=2}^{\infty} \Pr\{S_{n-1} \leq t - x\}\, d\mathsf{F}(x) \\ &= \mathsf{G}(t) + \int_{x=0}^{t} \sum_{n=1}^{\infty} \Pr\{S_n \leq t - x\}\, d\mathsf{F}(x) \\ &= \mathsf{G}(t) + \int_{x=0}^{t} m(t-x) d\mathsf{F}(x), \qquad t \geq 0. \end{aligned} \tag{5.111}$$

This is the *renewal equation* for delayed renewal processes and is a generalization of (5.55). It is shown to have a unique solution in [9], Section 11.1.

There is another useful integral equation very similar to (5.111) that arises from breaking up S_n as the sum of X_1 and $\widehat{S}_{n-1}$, where $\widehat{S}_{n-1} = X_2 + \cdots + X_n$. Letting $\widehat{m}(t)$ be the expected number of renewals in time t for an ordinary renewal process with interarrival CDF F, a similar argument to that above, starting with $\Pr\{S_n \leq t\} = \int_0^t \Pr\{\widehat{S}_{n-1} \leq t - x\}\, d\mathsf{G}(x)$ yields

$$m(t) = \mathsf{G}(t) + \int_{x=0}^{t} \widehat{m}(t-x) d\mathsf{G}(x). \tag{5.112}$$

This equation brings out the effect of the initial renewal interval clearly, and is useful in computation if one already knows $\widehat{m}(t)$.

Frequently, the most convenient way of dealing with $m(t)$ is through transforms. Following the same argument as that in (5.57), we get $L_m(r) = (1/r)L_{\mathsf{G}}(r) + L_m(r)L_{\mathsf{F}}(r)$. Solving, we get

$$L_m(r) = \frac{L_{\mathsf{G}}(r)}{r[1 - L_{\mathsf{F}}(r)]}. \tag{5.113}$$

We can find $m(t)$ from (5.113) by finding the inverse Laplace transform, using the same procedure as in Example 5.6.1. There is a second-order pole at $r = 0$ again, and, evaluating the residue, it is $1/L'_{\mathsf{F}}(0) = 1/\overline{X}_2$, which is not surprising in terms of Blackwell's theorem. We can also expand the numerator and denominator of (5.113) in a power series, as in (5.58). The inverse transform, corresponding to (5.59), is

$$m(t) = \frac{t}{\overline{X}} + \frac{\mathsf{E}\left[X_2^2\right]}{2\overline{X}} - \frac{\overline{X}_1}{\overline{X}} + \epsilon(t) \quad \text{for } t \to 0, \tag{5.114}$$

where $\lim_{t\to\infty} \epsilon(t) = 0$.

5.8.3 The equilibrium process

Consider an ordinary non-arithmetic renewal process with an inter-renewal interval X of CDF $\mathsf{F}(x)$. We have seen that the distribution of the interval from t to the next renewal approaches $\mathsf{F}_Y(y) = (1/\mathsf{E}[X]) \int_0^y [1 - \mathsf{F}(x)]dx$ as $t \to \infty$. This suggests that if we look at this renewal process starting at some very large t, we should see a delayed renewal process for which the CDF $\mathsf{G}(x)$ of the first renewal is equal to the residual life CDF $\mathsf{F}_Y(x)$ above and subsequent inter-renewal intervals should have the original CDF $\mathsf{F}(x)$ above. Thus it appears that such a delayed renewal process is the same as the original ordinary renewal process, except that it starts in 'steady state.' To verify this, we show that $m(t) = t/\overline{X}$ is a solution to (5.111) if $\mathsf{G}(t) = \mathsf{F}_Y(t)$. Substituting $(t-x)/\overline{X}$ for $m(t-x)$, the right-hand side of (5.111) is

$$\frac{\int_0^t [1 - \mathsf{F}(x)]dx}{\overline{X}_2} + \frac{\int_0^t (t - x)d\mathsf{F}(x)}{\overline{X}} = \frac{\int_0^t [1 - \mathsf{F}(x)]dx}{\overline{X}} + \frac{\int_0^t \mathsf{F}(x)dx}{\overline{X}} = \frac{t}{\overline{X}},$$

where we have used integration by parts for the first equality. This particular delayed renewal process is called the *equilibrium process*, since it starts off in steady state, and thus has no transients.

5.9 Summary

Sections 5.1–5.7 developed the central results about renewal processes that frequently appear in subsequent chapters. The chapter started with the SLLN, which was then used to establish the strong law for renewal processes, showing that the time average rate of renewals, $N(t)/t$, approaches $1/\overline{X}$ WP1 as $t \to \infty$. This, combined with the SLLN, is the basis for most subsequent results about time averages. Section 5.4 added a reward function $R(t)$ to the underlying renewal process. These reward functions are defined to depend only on the inter-renewal interval containing t, and are used to study many surprising aspects of renewal processes such as residual life, age, and duration. For all sample paths of a renewal process (except a subset of probability 0), the time-average reward for a given $R(t)$ is a constant, and that constant is the expected aggregate reward over an inter-renewal interval divided by the expected length of the inter-renewal interval.

The next topic, in Section 5.5, was that of stopping trials. These have obvious applications to situations where an experiment or game is played until some desired (or undesired) outcome (based on the results up to and including the given trial) occurs. This is a basic and important topic in its own right, but is also needed to understand both how the expected renewal rate $\mathsf{E}[N(t)]/t$ varies with time t and how renewal theory can be applied to queueing situations. Finally, we found that stopping rules were helpful in understanding G/G/1 queues, especially Little's theorem, and to derive and understand the Pollaczek–Khinchin expression for the expected delay in an M/G/1 queue.

This was followed, in Section 5.6, by an analysis of how $\mathsf{E}[N(t)]/t$ varies with t. This started by using Laplace transforms to get a complete solution of the ensemble average, $\mathsf{E}[N(t)]/t$, as a function of t, when the distribution of the inter-renewal interval has a rational Laplace transform. For the general case (where the Laplace transform is irrational or non-existent), the elementary renewal theorem shows that $\lim_{t\to\infty} \mathsf{E}[N(t)]/t = 1/\overline{X}$. The fact that the time average (WP1) and the limiting ensemble average are the same is not surprising, and the fact that the ensemble average has a limit is not surprising. These results are so fundamental to other results in probability, however, that they deserve to be understood.

Another fundamental result in Section 5.6 was Blackwell's renewal theorem, showing that the distribution of renewal epochs reaches a steady state as $t \to \infty$. The form of that steady state depends on whether the inter-renewal distribution is arithmetic (see (5.62)) or non-arithmetic (see (5.61)).

Section 5.7 tied together the results on rewards in Section 5.4 to those on ensemble averages in Section 5.6. Under some very minor restrictions imposed by the key renewal theorem, we found that, for non-arithmetic inter-renewal distributions, $\lim_{t\to\infty} \mathsf{E}[R(t)]$ is the same as the time-average value of reward.

Finally, all the results above were shown to apply to delayed renewal processes.

For further reading on renewal processes, see Feller [9], Ross [22], or Wolff [30]. Feller still appears to be the best source for deep understanding of renewal processes, but Ross and Wolff are somewhat more accessible.

5.10 Exercises

Exercise 5.1 The purpose of this exercise is to show that for an arbitrary renewal process, $N(t)$, the number of renewals in $(0, t]$ is a (non-defective) rv.

(a) Let $X_1, X_2, \dots$ be a sequence of IID inter-renewal rv s. Let $S_n = X_1 + \cdots + X_n$ be the corresponding renewal epochs for each $n \geq 1$. Assume that each X_i has a finite expectation $\overline{X} > 0$ and, for any given $t > 0$, use the WLLN to show that $\lim_{n\to\infty} \Pr\{S_n \leq t\} = 0$.

(b) Use (a) to show that $\lim_{n\to\infty} \Pr\{N(t) \geq n\} = 0$ for each $t > 0$ and explain why this means that $N(t)$ is a rv, i.e., is not defective.

(c) Now suppose that the X_i do not have a finite mean. Consider truncating each X_i to $\breve{X}_i$, where for any given $b > 0$, $\breve{X}_i = \min(X_i, b)$. Let $\breve{N}(t)$ be the renewal counting process for the inter-renewal intervals $\breve{X}_i$. Show that $\breve{N}(t)$ is non-defective for each $t > 0$. Show that $N(t) \leq \breve{N}(t)$ and thus that $N(t)$ is non-defective. Note: Large inter-renewal intervals create small values of $N(t)$, and thus $\mathsf{E}[X] = \infty$ has nothing to do with potentially large values of $N(t)$, so the argument here was purely technical.

Exercise 5.2 This exercise shows that, for an arbitrary renewal process, $N(t)$, the number of renewals in $(0, t]$, is non-defective and has finite expectation.

(a) Let the inter-renewal intervals have the CDF $\mathsf{F}_X(x)$, with, as usual, $\mathsf{F}_X(0) = 0$. Using whatever combination of mathematics and common sense is comfortable for you, show that for any ϵ, $0 < \epsilon < 1$, there is a $\delta > 0$ such that $\mathsf{F}_X(\delta) \leq 1 - \epsilon$. In other words, you are to show that a positive rv must lie in some range of positive values bounded away from 0 with positive probability.

(b) Show that $\Pr\{S_n \leq \delta\} \leq (1 - \epsilon)^n$.

(c) Show that $\mathsf{E}[N(\delta)] \leq 1/\epsilon$.

(d) For the ϵ, δ of (a), show that for every integer k, $\mathsf{E}[N(k\delta)] \leq k/\epsilon$ and thus that $\mathsf{E}[N(t)] \leq (t + \delta)/\epsilon\delta$ for any $t > 0$.

(e) Use the result here to show that $N(t)$ is non-defective.

Exercise 5.3 Let $\{X_i; i \geq 1\}$ be the inter-renewal intervals of a renewal process generalized to allow for inter-renewal intervals of size 0 and let $\Pr\{X_i = 0\} = \alpha$, $0 < \alpha < 1$. Let $\{Y_i;\ i \geq 1\}$ be the sequence of non-zero interarrival intervals. For example, if $X_1 = x_1 > 0,\ X_2 = 0,\ X_3 = x_3 > 0, \ldots$, then $Y_1 = x_1,\ Y_2 = x_3,\ \ldots$.

(a) Find the CDF of each Y_i in terms of that of the X_i.

(b) Find the PMF of the number of arrivals of the generalized renewal process at each epoch at which arrivals occur.

(c) Explain how to view the generalized renewal process as an ordinary renewal process with inter-renewal intervals $\{Y_i;\ i \geq 1\}$ and bulk arrivals at each renewal epoch.

(d) When a generalized renewal process is viewed as an ordinary renewal process with bulk arrivals, what is the distribution of the bulk arrivals? (The point of this part is to illustrate that bulk arrivals on an ordinary renewal process are considerably more general than generalized renewal processes.)

Exercise 5.4 Is it true for a renewal process that:

(a) $N(t) < n$ if and only if $S_n > t$?

(b) $N(t) \leq n$ if and only if $S_n \geq t$?

(c) $N(t) > n$ if and only if $S_n < t$?

Exercise 5.5 (This shows that convergence WP1 implies convergence in probability.) Let $\{Y_n;\ n \geq 1\}$ be a sequence of rv s that converges to 0 WP1. For any positive integers m and k, let

$$A(m, k) = \{\omega : |Y_n(\omega)| \leq 1/k \quad \text{for all } n \geq m\}.$$

(a) Show that if $\lim_{n\to\infty} Y_n(\omega) = 0$ for some given ω, then (for any given k) $\omega \in A(m, k)$ for some positive integer m.

(b) Show that for all $k \geq 1$

$$\Pr\Big\{\bigcup\nolimits_{m=1}^{\infty} A(m, k)\Big\} = 1.$$

(c) Show that, for all $m \geq 1$, $A(m,k) \subseteq A(m+1,\, k)$. Use this (plus (1.9)) to show that

$$\lim_{m\to\infty} \Pr\{A(m,k)\} = 1.$$

(d) Show that if $\omega \in A(m,k)$, then $|Y_m(\omega)| \leq 1/k$. Use this (plus (c)) to show that

$$\lim_{m\to\infty} \Pr\{|Y_m| > 1/k\} = 0.$$

Since $k \geq 1$ is arbitrary, this shows that $\{Y_n;\, n \geq 1\}$ converges in probabiity.

Exercise 5.6 A town starts a mosquito control program and the rv Z_n is the number of mosquitoes at the end of the nth year ($n = 0, 1, 2, \ldots$). Let X_n be the growth rate of the mosquito population in year n; i.e., $Z_n = X_n Z_{n-1};\; n \geq 1$. Assume that $\{X_n;\, n \geq 1\}$ is a sequence of IID rv s with the PMF $\Pr\{X{=}2\} = 1/2$; $\Pr\{X{=}1/2\} = 1/4$; $\Pr\{X{=}1/4\} = 1/4$. Suppose that Z_0, the initial number of mosquitoes, is some known constant and assume for simplicity and consistency that Z_n can take on non-integer values.

(a) Find $\mathsf{E}[Z_n]$ as a function of n and find $\lim_{n\to\infty} \mathsf{E}[Z_n]$.

(b) Let $W_n = \log_2 X_n$. Find $\mathsf{E}[W_n]$ and $\mathsf{E}\left[\log_2(Z_n/Z_0)\right]$ as a function of n.

(c) There is a constant α such that $\lim_{n\to\infty}(1/n)[\log_2(Z_n/Z_0)] = \alpha$ WP1. Find α and explain how this follows from the SLLN.

(d) Using (c), show that $\lim_{n\to\infty} Z_n = \beta$ WP1 for some β and evaluate β.

(e) Explain carefully how the result in (a) and the result in (d) are compatible. What you should learn from this problem is that the expected value of the log of a product of IID rv s might be more significant than the expected value of the product itself.

Exercise 5.7 In this exercise, you will find an explicit expression for $\{\omega : \lim_n Y_n = 0\}$. You need not be mathematically precise.

(a) Let $\{Y_n;\, n \geq 1\}$ be a sequence of rv s. Using the definition of convergence for a sequence of numbers, justify the following set equivalences:

$$\begin{aligned}\{\omega : \lim_n Y_n(\omega) = 0\} &= \bigcap\nolimits_{k=1}^{\infty} \{\omega : \text{there exists an } m \text{ such that } |Y_n(\omega)| \leq 1/k \\ &\qquad\qquad \text{for all } n \geq m\} \\ &= \bigcap\nolimits_{k=1}^{\infty} \bigcup\nolimits_{m=1}^{\infty} \{\omega : Y_n(\omega) \leq 1/k \text{ for all } n \geq m\} \\ &= \bigcap\nolimits_{k=1}^{\infty} \bigcup\nolimits_{m=1}^{\infty} \bigcap\nolimits_{n=m}^{\infty} \{\omega : Y_n(\omega) \leq 1/k\}.\end{aligned}$$

(b) Explain how this shows that $\{\omega : \lim_n Y_n(\omega) = 0\}$ must be an event.

(c) Use De Morgan's laws to show that the complement of the above equivalence is

$$\{\omega : \lim_n Y_n(\omega) = 0\}^{\mathsf{c}} = \bigcup\nolimits_{k=1}^{\infty} \bigcap\nolimits_{m=1}^{\infty} \bigcup\nolimits_{n=m}^{\infty} \{\omega : Y_n(\omega) > 1/k\}.$$

(d) Show that for $\{Y_n;\, n \geq 1\}$ to converge WP1, it is necessary and sufficient to satisfy

$$\Pr\left\{\bigcap\nolimits_{m=1}^{\infty} \bigcup\nolimits_{n=m}^{\infty} \{Y_n > 1/k\}\right\} = 0 \qquad \text{for all } k \geq 1.$$

(e) Show that for $\{Y_n;\, n \geq 1\}$ to converge WP1, it is necessary and sufficient to satisfy

$$\lim_{m\to\infty} \Pr\left\{\bigcup\nolimits_{n=m}^{\infty} \{Y_n > 1/k\}\right\} = 0 \qquad \text{for all } k \geq 1.$$

Hint: Use Exercise 5.8(a). Note: Part (e) provides an equivalent condition that is often useful in establishing convergence WP1. It also brings out quite clearly the difference between convergence WP1 and convergence in probability.

Exercise 5.8 (Borel–Cantelli lemma) Consider the event $\bigcap_{m\ge1}\bigcup_{n\ge m} A_n$, where $A_1, A_2, \ldots$ are arbitrary events.

(**a**) Show that

$$\lim_{m\to\infty} \Pr\Big\{\bigcup_{n\ge m} A_n\Big\} = 0 \qquad \Longleftrightarrow \qquad \Pr\Big\{\bigcap_{m\ge1}\bigcup_{n\ge m} A_n\Big\} = 0.$$

Hint: Apply (1.10).

(**b**) Show that if $\sum_{m=1}^{\infty} \Pr\{A_m\} < \infty$, then $\lim_{m\to\infty} \Pr\{\bigcup_{n\ge m} A_n\} = 0$. Hint: First explain why $\sum_{m=1}^{\infty} \Pr\{A_m\} < \infty$ implies that $\lim_{m\to\infty} \Pr\{A_n\} = 0$. This well-known result is called the Borel–Cantelli lemma.

(**c**) Show that if $\sum_{m=1}^{\infty} \Pr\{A_m\} < \infty$, then $\Pr\{\bigcap_m \bigcup_{n\ge m} A_n\} = 0$.
The set $\{\bigcap_m \bigcup_{n\ge m} A_n\}$ is often referred to as the set of ω that are contained in infinitely many of the A_n. Without trying to be precise about what this latter statement means, explain why it is a good way to think about $\{\bigcap_m \bigcup_{n\ge m} A_n\}$. Hint: Consider an ω that is contained in some finite number k of the sets A_n and argue that there must be an integer m such that $\omega \notin A_n$ for all $n > m$.

Exercise 5.9 Let $\{X_i;\ i\ge1\}$ be the inter-renewal intervals of a renewal process and assume that $\mathsf{E}[X_i] = \infty$. Let $b > 0$ be an arbitrary number and $\breve{X}_i$ be a truncated rv defined by $\breve{X}_i = X_i$ if $X_i \le b$ and $\breve{X}_i = b$ otherwise.

(**a**) Show that for any constant $M > 0$, there is a b sufficiently large so that $\mathsf{E}[\breve{X}_i] \ge M$.

(**b**) Let $\{\breve{N}(t);\ t\ge0\}$ be the renewal counting process with inter-renewal intervals $\{\breve{X}_i;\ i \ge 1\}$ and show that for all $t > 0$, $\breve{N}(t) \ge N(t)$.

(**c**) Show that for all sample functions $N(t,\omega)$, except a set of probability 0, $N(t,\omega)/t < 2/M$ for all sufficiently large t. Note: Since M is arbitrary, this means that $\lim N(t)/t = 0$ WP1.

Exercise 5.10 The first part of the proof of the SLLN showed that if $\mathsf{E}\left[X^4\right] < \infty$, then $\mathsf{E}\left[X^2\right] < \infty$. Here we generalize this to show that for $1 \le m < n$, if $\mathsf{E}[|X^n| < \infty]$, then $\mathsf{E}[|X^m| < \infty]$.

(**a**) For any $x > 0$, show that $x^m \le 1 + x^n$.

(**b**) For any rv X, show that if $\mathsf{E}[|X^n|] < \infty$, then $\mathsf{E}[|X^m|] \le 1 + \mathsf{E}[|X^n|]$.

Exercise 5.11 Let $Y(t) = S_{N(t)+1} - t$ be the residual life at time t of a renewal process. First consider a renewal process in which the interarrival time has density $\mathsf{f}_X(x) = e^{-x};\ x \ge 0$, and next consider a renewal process with density

$$\mathsf{f}_X(x) = \frac{3}{(x+1)^4}, \qquad x \ge 0.$$

For each of the above densities, use renewal–reward theory to find:

(**a**) the time average of $Y(t)$;

(**b**) the time average of $Y^2(t)$ (i.e., $\lim_{T\to\infty} (1/T) \int_0^T Y^2(t)dt$).

For the exponential density, verify your answers by finding $\mathsf{E}[Y(t)]$ and $\mathsf{E}\left[Y^2(t)\right]$ directly.

Exercise 5.12 Consider a variation of an M/G/1 queueing system in which there is no facility to save waiting customers. Assume customers arrive according to a Poisson process of rate λ. If the server is busy, the customer departs and is lost forever; if the server is not busy, the customer enters service with a service time CDF denoted by $\mathsf{F}_Y(y)$.

Successive service times (for those customers that are served) are IID and independent of arrival times. Assume that customer number 0 arrives and enters service at time $t = 0$.

(a) Show that the sequence of times $S_1, S_2, \ldots$ at which successive successful customers enter service are the renewal times of a renewal process. Show that each inter-renewal interval $X_i = S_i - S_{i-1}$ (where $S_0 = 0$) is the sum of two independent rv s, $Y_i + U_i$ where Y_i is the ith service time; find the probability density of U_i.

(b) Assume that a reward (actually a cost in this case) of one unit is incurred for each customer turned away. Sketch the expected reward function as a function of time for the sample function of inter-renewal intervals and service intervals shown below; the expectation is to be taken over those (unshown) arrivals of customers that must be turned away.

Y_1 Y_1 Y_1

$S_0 = 0$ S_1 S_2

(c) Let $\int_0^t R(\tau)d\tau$ denote the accumulated reward (i.e., cost) from 0 to t and find the limit as $t \to \infty$ of $(1/t)\int_0^t R(\tau)d\tau$. Explain (without any attempt to be rigorous or formal) why this limit exists with probability 1.

(d) In the limit of large t, find the expected reward from time t until the next renewal. Hint: Sketch this expected reward as a function of t for a given sample of inter-renewal intervals and service intervals; then find the time average.

(e) Now assume that the arrivals are deterministic, with the first arrival at time 0 and the nth arrival at time $n - 1$. Does the sequence of times $S_1, S_2, \ldots$ at which subsequent customers start service still constitute the renewal times of a renewal process? Draw a sketch of arrivals, departures, and service time intervals. Again find $\lim_{t\to\infty}(\int_0^t R(\tau)\,d\tau)/t$.

Exercise 5.13 Let $Z(t) = t - S_{N(t)}$ be the age of a renewal process and $Y(t) = S_{N(t)+1} - t$ be the residual life. Let $\mathsf{F}_X(x)$ be the CDF of the inter-renewal interval and find the following as a function of $\mathsf{F}_X(x)$:

(a) $\Pr\{Y(t){>}x \mid Z(t){=}s\}$;

(b) $\Pr\{Y(t){>}x \mid Z(t{+}x/2){=}s\}$;

(c) $\Pr\{Y(t){>}x \mid Z(t{+}x){>}s\}$ for a Poisson process.

Exercise 5.14 Let $\mathsf{F}_Z(z)$ be the fraction of time (over the limiting interval $(0, \infty)$) that the age of a renewal process is at most z. Show that $\mathsf{F}_Z(z)$ satisfies

$$\mathsf{F}_Z(z) = \frac{1}{\overline{X}} \int_{x=0}^{z} \Pr\{X > x\}\, dx \qquad \text{WP1.}$$

Hint: Follow the argument in Example 5.4.7.

Exercise 5.15 **(a)** Let J be a stopping rule and $\mathbb{I}_{\{J \geq n\}}$ be the indicator rv of the event $\{J \geq n\}$. Show that $J = \sum_{n \geq 1} \mathbb{I}_{\{J \geq n\}}$.

(b) Show that $\mathbb{I}_{J\geq 1} \geq \mathbb{I}_{J\geq 2} \geq \ldots$, i.e., show that for each $n \geq 1$, $\mathbb{I}_{J\geq n}(\omega) \geq \mathbb{I}_{J\geq n+1}(\omega)$ for each $\omega \in \Omega$ (except perhaps for a set of probability 0).

Exercise 5.16 **(a)** Use Wald's equality to compute the expected number of trials of a Bernoulli process up to and including the kth success.

(b) Use elementary means to find the expected number of trials up to and including the first success. Use this to find the expected number of trials up to and including the kth success. Compare with (a).

Exercise 5.17 A gambler with an initial finite capital of $d > 0$ dollars starts to play a dollar slot machine. At each play, either his dollar is lost or is returned with some additional number of dollars. Let X_i be his change of capital on the ith play. Assume that $\{X_i;\ i{=}1, 2, \ldots\}$ is a set of IID rv s taking on integer values $\{-1, 0, 1, \ldots\}$. Assume that $\mathsf{E}[X_i] < 0$. The gambler plays until losing all his money (i.e., the initial d dollars plus subsequent winnings).

(a) Let J be the number of plays until the gambler loses all his money. Is the WLLN sufficient to argue that $\lim_{n\to\infty} \Pr\{J > n\} = 0$ (i.e., that J is a rv) or is the SLLN necessary?

(b) Find $\mathsf{E}[J]$. Hint: The fact that there is only one possible negative outcome is important here.

Exercise 5.18 Let $\{X_i; i \geq 1\}$ be IID binary rv s with $P_X(0) = P_X(1) = 1/2$. Let J be a positive integer-valued rv defined on the above sample space of binary sequences and let $S_J = \sum_{i=1}^{J} X_i$. Find the simplest example you can in which J is not a stopping trial for $\{X_i;\ i \geq 1\}$ and where $\mathsf{E}[X]\,\mathsf{E}[J] \neq \mathsf{E}[S_J]$. Hint: Try letting J take on only the values 1 and 2.

Exercise 5.19 Let $J = \min\{n \mid S_n{\leq}b \text{ or } S_n{\geq}a\}$, where a is a positive integer, b is a negative integer, and $S_n = X_1 + X_2 + \cdots + X_n$. Assume that $\{X_i;\ i{\geq}1\}$ is a set of zero-mean IID rv s that can take on only the set of values $\{-1, 0, +1\}$, each with positive probability.

(a) Is J a stopping rule? Why or why not? Hint: The more difficult part of this is to argue that J is a rv (i.e., non-defective); you do not need to construct a proof of this, but try to argue why it must be true.

(b) What are the possible values of S_J?

(c) Find an expression for $\mathsf{E}[S_J]$ in terms of p, a, and b, where $p = \Pr\{S_J \geq a\}$.

(d) Find an expression for $\mathsf{E}[S_J]$ from Wald's equality. Use this to solve for p.

Exercise 5.20 Show that the interchange of expectation and sum in (5.34) is valid if $\mathsf{E}[J] < \infty$. Hint: First express the sum as $\sum_{n=1}^{k-1} X_n \mathbb{I}_{J\geq n} + \sum_{n=k}^{\infty}(X_n^+ + X_n^-)\mathbb{I}_{J\geq n}$ and then consider the limit as $k \to \infty$.

Exercise 5.21 Consider a G/G/1 queue with inter-arrival sequence $\{X_i;\ i \geq 1\}$ and service time sequence $\{V_i;\ i \geq 0\}$. Assume that $\mathsf{E}[X] < \infty$ and that $\mathsf{E}[V] < \mathsf{E}[X]$.

(a) Let $Z_n = \sum_{i=1}^{n}(X_i - V_{i-1})$ and let I_n be the sum of the intervals between 0 and $S_n = \sum_{i=1}^{n} X_i$ during which the server is empty. Show that $I_n \geq Z_n$ and explain the difference between them. Show that if arrival n enters an empty system, then $I_n = Z_n$.

(b) Assume that the inter-arrival distribution is bounded in the sense that for some $b < \infty$, $\Pr\{X > b\} = 0$. Show that the duration of each idle period for the G/G/1 queue has a duration less than or equal to b.

(c) Show that the number of idle intervals for the server up to time S_n is at least Z_n/b. Show that the number of renewals up to time S_n ($N^r(S_n)$) at which an arrival sees an empty system is at least (Z_n/b).

(d) Show that the renewal counting process satisfies $\lim_{t\to\infty} N^r(t)/t = a$ WP1 for some $a > 0$. Show that the mean number of arrivals per arrival to an empty system is $1/a < \infty$.

(e) Assume everything above except to remove the assumption that $\Pr\{X > b\} = 0$ for some b. Use the truncation method on X to establish (d) for this new case.

Exercise 5.22 A company has three promising approaches to the design of a new wireless application. Unknown to the company, the first approach requires two weeks, the second, four weeks, and the third eight weeks, but only the first will be successful. The company hires one programmer after another for the design, stopping when a design is successful. Let X_i be the time taken by the ith programmer, assuming that each programmer independently chooses an approach with equal probability (i.e., $\mathsf{p}_{X_i}(x) = 1/3$ for $x = 2$, 4, and 8). Let J be the number of the first programmer who is successful and let $T = \sum_{i=1}^{J} X_i$ be the time of the first success.

(a) Use Wald's equality to find $\mathsf{E}[T]$.

(b) Compute $\mathsf{E}[\sum_{i=1}^{J} X_i \mid J{=}n]$ and show that it is not equal to $\mathsf{E}[\sum_{i=1}^{n} X_i]$.

(c) Use (b) for a second derivation of $\mathsf{E}[T]$.

(d) Find $\mathsf{E}[T]$ if the company tells successive programmers which approaches have been tried unsuccessfully earlier.

Exercise 5.23 **(a)** Consider a renewal process for which the inter-renewal intervals have the PMF $p_X(1) = p_X(2) = 1/2$. Let $m(t) = \mathsf{E}[N(t)]$ and use elementary combinatorics to show that $m(1) = 1/2$, $m(2) = 5/4$, and $m(3) = 15/8$.

(b) Use elementary means to show that $\mathsf{E}\left[S_{N(1)}\right] = 1/2$ and $\mathsf{E}\left[S_{N(1)+1}\right] = 9/4$. Verify (5.37) in this case (i.e., for $t = 1$) and show that $N(1)$ is not a stopping trial. Note also that the expected duration, $\mathsf{E}\left[S_{N(1)+1} - S_{N(1)}\right]$ is not equal to $\overline{X}$.

(c) Generalize (a) so that $\Pr\{X = 1\} = 1 - p$ and $\Pr\{X = 2\} = p$. Let $W_n = N(n) - N(n-1)$, i.e., W_n is 1 if an arrival occurs at time n and is 0 if no arrival occurs at n. Show that $\overline{W}_n$ satisfies the difference equation $\overline{W}_n = 1 - p\overline{W}_{n-1}$ for $n \geq 1$, where by convention $\overline{W}_0 = 1$. Use this to show that

$$\overline{W}_n = \frac{1 - (-p)^{n+1}}{1 + p}. \tag{5.115}$$

From this, solve for $m(n)$ for $n \geq 1$ and verify the result in (a).

Exercise 5.24 Let $\{N(t);\ t > 0\}$ be a renewal counting process generalized to allow for inter-renewal intervals $\{X_i\}$ of duration 0. Let each X_i have the PMF $\Pr\{X_i = 0\} = 1-\epsilon$; $\Pr\{X_i = 1/\epsilon\} = \epsilon$.

(a) Sketch a typical sample function of $\{N(t);\ t > 0\}$. Note that $N(0)$ can be non-zero (i.e., $N(0)$ is the number of zero interarrival times that occur before the first non-zero interarrival time).

(b) Evaluate $\mathsf{E}[N(t)]$ as a function of t.

(c) Sketch $\mathsf{E}[N(t)]/t$ as a function of t.

(d) Evaluate $\mathsf{E}\left[S_{N(t)+1}\right]$ as a function of t (do this directly, and then use Wald's equality as a check on your work).

(e) Sketch the lower bound $\mathsf{E}[N(t)]/t \ge 1/\mathsf{E}[X] - 1/t$ on the same graph with (c).

(f) Sketch $\mathsf{E}\left[S_{N(t)+1} - t\right]$ as a function of t and find the time average of this quantity.

(g) Evaluate $\mathsf{E}\left[S_{N(t)}\right]$ as a function of t; verify that $\mathsf{E}\left[S_{N(t)}\right] \ne \mathsf{E}[X]\,\mathsf{E}[N(t)]$.

Exercise 5.25 Let $\{N(t);\ t > 0\}$ be a renewal counting process and let $m(t) = \mathsf{E}[N(t)]$ be the expected number of arrivals up to and including time t. Let $\{X_i; i \ge 1\}$ be the sequence of inter-renewal intervals and assume that $\mathsf{F}_X(0) = 0$.

(a) For all $x > 0$ and $t > x$ show that $\mathsf{E}[N(t)|X_1{=}x] = \mathsf{E}[N(t-x)] + 1$.

(b) Use (a) to show that $m(t) = \mathsf{F}_X(t) + \int_0^t m(t-x)\, d\mathsf{F}_X(x)$ for $t > 0$. This equation is the renewal equation, derived differently in (5.55).

(c) Suppose that X is an exponential rv of parameter λ. Evaluate $L_m(s)$ from (5.57); verify that the inverse Laplace transform is $\lambda t;\ t \ge 0$.

Exercise 5.26 **(a)** Let the inter-renewal interval of a renewal process have a second-order Erlang density, $\mathsf{f}_X(x) = \lambda^2 x \exp(-\lambda x)$. Evaluate the Laplace transform of $m(t) = \mathsf{E}[N(t)]$.

(b) Use this to evaluate $m(t)$ for $t \ge 0$. Verify that your answer agrees with (5.59).

(c) Evaluate the slope of $m(t)$ at $t = 0$ and explain why that slope is not surprising.

(d) View the renewals here as being the even numbered arrivals in a Poisson process of rate λ. Sketch $m(t)$ for the process here and show one half the expected number of arrivals for the Poisson process on the same sketch. Explain the difference between the two.

Exercise 5.27 **(a)** Let $N(t)$ be the number of arrivals in the interval $(0, t]$ for a Poisson process of rate λ. Show that the probability that $N(t)$ is even is $[1 + \exp(-2\lambda t)]/2$. Hint: Look at the power series expansion of $\exp(-\lambda t)$ and that of $\exp(\lambda t)$, and look at the sum of the two. Compare this with $\sum_{n \text{ even}} \Pr\{N(t) = n\}$.

(b) Let $\widetilde{N}(t)$ be the number of even numbered arrivals in $(0, t]$. Show that $\widetilde{N}(t) = N(t)/2 - \mathbb{I}_{\text{odd}}(t)/2$, where $\mathbb{I}_{\text{odd}}(t)$ is a rv that is 1 if $N(t)$ is odd and 0 otherwise.

(c) Use (a) and (b) to find $\mathsf{E}[\widetilde{N}(t)]$. Note that this is $m(t)$ for a renewal process with second-order Erlang inter-renewal intervals.

Exercise 5.28 **(a)** Consider a function $r(z) \ge 0$ defined as follows for $0 \le z < \infty$. For each integer $n \ge 1$ and each integer k, $1 \le k < n$, $r(z) = 1$ for $n + k/n \le z \le n + k/n + 2^{-n}$. For all other z, $r(z) = 0$. Sketch this function and show that $r(z)$ is not directly Riemann integrable.

(b) Evaluate the Riemann integral $\int_0^\infty r(z)dz$.

(c) Suppose $r(z)$ is decreasing, i.e., that $r(z) \ge r(y)$ for all $y > z > 0$. Show that if $r(z)$ is Riemann integrable, it is also directly Riemann integrable.

(d) Suppose $f(z) \geq 0$, defined for $z \geq 0$, is decreasing and Riemann integrable and that $f(z) \geq r(z)$ for all $z \geq 0$. Show that $r(z)$ is directly Riemann integrable.

(e) Let X be a non-negative rv for which $\mathsf{E}\left[X^2\right] < \infty$. Show that $x\,\mathsf{F}^{\mathsf{c}}_X(x)$ is directly Riemann integrable. Hint: Consider $y\mathsf{F}^{\mathsf{c}}_X(y) + \int_y^\infty \mathsf{F}_X(x)\,dx$ and use Figure 1.7 (or use integration by parts) to show that this expression is decreasing in y.

Exercise 5.29 Let $Z(t), Y(t), \widetilde{X}(t)$ denote the age, residual life, and duration at time t for a renewal counting process $\{N(t);\, t > 0\}$ in which the interarrival time has a density given by $f(x)$. Find the following probability densities; assume steady state:

(a) $\mathsf{f}_{Y(t)}(y \mid Z(t{+}s/2){=}s)$ for given $s > 0$;

(b) $\mathsf{f}_{Y(t),Z(t)}(y, z)$;

(c) $\mathsf{f}_{Y(t)}(y \mid \widetilde{X}(t){=}x)$;

(d) $\mathsf{f}_{Z(t)}(z \mid Y(t{-}s/2){=}s)$ for given $s > 0$;

(e) $\mathsf{f}_{Y(t)}(y \mid Z(t{+}s/2){\geq}s)$ for given $s > 0$.

Exercise 5.30 **(a)** Find $\lim_{t\to\infty}\{\mathsf{E}[N(t)] - t/\overline{X}\}$ for a renewal counting process $\{N(t);\, t > 0\}$ with inter-renewal times $\{X_i;\, i \geq 1\}$. Hint: Use Wald's equation.

(b) Evaluate your result for the case in which X is an exponential rv (you already know what the result should be in this case).

(c) Evaluate your result for a case in which $\mathsf{E}[X] < \infty$ and $\mathsf{E}\left[X^2\right] = \infty$. Explain (very briefly) why this does not contradict the elementary renewal theorem.

Exercise 5.31 Customers arrive at a bus stop according to a Poisson process of rate λ. Independently, buses arrive according to a renewal process with the inter-renewal interval CDF $\mathsf{F}_X(x)$. At the epoch of a bus arrival, all waiting passengers enter the bus and the bus leaves immediately. Let $R(t)$ be the number of customers waiting at time t.

(a) Draw a sketch of a sample function of $R(t)$.

(b) Given that the first bus arrives at time $X_1 = x$, find the expected number of customers picked up; then find $\mathsf{E}\left[\int_0^x R(t)dt\right]$, again given the first bus arrival at $X_1 = x$.

(c) Find $\lim_{t\to\infty}\left[\int_0^t R(\tau)d\tau\right]/t$ (WP1). Assuming that F_X is a non-arithmetic distribution, find $\lim_{t\to\infty}\mathsf{E}[R(t)]$. Interpret what these quantities mean.

(d) Use (c) to find the time-average expected wait per customer.

(e) Find the fraction of time that there are no customers at the bus stop. Hint: This part is independent of (a), (b), and (c); check your answer for $\mathsf{E}[X] \ll 1/\lambda$.

Exercise 5.32 Consider the same setup as in Exercise 5.31 except that now customers arrive according to a non-arithmetic renewal process independent of the bus arrival process. Let $1/\lambda$ be the expected inter-renewal interval for the customer renewal process. Assume that both renewal processes are in steady state (i.e., either we look only at $t \gg 0$, or we assume that they are equilibrium processes). Given that the nth customer arrives at time t, find the expected wait for customer n. Find the expected wait for customer n without conditioning on the arrival time.

Exercise 5.33 Let $\{N_1(t);\, t > 0\}$ be a Poisson counting process of rate λ. Assume that the arrivals from this process are switched on and off by arrivals from a non-arithmetic renewal counting process $\{N_2(t);\, t > 0\}$ (see figure below). The two processes are independent.

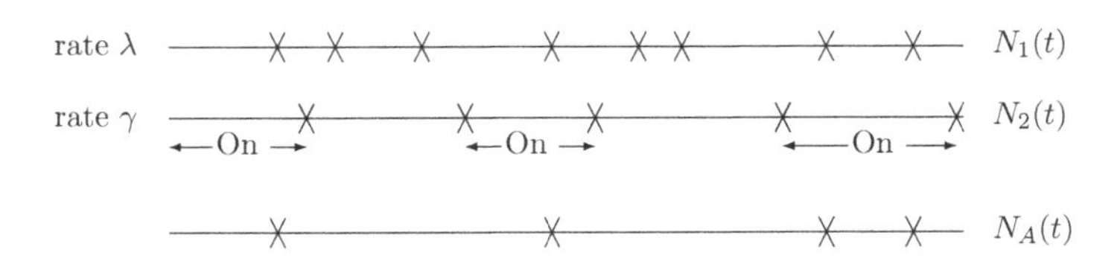

Let $\{N_A(t); t \geq 0\}$ be the switched process; i.e., $N_A(t)$ includes arrivals from $\{N_1(t); t > 0\}$ while $N_2(t)$ is even and excludes arrivals from $\{N_1(t); t > 0\}$ while $N_2(t)$ is odd.

(a) Is $N_A(t)$ a renewal counting process? Explain your answer and if you are not sure, look at several examples for $N_2(t)$.

(b) Find $\lim_{t\to\infty} [N_A(t)]/t$ and explain why the limit exists with probability 1. Hint: Use symmetry – i.e., look at $N_1(t) - N_A(t)$. To show why the limit exists, use the renewal–reward theorem. What is the appropriate renewal process to use here?

(c) Now suppose that $\{N_1(t); t>0\}$ is a non-arithmetic renewal counting process but not a Poisson process and let the expected inter-renewal interval be $1/\lambda$. For any given δ, find $\lim_{t\to\infty} \mathsf{E}[N_A(t+\delta) - N_A(t)]$ and explain your reasoning. Why does your argument in (b) fail to demonstrate a time average for this case?

Exercise 5.34 An M/G/1 queue has arrivals at rate λ and a service time distribution given by $\mathsf{F}_Y(y)$. Assume that $\lambda < 1/\mathsf{E}[Y]$. Epochs at which the system becomes empty define a renewal process. Let $\mathsf{F}_Z(z)$ be the CDF of the inter-renewal intervals and let $\mathsf{E}[Z]$ be the mean inter-renewal interval.

(a) Find the fraction of time that the system is empty as a function of λ and $\mathsf{E}[Z]$. State carefully what you mean by such a fraction.

(b) Apply Little's theorem, not to the system as a whole, but to the number of customers in the server (i.e., 0 or 1). Use this to find the fraction of time that the server is busy.

(c) Combine your results in (a) and (b) to find $\mathsf{E}[Z]$ in terms of λ and $\mathsf{E}[Y]$; give the fraction of time that the system is idle in terms of λ and $\mathsf{E}[Y]$.

(d) Find the expected duration of a busy period.

Exercise 5.35 Consider a sequence $X_1, X_2, \ldots$ of IID binary rv s with $\Pr\{X_n = 1\} = p_1$ and $\Pr\{X_n = 0\} = p_0 = 1 - p_1$. A *renewal* is said to occur at time $n \geq 2$ if $X_{n-1} = 0$ and $X_n = 1$.

(a) Show that $\{N(n); n > 0\}$ is a renewal counting process where $N(n)$ is the number of renewals up to and including time n.

(b) What is the probability that a renewal occurs at time n, $n \geq 2$?

(c) Find the expected inter-renewal interval; use Blackwell's theorem.

(d) Now change the definition of renewal so that a renewal occurs at time n if $X_{n-1} = 1$ and $X_n = 1$. Show that $\{N^*(n); n \geq 0\}$ is a delayed renewal counting process where N_n^* is the number of renewals up to and including n for this new definition of renewal.

(e) Find $\mathsf{E}[Y_i]$ for $i \geq 2$ for the case in (d).

(f) Find $\mathsf{E}[Y_1]$ for the case in (d). Hint: Show that $\mathsf{E}[Y_1|X_1 = 1] = 1 + \mathsf{E}[Y_2]$ and $\mathsf{E}[Y_1|X_1 = 0] = 1 + \mathsf{E}[Y_1]$.

(g) Looking at your results above for the strings (0,1) and (1,1), show that for an arbitrary string $\boldsymbol{a} = (a_1, \ldots, a_k)$, the arrival process of successive occurrences of the string is a renewal process if no proper suffix of $\boldsymbol{a}$ is a prefix of $\boldsymbol{a}$. Otherwise it is a delayed renewal process.

(h) Suppose a string $\boldsymbol{a} = (a_1, \ldots, a_k)$ of length k has no proper suffixes equal to a prefix. Show that the time to the first renewal satisfies

$$\mathsf{E}[Y_1] = \frac{1}{\prod_{\ell=1}^{k} p_{a_\ell}}.$$

(i) Suppose the string $\boldsymbol{a} = (a_1, \ldots, a_k)$ has at least one proper suffix equal to a prefix, and suppose i is the length of the longest such suffix. Show that the expected time until the first occurrence of $\boldsymbol{a}$ is given by

$$\mathsf{E}[Y_1] = \frac{1}{\prod_{\ell=1}^{k} p_{a_\ell}} + \mathsf{E}[U_i],$$

where $\mathsf{E}[U_i]$ is the expected time until the first occurrence of the string $(a_1, \ldots, a_i)$.

(j) Show that the expected time until the first occurrence of $\boldsymbol{a} = (a_1, \ldots, a_k)$ is given by

$$\mathsf{E}[Y_1] = \sum_{i=1}^{k} \frac{\mathbb{I}_i}{\prod_{\ell=1}^{i} p_{a_\ell}},$$

where, for $1 \le i \le k$, $\mathbb{I}_i$ is 1 if the prefix of $\boldsymbol{a}$ of length i is equal to the suffix of length i. Hint: Use (h) recursively. Also show that if $\boldsymbol{a}$ has a suffix of length i equal to the prefix of length i and also a suffix of length j equal to a prefix of length j where $j < i$, then the suffix of $(a_1, \ldots, a_i)$ of length j is also equal to the prefix of both $\boldsymbol{a}$ and $(a_1, \ldots, a_i)$ of length j.

(k) Use (i) to find, first, the expected time until the first occurrence of (1,1,1,1,1,1,0) and, second, that of (1,1,1,1,1,1). Use (4.31) to check the relationship between these answers.

Exercise 5.36 A large system is controlled by n identical computers. Each computer independently alternates between an operational state and a repair state. The duration of the operational state, from completion of one repair until the next need for repair, is a rv X with finite expected duration $\mathsf{E}[X]$. The time required to repair a computer is an exponentially distributed rv with density $\lambda e^{-\lambda t}$. All operating durations and repair durations are independent. Assume that all computers are in the repair state at time 0.

(a) For a single computer, say the ith, do the epochs at which the computer enters the repair state form a renewal process? If so, find the expected inter-renewal interval.

(b) Do the epochs at which it enters the operational state form a renewal process?

(c) Find the fraction of time over which the ith computer is operational and explain what you mean by fraction of time.

(d) Let $Q_i(t)$ be the probability that the ith computer is operational at time t and find $\lim_{t\to\infty} Q_i(t)$.

(e) The system is in failure mode at a given time if all computers are in the repair state at that time. Do the epochs at which system failure modes begin form a renewal process?

(f) Let $\Pr\{t\}$ be the probability that the the system is in failure mode at time t. Find $\lim_{t\to\infty} \Pr\{t\}$. Hint: Look at (d).

(g) For δ small, find the probability that the system enters failure mode in the interval $(t, t+\delta]$ in the limit as $t \to \infty$.

(h) Find the expected time between successive entries into failure mode.

(i) Next assume that the repair time of each computer has an arbitrary density rather than exponential, but has a mean repair time of $1/\lambda$. Do the epochs at which system failure modes begin form a renewal process?

(j) Repeat (f) for the assumption in (i).

Exercise 5.37 Let $\{N_1(t);\ t{>}0\}$ and $\{N_2(t);\ t{>}0\}$ be independent renewal counting processes. Assume that each has the same CDF $\mathsf{F}(x)$ for interarrival intervals and assume that a density $f(x)$ exists for the interarrival intervals.

(a) Is the counting process $\{N_1(t){+}N_2(t);\ t > 0\}$ a renewal counting process? Explain.

(b) Let $Y(t)$ be the interval from t until the first arrival (from either process) after t. Find an expression for the CDF of $Y(t)$ in the limit $t \to \infty$ (you may assume that time averages and ensemble averages are the same).

(c) Assume that a reward R of rate 1 unit per second starts to be earned whenever an arrival from process 1 occurs and ceases to be earned whenever an arrival from process 2 occurs. Assume that $\lim_{t\to\infty}(1/t)\int_0^t R(\tau)\,d\tau$ exists with probability 1 and find its numerical value.

(d) Let $Z(t)$ be the interval from t until the first time after t that $R(t)$ (as in (c)) changes value. Find an expression for $\mathsf{E}[Z(t)]$ in the limit $t \to \infty$. Hint: Make sure you understand why $Z(t)$ is not the same as $Y(t)$ in (b). You might find it easiest to first find the expectation of $Z(t)$ conditional on both the duration of the $\{N_1(t);\ t > 0\}$ interarrival interval containing t and the duration of the $\{N_2(t); t \geq 0\}$ interarrival interval containing t; draw pictures!

Exercise 5.38 This problem provides another way of treating ensemble averages for renewal–reward problems. Assume for notational simplicity that X is a continuous rv.

(a) Show that $\Pr\{\text{one or more arrivals in } (\tau, \tau+\delta)\} = m(\tau+\delta) - m(\tau) - o(\delta)$, where $o(\delta) \geq 0$ and $\lim_{\delta\to 0} o(\delta)/\delta = 0$.

(b) Show that $\Pr\{Z(t) \in [z, z+\delta),\ \widetilde{X}(t) \in (x, x+\delta)\}$ is equal to $[m(t-z) - m(t-z-\delta) - o(\delta)][\mathsf{F}_X(x+\delta) - \mathsf{F}_X(x)]$ for $x \geq z+\delta$.

(c) Assuming that $m'(\tau) = dm(\tau)/d\tau$ exists for all τ, show that the joint density of $Z(t), \widetilde{X}(t)$ is $\mathsf{f}_{Z(t),\widetilde{X}(t)}(z,x) = m'(t-z)\mathsf{f}_X(x)$ for $x > z$.

(d) Show that $\mathsf{E}[R(t)] = \int_{z=0}^{t}\int_{x=z}^{\infty} \mathsf{R}\,(z,x)\mathsf{f}_X(x)\,dx\, m'(t-z)\,dz$.

Exercise 5.39 In this problem, we show how to calculate the residual life distribution $Y(t)$ as a transient in t. Let $\mu(t) = dm(t)/dt$, where $m(t) = \mathsf{E}[N(t)]$, and let the interarrival distribution have the density $\mathsf{f}_X(x)$. Let $Y(t)$ have the density $\mathsf{f}_{Y(t)}(y)$.

(a) Show that these densities are related by the integral equation

$$\mu(t+y) = \mathsf{f}_{Y(t)}(y) + \int_{u=0}^{y} \mu(t+u)\mathsf{f}_X(y-u)\,du.$$

(b) Let $L_{\mu,t}(r) = \int_{y\geq 0} \mu(t+y)e^{-ry}dy$ and let $L_{Y(t)}(r)$ and $L_X(r)$ be the Laplace transforms of $\mathsf{f}_{Y(t)}(y)$ and $\mathsf{f}_X(x)$ respectively. Find $L_{Y(t)}(r)$ as a function of $L_{\mu,t}$ and L_X.

(c) Consider the inter-renewal density $\mathsf{f}_X(x) = (1/2)e^{-x} + e^{-2x}$ for $x \geq 0$ (as in Example 5.6.1). Find $L_{\mu,t}(r)$ and $L_{Y(t)}(r)$ for this example.

(d) Find $\mathsf{f}_{Y(t)}(y)$. Show that your answer reduces to that of (5.30) in the limit as $t \to \infty$.

(e) Explain how to go about finding $\mathsf{f}_{Y(t)}(y)$ in general, assuming that f_X has a rational Laplace transform.

Exercise 5.40 This problem is designed to give you an alternative way of looking at ensemble averages for renewal–reward problems. First we find an exact expression for $\Pr\{S_{N(t)} > s\}$. We find this for arbitrary s and t, $0 < s < t$.

(a) By breaking the event $\{S_{N(t)} > s\}$ into subevents $\{S_{N(t)} > s, N(t) = n\}$, explain each of the following steps:

$$\begin{aligned}
\Pr\{S_{N(t)} > s\} &= \sum_{n=1}^{\infty} \Pr\{t \geq S_n > s, S_{n+1} > t\} \\
&= \sum_{n=1}^{\infty} \int_{y=s}^{t} \Pr\{S_{n+1}{>}t \mid S_n{=}y\}\, d\mathsf{F}_{S_n}(y) \\
&= \int_{y=s}^{t} \mathsf{F}^{\mathsf{c}}_X(t{-}y)\, d\sum_{n=1}^{\infty} \mathsf{F}_{S_n}(y) \\
&= \int_{y=s}^{t} \mathsf{F}^{\mathsf{c}}_X(t{-}y)\, dm(y) \quad \text{where } m(y) = \mathsf{E}\left[N(y)\right].
\end{aligned}$$

(b) Show that for $0 < s < t < u$,

$$\Pr\{S_{N(t)} > s, S_{N(t)+1} > u\} = \int_{y=s}^{t} \mathsf{F}^{\mathsf{c}}_X(u-y)\, dm(y).$$

(c) Draw a two-dimensional sketch, with age and duration as the axes, and show the region of (age, duration) values corresponding to the event $\{S_N(t) > s,\ S_{N(t)+1} > u\}$.

(d) Assume that for large t, $dm(y)$ can be approximated (according to Blackwell) as $(1/\overline{X})dy$, where $\overline{X} = \mathsf{E}[X]$. Assuming that X also has a density, use the result in (b) and (c) to find the joint density of age and duration.

Exercise 5.41 Show that for a G/G/1 queue, the time-average wait in the system is the same as $\lim_{n\to\infty} \mathsf{E}[W_n]$. Hint: Consider an integer renewal counting process $\{M(n); n \geq 0\}$, where $M(n)$ is the number of renewals in the G/G/1 process of Section 5.5.3 that have occurred by the nth arrival. Show that this renewal process has a span of 1. Then consider $\{W_n; n \geq 1\}$ as a reward within this renewal process.

Exercise 5.42 If one extends the definition of renewal processes to include inter-renewal intervals of duration 0, with $\Pr\{X{=}0\} = \alpha$, show that the expected number of simultaneous renewals at a renewal epoch is $1/(1-\alpha)$, and that, for a non-arithmetic process, the probability of one or more renewals in the interval $(t, t+\delta]$ tends to $(1-\alpha)\delta/\mathsf{E}[X] + o(\delta)$ as $t \to \infty$.

Exercise 5.43 The purpose of this exercise is to show why the interchange of expectation and sum in the proof of Wald's equality is justified when $\mathsf{E}[J] < \infty$ but not otherwise. Let $X_1, X_2, \ldots$ be a sequence of IID rv s, each with the CDF F_X. Assume that $\mathsf{E}[|X|] < \infty$.

(a) Show that $S_n = X_1 + \cdots + X_n$ is a rv for each integer $n > 0$. Note: S_n is obviously a mapping from the sample space to the real numbers, but you must show that it is finite WP1. Hint: Recall the additivity axiom for the real numbers.

(b) Let J be a stopping trial for $X_1, X_2, \ldots$. Show that $S_J = X_1 + \cdots + X_J$ is a rv. Hint: Represent $\Pr\{S_J \leq A\}$ as $\sum_{n=1}^{\infty} \Pr\{J = n, S_n \leq A\}$.

(c) For the stopping trial J above, let $J^{(k)} = \min(J, k)$ be the stopping trial J truncated to integer k. Explain why the interchange of sum and expectation in the proof of Wald's equality is justified in this case, so $\mathsf{E}\left[S_{J^{(k)}}\right] = \overline{X}\mathsf{E}\left[J^{(k)}\right]$.

(d) In (d), (e), and (f), assume, in addition to the assumptions above, that $\mathsf{F}_X(0) = 0$, i.e., that the X_i are positive rv s. Show that $\lim_{k\to\infty} \mathsf{E}\left[S_{J^{(k)}}\right] < \infty$ if $\mathsf{E}[J] < \infty$ and $\lim_{k\to\infty} \mathsf{E}\left[S_{J^{(k)}}\right] = \infty$ if $\mathsf{E}[J] = \infty$.

(e) Show that

$$\Pr\left\{S_{J^{(k)}} > x\right\} \leq \Pr\{S_J > x\} \qquad \text{for all } k, x.$$

(f) Show that $\mathsf{E}[S_J] = \overline{X}\mathsf{E}[J]$ if $\mathsf{E}[J] < \infty$ and $\mathsf{E}[S_J] = \infty$ if $\mathsf{E}[J] = \infty$.

(g) Now assume that X has both negative and positive values with non-zero probability and let $X^+ = \max(0, X)$ and $X^- = \min(X, 0)$. Express S_J as $S_J^+ + S_J^-$, where $S_J^+ = \sum_{i=1}^{J} X_i^+$ and $S_J^- = \sum_{i=1}^{J} X_i^-$. Show that $\mathsf{E}[S_J] = \overline{X}\mathsf{E}[J]$ if $\mathsf{E}[J] < \infty$ and that $\mathsf{E}\left[S_j\right]$ is undefined otherwise.

Exercise 5.44 This is a very simple exercise designed to clarify confusion about the roles of past, present, and future in stopping rules. Let $\{X_n;\, n \geq 1\}$ be a sequence of IID binary rv s, each with the PMF $\mathsf{p}_X(1) = 1/2$, $\mathsf{p}_X(0) = 1/2$. Let J be a positive integer-valued rv that takes on the sample value n of the first trial for which $X_n = 1$. That is, for each $n \geq 1$,

$$\{J = n\} = \{X_1{=}0,\ X_2{=}0, \ldots, X_{n-1}{=}0,\ X_n{=}1\}.$$

(a) Use the definition of stopping trial, Definition 5.5.1 in the text, to show that J is a stopping trial for $\{X_n;\, n \geq 1\}$.

(b) Show that for any given n, the rv s X_n and $\mathbb{I}_{J=n}$ are *statistically dependent.*

(c) Show that for every $m > n$, X_n and $\mathbb{I}_{J=m}$ are *statistically dependent.*

(d) Show that for every $m < n$, X_n and $\mathbb{I}_{J=m}$ are *statistically independent.*

(e) Show that X_n and $\mathbb{I}_{J\geq n}$ are *statistically independent*. Give the simplest characterization you can of the event $\{J \geq n\}$.

(f) Show that X_n and $\mathbb{I}_{J>n}$ are *statistically dependent.*

Note: The results here are characteristic of most sequences of IID rv s. For most people, this requires some realignment of intuition, since $\{J \geq n\}$ is the union of $\{J = m\}$ for all $m \geq n$, and all of these events are highly dependent on X_n. The right way to think of this is that $\{J \geq n\}$ is the complement of $\{J < n\}$, which is determined by $X_1, \ldots, X_{n-1}$. Thus $\{J \geq n\}$ is also determined by $X_1, \ldots, X_{n-1}$ and is thus independent of X_n. The moral of the story is that thinking of stopping rules as rv s independent of the future is very tricky, even in totally obvious cases such as this.

Exercise 5.45 Assume a friend has developed an excellent program for finding the steady-state probabilities for finite-state Markov chains. More precisely, given the transition matrix [P], the program returns $\lim_{n\to\infty} P_{ii}^n$ for each i. Assume all chains are aperiodic.

(a) You want to find the expected time to first reach a given state k starting from a different state m for a Markov chain with transition matrix $[P]$. You modify the matrix to $[P']$ where $P'_{km} = 1$, $P'_{kj} = 0$ for $j \neq m$, and $P'_{ij} = P_{ij}$ otherwise. How do you find the desired first-passage time from the program output given $[P']$ as an input? Hint: The times at which a Markov chain enters any given state can be considered as renewals in a (perhaps delayed) renewal process.

(b) Using the same $[P']$ as the program input, how can you find the expected number of returns to state m before the first passage to state k?

(c) Suppose, for the same Markov chain $[P]$ and the same starting state m, you want to find the probability of reaching some given state n before the first passage to k. Modify $[P]$ to some $[P'']$ so that the above program with P'' as an input allows you to easily find the desired probability.

(d) Let $\Pr\{X(0) = i\} = Q_i$, $1 \leq i \leq \mathsf{M}$ be an arbitrary set of initial probabilities for the same Markov chain $[P]$ as above. Show how to modify $[P]$ to some $[P''']$ for which the steady-state probabilities allow you to easily find the expected time of the first passage to state k.

Exercise 5.46 Consider a ferry that carries cars across a river. The ferry holds an integer number k of cars and departs the dock when full. At that time, a new ferry immediately appears and begins loading newly arriving cars ad infinitum. The ferry business has been good, but customers complain about the long wait for the ferry to become full.

(a) Assume that cars arrive according to a renewal process. The IID interarrival times have mean $\overline{X}$, variance σ^2, and moment generating function $\mathsf{g}_X(r)$. Does the sequence of departure times of the ferries form a renewal process? Explain carefully.

(b) Find the expected time that a customer waits, starting from its arrival at the ferry terminal and ending at the departure of its ferry. Note 1: Part of the problem here is to give a reasonable definition of the expected customer waiting time. Note 2: It might be useful to consider $k = 1$ and $k = 2$ first.

(c) Is there a 'slow truck' phenomenon (a dependence on $\mathsf{E}\left[X^2\right]$) here? Give an intuitive explanation. Hint: Look at $k = 1$ and $k = 2$ again.

(d) In an effort to decrease waiting, the ferry managers institute a policy where no customer ever has to wait more than one hour. Thus, the first customer to arrive after a ferry departure waits for either one hour or the time at which the ferry is full, whichever comes first, and then the ferry leaves and a new ferry starts to accumulate new customers. Does the sequence of ferry departures form a renewal process under this new system?

Does the sequence of times at which each successive empty ferry is entered by its first customer form a renewal process? You can assume here that $t = 0$ is the time of the first arrival to the first ferry. Explain carefully.

(e) Give an expression for the expected waiting time of the first new customer to enter an empty ferry under this new strategy.

6 Countable-state Markov chains

6.1 Introductory examples

Markov chains with a countably-infinite state space (more briefly, *countable-state Markov chains*) exhibit some types of behavior not possible for chains with a finite state space. With the exception of the first example to follow and the section on branching processes, we label the states by the non-negative integers. This is convenient for modeling parameters such as customer arrivals in queueing systems, and provides a consistent notation for the general case.

The following two examples give some insight into the new issues posed by countable state spaces.

Example 6.1.1 Consider the familiar Bernoulli process $\{S_n = X_1 + \cdots + X_n;\ n \geq 1\}$, where $\{X_n;\ n \geq 1\}$ is an independent identically distributed (IID) binary sequence with $\mathsf{p}_X(1) = p$ and $\mathsf{p}_X(-1) = (1-p) = q$. The sequence $\{S_n;\ n \geq 1\}$ is a sequence of integer random variables (rv s), where $S_n = S_{n-1} + 1$ with probability p and $S_n = S_{n-1} - 1$ with probability q. This sequence can be modeled by the Markov chain in Figure 6.1.

Using the notation of Markov chains, P_{0j}^n is the probability of being in state j at the end of the nth transition, conditional on starting in state 0. The final state j is the number of positive transitions k less the number of negative transitions $n-k$, i.e., $j = 2k - n$. Thus, using the binomial formula,

$$P_{0j}^n = \binom{n}{k} p^k q^{n-k}, \qquad \text{where } k = \frac{j+n}{2}, \quad j+n \text{ even}. \tag{6.1}$$

All states in this Markov chain communicate with all other states, and are thus in the same class. The formula makes it clear that this class, i.e., the entire set of states in the Markov chain, is periodic with period 2. For n even, the state is even and for n odd, the state is odd.

What is more important than the periodicity, however, is what happens to the state probabilities for large n. As we saw in the Gaussian approximation to the binomial probability mass function (PMF) in (1.83),

$$P_{0j}^n \sim \frac{1}{\sqrt{2\pi npq}} \exp\left[\frac{-(k-np)^2}{2pqn}\right], \qquad \text{where } k = \frac{j+n}{2}, \quad j+n \text{ even}. \tag{6.2}$$

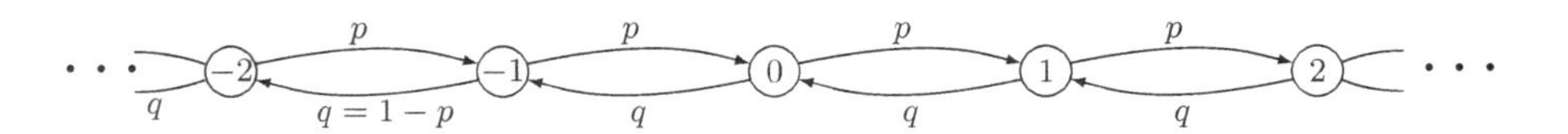

Figure 6.1 A Markov chain with a countable-state space modeling a Bernoulli process. If $p > 1/2$, then as time n increases, the state X_n becomes large with high probability, i.e., $\lim_{n\to\infty} \Pr\{X_n \geq j\} = 1$ for each integer j. Similarly, for $p < 1/2$, the state becomes highly negative.

In other words, P_{0j}^n, as a function of j, looks like a quantized form of the Gaussian density for large n. The significant terms of that distribution are close to $k = np$, i.e., to $j = n(2p - 1)$. For $p > 1/2$, the state increases with increasing n. Its distribution is centered at $n(2p - 1)$, but the distribution spreads out with $\sqrt{n}$. For $p < 1/2$, the state similarly decreases and spreads out. The most interesting case is $p = 1/2$, where the distribution remains centered at 0, but due to the spreading, the PMF approaches 0 as $1/\sqrt{n}$ for all j.

For this example, then, the probability of each state approaches 0 as $n \to \infty$, and this holds for all choices of p, $0 < p < 1$. If we attempt to define a steady-state probability to be 0 for each state, then these probabilities do not sum to 1, so they cannot be viewed as a steady-state distribution. Thus, for countable-state Markov chains, the notions of recurrence and steady-state probabilities will have to be modified from that with finite-state Markov chains. The same type of situation occurs whenever $\{S_n; n \geq 1\}$ is a sequence of sums of arbitrary IID integer-valued rv s.

Most countable-state Markov chains that are useful in applications are quite different from Example 6.1.1, and instead are quite similar to finite-state Markov chains. The following example bears a close resemblance to Example 6.1.1, but at the same time is a countable-state Markov chain that will keep reappearing in a large number of contexts. It is a special case of a birth–death process, which we study in Section 6.4.

Example 6.1.2 Figure 6.2 is similar to Figure 6.1 except that the negative states have been eliminated. A sequence of IID binary rv s $\{X_n; n \geq 1\}$, with $\mathsf{p}_X(1) = p$ and $\mathsf{p}_X(-1) = q = 1 - p$, controls the state transitions. Now, however, $S_n = \max(0, S_{n-1} + X_n)$, so that S_n is a non-negative rv. All states again communicate, and because of the self transition at state 0, the chain is aperiodic.

For $p > 1/2$, transitions to the right occur with higher frequency than transitions to the left. Thus, reasoning heuristically, we expect the state S_n at time n to drift to the right

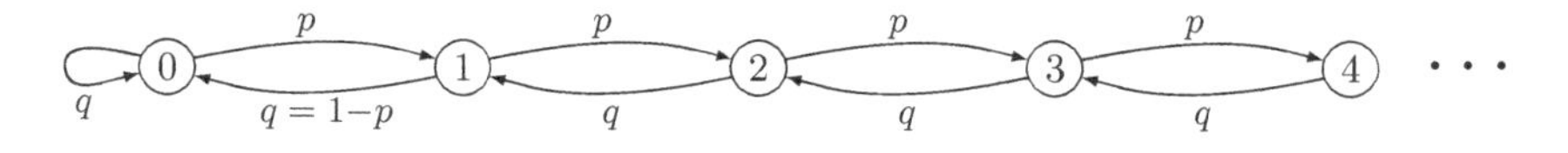

Figure 6.2 A Markov chain with a countable-state space. If $p > 1/2$, then as time n increases, the state X_n becomes large with high probability, i.e., $\lim_{n\to\infty} \Pr\{X_n \geq j\} = 1$ for each integer j.

with increasing n. Given $S_0 = 0$, the probability P_{0j}^n of being in state j at time n, should then tend to zero for any fixed j with increasing n. As in Example 6.1.1, we see that a steady state does not exist. In more poetic terms, the state wanders off into the wild blue yonder.

One way to understand this chain better is to look at what happens if the chain is truncated The truncation of Figure 6.2 to k states (i.e., $0, 1, \ldots, k-1$) is analyzed in Exercise 4.9. The solution there defines $\rho = p/q$ and shows that if $\rho \neq 1$, then $\pi_i = (1-\rho)\rho^i/(1-\rho^k)$ for each i, $0 \leq i < k$. For $\rho = 1$, $\pi_i = 1/k$ for each i. For $\rho < 1$, the limiting behavior as $k \to \infty$ is $\pi_i = (1-\rho)\rho^i$. Thus for $\rho < 1$ ($p < 1/2$), the steady-state probabilities for the truncated Markov chain approach a limit which we later interpret as the steady-state probabilities for the untruncated chain. For $\rho > 1$ ($p > 1/2$), on the other hand, the steady-state probabilities for the truncated case are geometrically decreasing from the right, and the states with significant probability keep moving to the right as k increases. Although the probability of each fixed state j approaches 0 as k increases, the truncated chain never resembles the untruncated chain.

Perhaps the most interesting case is that where $p = 1/2$. The nth-order transition probabilities, P_{0j}^n can be calculated exactly for this case (see Exercise 6.3) and are very similar to those of Example 6.1.1. In particular,

$$P_{0j}^n = \begin{cases} \binom{n}{(j+n)/2} 2^{-n} & \text{for } j \geq 0, (j+n) \text{ even}, \\ \binom{n}{(j+n+1)/2} 2^{-n} & \text{for } j \geq 0, (j+n) \text{ odd}, \end{cases} \tag{6.3}$$

$$\sim \sqrt{\frac{2}{\pi n}} \exp\left[\frac{-j^2}{2n}\right] \qquad \text{for } j \geq 0. \tag{6.4}$$

We see that P_{0j}^n for large n is approximated by the positive side of a quantized Gaussian distribution. It looks like the positive side of the PMF of (6.1) except that it is no longer periodic. For large n, P_{0j}^n is concentrated in a region of width $\sqrt{n}$ around $j = 0$, and the PMF goes to 0 as $1/\sqrt{n}$ for each j as $n \to \infty$.

Fortunately, the strange behavior of Figure 6.2 when $p \geq q$ is not typical of the Markov chains of interest for most applications. For typical countable-state Markov chains, a steady state does exist, and the steady-state probabilities of all but a finite number of states (the number depending on the chain and the application) can almost be ignored for numerical calculations.

6.2 First-passage times and recurrent states

The matrix approach used to analyze finite-state Markov chains does not generalize easily to the countable-state case. Fortunately, a combination of first-passage-time analysis and renewal theory is ideally suited for this purpose, especially for analyzing the long-term behavior of countable-state Markov chains. This section develops some of the properties of first-passage times, both from a state back to itself and from one state to another. This is used to define recurrence and transcience and also to develop the basic properties of recurrent classes of states.

In Section 6.3, we use renewal theory to see that there are two kinds of recurrent states – positive recurrent and null recurrent. For example, we will see that the states of the Markov chain in Figure 6.2 are positive recurrent for $p < 1/2$, null recurrent for $p = 1/2$, and transient for $p > 1/2$. We also use renewal theory to develop the needed general theorems about countable-state Markov chains.

The present section will not discuss positive recurrence and null recurrence, and some of the results will be rederived later using renewal theory. The approach here is valuable in that it is less abstract, and thus in some ways more insightful, than the renewal theory approach.

Definition 6.2.1 *The **first-passage-time probability**, $f_{ij}(n)$, from state i to j of a Markov chain, is the probability, conditional on $X_0 = i$, that n is the smallest $m \geq 1$ for which $X_m = j$. That is, for $n = 1$, $f_{ij}(1) = P_{ij}$. For $n \geq 2$,*

$$f_{ij}(n) = \Pr\{X_n{=}j, X_{n-1}{\neq}j, X_{n-2}{\neq}j, \ldots, X_1{\neq}j | X_0{=}i\}. \tag{6.5}$$

The distinction between $f_{ij}(n)$ and $P_{ij}^n = \Pr\{X_n = j | X_0 = i\}$ is that $f_{ij}(n)$ is the probability that the *first* entry to j (after time 0) occurs at time n, whereas P_{ij}^n is the probability that *any* entry to j occurs at time n, both conditional on starting in state i at time 0. The definition also applies for $i = j$; $f_{jj}(n)$ is thus the probability, given $X_0 = j$, that the first occurrence of state j after time 0 occurs at time n.

Since the transition probabilities are independent of time, $f_{kj}(n-1)$ is also the probability, given $X_1 = k$, that the first subsequent occurrence of state j occurs at time n. Thus we can calculate $f_{ij}(n)$ from the iterative relations

$$f_{ij}(n) = \sum_{k \neq j} P_{ik} f_{kj}(n-1), \quad n > 1, \qquad f_{ij}(1) = P_{ij}. \tag{6.6}$$

Note that this sum excludes $k = j$, since $P_{ij} f_{jj}(n-1)$ is the probability that state j occurs *first* at epoch 1 and *second* at epoch n. Recall from the Chapman–Kolmogorov equations that $P_{ij}^n = \sum_k P_{ik} P_{kj}^{n-1}$. Thus the only difference between the iterative expressions for $f_{ij}(n)$ and P_{ij}^n is the exclusion of $k = j$ in the expression for $f_{ij}(n)$.

With the iterative approach of (6.6), the first-passage-time probabilities $f_{ij}(n)$ for a given n must be calculated for all i before proceeding to calculate them for the next larger value of n. This also gives us $f_{jj}(n)$, although $f_{jj}(n)$ is not used in the iteration.

Let $\mathsf{F}_{ij}(n)$, for $n \geq 1$, be the probability, given $X_0 = i$, that state j occurs at some time between 1 and n inclusive. Thus,

$$\mathsf{F}_{ij}(n) = \sum_{m=1}^{n} f_{ij}(m). \tag{6.7}$$

For each i, j, $\mathsf{F}_{ij}(n)$ is non-decreasing in n and (since it is a probability) is upper bounded by 1. Thus $\mathsf{F}_{ij}(\infty)$ $\big($i.e., $\lim_{n\to\infty} \mathsf{F}_{ij}(n)\big)$ must exist, and is the probability, given $X_0 = i$, that state j will ever occur. If $\mathsf{F}_{ij}(\infty) = 1$, then, given $X_0 = i$, state j occurs eventually with probability 1. In this case, we can define a rv T_{ij}, conditional on $X_0 = i$, as the *first-passage time* from i to j. Then $f_{ij}(n)$ is the PMF of T_{ij} and $\mathsf{F}_{ij}(n)$ is the cumulative distribution function (CDF) of T_{ij}. If $\mathsf{F}_{ij}(\infty) < 1$, then T_{ij} is a defective rv, since, with some positive probability, there is no first passage to j.

The first-passage time T_{jj} from a state j back to itself is of particular importance. It has the PMF $f_{jj}(n)$ and the CDF $\mathsf{F}_{jj}(n)$. It is a rv (as opposed to a defective rv) if $\mathsf{F}_{jj}(\infty) = 1$, i.e., if the state eventually returns to state j with probability 1(WP1), given that it starts in state j. This suggests the following definition of recurrence.

Definition 6.2.2 *A state j in a countable-state Markov chain is* ***recurrent*** *if* $\mathsf{F}_{jj}(\infty) = 1$. *It is* transient *if* $\mathsf{F}_{jj}(\infty) < 1$.

Thus each state j in a countable-state Markov chain is either recurrent or transient, and is recurrent if and only if an eventual return to j (conditional on $X_0 = j$) occurs WP1. Equivalently, j is recurrent if and only if T_{jj}, the time of first return to j, is a rv. As shown shortly, this definition is consistent with the one in Chapter 4 for the special case of finite-state chains. For a countably-infinite state space, however, the earlier definition is not adequate. An example is provided by the case $p > 1/2$ in Figure 6.2. Here i and j communicate for all states i and j, but it is intuitively obvious (and shown in Exercise 6.2, and further explained in Section 6.4) that each state is transient.

If the initial state X_0 of a Markov chain is a recurrent state j, then T_{jj} is the integer time of the first recurrence of state j. At that recurrence, the Markov chain is in the same state j as it started in, and the discrete interval from T_{jj} to the next occurrence of state j, say $T_{jj,2}$, has the same distribution as T_{jj} and is clearly independent of T_{jj}. Similarly, the sequence of successive recurrence intervals, $T_{jj}, T_{jj,2}, T_{jj,3}, \ldots$ is a sequence of IID rv s. This sequence of recurrence intervals[1] is then the sequence of inter-renewal intervals of a renewal process, where each renewal interval has the distribution of T_{jj}. These inter-renewal intervals have the PMF $f_{jj}(n)$ and the CDF $\mathsf{F}_{jj}(n)$.

Since results about Markov chains depend heavily on whether states are recurrent or transient, we will look carefully at the probabilities $\mathsf{F}_{ij}(n)$. Substituting (6.6) into (6.7) and noting that $f_{ij}(n)$ in (6.6) has a different form for $n = 1$ than for $n > 1$, we obtain

$$\begin{aligned}\mathsf{F}_{ij}(n) = P_{ij} + \sum_{m=2}^{n} f_{ij}(m) &= P_{ij} + \sum_{m=2}^{n}\sum_{k\neq j} P_{ik} f_{kj}(m-1)\\ &= P_{ij} + \sum_{k\neq j} P_{ik}\mathsf{F}_{kj}(n-1); \quad n > 1; \qquad \mathsf{F}_{ij}(1) = P_{ij}.\end{aligned} \tag{6.8}$$

Since $\mathsf{F}_{ij}(n)$ is non-decreasing in n and upper bounded by 1, the limit $\mathsf{F}_{ij}(\infty)$ must exist. Similarly, $\sum_{k\neq j} P_{ik}\mathsf{F}_{kj}(n-1)$ is non-decreasing in n and upper bounded by 1, so it also has a limit, equal to $\sum_{k\neq j} P_{ik}\mathsf{F}_{kj}(\infty)$. Thus

$$\mathsf{F}_{ij}(\infty) = P_{ij} + \sum_{k\neq j} P_{ik}\mathsf{F}_{kj}(\infty). \tag{6.9}$$

[1] Note that in Chapter 5 the inter-renewal intervals were denoted $X_1, X_2, \ldots$, whereas here $X_0, X_1, \ldots$ is the sequence of states in the Markov chain and $T_{jj}, T_{jj,2}, \ldots$ is the sequence of inter-renewal intervals.

For any given j, (6.9) can be viewed as a set of linear equations in the variables $\mathsf{F}_{ij}(\infty)$ for each state i. This set of equations does not always have a unique solution. In fact, substituting x_{ij} for $F_{ij}(\infty)$, the equations

$$x_{ij} = P_{ij} + \sum_{k \neq j} P_{ik} x_{kj} \quad \text{all states } i \tag{6.10}$$

always have a solution in which $x_{ij} = 1$ for all i. If state j is transient, however, there is another solution in which x_{ij} is the true value of $\mathsf{F}_{ij}(\infty)$ and $\mathsf{F}_{jj}(\infty) < 1$. Exercise 6.1 shows that if (6.10) is satisfied by a set of non-negative numbers $\{x_{ij};\ 1 \leq i \leq J\}$, then $\mathsf{F}_{ij}(\infty) \leq x_{ij}$ for each i. In other words, $\mathsf{F}_{ij}(\infty)$ is the minimum x_{ij} over solutions to (6.10).

An interesting feature of (6.9) arises in the case $i = j$,

$$\mathsf{F}_{jj}(\infty) = P_{jj} + \sum_{k \neq j} P_{jk} \mathsf{F}_{kj}(\infty).$$

If j is recurrent, i.e., if $\mathsf{F}_{jj}(\infty) = 1$, then the right-hand side of this equation is also 1. Thus, for each $k \neq j$ for which $P_{jk} > 0$, it is necessary that $\mathsf{F}_{kj}(\infty) = 1$. In other words, if j is recurrent, then each possible transition out of j goes to a state from which an eventual return to j occurs WP1. This extends to the following lemma.

Lemma 6.2.3 *Assume that state j is recurrent and that a path exists from j to i. Then* $\mathsf{F}_{ij}(\infty) = 1$.

Proof We just saw that if j is recurrent and $P_{jk} > 0$, then $\mathsf{F}_{kj}(\infty) = 1$. Thus the lemma is proven for paths of length 1. Next consider any path of length 2, say j, k, ℓ, where $P_{jk} > 0$ and $P_{k\ell} > 0$. From (6.9),

$$\mathsf{F}_{kj}(\infty) = P_{kj} + \sum_{\ell \neq k} P_{k\ell} \mathsf{F}_{\ell j}(\infty).$$

Since $\mathsf{F}_{kj}(\infty) = 1$, the right-hand side of the above equation is also 1. Since $P_{k\ell} > 0$, it follows that $\mathsf{F}_{\ell j}(\infty)$ must be 1. This argument iterates over increasing path lengths. □

This lemma gives us some insight into the difference between recurrence for finite-state chains and countable-state chains. For the finite-state case, j is recurrent if there is a path back to j from each i accessible from j. Here, not only is a return path required, but an eventual return with probability 1 (i.e., $\mathsf{F}_{ij}(\infty) = 1$).

The following result, Lemma 6.2.4, is similar. The two lemmas each concern a state i that is accessible from a recurrent state j. Lemma 6.2.3 says that starting in i, an eventual return to j must occur and Lemma 6.2.4 says that starting in j, an eventual visit to i must occur. In other words, Lemma 6.2.4 says that if j is recurrent, then any state reachable from j must be reached eventually.

Lemma 6.2.4 *Assume that state j is recurrent and that a path exists from j to i. Then* $\mathsf{F}_{ji}(\infty) = 1$.

Proof Let T_{jj} be the first-passage time from state j to state j and let $T_{jj}, T_{jj,2}, T_{jj,3}, \dots$ be the sequence of successive recurrence intervals. Since j is recurrent, these successive recurrence intervals are IID rv s (i.e., non-defective), and thus, for each n, $T_{jj}^{(n)} = T_{jj} + T_{jj,2} + \cdots + T_{jj,n}$ must be a rv (see Exercise 1.13). This rv is the time of the nth recurrence of state j, and this nth recurrence must occur eventually since $T_{jj}^{(n)}$ is a rv.

By assumption, there is a path, say $j, k, \ell, \dots, h, i$ from j to i and this path is taken, starting at time 0, with probability $\alpha = P_{jk}P_{k\ell} \cdots P_{hi} > 0$. If this initial path is taken, then state i is visited before the first recurrence of state j. Thus the probability that the first recurrence of state j occurs before i is visited is at most $1 - \alpha$. In the same way, the given path from j to i can be taken starting at each recurrence of state j. It follows that the probability that the nth recurrence of state j occurs before the first passage to state i is at most $(1 - \alpha)^n$. Since the nth recurrence of state j occurs eventually, the probability that i is never visited is at most $(1 - \alpha)^n$. Since n is arbitrary, the probability that i is never visited is 0, so $\mathsf{F}_{ji}(\infty) = 1$. □

The above proof is a little cavalier in the sense that $\mathsf{F}_{ji}(\infty)$ is $\lim_{t\to\infty} \mathsf{F}_{ji}(t)$ and this limit has been exchanged with others without justification. Exercise 6.5 shows how to handle this mathematical detail.

We now get an unexpected dividend from Lemmas 6.2.3 and 6.2.4. If state j is recurrent and i is accessible from j, we now show that i is recurrent also. The reason is that, starting in state i, state j is eventually reached WP1, and then, from j, state i is eventually reached. Stating this in an equation,

$$\mathsf{F}_{ii}(t + \tau) \;\geq\; \mathsf{F}_{ij}(t)\mathsf{F}_{ji}(\tau).$$

Going to the limits $t \to \infty, \tau \to \infty$,

$$\lim_{n\to\infty} \mathsf{F}_{ii}(n) \;\geq\; \lim_{t\to\infty} \mathsf{F}_{ij}(t) \lim_{\tau\to\infty} \mathsf{F}_{ji}(\tau) \;=\; 1.$$

There is a slight peculiarity here in that we have shown that i is recurrent assuming only that j is recurrent and that $j \to i$. The resolution is that, from Lemma 6.2.3, if j is recurrent and $j \to i$, then $F_{ij}(\infty) = 1$, which implies that $i \to j$.

We can summarize these results in the following theorem.

Theorem 6.2.5 *If state j of a countable-state Markov chain is recurrent, then every state i in the same class as j is recurrent, i.e.,* $\mathsf{F}_{ii}(\infty) = 1$. *Also* $\mathsf{F}_{ji}(\infty) = 1$ *and* $\mathsf{F}_{ij}(\infty) = 1$

This says that either each state in a class is recurrent or each is transient. Furthermore, for a recurrent class, the first-passage time from each state to each other state is a rv.

We have defined a state j to be recurrent if $\mathsf{F}_{jj}(\infty) = 1$ and have seen that if j is recurrent, then the returns to state j given $X_0 = j$ form a renewal process. All of the results of renewal theory can then be applied to the random sequence of integer times at which j is entered. Several important results from renewal theory are then stated in the following theorem.

Theorem 6.2.6 *Let $\{N_{jj}(t);\, t \geq 0\}$ be the counting process for occurrences of state j up to time t in a Markov chain with $X_0 = j$. The following conditions are then equivalent:*

1. *State j is recurrent.*
2. $\lim_{t\to\infty} N_{jj}(t) = \infty$ *WP1.*
3. $\lim_{t\to\infty} \mathsf{E}\left[N_{jj}(t)\right] = \infty.$
4. $\lim_{t\to\infty} \sum_{1\le n\le t} P_{jj}^n = \infty.$

Proof First assume that j is recurrent, i.e., that $\mathsf{F}_{jj}(\infty) = 1$. This implies that the inter-renewal times between occurrences of j are IID rv s, and consequently $\{N_{jj}(t);\ t \ge 1\}$ is a renewal counting process. Recall from Lemma 5.3.2 that, whether or not the expected inter-renewal time $\mathsf{E}\left[T_{jj}\right]$ is finite, $\lim_{t\to\infty} N_{jj}(t) = \infty$ WP1 and $\lim_{t\to\infty} \mathsf{E}\left[N_{jj}(t)\right] = \infty$.

Next assume that state j is transient. In this case, the inter-renewal time T_{jj} is not a rv, so $\{N_{jj}(t);\ t \ge 0\}$ is not a renewal process. An eventual return to state j occurs only with probability $\mathsf{F}_{jj}(\infty) < 1$, and, since subsequent returns are independent, the total number of returns to state j is a geometric rv with mean $\mathsf{F}_{jj}(\infty)/[1 - \mathsf{F}_{jj}(\infty)]$. Thus the total number of returns is finite WP1 and the expected total number of returns is finite. This establishes the first three equivalences.

Finally, note that P_{jj}^n is the probability of a transition to state j at integer time n, and is thus equal to the expectation of a transition to j at integer time n (i.e., the number of transitions at time t is 1 with probability P_{jj}^n and 0 otherwise). Since $N_{jj}(t)$ is the sum of the number of transitions to j over times 1 to t, we have

$$\mathsf{E}\left[N_{jj}(t)\right] = \sum_{1\le n\le t} P_{jj}^n,$$

which establishes the final equivalence. □

6.3 Renewal theory applied to Markov chains

We have been looking at the counting process $\{N_{jj}(t);\ t > 0\}$, which is the number of returns to j by time t given $X_0 = j$. We also want to look at the counting process $\{N_{ij}(t);\ t > 0\}$, which is the number of visits to state j by time t given $X_0 = i$. If j is a recurrent state and i is in the same class, then Theorem 6.2.5 says that T_{ij}, the first-passage time from i to j, is a rv. Each subsequent interval between visits to j, i.e., each interval between subsequent recurrence times, is an IID rv with the distribution of T_{jj}. Thus $\{N_{ij}(t);\ t > 0\}$ is a delayed renewal process, which proves the following lemma.

Lemma 6.3.1 *Let* $\{N_{ij}(t);\ t \ge 0\}$ *be the counting process for transitions into state j up to time t for a Markov chain conditional on the initial state* $X_0 = i \ne j$. *Then if i and j are in the same recurrent class,* $\{N_{ij}(t);\ t \ge 0\}$ *is a delayed renewal process.*

6.3.1 Renewal theory and positive recurrence

Recall that a state j is recurrent if the first-passage time T_{jj} is a rv (i.e., if $\mathsf{F}_{jj}(\infty) = 1$), and it then follows that $\{N_{jj}(t);\ t > 0\}$ is a renewal counting process. It is possible, however, for $\mathsf{E}\left[T_{jj}\right]$ to be either finite or infinite. We have seen many positive rv s that have infinite expectation, but having a *first-passage time* that is a rv (i.e., finite WP1) but infinite in expectation might seem strange at first. An example of this phenomenon is the Markov chain of Figure 6.2 with $p = 1/2$, but it is tedious to verify this by direct calculation from $\mathsf{F}_{jj}(t)$. We will find better ways to calculate $\mathsf{E}\left[T_{jj}\right]$ later, and can then look at several examples.

In studying renewal processes, we found that the expected inter-renewal interval, in this case $\mathsf{E}\left[T_{jj}\right]$, is the reciprocal of the rate at which renewals occur, and that if $\mathsf{E}\left[T_{jj}\right] = \infty$, the renewal rate is 0. Thus it is not surprising that there is a large difference between the case where $\mathsf{E}\left[T_{jj}\right] = \infty$ and that where $\mathsf{E}\left[T_{jj}\right] < \infty$. This suggests the following definition.

Definition 6.3.2 *A state j in a countable-state Markov chain is* ***positive recurrent*** *if $\mathsf{F}_{jj}(\infty) = 1$ and $\overline{T}_{jj} < \infty$. It is* null recurrent *if $\mathsf{F}_{jj}(\infty) = 1$ and $\overline{T}_{jj} = \infty$.*

Each state of a Markov chain is thus classified as one of the following three types – positive recurrent, null recurrent, or transient. For the example of Figure 6.2, null recurrence lies on a boundary between positive recurrence and transience, which is often a good way to view null recurrence.

Assume that state j is recurrent and consider the renewal process $\{N_{jj}(t); t \geq 0\}$. The following theorem simply applies the limit theorems of renewal processes to this case.

Theorem 6.3.3 *If j is a recurrent state in a countable-state Markov chain, then the renewal counting process $N_{jj}(t)$ satisfies both*

$$\lim_{t\to\infty} N_{jj}(t)/t = 1/\overline{T}_{jj} \quad WP1, \tag{6.11}$$

and

$$\lim_{t\to\infty} \mathsf{E}\left[N_{jj}(t)/t\right] = 1/\overline{T}_{jj}. \tag{6.12}$$

Equations (6.11) and (6.12) are valid whether j is positive recurrent or null recurrent.

Proof The first equation, (6.11), is the strong law for renewal processes given in Theorem 5.3.1. The case where $\overline{T}_{jj} = \infty$ is treated in Exercise 5.9. The second equation is the elementary renewal theorem, Theorem 5.6.2. □

Equations (6.11) and (6.12) suggest that $1/\overline{T}_{jj}$ has some of the properties associated with a steady-state probability for state j. For a Markov chain consisting of a single class of states, all positive recurrent, we will strengthen this association further in Theorem 6.3.8 by showing that there is a unique *steady-state distribution*, $\{\pi_j, j \geq 0\}$, such that $\pi_j = 1/\overline{T}_{jj}$ for all j and such that $\pi_j = \sum_i \pi_i P_{ij}$ for all $j \geq 0$ and $\sum_j \pi_j = 1$. The following theorem starts this development by showing that (6.11) and (6.12) are independent of the starting state.

Theorem 6.3.4 *Let j be a recurrent state in a Markov chain and let i be any state in the same class as j. Given $X_0 = i$, let $N_{ij}(t)$ be the number of transitions into state j by time t and let $\overline{T}_{jj}$ be the expected recurrence time of state j (either finite or infinite). Then*

$$\lim_{t\to\infty} N_{ij}(t)/t = 1/\overline{T}_{jj} \quad WP1 \tag{6.13}$$

and

$$\lim_{t\to\infty} \mathsf{E}\left[N_{ij}(t)/t\right] = 1/\overline{T}_{jj}. \tag{6.14}$$

Proof Since i and j are recurrent and in the same class, Lemma 6.3.1 asserts that $\{N_{ij}(t);\ t \geq 0\}$ is a delayed renewal process for $j \neq i$. Thus (6.13) and (6.14) follow from Theorems 5.8.3 and 5.8.4. □

Theorem 6.3.5 *All states in the same class of a Markov chain are of the same type – either all positive recurrent, all null recurrent, or all transient.*

Proof Let j be a recurrent state. From Theorem 6.2.5, all states in a class are recurrent or all are transient. Suppose that j is positive recurrent, so that $1/\overline{T}_{jj} > 0$. Let i be in the same class as j, and consider the renewal–reward process on $\{N_{jj}(t);\ t \geq 0\}$ for which $R(t) = 1$ whenever the process is in state i (i.e., if $X_n = i$, then $R(t) = 1$ for $n \leq t < n+1$). The reward is 0 otherwise. Let $\mathsf{E}[\mathrm{R}]$ be the expected reward in an inter-renewal interval; this must be positive since i is accessible from j. From the strong law for renewal–reward processes, Theorem 5.4.5,

$$\lim_{t\to\infty} \frac{1}{t}\int_0^t R(\tau)d\tau = \frac{\mathsf{E}[\mathrm{R}]}{\overline{T}_{jj}} \qquad \text{WP1.}$$

The term on the left is the time-average number of transitions into state i, given $X_0 = j$, and this is $1/\overline{T}_{ii}$ from (6.13). Since $\mathsf{E}[\mathrm{R}] > 0$ and $\overline{T}_{jj} < \infty$, we have $1/\overline{T}_{ii} > 0$, so i is positive recurrent. Thus if one state is positive recurrent, the entire class is, completing the proof. □

If all of the states in a Markov chain are in a null-recurrent class, then $1/\overline{T}_{jj} = 0$ for each state, and one might think of $1/\overline{T}_{jj} = 0$ as a 'steady-state' probability for j in the sense that 0 is both the time-average rate of occurrence of j and the limiting probability of j. However, these 'probabilities' do not add up to 1, so a steady-state probability *distribution* does not exist. This appears rather paradoxical at first, but the example of Figure 6.2 with $p = 1/2$ will help clarify the situation. As time n increases (starting with $X_0 = i$, say), the rv X_n spreads out over more and more states around i, and thus is less likely to be in each individual state. For each j, $\lim_{n\to\infty} P_{ij}^n = 0$. Thus, $\sum_j \left(\lim_{n\to\infty} P_{ij}^n\right) = 0$. On the other hand, for every n, $\sum_j P_{ij}^n = 1$. This is one of those unusual examples where a limit and a sum cannot be interchanged.

6.3.2 Steady state

In Section 4.3.1, we defined the steady-state distribution of a finite-state Markov chain as a probability vector $\boldsymbol{\pi}$ that satisfies $\boldsymbol{\pi} = \boldsymbol{\pi}[P]$. In the same way, we define the steady-state distribution of a countable-state Markov chain as a PMF $\{\pi_i;\ i \geq 0\}$ that satisfies

$$\pi_j = \sum_i \pi_i P_{ij} \quad \text{for all } j, \qquad \pi_i \geq 0 \quad \text{for all } i, \qquad \sum_i \pi_i = 1. \tag{6.15}$$

Assume that a set of numbers $\{\pi_i;\ i \geq 0\}$ exists satisfying (6.15) and assume that this set is chosen as the initial PMF for the Markov chain, i.e., $\Pr\{X_0 = i\} = \pi_i$ for each $i \geq 0$. Then for each $j \geq 0$,

$$\Pr\{X_1 = j\} = \sum_i \Pr\{X_0 = i\}\, P_{ij} = \sum_i \pi_i P_{ij} = \pi_j.$$

In the same way, using induction on n,

$$\Pr\{X_n = j\} = \sum_i \Pr\{X_{n-1} = i\}\, P_{ij} = \sum_i \pi_i P_{ij} = \pi_j \tag{6.16}$$

for all j and all $n \geq 0$. The fact that $\Pr\{X_n = j\} = \pi_j$ for all $j \geq 0$ and $n \geq 0$ motivates the definition of steady-state distribution above. On the other hand, Theorem 6.3.3 showed that $1/\overline{T}_{jj}$ can be viewed as a 'steady-state' probability for state j, both in a time-average and a limiting ensemble-average sense. In this section, we bring these notions of steady state together and relate π_j to $1/\overline{T}_{jj}$.

Before doing this, it will be helpful to restrict our attention to Markov chains with a single class of states. For finite-state chains, it was sufficient to restrict attention to a single recurrent class, whereas now a new term is needed, since a Markov chain with a single class might be positive recurrent, null recurrent, or transient.

Definition 6.3.6 *An **irreducible class of states** C is a class with no exit, i.e., a class for which $P_{ij} = 0$ for all $i \in \mathsf{C}$ and $j \notin \mathsf{C}$. An **irreducible Markov chain** is a Markov chain for which all states are in a single irreducible class, i.e., for which all states communicate.*

The following results are largely restricted to irreducible Markov chains, since the relations between multiple classes are largely the same as for finite-state Markov chains. Between-class relations are largely separable from the properties of states within an irreducible class, and thus irreducible classes can be studied most efficiently by considering irreducible Markov chains.

The following lemma will start the process of associating steady-state probabilities $\{\pi_i;\ i \geq 0\}$ satisfying (6.15) with recurrence rates $1/\overline{T}_{jj}$.

Lemma 6.3.7 *Consider an irreducible countable-state Markov chain in which a PMF $\{\pi_i;\ i \geq 0\}$ satisfies (6.15). Then this chain is positive recurrent.*

Proof Consider using the given PMF $\{\pi_i;\ i \geq 0\}$ as the initial distribution of the Markov chain, i.e., $\Pr\{X_0{=}j\} = \pi_j$ for all $j \geq 0$. Then, as shown above, $\Pr\{X_n{=}j\} = \pi_j$ for all $n \geq 0$, $j \geq 0$. For any given j, let $\widetilde{N}_j(t)$ be the number of occurrences of state j from time 1 to t given the PMF $\{\pi_i;\ i \geq 0\}$ for X_0. Equating $\Pr\{X_n{=}j\} = \pi_j$ to the expectation of an occurrence of j at time n, we have

$$(1/t)\mathsf{E}\left[\widetilde{N}_j(t)\right] = (1/t)\sum_{1\leq n\leq t} \Pr\{X_n{=}j\} = \pi_j \qquad \text{for all integers } t \geq 1.$$

Conditioning this on the possible starting states i,

$$\pi_j = (1/t)\mathsf{E}\left[\widetilde{N}_j(t)\right] = (1/t)\sum_i \pi_i \mathsf{E}\left[N_{ij}(t)\right] \qquad \text{for all integers } t \geq 1. \tag{6.17}$$

For any given state i, let T_{ij} be the time (perhaps infinite) of the first occurrence of state j given $X_0 = i$. Conditional on $T_{ij} = m$, we have $N_{ij}(t) = 1 + \hat{N}_{jj}(m,t)$, where $\hat{N}_{jj}(m,t)$ is the number of recurrences of j from $m+1$ to t. Thus, for all $t > m$,

$$\mathsf{E}\left[N_{ij}(t) \mid T_{ij}{=}m\right] = 1 + \mathsf{E}\left[N_{jj}(t-m)\right] \leq 1 + \mathsf{E}\left[N_{jj}(t)\right]. \tag{6.18}$$

For $t < m$, the left-hand side of (6.18) is 0 and for $t = m$, it is 1. Thus (6.18) is valid for all m including $m = \infty$. Since the upper bound does not depend on m, $\mathsf{E}\left[N_{ij}(t)\right] \leq 1 + \mathsf{E}\left[N_{jj}(t)\right]$. Substituting this into (6.17) and summing over i, we get

$$\pi_j \leq (1/t)\sum_i \pi_i\left(1 + \mathsf{E}\left[N_{jj}(t)\right]\right) = (1/t) + (1/t)\mathsf{E}\left[N_{jj}(t)\right]. \tag{6.19}$$

From this, $\mathsf{E}\left[N_{jj}(t)\right] \geq t\pi_j - 1$. Since $\sum_i \pi_i = 1$, there must be some j for which π_j is strictly positive. For any given such j, $\lim_{t\to\infty} \mathsf{E}\left[N_{jj}(t)\right] = \infty$.

This shows (using Theorem 6.2.6) that state j is recurrent. It then follows from Theorem 6.3.3 that $\lim_{t\to\infty}(1/t)\mathsf{E}\left[N_{jj}(t)\right]$ exists and is equal to $1/\overline{T}_{jj}$ for all j. Taking this limit in (6.19),

$$\pi_j \leq \lim_{t\to\infty}(1/t)\mathsf{E}\left[N_{jj}(t)\right] = 1/\overline{T}_{jj}. \tag{6.20}$$

Since $\pi_j > 0$, $\overline{T}_{jj} < \infty$, so that state j is positive recurrent. From Theorem 6.3.5, all states are then positive recurrent. □

With this lemma, we can prove the following theorem. The theorem is particularly valuable for applications, since it says that *if* you can find steady-state probabilities, then you can forget about all the nuances and subtleties of countable-state chains; essentially the chain can be treated as a finite-state chain. More than that, the theorem provides a link between steady-state probabilities and expected first-passage times.

Theorem 6.3.8 *Assume an irreducible Markov chain with transition probabilities $\{P_{ij}\}$. If (6.15) has a solution, then the solution $\{\pi_i;\ i \geq 0\}$ is unique. Furthermore, the chain is positive recurrent and*

$$\pi_j = 1/\overline{T}_{jj} > 0 \qquad \text{for all } j \geq 0. \tag{6.21}$$

Conversely, if the chain is positive recurrent then (6.15) has a solution.

Proof First assume that $\{\pi_j;\ j \geq 0\}$ satisfies (6.15). From Lemma 6.3.7, each state j is positive recurrent, so that $\overline{T}_{jj} < \infty$ and $1/\overline{T}_{jj} > 0$ for all j. Taking the limit $t \to \infty$ in (6.19), we see that $\pi_j \leq 1/\overline{T}_{jj}$ for all j. We next develop a lower bound on π_j from (6.17). For each integer $m > 0$,

$$\pi_j \geq (1/t)\sum_{i=1}^{m} \pi_i \mathsf{E}\left[N_{ij}(t)\right] \qquad \text{for all integers } t \geq 1.$$

Using Theorem 6.3.4 to take the limit of this as $t \to \infty$,

$$\pi_j \geq \lim_{t\to\infty}(1/t)\mathsf{E}\left[N_{jj}(t)\right] = 1/\overline{T}_{jj}.$$

Combined with $\pi_j \leq 1/\overline{T}_{jj}$, this establishes (6.21).

Conversely, assume the chain is positive recurrent. We will choose $\pi_j = 1/\overline{T}_{jj}$ and then show that the resulting $\{\pi_i; i \geq 0\}$ is a PMF satisfying (6.15). Each π_j is positive since by definition of positive recurrence, $\overline{T}_{jj} < \infty$. We first show that $\sum_j \pi_j = 1$. For any given i, consider the renewal process $\{N_{ii}(t); t > 0\}$ and, for any j, let $R_{ij}(t)$ be the reward process assigning unit reward on each occurrence of state j. Let R_{ij} be the aggregate reward in the first inter-renewal interval. For $j = i$, note that $\mathrm{R}_{ii} = 1$ since a unit reward occurs at the end of the interval and cannot occur elsewhere. Since each step of the Markov process enters some state j and the first renewal occurs at time T_{ii}, we have

$$T_{ii} = \sum_{j=1}^{\infty} \mathrm{R}_{ij}.$$

Taking the expectation of each side,

$$\mathsf{E}[T_{ii}] = \mathsf{E}\left[\lim_{\ell\to\infty} \sum_{j=1}^{\ell} \mathrm{R}_{ij}\right] = \lim_{\ell\to\infty} \sum_{j=1}^{\ell} \mathsf{E}\left[\mathrm{R}_{ij}\right], \tag{6.22}$$

where the interchange of limit and expectation is warranted because the terms in the limit are non-decreasing and bounded. Using Theorem 5.4.5, $\mathsf{E}\left[\mathrm{R}_{ij}\right] = \pi_j/\pi_i$. Thus we can rewrite (6.22) as $1/\pi_i = \sum_j \pi_j/\pi_i$, which shows that $\sum_j \pi_j = 1$.

Finally we must show that $\pi_i = \sum_j \pi_j P_{ji}$. The proof is a variation on the one just used. We look again at the renewal process $N_{ii}(t)$ and define a reward function $R_{i(j,i)}(t)$ that assigns one unit of reward for each transition from state j to state i. A unit reward can occur only at the end of the first renewal and that unit reward goes to the state j from which the transition to i occurs. Thus, arguing as above,

$$\sum_{j=1}^{\infty} R_{i(j,i)} = 1, \qquad \sum_{j=1}^{\infty} \mathsf{E}\left[R_{i(j,i)}\right] = 1. \tag{6.23}$$

By looking at the delayed renewal process $\{N_{ij}(t); t \geq 0\}$, we see that $R_{i(j,i)}(t)$ is 1 whenever a renewal of state j occurs followed immediately by a transition to i. Thus the reward rate for $R_{i(j,i)}(t)$ is $\pi_j P_{ji}$. Applying Theorem 5.4.5, the expected reward over the first renewal of state i is $\pi_j P_{ji}/\pi_i$. Substituting this into (6.23) then completes the proof. □

In practice, it is usually easy to see whether a chain is irreducible. We shall also see by a number of examples that the steady-state distribution can often be calculated from (6.15). Theorem 6.3.8 then says that the calculated distribution is unique and that its existence guarantees that the chain is positive recurrent. It also shows that the expected inter-renewal interval T_{jj} for each state j is given simply by $1/\pi_j$. Finally, Exercise 6.7 provides a simple way to calculate $\overline{T}_{ij}$.

6.3.3 Blackwell's theorem applied to Markov chains

The results so far about steady state have ignored a large part of what we would intuitively view as a steady state, i.e., the notion that P_{ij}^n should approach π_j as $n \to \infty$, independent of i. We know from studying finite-state Markov chains that this is too much to hope for if an irreducible chain is periodic, but the discussion for countable-state chains has not yet discussed periodicity.

Consider applying Blackwell's theorem to the renewal process $\{N_{jj}(t); t \geq 0\}$ for a positive-recurrent chain. Recall that the period of a given state j in a Markov chain (whether the chain has a countable or finite number of states) is the greatest common divisor (gcd) of the set of integers $n > 0$ such that $P^n_{jj} > 0$. If this period is d, then $\{N_{jj}(t); t \geq 0\}$ is arithmetic with span d (i.e., renewals occur only at times that are multiples of d). From Blackwell's theorem in the arithmetic form of (5.64),

$$\lim_{n\to\infty} P^{nd}_{jj} = d/\overline{T}_{jj}. \tag{6.24}$$

As with finite-state chains, all states have the same period d for a periodic positive-recurrent chain, and, as in Theorem 4.2.9, the states can be partitioned into d classes and all transitions cycle from S_0 to S_1 to S_2 and so forth to S_{d-1} and then back to S_0. For the case where state j is aperiodic (i.e., $d = 1$), this says that

$$\lim_{n\to\infty} P^{n}_{jj} = 1/\overline{T}_{jj}. \tag{6.25}$$

Finally, Blackwell's theorem for delayed renewal processes, Theorem 5.8.5, asserts that if the states are aperiodic, then for all states, i and j,

$$\lim_{n\to\infty} P^{n}_{ij} = 1/\overline{T}_{jj}. \tag{6.26}$$

An aperiodic positive-recurrent Markov chain is defined to be an *ergodic* chain, and the interpretation of this is the same as with finite-state Markov chains. In particular, for ergodic chains, the steady-state probability π_j is, WP1, the time-average rate of that state. It is also the time- and ensemble-average rate of the state, and, from (6.26), the limiting probability of the state, independent of the starting state.

The independence from the starting state in the limit $n \to \infty$ in (6.26) is unfortunately not uniform in i and j. For example, the state can change by at most 1 per unit time in the example in Figure 6.2, and thus starting in a very high-numbered state i will significantly influence P^n_{ij} until n is very large.

6.3.4 Age of an arithmetic renewal process

Consider a renewal process $\{N(t); t > 0\}$ in which the inter-renewal rv s $\{W_j; j \geq 1\}$ are arithmetic with span 1. We will use a Markov chain to model the age of this process (see Figure 6.3). The probability that a renewal occurs at a particular integer time depends on the past only through the integer time back to the last renewal. The state of the

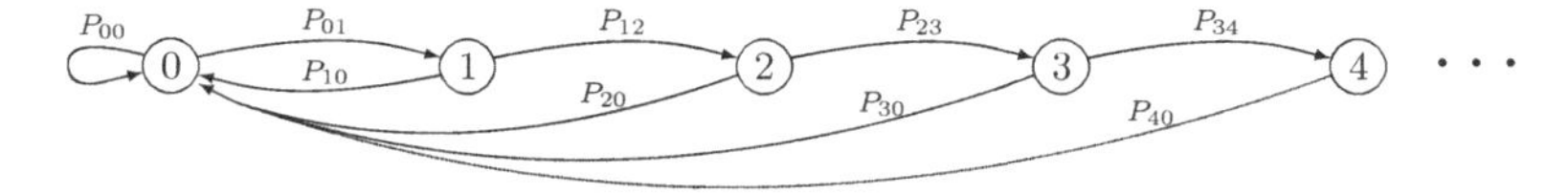

Figure 6.3 A Markov chain model of the age of a renewal process.

Markov chain during a unit interval will be taken as the age of the renewal process at the beginning of the interval. Thus, at each unit of time, the age either increases by 1 or a renewal occurs and the age decreases to 0 (i.e., if a renewal occurs at time t, the age at time t is 0).

$\Pr\{W > j\}$ is the probability that an inter-renewal interval lasts for more than j time units. We assume that $\Pr\{W > 0\} = 1$, so that each renewal interval lasts at least one time unit. The probability $P_{j,0}$ in the Markov chain is the probability that a renewal interval has duration $j + 1$, given that the interval exceeds j. Thus, for example, P_{00} is the probability that the renewal interval is equal to 1. We see that $P_{j,j+1}$ is $1 - P_{j,0}$, which is $\Pr\{W > j + 1\} / \Pr\{W > j\}$. We can then solve for the steady-state probabilities in the chain: for $j > 0$,

$$\pi_j = \pi_{j-1}P_{j-1,j} = \pi_{j-2}P_{j-2,j-1}P_{j-1,j} = \pi_0 P_{0,1}P_{1,2} \dots P_{j-1,j}.$$

The first equality above results from the fact that state j, for $j > 0$ can be entered only from state $j-1$. The subsequent equalities come from substituting in the same expression for π_{j-1}, then π_{j-2}, and so forth.

$$\pi_j = \pi_0 \frac{\Pr\{W > 1\}\Pr\{W > 2\}}{\Pr\{W > 0\}\Pr\{W > 1\}} \cdots \frac{\Pr\{W > j\}}{\Pr\{W > j - 1\}} = \pi_0 \Pr\{W > j\}. \tag{6.27}$$

We have canceled out all the cross terms above and used the fact that $\Pr\{W > 0\} = 1$. Another way to see that $\pi_j = \pi_0 \Pr\{W > j\}$ is to observe that state 0 occurs exactly once in each inter-renewal interval; state j occurs exactly once in those inter-renewal intervals of duration j or more.

Since the steady-state probabilities must sum to 1, (6.27) can be solved for π_0 as

$$\pi_0 = \frac{1}{\sum_{j=0}^{\infty} \Pr\{W > j\}} = \frac{1}{\mathsf{E}[W]}. \tag{6.28}$$

The second equality follows by expressing $\mathsf{E}[W]$ as the integral of the complementary CDF of W. As a check on (6.28) (and an easier way of deriving it), note that this is simply Theorem 6.3.8 applied to state 0. Combining (6.28) with (6.27), the steady-state probabilities for $j \geq 0$ are

$$\pi_j = \frac{\Pr\{W > j\}}{\mathsf{E}[W]}. \tag{6.29}$$

In terms of the renewal process, π_j is the probability that, at some large integer time, the age of the process will be j. Note that if the age of the process at an integer time is j, then the age increases toward $j + 1$ at the next integer time, at which point it either drops to 0 or continues to rise. Thus π_j can be interpreted as the fraction of time that the age of the process is between j and $j + 1$. Recall from (5.30) (and the fact that residual life and age are equally distributed) that the CDF of the time-average age is given by $\mathsf{F}_Z(j) = \int_0^j \Pr\{W > w\}\, dw / \mathsf{E}[W]$. Thus, the probability that the age is between j and $j+1$ is $\mathsf{F}_Z(j + 1) - \mathsf{F}_Z(j)$. Since W is an integer rv, this is $\Pr\{W > j\} / \mathsf{E}[W]$, in agreement with the result here.

The analysis here gives a new, and intuitively satisfying, explanation of why the age of a renewal process is so different from the inter-renewal time. The Markov chain shows the ever increasing loops that give rise to large expected age when the inter-renewal time is heavy tailed (i.e., has a CDF that goes to 0 slowly with increasing time). These loops can be associated with the isosceles triangles of Figure 5.8. The advantage here is that we can associate the states with steady-state probabilities if the chain is positive recurrent. Even when the Markov chain is null recurrent (i.e., the associated renewal process has infinite expected age), it seems easier to visualize the phenomenon of infinite expected age.

6.4 Birth–death Markov chains

A *birth–death Markov chain* is a Markov chain in which the state space is the set of non-negative integers; for all $i \geq 0$, the transition probabilities satisfy $P_{i,i+1} > 0$ and $P_{i+1,i} > 0$, and for all $|i - j| > 1$, $P_{ij} = 0$ (see Figure 6.4). A transition from state i to state $i + 1$ is regarded as a birth and one from state $i + 1$ to state i as a death. Thus the restriction on the transition probabilities means that only one birth or death can occur in one unit of time. Many applications of birth–death processes arise in queueing theory, where the state is the number of customers, births are customer arrivals, and deaths are customer departures. The restriction to only one arrival or departure at a time seems rather peculiar, but usually such a chain is a finely sampled approximation to a continuous-time process, and the time increments are then small enough that multiple arrivals or departures in a time increment are unlikely and can be ignored in the limit.

We denote $P_{i,i+1}$ by p_i and $P_{i,i-1}$ by q_i. Thus $P_{ii} = 1 - p_i - q_i$. There is an easy way to find the steady-state probabilities of these birth–death chains. In any sample function of the process, note that the number of transitions from state i to state $i + 1$ differs by at most 1 from the number of transitions from state $i + 1$ to state i. If the process starts to the left of state i and ends to the right, then one more $i \to i + 1$ transition occurs than $i + 1 \to i$, etc. Thus if we visualize a renewal–reward process with renewals on occurrences of state i and unit reward on transitions from state i to state $i+1$, the limiting time-average number of transitions per unit time is $\pi_i p_i$. Similarly, the limiting time-average number of transitions per unit time from state $i + 1$ to state i is $\pi_{i+1}q_{i+1}$. Since these two must be equal in the limit,

$$\pi_i p_i = \pi_{i+1} q_{i+1} \qquad \text{for } i \geq 0. \tag{6.30}$$

The intuition in (6.30) is simply that the rate at which downward transitions occur from state $i + 1$ to state i must equal the rate of upward transitions. Since this result is very

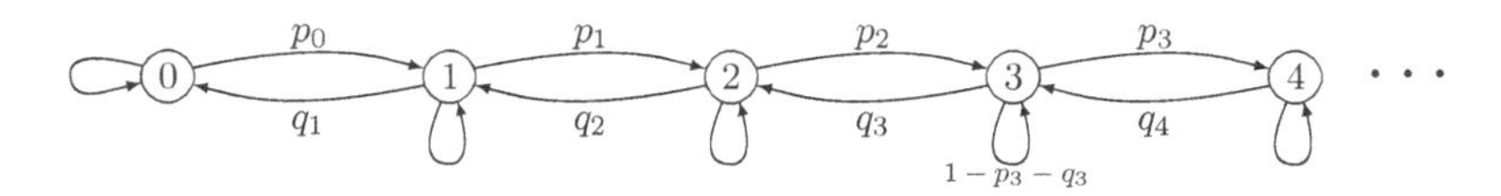

Figure 6.4 Birth–death Markov chain.

important, both here and in our later study of continuous-time birth–death processes, we show that (6.30) also results from using the steady-state equations in (6.15):

$$\pi_i = p_{i-1}\pi_{i-1} + (1 - p_i - q_i)\pi_i + q_{i+1}\pi_{i+1}, \quad i > 0; \tag{6.31}$$

$$\pi_0 = (1 - p_0)\pi_0 + q_1\pi_1. \tag{6.32}$$

From (6.32), $p_0\pi_0 = q_1\pi_1$. To see that (6.30) is satisfied for $i > 0$, we use induction on i, with $i = 0$ as the base. Thus assume, for a given i, that $p_{i-1}\pi_{i-1} = q_i\pi_i$. Substituting this in (6.31), we get $p_i\pi_i = q_{i+1}\pi_{i+1}$, thus completing the inductive proof.

It is convenient to define ρ_i as p_i/q_{i+1}. Then we have $\pi_{i+1} = \rho_i\pi_i$, and iterating this,

$$\pi_i = \pi_0 \prod_{j=0}^{i-1} \rho_j; \quad \pi_0 = \frac{1}{1 + \sum_{i=1}^{\infty} \prod_{j=0}^{i-1} \rho_j}. \tag{6.33}$$

If $\sum_{i\geq 1} \prod_{0\leq j<i} \rho_j < \infty$, then π_0 is positive and all the states are positive recurrent. If this sum of products is infinite, then no state is positive recurrent. If ρ_j is bounded below 1, say $\rho_j \leq 1 - \epsilon$ for some fixed $e > 0$ and all sufficiently large j, then this sum of products will converge and the states will be positive recurrent.

For the simple birth–death process of Figure 6.2, if we define $\rho = q/p$, then $\rho_j = \rho$ for all j. For $\rho < 1$, (6.33) simplifies to $\pi_i = \pi_0\rho^i$ for all $i \geq 0$, $\pi_0 = 1 - \rho$, and thus $\pi_i = (1 - \rho)\rho^i$ for $i \geq 0$. Exercise 6.2 shows how to find $\mathsf{F}_{ij}(\infty)$ for all i,j in the case where $\rho \geq 1$. We have seen that the simple birth–death chain of Figure 6.2 is transient if $\rho > 1$. This is not necessarily so in the case where self transitions exist, but the chain is still either transient or null recurrent. An example of this will arise in Exercise 7.3.

6.5 Reversible Markov chains

Many important Markov chains have the property that, in steady state, the sequence of states looked at backwards in time, i.e., $\ldots X_{n+1}, X_n, X_{n-1}, \ldots$, has the same probabilistic structure as the sequence of states running forward in time. This equivalence between the forward chain and the backward chain leads to a number of results that are intuitively quite surprising and that are quite difficult to derive by other means. We shall develop and study these results here and then redevelop them in Chapter 7 for Markov processes with discrete state spaces. This set of ideas, and its use in queueing and queueing networks, has been an active area of queueing research over many years. It leads to many simple results for systems that initially appear very complex. We only scratch the surface here and refer the interested reader to [17] for a more comprehensive treatment. Before going into reversibility, we describe the backward chain for an arbitrary Markov chain.

The defining characteristic of a Markov chain $\{X_n; n \geq 0\}$ is that for all $n \geq 0$

$$\Pr\{X_{n+1} \mid X_n, X_{n-1}, \ldots, X_0\} = \Pr\{X_{n+1} \mid X_n\}. \tag{6.34}$$

For homogeneous chains, which we have been assuming throughout, $\Pr\{X_{n+1} = j \mid X_n = i\} = P_{ij}$, independent of n. For any $k > 1$, we can extend (6.34) to get

$$\begin{aligned}
&\Pr\{X_{n+k}, X_{n+k-1}, \dots, X_{n+1} \mid X_n, X_{n-1}, \dots, X_0\} \\
&= \Pr\{X_{n+k} \mid X_{n+k-1}\} \Pr\{X_{n+k-1} \mid X_{n+k-2}\} \cdots \Pr\{X_{n+1} \mid X_n\} \\
&= \Pr\{X_{n+k}, X_{n+k-1}, \dots, X_{n+1} \mid X_n\}. \qquad (6.35)
\end{aligned}$$

By letting A^+ be any event defined on the states $X_{n+1}, \dots, X_{n+k}$ and letting A^- be any event defined on $X_0, \dots, X_{n-1}$, this can be written more succinctly as

$$\Pr\left\{A^+ \mid X_n, A^-\right\} = \Pr\left\{A^+ \mid X_n\right\}. \qquad (6.36)$$

This says that, given state X_n, any future event A^+ is statistically independent of any past event A^-. This result, namely that past and future are independent given the present state, is equivalent to (6.34) for defining a Markov chain, but it has the advantage of showing the symmetry between past and future. This symmetry is best brought out by multiplying both sides of (6.36) by $\Pr\left\{A^- \mid X_n\right\}$, obtaining

$$\Pr\left\{A^+, A^- \mid X_n\right\} = \Pr\left\{A^+ \mid X_n\right\} \Pr\left\{A^- \mid X_n\right\}. \qquad (6.37)$$

Much more broadly, any three ordered events, say $A^- \to X_0 \to A^+$, are said to satisfy the Markov property if $\Pr\left\{A^+ \mid X_0 A^-\right\} = \Pr\left\{A^+ \mid X_0\right\}$. This implies the more symmetric form $\Pr\left\{A^- A^+ \mid X_0)\right\} = \Pr\left\{A^- \mid X_0\right\} \Pr\left\{A^+ \mid X_0\right\}$. This symmetric form says that, conditional on the current state, the past and future states are statistically independent. Dividing both sides by $\Pr\left\{A^+ \mid X_n\right\}$ then yields

$$\Pr\left\{A^- \mid X_n, A^+\right\} = \Pr\left\{A^- \mid X_n\right\}. \qquad (6.38)$$

By letting A^- be X_{n-1} and A^+ be $X_{n+1}, X_{n+2}, \dots, X_{n+k}$, this becomes

$$\Pr\{X_{n-1} \mid X_n, X_{n+1}, \dots, X_{n+k}\} = \Pr\{X_{n-1} \mid X_n\}.$$

This is the equivalent form to (6.34) for the backward chain, and says that the backward chain is also a Markov chain. By Bayes' law, $\Pr\{X_{n-1} \mid X_n\}$ can be evaluated as

$$\Pr\{X_{n-1} \mid X_n\} = \frac{\Pr\{X_n \mid X_{n-1}\} \Pr\{X_{n-1}\}}{\Pr\{X_n\}}. \qquad (6.39)$$

Since the distribution of X_n can vary with n, $\Pr\{X_{n-1} \mid X_n\}$ can also depend on n. *Thus the backward Markov chain is not necessarily homogeneous.* This should not be surprising, since the forward chain was defined with some arbitrary distribution for the initial state at time 0. This initial distribution was not relevant for (6.34)–(6.36), but as soon as $\Pr\left\{A^- \mid X_n\right\}$ is introduced, the initial state implicitly becomes a part of each equation and destroys the symmetry between past and future. For a chain in steady state, i.e., a positive-recurrent chain for which the distribution of each X_n is the steady-state distribution satisfying (6.15), however, $\Pr\{X_n = j\} = \Pr\{X_{n-1} = j\} = \pi_j$ for all j, and we have

$$\Pr\{X_{n-1} = j \mid X_n = i\} = P_{ji}\pi_j / \pi_i \qquad \text{in steady state.} \qquad (6.40)$$

Thus the backward chain is homogeneous if the forward chain is in steady state. For a chain with steady-state probabilities $\{\pi_i;\ i \geq 0\}$, we define the backward transition probabilities P^*_{ij} as

$$\pi_i P^*_{ij} = \pi_j P_{ji}. \qquad (6.41)$$

From (6.39), the backward transition probability P^*_{ij}, for a Markov chain in steady state, is then equal to $\Pr\{X_{n-1} = j \mid X_n = i\}$, the probability that the previous state is j given that the current state is i.

Now consider a new Markov chain with transition probabilities $\{P^*_{ij}\}$. Over some segment of time for which both this new chain and the old chain are in steady state, the set of states generated by the new chain is statistically indistinguishable from the backward running sequence of states from the original chain. It is somewhat simpler, in talking about forward and backward running chains, to visualize Markov chains running in steady state from $t = -\infty$ to $t = +\infty$. If one is uncomfortable with this, one can also visualize starting the Markov chain at some very negative time with an initial distribution equal to the steady-state distribution.

Definition 6.5.1 *A Markov chain that has steady-state probabilities $\{\pi_i;\ i \geq 0\}$ is* ***reversible*** *if $P_{ij} = \pi_j P_{ji}/\pi_i$ for all i,j, i.e., if $P^*_{ij} = P_{ij}$ for all i,j.*

Thus the chain is reversible if, in steady state, the backward running sequence of states is statistically indistinguishable from the forward running sequence. Comparing (6.41) with the steady-state equations (6.30) that we derived for birth–death chains, we have proved the following important theorem.

Theorem 6.5.2 *Every birth–death chain with a steady-state probability distribution is reversible.*

We saw that for birth–death chains, the equation $\pi_i P_{ij} = \pi_j P_{ji}$ (which only had to be considered for $|i - j| \leq 1$) provided a very simple way of calculating the steady-state probabilities. Unfortunately, it appears that in general we must first calculate the steady-state probabilities in order to show that a chain is reversible. The following simple theorem gives us a convenient escape from this dilemma.

Theorem 6.5.3 *Assume that an irreducible Markov chain has transition probabilities $\{P_{ij}\}$. Suppose $\{\pi_i\}$ is a set of positive numbers summing to 1 and satisfying*

$$\pi_i P_{ij} = \pi_j P_{ji}, \qquad \text{all } i,j. \tag{6.42}$$

Then, first, $\{\pi_i;\ i \geq 0\}$ is the steady-state distribution for the chain, and, second, the chain is reversible.

Proof Given a solution to (6.42) for all i and j, we can sum (6.42) over i for each j.

$$\sum_i \pi_i P_{ij} = \pi_j \sum_i P_{ji} = \pi_j. \tag{6.43}$$

Thus the solution to (6.42), along with the constraints $\pi_i > 0$, $\sum_i \pi_i = 1$, satisfies the steady-state equations, (6.15). From Theorem 6.3.8, this is the unique steady-state distribution. Since (6.42) is satisfied, the chain is also reversible. □

It is often possible, sometimes by using an educated guess, to find a solution to (6.42). If this is successful, then we are assured both that the chain is reversible and that the actual steady-state probabilities have been found.

Note that the theorem applies to periodic chains as well as to aperiodic chains. If the chain is periodic, then the steady-state probabilities have to be interpreted as average values over the period, but Theorem 6.3.8 shows that (6.43) still has a unique solution (assuming a positive-recurrent chain). On the other hand, for a chain with period $d > 1$, there are d subclasses of states and the sequence $\{X_n\}$ must rotate between these classes in a fixed order. For this same order to be followed in the backward chain, the only possibility is $d = 2$. Thus periodic chains with periods other than 2 cannot be reversible.

There are several simple tests that can be used to show that an irreducible chain is not reversible. First, the steady-state probabilities must satisfy $\pi_i > 0$ for all i, and thus, if $P_{ij} > 0$ but $P_{ji} = 0$ for some i, j, then (6.42) cannot be satisfied and the chain is not reversible. Second, consider any set of three states, i, j, k. If $P_{ji}P_{ik}P_{kj}$ is unequal to $P_{jk}P_{ki}P_{ij}$, then the chain cannot be reversible. To see this, note that (6.42) requires that

$$\pi_i = \pi_j P_{ji}/P_{ij} = \pi_k P_{ki}/P_{ik}.$$

Thus, $\pi_j P_{ji} P_{ik} = \pi_k P_{ki} P_{ij}$. Equation (6.42) also requires that $\pi_j P_{jk} = \pi_k P_{kj}$. Taking the ratio of these equations, we see that $P_{ji}P_{ik}P_{kj} = P_{jk}P_{ki}P_{ij}$. Thus if this equation is not satisfied, the chain cannot be reversible. In retrospect, this result is not surprising. What it says is that for any cycle of three states, the probability of three transitions going around the cycle in one direction must be the same as the probability of going around the cycle in the opposite (and therefore backwards) direction.

It is also true (see [22] for a proof) that a necessary and sufficient condition for a chain to be reversible is that the product of transition probabilities around any cycle of arbitrary length must be the same as the product of transition probabilities going around the cycle in the opposite direction. This does not seem to be a widely useful way to demonstrate reversibility.

There is another result, generalizing Theorem 6.5.3, for finding the steady-state probabilities of an arbitrary Markov chain and simultaneously finding the transition probabilities of the backward chain.

Theorem 6.5.4 *Assume that an irreducible Markov chain has transition probabilities $\{P_{ij}\}$. Suppose $\{\pi_i\}$ is a set of positive numbers summing to 1 and that $\{P^*_{ij}\}$ is a set of transition probabilities satisfying*

$$\pi_i P_{ij} = \pi_j P^*_{ji}, \quad \text{all } i, j. \tag{6.44}$$

Then $\{\pi_i\}$ is the steady-state distribution and $\{P^*_{ij}\}$ is the set of transition probabilities for the backward chain.

Proof Summing (6.44) over i, we get the steady-state equations for the Markov chain, so the fact that the given $\{\pi_i\}$ satisfy these equations asserts that they are the steady-state probabilities. Equation (6.44) then asserts that $\{P^*_{ij}\}$ is the set of transition probabilities for the backward chain. □

The following two sections illustrate some important applications of reversibility.

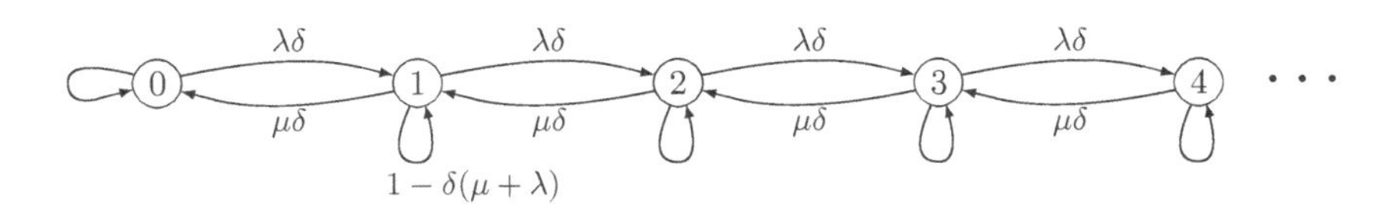

Figure 6.5 Sampled-time approximation to M/M/1 queue for time increment δ.

6.6 The M/M/1 sampled-time Markov chain

The M/M/1 Markov chain is a sampled-time model of the M/M/1 queueing system. Recall that the M/M/1 queue has Poisson arrivals at some rate λ and IID exponential service times at some rate μ. We assume throughout this section that $\lambda < \mu$ (this is required to make the states positive-recurrent). For some given small increment of time δ, we visualize observing the state of the system at the sample times $n\delta$. As indicated in Figure 6.5, the probability of an arrival in the interval from $(n-1)\delta$ to $n\delta$ is modeled as $\lambda\delta$, independent of the state of the chain at time $(n-1)\delta$ and thus independent of all prior arrivals and departures. Thus the arrival process, viewed as arrivals in subsequent intervals of duration δ, is Bernoulli, thus approximating the Poisson arrivals. This is a sampled-time approximation to the Poisson arrival process of rate λ for a continuous-time M/M/1 queue.

When the system is non-empty (i.e., the state of the chain is 1 or more), the probability of a departure in the interval $(n-1)\delta$ to $n\delta$ is $\mu\delta$, thus modeling the exponential service times. When the system is empty, of course, departures cannot occur.

Note that in our sampled-time model, there can be at most one arrival or departure in an interval $(n-1)\delta$ to $n\delta$. As in the Poisson process, the probability of more than one arrival, more than one departure, or both an arrival and a departure in an increment δ is of order δ^2 for the actual continuous-time M/M/1 system being modeled. Thus, for δ very small, we expect the sampled-time model to be relatively good. At any rate, if $\delta \leq 1/(\mu + \lambda)$, the self transitions have non-negative probability and the model can be analyzed with no further approximations.

Since this chain is a birth–death chain, we can use (6.33) to determine the steady-state probabilities; they are

$$\pi_i = \pi_0 \rho^i \; ; \; \rho = \lambda/\mu < 1.$$

Setting the sum of the π_i to 1, we find that $\pi_0 = 1 - \rho$, so

$$\pi_i = (1-\rho)\rho^i \,, \qquad \text{all } i \geq 0. \tag{6.45}$$

Thus the steady-state probabilities exist and the chain is a birth–death chain, so from Theorem 6.5.2, it is reversible. We now exploit the consequences of reversibility to find some rather surprising results about the M/M/1 chain in steady state. Figure 6.6 illustrates a sample path of arrivals and departures for the chain. To avoid the confusion associated with the backward chain evolving backward in time, we refer to the original chain as the chain moving to the right and to the backward chain as the chain moving to the left.

There are two types of correspondence between the right-moving and the left-moving chain:

1. The left-moving chain has the same Markov chain description as the right-moving chain, and thus can be viewed as an M/M/1 chain in its own right. We still label the sampled-time intervals from left to right, however, so that the left-moving chain makes transitions from X_{n+1} to X_n to X_{n-1} etc. Thus, for example, if $X_n = i$ and $X_{n-1} = i + 1$, the left-moving chain is viewed as having an arrival in the interval from $n\delta$ to $(n-1)\delta$. The sequence of such downward transitions moving to the left is a Bernoulli process.
2. Each sample function $\ldots x_{n-1}, x_n, x_{n+1} \ldots$ of the right-moving chain corresponds to the same sample function $\ldots x_{n+1}, x_n, x_{n-1} \ldots$ of the left-moving chain, where $X_{n-1} = x_{n-1}$ is to the left of $X_n = x_n$ for both chains. With this correspondence, an arrival to the right-moving chain in the interval $(n-1)\delta$ to $n\delta$ is a departure from the left-moving chain in the interval $n\delta$ to $(n-1)\delta$, and a departure from the right-moving chain is an arrival to the left-moving chain. Using this correspondence, each event in the left-moving chain corresponds to some event in the right-moving chain.

In each of the properties of the M/M/1 chain to be derived below, a property of the left-moving chain is developed through correspondence 1 above, and then that property is translated into a property of the right-moving chain by correspondence 2.

Property 1 Since the arrival process of the right-moving chain is Bernoulli, the arrival process of the left-moving chain is also Bernoulli (by correspondence 1). Looking at a sample function x_{n+1}, x_n, x_{n-1} of the left-moving chain (i.e., using correspondence 2), an arrival in the interval $n\delta$ to $(n-1)\delta$ of the left-moving chain is a departure in the interval $(n-1)\delta$ to $n\delta$ of the right-moving chain. Since the arrivals in successive increments of the left-moving chain are independent and have probability $\lambda\delta$ in each increment δ, we conclude that departures in the right-moving chain are similarly Bernoulli (see Figure 6.6).

The fact that the departure process is Bernoulli with departure probability $\lambda\delta$ in each increment is surprising. Note that the probability of a departure in the interval $(n\delta - \delta, n\delta]$ is $\mu\delta$ conditional on $X_{n-1} \geq 1$ and is 0 conditional on $X_{n-1} = 0$. Since $\Pr\{X_{n-1} \geq 1\} = 1 - \Pr\{X_{n-1} = 0\} = \rho$, we see that the unconditional probability of a departure in the interval $(n\delta - \delta, n\delta]$ is $\rho\mu\delta = \lambda\delta$ as asserted above. The fact that successive departures are independent is much harder to derive without using reversibility (see Exercise 6.15).

Property 2 In the original (right-moving) chain, arrivals in the time increments after $n\delta$ are independent of X_n. Thus, for the left-moving chain, arrivals in time increments to the left of $n\delta$ are independent of the state of the chain at $n\delta$. From the correspondence between sample paths, however, a left chain arrival is a right chain departure, so that for the right-moving chain, departures in the time increments prior to $n\delta$ are independent of X_n, which is equivalent to saying that the state X_n is independent of the prior departures. This means that if one observes the departures prior to time $n\delta$, one obtains no information about the state of the chain at $n\delta$. This is again a surprising result. To make it seem more plausible, note that an unusually large number of departures in an interval from

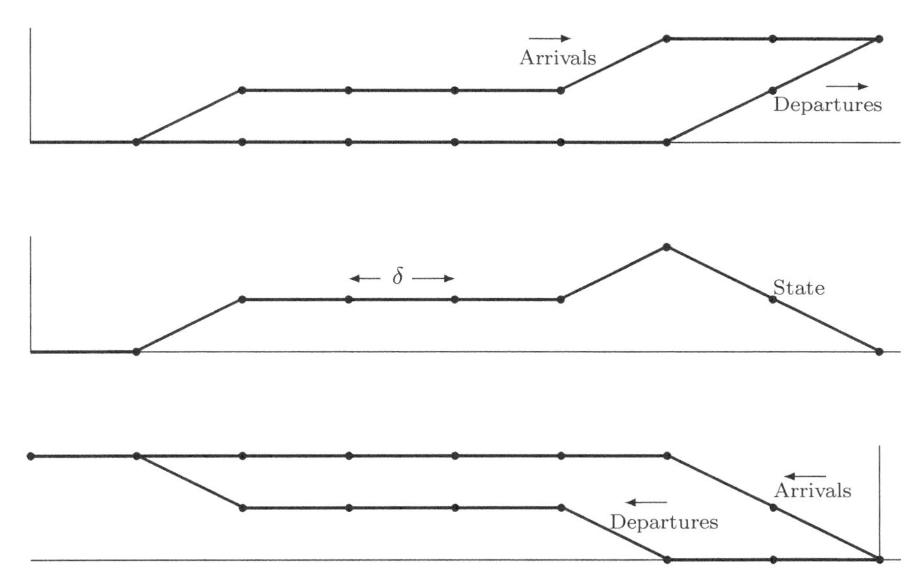

Figure 6.6 Sample function of M/M/1 chain over a busy period and corresponding arrivals and departures for right- and left-moving chains. Arrivals and departures are viewed as occurring between the sample times, and an arrival in the left-moving chain between times $n\delta$ and $(n+1)\delta$ corresponds to a departure in the right-moving chain between $(n+1)\delta$ and $n\delta$.

$(n-m)\delta$ to $n\delta$ indicates that a large number of customers were probably in the system at time $(n-m)\delta$, but it does not appear to say much (and in fact it says exactly nothing) about the number remaining at $n\delta$.

The following theorem summarizes these results.

Theorem 6.6.1 (Burke's theorem for sampled-time) *Given an M/M/1 Markov chain in steady state with $\lambda < \mu$,*

(a) the departure process is Bernoulli with departure probability $\lambda\delta$ per increment,
(b) the state X_n at any time $n\delta$ is independent of departures prior to $n\delta$.

The proof of Burke's theorem above did not use the fact that the departure probability is the same for all states except state 0. Thus these results remain valid for any birth–death chain with Bernoulli arrivals that are independent of the current state (i.e., for which $P_{i,i+1} = \lambda\delta$ for all $i \geq 0$). One important example of such a chain is the sampled time approximation to an M/M/m queue. Here there are m servers, and the probability of departure from state i in an increment δ is $\mu i\delta$ for $i \leq m$ and $\mu m\delta$ for $i > m$. For the states to be recurrent, and thus for a steady state to exist, λ must be less than μm. Subject to this restriction, properties (a) and (b) above are valid for sampled-time M/M/m queues.

6.7 Branching processes

Branching processes provide a simple model for studying the population of various types of individuals from one generation to the next. The individuals could be photons in a photo-multiplier, particles in a cloud chamber, micro-organisms, insects, or branches in a data structure.

Let X_n be the number of individuals in generation n of some population. Each of these X_n individuals, independently of each other, produces a random number of offspring, and these offspring collectively make up generation $n+1$. More precisely, a *branching process* is a Markov chain $\{X_n; n \geq 1\}$ in which the X_n models the number of individuals in generation n and successive X_n evolve as follows: for the individuals $\{1, 2, ..., X_n\}$ in generation n, let $Y_{k,n}$ be the number of offspring of individual k. The rv s $Y_{k,n}$ are defined to be IID over k and n, with a PMF $p_j = \Pr\{Y_{k,n} = j\}$. The state at time $n+1$, namely the number of individuals in generation $n+1$, is

$$X_{n+1} = \sum_{k=1}^{X_n} Y_{k,n}. \tag{6.46}$$

Assume a given distribution (perhaps deterministic) for the initial state X_0. The transition probability, $P_{ij} = \Pr\{X_{n+1} = j \mid X_n = i\}$, is just the probability that $Y_{1,n} + Y_{2,n} + \cdots + Y_{i,n} = j$. The zero state (i.e., the state in which there are *no* individuals) is a trapping state (i.e., $P_{00} = 1$) since no future offspring can arise in this case.

One of the most important issues about a branching process is the probability that the population dies out eventually. Naturally, if p_0 (the probability that an individual has no offspring) is zero, then each generation must be at least as large as the generation before, and the population cannot die out unless $X_0 = 0$. We assume in what follows that $p_0 > 0$ and $X_0 > 0$. Recall that $\mathsf{F}_{ij}(n)$ was defined as the probability, given $X_0 = i$, that state j is entered between times 1 and n. From (6.8), this satisfies the iterative relation

$$\mathsf{F}_{ij}(n) = P_{ij} + \sum_{k \neq j} P_{ik} \mathsf{F}_{kj}(n-1), n > 1; \quad \mathsf{F}_{ij}(1) = P_{ij}. \tag{6.47}$$

The probability that the process dies out by time n or before, given $X_0 = i$, is thus $\mathsf{F}_{i0}(n)$. For the nth generation to die out, starting with an initial population of i individuals, the descendants of each of those i individuals must die out. Since each individual generates descendants independently, we have $\mathsf{F}_{i0}(n) = [\mathsf{F}_{10}(n)]^i$ for all i and n. Because of this relationship, it is sufficient to find $\mathsf{F}_{10}(n)$, which can be determined from (6.47). Observe that P_{1k} is just p_k, the probability that an individual will have k offspring. Thus, (6.47) becomes

$$\mathsf{F}_{10}(n) = p_0 + \sum_{k=1}^{\infty} p_k [\mathsf{F}_{10}(n-1)]^k = \sum_{k=0}^{\infty} p_k [\mathsf{F}_{10}(n-1)]^k. \tag{6.48}$$

Let $h(z) = \sum_k p_k z^k$ be the z transform of the number of an individual's offspring. Then (6.48) can be written as

$$\mathsf{F}_{10}(n) = h(\mathsf{F}_{10}(n-1)). \tag{6.49}$$

This iteration starts with $\mathsf{F}_{10}(1) = p_0$. Figure 6.7 shows a graphical construction for evaluating $\mathsf{F}_{10}(n)$. Having found $\mathsf{F}_{10}(n)$ as an ordinate on the graph for a given value of n, we find the same value as an abscissa by drawing a horizontal line over to the straight line of slope 1; we then draw a vertical line back to the curve $h(z)$ to find $h(\mathsf{F}_{10}(n)) = \mathsf{F}_{10}(n+1)$.

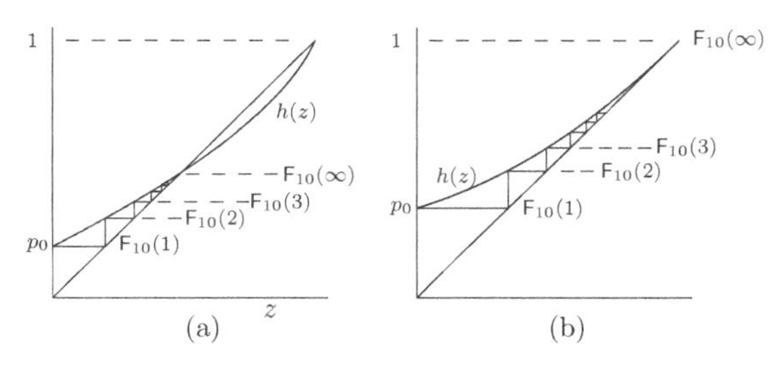

Figure 6.7 Graphical construction to find the probability that a population dies out. Here $\mathsf{F}_{10}(n)$ is the probability that a population starting with one member at generation 0 dies out by generation n or before. Thus $\mathsf{F}_{10}(\infty)$ is the probability that the population ever dies out. In (a) $\mathsf{F}_{1,0}(\infty) < 1$, whereas in (b) $\mathsf{F}_{1,0}(\infty) = 1$.

For the two subfigures shown, it can be seen that $\mathsf{F}_{10}(\infty)$ is equal to the smallest root of the equation $h(z) - z = 0$. We next show that these two figures are representative of all possibilities. Since $h(z)$ is a z transform, we know that $h(1) = 1$, so that $z = 1$ is one root of $h(z) - z = 0$. Also, $h'(1) = \overline{Y}$, where $\overline{Y} = \sum_k kp_k$ is the expected number of an individual's offspring. If $\overline{Y} > 1$, as in Figure 6.7(a), then $h(z) - z$ is negative for z slightly smaller than 1. Also, for $z = 0$, $h(z) - z = h(0) = p_0 > 0$. Since $h''(z) \geq 0$, there is exactly one root of $h(z) - z = 0$ for $0 < z < 1$, and that root is equal to $\mathsf{F}_{10}(\infty)$. By the same type of analysis, it can be seen that if $\overline{Y} \leq 1$, as in Figure 6.7(b), then there is no root of $h(z) - z = 0$ for $z < 1$, and $\mathsf{F}_{10}(\infty) = 1$.

As we saw earlier, $\mathsf{F}_{i0}(\infty) = [\mathsf{F}_{10}(\infty)]^i$, so that for any initial population size, there is a probability strictly between 0 and 1 that successive generations eventually die out for $\overline{Y} > 1$, and probability 1 that successive generations eventually die out for $\overline{Y} \leq 1$. Since state 0 is accessible from all i, but $\mathsf{F}_{0i}(\infty) = 0$, it follows from Lemma 6.2.4 that all states other than state 0 are transient.

We next evaluate the expected number of individuals in a given generation. Conditional on $X_{n-1} = i$, (6.46) shows that the expected value of X_n is $i\overline{Y}$. Taking the expectation over X_{n-1}, we have

$$\mathsf{E}[X_n] = \overline{Y}\mathsf{E}\left[X_{n-1}\right]. \tag{6.50}$$

Iterating this equation, we get

$$\mathsf{E}[X_n] = \overline{Y}^n\mathsf{E}[X_0]. \tag{6.51}$$

Thus, if $\overline{Y} > 1$, the expected number of individuals in a generation increases exponentially with n, and $\overline{Y}$ gives the rate of growth. Physical processes do not grow exponentially forever, so branching processes are appropriate models of such physical processes only over some finite range of population. Even more important, the model here assumes that the number of offspring of a single member is independent of the total population, which is highly questionable in many areas of population growth. The advantage of an oversimplified model such as this is that it explains what would happen under these idealized conditions, thus providing insight into how the model should be changed for more realistic scenarios.

It is important to realize that, for branching processes, the mean number of individuals is not a good measure of the actual number of individuals. For $\overline{Y} = 1$ and $X_0 = 1$, the expected number of individuals in each generation is 1, but the probability that $X_n = 0$ approaches 1 with increasing n; this means that as n gets large, the nth generation contains a large number of individuals with a very small probability and contains no individuals with a very large probability. For $\overline{Y} > 1$, we have just seen that there is a positive probability that the population dies out, but the expected number is growing exponentially.

A surprising result, which is derived from the theory of martingales in Chapter 9, is that if $X_0 = 1$ and $\overline{Y} > 1$, then the sequence of rv s $X_n/\overline{Y}^n$ has a limit with probability 1. This limit is a rv; it has the value 0 with probability $\mathsf{F}_{10}(\infty)$, and has larger values with some given distribution. Intuitively, for large n, X_n is either 0 or very large. If it is very large, it tends to grow in an orderly way, increasing by a multiple of $\overline{Y}$ in each subsequent generation.

6.8 Round-robin service and processor sharing

Typical queueing systems have one or more servers who each serve customers in first come, first served (FCFS) order, serving one customer completely while other customers wait. These typical systems have larger average delay than necessary. For example, if two customers with service requirements of 10 and 1 units respectively are waiting when a single server becomes empty, then serving the first before the second results in departures at times 10 and 11, for an average delay of 10.5. Serving the customers in the opposite order results in departures at times 1 and 11, for an average delay of 6. Supermarkets have recognized this for years and have special express checkout lines for customers with small service requirements.

Giving priority to customers with small service requirements, however, has some disadvantages; first, customers with high service requirements can feel discriminated against, and second, it is not always possible to determine the service requirements of customers before they are served. The following alternative to priorities is popular in both the computer and data network industries. When a processor in a computer system has many jobs to accomplish, it often serves these jobs on a time-shared basis, spending a small increment of time on one, then the next, and so forth. In data networks, particularly high-speed networks, messages are broken into small fixed-length packets, and then the packets from different messages can be transmitted on an alternating basis between messages.

A *round-robin* service system is a system in which, if there are m customers in the system, say $c_1, c_2, \ldots, c_m$, then c_1 is served for an incremental interval δ, followed by c_2 being served for an interval δ, and so forth up to c_m. After c_m is served for an interval δ, the server returns and starts serving c_1 for an interval δ again. Thus the customers are served in a cyclic, or 'round-robin' order, each getting a small increment of service on each visit from the server. When a customer's service is completed, the customer leaves the system, m is reduced, and the server continues rotating through the now reduced cycle as before. When a new customer arrives, m is increased and the new customer must be inserted into the cycle of existing customers in a way to be discussed later.

Processor sharing is the limit of round-robin service as the increment δ goes to zero. Thus, with processor sharing, if m customers are in the system, all are being served simultaneously, but each is being served at $1/m$ times the basic server rate. For the example of two customers with service requirement 1 and 10, each customer is initially served at rate $1/2$, so one customer departs at time 2. At that time, the remaining customer is served at rate 1 and departs at time 11. For round-robin service with an increment of 1, the customer with unit service requirement departs at either time 1 or 2, depending on the initial order of service. With other increments of service, the results are slightly different.

We first analyze round-robin service and then go to the processor-sharing limit as $\delta \to 0$. As the above example suggests, the results are somewhat cleaner in the limiting case, but more realistic in the round-robin case. Round-robin service provides a good example of the use of backward transition probabilities to find the steady-state distribution of a Markov chain. The techniques used here are quite similar to those used in the next chapter to analyze queueing networks.

Assume a Bernoulli arrival process in which the probability of an arrival in an interval δ is $\lambda\delta$. Assume that the ith arriving customer has a service requirement W_i. The rv s W_i, $i \geq 1$, are IID and independent of the arrival epochs. Thus, in terms of the arrival process and the service requirements, this is the same as an M/G/1 queue (see Section 5.5.5), but with M/G/1 queues, the server serves each customer completely before going on to the next customer. We shall find that the round-robin service here avoids the 'slow truck effect' identified with the M/G/1 queue.

For simplicity, assume that W_i is arithmetic with span δ, taking on only values that are positive integer multiples of δ. Let $f(j) = \Pr\{W_i = j\delta\}$, $j \geq 1$ and let $\overline{F}(j) = \Pr\{W_i > j\delta\}$. Note that if a customer has already received j increments of service, then the probability that that customer will depart after 1 more increment is $f(j+1)/\overline{F}(j)$. This probability of departure on the next service increment after the jth is denoted by

$$g(j) = f(j+1)/\overline{F}(j); \; j \geq 1. \tag{6.52}$$

The state $\boldsymbol{s}$ of a round-robin system can be expressed as the number, m, of customers in the system, along with an ordered listing of how many service increments each of those m customers have received, i.e.,

$$\boldsymbol{s} = (m, z_1, z_2, \ldots, z_m), \tag{6.53}$$

where $z_1\delta$ is the amount of service already received by the customer at the front of the queue, $z_2\delta$ is the service already received by the next customer in order, etc. In the special case of an idle queue, $\boldsymbol{s} = (0)$, which we denote as ϕ.

Given that the state X_n at time $n\delta$ is $\boldsymbol{s} \neq \phi$, the state X_{n+1} at time $n\delta + \delta$ evolves as follows:

- a new arrival enters with probability $\lambda\delta$ and is placed at the front of the queue;
- the customer at the front of the queue receives an increment δ of service;

- the customer departs if service is complete;
- otherwise, the customer goes to the back of the queue.

It can be seen that the state transition depends, first, on whether a new arrival occurs (an event of probability $\lambda\delta$), and, second, on whether a departure occurs. If neither an arrival nor a departure occurs, then the queue simply rotates. The new state is $s' = r(s)$, where the rotation operator $r(s)$ is defined by $r(s) = (m, z_2, \ldots, z_m, z_1+1)$. If a departure but no arrival occurs, then the customer at the front of the queue receives its last unit of service and departs. The new state is $s' = d(s)$, where the departure operator $d(s)$ is defined by $d(s) = (m-1, z_2, \ldots, z_m)$.

If an arrival occurs, the new customer receives one unit of service and goes to the back of the queue if more than one unit of service is required. In this case, the new state is $s' = a(s)$, where the arrival operator $a(s)$ is defined by $a(s) = (m+1, z_1, z_2, \ldots, z_m, 1)$. If only one unit of service is required by a new arrival, the arrival departs and $s' = s$. In the special case of an empty queue, $s = \phi$, the state is unchanged if either no arrival occurs or an arrival requiring one increment of service arrives. Otherwise, the new state is $s = (1, 1)$, i.e., the one customer in the system has received one increment of service.

We next find the probability of each transition for $s \neq \phi$. The probability of no arrival is $1 - \lambda\delta$. Given no arrival, and given a non-empty system, $s \neq \phi$, the probability of a departure is $g(z_1) = f(z_1+1)/\overline{F}(z_1)$, i.e., the probability that one more increment of service allows the customer at the front of the queue to depart. Thus the probability of a departure is $(1-\lambda\delta)g(z_1)$ and the probability of a rotation is $(1-\lambda\delta)[1-g(z_1)]$. Finally, the probability of an arrival is $\lambda\delta$, and given an arrival, the new arrival will leave the system after one unit of service with probability $g(0) = f(1)$. Thus the probability of an arrival and no departure is $\lambda\delta[1-f(1)]$ and the probability of an unchanged system is $\lambda\delta f(1)$. To summarize, for $s \neq \phi$,

$$\begin{aligned}
P_{s,r(s)} &= (1-\lambda\delta)[1-g(z_1)], & r(s) &= (m, z_2, \ldots, z_m, z_1+1);\\
P_{s,d(s)} &= (1-\lambda\delta)g(z_1), & d(s) &= (m-1, z_2, \ldots, z_m);\\
P_{s,a(s)} &= \lambda\delta[1-f(1)], & a(s) &= (m+1, z_1, z_2, \ldots, z_m, 1);\\
P_{s,s} &= \lambda\delta f(1). & &
\end{aligned} \tag{6.54}$$

For the special case of the idle state, $P_{\phi,\phi} = (1-\lambda\delta)+\lambda\delta f(1)$ and $P_{\phi,(1,1)} = \lambda\delta(1-f(1))$.

We now find the steady-state distribution for this Markov chain by looking at the backward Markov chain. We will hypothesize backward transition probabilities, and then use Theorem 6.5.4 to verify that the hypothesis is correct. Consider the backward transitions corresponding to each of the forward transitions in (6.54). A rotation in forward time causes the elements $z_1, \ldots, z_m$ in the state $s = (m, z_1, \ldots, z_m)$ to rotate left, and the left-most element (corresponding to the front of the queue) is incremented while rotating to the right end. The backward transition from $r(s)$ to s corresponds to the elements $z_2, \ldots, z_m, z_1+1$ rotating to the right, with the right-most element being decremented while rotating to the left end. If we view the transitions in backward time as a kind of round-robin system, we see that the rotation is in the opposite direction from the forward time system.

In the backward time system, we view the numbers $z_1, \ldots, z_m$ in the state as the remaining service required before the corresponding customers can depart. Thus, these numbers decrease in the backward-moving system. Also, since the customer rotation in the backward-moving system is opposite to that in the forward-moving system, z_m is the remaining service of the customer at the front of the queue, and z_1 is the remaining service of the customer at the back of the queue. We also view departures in forward time as arrivals in backward time. Thus the backward transition from $d(\boldsymbol{s}) = (m-1, z_2, \ldots, z_m)$ to $\boldsymbol{s} = (m, z_1, \ldots, z_m)$ corresponds to an arrival requiring $z_1 + 1$ units of service; the arrival goes to the front of the queue, receives one increment of service, and then goes to the back of the queue with z_1 increments of remaining service.

The most desirable result we could now hope for is that the arrivals in backward time are Bernoulli. This is a reasonable hypothesis to make, partly because it is plausible, and partly because it is easy to check via Theorem 6.5.4. Fortunately, we shall find that it is valid. According to this hypothesis, the backward transition probability $P^*_{r(\boldsymbol{s}),\boldsymbol{s}}$ is given by $1 - \lambda\delta$; i.e., given that X_{n+1} is $r(\boldsymbol{s}) = (m, z_2, \ldots, z_m, z_1 + 1)$, and given that there is no arrival in the backward system at time $(n+1)\delta$, then the only possible state at time n is $\boldsymbol{s} = (m, z_1, \ldots, z_m)$. Next consider a backward transition from $d(\boldsymbol{s}) = (m-1, z_2, \ldots, z_n)$ to $\boldsymbol{s} = (m, z_1, z_2, \ldots, z_m)$. This corresponds to an arrival in the backward moving system; the arrival requires z_1+1 increments of service, one of which is provided immediately, leaving the arrival at the back of the queue with z_1 required increments of service remaining. The probability of this transition is $P^*_{d(\boldsymbol{s}),\boldsymbol{s}} = \lambda\delta f(z_1+1)$. Calculating the other backward transitions in the same way, the hypothesized backward transition probabilities are given by

$$P^*_{r(\boldsymbol{s}),\boldsymbol{s}} = 1 - \lambda\delta; \qquad P^*_{d(\boldsymbol{s}),\boldsymbol{s}} = \lambda\delta f(z_1 + 1);$$
$$P^*_{a(\boldsymbol{s}),\boldsymbol{s}} = 1 - \lambda\delta; \qquad P^*_{\boldsymbol{s},\boldsymbol{s}} = \lambda\delta f(1). \tag{6.55}$$

One should view (6.55) as an hypothesis for the backward transition probabilities. The arguments leading up to (6.55) are simply motivation for this hypothesis. If the hypothesis is correct, we can combine (6.54) and (6.55) to express the steady-state equations of Theorem 6.5.4 (for $\boldsymbol{s} \neq \phi$) as

$$\pi_{\boldsymbol{s}} P_{\boldsymbol{s},r(\boldsymbol{s})} = \pi_{r(\boldsymbol{s})} P^*_{r(\boldsymbol{s}),\,\boldsymbol{s}}, \qquad (1-\lambda\delta)[1-g(z_1)]\pi_{\boldsymbol{s}} = (1-\lambda\delta)\pi_{r(\boldsymbol{s})}; \tag{6.56}$$

$$\pi_{\boldsymbol{s}} P_{\boldsymbol{s},d(\boldsymbol{s})} = \pi_{d(\boldsymbol{s})} P^*_{d(\boldsymbol{s}),\,\boldsymbol{s}}, \qquad (1-\lambda\delta)g(z_1)\pi_{\boldsymbol{s}} = \lambda\delta f(z_1+1)\pi_{d(\boldsymbol{s})}; \tag{6.57}$$

$$\pi_{\boldsymbol{s}} P_{\boldsymbol{s},a(\boldsymbol{s})} = \pi_{a(\boldsymbol{s})} P^*_{a(\boldsymbol{s}),\,\boldsymbol{s}}, \qquad \lambda\delta[1-f(1)]\pi_{\boldsymbol{s}} = (1-\lambda\delta)\pi_{a(\boldsymbol{s})}; \tag{6.58}$$

$$\pi_{\boldsymbol{s}} P_{\boldsymbol{s},\boldsymbol{s}} = \pi_{\boldsymbol{s}} P^*_{\boldsymbol{s},\boldsymbol{s}}, \qquad \lambda\delta f(1)\pi_{\boldsymbol{s}} = \lambda\delta f(1)\pi_{\boldsymbol{s}}. \tag{6.59}$$

We next show that (6.57), applied repeatedly, will allow us to solve for $\pi_{\boldsymbol{s}}$ (if λ is small enough for the states to be positive recurrent). Verifying that the solution also satisfies (6.56) and (6.58) will then verify the hypothesis. Since $f(z_1+1)/g(z_1)$ is $\overline{F}(z_1)$ from (6.52), we have

$$\pi_{\boldsymbol{s}} = \frac{\lambda\delta}{1-\lambda\delta}\,\overline{F}(z_1)\pi_{d(\boldsymbol{s})}. \tag{6.60}$$

For $m > 1$, $d(\mathbf{s}) = (m-1, z_2, \ldots, z_m)$, so we can apply (6.60) to $\pi_{d(\mathbf{s})}$, and substitute the result back into (6.60), yielding

$$\pi_{\mathbf{s}} = \left(\frac{\lambda\delta}{1-\lambda\delta}\right)^2 \overline{F}(z_1)\overline{F}(z_2)\pi_{d(d(\mathbf{s}))}, \tag{6.61}$$

where $d(d(\mathbf{s})) = (m-2, z_3, \ldots, z_m)$. Applying (6.60) repeatedly to $\pi_{d(d(\mathbf{s}))}, \pi_{d(d(d(\mathbf{s})))}$, and so forth, we eventually get

$$\pi_{\mathbf{s}} = \left(\frac{\lambda\delta}{1-\lambda\delta}\right)^m \left(\prod_{j=1}^{m} \overline{F}(z_j)\right)\pi_\phi. \tag{6.62}$$

Before this can be accepted as a steady-state distribution, we must verify that it satisfies (6.56) and (6.58). The left-hand side of (6.56) is $(1-\lambda\delta)[1-g(z_1)]\pi_{\mathbf{s}}$, and, from (6.52), $1-g(z_1) = [\overline{F}(z_1) - f(z_1+1)]/\overline{F}(z_1) = \overline{F}(z_1+1)/(z_1)$. Thus using (6.62), the left-hand side of (6.56) is

$$(1-\lambda\delta)\,\frac{\overline{F}(z_1+1)}{\overline{F}(z_1)}\left(\frac{\lambda\delta}{1-\lambda\delta}\right)^m\left(\prod_{j=1}^{m}\overline{F}(z_j)\right)\pi_\phi = (1-\lambda\delta)\left(\frac{\lambda\delta}{1-\lambda\delta}\right)^m\left(\prod_{j=2}^{m}\overline{F}(z_j)\right)\overline{F}(z_1+1)\pi_\phi.$$

This is equal to $(1-\lambda\delta)\pi_{r(\mathbf{s})}$, verifying (6.56). Equation (6.58) is verified in the same way. We now have to find whether there is a solution for π_ϕ such that these probabilities sum to 1. First define $P_m = \sum_{z_1,\ldots,z_m} \pi(m, z_1, \ldots, z_m)$. This is the probability of m customers in the system. Whenever a new customer enters the system, it receives one increment of service immediately, so each $z_i \geq 1$. Using the hypothesized solution in (6.62),

$$P_m = \left(\frac{\lambda\delta}{1-\lambda\delta}\right)^m \left(\prod_{j=1}^{m}\sum_{i=1}^{\infty}\overline{F}(i)\right)\pi_\phi. \tag{6.63}$$

Since $\overline{F}(i) = \Pr\{W > i\delta\}$, since W is arithmetic with span δ, and since the mean of a non-negative random variable is the integral of its complementary CDF, we have

$$\delta\sum_{i=1}^{\infty}\overline{F}(i) = \mathsf{E}[W] - \delta, \tag{6.64}$$

$$P_m = \left(\frac{\lambda}{1-\lambda\delta}\right)^m \left(\mathsf{E}[W] - \delta\right)^m \pi_\phi. \tag{6.65}$$

Defining $\rho = [\lambda/(1-\lambda\delta)]\{\mathsf{E}[W] - \delta\}$, we see $P_m = \rho^m\pi_\phi$. If $\rho < 1$, then $\pi_\phi = 1-\rho$, and

$$P_m = (1-\rho)\rho^m, \quad m \geq 0. \tag{6.66}$$

The condition $\rho < 1$ is required for the states to be positive recurrent. The expected number of customers in the system for a round-robin queue is $\sum_m mP_m = \rho/(1-\rho)$, and, using Little's theorem, Theorem 5.5.9, the expected delay is $\rho/[\lambda(1-\rho)]$. In using Little's theorem here, however, we are viewing the time a customer spends in the system as starting when the number m in the state increases; i.e., if a customer arrives at time

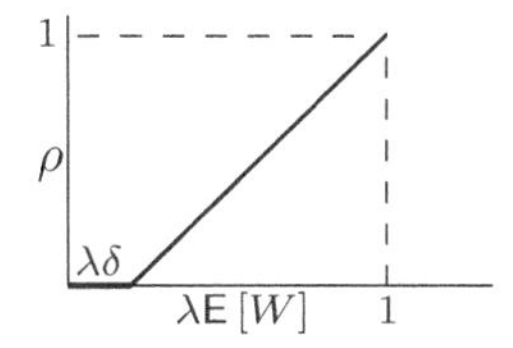

Figure 6.8 ρ as a function of $\lambda\mathsf{E}[W]$ for given $\lambda\delta$.

$n\delta$, it goes to the front of the queue and receives one increment of service, and then, assuming it needs more than one increment, the number m in the state increases at time $(n+1)\delta$. Thus the actual expected delay, including the original δ when the customer is being served but not counted in the state, is $\delta + \rho/[\lambda(1-\rho)]$.

The relation between ρ and $\lambda\mathsf{E}[W]$ is shown in Figure 6.8, and it is seen that $\rho < 1$ for $\lambda\mathsf{E}[W] < 1$. The extreme case where $\lambda\delta = \lambda\mathsf{E}[W]$ is the case for which each customer requires exactly one unit of service. Since at most one customer can arrive per time increment, the state always remains at $\boldsymbol{s} = \phi$, and the delay is δ, i.e., the original increment of service received when a customer arrives.

Note that (6.66) is the same as the distribution of customers in the system for the M/M/1 Markov chain in (6.45), except for the anomaly in the definition of ρ here. We then have the surprising result that if round-robin queueing is used rather than FCFS, then the distribution of the number of customers in the system is approximately the same as that for an M/M/1 queue. In other words, the slow truck effect associated with the M/G/1 queue has been eliminated.

Another remarkable feature of round-robin systems is that one can also calculate the expected delay for a customer conditional on the required service of that customer. This is done in Exercise 6.18, and it is found that the expected delay is linear in the required service.

Next we look at processor sharing by going to the limit as $\delta \to 0$. We first eliminate the assumption that the service requirement distribution is arithmetic with span δ. Assume that the server always spends an increment of time δ on the customer at the front of the queue, and if service is finished before the interval of length δ ends, the server is idle until the next sample time. The analysis of the steady-state distribution above is still valid if we define $\overline{F}(j) = \Pr\{W > j\delta\}$, and $f(j) = \overline{F}(j) - \overline{F}(j+1)$. In this case $\delta \sum_{i=1}^{\infty} \overline{F}(i)$ lies between $\mathsf{E}[W] - \delta$ and $\mathsf{E}[W]$. As $\delta \to 0$, $\rho = \lambda\mathsf{E}[W]$, and distribution of time in the system becomes identical to that of the M/M/1 system.

6.9 Summary

This chapter extended the finite-state Markov chain results of Chapter 4 to the case of countably-infinite state spaces. It also provided an excellent example of how renewal processes can be used for understanding other kinds of processes. In Section 6.2, the first-passage-time rv s were used to construct renewal processes with renewals on successive transitions to a given state. In Section 6.3, these renewal processes were used in place of the eigenvectors and eigenvalues of Chapter 4 to rederive the basic properties of Markov chains. The central result of this was Theorem 6.3.8, which showed that, for an irreducible chain, the states are positive recurrent if and only if the steady-state equations, (6.15), have a solution. Also if (6.15) has a solution, it is positive and unique. We also showed that these steady-state probabilities are, WP1, time-averages for sample paths, and that, for an ergodic chain, they are limiting probabilities independent of the starting state.

We found that the major complications that result from countable-state spaces are, first, different kinds of transient behavior, and, second, the possibility of null-recurrent states. For finite-state Markov chains, a state is transient only if it can reach some other state from which it cannot return. For countably-infinite chains, there is also the case, as in Figure 6.2 for $p > 1/2$, where the state just wanders away, never to return. Null recurrence is a limiting situation where the state wanders away and returns WP1, but with an infinite expected time. There is not much engineering significance to null recurrence; it is highly sensitive to modeling details over the entire infinite set of states. One usually uses countably-infinite chains to simplify models; for example, if a buffer is very large and we do not expect it to overflow, we assume it is infinite. Finding out, then, that the chain is transient or null recurrent simply means that the modeling assumption is not very good.

We next studied birth–death Markov chains and reversibility. Birth–death chains are widely used in queueing theory as sample-time approximations for systems with Poisson arrivals and various generalizations of exponentially distributed service times. Equation (6.33) gives their steady-state probabilities if positive recurrent, and shows the conditions under which they are positive recurrent. We showed that these chains are reversible if they are positive recurrent.

Theorems 6.5.3 and 6.5.4 provide a simple way to find the steady-state distribution of reversible chains and also of chains where the backward chain behavior could be hypothesized or deduced. We used reversibility to show that M/M/1 and M/M/m Markov chains satisfy Burke's theorem for sampled time – namely that the departure process is Bernoulli, and that the state at any time is independent of departures before that time.

Branching processes were introduced in Section 6.7 as a model to study the growth of various kinds of elements that reproduce. In general, for these models (assuming $p_0 > 0$), there is one trapping state and all other states are transient. Figure 6.7 showed how to find the probability that the trapping state is entered by the nth generation, and also the probability that it is entered eventually. If the expected number of offspring of an element is at most 1, then the population dies out WP1, and otherwise, the population dies out with some given probability q, and grows without bound with probability $1-q$.

Round-robin queueing was then used as a more complex example of how to use the backward process to deduce the steady-state distribution of a rather complicated Markov chain; this also gave us added insight into the behavior of queueing systems and allowed us to show that, in the processor-sharing limit, the distribution of number of customers is the same as that in an M/M/1 queue.

For further reading on Markov chains with countably-infinite state spaces, see [9], [22], or [30]. Feller [9] is particularly complete, but Ross [22] and Wolff [30] are somewhat more accessible. Harris [15] is the standard reference on branching processes and Kelly [17] is the standard reference on reversibility. The material on round-robin systems is from [32] and is generalized there.

6.10 Exercises

Exercise 6.1 Let $\{P_{ij};\ i,j \geq 0\}$ be the set of transition probabilities for a countable-state Markov chain. For each i, j, let $\mathsf{F}_{ij}(n)$ be the probability that state j occurs sometime between time 1 and n inclusive, given $X_0 = i$. For some given j, assume that $\{x_i;\ i \geq 0\}$ is a set of non-negative numbers satisfying $x_i = P_{ij} + \sum_{k \neq j} P_{ik} x_k$ for all $i \geq 0$. Show that $x_i \geq \mathsf{F}_{ij}(n)$ for all n and i, and hence that $x_i \geq \mathsf{F}_{ij}(\infty)$ for all i. Hint: Use induction.

Exercise 6.2 **(a)** For the Markov chain in Figure 6.2, show that, for $p \geq 1/2$, $\mathsf{F}_{00}(\infty) = 2(1-p)$ and show that $\mathsf{F}_{i0}(\infty) = [(1-p)/p]^i$ for $i \geq 1$. Hint: First show that this solution satisfies (6.10) and then show that (6.10) has no smaller solution (Exercise 6.1 shows that $F_{i0}(\infty)$ is the smallest solution to (6.10)). Note that you have shown that the chain is transient for $p > 1/2$ and that it is recurrent for $p = 1/2$.

(b) Under the same conditions as (a), show that $\mathsf{F}_{ij}(\infty)$ equals $2(1-p)$ for $i = j$, equals $[(1-p)/p]^{i-j}$ for $i > j$, and equals 1 for $i < j$.

Exercise 6.3 **(a)** Show that the nth-order transition probabilities, starting in state 0, for the Markov chain in Figure 6.2 satisfy

$$P^n_{0j} = pP^{n-1}_{0,\,j-1} + qP^{n-1}_{0,\,j+1} \quad j \neq 0; \qquad P^n_{00} = qP^{n-1}_{00} + qP^{n-1}_{01}.$$

Hint: Use the Chapman–Kolmogorov equality, (4.7).

(b) For $p = 1/2$, use this equation to calculate P^n_{0j} iteratively for $n = 1, 2, 3, 4$. Verify (6.3) for these values and then use induction to verify (6.3) in general. Note: This becomes an absolute mess for $p \neq 1/2$, so do not attempt this in general.

(c) As a more interesting approach, which brings out the relationship of Figures 6.2 and 6.1, note that (6.3), with $j + n$ even, is the probability that $S_n = j$ for the chain in Figure 6.1 and (6.3) with $j + n$ odd is the probability that $S_n = -j - 1$ for the chain in Figure 6.1. By viewing each transition over the self loop at state 0 as a sign reversal for the chain in Figure 6.1, explain why this surprising result is true. (Again, this does not work for $p \neq 1/2$, since the sign reversals also reverse the $+1$, -1 transitions.)

Exercise 6.4 Let j be a transient state in a Markov chain and let j be accessible from i. Show that i is transient also. Interpret this as a form of Murphy's law (if something bad can happen, it will, where the bad thing is the lack of an eventual return). Note: Give a direct demonstration rather than using Theorem 6.2.5.

Exercise 6.5 This exercise treats a mathematical detail from the proof of Lemma 6.2.4. Assume that j is recurrent, there is a path of probability $\alpha > 0$ from j to i, and $n \geq 1$ is an integer. Let T_{ji} be the possibly defective rv giving the first-passage time from j to i.

Verify the following equations for any integer $t > 0$:

$$\begin{aligned}
\Pr\left\{T_{ji} > t\right\} &= \Pr\left\{T_{ji} > t, T_{jj}^{(n)} > t\right\} + \Pr\left\{T_{ji} > t, T_{jj}^{(n)} \le t\right\} \\
&\le \Pr\left\{T_{jj}^{(n)} > t\right\} + \Pr\left\{T_{ji} > T_{jj}^{(n)}\right\} \\
&\le \Pr\left\{T_{jj}^{(n)} > t\right\} + (1-\alpha)^n; \\
\Pr\left\{T_{ji} = \infty\right\} &\le (1-\alpha)^n.
\end{aligned}$$

Exercise 6.6 (Renewal theory proof that a class is all recurrent or all transient) Assume that state j in a countable-state Markov chain is recurrent and i communicates with j. More specifically, assume that there is a path of length m from i to j and one of length k from j to i.

(a) Show that for any $n > 0$, $P_{ii}^{m+n+k} \ge P_{ij}^m P_{jj}^n P_{ji}^k$.

(b) By summing over n, show that state i is also recurrent. Hint: Use Theorem 6.2.6.

(c) Explain why this shows that all states in a class are recurrent or all are transient.

Exercise 6.7 Consider an irreducible Markov chain that is positive recurrent. Recall the technique used to find the expected first-passage time from j to i, i.e., $\overline{T}_{ji}$, in Section 4.5. The state i was turned into a trapping state by turning all the transitions out of i into a single transition $P_{ii} = 1$. Here, in order to preserve the positive recurrence, we instead move all transitions out of state i into the single transition $P_{ij} = 1$.

(a) Use Figure 6.2 to illustrate that the above strategy can turn an irreducible chain into a reducible chain. Also explain why states i and j are still positive recurrent and still in the same class.

(b) Let $\{\pi'_k;\, k \ge 0\}$ be the steady-state probabilities for the positive-recurrent class in the modified Markov chain. Show that the expected first passage time $\overline{T}'_{ji}$ from j to i in the modified Markov chain is $(1/\pi'_i) - 1$.

(c) Show that the expected first passage time from j to i is the same in the modified and unmodified chains.

(d) Show by example that after the modification above, two states i and j that were not positive recurrent before the modification can become positive recurrent and the above technique can again be used to find the expected first-passage time.

Exercise 6.8 Let $\{X_n;\, n \ge 0\}$ be a branching process with $X_0 = 1$. Let $\overline{Y}$ and σ^2 be the mean and variance respectively of the number of offspring of an individual.

(a) Argue that $\lim_{n\to\infty} X_n$ exists WP1 and either has the value 0 (with probability $\mathsf{F}_{10}(\infty)$) or the value ∞ (with probability $1 - \mathsf{F}_{10}(\infty)$).

(b) Show that $\mathsf{VAR}[X_n] = \sigma^2 \overline{Y}^{n-1}(\overline{Y}^n - 1)/(\overline{Y} - 1)$ for $\overline{Y} \ne 1$ and $\mathsf{VAR}[X_n] = n\sigma^2$ for $\overline{Y} = 1$.

Exercise 6.9 There are n states and for each pair of states i and j, a positive number $d_{ij} = d_{ji}$ is given. A particle moves from state to state in the following manner. Given that the particle is in any state i, it will next move to any $j \ne i$ with probability P_{ij} given by

$$P_{ij} = \frac{d_{ij}}{\sum_{j \ne i} d_{ij}}.$$

Assume that $P_{ii} = 0$ for all i. Show that the sequence of positions is a reversible Markov chain and find the limiting probabilities.

Exercise 6.10 Consider a reversible Markov chain with transition probabilities P_{ij} and limiting probabilities π_i. Also consider the same chain truncated to the states $0, 1, \ldots, M$. That is, the transition probabilities $\{P'_{ij}\}$ of the truncated chain are

$$P'_{ij} = \begin{cases} \dfrac{P_{ij}}{\sum_{k=0}^{m} P_{ik}}, & 0 \le i,j \le M, \\ 0, & \text{elsewhere.} \end{cases}$$

Show that the truncated chain is also reversible and has limiting probabilities given by

$$\overline{\pi}_i = \frac{\pi_i \sum_{j=0}^{M} P_{ij}}{\sum_{k=0}^{M} \pi_i \sum_{m=0}^{M} P_{km}}.$$

Exercise 6.11 A Markov chain (with states $\{0, 1, 2, \ldots, J-1\}$, where J is either finite or infinite) has transition probabilities $\{P_{ij};\ i,j \ge 0\}$. Assume that $P_{0j} > 0$ for all $j > 0$ and $P_{j0} > 0$ for all $j > 0$. Also assume that for all i, j, k, we have $P_{ij}P_{jk}P_{ki} = P_{ik}P_{kj}P_{ji}$.

(a) Assuming also that all states are positive recurrent, show that the chain is reversible and find the steady-state probabilities $\{\pi_i\}$ in simplest form.

(b) Find a condition on $\{P_{0j};\ j \ge 0\}$ and $\{P_{j0};\ j \ge 0\}$ that is sufficient to ensure that all states are positive recurrent.

Exercise 6.12 **(a)** Use the birth and death model described in Figure 6.4 to find the steady-state PMF for the number of customers in the system (queue plus service facility) for the following queues:

(i) M/M/1 with arrival probability $\lambda\delta$, service completion probability $\mu\delta$;

(ii) M/M/m with arrival probability $\lambda\delta$, service completion probability $i\mu\delta$ for i servers busy, $1 \le i \le m$;

(iii) M/M/∞ with arrival probability $\lambda\delta$, service probability $i\mu\delta$ for i servers. Assume δ so small that $i\mu\delta < 1$ for all i of interest.

Assume the system is positive recurrent.

(b) For each of the queues above give necessary conditions (if any) for the states in the chain to be (i) transient, (ii) null recurrent, iii) positive recurrent.

(c) For each of the queues find:

$L =$ (steady-state) mean number of customers in the system;

$L^q =$ (steady-state) mean number of customers in the queue;

$W =$ (steady-state) mean waiting time in the system;

$W^q =$ (steady-state) mean waiting time in the queue.

Exercise 6.13 **(a)** Given that an arrival occurs in the interval $(n\delta, (n+1)\delta)$ for the sampled-time M/M/1 model in Figure 6.5, find the conditional PMF of the state of the system at time $n\delta$ (assume n is arbitrarily large and assume positive recurrence).

(b) For the same model, again in steady state but not conditioned on an arrival in $(n\delta, (n+1)\delta)$, find the probability $Q(i,j) (i \ge j > 0)$ that the system is in state i at $n\delta$ and that $i-j$ departures occur before the next arrival.

(c) Find the expected number of customers seen in the system by the first arrival after time $n\delta$. Note: The purpose of this exercise is to make you cautious about the meaning of 'the state seen by a random arrival'.

Exercise 6.14 Find the backward transition probabilities for the Markov chain model of age in Figure 6.3. Draw the graph for the backward Markov chain, and interpret it as a model for residual life.

Exercise 6.15 Consider the sampled-time approximation to the M/M/1 queue in Figure 6.5.

(a) Give the steady-state probabilities for this chain (no explanations or calculations required – just the answer).

In (b)–(g) do not use reversibility and do not use Burke's theorem. Let X_n be the state of the system at time $n\delta$ and let D_n be a rv taking on the value 1 if a departure occurs between $n\delta$ and $(n+1)\delta$, and the value 0 if no departure occurs. Assume that the system is in steady state at time $n\delta$.

(b) Find $\Pr\{X_n = i, D_n = j\}$ for $i \geq 0, j = 0, 1$.

(c) Find $\Pr\{D_n = 1\}$.

(d) Find $\Pr\{X_n = i \mid D_n = 1\}$ for $i \geq 0$.

(e) Find $\Pr\{X_{n+1} = i \mid D_n = 1\}$ and show that X_{n+1} is statistically independent of D_n. Hint: Use (d); also show that $\Pr\{X_{n+1} = i\} = \Pr\{X_{n+1} = i \mid D_n = 1\}$ for all $i \geq 0$ is sufficient to show independence.

(f) Find $\Pr\{X_{n+1} = i, D_{n+1} = j \mid D_n\}$ and show that the pair of variables (X_{n+1}, D_{n+1}) is statistically independent of D_n.

(g) For each $k > 1$, find $\Pr\{X_{n+k} = i, D_{n+k} = j \mid D_{n+k-1}, D_{n+k-2}, \ldots, D_n\}$ and show that the pair (X_{n+k}, D_{n+k}) is statistically independent of $(D_{n+k-1}, D_{n+k-2}, \ldots, D_n)$. Hint: Use induction on k; as a substep, find $\Pr\{X_{n+k} = i \mid D_{n+k-1} = 1, D_{n+k-2}, \ldots, D_n\}$ and show that X_{n+k} is independent of $D_{n+k-1}, D_{n+k-2}, \ldots, D_n$.

(h) What do your results mean relative to Burke's theorem.

Exercise 6.16 Let $\{X_n, n \geq 1\}$ denote a positive recurrent Markov chain having a countable-state space. Now consider a new stochastic process $\{Y_n, n \geq 0\}$ that only accepts values of the Markov chain that are between 0 and some integer m. For instance, if $m = 3$ and $X_1 = 1, X_2 = 3, X_3 = 5, X_4 = 6, X_5 = 2$, then $Y_1 = 1, Y_2 = 3, Y_3 = 2$.

(a) Is $\{Y_n, n \geq 0\}$ a Markov chain? Explain briefly.

(b) Let p_j denote the proportion of time that $\{X_n, n \geq 1\}$ is in state j. If $p_j > 0$ for all j, what proportion of time is $\{Y_n, n \geq 0\}$ in each of the states $0, 1, \ldots, m$?

(c) Suppose $\{X_n\}$ is null recurrent and let $p_i(m), i = 0, 1, \ldots, m$ denote the long-run proportions for $\{Y_n, n \geq 0\}$. Show that for $j \neq i$, $p_j(m) = p_i(m)$ E[time the X process spends in j between returns to i].

Exercise 6.17 Verify that (6.58) is satisfied by the hypothesized solution to π in (6.62). Also show that the equations involving the idle state ϕ are satisfied.

Exercise 6.18 Replace the state $\mathbf{m} = (m, z_1, \ldots, z_m)$ in Section 6.8 with an expanded state $\mathbf{m} = (m, z_1, w_1, z_2, w_2, \ldots, z_m, w_m)$, where m and $\{z_i;\ 1 \leq i \leq m\}$ are as before and $w_1, w_2, \ldots, w_m$ are the original service requirements of the m customers.

(a) Hypothesizing the same backward round-robin system as in Section 6.8, find the backward transition probabilities and give the equations corresponding to (6.56)–(6.59) for the expanded state description.

(b) Solve the resulting equations to show that

$$\pi_{\mathbf{m}} = \pi + \phi\left(\frac{\lambda\delta}{1-\lambda\delta}\right)^m \prod_{j=1}^{m} f(w_j).$$

(c) Show that the probability that there are m customers in the system, and that those customers have original service requirements given by $w_1, \dots, w_m$, is

$$\Pr\{m, w_1, \dots, w_m\} = \pi_\phi\left(\frac{\lambda\delta}{1-\lambda\delta}\right)^m \prod_{j=1}^{m} (w_j - 1)f(w_j).$$

(d) Given that a customer has original service requirement w, find the expected time that customer spends in the system.

7 Markov processes with countable-state spaces

7.1 Introduction

Recall that a Markov chain is a discrete-time process $\{X_n;\ n \geq 0\}$ for which the state at each time $n \geq 1$ is an integer-valued random variable (rv) that depends on $X_0, \ldots, X_{n-1}$ only through X_{n-1}. A *countable-state Markov process*[1] $\{X(t);\ t \geq 0\}$ (Markov process for short) is a continuous-time process for which, first, the interval between state changes is exponential at a rate depending only on the current state $X(t)$, and, second, the sequence of successive distinct states forms a discrete-time Markov chain.

To be more specific, let $X_0{=}i,\ X_1{=}j,\ X_2{=}k, \ldots$ denote a sample path of the sequence of states in the Markov chain (henceforth called the *embedded Markov chain*). Then the *holding interval* U_n between the time that state $X_{n-1} = \ell$ is entered and that state X_n is entered is a non-negative exponential rv with parameter ν_ℓ, i.e., for all $u \geq 0$,

$$\Pr\{U_n \leq u \mid X_{n-1} = \ell\} = 1 - \exp(-\nu_\ell u). \tag{7.1}$$

Furthermore, U_n, conditional on X_{n-1}, is jointly independent of X_m for all $m \neq n-1$ and of U_m for all $m \neq n$.

If we visualize starting this process at time 0 in state $X_0 = i$, then the first transition of the embedded Markov chain enters state $X_1 = j$ with the transition probability P_{ij} of the embedded chain. This transition occurs at time U_1, where U_1 is independent of X_1 and exponential with rate ν_i. Next, conditional on $X_1 = j$, the next transition enters state $X_2 = k$ with the transition probability P_{jk}. This transition occurs after an interval U_2, i.e., at time $U_1 + U_2$, where U_2 is independent of X_2 and exponential with rate ν_j. Subsequent transitions occur similarly, with the new state, say $X_n = i$, determined from the old state, say $X_{n-1} = \ell$, via $P_{\ell i}$, and the new holding interval U_n determined via the exponential rate ν_ℓ. Figure 7.1 illustrates the statistical dependencies between the rv s $\{X_n;\ n \geq 0\}$ and $\{U_n;\ n \geq 1\}$.

The epochs at which successive transitions occur are denoted $S_1, S_2, \ldots$, so we have $S_1 = U_1$, $S_2 = U_1 + U_2$, and in general $S_n = \sum_{m=1}^{n} U_m$ for $n \geq 1$ with $S_0 = 0$. The state of a Markov process at any time $t > 0$ is denoted by $X(t)$ and is given by

$$X(t) = X_n \quad \text{for } S_n \leq t < S_{n+1} \qquad \text{for each } n \geq 0.$$

[1] These processes are often called *continuous-time* Markov chains.

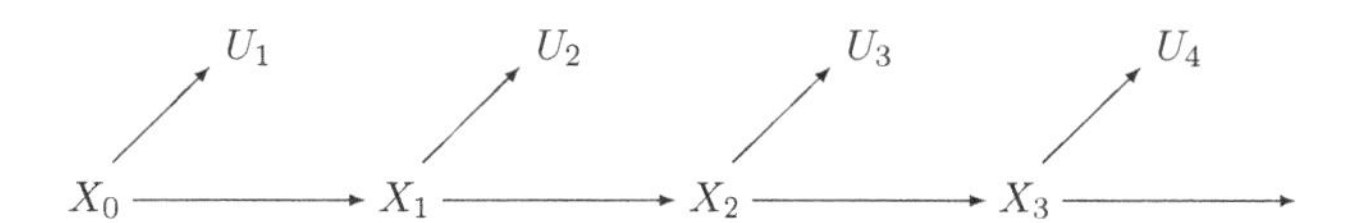

Figure 7.1 The statistical dependencies between the rv s of a Markov process. Each holding interval U_n, conditional on the current state X_{n-1}, is independent of all other states and holding intervals.

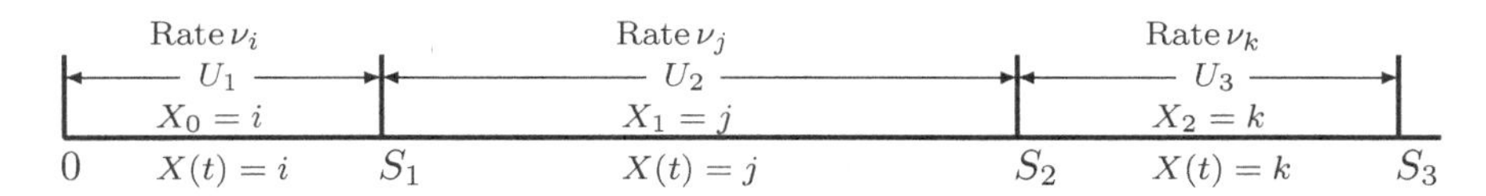

Figure 7.2 The relationship of the holding intervals $\{U_n;\ n \geq 1\}$ and the epochs $\{S_n;\ n \geq 1\}$ at which state changes occur. The state $X(t)$ of the Markov process and the corresponding state of the embedded Markov chain are also illustrated. Note that if $X_n = i$, then $X(t) = i$ for $S_n \leq t < S_{n+1}$

This defines a stochastic process $\{X(t);\ t \geq 0\}$ in the sense that each sample point $\omega \in \Omega$ maps to a sequence of sample values of $\{X_n;\ n \geq 0\}$ and $\{S_n;\ n \geq 1\}$, and thus into a sample function of $\{X(t);\ t \geq 0\}$. This stochastic process is what is usually referred to as a Markov process, but it is often simpler to view $\{X_n;\ n \geq 0\}, \{S_n;\ n \geq 1\}$ as a characterization of the process. Figure 7.2 illustrates the relationship between all these quantities.

This can be summarized in the following definition.

Definition 7.1.1 *A **countable-state Markov process** $\{X(t);\ t \geq 0\}$ is a stochastic process mapping each non-negative real number t to the non-negative integer-valued rv $X(t)$ in such a way that for each $t \geq 0$,*

$$X(t) = X_n \quad \textit{for } S_n \leq t < S_{n+1}; \qquad S_0 = 0;\ S_n = \sum_{m=1}^{n} U_m \quad \textit{for } n \geq 1, \tag{7.2}$$

where $\{X_n;\ n \geq 0\}$ is a Markov chain with a countably infinite or finite state space and each U_n, given $X_{n-1} = i$, is exponential with rate $\nu_i > 0$ and is conditionally independent of all other U_m and X_m.

The tacit assumptions that the state space is the set of non-negative integers and that the process starts at $t = 0$ are taken for notational simplicity.

We assume throughout this chapter (except in a few places where specified otherwise) that the embedded Markov chain has no self transitions, i.e., $P_{ii} = 0$ for all states i. One reason for this is that such transitions are invisible in $\{X(t);\ t \geq 0\}$. Another is that with this assumption, the sample functions of $\{X(t);\ t \geq 0\}$ and the joint sample functions of $\{X_n;\ n \geq 0\}$ and $\{U_n;\ n \geq 1\}$ uniquely specify each other.

We are not interested for the moment in exploring the probability distribution of $X(t)$ for given values of t, but one important feature of this distribution is that for any states i, j, any integer $\ell > 0$, and any times $t > \tau > s_1 > s_2 > \cdots > s_\ell > 0$,

$$\Pr\{X(t)=j \mid X(\tau)=i, \{X(s_m) = x(s_m);\ m \leq \ell\}\} = \Pr\{X(t-\tau)=j \mid X(0)=i\}. \tag{7.3}$$

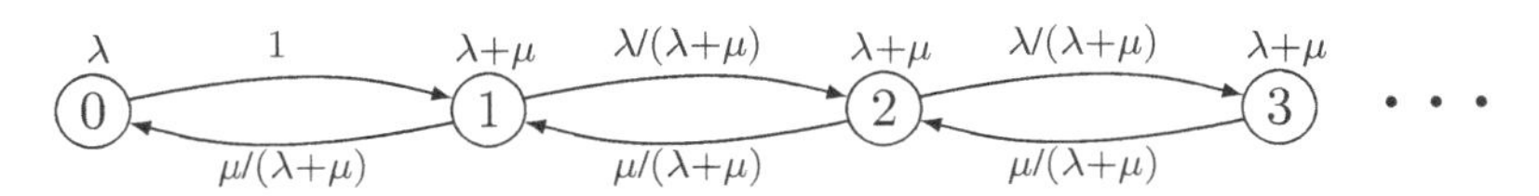

Figure 7.3 The embedded Markov chain for an M/M/1 queue. Each node i is labeled with the corresponding rate ν_i of the exponentially distributed holding interval to the next transition. Each transition, say i to j, is labeled with the corresponding transition probability P_{ij} in the embedded Markov chain.

This property arises because of the memoryless property of the exponential distribution. That is, if $X(\tau) = i$, it makes no difference how long the process has been in state i before τ; the time to the next transition is still exponential with rate ν_i and subsequent states and holding intervals are determined as if the process starts in state i at time 0. This will be seen more clearly in the following exposition. This property is the reason why these processes are called Markov, and is often taken as the defining property of Markov processes.

Example 7.1.2 (The M/M/1 queue) An M/M/1 queue has Poisson arrivals at a rate denoted by λ and has a single server with an exponential service distribution of rate $\mu > \lambda$ (see Figure 7.3). The service times are independent of each other and also of the arrivals. The state $X(t)$ of the queue is the total number of customers either in the queue or in service. The process $\{X(t); t > 0\}$ is a Markov process with the following parameters.

For $X(t) = 0$, the time to the next transition is the time until the next arrival, so $\nu_0 = \lambda$. When $X(t) = i$, $i \geq 1$, the server is busy and the time to the next transition is the time until either a new arrival occurs or a departure occurs. Thus $\nu_i = \lambda + \mu$. For the embedded Markov chain, $P_{01} = 1$ since only arrivals are possible in state 0, and they increase the state to 1. In the other states, $P_{i,i-1} = \mu/(\lambda{+}\mu)$ and $P_{i,i+1} = \lambda/(\lambda{+}\mu)$.

The embedded Markov chain is a birth–death chain, and its steady-state probabilities can be calculated easily using (6.30). The result is

$$\begin{aligned} \pi_0 &= \frac{1-\rho}{2}, && \text{where } \rho = \frac{\lambda}{\mu}, \\ \pi_n &= \frac{1-\rho^2}{2}\rho^{n-1} && \text{for } n \geq 1. \end{aligned} \tag{7.4}$$

Note that if $\lambda << \mu$, then π_0 and π_1 are each close to 1/2 (i.e., the embedded chain mostly alternates between states 0 and 1, and higher-ordered states are rarely entered), whereas because of the large holding interval in state 0, the process spends most of its time in state 0 waiting for arrivals. The steady-state probability π_i of state i in the embedded chain is the long-term fraction of the total transitions that go to state i. We will shortly learn how to find the long-term *fraction of time* spent in state i as opposed to this fraction of transitions, but for now we return to the general study of Markov processes.

The evolution of a Markov process can be visualized in several ways. We have already looked at the first, in which for each state $X_{n-1} = i$ in the embedded chain, the next state X_n is determined by the probabilities $\{P_{ij};\, j \geq 0\}$ of the embedded Markov chain, and the holding interval U_n is independently determined by the exponential distribution with rate ν_i.

For a second viewpoint, suppose an independent Poisson process of rate $\nu_i > 0$ is associated with each state i. When the Markov process enters a given state i, the next transition occurs at the next arrival epoch in the Poisson process for state i. At that epoch, a new state is chosen according to the transition probabilities P_{ij}. Since the choice of next state, given state i, is independent of the interval in state i, this view describes the same process as the first view.

For a third visualization, suppose, for each pair of states i and j, that an independent Poisson process of rate $\nu_i P_{ij}$ is associated with a possible transition to j conditional on being in i. When the Markov process enters a given state i, both the time of the next transition and the choice of the next state are determined by the set of i to j Poisson processes over all possible next states j. The transition occurs at the epoch of the first arrival, for the given i, to any of the i to j processes, and the next state is the j for which that first arrival occurs. Since such a collection of independent Poisson processes is equivalent to a single process of rate ν_i followed by an independent selection according to the transition probabilities P_{ij}, this view again describes the same process as the other views.

It is convenient in this third visualization to define the rate from any state i to any other state j as

$$q_{ij} = \nu_i P_{ij}.$$

If we sum over j, we see that ν_i and P_{ij} are also uniquely determined by $\{q_{ij};\, i,j \geq 0\}$ as

$$\nu_i = \sum_j q_{ij}; \qquad P_{ij} = q_{ij}/\nu_i. \tag{7.5}$$

This means that the fundamental characterization of the Markov process in terms of the P_{ij} and the ν_i can be replaced by a characterization in terms of the set of transition rates q_{ij}. In many cases, this is a more natural approach. For the M/M/1 queue, for example, $q_{i,i+1}$ is simply the arrival rate λ. Similarly, $q_{i,i-1}$ is the departure rate μ when there are customers to be served, i.e., when $i > 0$. Figure 7.4 shows Figure 7.3 incorporating this notational simplification.

Note that the interarrival density for the Poisson process, from any given state i to other state j, is given by $q_{ij} \exp(-q_{ij}x)$. On the other hand, given that the process is in

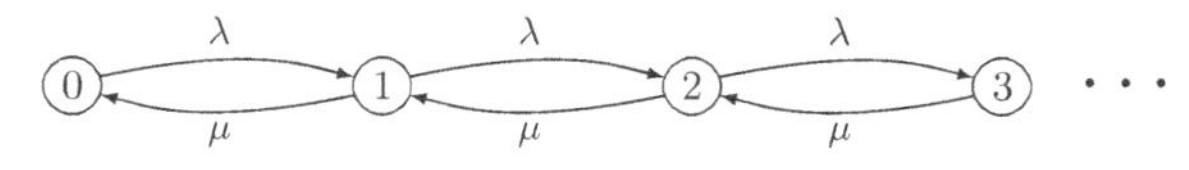

Figure 7.4 The Markov process for an M/M/1 queue. Each transition (i,j) is labeled with the corresponding transition rate q_{ij}.

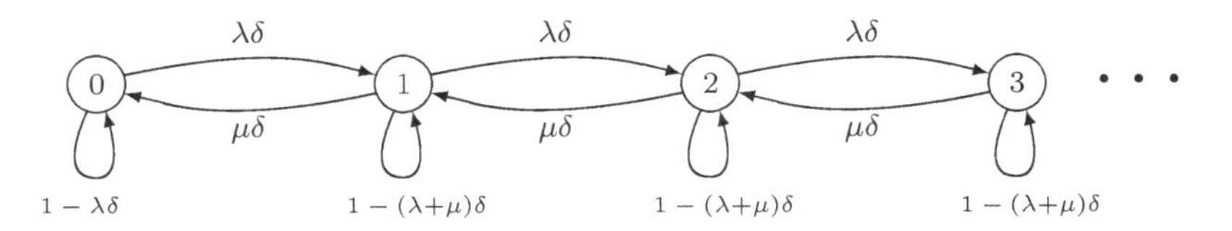

Figure 7.5 Approximating an M/M/1 queue by a sampled-time Markov chain.

state i, the probability density for the interval until the next transition, whether conditioned on the next state or not, is $\nu_i \exp(-\nu_i x)$, where $\nu_i = \sum_j q_{ij}$. One might argue, incorrectly, that, conditional on the next transition being to state j, the time to that transition has density $q_{ij} \exp(-q_{ij}x)$. Exercise 7.1 uses an M/M/1 queue to provide a guided explanation of why this argument is incorrect.

7.1.1 The sampled-time approximation to a Markov process

As yet another way to visualize a Markov process, consider approximating the process by viewing it only at times separated by a given increment size δ. The Poisson processes above are then approximated by Bernoulli processes where the transition probability from i to j in the sampled-time chain is defined to be $q_{ij}\delta$ for all $j \neq i$.

The Markov process is then approximated by a Markov chain. Since each δq_{ij} decreases with decreasing δ, there is an increasing probability of no transition out of any given state in the time increment δ. These must be modeled with self transition probabilities, say $P_{ii}(\delta)$ which must satisfy

$$P_{ii}(\delta) = 1 - \sum_j q_{ij}\delta = 1 - \nu_i\delta \qquad \text{for each } i \geq 0.$$

This is illustrated in Figure 7.5 for the M/M/1 queue. Recall that this sampled-time M/M/1 Markov chain was analyzed in Section 6.6 and the steady-state probabilities were shown to be

$$p_i(\delta) = (1-\rho)\rho^i \qquad \text{for all } i \geq 0 \quad \text{where } \rho = \lambda/\mu. \tag{7.6}$$

These steady-state probabilities are denoted by $p_i(\delta)$ to avoid confusion with the steady-state probabilities for the embedded chain. As discussed in detail later, the steady-state probabilities in (7.6) do not depend on δ, so long as δ is small enough that the self transition probabilities are non-negative.

This sampled-time approximation is an approximation in two ways. First, transitions occur only at integer multiples of the increment δ, and, second, $q_{ij}\delta$ is an approximation to $\Pr\{X(\delta){=}j \mid X(0){=}i\}$. From (7.3), $\Pr\{X(\delta){=}j \mid X(0) = i\} = q_{ij}\delta + o(\delta)$, so this second approximation is increasingly good as $\delta \to 0$.

As observed above, the steady-state probabilities for the sampled-time approximation to an M/M/1 queue do not depend on δ. As seen later, whenever the embedded chain is positive recurrent and a sampled-time approximation exists for a Markov process, then the steady-state probability of each state i is independent of δ and represents the limiting fraction of time spent in state i with probability 1.

Figure 7.6 Approximating a generic Markov process by its sampled-time Markov chain.

Figure 7.6 illustrates the sampled-time approximation of a generic Markov process. Note that $P_{ii}(\delta)$ is equal to $1 - \delta\nu_i$ for each i in any such approximation, and thus it is necessary for δ to be small enough to satisfy $\nu_i\delta \leq 1$ for all i. For a finite state space, this is satisfied for any $\delta \leq [\max_i \nu_i]^{-1}$. For a countably-infinite state space, however, the sampled-time approximation requires the existence of some finite B such that $\nu_i \leq B$ for all i. The consequences of having no such bound are explored in the next section.

7.2 Steady-state behavior of irreducible Markov processes

Definition 7.2.1 *An **irreducible Markov process** is a Markov process for which the embedded Markov chain is irreducible (i.e., all states are in the same class).*

The analysis in this chapter is restricted almost entirely to irreducible Markov processes. The reason for this restriction is not that Markov processes with multiple classes of states are unimportant, but rather that they can usually be best understood by first looking at the embedded Markov chains for the various classes making up that overall chain.

We will ask the same types of steady-state questions for Markov processes as we asked about Markov chains. In particular, under what conditions is there a set of steady-state probabilities, $p_0, p_1, \ldots$ with the property that for any given starting state $X(0) = i$, the limiting fraction of time spent in any given state j is p_j with probability 1 (WP1)? Do the probabilities $\{p_j;\, j \geq 0\}$ also have the property that $p_j = \lim_{t\to\infty} \Pr\{X(t) = j \mid X_0 = i\}$?

We will find that simply having a positive-recurrent embedded Markov chain is not quite enough to ensure that such a set of probabilities exists. It is also necessary for the steady-state probabilities $\{\pi_i;\, i \geq 0\}$ of the embedded-chain and the holding-interval parameters $\{\nu_i;\, i \geq 0\}$ to satisfy $\sum_i \pi_i/\nu_i < \infty$. We will interpret this latter condition as asserting that the limiting long-term rate at which transitions occur is strictly positive. Finally, we will show that when these conditions are satisfied, the steady-state probabilities for the process are related to those of the embedded chain by

$$p_j = \frac{\pi_j/\nu_j}{\sum_k \pi_k/\nu_k}. \tag{7.7}$$

Definition 7.2.2 *Given a Markov process with a positive-recurrent embedded chain with steady-state probabilities $\{\pi_i : i \geq 0\}$ and given that $\sum_k \pi_k/\nu_k < \infty$, the*

***steady-state process probabilities**, $p_0, p_1, \ldots$ are the numbers satisfying (7.7) where $\{\nu_i;\ i \geq 0\}$ are the holding-interval rates.*

As one might guess, the appropriate approach to answering the above questions comes from applying renewal theory to various renewal processes associated with the Markov process. Many of the needed results for this have already been developed in looking at the steady-state behavior of countable-state Markov chains.

We start with a very technical lemma that will perhaps appear obvious, and the reader is welcome to ignore the proof until perhaps questioning the issue later. The lemma is not restricted to irreducible processes, although we only use it in that case.

Lemma 7.2.3 *Consider a Markov process for which the embedded chain starts in some given state i. Then the holding intervals, U_1, U_2, ... are all rv s. Let $M_i(t)$ be the number of transitions made by the process up to and including time t. Then WP1,*

$$\lim_{t\to\infty} M_i(t) = \infty. \tag{7.8}$$

Proof The first holding interval U_1 is exponential with rate $\nu_i > 0$, so it is clearly a rv (i.e., non-defective). In general, the state after transition $(n-1)$ has the probability mass function (PMF) P_{ij}^{n-1}, so the complementary cumulative distribution function of U_n is

$$\begin{aligned}\Pr\{U_n > u\} &= \lim_{k\to\infty} \sum_{j=1}^{k} P_{ij}^{n-1} \exp(-\nu_j u). \\ &\leq \sum_{j=1}^{k} P_{ij}^{n-1} \exp(-\nu_j u) + \sum_{j=k+1}^{\infty} P_{ij}^{n+1} \qquad \text{for every } k.\end{aligned}$$

For each k, the first sum above approaches 0 with increasing u. Since the second sum approaches 0 with increasing k, the limit as $u \to \infty$ must be 0 and U_n is a rv.

It follows that each $S_n = U_1 + \cdots + U_n$ is also a rv. Since $S_1, S_2, \ldots$ are the arrival epochs of an arrival process, we have the set equality $\{S_n \leq t\} = \{M_i(t) \geq n\}$ for each choice of n. Since S_n is a rv, we have $\lim_{t\to\infty} \Pr\{S_n \leq t\} = 1$ for each n. Thus $\lim_{t\to\infty} \Pr\{M_i(t) \geq n\} = 1$ for all n. This means that the set of sample points ω for which $\lim_{t\to\infty} M_i(t,\omega) < n$ has probability 0 for all n, and thus $\lim_{t\to\infty} M_i(t,\omega) = \infty$ WP1. □

7.2.1 Renewals on successive entries to a given state

For an irreducible Markov process with $X_0 = i$, let $M_{ij}(t)$ be the number of transitions into state j over the interval $(0, t]$. We want to find when this is a delayed renewal counting process. It is clear that the sequence of epochs at which state j is entered form renewal points, since they form renewal points in the embedded Markov chain and the holding intervals between transitions depend only on the current state. The questions are whether the first entry to state j must occur within some finite time, and whether recurrences to j must occur within finite time. The following lemma answers these questions for the case where the embedded chain is recurrent (either positive recurrent or null recurrent).

Lemma 7.2.4 *Consider a Markov process with an irreducible recurrent embedded chain* $\{X_n;\, n \geq 0\}$. *Given* $X_0 = i$, *let* $\{M_{ij}(t);\, t \geq 0\}$ *be the number of transitions into a given state* j *in the interval* $(0, t]$. *Then* $\{M_{ij}(t);\, t \geq 0\}$ *is a delayed (or ordinary) renewal counting process.*

Proof Given $X_0 = i$, let $N_{ij}(n)$ be the number of transitions into state j that occur in the embedded Markov chain by the nth transition of the embedded chain. From Lemma 6.3.1, $\{N_{ij}(n);\, n \geq 0\}$ is a delayed renewal process, so from Lemma 5.8.2, $\lim_{n\to\infty} N_{ij}(n) = \infty$ WP1. Note that $M_{ij}(t) = N_{ij}(M_i(t))$, where $M_i(t)$ is the total number of state transitions (between all states) in the interval $(0, t]$. Thus, WP1,

$$\lim_{t\to\infty} M_{ij}(t) = \lim_{t\to\infty} N_{ij}(M_i(t)) = \lim_{n\to\infty} N_{ij}(n) = \infty,$$

where we have used Lemma 7.2.3, which asserts that $\lim_{t\to\infty} M_i(t) = \infty$ WP1.

It follows that the interval W_1 until the first transition to state j, and the subsequent interval W_2 until the next transition to state j, are both finite WP1. Subsequent intervals have the same distribution as W_2, and all intervals are independent, so $\{M_{ij}(t);\, t \geq 0\}$ is a delayed renewal process with inter-renewal intervals $\{W_k;\, k \geq 1\}$. If $i = j$, then all W_k are identically distributed and we have an ordinary renewal process, completing the proof. □

The inter-renewal intervals $W_2, W_3, \ldots$ for $\{M_{ij}(t);\, t \geq 0\}$ above are well-defined non-negative independent identically distributed (IID) rv s whose distribution depends on j but not i. They either have an expectation as a finite number or have an infinite expectation. In either case, this expectation is denoted as $\mathsf{E}[W(j)] = \overline{W}(j)$. This is the mean time between successive entries to state j, and we will see later that in some cases this mean time can be infinite.

7.2.2 The limiting fraction of time in each state

In order to study the fraction of time spent in state j, we define a delayed renewal–reward process, based on $\{M_{ij}(t);\, t \geq 0\}$, for which unit reward is accumulated whenever the process is in state j. That is (given $X_0 = i$), $R_{ij}(t) = 1$ when $X(t) = j$ and $R_{ij}(t) = 0$ otherwise (see Figure 7.7). Given that the $(n-1)$th transition of the embedded chain enters state j, the interval U_n is exponential with rate ν_j, so $\mathsf{E}[U_n \mid X_{n-1}{=}j] = 1/\nu_j$.

Let $p_j(i)$ be the limiting time-average fraction of time spent in state j for $X_0 = i$. We will see later that such a limit exists WP1, that the limit does not depend on i, and that

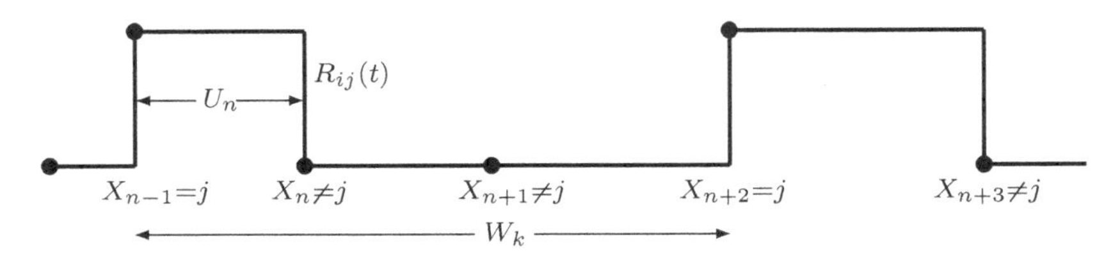

Figure 7.7 The delayed renewal–reward process $\{R_{ij}(t);\ t \geq 0\}$ for time in state j. The reward is 1 whenever the process is in state j, i.e., $R_{ij}(t) = \mathbb{I}_{\{X(t)=j\}}$. A renewal occurs on each entry to state j, so the reward starts at each such entry and continues until a state transition, assumed to enter a state other than j. The reward then ceases until the next renewal, i.e., the next entry to state j. The figure illustrates the kth inter-renewal interval, of duration W_k, which is assumed to start on the $(n-1)$th state transition. The expected interval over which a reward is accumulated is $1/\nu_j$ and the expected duration of the inter-renewal interval is $\overline{W}(j)$.

it is equal to the steady-state probability p_j in (7.7). Since $\overline{U}(j) = 1/\nu_j$, Theorems 5.4.5 and 5.8.6, for ordinary and delayed renewal–reward processes respectively, state that[2]

$$p_j(i) = \lim_{t\to\infty} \frac{\int_0^t R_{ij}(\tau)d\tau}{t} \qquad \text{WP1} \tag{7.9}$$

$$= \frac{\overline{U}(j)}{\overline{W}(j)} = \frac{1}{\nu_j \overline{W}(j)}. \tag{7.10}$$

This shows that the limiting time average, $p_j(i)$, exists WP1 and is independent of the starting state i. We show later that it is the steady-state process probability given by (7.7).
We can also investigate the limit, as $t \to \infty$, of the probability that $X(t) = j$. This is equal to $\lim_{t\to\infty} \mathsf{E}[R(t)]$ for the renewal–reward process above. Because of the exponential holding intervals, the inter-renewal times are non-arithmetic, and from Blackwell's theorem, in the form of (5.108),

$$\lim_{t\to\infty} \Pr\{X(t) = j\} = \frac{1}{\nu_j \overline{W}(j)} = p_j(i). \tag{7.11}$$

We summarize these results in the following lemma.

Lemma 7.2.5 *Consider an irreducible Markov process with a recurrent embedded Markov chain starting in $X_0 = i$. Then WP1, the limiting time average in state j is given by $p_j(i) = 1/[\nu_j \overline{W}(j)]$. This is also the limit, as $t \to \infty$, of* $\Pr\{X(t) = j\}$.

7.2.3 Finding $\{p_j(i);\ j \geq 0\}$ in terms of $\{\pi_j;\ j \geq 0\}$

Next we must express the mean inter-renewal time, $\overline{W}(j)$, in terms of more accessible quantities that essentially allow us to show that $p_j(i) = p_j$, where p_j is the steady-state process probability of Definition 7.2.2. We assume that the embedded chain is not only

[2] Theorems 5.4.5 and 5.8.6 do not cover the case where $\overline{W}(j) = \infty$, but, since the expected reward per renewal interval is finite, it is not hard to verify (7.9) in that special case.

recurrent but also positive recurrent with steady-state probabilities $\{\pi_j;\ j \geq 0\}$. We continue to assume a given starting state $X_0 = i$. Applying the strong law for delayed renewal processes (Theorem 5.8.3) to the Markov process,

$$\lim_{t\to\infty} M_{ij}(t)/t = 1/\overline{W}(j) \qquad \text{WP1}. \tag{7.12}$$

As before, $M_{ij}(t) = N_{ij}(M_i(t))$. Since $\lim_{t\to\infty} M_i(t) = \infty$ WP1,

$$\lim_{t\to\infty} \frac{M_{ij}(t)}{M_i(t)} = \lim_{t\to\infty} \frac{N_{ij}(M_i(t))}{M_i(t)} = \lim_{n\to\infty} \frac{N_{ij}(n)}{n} = \pi_j \qquad \text{WP1}. \tag{7.13}$$

In the last step, we applied the same strong law to the embedded chain. Combining (7.12) and (7.13), the following equalities hold WP1:

$$\begin{aligned} \frac{1}{\overline{W}(j)} &= \lim_{t\to\infty} \frac{M_{ij}(t)}{t} \\ &= \lim_{t\to\infty} \frac{M_{ij}(t)}{M_i(t)} \frac{M_i(t)}{t} \\ &= \pi_j \lim_{t\to\infty} \frac{M_i(t)}{t}. \end{aligned} \tag{7.14}$$

This tells us that $\overline{W}(j)\pi_j$ is the same for all j. Also, since $\pi_j > 0$ for a positive-recurrent chain, it tells us that if $\overline{W}(j) < \infty$ for one state j, it is finite for all states. As seen later in Example 7.2.8, however, it is possible to have $\overline{W}_j = \infty$ for all j. These expected recurrence times are finite if and only if $\lim_{t\to\infty} M_i(t)/t > 0$. Finally, it says implicitly that $\lim_{t\to\infty} M_i(t)/t$ exists WP1 and has the same value for all starting states i.

There is relatively little left to do, and the following theorem does most of it.

Theorem 7.2.6 *Consider an irreducible Markov process with a positive-recurrent embedded Markov chain. Let $\{\pi_j;\ j \geq 0\}$ be the steady-state probabilities of the embedded chain and let $X_0 = i$ be the starting state. Then, WP1, the limiting time-average fraction of time spent in each state j is equal to the steady-state process probability defined in (7.7), i.e.,*

$$p_j(i) \;=\; p_j \;=\; \frac{\pi_j/\nu_j}{\sum_k \pi_k/\nu_k}. \tag{7.15}$$

The expected time between returns to state j is

$$\overline{W}(j) \;=\; \frac{\sum_k \pi_k/\nu_k}{\pi_j} \;=\; \frac{1}{p_j\nu_j}, \tag{7.16}$$

and the limiting rate at which transitions take place is independent of the starting state and given by

$$\lim_{t\to\infty} \frac{M_i(t)}{t} = \frac{1}{\sum_k \pi_k/\nu_k} \qquad \textit{WP1}. \tag{7.17}$$

Discussion Recall that $p_j(i)$ was defined as a time average WP1, and we saw earlier that this time average exists with a value independent of i. The theorem states that this time average (and the limiting ensemble average) is given by the steady-state process probabilities in (7.7). Thus, after the proof, we can stop distinguishing these quantities.

At a superficial level, the theorem is almost obvious from what we have done. In particular, substituting (7.14) into (7.9), we see that

$$p_j(i) = \frac{\pi_j}{\nu_j} \lim_{t\to\infty} \frac{M_i(t)}{t} \qquad \text{WP1.} \tag{7.18}$$

Since $p_j(i) = \lim_{t\to\infty} \Pr\{X(t) = j\}$, and since $X(t)$ is in some state at all times, we would conjecture (and perhaps insist if we did not read on) that $\sum_j p_j(i) = 1$. Adding that condition to normalize (7.18), we get (7.15), and (7.16) and (7.17) follow immediately. The trouble is that if $\sum_j \pi_j/\nu_j = \infty$, then (7.15) says that $p_j = 0$ for all j, and (7.17) says that $\lim M_i(t)/t = 0$, i.e., the process 'gets tired' with increasing t and the rate of transitions decreases toward 0. The rather technical proof to follow deals with these limits more carefully.

Proof We have seen in (7.14) that $\lim_{t\to\infty} M_i(t)/t$ is equal to a constant, say α, WP1 and that this constant is the same for all starting states i. We first consider the case where $\alpha > 0$. In this case, from (7.14), $\overline{W}(j) < \infty$ for all j. Choosing any given j and any positive integer ℓ, consider a renewal–reward process with renewals on transitions to j and a reward $R^{\ell}_{ij}(t) = 1$ when $X(t) \le \ell$. This reward is independent of j and equal to $\sum_{k=1}^{\ell} R_{ik}(t)$. Thus, from (7.9), we have

$$\lim_{t\to\infty} \frac{\int_0^t R^{\ell}_{ij}(\tau)d\tau}{t} = \sum_{k=1}^{\ell} p_k(i). \tag{7.19}$$

If we let $\mathsf{E}[\mathsf{R}^{\ell}_j]$ be the expected reward over a renewal interval, then, from Theorem 5.8.6,

$$\lim_{t\to\infty} \frac{\int_0^t R^{\ell}_{ij}(\tau)d\tau}{t} = \frac{\mathsf{E}\left[\mathsf{R}^{\ell}_j\right]}{\overline{W}(j)}. \tag{7.20}$$

Note that $\mathsf{E}[\mathsf{R}^{\ell}_j]$ above is non-decreasing in ℓ and goes to the limit $\overline{W}(j)$ as $\ell \to \infty$. Thus, combining (7.19) and (7.20), we see that

$$\lim_{\ell\to\infty} \sum_{k=1}^{\ell} p_k(i) = 1.$$

With this added relation, (7.15), (7.16), and (7.17) follow as in the discussion. This completes the proof for the case where $\alpha > 0$.

For the remaining case, where $\lim_{t\to\infty} M_i(t)/t = \alpha = 0$, (7.14) shows that $\overline{W}(j) = \infty$ for all j and (7.18) then shows that $p_j(i) = 0$ for all j. We give a guided proof in Exercise 7.6 that, for $\alpha = 0$, we must have $\sum_i \pi_i/\nu_i = \infty$. It follows that (7.15), (7.16), and (7.17) are all satisfied. □

This has been quite a difficult proof for something that might seem almost obvious for simple examples. However, the fact that these time averages are valid over all sample points WP1 is not obvious and the fact that $\pi_j\overline{W}(j)$ is independent of j is certainly not obvious.

The most subtle thing here, however, is that if $\sum_i \pi_i/\nu_i = \infty$, then $p_j = 0$ for all states j. This is strange because the time-average state probabilities do not add to 1, and also strange because the embedded Markov chain continues to make transitions, and these transitions, in steady state for the Markov chain, occur with the probabilities π_i. Example 7.2.8 and Exercise 7.3 give some insight into this. Some added insight can be gained by looking at the embedded Markov chain starting in steady state, i.e., with probabilities $\{\pi_i; i \geq 0\}$. Given $X_0 = i$, the expected time to a transition is $1/\nu_i$, so the unconditional expected time to a transition is $\sum_i \pi_i/\nu_i$, which is infinite for the case under consideration. This is not a phenomenon that can be easily understood intuitively, but Example 7.2.8 and Exercise 7.3 will help.

7.2.4 Solving for the steady-state process probabilities directly

We continue to assume that the embedded chain is positive recurrent with steady-state probabilities $\{\pi_i; i \geq 0\}$ and again return to the case of application interest in which $\sum_k \pi_k/\nu_k < \infty$. We have seen that a Markov process can be specified in terms of the time transitions $q_{ij} = \nu_i P_{ij}$, and it is useful to express the steady-state equations for p_j directly in terms of q_{ij} rather than indirectly in terms of the embedded chain. As a useful prelude to this, we first express the π_j in terms of the p_j. Denote $\sum_k \pi_k/\nu_k$ as $\beta < \infty$. Then, from (7.15), $p_j = \pi_j/\nu_j\beta$, so $\pi_j = p_j\nu_j\beta$. Expressing this along with the normalization $\sum_k \pi_k = 1$, we obtain

$$\pi_i = \frac{p_i\nu_i}{\sum_k p_k\nu_k}. \tag{7.21}$$

Thus, $\beta = 1/\sum_k p_k\nu_k$, so

$$\sum_k \pi_k/\nu_k = \frac{1}{\sum_k p_k\nu_k}. \tag{7.22}$$

We can now substitute π_i as given by (7.21) into the steady-state equations for the embedded Markov chain, i.e., $\pi_j = \sum_i \pi_i P_{ij}$ for all j, obtaining

$$p_j\nu_j = \sum_i p_i\nu_i P_{ij}$$

for each state j. Since $\nu_i P_{ij} = q_{ij}$,

$$p_j\nu_j = \sum_i p_i q_{ij}, \qquad \sum_i p_i = 1. \tag{7.23}$$

This set of equations is known as the steady-state equations for the Markov process. The normalization condition $\sum_i p_i = 1$ is a consequence of (7.22) and also of (7.15). Equation (7.23) has the intuitive interpretation that the term $p_j\nu_j$ is the steady-state rate at which transitions occur out of state j and the term $\sum_i p_i q_{ij}$ is the rate at which transitions occur into state j. Since the total number of entries to j must differ by at most 1 from the exits from j for each sample path, this equation is not surprising.

The embedded chain is positive recurrent, so its steady-state equations have a unique solution with all $\pi_i > 0$. Thus (7.23) also has a unique solution with all $p_i > 0$ under the added condition that $\sum_i \pi_i/\nu_i < \infty$. However, we would like to solve (7.23) directly without worrying about the embedded chain.

If we find a solution to (7.23), however, and if $\sum_i p_i \nu_i < \infty$ in that solution, then the corresponding set of π_i from (7.21) must be the unique steady-state solution for the embedded chain. Thus the solution for p_i must be the corresponding steady-state solution for the Markov process. This is summarized in the following theorem.

Theorem 7.2.7 *Assume an irreducible Markov process and let $\{p_i;\ i \geq 0\}$ be a solution to (7.23). If $\sum_i p_i \nu_i < \infty$, then, first, that solution is unique, second, each p_i is positive, and, third, the embedded Markov chain is positive recurrent with steady-state probabilities satisfying (7.21). Also, if the embedded chain is positive recurrent, and $\sum_i \pi_i/\nu_i < \infty$ then the set of p_i satisfying (7.15) is the unique solution to (7.23).*

7.2.5 The sampled-time approximation again

For an alternative view of the probabilities $\{p_i;\ i \geq 0\}$, consider the special case (but the typical case) where the transition rates $\{\nu_i;\ i \geq 0\}$ are bounded. Consider the sampled-time approximation to the process for a given increment size $\delta \leq [\max_i \nu_i]^{-1}$ (see Figure 7.6). Let $\{p_i(\delta);\ i \geq 0\}$ be the set of steady-state probabilities for the sampled-time chain, assuming that they exist. These steady-state probabilities satisfy

$$p_j(\delta) = \sum_{i \neq j} p_i(\delta) q_{ij} \delta + p_j(\delta)(1 - \nu_j \delta); \quad p_j(\delta) \geq 0, \quad \sum_j p_j(\delta) = 1. \tag{7.24}$$

The first equation simplifies to $p_j(\delta)\nu_j = \sum_{i \neq j} p_i(\delta) q_{ij}$, which is the same as (7.23). It follows that the steady-state probabilities $\{p_i;\ i \geq 0\}$ for the process are the same as the steady-state probabilities $\{p_i(\delta);\ i \geq 0\}$ for the sampled-time approximation. Note that this is not an approximation; $p_i(\delta)$ is exactly equal to p_i for all values of $\delta \leq 1/\sup_i \nu_i$. We shall see later that the dynamics of a Markov process are not quite so well modeled by the sampled time approximation except in the limit $\delta \to 0$.

7.2.6 Pathological cases

Example 7.2.8 (Zero transition rate) Consider the Markov process with a positive-recurrent embedded chain in Figure 7.8. This models a variation of an M/M/1 queue in which the server becomes increasingly rattled and slow as the queue builds up, and the customers become almost equally discouraged about entering. The downward drift in the transitions is more than overcome by the slow-down in large numbered states, and it is easily verified that $\sum_i \pi_i/\nu_i = \infty$. Transitions continue to occur, but the number of transitions per unit time goes to 0 with increasing time. Although the embedded chain has a steady-state solution, the process cannot be viewed as having any sort of steady state. Exercise 7.3 gives some added insight into this type of situation.

It is also possible for (7.23) to have a solution for $\{p_i;\ i \geq 0\}$ with $\sum_i p_i = 1$, but $\sum_i p_i \nu_i = \infty$. This is not possible for a positive-recurrent embedded chain, but is possible both if the embedded Markov chain is transient and if it is null recurrent. A

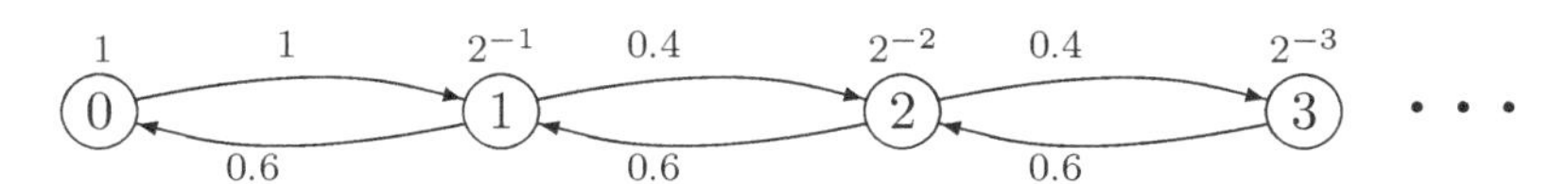

Figure 7.8 The Markov process for a variation on M/M/1, where arrivals and services get slower with increasing state. Each node i has a rate $\nu_i = 2^{-i}$. The embedded chain transition probabilities are $P_{i,i+1} = 0.4$ for $i \geq 1$ and $P_{i,i-1} = 0.6$ for $i \geq 1$, thus ensuring that the embedded Markov chain is positive recurrent. Note that $q_{i,i+1} > q_{i+1,i}$, thus ensuring that the Markov process drifts to the right.

transient chain means that there is a positive probability that the embedded chain will *never* return to a state after leaving it, and thus there can be no sensible kind of steady-state behavior for the process. These processes are characterized by arbitrarily large transition rates from the various states, and these allow the process to transit through an infinite number of states in a finite time.

Processes for which there is a non-zero probability of passing through an infinite number of states in a finite time are called *irregular*. Exercises 7.7 and 7.8 give some insight into irregular processes. Exercise 7.9 gives an example of a process that is not irregular, but for which (7.23) has a solution with $\sum_i p_i = 1$ and the embedded Markov chain is null recurrent. We restrict our attention in what follows to irreducible Markov chains for which (7.23) has a solution, $\sum p_i = 1$, and $\sum p_i \nu_i < \infty$. This is slightly more restrictive than necessary, but processes for which $\sum_i p_i \nu_i = \infty$ (see Exercise 7.9) are not very robust.

7.3 The Kolmogorov differential equations

Let $P_{ij}(t)$ be the probability that a Markov process $\{X(t);\, t \geq 0\}$ is in state j at time t given that $X(0) = i$,

$$P_{ij}(t) = \Pr\{X(t)=j \mid X(0)=i\}. \tag{7.25}$$

$P_{ij}(t)$ is analogous to the nth-order transition probabilities P_{ij}^n for Markov chains. We have already seen that $\lim_{t\to\infty} P_{ij}(t) = p_j$ for the case where the embedded chain is positive recurrent and $\sum_i \pi_i/\nu_i < \infty$. Here we want to find the transient behavior, and we start by deriving the Chapman–Kolmogorov equations for Markov processes. Let s and t be arbitrary times, $0 < s < t$. By including the state at time s, we can rewrite (7.25) as

$$\begin{aligned} P_{ij}(t) &= \sum_k \Pr\{X(t)=j,\, X(s)=k \mid X(0)=i\} \\ &= \sum_k \Pr\{X(s)=k \mid X(0)=i\} \Pr\{X(t)=j \mid X(s)=k\}, \quad \text{all } i,j, \end{aligned} \tag{7.26}$$

where we have used the Markov property, (7.3). Given that $X(s) = k$, the residual time until the next transition after s is exponential with rate ν_k, and thus the process starting at time s in state k is statistically identical to that starting at time 0 in state k. Thus, for any s, $0 \le s \le t$, we have

$$\Pr\{X(t)=j \mid X(s)=k\} = P_{kj}(t-s).$$

Substituting this into (7.26), we have the *Chapman–Kolmogorov equations* for a Markov process,

$$P_{ij}(t) = \sum_k P_{ik}(s)P_{kj}(t-s). \tag{7.27}$$

These equations correspond to (4.7) for Markov chains. We now use these equations to derive two types of sets of differential equations for $P_{ij}(t)$. The first are called the *Kolmogorov forward differential equations*, and the second the *Kolmogorov backward differential equations*. The forward equations are obtained by letting s approach t from below, and the backward equations are obtained by letting s approach 0 from above. First we derive the forward equations.

For $t-s$ small and positive, $P_{kj}(t-s)$ in (7.27) can be expressed as $(t{-}s)q_{kj} + o(t{-}s)$ for $k \neq j$. Similarly, $P_{jj}(t{-}s)$ can be expressed as $1{-}(t{-}s)\nu_j{+}o(s)$. Thus (7.27) becomes

$$P_{ij}(t) = \sum_{k\neq j} [P_{ik}(s)(t-s)q_{kj}] + P_{ij}(s)[1-(t-s)\nu_j] + o(t-s). \tag{7.28}$$

We want to express this, in the limit $s \to t$, as a differential equation. To do this, subtract $P_{ij}(s)$ from both sides and divide by $t-s$:

$$\frac{P_{ij}(t)-P_{ij}(s)}{t-s} = \sum_{k\neq j} \big(P_{ik}(s)q_{kj}\big) - P_{ij}(s)\nu_j + \frac{o(s)}{s}. \tag{7.29}$$

Taking the limit as $s \to t$ from below,[3] we get the Kolmogorov forward equations,

$$\frac{dP_{ij}(t)}{dt} = \sum_{k\neq j} \big(P_{ik}(t)q_{kj}\big) - P_{ij}(t)\nu_j. \tag{7.30}$$

The first term on the right-hand side of (7.30) is the rate at which transitions occur into state j at time t and the second term is the rate at which transitions occur out of state j. Thus the difference of these terms is the net rate at which transitions occur into j, which is the rate at which $P_{ij}(t)$ is increasing at time t.

Example 7.3.1 (A queueless M/M/1 queue) Consider the following two-state Markov process where $q_{01} = \lambda$ and $q_{10} = \mu$:

[3] We have assumed that the sum and the limit in (7.29) can be interchanged. This is certainly valid if the state space is finite, which is the only case we analyze in what follows.

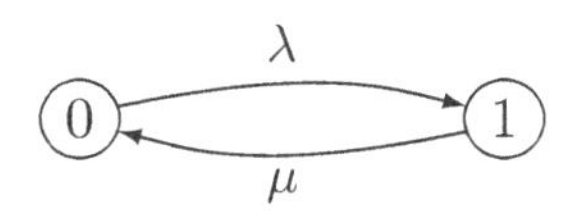

This can be viewed as a model for an M/M/1 queue with no storage for waiting customers. When the system is empty (state 0), memoryless customers arrive at rate λ, and when the server is busy, an exponential server operates at rate μ, with the system returning to state 0 when service is completed.

To find $P_{01}(t)$, the probability of state 1 at time t conditional on state 0 at time 0, we use the Kolmogorov forward equations for $P_{01}(t)$, getting

$$\frac{dP_{01}(t)}{dt} \;=\; P_{00}(t)q_{01} - P_{01}(t)\nu_1 \;=\; P_{00}(t)\lambda - P_{01}(t)\mu.$$

Using the fact that $P_{00}(t) = 1 - P_{01}(t)$, this becomes

$$\frac{dP_{01}(t)}{dt} = \lambda - P_{01}(t)(\lambda + \mu).$$

Using the boundary condition $P_{01}(0) = 0$, the solution is

$$P_{01}(t) = \frac{\lambda}{\lambda + \mu}\left[1 - e^{-(\lambda+\mu)t}\right]. \tag{7.31}$$

Thus $P_{01}(t)$ is 0 at $t = 0$ and increases as $t \to \infty$ to its steady-state value in state 1, which is $\lambda/(\lambda + \mu)$.

In general, for any given starting state i in a Markov process with M states, (7.30) provides a set of M simultaneous linear differential equations, one for each j, $1 \leq j \leq$ M. As we saw in the example, one of these is redundant because $\sum_{j=1}^{\mathsf{M}} P_{ij}(t) = 1$. This leaves M $-$ 1 simultaneous linear differential equations to be solved.

For more than two or three states, it is more convenient to express (7.30) in matrix form. Let $[P(t)]$ (for each $t > 0$) be an M $\times$ M matrix whose i,j element is $P_{ij}(t)$. Let $[Q]$ be an M $\times$ M matrix whose i,j element is q_{ij} for each $i \neq j$ and $-\nu_j$ for $i = j$. Then (7.30) becomes

$$\frac{d[P(t)]}{dt} = [P(t)][Q]. \tag{7.32}$$

For Example 7.3.1, $P_{ij}(t)$ can be calculated for each i,j as in (7.31), resulting in

$$[P(t)] = \begin{bmatrix} \frac{\mu}{\lambda+\mu} + \frac{\lambda}{\lambda+\mu}e^{-(\lambda+\mu)t} & \frac{\lambda}{\lambda+\mu} - \frac{\lambda}{\lambda+\mu}e^{-(\lambda+\mu)t} \\ \frac{\mu}{\lambda+\mu} + \frac{\lambda}{\lambda+\mu}e^{-(\lambda+\mu)t} & \frac{\lambda}{\lambda+\mu} - \frac{\lambda}{\lambda+\mu}e^{-(\lambda+\mu)t} \end{bmatrix} \qquad [Q] = \begin{bmatrix} -\lambda & \lambda \\ \mu & -\mu \end{bmatrix}.$$

In order to provide some insight into the general solution of (7.32), we go back to the sampled-time approximation of a Markov process. With an increment of size δ between samples, the probability of a transition from i to j, $i \neq j$, is $q_{ij}\delta + o(\delta)$, and the probability of remaining in state i is $1 - \nu_i\delta + o(\delta)$. Thus, in terms of the matrix $[Q]$ of transition rates, the transition probability matrix in the sampled-time model is $[I] + \delta[Q]$, where $[I]$ is the identity matrix. We denote this matrix by $[W_\delta] = [I] + \delta[Q]$. Note that λ is an eigenvalue of $[Q]$ if and only if $1 + \lambda\delta$ is an eigenvalue of $[W_\delta]$. Also the eigenvectors of these corresponding eigenvalues are the same. That is, if $\boldsymbol{\nu}$ is a right eigenvector of $[Q]$ with eigenvalue λ, then $\boldsymbol{\nu}$ is a right eigenvector of $[W_\delta]$ with eigenvalue $1 + \lambda\delta$, and conversely. Similarly, if $\boldsymbol{p}$ is a left eigenvector of $[Q]$ with eigenvalue λ, then $\boldsymbol{p}$ is a left eigenvectorof $[W_\delta]$ with eigenvalue $1 + \lambda\delta$, and conversely.

We saw in Section 7.2.5 that the steady-state probability vector $\boldsymbol{p}$ of a Markov process is the same as that of any sampled-time approximation. We have now seen that, in addition, *all* the eigenvectors are the same and the eigenvalues are simply related. Thus study of these differential equations can be largely replaced by studying the sampled-time approximation.

The following theorem uses our knowledge of the eigenvalues and eigenvectors of transition matrices such as $[W_\delta]$ in Section 4.4, to be more specific about the properties of $[Q]$.

Theorem 7.3.2 *Consider an irreducible finite-state Markov process with* M *states. Then the matrix* $[Q]$ *for that process has an eigenvalue* λ *equal to* 0*. That eigenvalue has a right eigenvector* $\boldsymbol{e} = (1, 1, \dots, 1)^{\mathsf{T}}$ *which is unique within a scale factor. It has a left eigenvector* $\boldsymbol{p} = (p_1, \dots, p_{\mathsf{M}})$ *that is positive, sums to* 1*, satisfies (7.23), and is unique within a scale factor. All the other eigenvalues of* $[Q]$ *have strictly negative real parts.*

Proof Since all M states communicate, the sampled-time chain is recurrent. From Theorem 4.4.2, $[W_\delta]$ has a unique eigenvalue $\lambda = 1$. The corresponding right eigenvector is $\boldsymbol{e}$ and the left eigenvector is the steady-state probability vector $\boldsymbol{p}$ as given in (4.8). Since $[W_\delta]$ is recurrent, the components of $\boldsymbol{p}$ are strictly positive. From the equivalence of (7.23) and (7.24), $\boldsymbol{p}$, as given by (7.23), is the steady-state probability vector of the process. Each eigenvalue λ_δ of $[W_\delta]$ corresponds to an eigenvalue λ of $[Q]$ with the correspondence $\lambda_\delta = 1 + \lambda\delta$, i.e., $\lambda = (\lambda_\delta - 1)/\delta$. Thus the eigenvalue 1 of $[W_\delta]$ corresponds to the eigenvalue 0 of $[Q]$. Since $|\lambda_\delta| \leq 1$ and $\lambda_\delta \neq 1$ for all other eigenvalues, the other eigenvalues of $[Q]$ all have strictly negative real parts, completing the proof. □

We complete this section by deriving the Komogorov backward equations. For s small and positive, the Chapman–Kolmogorov equations in (7.27) become

$$
\begin{aligned}
P_{ij}(t) &= \sum_k P_{ik}(s)P_{kj}(t-s) \\
&= \sum_{k \neq i} s q_{ik} P_{kj}(t-s) + (1 - s\nu_i)P_{ij}(t-s) + o(s).
\end{aligned}
$$

Subtracting $P_{ij}(t-s)$ from both sides and dividing by s,

$$
\begin{aligned}
\frac{P_{ij}(t) - P_{ij}(t-s)}{s} &= \sum_{k \neq i} q_{ik} P_{kj}(t-s) - \nu_i P_{ij}(t-s) + \frac{o(s)}{s} \\
\frac{dP_{ij}(t)}{dt} &= \sum_{k \neq i} q_{ik} P_{kj}(t) - \nu_i P_{ij}(t). \qquad (7.33)
\end{aligned}
$$

In matrix form, this is expressed as

$$\frac{d[P(t)]}{dt} = [Q][P(t)]. \tag{7.34}$$

By comparing (7.34) and (7.32), we see that $[Q][P(t)] = [P(t)][Q]$, i.e., that the matrices $[Q]$ and $[P(t)]$ commute. Simultaneous linear differential equations appear in so many applications that we leave the further exploration of these forward and backward equations as simple differential equation topics rather than topics that have special properties for Markov processes.

7.4 Uniformization

Up until now, we have discussed Markov processes under the assumption that $q_{ii} = 0$ (i.e., no transitions from a state into itself are allowed). We now consider what happens if this restriction is removed. Suppose we start with some Markov process defined by a set of transition rates q_{ij} with $q_{ii} = 0$, and we modify this process by some arbitrary choice of $q_{ii} \geq 0$ for each state i. This modification changes the embedded Markov chain, since ν_i is increased from $\sum_{k \neq i} q_{ik}$ to $\sum_{k \neq i} q_{ik} + q_{ii}$. From (7.5), P_{ij} is changed to q_{ij}/ν_i for the new value of ν_i for each i, j. Thus the steady-state probabilities π_i for the embedded chain are changed. The Markov process $\{X(t);\ t \geq 0\}$ is not changed, since a transition from i into itself does not change $X(t)$ and does not change the distribution of the time until the next transition to a different state. The steady-state probabilities for the process still satisfy

$$p_j \nu_j = \sum_k p_k q_{kj}, \qquad \sum_i p_i = 1. \tag{7.35}$$

The addition of the new term q_{jj} increases ν_j by q_{jj}, thus increasing $p_j \nu_j$ by $p_j q_{jj}$. Similarly $\sum_k p_k q_{kj}$ is increased by $p_j q_{jj}$, so that the solution is unchanged (as we already determined it must be).

A particularly convenient way to add self transitions is to add them in such a way as to make the transition rate ν_j the same for all states. Assuming that the transition rates $\{\nu_i;\ i \geq 0\}$ are bounded, we define ν^* as $\sup_j \nu_j$ for the original transition rates. Then we set $q_{jj} = \nu^* - \sum_{k \neq j} q_{jk}$ for each j. With this addition of self transitions, all transition rates become ν^*. From (7.21), we see that the new steady-state probabilities, π_i^*, in the embedded Markov chain become equal to the steady-state process probabilities, p_i. Naturally, we could also choose any ν greater than ν^* and increase each q_{jj} to make all transition rates equal to that value of ν. When the transition rates are all changed to be ν^*, the resulting embedded chain is called a *uniformized chain* and the Markov process is called the *uniformized process*. The uniformized process is the same as the original process, except that quantities like the number of transitions over some interval are different because of the self transitions.

Assuming that all transition rates are made equal to ν^*, the new transition probabilities in the embedded chain become $P_{ij}^* = q_{ij}/\nu^*$. Let $N(t)$ be the total number of transitions that occur from 0 to t in the uniformized process. Since the rate of transitions is the same from all states and the inter-transition intervals are independent and identically exponentially distributed, $N(t)$ is a Poisson counting process of rate ν^*. Also, $N(t)$ is independent of the sequence of transitions in the embedded uniformized Markov chain. Thus, given that $N(t) = n$, the probability that $X(t) = j$ given that $X(0) = i$ is just the probability that the embedded chain goes from i to j in n steps, i.e., P_{ij}^{*n}. This gives us another formula for calculating $P_{ij}(t)$, (i.e., the probability that $X(t) = j$ given that $X(0) = i$):

$$P_{ij}(t) = \sum_{n=0}^{\infty} P_{ij}^{*n} \frac{e^{-\nu^* t}(\nu^* t)^n}{n!}. \tag{7.36}$$

Another situation where the uniformized process is useful is in extending Markov decision theory to Markov processes, but we do not pursue this.

7.5 Birth–death processes

Birth–death processes are very similar to the birth–death Markov chains that we studied earlier. Here transitions occur only between neighboring states, so it is convenient to define λ_i as $q_{i,i+1}$ and μ_i as $q_{i,i-1}$ (see Figure 7.9). Since the number of transitions from i to $i+1$ is within 1 of the number of transitions from $i+1$ to i for every sample path, we conclude that

$$p_i \lambda_i = p_{i+1} \mu_{i+1}. \tag{7.37}$$

This can also be obtained inductively from (7.23) using the same argument that we used earlier for birth–death Markov chains.

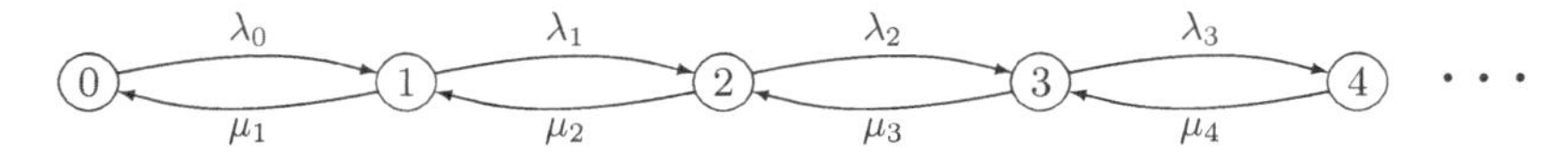

Figure 7.9 Birth–death process.

Define ρ_i as λ_i/μ_{i+1}. Then applying (7.37) iteratively, we obtain the steady-state equations

$$p_i = p_0 \prod_{j=0}^{i-1} \rho_j\,, \quad i \geq 1. \tag{7.38}$$

We can solve for p_0 by substituting (7.38) into $\sum_i p_i$, yielding

$$p_0 = \frac{1}{1 + \sum_{i=1}^{\infty} \prod_{j=0}^{i-1} \rho_j}. \tag{7.39}$$

7.5.1 The M/M/1 queue again

For the M/M/1 queue, the state of the Markov process is the number of customers in the system (i.e., customers either in queue or in service). The transitions from i to $i+1$

correspond to arrivals, and since the arrival process is Poisson of rate λ, we have $\lambda_i = \lambda$ for all $i \geq 0$. The transitions from i to $i-1$ correspond to departures, and since the service time distribution is exponential with parameter μ, say, we have $\mu_i = \mu$ for all $i \geq 1$. Thus, (7.39) simplifies to $p_0 = 1 - \rho$, where $\rho = \lambda/\mu$ and thus

$$p_i = (1-\rho)\rho^i, \qquad i \geq 0. \tag{7.40}$$

We assume that $\rho < 1$, which is required for positive recurrence. The probability that there are i or more customers in the system in steady state is then given by $\Pr\{X(t) \geq i\} = \rho^i$ and the expected number of customers in the system is given by

$$\mathsf{E}[X(t)] = \sum_{i=1}^{\infty} \Pr\{X(t) \geq i\} = \frac{\rho}{1-\rho}. \tag{7.41}$$

The expected time that a customer spends in the system in steady state can now be determined by Little's formula (Theorem 5.5.9):

$$\mathsf{E}\big[\text{system time}\big] = \frac{\mathsf{E}[X(t)]}{\lambda} = \frac{\rho}{\lambda(1-\rho)} = \frac{1}{\mu-\lambda}. \tag{7.42}$$

The expected time that a customer spends in the queue (i.e., before entering service) is just the expected system time less the expected service time, so

$$\mathsf{E}\big[\text{queueing time}\big] = \frac{1}{\mu-\lambda} - \frac{1}{\mu} = \frac{\rho}{\mu-\lambda}. \tag{7.43}$$

Finally, the expected number of customers in the queue can be found by applying Little's formula to (7.43),

$$\mathsf{E}\big[\text{number in queue}\big] = \frac{\lambda\rho}{\mu-\lambda}. \tag{7.44}$$

Note that the expected numbers of customers in the system and in the queue depend only on ρ, so that if the arrival rate and service rate are both speeded up by the same factor, these expected values are unchanged. The expected system time and queueing time, however, would decrease by the factor of the rate increases. Note also that as ρ approaches 1, all these quantities approach infinity as $1/(1-\rho)$. At the value $\rho = 1$, the embedded Markov chain becomes null recurrent and the steady-state probabilities (both $\{\pi_i;\ i \geq 0\}$ and $\{p_i;\ i \geq 0\}$) can be viewed as being all 0 or as failing to exist.

One final result about the M/M/1 queue was established in Section 2.3.3. This showed that in steady state, the system time is an exponential rv of rate $\mu - \lambda$.

7.5.2 Other birth–death systems

There are many types of queueing systems that can be modeled as birth–death processes. For example, the arrival rate could vary with the number in the system and the service rate could vary with the number in the system. All of these systems can be analyzed in steady state in the same way, but (7.38) and (7.39) can become quite messy in these more complex systems. As an example, we analyze the M/M/m system. Here there are m servers, each with exponentially distributed service times with parameter μ. When

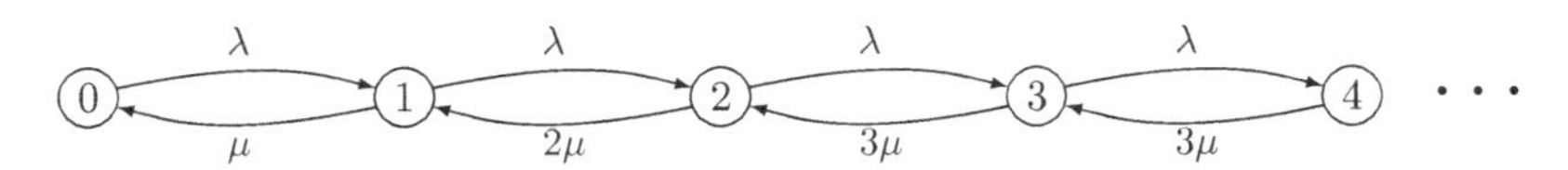

Figure 7.10 M/M/m queue for $m = 3$.

i customers are in the system, there are i servers working for $i < m$ and all m servers are working for $i \geq m$. With i servers working, the probability of a departure in an incremental time δ is $i\mu\delta$, so that μ_i is $i\mu$ for $i < m$ and $m\mu$ for $i \geq m$ (see Figure 7.10).

Define $\rho = \lambda/(m\mu)$. Then in terms of our general birth–death process notation, $\rho_i = m\rho/(i+1)$ for $i < m$ and $\rho_i = \rho$ for $i \geq m$. From (7.38), we have

$$p_i = p_0 \frac{m\rho}{1} \frac{m\rho}{2} \cdots \frac{m\rho}{i} = \frac{p_0 (m\rho)^i}{i!}, \qquad i \leq m; \tag{7.45}$$

$$p_i = \frac{p_0 \rho^i m^m}{m!}, \qquad i \geq m. \tag{7.46}$$

We can find p_0 by summing p_i and setting the result equal to 1; a solution exists if $\rho < 1$. Nothing simplifies much in this sum, except that $\sum_{i \geq m} p_i = p_0 (\rho m)^m / [m!(1-\rho)]$, and the solution is

$$p_0 = \left[\frac{(m\rho)^m}{m!(1-\rho)} + \sum_{i=0}^{m-1} \frac{(m\rho)^i}{i!} \right]^{-1}. \tag{7.47}$$

7.6 Reversibility for Markov processes

In Section 6.5 on reversibility for Markov chains, (6.44) showed that the backward transition probabilities P_{ij}^* in steady state satisfy

$$\pi_i P_{ij}^* = \pi_j P_{ji}. \tag{7.48}$$

These equations are then valid for the embedded chain of a Markov process. Next, consider backward transitions in the process itself. Given that the process is in state i, the probability of a transition in an increment δ of time is $\nu_i \delta + o(\delta)$, and transitions in successive increments are independent. Thus, if we view the process running backward in time, the probability of a transition in each increment δ of time is also $\nu_i \delta + o(\delta)$ with independence between increments. Thus, going to the limit $\delta \to 0$, the distribution of the time backward to a transition is exponential with parameter ν_i. This means that the process running backwards is again a Markov process with transition probabilities P_{ij}^* and transition rates ν_i. Figure 7.11 helps to illustrate this.

Since the steady-state probabilities $\{p_i; i \geq 0\}$ for the Markov process are determined by

$$p_i = \frac{\pi_i / \nu_i}{\sum_k \pi_k / \nu_k}, \tag{7.49}$$

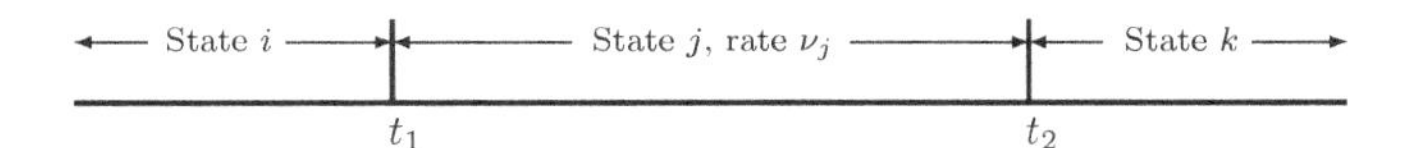

Figure 7.11 The forward process enters state j at time t_1 and departs at t_2. The backward process enters state j at time t_2 and departs at t_1. In any sample function, as illustrated, the interval in a given state is the same in the forward and backward process. Given $X(t) = j$, the time forward to the next transition and the time backward to the previous transition are each exponential with rate ν_j.

and since $\{\pi_i;\ i \geq 0\}$ and $\{\nu_i;\ i \geq 0\}$ are the same for the forward and backward processes, we see that the steady-state probabilities in the backward Markov process are the same as the steady-state probabilities in the forward process. This result can also be seen by the correspondence between sample functions in the forward and backward processes.

The *transition rates* in the backward process are defined by $q^*_{ij} = \nu_i P^*_{ij}$. Using (7.48), we have

$$q^*_{ij} = \nu_i P^*_{ij} = \frac{\nu_i \pi_j P_{ji}}{\pi_i} = \frac{\nu_i \pi_j q_{ji}}{\pi_i \nu_j}. \tag{7.50}$$

From (7.49), we note that $p_j = \alpha\pi_j/\nu_j$ and $p_i = \alpha\pi_i/\nu_i$ for the same value of α. Thus the ratio of π_j/ν_j to π_i/ν_i is p_j/p_i. This simplifies (7.50) to $q^*_{ij} = p_j q_{ji}/p_i$, and

$$p_i q^*_{ij} = p_j q_{ji}. \tag{7.51}$$

This equation can be used as an alternative definition of the backward transition rates. To interpret this, let δ be a vanishingly small increment of time and assume the process is in steady state at time t. Then $\delta p_j q_{ji} \approx \Pr\{X(t) = j\}\Pr\{X(t+\delta) = i \mid X(t) = j\}$, whereas $\delta p_i q^*_{ij} \approx \Pr\{X(t+\delta) = i\}\Pr\{X(t) = j \mid X(t+\delta) = i\}$.

A Markov process is defined to be *reversible* if $q^*_{ij} = q_{ij}$ for all i, j. If the embedded Markov chain is reversible (i.e., $P^*_{ij} = P_{ij}$ for all i, j), then one can repeat the above steps using P_{ij} and q_{ij} in place of P^*_{ij} and q^*_{ij} to see that $p_i q_{ij} = p_j q_{ji}$ for all i, j. Thus, if the embedded chain is reversible, the process is also. Similarly, if the Markov process is reversible, the above argument can be reversed to see that the embedded chain is reversible. Thus, we have the following useful lemma.

Lemma 7.6.1 *Assume that steady-state probabilities $\{p_i;\ i \geq 0\}$ exist in an irreducible Markov process (i.e., (7.23) has a solution and $\sum p_i \nu_i < \infty$). Then the Markov process is reversible if and only if the embedded chain is reversible.*

One can find the steady-state probabilities of a reversible Markov process and simultaneously show that it is reversible by the following useful theorem (which is directly analogous to Theorem 6.5.3).

Theorem 7.6.2 *For an irreducible Markov process, assume that $\{p_i;\ i \geq 0\}$ is a set of non-negative numbers summing to 1, satisfying $\sum_i p_i \nu_i \leq \infty$, and satisfying*

$$p_i q_{ij} = p_j q_{ji} \qquad \text{for all } i, j. \tag{7.52}$$

Then $\{p_i;\ i \geq 0\}$ is the set of steady-state probabilities for the process, $p_i > 0$ for all i, the process is reversible, and the embedded chain is positive recurrent.

Proof Summing (7.52) over i, we obtain

$$\sum_i p_i q_{ij} = p_j \nu_j \quad \text{for all } j.$$

These, along with $\sum_i p_i = 1$ are the steady-state equations for the process. These equations have a solution, and by Theorem 7.2.7, $p_i > 0$ for all i, the embedded chain is positive recurrent, and $p_i = \lim_{t\to\infty} \Pr\{X(t) = i\}$. Comparing (7.52) with (7.51), we see that $q_{ij} = q^*_{ij}$, so the process is reversible. □

There are many irreducible Markov processes that are not reversible but for which the backward process has interesting properties that can be deduced, at least intuitively, from the forward process. Jackson networks (to be studied shortly) and many more complex networks of queues fall into this category. The following simple theorem allows us to use whatever combination of intuitive reasoning and wishful thinking we desire to guess both the transition rates q^*_{ij} in the backward process and the steady-state probabilities, and to then verify rigorously that the guess is correct. One might think that guessing is somehow unscientific, but in fact, the art of educated guessing and intuitive reasoning leads to much of the best research.

Theorem 7.6.3 *For an irreducible Markov process, assume that a set of positive numbers $\{p_i;\, i \geq 0\}$ satisfy $\sum_i p_i = 1$ and $\sum_i p_i \nu_i < \infty$. Also assume that a set of non-negative numbers $\{q^*_{ij}\}$ satisfy the two sets of equations*

$$\sum_j q_{ij} = \sum_j q^*_{ij} \quad \textit{for all } i, \tag{7.53}$$

$$p_i q_{ij} = p_j q^*_{ji} \quad \textit{for all } i, j. \tag{7.54}$$

Then $\{p_i\}$ is the set of steady-state probabilities for the process, $p_i > 0$ for all i, the embedded chain is positive recurrent, and $\{q^*_{ij}\}$ is the set of transition rates in the backward process.

Proof Sum (7.54) over i. Using the fact that $\sum_j q_{ij} = \nu_i$ and using (7.53), we obtain

$$\sum_i p_i q_{ij} = p_j \nu_j \quad \text{for all } j. \tag{7.55}$$

These, along with $\sum_i p_i = 1$, are the steady-state equations for the process. These equations thus have a solution, and by Theorem 7.2.7, $p_i > 0$ for all i, the embedded chain is positive recurrent, and $p_i = \lim_{t\to\infty} \Pr\{X(t) = i\}$. Finally, q^*_{ij} as given by (7.54) is the backward transition rate as given by (7.51) for all i, j. □

We see that Theorem 7.6.2 is just a special case of Theorem 7.6.3 in which the guess about q^*_{ij} is that $q^*_{ij} = q_{ij}$.

Birth–death processes are all reversible if the steady-state probabilities exist. To see this, note that (7.37) (the equation to find the steady-state probabilities) is just (7.52) applied to the special case of birth–death processes. Due to the importance of this, we state it as a theorem.

Theorem 7.6.4 *For a birth–death process, if there is a solution $\{p_i; i \geq 0\}$ to (7.37) with $\sum_i p_i = 1$ and $\sum_i p_i \nu_i < \infty$, then the process is reversible, and the embedded chain is positive recurrent and reversible.*

Since the M/M/1 queueing process is a birth–death process, it is also reversible. Burke's theorem, which was given as Theorem 6.6.1 for sampled-time M/M/1 queues, can now be established for continuous-time M/M/1 queues. Note that the theorem here contains an extra part, part (c).

Theorem 7.6.5 (Burke's theorem) *Given an M/M/1 queueing system in steady state with $\lambda < \mu$,*

(a) the departure process is Poisson with rate λ,
(b) the state $X(t)$ at any time t is independent of departures prior to t, and
(c) for first come, first served (FCFS) service, given that a customer departs at time t, the arrival time of that customer is independent of the departures prior to t.

Proof The proofs of parts (a) and (b) are the same as the proof of Burke's theorem for sampled time, Theorem 6.6.1, and thus will not be repeated. For part (c), note that with FCFS service, the mth customer to arrive at the system is also the mth customer to depart. Figure 7.12 illustrates that the association between arrivals and departures is the same in the backward system as in the forward system (even though the customer ordering is reversed in the backward system). In the forward, right moving system, let τ be the epoch of some given arrival. The customers arriving after τ wait behind the given arrival in the queue, and have no effect on the given customer's service. Thus the interval from τ to the given customer's service completion is independent of arrivals after τ.

Since the backward, left moving, system is also an M/M/1 queue, the interval from a given backward arrival, say at epoch t, moving left until the corresponding departure, is independent of arrivals to the left of t. From the correspondence between sample functions in the right moving and left moving systems, given a departure at epoch t in the right moving system, the departures before time t are independent of the arrival epoch of the given customer departing at t; this completes the proof. □

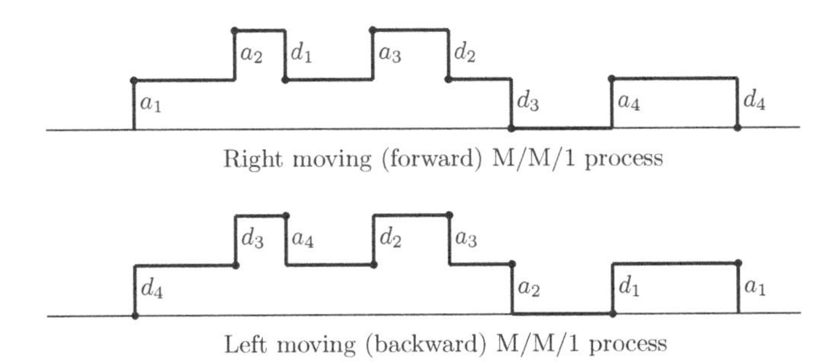

Figure 7.12 FCFS arrivals and departures in right and left moving M/M/1 processes.

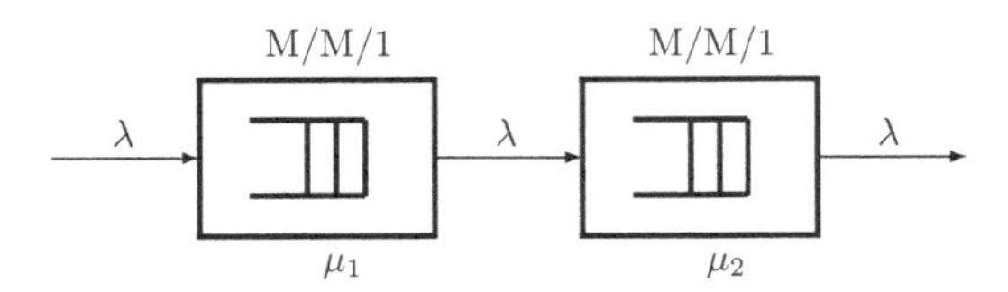

Figure 7.13 A tandem queueing system. Assuming that $\lambda < \mu_1$ and $\lambda < \mu_2$, the departures from each queue are Poisson of rate λ.

Part (c) of Burke's theorem does not apply to sampled-time M/M/1 queues because the sampled-time model does not allow for both an arrival and departure in the same increment of time.

Note that the proof of Burke's theorem (including parts (a) and (b) from Section 6.6) does not make use of the fact that the transition rate $q_{i,i-1} = \mu$ for $i \geq 1$ in the M/M/1 queue. Thus Burke's theorem remains true for any birth–death Markov process in steady state for which $q_{i,i+1} = \lambda$ for all $i \geq 0$. For example, parts (a) and (b) are valid for M/M/m queues; part (c) is also valid (see [30]), but the argument here is not adequate since the first customer to enter the system might not be the first to depart.

We next show how Burke's theorem can be used to analyze a tandem pair of queues. As illustrated in Figure 7.13, the first queueng system is M/M/1 with Poisson arrivals at rate λ and IID exponential service times at rate $\mu_1 > \lambda$. The departures from this queueing system are the arrivals to a second queueing system. Departures from system 1 instantaneously enter queueing system 2.

Queueing system 2 has a single server with IID exponential service times at rate $\mu_2 > \lambda$. The successive service times at system 2 are assumed to be independent of the arrival times and service times at queueing system 1. Since Burke's theorem shows that the departures from the first system are Poisson with rate λ, the arrivals to the second queue are Poisson with rate λ. The arrival times at the second queueing system are functions of the arrival and service times at system 1, and are thus independent of the service times at system 2. Thus the second system is also M/M/1.

Let $X(t)$ be the state of queueing system 1 and $Y(t)$ be the state of queueing system 2. Since $X(t)$ at time t is independent of the departures from system 1 prior to t, $X(t)$ is independent of the arrivals to system 2 prior to time t. Since $Y(t)$ depends only on the arrivals to system 2 prior to t and on the service times that have been completed prior to t, we see that $X(t)$ is independent of $Y(t)$. This leaves a slight nit-picking question about what happens at the instant of a departure from system 1. We have considered the state $X(t)$ at the instant of a departure to be the number of customers remaining in system 1 not counting the departing customer. Also the state $Y(t)$ is the state in system 2 including the new arrival at instant t. The state $X(t)$ then is independent of the departures up to and including t, so that $X(t)$ and $Y(t)$ are still independent.

Next assume that both systems use FCFS service. Consider a customer that leaves system 1 at time t. The time at which that customer arrived at system 1, and thus the waiting time in system 1 for that customer, is independent of the departures prior to t. This means that the state of system 2 immediately before the given customer arrives at time t is independent of the time the customer spent in system 1. It therefore follows that

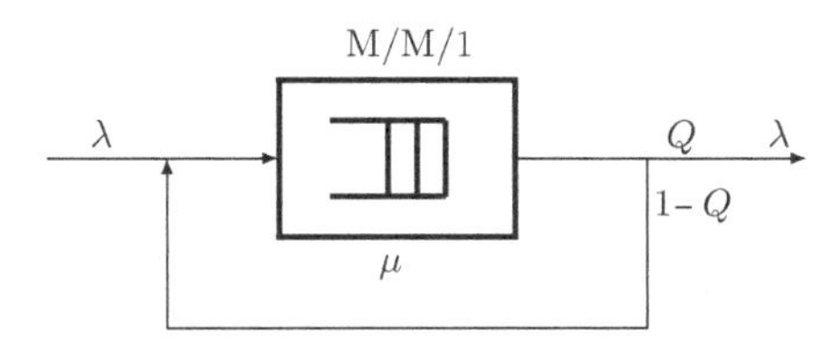

Figure 7.14 A queue with feedback. Assuming that $\mu > \lambda/Q$, the exogenous output is Poisson of rate λ.

Figure 7.15 Sample path of arrivals and departures for queue with feedback.

the time that the customer spends in system 2 is independent of the time spent in system 1. Thus the total system time that a customer spends in both system 1 and system 2 is the sum of two independent rv s.

This same argument can be applied to more than two queueing systems in tandem. It can also be applied to more general networks of queues, each with single servers with exponentially distributed service times. The restriction here is that there cannot be any cycle of queueing systems where departures from each queue in the cycle can enter the next queue in the cycle. The problem posed by such cycles can be seen easily in the following example of a single queueing system with feedback (see Figure 7.14).

We assume that the queueing system in Figure 7.14 has a single server with IID exponential service times that are independent of arrival times. The exogenous arrivals from outside the system are Poisson with rate λ. With probability Q, the departures from the queue leave the entire system, and, alternatively, with probability $1 - Q$, they return instantaneously to the input of the queue. Successive choices between leaving the system and returning to the input are IID and independent of exogenous arrivals and of service times. Figure 7.15 shows a sample function of the arrivals and departures in the case in which the service rate μ is very much greater than the exogenous arrival rate λ. Each exogenous arrival spawns a geometrically distributed set of departures and simultaneous reentries. Thus the overall arrival process to the queue, counting both exogenous arrivals and feedback from the output, is not Poisson. Note, however, that if we look at the Markov process description, the departures that are fed back to the input correspond to self loops from one state to itself. Thus the Markov process is the same as one without the self loops with a service rate equal to μQ. Thus, from Burke's theorem, the exogenous departures are Poisson with rate λ. Also the steady-state distribution of $X(t)$ is $\Pr\{X(t) = i\} = (1 - \rho)\rho^i$, where $\rho = \lambda/(\mu Q)$ (assuming, of course, that $\rho < 1$).

The tandem queueing system of Figure 7.13 can also be regarded as a combined Markov process in which the state at time t is the pair $(X(t), Y(t))$. The transitions in this process correspond to, first, exogenous arrivals in which $X(t)$ increases, second, exogenous departures in which $Y(t)$ decreases, and, third, transfers from system 1 to system 2 in which $X(t)$ decreases and $Y(t)$ simultaneously increases. The combined process is not reversible since there is no transition in which $X(t)$ increases and $Y(t)$ simultaneously decreases. In the next section, we show how to analyze these combined Markov processes for more general networks of queues.

7.7 Jackson networks

In many queueing situations, a customer has to wait in a number of different queues before completing the desired transaction and leaving the system. For example, when we go to the registry of motor vehicles to get a driver's license, we must wait in one queue to have the application processed, in another queue to pay for the license, and in yet a third queue to obtain a photograph for the license. In a multiprocessor computer facility, a job can be queued waiting for service at one processor, then go to wait for another processor, and so forth; frequently the same processor is visited several times before the job is completed. In a data network, packets traverse multiple intermediate nodes; at each node they enter a queue waiting for transmission to other nodes.

Such systems are modeled by a network of queues, and Jackson networks are perhaps the simplest models of such networks. In such a model, we have a network of k interconnected queueing systems which we call nodes. Each of the k nodes receives customers (i.e., tasks or jobs) both from outside the network (exogenous inputs) and from other nodes within the network (endogenous inputs). It is assumed that the exogenous inputs to each node i form a Poisson process of rate r_i and that these Poisson processes are independent of each other. For analytical convenience, we regard this as a single Poisson input process of rate λ_0, with each input independently going to each node i with probability $Q_{0i} = r_i/\lambda_0$.

Each node i contains a single server, and the successive service times at node i are IID rv s with an exponentially distributed service time of rate μ_i. The service times at each node are also independent of the service times at all other nodes and independent of the exogenous arrival times at all nodes. When a customer completes service at a given node i, that customer is routed to node j with probability Q_{ij} (see Figure 7.16). It is also possible for the customer to depart from the network entirely (called an exogenous departure), and this occurs with probability $Q_{i0} = 1 - \sum_{j\geq 1} Q_{ij}$. For a customer departing from node i, the next node j is a random variable with PMF $\{Q_{ij}, 0 \leq j \leq k\}$.

Successive choices of the next node for customers at node i are IID, independent of the customer routing at other nodes, independent of all service times, and independent of the exogenous inputs. Notationally, we are regarding the outside world as a fictitious node 0 from which customers appear and to which they disappear.

When a customer is routed from node i to node j, it is assumed that the routing is instantaneous; thus at the epoch of a departure from node i, there is a simultaneous

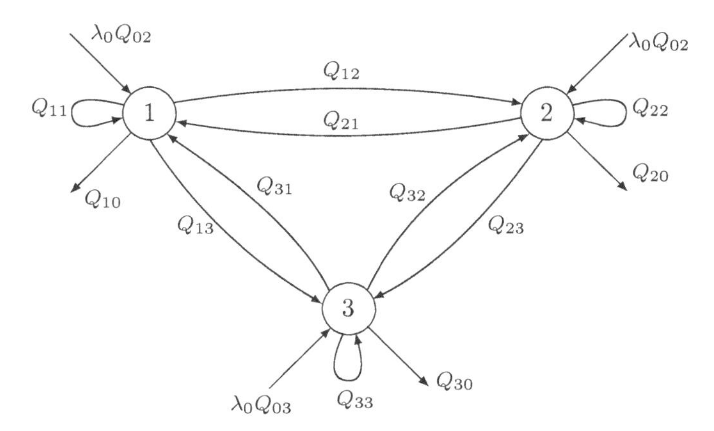

Figure 7.16 A Jackson network with three nodes. Given a departure from node i, the probability that departure goes to node j (or, for $j = 0$, departs the system) is Q_{ij}. Note that a departure from node i can reenter node i with probability Q_{ii}. The overall exogenous arrival rate is λ_0, and, conditional on an arrival, the probability the arrival enters node i is Q_{0i}.

endogenous arrival at node j. Thus a node j receives Poisson exogenous arrivals from outside the system at rate $\lambda_0 Q_{0j}$ and receives endogenous arrivals from other nodes according to the probabilistic rules just described. We can visualize these combined exogenous and endogenous arrivals as being served in FCFS fashion, but it really makes no difference in which order they are served, since the customers are statistically identical and simply give rise to service at node j at rate μ_j whenever there are customers to be served.

The Jackson queueing network, as just defined, is fully described by the exogenous input rate λ_0, the service rates $\{\mu_i\}$, and the routing probabilities $\{Q_{ij};\, 0 \leq i,j \leq k\}$. The network as a whole is a Markov process in which the state is a vector $\boldsymbol{m} = (m_1, m_2, \ldots, m_k)$, where $m_i, 1 \leq i \leq k$, is the number of customers at node i. State changes occur upon exogenous arrivals to the various nodes, exogenous departures from the various nodes, and departures from one node that enter another node. In a vanishingly small interval δ of time, given that the state at the beginning of that interval is $\boldsymbol{m}$, an exogenous arrival at node j occurs in the interval with probability $\lambda_0 Q_{0j}\delta$ and changes the state to $\boldsymbol{m}' = \boldsymbol{m} + \boldsymbol{e}_j$, where $\boldsymbol{e}_j$ is a unit vector with a one in position j. If $m_i > 0$, an exogenous departure from node i occurs in the interval with probability $\mu_i Q_{i0}\delta$ and changes the state to $\boldsymbol{m}' = \boldsymbol{m} - \boldsymbol{e}_i$. Finally, if $m_i > 0$, a departure from node i entering node j occurs in the interval with probability $\mu_i Q_{ij}\delta$ and changes the state to $\boldsymbol{m}' = \boldsymbol{m} - \boldsymbol{e}_i + \boldsymbol{e}_j$. Thus, the transition rates are given by

$$
\begin{aligned}
q_{\boldsymbol{m},\boldsymbol{m}'} &= \lambda_0 Q_{0j} && \text{for } \boldsymbol{m}' = \boldsymbol{m} + \boldsymbol{e}_j, \quad 1 \leq i \leq k, && (7.56)\\
&= \mu_i Q_{i0} && \text{for } \boldsymbol{m}' = \boldsymbol{m} - \boldsymbol{e}_i, \quad m_i > 0, \quad 1 \leq i \leq k, && (7.57)\\
&= \mu_i Q_{ij} && \text{for } \boldsymbol{m}' = \boldsymbol{m} - \boldsymbol{e}_i + \boldsymbol{e}_j, \quad m_i > 0, \quad 1 \leq i,j \leq k, && (7.58)\\
&= 0 && \text{for all other choices of } \boldsymbol{m}'.
\end{aligned}
$$

Note that a departure from node i that reenters node i causes a transition from state $\boldsymbol{m}$ back into state $\boldsymbol{m}$; we disallowed such transitions in Sections 7.1 and 7.2, but showed that they caused no problems in our discussion of uniformization. It is convenient to allow these self transitions here, partly for the added generality and partly to illustrate that the single-node network with feedback of Figure 7.14 is an example of a Jackson network.

Our objective is to find the steady-state probabilities $p(\boldsymbol{m})$ for this type of process, and our plan of attack is in accordance with Theorem 7.6.3; i.e., we shall guess a set of transition rates for the backward Markov process, use these to guess $p(\boldsymbol{m})$, and then verify that the guesses are correct. Before making these guesses, however, we must find out a little more about how the system works, so as to guide the guesswork. Let us define λ_i for each i, $1 \leq i \leq k$, as the time-average overall rate of arrivals to node i, including both exogenous and endogenous arrivals. Since λ_0 is the rate of exogenous inputs, we can interpret λ_i/λ_0 as the expected number of visits to node i per exogenous input. The endogenous arrivals to node i are not necessarily Poisson, as the example of a single queue with feedback shows, and we are not even sure at this point that such a time-average rate exists in any reasonable sense. However, let us assume for the time being that such rates exist and that the time-average rate of departures from each node equals the time-average rate of arrivals (i.e., the queue sizes do not grow linearly with time). Then these rates must satisfy the equation

$$\lambda_j = \sum_{j=0}^{k} \lambda_i Q_{ij}, \qquad 1 \leq j \leq k. \tag{7.59}$$

To see this, note that $\lambda_0 Q_{0j}$ is the rate of exogenous arrivals to j. Also λ_i is the time-average rate at which customers depart from queue i, and $\lambda_i Q_{ij}$ is the rate at which customers go from node i to node j. Thus, the right-hand side of (7.59) is the sum of the exogenous and endogenous arrival rates to node j. Note the distinction between the time-average rate of customers going from i to j in (7.59) and the rate $q_{\boldsymbol{m},\boldsymbol{m}'} = \mu_i Q_{ij}$ for $\boldsymbol{m}' = \boldsymbol{m} - \boldsymbol{e}_i + \boldsymbol{e}_j, m_i > 0$ in (7.58). The rate in (7.58) is conditioned on a state $\boldsymbol{m}$ with $m_i > 0$, whereas that in (7.59) is the overall time-average rate, averaged over all states.

Note that $\{Q_{ij};\ 0 \leq i,j \leq k\}$ forms a stochastic matrix and (7.59) is formally equivalent to the equations for steady-state probabilities (except that steady-state probabilities sum to 1). The usual equations for steady-state probabilities include an equation for $j = 0$, but that equation is redundant. Thus we know that, if there is a path between each pair of nodes (including the fictitious node 0), then (7.59) has a solution for $\{\lambda_i;\ 0 \leq i \leq k\}$, and that solution is unique within a scale factor. The known value of λ_0 determines this scale factor and makes the solution unique. Note that we do not have to be careful at this point about whether these rates are time averages in any particular sense, since this will be verified later; we do have to make sure that (7.59) has a solution, however, since it will appear in our solution for $p(\boldsymbol{m})$. Thus we assume in what follows that a path exists between each pair of nodes, and thus that (7.59) has a unique solution as a function of λ_0.

We now make the final necessary assumption about the network, which is that $\mu_i > \lambda_i$ for each node i. This will turn out to be required in order to make the process positive recurrent. We also define ρ_i as λ_i/μ_i. We shall find that, even though the inputs to an individual node i are not Poisson in general, there is a steady-state distribution for the number of customers at i, and that distribution is the same as that of an M/M/1 queue with the parameter ρ_i.

Now consider the backward time process. We have seen that only three kinds of transitions are possible in the forward process. First, there are transitions from $\boldsymbol{m}$ to $\boldsymbol{m}' = \boldsymbol{m} + \boldsymbol{e}_j$ for any j, $1 \leq j \leq k$. Second, there are transitions from $\boldsymbol{m}$ to $\boldsymbol{m} - \boldsymbol{e}_i$ for any i, $1 \geq i \leq k$, such that $m_i > 0$. Third, there are transitions from $\boldsymbol{m}$ to $\boldsymbol{m}' = \boldsymbol{m} - \boldsymbol{e}_i + \boldsymbol{e}_j$ for $1 \leq i,j \leq k$ with $m_i > 0$. Thus in the backward process, transitions from $\boldsymbol{m}'$ to $\boldsymbol{m}$ are possible only for the $\boldsymbol{m}, \boldsymbol{m}'$ pairs above. Corresponding to each arrival in the forward process, there is a departure in the backward process; for each forward departure, there is a backward arrival; and for each forward passage from i to j, there is a backward passage from j to i.

We now make the conjecture that the backward process is itself a Jackson network with Poisson exogenous arrivals at rates $\{\lambda_0 Q^*_{0j}\}$, service times that are exponential with rates $\{\mu_i\}$, and routing probabilities $\{Q^*_{ij}\}$. The backward routing probabilities $\{Q^*_{ij}\}$ must be chosen to be consistent with the transition rates in the forward process. Since each transition from i to j in the forward process must correspond to a transition from j to i in the backward process, we should have

$$\lambda_i Q_{ij} = \lambda_j Q^*_{ji}, \quad 0 \leq i,j \leq k. \tag{7.60}$$

Note that $\lambda_i Q_{ij}$ represents the rate at which forward transitions go from i to j, and λ_i represents the rate at which forward transitions *leave* node i. Equation (7.60) takes advantage of the fact that λ_i is also the rate at which forward transitions *enter* node i, and thus the rate at which backward transitions *leave* node i. Using the conjecture that the backward time system is a Jackson network with routing probabilities $\{Q^*_{ij};\ 0 \leq i,j \leq k\}$, we can write down the backward transition rates in the same way as (7.56)–(7.58):

$$\begin{aligned} q^*_{\boldsymbol{m},\boldsymbol{m}'} &= \lambda_0 Q^*_{0j} && \text{for } \boldsymbol{m}' = \boldsymbol{m} + \boldsymbol{e}_j, && (7.61)\\ &= \mu_i Q^*_{i0} && \text{for } \boldsymbol{m}' = \boldsymbol{m} - \boldsymbol{e}_i,\ m_i > 0,\ 1 \leq i \leq k, && (7.62)\\ &= \mu_i Q^*_{ij} && \text{for } \boldsymbol{m}' = \boldsymbol{m} - \boldsymbol{e}_i + \boldsymbol{e}_j,\ \ m - i > 0,\ \ 1 \leq i,\ \ j \leq k. && (7.63) \end{aligned}$$

If we substitute (7.60) into (7.61)–(7.63), we obtain

$$\begin{aligned} q^*_{\boldsymbol{m},\boldsymbol{m}'} &= \lambda_j Q_{j0} && \text{for } \boldsymbol{m}' = \boldsymbol{m} + \boldsymbol{e}_j, \quad 1 \leq j \leq k, && (7.64)\\ &= (\mu_i/\lambda_i)\lambda_0 Q_{0i} && \text{for } \boldsymbol{m}' = \boldsymbol{m} - \boldsymbol{e}_i,\ m_i > 0, \quad 1 \leq i \leq k, && (7.65)\\ &= (\mu_i/\lambda_i)\lambda_j Q_{ji} && \text{for } \boldsymbol{m}' = \boldsymbol{m} - \boldsymbol{e}_i + \boldsymbol{e}_j, \quad m_i > 0,\ 1 \leq i,j \leq k. && (7.66) \end{aligned}$$

This gives us our hypothesized backward transition rates in terms of the parameters of the original Jackson network. To use Theorem 7.6.3, we must verify that there is a set of positive numbers, $p(\boldsymbol{m})$, satisfying $\sum_{\boldsymbol{m}} p(\boldsymbol{m}) = 1$ and $\sum_{\boldsymbol{m}} \nu_{\boldsymbol{m}} p_{\boldsymbol{m}} < \infty$, and a set of non-negative numbers $q^*_{\boldsymbol{m}',\boldsymbol{m}}$ satisfying the following two sets of equations:

$$p(\boldsymbol{m}) q_{\boldsymbol{m},\boldsymbol{m}'} = p(\boldsymbol{m}') q^*_{\boldsymbol{m}',\boldsymbol{m}} \qquad \text{for all } \boldsymbol{m}, \boldsymbol{m}'; \tag{7.67}$$

$$\sum_{m'} q_{m,m'} = \sum_{m'} q^*_{m,m'} \qquad \text{for all } m. \tag{7.68}$$

We verify (7.67) by substituting (7.56)–(7.58) on the left-hand side of (7.67) and (7.64)–(7.66) on the right-hand side. Recalling that ρ_i is defined as λ_i/μ_i, and canceling out common terms on each side, we have

$$p(\boldsymbol{m}) = p(\boldsymbol{m}')/\rho_j \qquad \text{for } \boldsymbol{m}' = \boldsymbol{m} + \boldsymbol{e}_j, \tag{7.69}$$

$$p(\boldsymbol{m}) = p(\boldsymbol{m}')\rho_i \qquad \text{for } \boldsymbol{m}' = \boldsymbol{m} - \boldsymbol{e}_i, m_i > 0, \tag{7.70}$$

$$p(\boldsymbol{m}) = p(\boldsymbol{m}')\rho_i/\rho_j \qquad \text{for } \boldsymbol{m}' = \boldsymbol{m} - \boldsymbol{e}_i + \boldsymbol{e}_j, \qquad m_i > 0. \tag{7.71}$$

Looking at the case $\boldsymbol{m}' = \boldsymbol{m} - \boldsymbol{e}_i$, and using this equation repeatedly to get from state $(0, 0, \dots, 0)$ up to an arbitrary $\boldsymbol{m}$, we obtain

$$p(\boldsymbol{m}) = p(0, 0, \dots, 0) \prod_{i=1}^{k} \rho_i^{m_i}. \tag{7.72}$$

It is easy to verify that (7.72) satisfies (7.69)–(7.71) for all possible transitions. Summing over all $\boldsymbol{m}$ to solve for $p(0, 0, \dots, 0)$, we get

$$\begin{aligned} 1 = \sum_{m_1, m_2, \dots, m_k} p(\boldsymbol{m}) &= p(0, 0, \dots, 0) \sum_{m_1} \rho_1^{m_1} \sum_{m_2} \rho_2^{m_2} \dots \sum_{m_k} \rho_k^{m_k} \\ &= p(0, 0, \dots, 0)(1-\rho_1)^{-1}(1-\rho_2)^{-1} \cdots (1-\rho_k)^{-1}. \end{aligned}$$

Thus, $p(0, 0, \dots, 0) = (1-\rho_1)(1-\rho_2)\cdots(1-\rho_k)$, and substituting this in (7.72), we get

$$p(\boldsymbol{m}) = \prod_{i=1}^{k} p_i(m_i) = \prod_{i=1}^{k} \left[(1-\rho_i)\rho_i^{m_i}\right], \tag{7.73}$$

where $p_i(m) = (1-\rho_i)\rho_i^m$ is the steady-state distribution of a single M/M/1 queue. Now that we have found the steady-state distribution implied by our assumption about the backward process being a Jackson network, our remaining task is to verify (7.68)

To verify (7.68), i.e., $\sum_{m'} q_{m,m'} = \sum_{m'} q^*_{m,m'}$, first consider the right-hand side. Using (7.61) to sum over all $\boldsymbol{m}' = \boldsymbol{m} + \boldsymbol{e}_j$, then (7.62) to sum over $\boldsymbol{m}' = \boldsymbol{m} - \boldsymbol{e}_i$ (for i such that $m_i > 0$), and finally (7.63) to sum over $\boldsymbol{m}' = \boldsymbol{m} - \boldsymbol{e}_i + \boldsymbol{e}_j$ (again for i such that $m_i > 0$), we get

$$\sum_{m'} q^*_{m,m'} = \sum_{j=1}^{k} \lambda_0 Q^*_{0j} + \sum_{i:m_i>0} \mu_i Q^*_{i0} + \sum_{i:m_i>0} \mu_i \sum_{j=1}^{k} Q^*_{ij}. \tag{7.74}$$

Using the fact Q^* is a stochastic matrix,

$$\sum_{m'} q^*_{m,m'} = \lambda_0 + \sum_{i:m_i>0} \mu_i. \tag{7.75}$$

The left-hand side of (7.68) can be summed in the same way to get the result on the right-hand side of (7.75), but we can see that this must be the result by simply observing that λ_0 is the rate of exogenous arrivals and $\sum_{i:m_i>0} \mu_i$ is the overall rate of service completions in state $\boldsymbol{m}$. Note that this also verifies that $\nu_{\boldsymbol{m}} = \sum_{\boldsymbol{m}'} q_{\boldsymbol{m},\boldsymbol{m}'} \geq \lambda_0 + \sum_i \mu_i$, and since $\nu_{\boldsymbol{m}}$ is bounded, $\sum_{\boldsymbol{m}} \nu_{\boldsymbol{m}} p(\boldsymbol{m}) < \infty$. Since all the conditions of Theorem 7.6.3 are satisfied, $p(\boldsymbol{m})$, as given in (7.73), gives the steady-state probabilities for the Jackson network. This also verifies that the backward process is a Jackson network, and hence the exogenous departures are Poisson and independent.

Although the exogenous arrivals and departures in a Jackson network are Poisson, the endogenous processes of customers traveling from one node to another are typically not Poisson if there are feedback paths in the network. Also, although (7.73) shows that the numbers of customers at the different nodes are independent rv s at any given time in steady state, it is not generally true that the number of customers at one node at one time is independent of the number of customers at another node at another time.

There are many generalizations of the reversibility arguments used above, and many network situations in which the nodes have independent states at a common time. We discuss just two of them here and refer the reader to Kelly [17] for a complete treatment.

For the first generalization, assume that the service time at each node depends on the number of customers at that node, i.e., μ_i is replaced by μ_{i,m_i}. Note that this includes the M/M/m type of situation in which each node has several independent exponential servers. With this modification, the transition rates in (7.57) and (7.58) are modified by replacing μ_i with μ_{i,m_i}. The hypothesized backward transition rates are modified in the same way, and the only effect of these changes is to replace ρ_i and ρ_j for each i and j in (7.69)–(7.71) with $\rho_{i,m_i} = \lambda_i/\mu_{i,m_i}$ and $\rho_{j,m_j} = \lambda_j/\mu_{j,m_j}$. With this change, (7.72) becomes

$$p(\boldsymbol{m}) = \prod_{i=1}^{k} p_i(m_i) = \prod_{i=1}^{k} p_i(0) \prod_{j=0}^{m_i} \rho_{i,j}, \tag{7.76}$$

$$p_i(0) = \left[1 + \sum_{m=1}^{\infty} \prod_{j=0}^{m-1} \rho_{i,j}\right]^{-1}. \tag{7.77}$$

Thus, $p(\boldsymbol{m})$ is given by the product distribution of k individual birth–death systems.

7.7.1 Closed Jackson networks

The second generalization is to a network of queues with a fixed number M of customers in the system and with no exogenous inputs or outputs. Such networks are called *closed Jackson* networks, whereas the networks analyzed above are often called *open Jackson* networks. Suppose a k-node closed network has routing probabilities Q_{ij}, $1 \leq i,j \leq k$, where $\sum_j Q_{ij} = 1$, and has exponential service times of rate μ_i (this can be generalized to μ_{i,m_i} as above). We make the same assumptions as before about independence of service variables and routing variables, and assume that there is a path between each pair of nodes. Since $\{Q_{ij};\ 1 \leq i,j \leq k\}$ forms an irreducible stochastic matrix, there is a one-dimensional set of solutions to the steady-state equations

$$\lambda_j = \sum_i \lambda_i Q_{ij}, \qquad 1 \leq j \leq k. \tag{7.78}$$

We interpret λ_i as the time-average rate of transitions that go into node i. Since this set of equations can only be solved within an unknown multiplicative constant, and since this constant can only be determined at the end of the argument, we define $\{\pi_i;\ 1 \le i \le k\}$ as the particular solution of (7.78) satisfying

$$\pi_j = \sum_i \pi_i Q_{ij};\ 1 \le j \le k, \quad \sum_i \pi_i = 1. \tag{7.79}$$

Thus, for all i, $\lambda_i = \alpha \pi_i$, where α is some unknown constant. The state of the Markov process is again taken as $\boldsymbol{m} = (m_1, m_2, \dots, m_k)$ with the condition $\sum_i m_i = M$. The transition rates of the Markov process are the same as for open networks, except that there are no exogenous arrivals or departures; thus (7.56)–(7.58) are replaced by

$$q_{\boldsymbol{m},\boldsymbol{m}'} = \mu_i Q_{ij} \quad \text{for } \boldsymbol{m}' = \boldsymbol{m} - \boldsymbol{e}_i + \boldsymbol{e}_j, \quad m_i > 0,\ 1 \le i,j \le k. \tag{7.80}$$

We hypothesize that the backward time process is also a closed Jackson network, and as before, we conclude that if the hypothesis is true, the backward transition rates should be

$$q^*_{\boldsymbol{m},\boldsymbol{m}'} = \mu_i Q^*_{ij} \quad \text{for } \boldsymbol{m}' = \boldsymbol{m} - \boldsymbol{e}_i + \boldsymbol{e}_j, \quad m_i > 0,\ 1 \le i,j \le k, \tag{7.81}$$

where

$$\lambda_i Q_{ij} = \lambda_j Q^*_{ji} \qquad \text{for } 1 \le i,j \le k. \tag{7.82}$$

In order to use Theorem 7.6.3 again, we must verify that a PMF $p(\boldsymbol{m})$ exists satisfying $p(\boldsymbol{m})q_{\boldsymbol{m},\boldsymbol{m}'} = p(\boldsymbol{m}')q^*_{\boldsymbol{m}',\boldsymbol{m}}$ for all possible states and transitions, and we must also verify that $\sum_{\boldsymbol{m}'} q_{\boldsymbol{m},\boldsymbol{m}'} = \sum_{\boldsymbol{m}'} q^*_{\boldsymbol{m},\boldsymbol{m}'}$ for all possible $\boldsymbol{m}$. This latter verification is virtually the same as before and is left as an exercise. The former verification, with the use of (7.72), (7.73), and (7.74), becomes

$$p(\boldsymbol{m})(\mu_i/\lambda_i) = p(\boldsymbol{m}')(\mu_j/\lambda_j) \quad \text{for } \boldsymbol{m}' = \boldsymbol{m} - \boldsymbol{e}_i + \boldsymbol{e}_j, \qquad m_i > 0. \tag{7.83}$$

Using the open network solution to guide our intuition, we see that the following choice of $p(\boldsymbol{m})$ satisfies (7.83) for all possible $\boldsymbol{m}$ (i.e., all $\boldsymbol{m}$ such that $\sum_i m_i = M$):

$$p(\boldsymbol{m}) = A \prod_{i=1}^{k} (\lambda_i/\mu_i)^{m_i} \quad \text{for } \boldsymbol{m} \text{ such that } \sum_i m_i = M. \tag{7.84}$$

The constant A is a normalizing constant, chosen to make $p(\boldsymbol{m})$ sum to unity. The problem with (7.84) is that we do not know λ_i (except within a multiplicative constant independent of i). Fortunately, however, if we substitute π_i/α for λ_i, we see that α is raised to the power $-M$, independent of the state $\boldsymbol{m}$. Thus, letting $A' = A\alpha^{-M}$, our solution becomes

$$p(\boldsymbol{m}) = A' \prod_{i=1}^{k} (\pi_i/\mu_i)^{m_i} \quad \text{for } \boldsymbol{m} \text{ such that } \sum_i m_i = M, \tag{7.85}$$

$$\frac{1}{A'} = \sum_{\boldsymbol{m}:\ \sum_i m_i = M} \ \prod_{i=1}^{k} \left(\frac{\pi_i}{\mu_i}\right)^{m_i}. \tag{7.86}$$

Note that the steady-state distribution of the closed Jackson network has been found without solving for the time-average transition rates. Note also that the steady-state distribution looks very similar to that for an open network; i.e., it is a product distribution over the nodes with a geometric type distribution within each node. This is somewhat misleading, however, since the constant A' can be quite difficult to calculate. It is surprising at first that the parameter of the geometric distribution can be changed by a constant multiplier in (7.85) and (7.86) (i.e., π_i could be replaced with λ_i) and the solution does not change; the important quantity is the relative values of π_i/μ_i from one value of i to another rather than the absolute value.

In order to find λ_i (and this is important, since it says how quickly the system is doing its work), note that $\lambda_i = \mu_i \Pr\{m_i > 0\}$). Solving for $\Pr\{m_i > 0\}$ requires finding the constant A' in (7.80). In fact, the major difference between open and closed networks is that the relevant constants for closed networks are tedious to calculate (even by computer) for large networks and large M.

7.8 Semi-Markov processes

Semi-Markov processes are generalizations of Markov processes in which the time intervals between transitions have arbitrary distributions rather than exponential distributions. To be specific, there is an embedded Markov chain, $\{X_n;\ n \geq 0\}$ with a finite or countably infinite state space, and a sequence $\{U_n;\ n \geq 1\}$ of holding intervals between state transitions. The epochs at which state transitions occur are then given, for $n \geq 1$, as $S_n = \sum_{m=1}^{n} U_n$. The process starts at time 0 with S_0 defined to be 0. The semi-Markov process is then the continuous-time process $\{X(t);\ t \geq 0\}$ where, for each $n \geq 0$, $X(t) = X_n$ for t in the interval $S_n \leq X_n < S_{n+1}$. Initially, $X_0 = i$ where i is any given element of the state space.

The holding intervals $\{U_n;\ n \geq 1\}$ are non-negative rv s that depend only on the current state X_{n-1} and the next state X_n. More precisely, given $X_{n-1} = j$ and $X_n = k$, say, the interval U_n is independent of $\{U_m;\ m < n\}$ and independent of $\{X_m;\ m < n-1\}$ The conditional CDF for such an interval U_n is denoted by $G_{jk}(u)$, i.e.,

$$\Pr\{U_n \leq u \mid X_{n-1} = j, X_n = k\} = G_{jk}(u). \tag{7.87}$$

The dependencies between the rv s $\{X_n;\ n \geq 0\}$ and $\{U_n;\ n \geq 1\}$ are illustrated in Figure 7.17.

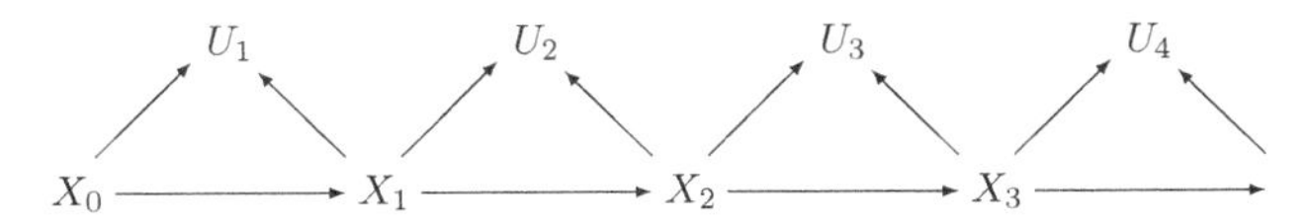

Figure 7.17 The statistical dependencies between the rv s of a semi-Markov process. Each holding interval U_n, conditional on the current state X_{n-1} and next state X_n, is independent of all other states and holding intervals. Note that, conditional on X_n, the holding intervals $U_n, U_{n-1}, \ldots$ are statistically independent of $U_{n+1}, X_{n+2}, \ldots$.

The conditional mean of U_n, conditional on $X_{n-1} = j, X_n = k$, is denoted $\overline{U}(j,k)$, i.e.,

$$\overline{U}(j,k) = \mathsf{E}\left[U_n \mid X_{n-1} = j, X_n = k\right] = \int_{u \geq 0} [1 - G_{jk}(u)]du. \tag{7.88}$$

A semi-Markov process evolves in essentially the same way as a Markov process. Given an initial state, $X_0 = i$ at time 0, a new state $X_1 = j$ is selected according to the embedded chain with probability P_{ij}. Then $U_1 = S_1$ is selected using the distribution $G_{ij}(u)$. Next a new state $X_2 = k$ is chosen according to the probability P_{jk}; then, given $X_1 = j$ and $X_2 = k$, the interval U_2 is selected with CDF $G_{jk}(u)$. Successive state transitions and transition times are chosen in the same way.

The steady-state behavior of semi-Markov processes can be analyzed in virtually the same way as that of Markov processes. We outline this in what follows, and often omit proofs where they are the same as the corresponding proof for Markov processes. First, since the holding intervals, U_n, are rv s, the transition epochs, $S_n = \sum_{m=1}^{n} U_n$, are also rv s. The following lemma then follows in the same way as Lemma 7.2.3 for Markov processes.

Lemma 7.8.1 *Let $M_i(t)$ be the number of transitions in a semi-Markov process in the interval $(0,t]$ for some given initial state $X_0 = i$. Then $\lim_{t\to\infty} M_i(t) = \infty$ WP1.*

In what follows, we assume that the embedded Markov chain is irreducible and positive recurrent. We want to find the limiting fraction of time that the process spends in any given state, say j. We will find that this limit exists WP1, and will find that it depends only on the steady-state probabilities of the embedded Markov chain and on the expected holding interval in each state. This is given by

$$\overline{U}(j) = \mathsf{E}\left[U_n \mid X_{n-1} = j\right] = \sum_k P_{jk}\mathsf{E}\left[U_n \mid X_{n-1} = j, X_n = k\right] = \sum_k P_{jk}\overline{U}(j,k), \tag{7.89}$$

where $\overline{U}(j,k)$ is given in (7.88). The steady-state probabilities $\{\pi_i;\ i \geq 0\}$ for the embedded chain tell us the fraction of transitions that enter any given state i. Since $\overline{U}(i)$ is the expected holding interval in i per transition into i, we would guess that the fraction of time spent in state i should be proportional to $\pi_i\overline{U}(i)$. Normalizing, we would guess that the time-average probability of being in state i should be

$$p_j = \frac{\pi_j\overline{U}(j)}{\sum_k \pi_k\overline{U}(k)}. \tag{7.90}$$

Identifying the mean holding interval, $\overline{U}_j$, with $1/\nu_j$, this is the same result that we established for the Markov process case. Using the same arguments, we find this is valid for the semi-Markov case. It is valid both in the conventional case where each p_j is positive and $\sum_j p_j = 1$, and also in the case where $\sum_k \pi_k\overline{U}(k) = \infty$, where each $p_j = 0$. The analysis is based on the fact that successive transitions to some given state, say j, given $X_0 = i$, form a delayed renewal process.

Lemma 7.8.2 *Consider a semi-Markov process with an irreducible recurrent embedded chain $\{X_n; n \geq 0\}$. Given $X_0 = i$, let $\{M_{ij}(t); t \geq 0\}$ be the number of transitions into* *a given state j in the interval $(0,t]$. Then $\{M_{ij}(t); t > 0\}$ is a delayed renewal process (or, if $j = i$, is an ordinary renewal process).*

This is the same as Lemma 7.2.4, but it is not quite so obvious that successive intervals between visits to state j are statistically independent. This can be seen, however, from Figure 7.17, which makes it clear that, given $X_n = j$, the future holding intervals, $U_n, U_{n+1}, \ldots$ are independent of the past intervals $U_{n-1}, U_{n-2}, \ldots$.

Next, using the same renewal–reward process as in Lemma 7.2.5, assigning reward 1 whenever $X(t) = j$, we define W_n as the interval between the $(n-1)$th and the nth entry to state j and get the following lemma.

Lemma 7.8.3 *Consider a semi-Markov process with an irreducible, recurrent, embedded Markov chain starting in $X_0 = i$. Then with probability 1, the limiting time average in state j is given by $p_j(i) = \overline{U}_j/[\overline{W}(j)]$.*

This lemma has omitted any assertion about the limiting ensemble probability of state j, i.e., $\lim_{t\to\infty} \Pr\{X(t) = j\}$. This follows easily from Blackwell's theorem, but depends on whether the successive entries to state j, i.e., $\{W_n; n \geq 1\}$, are arithmetic or non-arithmetic. This is explored in Exercise 7.33. The lemma shows (as expected) that the limiting time average in each state is independent of the starting state, so we henceforth replace $p_j(i)$ with p_j.

Next, let $M_i(t)$ be the total number of transitions in the semi-Markov process up to and including time n, given $X_0 = i$. This is not a renewal counting process, but, as with Markov processes, it provides a way to combine the time-average results for all states j. The following theorem is the same as that for Markov processes, except for the omission of ensemble-average results.

Theorem 7.8.4 *Consider a semi-Markov process with an irreducible, positive-recurrent, embedded Markov chain. Let $\{\pi_j; j \geq 0\}$ be the steady-state probabilities of the embedded chain and let $X_0 = i$ be the starting state. Then, WP1, the limiting time-average fraction of time spent in any arbitrary state j is given by*

$$p_j = \frac{\pi_j \overline{U}(j)}{\sum_k \pi_k \overline{U}(j)}. \tag{7.91}$$

The expected time between returns to state j is

$$\overline{W}(j) = \frac{\sum_k \pi_k \overline{U}(k)}{\pi_j}, \tag{7.92}$$

and the rate at which transitions take place is independent of X_0 and given by

$$\lim_{t\to\infty} \frac{M_i(t)}{t} = \frac{1}{\sum_k \pi_k \overline{U}(k)} \qquad \textit{WP1}. \tag{7.93}$$

For a semi-Markov process, knowing the steady-state probability of $X(t) = j$ for large t does not completely specify the steady-state behavior. Another important steady-state question is to determine the fraction of time involved in i to j transitions. To make this notion precise, define $Y(t)$ as the residual time until the next transition after time t (i.e.,

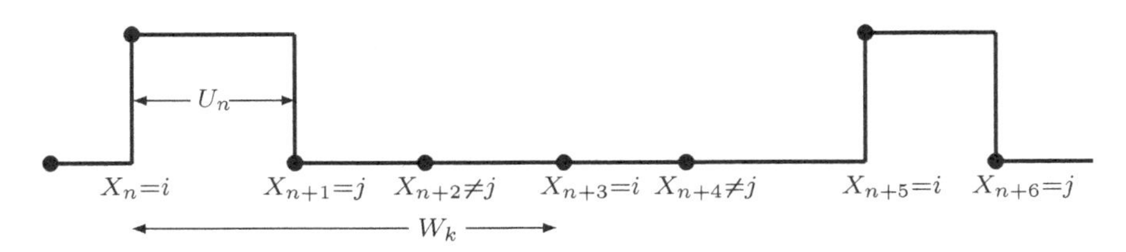

Figure 7.18 The renewal–reward process for i to j transitions. The expected value of U_n if $X_n = i$ and $X_{n+1} = j$ is $\overline{U}(i,j)$ and the expected interval between entries to i is $\overline{W}(i)$.

$t + Y(t)$ is the epoch of the next transition after time t). We want to determine the fraction of time t over which $X(t) = i$ and $X(t + Y(t)) = j$. Equivalently, for a non-arithmetic process, we want to determine $\Pr\{X(t) = i, X(t + Y(t) = j)\}$ in the limit as $t \to \infty$. Call this limit $Q(i,j)$.

Consider a renewal process, starting in state i and with renewals on transitions to state i. Define a reward $R(t) = 1$ for $X(t) = i$, $X(t + Y(t)) = j$ and $R(t) = 0$ otherwise (see Figure 7.18). That is, for each n such that $X(S_n) = i$ and $X(S_{n+1}) = j$, $R(t) = 1$ for $S_n \le t < S_{n+1}$. The expected reward in an inter-renewal interval is then $P_{ij}\overline{U}(i,j)$. It follows that $Q(i,j)$ is given by

$$Q(i,j) = \lim_{t \to \infty} \frac{\int_0^t R(\tau)d\tau}{t} = \frac{P_{ij}\overline{U}(i,j)}{\overline{W}(i)} = \frac{p_i P_{ij}\overline{U}(i,j)}{\overline{U}(i)}. \tag{7.94}$$

7.8.1 Example – the M/G/1 queue

As one example of a semi-Markov chain, consider an M/G/1 queue. Rather than the usual interpretation in which the state of the system is the number of customers in the system, we view the state of the system as changing only at departure times; the new state at a departure time is the number of customers left behind by the departure. This state then remains fixed until the next departure. New customers still enter the system according to the Poisson arrival process, but these new customers are not considered as part of the state until the next departure time. The number of customers in the system at arrival epochs does not in general constitute a 'state' for the system, since the age of the current service is also necessary as part of the statistical characterization of the process.

One purpose of this example is to illustrate that it is often more convenient to visualize the transition interval $U_n = S_n - S_{n-1}$ as being chosen first and the new state X_n as being chosen second rather than choosing the state first and the transition time second. For the M/G/1 queue, first suppose that the state is some $i > 0$. In this case, service begins on the next customer immediately after the old customer departs. Thus, U_n, conditional on $X_n = i$ for $i > 0$, has the distribution of the service time, say $G(u)$. The mean interval until a state transition occurs is

$$\overline{U}(i) = \int_0^\infty [1 - G(u)]du, \quad i > 0. \tag{7.95}$$

Given the interval u for a transition from state $i > 0$, the number of arrivals in that period is a Poisson rv with mean λu, where λ is the Poisson arrival rate. Since the next state j is the old state i, plus the number of new arrivals, minus the single departure,

$$\Pr\{X_{n+1} = j \mid X_n = i, U_n = u\} = \frac{(\lambda u)^{j+i+1} \exp(-\lambda u)}{(j - i + 1)!} \tag{7.96}$$

for $j \geq i - 1$. For $j < i - 1$, the probability above is 0. The unconditional probability P_{ij} of a transition from i to j can then be found by multiplying the right-hand side of (7.96) by the probability density $g(u)$ of the service time and integrating over u:

$$P_{ij} = \int_0^\infty \frac{G(u)(\lambda u)^{j-i+1} \exp(-\lambda u)}{(j - i + 1)} du, \quad j \geq i - 1, i > 0. \tag{7.97}$$

For the case $i = 0$, the server must wait until the next arrival before starting service. Thus the expected time from entering the empty state until a service completion is

$$\overline{U}(0) = (1/\lambda) + \int_0^\infty [1 - G(u)]du. \tag{7.98}$$

We can evaluate P_{0j} by observing that the departure of that first arrival leaves j customers in this system if and only if j customers arrive during the service time of that first customer; i.e., the new state does not depend on how long the server waits for a new customer to serve, but only on the arrivals while that customer is being served. Letting $g(u)$ be the density of the service time,

$$P_{0j} = \int_0^\infty \frac{g(u)(\lambda u)^j \exp(-\lambda u)}{j!} du, \quad j \geq 0. \tag{7.99}$$

7.9 Summary

We have seen that Markov processes with countable-state spaces are remarkably similar to Markov chains with countable-state spaces. In fact, we initially defined these processes in terms of an embedded Markov chain along with an exponential holding interval of some given rate ν_j for each state j. Markov processes can be equivalently described in terms of a set of transition rates $\{q_{ij};\ i,j \geq 0\}$ between each pair of states.

As with countable-state Markov chains, we found that renewal theory is the appropriate tool to analyze the main properties of these Markov processes. The main result from this analysis is Theorem 7.2.6. This shows that if the embedded chain is positive recurrent, then the time-average fraction of time spent in each state j is $p_j = \pi_j/(\nu_j \sum_k \pi_k/\nu_k)$ WP1. It also shows that p_j is the probability, as $t \to \infty$, of being in state j, independent of the starting state. Finally, the expected recurrence time of state j is $\overline{W}_j = 1/(p_j\nu_j)$. Note that if $\sum_k \pi_k/\nu_k < \infty$, then $p_j > 0$ for each k and $\sum_k p_k = 1$. This is the usual situation and in this case p_j is in every sense viewed as a steady-state probability in time.

In the special case where $\sum_k \pi_k/\nu_k = \infty$ we have $p_j = 0$ for all j. In this case, the expected time between state transitions is infinite if the embedded chain is in steady state. Thus the process cannot really be viewed as having a steady state.

The steady-state process probabilities can also be calculated by the steady-state equations $\nu_j p_j = \sum_i p_i q_{ij}$, $\sum_i p_i = 1$. This has a unique solution if the embedded chain is positive recurrent. There is a special case in which these equations have a solution but $\sum_j p_j \nu_j = \infty$. In this case, the embedded chain is not positive recurrent, the solution $\{p_j; j \geq 0\}$ cannot be interpreted as a steady state, and the embedded chain is either transient or null recurrent.

Section 7.3 developed the Kolmogorov backward and forward differential equations for the transient probabilities $P_{ij}(t)$ of being in state j at time t given state i at time 0. We showed that for finite-state processes, these equations can be solved by finding the eigenvalues and eigenvectors of the transition rate matrix Q. There are close analogies between this analysis and the algebraic treatment of finite-state Markov chains in Chapter 4, and Exercise 6.7 shows how the transients of the process are related to the transients of the sampled time approximation.

For irreducible processes with bounded transition rates, uniformization was introduced as a way to simplify the structure of the process. The addition of self transitions does not change the process itself, but can be used to adjust the transition rates ν_i to be the same for all states. This changes the embedded Markov chain, and the steady-state probabilities for the embedded chain become the same as those for the process. The epochs at which transitions occur then form a Poisson process which is independent of the set of states entered. This yields a separation between the transition epochs and the sequence of states.

The next two sections analyzed birth–death processes and reversibility. The results about birth–death Markov chains and reversibility for Markov chains carried over almost without change to Markov processes. These results are central in queueing theory, and Burke's theorem helped us to analyze simple queueing networks with no feedback. We also saw how feedback complicates the problem.

Jackson networks were next discussed. These are important in their own right and provide a good example of how one can solve complex queueing problems by studying the reverse time process and making educated guesses about the steady-state behavior. The somewhat startling result here is that in steady state, and at a fixed time, the number of customers at each node is independent of the number at each other node and satisfies the same distribution as for an M/M/1 queue. Also the exogenous departures from the network are Poisson and independent from node to node. We emphasized that the number of customers at one node at one time is often dependent on the number at other nodes at other times. The above independence holds only when all nodes are viewed at the same time.

Finally, semi-Markov processes were introduced. Renewal theory again provided the key to analysis of these systems. Theorem 7.8.4 showed how to find the steady-state probabilities of these processes, and it was shown that these probabilities could be interpreted both as time averages and, in the case of non-arithmetic transition times, as limiting probabilities in time.

For further reading on Markov processes, see [17], [22], [30], and [9].

7.10 Exercises

Exercise 7.1 Consider an M/M/1 queue as represented in Figure 7.4. Assume throughout that $X_0 = i$, where $i > 0$. The purpose of this exercise is to understand the relationship between the holding interval until the next state transition and the interval until the next arrival to the M/M/1 queue. Your explanations in the following parts can and should be very brief.

(a) Explain why the expected holding interval $\mathsf{E}[U_1|X_0 = i]$ until the next state transition is $1/(\lambda + \mu)$.

(b) Explain why the expected holding interval U_1, conditional on $X_0 = i$ and $X_1 = i + 1$, is

$$\mathsf{E}[U_1|X_0 = i, X_1 = i + 1] = 1/(\lambda + \mu).$$

Show that $\mathsf{E}[U_1|X_0 = i, X_1 = i - 1]$ is the same.

(c) Let V be the time of the first arrival after time 0 (this may occur either before or after the time W of the first departure). Show that

$$\mathsf{E}[V|X_0 = i, X_1 = i + 1] = \frac{1}{\lambda + \mu},$$

$$\mathsf{E}[V|X_0 = i, X_1 = i - 1] = \frac{1}{\lambda + \mu} + \frac{1}{\lambda}.$$

Hint: In the second equation, use the memorylessness of the exponential rv and the fact that V under this condition is the time to the first departure plus the remaining time to an arrival.

(d) Use your solution to (c) plus the probability of moving up or down in the Markov chain to show that $\mathsf{E}[V] = 1/\lambda$. Note: You already know that $\mathsf{E}[V] = 1/\lambda$. The purpose here is to show that your solution to (c) is consistent with that fact.

Exercise 7.2 Consider a Markov process for which the embedded Markov chain is a birth–death chain with transition probabilities $P_{i,\,i+1} = 2/5$ for all $i \geq 0$, $P_{i,\,i-1} = 3/5$ for all $i \geq 1$, $P_{01} = 1$, and $P_{ij} = 0$ otherwise.

(a) Find the steady-state probabilities $\{\pi_i;\ i \geq 0\}$ for the embedded chain.

(b) Assume that the transition rate ν_i out of state i, for $i \geq 0$, is given by $\nu_i = 2^i$. Find the transition rates $\{q_{ij}\}$ between states and find the steady-state probabilities $\{p_i\}$ for the Markov process. Explain heuristically why $\pi_i \neq p_i$.

(c) Explain why there is no sampled-time approximation for this process. Then truncate the embedded chain to states 0 to m and find the steady-state probabilities for the sampled-time approximation to the truncated process.

(d) Show that as $m \to \infty$, the steady-state probabilities for the sequence of sampled-time approximations approach the probabilities p_i in (b).

Exercise 7.3 Consider a Markov process for which the embedded Markov chain is a birth–death chain with transition probabilities $P_{i,i+1} = 2/5$ for all $i \geq 1$, $P_{i,i-1} = 3/5$ for all $i \geq 1$, $P_{01} = 1$, and $P_{ij} = 0$ otherwise.

(a) Find the steady-state probabilities $\{\pi_i;\ i \geq 0\}$ for the embedded chain.

(b) Assume that the transition rate out of state i, for $i \geq 0$, is given by $\nu_i = 2^{-i}$. Find the transition rates $\{q_{ij}\}$ between states and show that there is no probability vector solution $\{p_i; i \geq 0\}$ to (7.23).

(c) Argue that the expected time between visits to any given state i is infinite. Find the expected number of transitions between visits to any given state i. Argue that, starting from any state i, an eventual return to state i occurs with probability 1.

(d) Consider the sampled-time approximation of this process with $\delta = 1$. Draw the graph of the resulting Markov chain and argue why it must be null recurrent.

Exercise 7.4 Consider the Markov process for the M/M/1 queue, as given in Figure 7.4.

(a) Find the steady-state process probabilities (as a function of $\rho = \lambda/\mu$) from (7.15) and also as the solution to (7.23). Verify that the two solutions are the same.

(b) For the remaining parts of the exercise, assume that $\rho = 0.01$, thus ensuring (for aiding intuition) that states 0 and 1 are much more probable than the other states. Assume that the process has been running for a very long time and is in steady state. Explain in your own words the difference between π_1 (the steady-state probability of state 1 in the embedded chain) and p_1 (the steady-state probability that the process is in state 1). More explicitly, what experiments could you perform (repeatedly) on the process to measure π_1 and p_1.

(c) Now suppose you want to start the process in steady state. Show that it is impossible to choose initial probabilities so that both the process and the embedded chain start in steady state. Which version of steady state is closest to your intuitive view? (There is no correct answer here, but it is important to realize that the notion of steady state is not quite as simple as you might imagine.)

(d) Let $M(t)$ be the number of transitions (counting both arrivals and departures) that take place by time t in this Markov process and assume that the embedded Markov chain starts in steady state at time 0. Let $U_1, U_2, \ldots$ be the sequence of holding intervals between transitions (with U_1 being the time to the first transition). Show that these rv s are identically distributed. Show by example that they are not independent (i.e., $M(t)$ is not a renewal process).

Exercise 7.5 Consider the Markov process illustrated below. The transitions are labeled by the rate q_{ij} at which those transitions occur. The process can be viewed as a single-server queue where arrivals become increasingly discouraged as the queue lengthens. The word *time average* below refers to the limiting time average over each sample path of the process, except for a set of sample paths of probability 0.

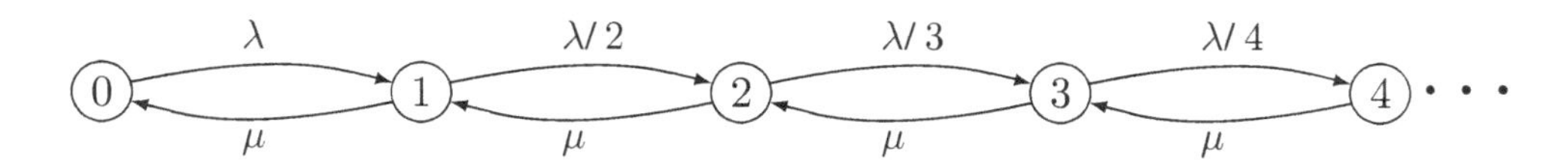

(a) Find the time-average fraction of time p_i spent in each state $i > 0$ in terms of p_0 and then solve for p_0. Hint: First find an equation relating p_i to p_{i+1} for each i. It also may help to recall the power series expansion of e^x.

(b) Find a closed form solution to $\sum_i p_i \nu_i$, where ν_i is the rate at which transitions out of state i occur. Show that the embedded chain is positive recurrent for all choices of $\lambda > 0$ and $\mu > 0$ and explain intuitively why this must be so.

(c) For the embedded Markov chain corresponding to this process, find the steady-state probabilities π_i for each $i \geq 0$ and the transition probabilities P_{ij} for each i, j.

(d) For each i, find both the time-average interval and the time-average number of overall state transitions between successive visits to i.

Exercise 7.6 (Detail from proof of Theorem 7.2.6) **(a)** Let U_n be the nth holding interval for a Markov process starting in steady state, with $\Pr\{X_0 = i\} = \pi_i$. Show that $\mathsf{E}[U_n] = \sum_k \pi_k/\nu_k$ for each integer n.

(b) Let S_n be the epoch of the nth transition. Show that $\Pr\{S_n \geq n\beta\} \leq (\sum_k \pi_k/\nu_k)/\beta$ for all $\beta > 0$.

(c) Let $M(t)$ be the number of transitions of the Markov process up to time t, given that X_0 is in steady state. Show that $\Pr\{M(n\beta) \geq n\} \geq 1 - (\sum_k \pi_k/\nu_k)/\beta$.

(d) Show that if $\sum_k \pi_k/\nu_k$ is finite, then $\lim_{t\to\infty} M(t)/t = 0$ WP1 is impossible. (Note that this is equivalent to showing that $\lim_{t\to\infty} M(t)/t = 0$ WP1 implies $\sum_k \pi_k/\nu_k = \infty$.)

(e) Let $M_i(t)$ be the number of transitions by time t starting in state i. Show that if $\sum_k \pi_k/\nu_k$ is finite, then $\lim_{t\to\infty} M_i(t)/t = 0$ WP1 is impossible.

Exercise 7.7 **(a)** Consider the process in the figure below. The process starts at $X(0) = 1$, and for all $i \geq 1$, $P_{i,i+1} = 1$ and $\nu_i = i^2$ for all i. Let S_n be the epoch when the nth transition occurs. Show that

$$\mathsf{E}[S_n] = \sum_{i=1}^{n} i^{-2} < 2 \text{ for all } n.$$

Hint: Upper bound the sum from $i = 2$ by integrating x^{-2} from $x = 1$.

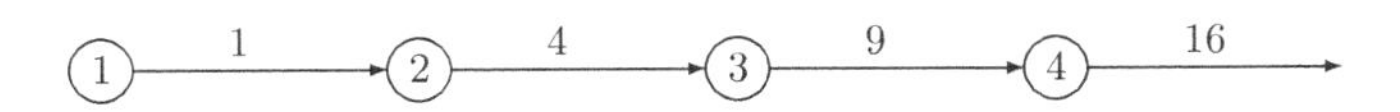

(b) Use the Markov inequality to show that $\Pr\{S_n > 4\} \leq 1/2$ for all n. Show that the probability of an infinite number of transitions by time 4 is at least $1/2$.

Exercise 7.8 **(a)** Consider a Markov process with the set of states $\{0, 1, \ldots\}$ in which the transition rates $\{q_{ij}\}$ between states are given by $q_{i,i+1} = (3/5)2^i$ for $i \geq 0$, $q_{i,i-1} = (2/5)2^i$ for $i \geq 1$, and $q_{ij} = 0$ otherwise. Find the transition rate ν_i out of state i for each $i \geq 0$ and find the transition probabilities $\{P_{ij}\}$ for the embedded Markov chain.

(b) Find a solution $\{p_i;\ i \geq 0\}$ with $\sum_i p_i = 1$ to (7.23).

(c) Show that all states of the embedded Markov chain are transient.

(d) Explain in your own words why your solution to (b) is not in any sense a set of steady-state probabilities.

Exercise 7.9 Let $q_{i,i+1} = 2^{i-1}$ for all $i \geq 0$ and let $q_{i,i-1} = 2^{i-1}$ for all $i \geq 1$. All other transition rates are 0.

(a) Solve the steady-state equations and show that $p_i = 2^{-i-1}$ for all $i \geq 0$.

(b) Find the transition probabilities for the embedded Markov chain and show that the chain is null recurrent.

(c) For any state i, consider the renewal process for which the Markov process starts in state i and renewals occur on each transition to state i. Show that, for each $i \geq 1$, the expected inter-renewal interval is equal to 2. Hint: Use renewal–reward theory.

(d) Show that the expected number of transitions between each entry into state i is infinite. Explain why this does *not* mean that an infinite number of transitions can occur in a finite time.

Exercise 7.10 **(a)** Consider the two-state Markov process of Example 7.3.1 with $q_{01} = \lambda$ and $q_{10} = \mu$. Find the eigenvalues and eigenvectors of the transition rate matrix $[Q]$.

(b) If $[Q]$ has M distinct eigenvalues, the differential equation $d[P(t)]/dt = [Q][P(t)]$ can be solved by the equation

$$[P(t)] = \sum_{i=1}^{\mathsf{M}} \boldsymbol{\nu}_i e^{t\lambda_i} \boldsymbol{p}^{\mathsf{T}}{}_i,$$

where $\boldsymbol{p}_i$ and $\boldsymbol{\nu}_i$ are the left and right eigenvectors of eigenvalue λ_i. Show that this equation gives the same solution as that given for Example 7.3.1.

Exercise 7.11 Consider the three-state Markov process below; the number given on edge (i,j) is q_{ij}, the transition rate from i to j. Assume that the process is in steady state.

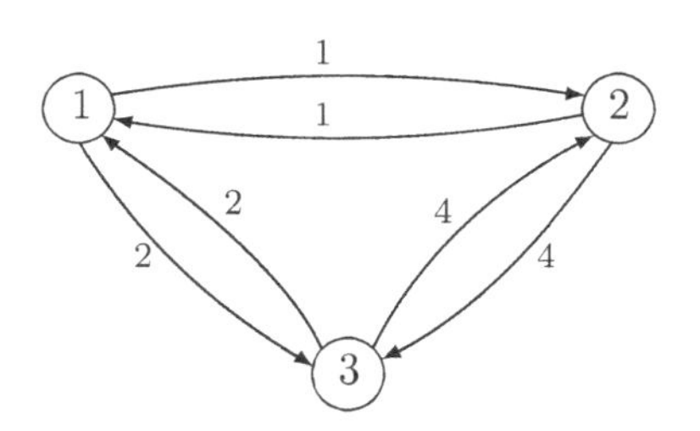

(a) Is this process reversible?

(b) Find p_i, the time-average fraction of time spent in state i for each i.

(c) Given that the process is in state i at time t, find the mean delay from t until the process leaves state i.

(d) Find π_i, the time-average fraction of all transitions that go into state i for each i.

(e) Suppose the process is in steady state at time t. Find the steady-state probability that the next state to be entered is state 1.

(f) Given that the process is in state 1 at time t, find the mean delay until the process first returns to state 1.

(g) Consider an arbitrary irreducible finite-state Markov process in which $q_{ij} = q_{ji}$ for all i,j. Either show that such a process is reversible or find a counter example.

Exercise 7.12 **(a)** Consider an M/M/1 queueing system with arrival rate λ, service rate μ, $\mu > \lambda$. Assume that the queue is in steady state. Given that an arrival occurs at time t, find the probability that the system is in state i immediately *after* time t.

(b) Assuming FCFS service, and conditional on i customers in the system immediately after the above arrival, characterize the time until the above customer departs as a sum of rv s.

(c) Find the unconditional probability density of the time until the above customer departs. Hint: You know (from splitting a Poisson process) that the sum of a geometrically distributed number of IID exponentially distributed rv s is exponentially distributed. Use the same idea here.

Exercise 7.13 **(a)** Consider an M/M/1 queue in steady state. Assume $\rho = \lambda/\mu < 1$. Find the probability $Q(i,j)$ for $i \geq j > 0$ that the system is in state i at time t and that $i - j$ departures occur before the next arrival.

(b) Find the PMF of the state immediately before the first arrival after time t.

(c) There is a well-known queueing principle called PASTA, standing for 'Poisson arrivals see time averages'. Given your results above, give a more precise statement of what that principle means in the case of the M/M/1 queue.

Exercise 7.14 A small bookie shop has room for at most two customers. Potential customers arrive at a Poisson rate of ten customers per hour; they enter if there is room and are turned away, never to return, otherwise. The bookie serves the admitted customers in order, requiring an exponentially distributed time of mean 4 minutes per customer.

(a) Find the steady-state distribution of the number of customers in the shop.

(b) Find the rate at which potential customers are turned away.

(c) Suppose the bookie hires an assistant; the bookie and assistant, working together, now serve each customer in an exponentially distributed time of mean 2 minutes, but there is only room for one customer (i.e., the customer being served) in the shop. Find the new rate at which customers are turned away.

Exercise 7.15 This exercise explores a continuous time version of a simple branching process.

Consider a population of primitive organisms which do nothing but procreate and die. In particular, the population starts at time 0 with one organism. This organism has an exponentially distributed lifespan T_0 with rate μ (i.e., $\Pr\{T_0 \geq \tau\} = e^{-\mu\tau}$). While this organism is alive, it gives birth to new organisms according to a Poisson process of rate λ. Each of these new organisms, while alive, gives birth to yet other organisms. The lifespan and birthrate for each of these new organisms are IID to those of the first organism. All these and subsequent organisms give birth and die in the same way, again independently of all other organisms.

(a) Let $X(t)$ be the number of (live) organisms in the population at time t. Show that $\{X(t);\ t \geq 0\}$ is a Markov process and specify the transition rates between the states.

(b) Find the embedded Markov chain $\{X_n;\ n \geq 0\}$ corresponding to the Markov process in (a). Find the transition probabilities for this Markov chain.

(c) Explain why the Markov process and Markov chain above are not irreducible. Note: The majority of results you have seen for Markov processes assume the process is irreducible, so be careful not to use those results in this exercise.

(d) For purposes of analysis, add an additional transition of rate λ from state 0 to state 1. Show that the Markov process and the embedded chain are irreducible. Find the values of λ and μ for which the modified chain is positive recurrent, null recurrent, and transient.

(e) Assume that $\lambda < \mu$. Find the steady-state process probabilities for the modified Markov process.

(f) Find the mean recurrence time between visits to state 0 for the modified Markov process.

(g) Find the mean time $\overline{T}$ for the population in the original Markov process to die out. Note: We have seen before that removing transitions from a Markov chain or process to create a trapping state can make it easier to find mean recurrence times. This is an example of the opposite, where adding an exit from a trapping state makes it easy to find the recurrence time.

Exercise 7.16 Consider the job sharing computer system illustrated below. Incoming jobs arrive from the left in a Poisson stream. Each job, independently of other jobs, requires preprocessing in system 1 with probability Q. Jobs in system 1 are served FCFS and the service times for successive jobs entering system 1 are IID with an exponential distribution of mean $1/\mu_1$. The jobs entering system 2 are also served FCFS and successive service times are IID with an exponential distribution of mean $1/\mu_2$. The service times in the two systems are independent of each other and of the arrival times. Assume that $\mu_1 > \lambda Q$ and that $\mu_2 > \lambda$. Assume that the combined system is in steady state.

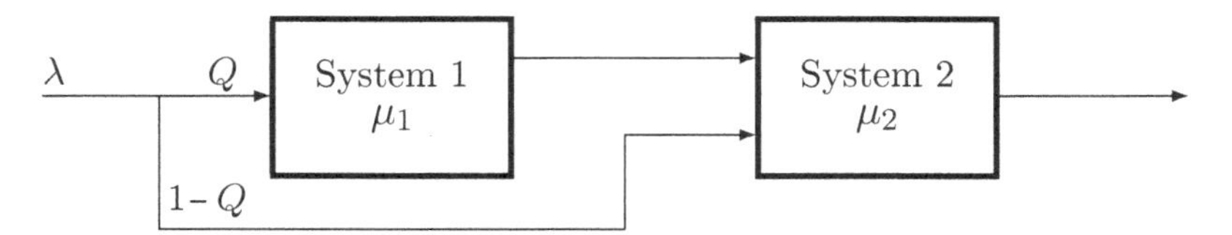

(a) Is the input to system 1 Poisson? Explain.

(b) Are each of the two input processes coming into system 2 Poisson? Explain.

(c) Give the joint steady-state PMF of the number of jobs in the two systems. Explain briefly.

(d) What is the probability that the first job to leave system 1 after time t is the same as the first job that entered the entire system after time t?

(e) What is the probability that the first job to leave system 2 after time t both passed through system 1 and arrived at system 1 after time t?

Exercise 7.17 Consider the following combined queueing system. The first queue system is M/M/1 with service rate μ_1. The second queue system has IID exponential service times with rate μ_2. Each departure from system 1 leaves the entire system with probability $1 - Q_1$ and enters system 2 with the remaining probability Q_1. System 2 has an additional Poisson input of rate λ_2, independent of inputs and outputs from the first system. Each departure from the second system independently leaves the combined system with probability Q_2 and reenters system 2 with probability $1 - Q_2$. For (a), (b), (c) assume that $Q_2 = 1$ (i.e., there is no feedback).

(a) Characterize the process of departures from system 1 that enter system 2 and characterize the overall process of arrivals to system 2.

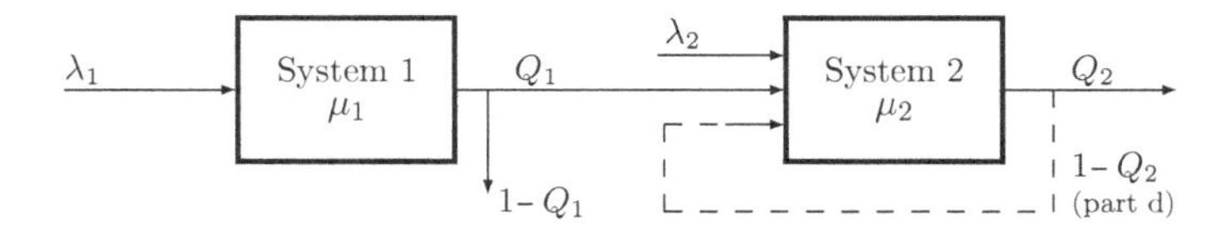

(b) Assuming FCFS service in each system, find the steady-state distribution of time that a customer spends in each system.

(c) For a customer that goes through both systems, show why the time in each system is independent of that in the other; find the distribution of the combined system time for such a customer.

(d) Now assume that $Q_2 < 1$. Is the departure process from the combined system Poisson? Which of the three input processes to system 2 are Poisson? Which of the input processes are independent? Explain your reasoning, but do not attempt formal proofs.

Exercise 7.18 Suppose a Markov chain with transition probabilities $\{P_{ij}\}$ is reversible. Suppose we change the transition probabilities out of state 0 from $\{P_{0j}; j \geq 0\}$ to $\{P'_{0j}; j \geq 0\}$. Assuming that all P_{ij} for $i \neq 0$ are unchanged, what is the most general way in which we can choose $\{P'_{0j}; j \geq 0\}$ so as to maintain reversibility? Your answer should be explicit about how the steady-state probabilities $\{\pi_i; i \geq 0\}$ are changed. Your answer should also indicate what this problem has to do with uniformization of reversible Markov processes, if anything. Hint: Given P_{ij} for all i, j, a single additional parameter will suffice to specify P'_{0j} for all j.

Exercise 7.19 Consider the closed queueing network in the figure below. There are three customers who are doomed forever to cycle between node 1 and node 2. Both nodes use FCFS service and have exponential IID service times. The service times at one node are also independent of those at the other node and are independent of the customer being served. The server at node i has mean service time $1/\mu_i$, $i = 1, 2$. Assume to be specific that $\mu_2 < \mu_1$.

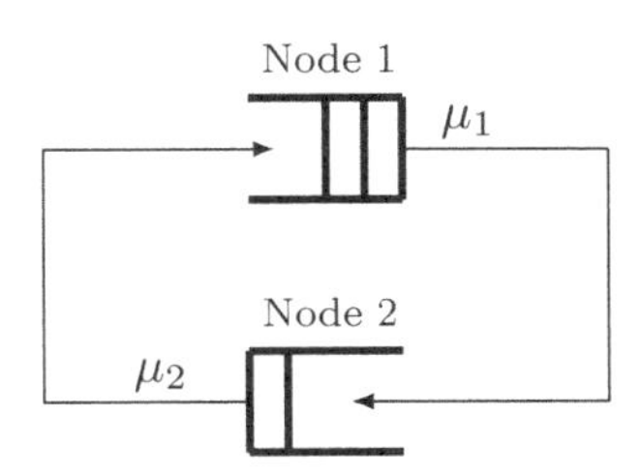

(a) The system can be represented by a four-state Markov process. Draw its graphical representation and label it with the individual states and the transition rates between them.

(b) Find the steady-state probability of each state.

(c) Find the time-average rate at which customers leave node 1.

(d) Find the time-average rate at which a given customer cycles through the system.

(e) Is the Markov process reversible? Suppose that the backward Markov process is interpreted as a closed queueing network. What does a departure from node 1 in the forward process correspond to in the backward process? Can the transitions of a single customer in the forward process be associated with transitions of a single customer in the backward process?

Exercise 7.20 Consider an M/G/1 queueing system with last come, first served (LCFS) preemptive resume service. That is, customers arrive according to a Poisson process of rate λ. A newly arriving customer interrupts the customer in service and enters service itself. When a customer is finished, it leaves the system and the customer that had been interrupted by the departing customer resumes service from where it had left off. For example, if customer 1 arrives at time 0 and requires 2 units of service, and customer 2 arrives at time 1 and requires 1 unit of service, then customer 1 is served from time 0 to 1; customer 2 is served from time 1 to 2 and leaves the system, and then customer 1 completes service from time 2 to 3. Let X_i be the service time required by the ith customer; the X_i are IID rv s with expected value $\mathsf{E}[X]$; they are independent of customer arrival times. Assume $\lambda\,\mathsf{E}[X] < 1$.

(a) Find the mean time between busy periods (i.e., the time until a new arrival occurs after the system becomes empty).

(b) Find the time-average fraction of time that the system is busy.

(c) Find the mean duration, $\mathsf{E}[B]$, of a busy period. Hint: Use (a) and (b).

(d) Explain briefly why the customer that starts a busy period remains in the system for the entire busy period; use this to find the expected system time of a customer given that that customer arrives when the system is empty.

(e) Is there any statistical dependence between the system time of a given customer (i.e., the time from the customer's arrival until departure) and the number of customers in the system when the given customer arrives?

(f) Show that a customer's expected system time is equal to $\mathsf{E}[B]$. Hint: Look carefully at your answers to (d) and (e).

(g) Let C be the expected system time of a customer conditional on the service time X of that customer being 1. Find (in terms of C) the expected system time of a customer conditional on $X = 2$. (Hint: Compare a customer with $X = 2$ to two customers with $X = 1$ each.) Repeat for arbitrary $X = x$.

(h) Find the constant C. Hint: Use (f) and (g); do not do any tedious calculations.

Exercise 7.21 Consider a queueing system with two classes of customers, type A and type B. Type A customers arrive according to a Poisson process of rate λ_A and customers of type B arrive according to an independent Poisson process of rate λ_B. The queue has a FCFS server with exponential IID service times of rate $\mu > \lambda_A + \lambda_B$. Characterize the departure process of class A customers; explain carefully. Hint: Consider the combined arrival process and be judicious about how to select between A and B types of customers.

Exercise 7.22 Consider a preemptive resume LCFS queueing system with two classes of customers. Type A customer arrivals are Poisson with rate λ_A and Type B customer arrivals are Poisson with rate λ_B. The service time for type A customers is exponential with rate μ_A and that for type B is exponential with rate μ_B. Each service time is independent of all other service times and of all arrival epochs.

Define the 'state' of the system at time t by the string of customer types in the system at t, in order of arrival. Thus state AB means that the system contains two customers, one of type A and the other of type B; the type B customer arrived later, so is in service. The set of possible states arising from transitions out of AB is as follows:

ABA if another type A arrives.
ABB if another type B arrives.
A if the customer in service (B) departs.

Note that whenever a customer completes service, the next most recently arrived resumes service, so the state changes by eliminating the final element in the string.

(a) Draw a graph for the states of the process, showing all states with two or fewer customers and a couple of states with three customers (label the empty state as E). Draw an arrow, labeled by the rate, for each state transition. Explain why these are states in a Markov process.

(b) Is this process reversible? Explain. Assume positive recurrence. Hint: If there is a transition from one state S to another state S', how is the number of transitions from S to S' related to the number from S' to S?

(c) Characterize the process of type A departures from the system (i.e., are they Poisson, do they form a renewal process, at what rate do they depart etc.?)

(d) Express the steady-state probability $\Pr\{A\}$ of state A in terms of the probability of the empty state $\Pr\{E\}$. Find the probability $\Pr\{AB\}$ and the probability $\Pr\{ABBA\}$ in terms of $\Pr\{E\}$. Use the notation $\rho_A = \lambda_A/\mu_A$ and $\rho_B = \lambda_B/\mu_B$.

(e) Let Q_n be the probability of n customers in the system, as a function of $Q_0 = \Pr\{E\}$. Show that $Q_n = (1-\rho)\rho^n$ where $\rho = \rho_A + \rho_B$.

Exercise 7.23 **(a)** Generalize Exercise 7.22 to the case in which there are m types of customers, each with independent Poisson arrivals and each with independent exponential service times. Let λ_i and μ_i be the arrival rate and service rate respectively of the ith user. Let $\rho_i = \lambda_i/\mu_i$ and assume that $\rho = \rho_1 + \rho_2 + \cdots + \rho_m < 1$. In particular, show, as before, that the probability of n customers in the system is $Q_n = (1-\rho)\rho^n$ for $0 \le n < \infty$.

(b) View the customers in (a) as a single type of customer with Poisson arrivals of rate $\lambda = \sum_i \lambda_i$ and with a service density $\sum_i (\lambda_i/\lambda)\mu_i \exp(-\mu_i x)$. Show that the expected service time is ρ/λ. Note that you have shown that, if a service distribution can be represented as a weighted sum of exponentials, then the distribution of customers in the system for LCFS service is the same as for the M/M/1 queue with equal mean service time.

Exercise 7.24 Consider a k-node Jackson type network with the modification that each node i has s servers rather than one server. Each server at i has an exponentially distributed service time with rate μ_i. The exogenous input rate to node i is $\rho_i = \lambda_0 Q_{0i}$ and each output from i is switched to j with probability Q_{ij} and switched out of the system with probability Q_{i0} (as in the text). Let λ_i, $1 \le i \le k$, be the solution, for given λ_0, to

$$\lambda_j = \sum_{i=0}^{k} \lambda_i Q_{ij},$$

$1 \le j \le k$, and assume that $\lambda_i < s\mu_i$; $1 \le i \le k$. Show that the steady-state probability of state $\boldsymbol{m}$ is

$$\Pr\{\boldsymbol{m}\} = \prod_{i=1}^{k} p_i(m_i),$$

where $p_i(m_i)$ is the probability of state m_i in an (M, M, s) queue. Hint: Simply extend the argument in the text to the multiple server case.

Exercise 7.25 Suppose a Markov process with the set of states A is reversible and has steady-state probabilities p_i, $i \in A$. Let B be a subset of A and assume that the process is changed by setting $q_{ij} = 0$ for all $i \in B, j \notin B$. Assuming that the new process (starting in B and remaining in B) is irreducible, show that the new process is reversible and find its steady-state probabilities.

Exercise 7.26 Consider a queueing system with two classes of customers. Type A customer arrivals are Poisson with rate λ_A and type B customer arrivals are Poisson with rate λ_B. The service time for type A customers is exponential with rate μ_A and that for type B is exponential with rate μ_B. Each service time is independent of all other service times and of all arrival epochs.

(a) First assume there are infinitely many identical servers, and each new arrival immediately enters an idle server and begins service. Let the state of the system be (i,j), where i and j are the numbers of type A and B customers respectively in service. Draw a graph of the state transitions for $i \le 2, j \le 2$. Find the steady-state PMF, $\{p(i,j);\ i,j \ge 0\}$, for the Markov process. Hint: Note that the type A and type B customers do not interact.

(b) Assume for the rest of the exercise that there is some finite number m of servers. Customers who arrive when all servers are occupied are turned away. Find the steady-state PMF, $\{p(i,j);\ i,j \ge 0, i+j \le m\}$, in terms of $p(0,0)$ for this Markov process. Hint: Combine (a) with the result of Exercise 7.25.

(c) Let Q_n be the probability that there are n customers in service at some given time in steady state. Show that $Q_n = p(0,0)\rho_n/n!$ for $0 \le n \le m$, where $\rho = \rho_A + \rho_B$, $\rho_A = \lambda_A/\mu_A$, and $\rho_B = \lambda_B/\mu_B$. Solve for $p(0,0)$.

Exercise 7.27 **(a)** Generalize Exercise 7.26 to the case in which there are K types of customers, each with independent Poisson arrivals and each with independent exponential service times. Let λ_k and μ_k be the arrival rate and service rate respectively for the kth user type, $1 \le k \le K$. Let $\rho_k = \lambda_k/\mu_k$ and $\rho = \rho_1+\rho_2+\cdots+\rho_K$. In particular, show, as before, that the probability of n customers in the system is $Q_n = p(0,\ldots,0)\rho_n/n!$ for $0 \le n \le m$.

(b) View the customers in (a) as a single type of customer with Poisson arrivals of rate $\lambda = \sum_k \lambda_k$ and with a service density $\sum_k(\lambda_k/\lambda)\mu_k \exp(-\mu_k x)$. Show that the expected service time is ρ/λ. Note that what you have shown is that, if a service distribution can be represented as a weighted sum of exponentials, then the distribution of customers in the system is the same as for the M/M/m/m queue with equal mean service time.

Exercise 7.28 Consider a sampled-time M/D/m/m queueing system. The arrival process is Bernoulli with probability $\lambda << 1$ of arrival in each time unit. There are m servers; each arrival enters a server if a server is not busy and otherwise the arrival is discarded. If an arrival enters a server, it keeps the server busy for d units of time and then departs; d is some integer constant and is the same for each server.

Let n, $0 \le n \le m$ be the number of customers in service at a given time and let x_i be the number of time units that the ith of those n customers (in order of arrival) has been in service. Thus the state of the system can be taken as $(n,\boldsymbol{x}) = (n, x_1, x_2, \dots, x_n)$, where $0 \le n \le m$ and $1 \le x_1 < x_2 < \cdots < x_n \le d$.

Let $A(n,\boldsymbol{x})$ denote the next state if the present state is $(n,\boldsymbol{x})$ and a new arrival enters service. That is,

$$A(n,\boldsymbol{x}) = (n+1, 1, x_1+1, x_2+1, \dots, x_n+1) \qquad \text{for } n < m \text{ and } x_n < d, \tag{7.100}$$

$$A(n,\boldsymbol{x}) = (n, 1, x_1+1, x_2+1, \dots, x_{n-1}+1) \qquad \text{for } n \le m \text{ and } x_n = d. \tag{7.101}$$

That is, the new customer receives one unit of service by the next state time, and all the old customers receive one additional unit of service. If the oldest customer has received d units of service, then it leaves the system by the next state time. Note that it is possible for a customer with d units of service at the present time to leave the system and be replaced by an arrival at the present time (i.e., (7.101) with $n = m$, $x_n = d$). Let $B(n,\boldsymbol{x})$ denote the next state if either no arrival occurs or if a new arrival is discarded:

$$B(n,\boldsymbol{x}) = (n, x_1+1, x_2+1, \dots, x_n+1) \qquad \text{for } x_n < d, \tag{7.102}$$

$$B(n,\boldsymbol{x}) = (n-1, x_1+1, x_2+1, \dots, x_{n-1}+1) \quad \text{for } x_n = d. \tag{7.103}$$

(a) Hypothesize that the *backward* chain for this system is also a sampled-time M/D/m/m queueing system, but that the state $(n, x_1, \dots, x_n)(0 \le n \le m, 1 \le x_1 < x_2 < \dots < x_n \le d)$ has a different interpretation: n is the number of customers as before, but x_i is now the remaining service required by customer i. Explain how this hypothesis leads to the following steady-state equations:

$$\lambda\pi_{n,\boldsymbol{x}} = (1-\lambda)\pi_{A(n,\boldsymbol{x})}, \qquad n < m, x_n < d, \tag{7.104}$$

$$\lambda\pi_{n,\boldsymbol{x}} = \lambda\pi_{A(n,\boldsymbol{x})}, \qquad n \le m, x_n = d, \tag{7.105}$$

$$(1-\lambda)\pi_{n,\boldsymbol{x}} = \lambda\pi_{B(n,\boldsymbol{x})}, \qquad n \le m, x_n = d, \tag{7.106}$$

$$(1-\lambda)\pi_{n,\boldsymbol{x}} = (1-\lambda)\pi_{B(n,\boldsymbol{x})}, \qquad n \le m, x_n < d. \tag{7.107}$$

(b) Using this hypothesis, find $\pi_{n,\boldsymbol{x}}$ in terms of π_0, where π_0 is the probability of an empty system. Hint: Use (7.106) and (7.107); your answer should depend on n, but not $\boldsymbol{x}$.

(c) Verify that the above hypothesis is correct.

(d) Find an expression for π_0.

(e) Find an expression for the steady-state probability that an arriving customer is discarded.

Exercise 7.29 A taxi alternates between three locations. When it reaches location 1 it is equally likely to go next to either location 2 or location 3. When it reaches location 2 it will next go to location 1 with probability 1/3 and to location 3 with probability 2/3. From location 3 it always goes to location 1. The mean time between locations i and j are $t_{12} = 20$, $t_{13} = 30$, $t_{23} = 30$. Assume exponential service and assume $t_{ij} = t_{ji}$.

(a) Find the (limiting) probability that the taxi's most recent stop was at location i, $i = 1, 2, 3$.

(b) What is the (limiting) probability that the taxi is heading for location 2?

(c) What fraction of time is the taxi traveling from location 2 to location 3? Note: Upon arrival at a location the taxi immediately departs.

Exercise 7.30 (Semi-Markov continuation of Exercise 7.5) **(a)** Assume that the Markov process in Exercise 7.5 is changed in the following way: whenever the process enters state 0, the time spent before leaving state 0 is now a *uniformly distributed* rv, taking values from 0 to $2/\lambda$. All other transitions remain the same. For this new process, determine whether the successive epochs of entry to state 0 form renewal epochs, whether the successive epochs of exit from state 0 form renewal epochs, and whether the successive entries to any other given state i form renewal epochs.

(b) For each i, find both the time-average interval and the time-average number of overall state transitions between successive visits to i.

(c) Is this modified process a Markov process in the sense that $\{X(t) = i \mid X(\tau) = j, X(s) = k\} = \Pr\{X(t) = i \mid X(\tau) = j\}$ for all $0 < s < \tau < t$ and all i, j, k? Explain.

Exercise 7.31 Consider an M/G/1 queueing system with Poisson arrivals of rate λ and expected service time $\mathsf{E}[X]$. Let $\rho = \lambda\mathsf{E}[X]$ and assume $\rho < 1$. Consider a semi-Markov process model of the M/G/1 queueing system in which transitions occur on departures from the queueing system and the state is the number of customers immediately following a departure.

(a) Suppose a colleague has calculated the steady-state probabilities $\{p_i\}$ of being in state i for each $i \geq 0$. For each $i \geq 0$, find the steady-state probability π of state i in the embedded Markov chain. Give your solution as a function of ρ, π_i, and p_0.

(b) Calculate p_0 as a function of ρ.

(c) Find π_i as a function of ρ and p_i.

(d) Is p_i the same as the steady-state probability that the queueing system contains i customers at a given time? Explain carefully.

Exercise 7.32 Consider an M/G/1 queue in which the arrival rate is λ and the service time distribution is uniform $(0, 2W)$ with $\lambda W < 1$. Define a semi-Markov chain following the framework for the M/G/1 queue in Section 7.8.1.

(a) Find P_{0j} for $j \geq 0$.

(b) Find P_{ij} for $i > 0; j \geq i - 1$.

Exercise 7.33 Consider a semi-Markov process for which the embedded Markov chain is irreducible and positive recurrent. Assume that the distribution of inter-renewal intervals for one state j is arithmetic with span d. Show that the distribution of inter-renewal intervals for all states is arithmetic with the same span.

8 Detection, decisions, and hypothesis testing

Detection, decision making, and hypothesis testing are different names for the same procedure. The word *detection* refers to the effort to decide whether some phenomenon is present or not in a given situation. For example, a radar system attempts to detect whether or not a target is present; a quality control system attempts to detect whether a unit is defective; a medical test detects whether a given disease is present. The meaning has been extended in the communication field to detect which one, among a finite set of mutually exclusive possible transmited signals, has been transmitted. *Decision making* is, again, the process of choosing between a number of mutually exclusive alternatives. *Hypothesis testing* is the same, except the mutually exclusive alternatives are called hypotheses. We usually use the word hypotheses for these alternatives in what follows, since the word seems to conjure up the appropriate intuitive images.

These problems will usually be modeled by a generic type of probability model. Each such model is characterized by a discrete random variable (rv) X called the hypothesis rv and a random vector (rv) $\boldsymbol{Y}$ called the observation rv. The observation might also be one-dimensional, i.e., an ordinary rv. The sample values of X are called hypotheses; it makes no difference what these hypotheses are called, so we usually number them, $0, 1, \ldots, \mathsf{M} - 1$. When the experiment is performed, the resulting sample point ω maps into a sample value $x \in \{0, 1, \mathsf{M}-1\}$ for X, and into a sample value $\boldsymbol{y}$ for $\boldsymbol{Y}$. The decision maker observes $\boldsymbol{y}$ (but not x) and maps $\boldsymbol{y}$ into a decision $\hat{x}(\boldsymbol{y})$. The decision is correct if $\hat{x}(\boldsymbol{y}) = x$.

The probability $p_x = \mathsf{p}_X(x)$ of hypothesis x is referred to as the a priori probability of hypothesis x. The probability model is completed by the conditional distribution of $\boldsymbol{Y}$, conditional on each sample value of X. These conditional distributions are called likelihoods in the terminology of hypothesis testing. In most of the situations we consider, these conditional distributions are represented either by a probability mass function (PMF) or by joint probability densities over $\mathbb{R}^n$ for a given $n \geq 1$. To avoid repeating everything, probability densities are usually assumed. Arguments that cannot be converted to discrete observations simply by replacing probability density functions (PDFs) by PMFs and changing integrals to sums will be discussed as they arise. There are also occasional comments about observations that cannot be described by PDFs or PMFs.

As with any probability model representing a real-world phenomenon, the rv s might model quantities that are actually random, or quantities such as coin tosses that might be viewed as either random or deterministic, or quantities such as physical constants that are deterministic but unknown. In addition, the model might be chosen for its simplicity, or for its similarity to some better-understood phenomenon, or for its faithfulness to some aspect of the real-world phenomenon.

Classical statisticians[1] are uncomfortable with the use of completely probabilistic models to study hypothesis testing, particularly when the 'correct' hypothesis is not obviously random with known probabilities. They have no objection to a separate probability model for the observation under each hypothesis, but are unwilling to choose a priori probabilities for the hypotheses. This is partly a practical matter, since statisticians design experiments to gather data and make decisions that often have considerable political and commercial importance. The use of a priori probabilities could be viewed as biasing these decisions and thus losing the appearance of impartiality.

The approach in this text, as pointed out frequently before, is to use a variety of probability models to gain insight and understanding about real-world phenomena. If we assume a variety of a priori probabilities and then see how the results depend on those choices, we often learn more than if we refuse to consider a priori probabilities at all. Another very similar approach that we often use is to consider a complete probabilistic model, but to assume that the observer does not know the a priori probabilities and makes a decision not based on them.[2] This is illustrated in the development of the Neyman–Pearson criterion in Section 8.4.

Before discussing how to make decisions, it is important to understand when and why decisions must be made. As an example, suppose we conclude, on the basis of an observation, that hypothesis 0 is correct with probability 2/3 and hypothesis 1 with probability 1/3. Simply making a decision on hypothesis 0 and forgetting about the probabilities throws away much of the information that has been gathered. The issue, however, is that sometimes choices must be made. In a communication system, the recipient wants to receive the message (perhaps with an occasional error) rather than a set of probabilities. In a control system, the controls must occasionally take action. Similarly managers must occasionally choose between courses of action, between products, and between people to hire. In a sense, it is by making decisions (and, in Chapter 10, by making estimates) that we return from the world of mathematical probability models to the world being modeled.

8.1 Decision criteria and the maximum a posteriori probability (MAP) criterion

There are a number of possible criteria for making decisions, and initially we concentrate on maximizing the probability of making correct decisions. For each hypothesis x, let p_X be the a priori probability that $X = x$ and let $\mathsf{f}_{Y|X}(\boldsymbol{y} \mid x)$ be the conditional probability density (called a likelihood) that $\boldsymbol{Y} = \boldsymbol{y}$ conditional on $X = x$. If $\mathsf{f}_{Y}(\boldsymbol{y}) > 0$, then the probability that $X = x$, conditional on $\boldsymbol{Y} = \boldsymbol{y}$, is given by Bayes' law as

[1] Statisticians have argued since the time of Bayes about the 'validity' of choosing a priori probabilities for hypotheses to be tested. Bayesian statisticians are comfortable with this practice and non-Bayesian or classical statisticians are not.

[2] Note that any decision procedure not using a priori probabilities can be made whether one believes the a priori probabilities exist but are unknown or do not exist at all.

$$\mathsf{p}_{X|Y}(x \mid \boldsymbol{y}) = \frac{p_x \mathsf{f}_{Y|X}(\boldsymbol{y} \mid x)}{\mathsf{f}_Y(\boldsymbol{y})}, \qquad \text{where } \mathsf{f}_Y(\boldsymbol{y}) = \sum_{x=0}^{\mathsf{M}-1} p_x \mathsf{f}_{Y|X}(\boldsymbol{y} \mid x). \tag{8.1}$$

Whether $\boldsymbol{Y}$ is discrete or continuous, the set of sample values where $\mathsf{p}_Y(\boldsymbol{y}) = 0$ or $\mathsf{f}_Y(\boldsymbol{y}) = 0$ is an event of zero probability. Thus we ignore this event in what follows and simply assume that $\mathsf{p}_Y(\boldsymbol{y}) > 0$ or $\mathsf{f}_Y(\boldsymbol{y}) > 0$ for all sample values.

The decision maker observes $\boldsymbol{y}$ and must choose one hypothesis, say $\hat{x}(\boldsymbol{y})$, from the set of possible hypotheses. The probability that hypothesis x is correct (i.e., the probability that $X = x$) conditional on observation $\boldsymbol{y}$ is given in (8.1). Thus, in order to maximize the probability of choosing correctly, the observer should choose that x for which $\mathsf{p}_{X|Y}(x \mid \boldsymbol{y})$ is maximized. Writing this as an equation,

$$\hat{x}(\boldsymbol{y}) = \arg\max_x \left[\mathsf{p}_{X|Y}(x \mid \boldsymbol{y})\right] \qquad \text{(MAP rule)}, \tag{8.2}$$

where $\arg\max_x$ means the argument $x \in \{0, \ldots, \mathsf{M}-1\}$ that maximizes the function. The conditional probability $\mathsf{p}_{X|Y}(x \mid \boldsymbol{y})$ is called an a posteriori probability, and thus the decision rule in (8.2) is called a MAP rule. Since we want to discuss other rules as well, we often denote the $\hat{x}(\boldsymbol{y})$ given in (8.2) as $\hat{x}_{\text{MAP}}(\boldsymbol{y})$.

An equivalent representation of (8.2) is obtained by substituting (8.1) into (8.2) and observing that $\mathsf{f}_Y(\boldsymbol{y})$ is the same for all hypotheses. Thus,

$$\hat{x}_{\text{MAP}}(\boldsymbol{y}) = \arg\max_x \left[p_x \mathsf{f}_{Y|X}(\boldsymbol{y} \mid x)\right]. \tag{8.3}$$

It is possible for the maximum in (8.3) to be achieved by several hypotheses. Each such hypothesis will maximize the probability of correct decision, but there is a need for a tie-breaking rule to actually make a decision. Ties could be broken in a random way (and this is essentially necessary in Section 8.4), but then the decision is more than a function of the observation. In many situations, it is useful for decisions to be (deterministic) functions of the observations.

Definition 8.1.1 *A **test** is a decision $\hat{x}(\boldsymbol{y})$ that is a (deterministic) function of the observation.*

In order to fully define the MAP rule in (8.3) as a test, we arbitrarily choose the largest maximizing x. An arbitrary test A, i.e., an arbitrary deterministic rule for choosing an hypothesis from an observation $\boldsymbol{y}$, can be viewed as a function, say $\hat{x}_A(\boldsymbol{y})$, mapping the set of observations to the set of hypotheses, $\{0, 1, \ldots, \mathsf{M}-1\}$. For any test A, then, $\mathsf{p}_{X|Y}(\hat{x}_A \mid \boldsymbol{y})$ is the probability that $\hat{x}_A(\boldsymbol{y})$ is the correct decision when test A is used on observation $\boldsymbol{y}$. Since $x_{\text{MAP}}(\boldsymbol{y})$ maximizes the probability of correct decision, we have

$$\mathsf{p}_{X|Y}\big(\hat{x}_{\text{MAP}}(\boldsymbol{y}) \mid \boldsymbol{y}\big) \ge \mathsf{p}_{X|Y}\big(\hat{x}_A(\boldsymbol{y}) \mid \boldsymbol{y}\big) \qquad \text{for all } A \text{ and } \boldsymbol{y}. \tag{8.4}$$

For simplicity of notation, we assume in what follows that the observation $\boldsymbol{Y}$, conditional on each hypothesis, has a joint probability density. Then $\boldsymbol{Y}$ also has a marginal density $\mathsf{f}_Y(\boldsymbol{y})$. Averaging (8.4) over observations,

$$\int \mathsf{f}_Y(\boldsymbol{y}) \mathsf{p}_{X|Y}\big(\hat{x}_{\text{MAP}}(\boldsymbol{y}) | \boldsymbol{y}\big)\, d\boldsymbol{y} \ge \int \mathsf{f}_Y(\boldsymbol{y}) \mathsf{p}_{X|Y}\big(\hat{x}_A(\boldsymbol{y}) | \boldsymbol{y}\big)\, d\boldsymbol{y}. \tag{8.5}$$

The quantity on the left is the overall probability of correct decision using $\hat{x}_{\text{MAP}}$, and that on the right is the overall probability of correct decision using $\hat{x}_A$. The above results are very simple, but also important and fundamental. We summarize them in the following theorem, which applies equally for observations with a PDF or PMF.

Theorem 8.1.2 *Assume that* $\boldsymbol{Y}$ *has a joint PDF or PMF conditional on each hypothesis. The MAP rule, given in (8.2) (or equivalently in (8.3)), maximizes the probability of correct decision conditional on each observed sample value* $\boldsymbol{y}$. *It also maximizes the overall probability of correct decision given in (8.5).*

Proof[3] The only questionable issue about the theorem is whether the integrals in (8.5) exist whenever $\boldsymbol{Y}$ has a density. Using (8.1), the integrands in (8.5) can be rewritten as

$$\int p_{\hat{x}_{\text{MAP}}(\boldsymbol{y})} \mathsf{f}_{\boldsymbol{Y}|X}\big(\boldsymbol{y} \mid \hat{x}_{\text{MAP}}(\boldsymbol{y})\big)\, d\boldsymbol{y} \ \geq\ \int p_{\hat{x}_{\boldsymbol{A}}(\boldsymbol{y})} \mathsf{f}_{\boldsymbol{Y}|X}\big(\boldsymbol{y} \mid \hat{x}_{\boldsymbol{A}}(\boldsymbol{y})\big)\, d\boldsymbol{y}. \tag{8.6}$$

The right-hand integrand of (8.6) can be expressed in a less obscure way by noting that the function $\hat{x}_{\boldsymbol{A}}(\boldsymbol{y})$ is specified by defining, for each hypothesis ℓ, the set A_ℓ of observations $\boldsymbol{y} \in \mathbb{R}^n$ for which $\hat{x}_{\boldsymbol{A}}(\boldsymbol{y}) = \ell$, i.e., $A_\ell = \{\boldsymbol{y} : \hat{x}_{\boldsymbol{A}}(\boldsymbol{y}) = \ell\}$. Using these sets, the integral on the right is given by

$$\begin{aligned}
\int p_{\hat{x}_{\boldsymbol{A}}(\boldsymbol{y})} \mathsf{f}_{\boldsymbol{Y}|X}\big(\boldsymbol{y} \mid \hat{x}_{\boldsymbol{A}}(\boldsymbol{y})\big)\, d\boldsymbol{y} &= \sum_\ell p_\ell \int_{\boldsymbol{y}\in A_\ell} \mathsf{f}_{\boldsymbol{Y}|X}(\boldsymbol{y} \mid \ell)\, d\boldsymbol{y} \\
&= \sum_\ell p_\ell \Pr\{A_\ell \mid X = \ell\} \qquad (8.7)\\
&= \sum_\ell p_\ell \Pr\left\{\hat{X}_A = \ell \mid X = \ell\right\}, \qquad (8.8)
\end{aligned}$$

where $\hat{X}_A = \hat{x}_A(\boldsymbol{Y})$. The integral above exists if the likelihoods are measurable functions and if the sets A_ℓ are measurable sets. This is virtually no restriction and is henceforth assumed without further question. The integral on the left-hand side of (8.6) is similarly $\sum_\ell p_\ell \Pr\{\hat{X}_{\text{MAP}} = \ell \mid X = \ell\}$, which exists if the likelihoods are measurable functions. □

Before discussing the implications and use of the MAP rule, we review the assumptions that have been made. First, we assumed a probability experiment in which all probabilities are known, and in which, for each performance of the experiment, one and only one hypothesis is correct. This conforms very well to a communication model in which a transmitter sends one of a set of possible signals, and the receiver, given signal plus noise, makes a decision on the transmitted signal. It does not always conform well to a scientific experiment attempting to verify the existence of some new phenomenon; in such situations, there is often no sensible way to model a priori probabilities. In Section 8.4, we find ways to avoid depending on a priori probabilities.

The next assumption was that maximizing the probability of correct decision is an appropriate decision criterion. In many situations, the costs of right and wrong decisions

[3] This proof deals with mathematical details and involves both elementary notions of measure theory and conceptual difficulty; readers may safely omit it or postpone the proof until learning measure theory.

are highly asymmetric. For example, when testing for a treatable but deadly disease, making an error when the disease is present is far more dangerous than making an error when the disease is not present. In Section 8.3 we define a minimum-cost formulation which allows us to treat these asymmetric cases.

The MAP rule can be extended to broader assumptions than the existence of a PMF or PDF. This is carried out in Exercise 8.12 but requires a slight generalization of the notion of a MAP rule at each sample observation $\boldsymbol{y}$.

In the next three sections, we restrict attention to the case of binary hypotheses. This allows us to understand most of the important ideas but simplifies the notation and details considerably. In Section 8.5, we again consider an arbitrary number of hypotheses.

8.2 Binary MAP detection

Assume a probability model in which the hypothesis rv X is binary with $\mathsf{p}_X(0) = p_0 > 0$ and $\mathsf{p}_X(1) = p_1 > 0$. Let $\boldsymbol{Y}$ be an n-rv $\boldsymbol{Y} = (Y_1, \ldots, Y_n)^\mathsf{T}$ whose conditional probability density, $\mathsf{f}_{Y|X}(\boldsymbol{y} \mid \ell)$, is initially assumed to be finite and non-zero for all $y \in \mathbb{R}^n$ and $\ell \in \{0, 1\}$. The marginal density of $\boldsymbol{Y}$ is given by $\mathsf{f}_Y(\boldsymbol{y}) = p_0 \mathsf{f}_{Y|X}(\boldsymbol{y} \mid 0) + p_1 \mathsf{f}_{Y|X}(\boldsymbol{y} \mid 1) > 0$. The a posteriori probability of $X = 0$ or $X = 1$ is given by

$$\mathsf{p}_{X|\boldsymbol{Y}}(\ell \mid \boldsymbol{y}) = \frac{p_\ell \mathsf{f}_{Y|X}(\boldsymbol{y} \mid l)}{\mathsf{f}_Y(\boldsymbol{y})}. \tag{8.9}$$

Writing out (8.2) explicitly for this case,

$$\frac{p_1 \mathsf{f}_{Y|X}(\boldsymbol{y} \mid 1)}{\mathsf{f}_Y(\boldsymbol{y})} \quad \mathop{\gtrless}_{\hat{x}(\boldsymbol{y})=0}^{\hat{x}(\boldsymbol{y})=1} \quad \frac{p_0 \mathsf{f}_{Y|X}(\boldsymbol{y} \mid 0)}{\mathsf{f}_Y(\boldsymbol{y})}. \tag{8.10}$$

This 'equation' indicates that the decision is 1 if the left-hand side is greater than or equal to the right, and is 0 if the left-hand side is less than the right. Choosing the decision to be 1 when equality holds is arbitrary and does not affect the probability of being correct. Canceling $\mathsf{f}_Y(\boldsymbol{y})$ and rearranging,

$$\Lambda(\boldsymbol{y}) \;=\; \frac{\mathsf{f}_{Y|X}(\boldsymbol{y} \mid 1)}{\mathsf{f}_{Y|X}(\boldsymbol{y} \mid 0)} \quad \mathop{\gtrless}_{\hat{x}(\boldsymbol{y})=0}^{\hat{x}(\boldsymbol{y})=1} \quad \frac{p_0}{p_1} = \eta. \tag{8.11}$$

The ratio $\Lambda(\boldsymbol{y}) = \mathsf{f}_{Y|X}(\boldsymbol{y} \mid 1)/\mathsf{f}_{Y|X}(\boldsymbol{y} \mid 0)$ is called the *likelihood ratio* for a binary decision problem. It is a function only of $\boldsymbol{y}$ and not the a priori probabilities.[4] The detection rule in (8.11) is called a *threshold rule* and its right-hand side, $\eta = p_0/p_1$ is called the *threshold*, which in this case is a function only of the a priori probabilities.

Note that if the a priori probability p_0 is increased in (8.11), then the threshold increases, and the set of $\boldsymbol{y}$ for which hypothesis 0 is chosen increases; this corresponds

[4] For non-Bayesians, the likelihood ratio is one rv in the model for $X = 0$ and another rv in the model for $X = 1$, but is not a rv overall.

to our intuition – the greater our initial conviction that X is 0, the stronger the evidence required to change our minds.

Next consider a slightly more general binary case in which $\mathsf{f}_{Y|X}(y \mid \ell)$ might be 0 (although, as discussed before $\mathsf{f}_{Y}(y) > 0$ and thus at least one of the likelihoods must be positive for each $\boldsymbol{y}$). In this case, the likelihood ratio can be either 0 or ∞ depending on which of the two likelihoods is 0. This more general case does not require any special care except for the development of the error curve in Section 8.4. Thus we include this generalization in what follows.

We will look later at a number of other detection rules, including maximum likelihood (ML), minimum-cost, and Neyman–Pearson. These are also essentially threshold rules that differ from the MAP test in the choice of the threshold η (and perhaps in the tie-breaking rule when the likelihood ratio equals the threshold). In general, a binary *threshold rule* is a decision rule between two hypotheses in which the likelihood ratio is compared to a threshold, where the threshold is a real number not dependent on the observation. As discussed in Section 8.3, there are also situations for which threshold rules are inappropriate.

An important special case of (8.11) is that in which $p_0 = p_1$. In this case $\eta = 1$. The rule is then $\hat{x}(\boldsymbol{y}) = 1$ for $\boldsymbol{y}$ such that $\mathsf{f}_{Y|X}(y \mid 1) \geq \mathsf{f}_{Y|X}(y \mid 0)$ (and $\hat{x}(\boldsymbol{y}) = 0$ otherwise). This is called a *ML rule or test*. The ML test is often used when p_0 and p_1 are unknown, as discussed later.

We now find the probability of error, $\Pr\{e_\eta \mid X{=}x\}$, under each hypothesis[5] x using a threshold at η. From this we can also find the overall probability of error $\Pr\{e_\eta\} = p_0\Pr\{e_\eta \mid X{=}0\} + p_1\Pr\{e_\eta \mid X{=}1\}$, where $\eta = p_0/p_1$.

Note that (8.11) partitions the space of observed sample values into two regions. The region $A_1 = \{\boldsymbol{y} : \Lambda(\boldsymbol{y}) \geq \eta\}$ specifies where $\hat{x} = 1$ and the region $A_0 = \{\boldsymbol{y} : \Lambda(\boldsymbol{y}) < \eta\}$ specifies where $\hat{x} = 0$. For $\mathrm{X} = 0$, an error occurs if and only if $\boldsymbol{y}$ is in A_0; for $X = 1$, an error occurs if and only if $\boldsymbol{y}$ is in A_1. Thus,

$$\Pr\{e_\eta \mid X{=}0\} = \int_{\boldsymbol{y} \in A_1} \mathsf{f}_{Y|X}(\boldsymbol{y} \mid 0)\, d\boldsymbol{y}; \tag{8.12}$$

$$\Pr\{e_\eta \mid X{=}1\} = \int_{\boldsymbol{y} \in A_0} \mathsf{f}_{Y|X}(\boldsymbol{y} \mid 1)\, d\boldsymbol{y}. \tag{8.13}$$

In several of the examples to follow, the error probability can be evaluated in a simpler way by working directly with the likelihood ratio. Since $\Lambda(\boldsymbol{y})$ is a function of the observed sample value $\boldsymbol{y}$, we can define the likelihood ratio rv $\Lambda(\boldsymbol{Y})$ in the usual way, i.e., for every sample point ω, $\boldsymbol{Y}(\omega)$ is the sample value for $\boldsymbol{Y}$, and $\Lambda(\boldsymbol{Y}(\omega))$ is the sample

[5] In the radar field, $\Pr\{e \mid X{=}0\}$ is called the probability of false alarm, and $\Pr\{e \mid X{=}1\}$ (for any given test) is called the probability of a miss. Also $1 - \Pr\{e \mid X{=}1\}$ is called the probability of detection. In statistics, $\Pr\{e \mid X{=}1\}$ is called the probability of error of the second kind, and $\Pr\{e \mid X{=}0\}$ is the probability of error of the first kind. I feel that all this terminology conceals the fundamental symmetry between the two hypotheses. The names 0 and 1 of the hypotheses could be interchanged, or they could be renamed, for example, as Alice and Bob or Apollo and Zeus, with only minor complications in notation.

value of $\Lambda(\boldsymbol{Y})$. In the same way, $\hat{x}(\boldsymbol{Y}(\omega))$ (or more briefly $\hat{X}$) is the decision rv. In these terms, (8.11) states that

$$\hat{X} = 1 \quad \text{if and only if} \quad \Lambda(\boldsymbol{Y}) \geq \eta. \tag{8.14}$$

Thus,

$$\Pr\{e_\eta \mid X{=}0\} = \Pr\left\{\hat{X}{=}1 \mid X{=}0\right\} = \Pr\{\Lambda(\boldsymbol{Y}) \geq \eta \mid X{=}0\}; \tag{8.15}$$

$$\Pr\{e_\eta \mid X{=}1\} = \Pr\left\{\hat{X}{=}0 \mid X{=}1\right\} = \Pr\{\Lambda(\boldsymbol{Y}) < \eta \mid X{=}1\}. \tag{8.16}$$

This means that if the one-dimensional quantity $\Lambda(\boldsymbol{y})$ can be found easily from the observation $\boldsymbol{y}$, then it can be used, with no further reference to $\boldsymbol{y}$, both to perform any threshold test and to find the resulting error probability.

8.2.1 Sufficient statistics I

The previous section suggests that binary hypothesis testing can often be separated into two parts, first finding the likelihood ratio for a given observation, and then doing everything else on the basis of the likelihood ratio. Sometimes it is simpler to find a quantity that is a one-to-one function of the likelihood ratio (the log of the likelihood ratio is a common choice) and sometimes it is simpler to calculate some intermediate quantity from which the likelihood ratio can be found. Both of these are called *sufficient statistics*, as defined more precisely below.

Definition 8.2.1 *For binary hypothesis testing, a* ***sufficient statistic*** *is any function $v(\boldsymbol{y})$ of the observation $\boldsymbol{y}$ from which the likelihood ratio can be calculated, i.e., $v(\boldsymbol{y})$ is a sufficient statistic if a function $u(v)$ exists such that $\Lambda(\boldsymbol{y}) = u(v(\boldsymbol{y}))$ for all $\boldsymbol{y}$.*

For example, $\boldsymbol{y}$ itself, $\Lambda(\boldsymbol{y})$, and any one-to-one function of $\Lambda(\boldsymbol{y})$ are sufficient statistics. For vector or process observations in particular, $\Lambda(\boldsymbol{Y})$ (or any one-to-one function of $\Lambda(\boldsymbol{Y})$) is often simpler to work with than $\boldsymbol{Y}$ itself, since $\Lambda(\boldsymbol{Y})$ is one-dimensional rather than many-dimensional. As indicated by these examples, $v(\boldsymbol{y})$ can be one-dimensional or multi-dimensional.

An important example of a sufficient statistic is the log-likelihood ratio, $\mathrm{LLR}(\boldsymbol{Y}) = \ln[\Lambda(\boldsymbol{Y})]$. We will often find that the LLR is more convenient to work with than $\Lambda(\boldsymbol{Y})$. We next look at some widely used examples of binary MAP detection where the LLR and other sufficient statistics are useful. We will then develop additional properties of sufficient statistics.

8.2.2 Binary detection with a one-dimensional observation

Example 8.2.2 (Detection of antipodal signals in Gaussian noise) First we look at a simple abstraction of a common digital communication system in which a single binary digit is transmitted and that digit plus Gaussian noise is received. The observation is then taken to be $Y = X + Z$, where $Z \sim \mathcal{N}(0, \sigma^2)$ is Gaussian and X (the transmitted rv and the hypothesis) is independent of Z and binary with possible values $b > 0$ and $-b$. We could visualize mapping the binary digits $\{1, 0\}$ to $\{b, -b\}$, but it is simpler to view the hypothesis directly as having values b and $-b$.

The receiver must detect, from the observed sample value y of Y, whether the sample value of X is b or $-b$.

We will see in subsequent examples that the approach here can be used if the binary hypothesis is a choice between two vectors in the presence of a Gaussian vector or a choice between two waveforms in the presence of a Gaussian process. In fact, not only can the same approach be used, but the problem essentially reduces to the single-dimensional case here.

Conditional on $X = b$, the observation is $Y \sim \mathrm{N}\,(b, \sigma^2)$ and, conditional on $X = -b$, $Y \sim \mathrm{N}\,(-b, \sigma^2)$, i.e.,

$$\mathsf{f}_{Y|X}(y \mid b) = \frac{1}{\sqrt{2\pi\sigma^2}} \exp\left[\frac{-(y-b)^2}{(2\sigma^2)}\right], \qquad \mathsf{f}_{Y|X}(y \mid -b) = \frac{1}{\sqrt{2\pi\sigma^2}} \exp\left[\frac{-(y+b)^2}{(2\sigma^2)}\right].$$

The likelihood ratio is the ratio of $\mathsf{f}(y \mid b)$ to $\mathsf{f}(y \mid -b)$, given by

$$\Lambda(y) = \exp\left[\frac{(y+b)^2 - (y-b)^2}{2\sigma^2}\right] = \exp\left[\frac{2yb}{\sigma^2}\right]. \tag{8.17}$$

Substituting this into (8.11), with $p_0 = \mathsf{p}_X(-b)$ and $p_1 = \mathsf{p}_X(b)$,

$$\Lambda(y) = \exp\left[\frac{2yb}{\sigma^2}\right] \underset{\hat{x}(y)=-b}{\overset{\hat{x}(y)=b}{\stackrel{\geq}{<}}} \frac{p_0}{p_1} = \eta. \tag{8.18}$$

This is further simplified by taking the logarithm, yielding

$$\mathrm{LLR}(y) = \left[\frac{2yb}{\sigma^2}\right] \underset{\hat{x}(y)=-b}{\overset{\hat{x}(y)=b}{\stackrel{\geq}{<}}} \ln \eta. \tag{8.19}$$

This can be rewritten as a threshold rule on y directly,

$$y \underset{\hat{x}(y)=-b}{\overset{\hat{x}(y)=b}{\stackrel{\geq}{<}}} \frac{\sigma^2 \ln \eta}{2b}. \tag{8.20}$$

In the ML case ($p_1 = p_0$), the threshold η is 1, so $\ln \eta = 0$. Thus, as illustrated in Figure 8.1, the ML rule maps $y \geq 0$ into $\hat{x} = b$ and $y < 0$ into $\hat{x} = -b$.

Still assuming ML, an error occurs, conditional on $X = -b$, if $Y \geq 0$. This is the same as the probability that the normalized Gaussian rv $(Y + b)/\sigma$ exceeds b/σ. This in turn is $Q(b/\sigma)$ where $Q(u)$ is the complementary cumulative distribution function (CDF) of a normalized Gaussian rv,

$$Q(u) = \int_u^\infty \frac{1}{2\pi} \exp\left(\frac{-z^2}{2}\right) dz. \tag{8.21}$$

It is easy to see (especially with the help of Figure 8.1) that with ML, the probability of error conditional on $X = b$ is the same, so

$$\Pr\{e \,|\, X = b\} = \Pr\{e \,|\, X = -b\} = \Pr\{e\} = Q(b/\sigma). \tag{8.22}$$

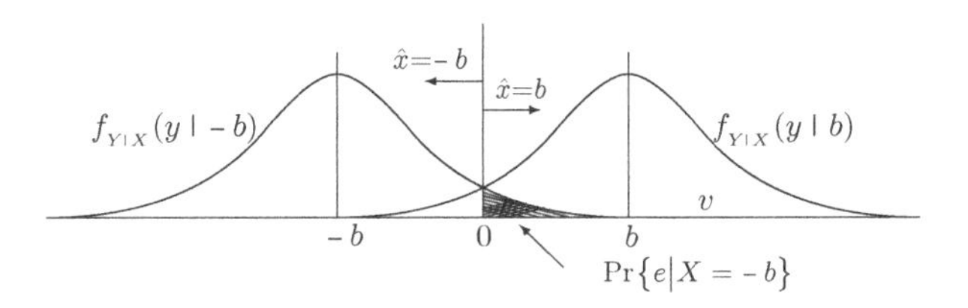

Figure 8.1 Binary hypothesis testing for antipodal signals, $-b$ and $+b$ with $\eta = 1$. The figure also illustrates the error probability given $X = -b$ and using ML detection ($\eta = 1$). It is clear from the symmetry between the two curves that $\Lambda(y) = 1$ at $y = 0$.

It can be seen that with ML detection, the error probability depends only on the ratio b/σ, which we denote as γ. The reason for this dependence on γ alone can be seen by dimensional analysis. That is, if the signal amplitude and the noise standard deviation are measured in a different system of units, the error probability would not change. We view γ^2 as a signal-to-noise ratio, i.e., the signal value squared (which can be interpreted as signal energy) divided by σ^2, which can be interpreted in this context as noise energy.

It is now time to look at what happens when $\eta = p_0/p_1$ is not 1. Using (8.20) to find the error probability as before, and using $\gamma = b/\sigma$,

$$\Pr\{e_\eta|X = -b\} = Q\left(\gamma + \frac{\ln \eta}{2\gamma}\right), \tag{8.23}$$

$$\Pr\{e_\eta|X = b\} = Q\left(\gamma - \frac{\ln \eta}{2\gamma}\right). \tag{8.24}$$

If $\ln \eta > 0$, i.e., if $p_0 > p_1$, then, for example, if $y = 0$, the observation alone would provide no evidence whether $X = -b$ or $X = b$, but we would choose $\hat{x}(y) = -b$ since it is more likely a priori. This gives an intuitive explanation why the threshold is moved to the right if $\ln \eta > 0$ and to the left if $\ln \eta < 0$.

Example 8.2.3 (Binary detection with a Gaussian noise rv) This is a very small generalization of the previous example. The binary rv X, instead of being antipodal (i.e., $X = \pm b$) is now arbitrary, taking on either the arbitrary value a or b, where $b > a$. The a priori probabilities are denoted by p_0 and p_1 respectively. As before, the observation (at the receiver) is $Y = X + Z$, where $Z \sim \mathrm{N}(0, \sigma^2)$ and X and Z are independent.

Conditional on $X = b$, $Y \sim \mathrm{N}(b, \sigma^2)$ and, conditional on $X = a$, $Y \sim \mathrm{N}(a, \sigma^2)$.

$$\mathsf{f}_{Y|X}(y \mid b) = \frac{1}{\sqrt{2\pi\sigma^2}} \exp\left[\frac{-(y-b)^2}{2\sigma^2}\right], \qquad \mathsf{f}_{Y|X}(y \mid a) = \frac{1}{\sqrt{2\pi\sigma^2}} \exp\left[\frac{-(y-a)^2}{2\sigma^2}\right].$$

The likelihood ratio is then

$$\begin{aligned}\Lambda(y) &= \exp\left[\frac{(y-a)^2 - (y-b)^2}{2\sigma^2}\right] = \exp\left[\frac{2(b-a)y + (a^2 - b^2)}{2\sigma^2}\right] \\ &= \exp\left[\left(\frac{b-a}{\sigma^2}\right)\left(y - \frac{b+a}{2}\right)\right].\end{aligned} \tag{8.25}$$

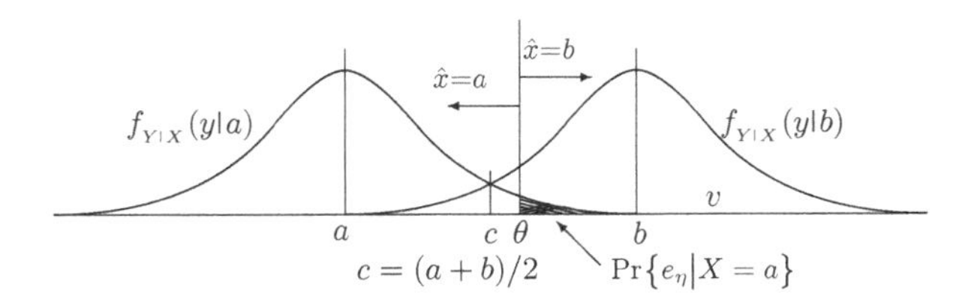

Figure 8.2 Binary hypothesis testing for arbitrary signals, a, b, for $b > a$. This is essentially the same as Figure 8.1 except the midpoint has been shifted to $(a+b)/2$ and the threshold η is assumed here to be greater than 1 (see (8.28)).

Substituting this into (8.11), we have

$$\exp\left[\left(\frac{b-a}{\sigma^2}\right)\left(y - \frac{b+a}{2}\right)\right] \underset{\hat{x}(y)=a}{\overset{\hat{x}(y)=b}{\gtrless}} \frac{p_0}{p_1} = \eta. \tag{8.26}$$

This is further simplified by taking the logarithm, yielding

$$\text{LLR}(y) = \left[\left(\frac{b-a}{\sigma^2}\right)\left(y - \frac{b+a}{2}\right)\right] \underset{\hat{x}(y)=a}{\overset{\hat{x}(y)=b}{\gtrless}} \ln(\eta). \tag{8.27}$$

Solving for y, (8.27) can be rewritten as a threshold rule on y directly,

$$y \underset{\hat{x}(y)=a}{\overset{\hat{x}(y)=b}{\gtrless}} \frac{\sigma^2 \ln \eta}{b-a} + \frac{b+a}{2}.$$

This says that the MAP rule simply compares y to a threshold $\sigma^2 \ln \eta/(b-a) + (b+a)/2$. Denoting this threshold for Y as θ, the MAP rule is

$$y \underset{\hat{x}(y)=a}{\overset{\hat{x}(y)=b}{\gtrless}} \theta, \qquad \text{where } \theta = \frac{\sigma^2 \ln \eta}{b-a} + \frac{b+a}{2}. \tag{8.28}$$

In the ML case ($p_1 = p_0$), the threshold η for Λ is 1 and the threshold θ for y is the midpoint[6] between a and b (i.e., $\theta = (b+a)/2$). For the MAP case, if η is larger or smaller than 1, θ is respectively larger or smaller than $(b+a)/2$ (see Figure 8.2).

From (8.28), $\Pr\{e_\eta \mid X{=}a\} = \Pr\{Y \geq \theta \mid X{=}a\}$. Given $X = a$, $Y \sim \mathrm{N}(a, \sigma^2)$, so, given $X = a$, $(Y-a)/\sigma$ is a normalized Gaussian variable:

$$\Pr\{Y \geq \theta \mid X{=}a\} = \Pr\left\{\frac{Y-a}{\sigma} \geq \frac{\theta-a}{\sigma} \mid X{=}a\right\} = Q\left(\frac{\theta-a}{\sigma}\right). \tag{8.29}$$

Replacing θ in (8.29) by its value in (8.28),

$$\Pr\{e_\eta \mid X{=}a\} = Q\left(\frac{\sigma \ln \eta}{b-a} + \frac{b-a}{2\sigma}\right). \tag{8.30}$$

[6] At this point, we see that this example is really the same as Example 8.2.2 with a simple linear shift of X and Y. Thus this example is redundant, but all these equations might be helpful to make this clear.

We evaluate $\Pr\{e_\eta \mid X{=}b\} = \Pr\{Y < \theta \mid X{=}b\}$ in the same way. Given $X = b$, Y is $\mathrm{N}\ (b, \sigma^2)$, so

$$\Pr\{Y < \theta \mid X{=}b\} = \Pr\left\{\frac{Y-b}{\sigma} < \frac{\theta - b}{\sigma} \mid X{=}b\right\} = 1 - Q\left(\frac{\theta - b}{\sigma}\right).$$

Using (8.28) for θ and noting that $Q(x) = 1 - Q(-x)$ for any x,

$$\Pr\{e_\eta \mid X{=}b\} = Q\left(\frac{-\sigma \ln \eta}{b-a} + \frac{b-a}{2\sigma}\right). \tag{8.31}$$

Note that (8.30) and (8.31) are functions only of $(b-a)/\sigma$ and η. That is, only the distance between b and a is relevant, rather than their individual values, and it is only this distance relative to σ that is relevant. This should be intuitively clear from Figure 8.2. If we define $\gamma = (b-a)/(2\sigma)$, then (8.30) and (8.31) simplify to

$$\Pr\{e_\eta \mid X{=}a\} = Q\left(\frac{\ln \eta}{2\gamma} + \gamma\right), \qquad \Pr\{e_\eta \mid X{=}b\} = Q\left(\frac{-\ln \eta}{2\gamma} + \gamma\right). \tag{8.32}$$

For ML detection, $\eta = 1$, and this simplifies further to

$$\Pr\{e \mid X{=}a\} = \Pr\{e \mid X{=}b\} = \Pr\{e\} = Q(\gamma). \tag{8.33}$$

As expected, the solution here is essentially the same as the antipodal case of the previous example. The only difference is that in the first example, the midpoint between the signals is 0 and in the present case it is arbitrary. Since this arbitrary offset is known at the receiver, it has nothing to do with the error probability. We still interpret γ^2 as a signal-to-noise ratio, and the energy in the offset is wasted (for the purpose of detection).

8.2.3 Binary MAP detection with vector observations

In this section, we consider the same basic problem as in the last section, except that here the observation consists of the sample values of n rv s instead of 1. There is a binary hypothesis with a priori probabilities $\mathsf{p}_X(0) = p_0 > 0$ and $\mathsf{p}_X(1) = p_1 > 0$. There are n observation rv s which we view as an n-**rv** $\boldsymbol{Y} = (Y_1, \ldots, Y_n)^\mathsf{T}$. Given a sample value $\boldsymbol{y}$ of $\boldsymbol{Y}$, we use the MAP rule to select the most probable hypothesis conditional on $\boldsymbol{y}$.

It is important to understand that the sample point ω resulting from an experiment leads to a sample value $X(\omega)$ for X and sample values $Y_1(\omega), \ldots, Y_n(\omega)$ for $Y_1, \ldots, Y_n$. When testing an hypothesis, one often performs many subexperiments, corresponding to the multiple observations $y_1, \ldots, y_n$. However, the sample value of the hypothesis (which is not observed) is constant over these subexperiments.

The analysis of this n-**rv** observation is virtually identical to that for a single observation rv (except for the examples which are based explicitly on particular probability densities). Throughout this section, we assume that the conditional joint distribution of $\boldsymbol{Y}$ conditional on either hypothesis has a density $\mathsf{f}_{\boldsymbol{Y}|X}(\boldsymbol{y}|\ell)$ that is positive over $\mathbb{R}^n$. Then, exactly as in Section 8.2.2,

$$\Lambda(\boldsymbol{y}) \quad = \quad \frac{\mathsf{f}_{\boldsymbol{Y}|X}(\boldsymbol{y} \mid 1)}{\mathsf{f}_{\boldsymbol{Y}|X}(\boldsymbol{y} \mid 0)} \quad \mathop{\gtrless}_{\hat{x}(\boldsymbol{y})=0}^{\hat{x}(\boldsymbol{y})=1} \quad \frac{p_0}{p_1} = \eta. \tag{8.34}$$

Here $\Lambda(\boldsymbol{y})$ is the likelihood ratio for the observed sample value $\boldsymbol{y}$. MAP detection simply compares $\Lambda(\boldsymbol{y})$ to the threshold η. The MAP principle applies here, i.e., the rule in (8.34) minimizes the error probability for each sample value $\boldsymbol{y}$ and thus also minimizes the overall error probability.

Extending the observation space from rv s to n-**rv** s becomes more interesting if we constrain the observations to be conditionally independent given the hypothesis. That is, we now assume that the joint density of observations given the hypothesis satisfies

$$\mathsf{f}_{\boldsymbol{Y}|X}(\boldsymbol{y} \mid x) = \prod_{i=1}^{n} \mathsf{f}_{Y_j|X}(y_i \mid x) \qquad \text{for all } \boldsymbol{y} \in \mathbb{R}^n,\ x \in \{0,1\}. \tag{8.35}$$

In this case, the likelihood ratio is given by

$$\Lambda(\boldsymbol{y}) = \prod_{j=1}^{n} \frac{\mathsf{f}_{Y_j|X}(y_j \mid 1)}{\mathsf{f}_{Y_j|X}(y_j \mid 0)}. \tag{8.36}$$

The MAP test then takes on a more attractive form if we take the logarithm of each side in (8.36). The logarithm of $\Lambda(\boldsymbol{y})$ is then a sum of n terms,

$$\mathrm{LLR}(\boldsymbol{y}) \;=\; \sum_{j=1}^{n} \mathrm{LLR}_j(y_j), \qquad \text{where } \mathrm{LLR}_j(y_j) \;=\; \ln \frac{\mathsf{f}_{Y_j|X}(y_j \mid 1)}{\mathsf{f}_{Y_j|X}(y_j \mid 0)}. \tag{8.37}$$

The test in (8.34) is then expressed as

$$\mathrm{LLR}(\boldsymbol{y}) \;=\; \sum_{j=1}^{n} \mathrm{LLR}_j(y_j) \;\mathop{\gtrless}_{\hat{x}(\boldsymbol{y})=0}^{\hat{x}(\boldsymbol{y})=1}\; \ln \eta. \tag{8.38}$$

Note that the rv s $\mathrm{LLR}_1(Y_1), \ldots, \mathrm{LLR}_n(Y_n)$ are conditionally independent given $X = 0$ or $X = 1$. This form becomes even more attractive if for each hypothesis, the rv s $\mathrm{LLR}_j(Y_j)$ are conditionally identically distributed over j. Chapter 9 analyzes this case as two random walks, one conditional on $X = 0$ and the other on $X = 1$. This case is then extended to sequential decision theory, where, instead of making a decision after some fixed number of observations, part of the decision process includes deciding after each observation whether to make a decision or continue making more observations.

Here, however, we use the more general formulation of (8.38) to study the detection of a vector signal in Gaussian noise. The example also applies to various hypothesis testing problems in which multiple noisy measurements are taken to distinguish between two hypotheses.

Example 8.2.4 (Binary detection of vector signals in Gaussian noise) Figure 8.3 illustrates the transmission of a single binary digit in a communication system. If 0 is to be transmitted, it is converted to a real vector $\boldsymbol{a} = (a_1, \ldots, a_n)^{\mathsf{T}}$ and similarly 1 is converted to a vector $\boldsymbol{b}$. Readers not familiar with the signal space view of digital communication can simply view this as an abstraction of converting 0 into one waveform and 1 into another. The transmitted signal is then $\boldsymbol{X}$, where $\boldsymbol{X} = \boldsymbol{a}$ or $\boldsymbol{b}$, and we regard these directly

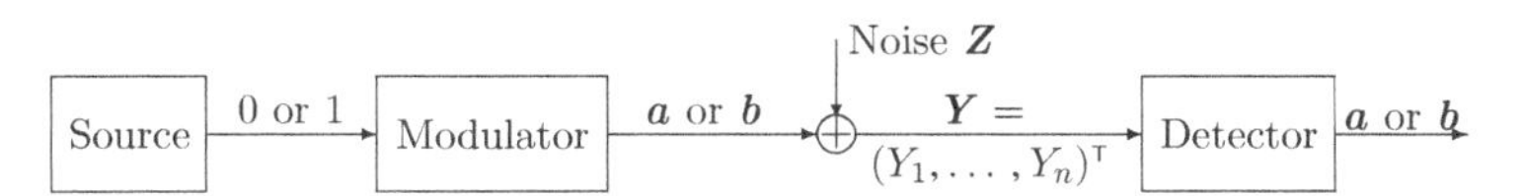

Figure 8.3 The source transmits a binary digit, 0 or 1. The binary digit 0 is mapped into the n-vector $\boldsymbol{a}$, and 1 is mapped into the n-vector $\boldsymbol{b}$. We view the two hypotheses as $\boldsymbol{X} = \boldsymbol{a}$ and $\boldsymbol{X} = \boldsymbol{b}$ (although the hypotheses could equally well be viewed as 0 and 1). After addition of IID Gaussian noise, $\mathcal{N}(0, \sigma^2[I_n])$, the detector chooses $\hat{x}$ to be $\boldsymbol{a}$ or $\boldsymbol{b}$.

as the two hypotheses. The receiver then observes $\boldsymbol{Y} = \boldsymbol{X} + \boldsymbol{Z}$, where $\boldsymbol{Z} = (Z_1, \dots, Z_n)^{\mathsf{T}}$ is $\mathrm{N}\ (0, \sigma^2[I_n])$ and independent of X.

Given $\boldsymbol{X} = \boldsymbol{b}$, the jth observation, Y_j, is $\mathrm{N}\ (b_j, \sigma^2)$ and the n observations are conditionally independent. Similarly, given $\boldsymbol{X} = \boldsymbol{a}$, the observations are independent, $\mathrm{N}\ (a_j, \sigma^2)$. From (8.38), the LLR for an observation vector $\boldsymbol{y}$ is the sum of the individual LLRs given in (8.27), i.e.,

$$\mathrm{LLR}(\boldsymbol{y}) = \sum_{j=1}^{n} \mathrm{LLR}(y_j), \qquad \text{where } \mathrm{LLR}(y_j) = \left(\frac{b_j - a_j}{\sigma^2}\right)\left(y_j - \frac{b_j + a_j}{2}\right). \tag{8.39}$$

Expressing this sum in vector notation,

$$\mathrm{LLR}(\boldsymbol{y}) = \frac{(\boldsymbol{b} - \boldsymbol{a})^{\mathsf{T}}}{\sigma^2}\left(\boldsymbol{y} - \frac{\boldsymbol{b} + \boldsymbol{a}}{2}\right). \tag{8.40}$$

The MAP test is then

$$\mathrm{LLR}(\boldsymbol{y}) = \frac{(\boldsymbol{b} - \boldsymbol{a})^{\mathsf{T}}}{\sigma^2}\left(\boldsymbol{y} - \frac{\boldsymbol{b} + \boldsymbol{a}}{2}\right) \underset{\hat{x}(\boldsymbol{y})=\boldsymbol{a}}{\overset{\hat{x}(\boldsymbol{y})=\boldsymbol{b}}{\gtrless}} \ln \eta. \tag{8.41}$$

This test involves the observation $\boldsymbol{y}$ only in terms of the inner product $(\boldsymbol{b} - \boldsymbol{a})^{\mathsf{T}}\boldsymbol{y}$. Thus $(\boldsymbol{b} - \boldsymbol{a})^{\mathsf{T}}\boldsymbol{y}$ is a sufficient statistic and the detector can perform a MAP test simply by calculating this scalar quantity and then calculating $\mathrm{LLR}(\boldsymbol{y})$ without further reference to $\boldsymbol{y}$.

This is interpreted in Figure 8.4 for the special case of ML, where $\ln \eta = 0$. Note that one point on the threshold is $\boldsymbol{y} = (\boldsymbol{a} + \boldsymbol{b})/2$. The other points on the threshold are those for which $\boldsymbol{y} - (\boldsymbol{b} + \boldsymbol{a})/2$ is orthogonal to $\boldsymbol{b} - \boldsymbol{a}$. As illustrated for two dimensions, this is the line through $(\boldsymbol{a} + \boldsymbol{b})/2$ that is perpendicular to the line joining $\boldsymbol{a}$ and $\boldsymbol{b}$. For n dimensions, the set of points orthogonal to $\boldsymbol{b} - \boldsymbol{a}$ is an $(n - 1)$-dimensional hyperplane. Thus $\boldsymbol{y}$ is on the threshold if $\boldsymbol{y} - (\boldsymbol{b} + \boldsymbol{a})/2$ is in that $(n - 1)$-dimensional hyperplane.

The most fundamental way of viewing (8.41) is to view it in a different coordinate basis. That is, view the observation $\boldsymbol{y}$ as a point in n-dimensional space in which one of the basis vectors is the normalization of $\boldsymbol{b} - \boldsymbol{a}$, i.e., $(\boldsymbol{b} - \boldsymbol{a})/\|\boldsymbol{b} - \boldsymbol{a}\|$, where

$$\|\boldsymbol{b} - \boldsymbol{a}\| = \sqrt{(\boldsymbol{b} - \boldsymbol{a})^{\mathsf{T}}(\boldsymbol{b} - \boldsymbol{a})}, \tag{8.42}$$

Thus $(\boldsymbol{b} - \boldsymbol{a})/\|\boldsymbol{b} - \boldsymbol{a}\|$ is the vector $\boldsymbol{b} - \boldsymbol{a}$ normalized to unit length.

The two hypotheses can then only be distinguished by the component of the observation vector in this direction, i.e., by $(\boldsymbol{b} - \boldsymbol{a})^{\mathsf{T}}\boldsymbol{y}/\|\boldsymbol{b} - \boldsymbol{a}\|$. This is what (8.41) says, but we now see that this is very intuitive geometrically. The measurements in orthogonal directions only measure noise in those directions. Because the noise is IID, the noise in these

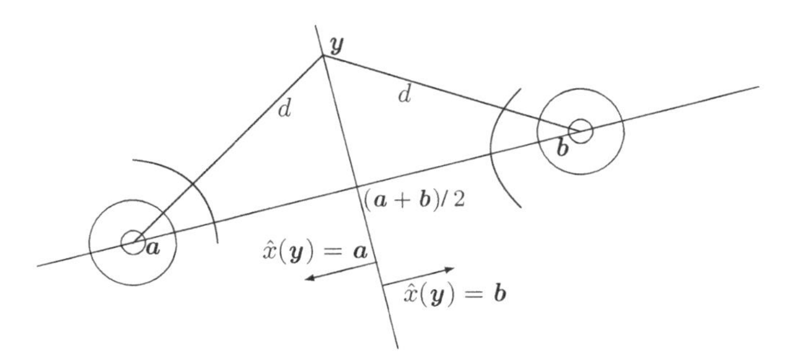

Figure 8.4 ML decision regions for binary signals in IID Gaussian noise. Note that, conditional on $X = a$, the equiprobable contours of Y form concentric spheres around a. Similarly, the equiprobable contours conditional on $X = b$ are concentric spheres around b. The two sets of spheres have the same radii for the same value of probability density. Thus all points on the perpendicular bisector between a and b are equidistant from a and b and are thus equiprobable for each hypothesis, i.e., they lie on the ML threshold boundary.

directions is independent of both the signal and the noise in the direction of interest, and thus can be ignored. This is often called the theorem of irrelevance.

When (8.41) is rewritten to express the detection rule in terms of the signal and noise in this dimension, it becomes

$$\text{LLR}(y) = \frac{\|\boldsymbol{b}-\boldsymbol{a}\|}{\sigma^2}\left(\frac{(\boldsymbol{b}-\boldsymbol{a})^{\mathsf{T}}\boldsymbol{y}}{\|\boldsymbol{b}-\boldsymbol{a}\|} - \frac{(\boldsymbol{b}-\boldsymbol{a})^{\mathsf{T}}(\boldsymbol{b}+\boldsymbol{a})/2}{\|\boldsymbol{b}-\boldsymbol{a}\|}\right) \mathop{\gtrless}_{\hat{x}(y)=\boldsymbol{a}}^{\hat{x}(y)=\boldsymbol{b}} \ln\eta. \tag{8.43}$$

If we let v be $(\boldsymbol{b}-\boldsymbol{a})^{\mathsf{T}}\boldsymbol{y}/\|\boldsymbol{b}-\boldsymbol{a}\|$, then v is the component of $\boldsymbol{y}$ in the signal direction, normalized so the noise in this direction is $\mathcal{N}(0,\sigma^2)$. Since v is multipied by the distance between $\boldsymbol{a}$ and $\boldsymbol{b}$, we see that this is the LLR for the one-dimensional detection problem in the $\boldsymbol{b}-\boldsymbol{a}$ direction.

Since this multi-dimensional binary detection problem has now been reduced to a one-dimensional problem with signal difference $\|\boldsymbol{b}-\boldsymbol{a}\|$ and noise $(0,\sigma^2)$, we can simply write down the error probability as found in (8.32):

$$\Pr\{e_\eta \mid X{=}\boldsymbol{a}\} = Q\left(\frac{\ln\eta}{2\gamma}+\gamma\right), \qquad \Pr\{e_\eta \mid X{=}\boldsymbol{b}\} = Q\left(\frac{-\ln\eta}{2\gamma}+\gamma\right), \tag{8.44}$$

where $\gamma = \|\boldsymbol{b}-\boldsymbol{a}\|/(2\sigma)$. This result is derived in a more prosaic way in Exercise 8.1.

Example 8.2.4 has shown that the LLR for detection between two n-vectors $\boldsymbol{a}$ and $\boldsymbol{b}$ in IID Gaussian noise is a function only of the magnitude of the received vector in the direction $\boldsymbol{b}-\boldsymbol{a}$. The error probabilities depend only on the signal-to-noise ratio γ^2 and the threshold η. When this problem is viewed in a coordinate system where $(\boldsymbol{b}-\boldsymbol{a})/\|\boldsymbol{b}-\boldsymbol{a}\|$ is a basis vector, the problem reduces to the one-dimensional case solved in Example 8.2.3. If the vectors are then translated so that the midpoint, $(\boldsymbol{a}+\boldsymbol{b})/2$, is at the origin, the problem further reduces to Example 8.2.2.

When we think about the spherical symmetry of IID Gaussian rv s, these results become unsurprising. However, both in the binary communication case, where vector signals are selected in the context of a more general situation, and in the hypothesis testing example where repeated tests must be done, we should consider the mechanics of reducing the vector problem to the one-dimensional case, i.e., to the problem of computing $(\boldsymbol{b}-\boldsymbol{a})^{\mathsf{T}}\boldsymbol{y}$.

In the communication example, $(\boldsymbol{b}-\boldsymbol{a})^{\mathsf{T}}\boldsymbol{y}$ is often called the correlation between the two vectors $(\boldsymbol{b}-\boldsymbol{a})$ and $\boldsymbol{y}$, and a receiver implementing this correlation is often called a correlation receiver. This operation is often done by creating a digital filter with impulse response $(b_n-a_n),\dots,(b_1-a_1)$ (i.e., by $\boldsymbol{b}-\boldsymbol{a}$ reversed in component order). If the received signal $\boldsymbol{y}$ is passed through this filter, then the output at time n is $(\boldsymbol{b}-\boldsymbol{a})^{\mathsf{T}}\boldsymbol{y}$. A receiver that implements $(\boldsymbol{b}-\boldsymbol{a})^{\mathsf{T}}\boldsymbol{y}$ in this way is called a matched-filter receiver. Thus correlation receivers and matched-filter receivers perform essentially the same function.

Example 8.2.5 (Gaussian non-IID noise) We consider Figure 8.3 again, but now generalize the noise to be $\mathrm{N}(0,[K_Z])$, where $[K_Z]$ is non-singular. From (3.24), the likelihoods are then:

$$p_{Y|X}(y \mid \boldsymbol{b}) = \frac{\exp\left[-\frac{1}{2}(y-\boldsymbol{b})^{\mathsf{T}}[K_Z^{-1}](y-\boldsymbol{b})\right]}{(2\pi)^{n/2}\sqrt{\det[K_Z]}}; \tag{8.45}$$

$$p_{Y|X}(y \mid \boldsymbol{a}) = \frac{\exp\left[-\frac{1}{2}(y-\boldsymbol{a})^{\mathsf{T}}[K_Z^{-1}](y-\boldsymbol{a})\right]}{(2\pi)^{n/2}\sqrt{\det[K_Z]}}. \tag{8.46}$$

The LLR is

$$\mathrm{LLR}(y) = \frac{1}{2}(y-\boldsymbol{a})^{\mathsf{T}}[K_Z^{-1}](y-\boldsymbol{a}) - \frac{1}{2}(y-\boldsymbol{b})^{\mathsf{T}}[K_Z^{-1}](y-\boldsymbol{b}) \tag{8.47}$$

$$= (\boldsymbol{b}-\boldsymbol{a})^{\mathsf{T}}[K_Z^{-1}]y + \frac{1}{2}\boldsymbol{a}^{\mathsf{T}}[K_Z^{-1}]\boldsymbol{a} - \frac{1}{2}\boldsymbol{b}^{\mathsf{T}}[K_Z^{-1}]\boldsymbol{b}. \tag{8.48}$$

This can be rewritten as

$$\mathrm{LLR}(\boldsymbol{y}) = (\boldsymbol{b}-\boldsymbol{a})^{\mathsf{T}}[K_Z^{-1}]\left[\boldsymbol{y} - \frac{\boldsymbol{b}+\boldsymbol{a}}{2}\right]. \tag{8.49}$$

The quantity $(\boldsymbol{b}-\boldsymbol{a})^{\mathsf{T}}[K_Z^{-1}]y$ is a sufficient statistic, and is simply a linear combination of the measurement variables $y_1,\dots,y_n$. Conditional on $\boldsymbol{X}=\boldsymbol{a}$, $\boldsymbol{Y}=\boldsymbol{a}+\boldsymbol{Z}$, so from (8.49),

$$\mathrm{E}\,[\mathrm{LLR}(\boldsymbol{Y} \mid \boldsymbol{X}{=}\boldsymbol{a})] = -(\boldsymbol{b}-\boldsymbol{a})^{\mathsf{T}}[K_Z^{-1}](\boldsymbol{b}-\boldsymbol{a})/2.$$

Defining γ as

$$\gamma = \sqrt{\frac{(\boldsymbol{b}-\boldsymbol{a})^{\mathsf{T}}}{2}[K_Z^{-1}]\frac{(\boldsymbol{b}-\boldsymbol{a})}{2}}, \tag{8.50}$$

we see that $\mathsf{E}\,[\mathrm{LLR}(\boldsymbol{Y} \mid \boldsymbol{X}{=}\boldsymbol{a})] = -2\gamma^2$. Similarly (see Exercise 8.2 for details),

$$\mathsf{VAR}\,[\mathrm{LLR}(\boldsymbol{Y} \mid \boldsymbol{X}{=}\boldsymbol{a})] = 4\gamma^2.$$

Then, as before, the conditional distribution of the log-likelihood ratio is given by

$$\text{Given } \boldsymbol{X}{=}\boldsymbol{a}, \;\; \text{LLR}(\boldsymbol{Y}) \sim \mathrm{N}\,(-2\gamma^2, 4\gamma^2). \tag{8.51}$$

In the same way,

$$\text{Given } \boldsymbol{X}{=}\boldsymbol{b}, \;\; \text{LLR}(\boldsymbol{Y}) \sim \mathrm{N}\,(2\gamma^2, 4\gamma^2). \tag{8.52}$$

The probability of error is then

$$\Pr\{e_\eta \mid \boldsymbol{X}{=}\boldsymbol{a}\} = Q\left(\frac{\ln \eta}{2\gamma} + \gamma\right), \quad \Pr\{e_\eta \mid \boldsymbol{X}{=}\boldsymbol{b}\} = Q\left(\frac{-\ln \eta}{2\gamma} + \gamma\right). \tag{8.53}$$

Note that the previous two examples are special cases of this more general result. The following theorem summarizes this.

Theorem 8.2.6 *Let the observed* **rv** $\boldsymbol{Y}$ *be given by* $\boldsymbol{Y} = \boldsymbol{a} + \boldsymbol{Z}$ *under* $\boldsymbol{X} = \boldsymbol{a}$ *and by* $\boldsymbol{Y} = \boldsymbol{b} + \boldsymbol{Z}$ *under* $\boldsymbol{X} = \boldsymbol{b}$ *and let* $\boldsymbol{Z} \sim \mathrm{N}\,(0, [K_{\boldsymbol{Z}}])$, *where* $[K_{\boldsymbol{Z}}]$ *is non-singular and* $\boldsymbol{Z}$ *is independent of* $\boldsymbol{X}$. *Then the distribution of the conditional LLRs is given by (8.51), (8.52) and the conditional error probabilities by (8.53).*

The definition of γ in (8.50) almost seems pulled out of a hat. It provides us with the general result in Theorem 8.2.6 very easily, but does not provide much insight. If we change the coordinate axes to an orthonormal expansion of the eigenvectors of $[K_{\boldsymbol{Z}}]$, then the noise components are independent and the LLR can be expressed as a sum of terms as in (8.39). The signals terms $b_j - a_j$ must be converted to this new coordinate system. We then see that signal components in the directions of small noise variance contribute more to the LLR than those in the directions of large noise variances. We can then easily see that with a limit on overall signal energy, $\sum_j (b_j - a_j)^2$, one achieves the smallest error probabilities by putting all the signal energy where the noise is smallest. This is not surprising, but it is reassuring that the theory shows this so easily.

The emphasis so far has been on Gaussian examples. For variety, the next example looks at finding the rate of a Poisson process.

Example 8.2.7 Consider a Poisson process for which the arrival rate λ is either λ_0 or λ_1, where $\lambda_0 > \lambda_1$. Let p_ℓ be the a priori probability that the rate is λ_ℓ. Suppose we observe the first n interarrival intervals, $Y_1, \ldots Y_n$, and make a MAP decision about the arrival rate from the sample values $y_1, \ldots, y_n$.

The conditional probability densities for the observations $Y_1, \ldots, Y_n$ are given by

$$\mathsf{f}_{\boldsymbol{Y}|X}(\boldsymbol{y} \mid x) = \prod_{j=1}^{n} \lambda_x e^{-\lambda_x y_j} \qquad \text{for } \boldsymbol{y} \geq 0.$$

The LLR is then

$$\text{LLR}(\boldsymbol{y}) = n \ln(\lambda_1/\lambda_0) + \sum_{j=1}^{n} (\lambda_0 - \lambda_1) y_j.$$

The MAP test in (8.38) is then

$$n\ln(\lambda_1/\lambda_0) + (\lambda_0 - \lambda_1)\sum_{j=1}^{n} y_j \quad \mathop{\gtrless}_{\hat{x}(y)=0}^{\hat{x}(y)=1} \quad \ln \eta. \tag{8.54}$$

Note that the test depends on $\boldsymbol{y}$ only through the epoch $s_n = \sum_j y_j$ of the nth arrival, and thus s_n is a sufficient statistic. This role of s_n should not be surprising, since we know that, under each hypothesis, the first $n-1$ arrivals are uniformly distributed conditional on the nth arrival time. With a little thought, one can see that (8.54) is also valid when $\lambda_0 < \lambda_1$.

8.2.4 Sufficient statistics II

The above examples have not only solved several hypothesis testing problems, but have also illustrated how to go about finding a sufficient statistic $v(\boldsymbol{y})$. One starts by writing an equation for the likelihood ratio (or LLR) and then one simplifies it. In each example, a relatively simple function $v(\boldsymbol{y})$ appeared from which $\Lambda(\boldsymbol{y})$ could be calculated, i.e., a function $v(\boldsymbol{y})$ appeared with the property that $\Lambda(\boldsymbol{y}) = u\big(v(\boldsymbol{y})\big)$ for some function u. This 'procedure' for finding a sufficient statistic was rather natural in the examples, but, as we have seen, sufficient statistics are not unique. For example, as we have seen, $\boldsymbol{y}$, $\Lambda(\boldsymbol{y})$, and $\mathrm{LLR}(\boldsymbol{y})$ are each sufficient statistics for any hypothesis testing problem. The sufficient statistics found in the examples, however, were noteworthy because of their simplicity and their ability to reduce dimensionality.

There are many other properties of sufficient statistics, and in fact there are a number of useful equivalent definitions of a sufficient statistic. Most of these apply both to binary and M-ary ($\mathsf{M} > 2$) hypotheses and most apply to both discrete and continuous observations. To avoid as many extraneous details as possible, however, we continue focusing on the binary case and restrict ourselves initially to discrete observations. When this initial discussion is extended shortly to observations with densities, we will find some strange phenomena that must be thought through carefully. When we extend the analysis to M-ary hypotheses in Section 8.5.1, we will find that almost no changes are needed.

Discrete observations

In this subsection, assume that the observation $\boldsymbol{Y}$ is discrete. It makes no difference whether $\boldsymbol{Y}$ is a **rv**, a rv, or simply a partition of the sample space. The hypothesis is binary with a priori probabilities $\{p_0, p_1\}$, although much of the discussion does not depend on the a priori probabilities.

Theorem 8.2.8 *Let $V = v(\boldsymbol{Y})$ be a function of $\boldsymbol{Y}$ for a binary hypothesis X with a discrete observation $\boldsymbol{Y}$. The following (for all sample values $\boldsymbol{y}$) are equivalent conditions for $v(\boldsymbol{y})$ to be a sufficient statistic:*

1. *A function u exists such that $\Lambda(\boldsymbol{y}) = u\big(v(\boldsymbol{y})\big)$.*

2. *For any given positive a priori probabilities, the a posteriori probabilities satisfy*

$$\mathsf{p}_{X|Y}(x \mid \boldsymbol{y}) = \mathsf{p}_{X|V}\big(x \mid v(\boldsymbol{y})\big). \tag{8.55}$$

3. *The likelihood ratio of* $\boldsymbol{y}$ *is the same as that of* $v(\boldsymbol{y})$, *i.e.,*

$$\Lambda(\boldsymbol{y}) = \frac{\mathsf{p}_{V|X}\big(v(\boldsymbol{y}) \mid 1\big)}{\mathsf{p}_{V|X}\big(v(\boldsymbol{y}) \mid 0\big)}. \tag{8.56}$$

Proof We show that condition 1 implies condition 2 implies condition 3 implies condition 1, which will complete the proof. To demonstrate condition 2 from condition 1, start with Bayes' law,

$$\begin{aligned}
\mathsf{p}_{X|Y}(0 \mid \boldsymbol{y}) &= \frac{p_0 \mathsf{p}_{Y|X}(\boldsymbol{y} \mid 0)}{p_0 \mathsf{p}_{Y|X}(\boldsymbol{y} \mid 0) + p_1 \mathsf{p}_{Y|X}(\boldsymbol{y} \mid 1)} \\
&= \frac{p_0}{p_0 + p_1 \Lambda(\boldsymbol{y})} = \frac{p_0}{p_0 + p_1 u\big(v(\boldsymbol{y})\big)},
\end{aligned}$$

where, in the second step we divided numerator and denominator by $\mathsf{p}_{Y|X}(\boldsymbol{y} \mid 0)$ and then used condition 1. This shows that $\mathsf{p}_{X|Y}(0 \mid \boldsymbol{y})$ is a function of $\boldsymbol{y}$ only through $v(\boldsymbol{y})$, i.e., that $\mathsf{p}_{X|Y}(0 \mid \boldsymbol{y})$ is the same for all $\boldsymbol{y}$ with a common value of $v(\boldsymbol{y})$. Thus $\mathsf{p}_{X|Y}(0 \mid \boldsymbol{y})$ is the conditional probability of $X = 0$ given $v(\boldsymbol{y})$. This establishes (8.55) for $X = 0$ (Exercise 8.4 spells this out in more detail). The case $X = 1$ follows since $\mathsf{p}_{X|Y}(1 \mid \boldsymbol{y}) = 1 - \mathsf{p}_{X|Y}(0 \mid \boldsymbol{y})$.

Next we show that condition 2 implies condition 3. Taking the ratio of (8.55) for $X = 1$ to that for $X = 0$,

$$\frac{\mathsf{p}_{X|Y}(1 \mid \boldsymbol{y})}{\mathsf{p}_{X|Y}(0 \mid \boldsymbol{y})} = \frac{\mathsf{p}_{X|V}\big(1 \mid v(\boldsymbol{y})\big)}{\mathsf{p}_{X|V}\big(0 \mid v(\boldsymbol{y})\big)}.$$

Applying Bayes' law to each numerator and denominator above and canceling $\mathsf{p}_Y(\boldsymbol{y})$ from both terms on the left, $\mathsf{p}_V\big(v(\boldsymbol{y})\big)$ from both terms on the right, and p_0 and p_1 throughout, we get (8.56).

Finally, going from condition 3 to condition 1 is obvious, since the right-hand side of (8.56) is a function of $\boldsymbol{y}$ only through $v(\boldsymbol{y})$. □

We next show that condition 2 in the above theorem means that the triplet $\boldsymbol{Y} \to V \to X$ is Markov. Recall from (6.37) that three ordered rv s, $\boldsymbol{Y} \to V \to X$ are said to be Markov if the joint PMF can be expressed as

$$\mathsf{p}_{YVX}(\boldsymbol{y}, v, x) = \mathsf{p}_Y(\boldsymbol{y})\, \mathsf{p}_{V|Y}(v \mid \boldsymbol{y})\, \mathsf{p}_{X|V}(x \mid v). \tag{8.57}$$

For the situation here, V is a function of $\boldsymbol{Y}$. Thus $\mathsf{p}_{V|Y}(v(\boldsymbol{y}) \mid \boldsymbol{y}) = 1$ and similarly, $\mathsf{p}_{YVX}(\boldsymbol{y}, v(\boldsymbol{y}), x) = \mathsf{p}_{YX}(\boldsymbol{y}, x)$. Thus the Markov property in (8.57) simplifies to

$$\mathsf{p}_{YX}(\boldsymbol{y}, x) = \mathsf{p}_Y(\boldsymbol{y})\; \mathsf{p}_{X|V}\big(x \mid v(\boldsymbol{y})\big). \tag{8.58}$$

Dividing both sides by $\mathsf{p}_Y(\boldsymbol{y})$ (which is positive for all $\boldsymbol{y}$ by convention), we get (8.55), showing that condition 2 of Theorem 8.2.8 holds if and only if $\boldsymbol{Y} \to V \to X$ is Markov.

Assuming that $\boldsymbol{Y} \to V \to X$ satisfies the Markov property we recall (see (6.38)) that $X \to V \to \boldsymbol{Y}$ also satisfies the Markov property. Thus

$$\mathsf{p}_{XV\boldsymbol{Y}}(x, v, \boldsymbol{y}) = \mathsf{p}_X(x)\, \mathsf{p}_{V|X}(v \mid x)\, \mathsf{p}_{\boldsymbol{Y}|V}(\boldsymbol{y} \mid v). \tag{8.59}$$

This is 0 for $v, \boldsymbol{y}$ such that $v \neq v(\boldsymbol{y})$. For all $\boldsymbol{y}$ and $v = v(\boldsymbol{y})$,

$$\mathsf{p}_{X\boldsymbol{Y}}(x, \boldsymbol{y}) = \mathsf{p}_X(x)\, \mathsf{p}_{V|X}(v(\boldsymbol{y}) \mid x)\, \mathsf{p}_{\boldsymbol{Y}|V}(\boldsymbol{y} \mid v(\boldsymbol{y})). \tag{8.60}$$

The relationship in (8.60) might seem almost intuitively obvious. It indicates that, conditional on v, the hypothesis says nothing more about $\boldsymbol{y}$. This intuition is slightly confused, since the notion of a sufficient statistic is that $\boldsymbol{y}$, given v, says nothing more about X. The equivalence of these two viewpoints lies in the symmetric form of the Markov relation, which says that X and $\boldsymbol{Y}$ are conditionally independent for any given $V = v$.

There are two other equivalent definitions of a sufficient statistic (also generalizable to multiple hypotheses and continuous observations) that are popular among statisticians. Their equivalence to each other is known as the Fisher–Neyman factorization theorem. This theorem, generalized to equivalence with the definitions already given, follows.

Theorem 8.2.9 (Fisher–Neyman factorization) *Let $V = v(\boldsymbol{Y})$ be a function of $\boldsymbol{Y}$ for a binary hypothesis X with a discrete observation $\boldsymbol{Y}$. Then (8.61) below is a necessary and sufficient condition for $v(\boldsymbol{y})$ to be a sufficient statistic.*

$$\mathsf{p}_{\boldsymbol{Y}|VX}(\boldsymbol{y} \mid v, x) = \mathsf{p}_{\boldsymbol{Y}|V}(\boldsymbol{y} \mid v) \qquad \textit{for all } x, v \textit{ such that } \mathsf{p}_{XV}(x, v) > 0. \tag{8.61}$$

Another necessary and sufficient condition is that functions $\phi(\boldsymbol{y})$ and $\widetilde{u}_x\big(v(\boldsymbol{y})\big)$ exist such that

$$\mathsf{p}_{\boldsymbol{Y}|X}(\boldsymbol{y} \mid x) = \phi(\boldsymbol{y})\, \widetilde{u}_x\big(v(\boldsymbol{y})\big). \tag{8.62}$$

Proof Equation (8.61) is simply the conditional PMF form for (8.59), so it is equivalent to condition 2 in Theorem 8.2.8.

Assume the likelihoods satisfy (8.62). Then taking the ratio of (8.62) for $x = 1$ to that for $x = 0$, the term $\phi(\boldsymbol{y})$ cancels out in the ratio, leaving $\Lambda(\boldsymbol{y}) = \widetilde{u}_1(v(\boldsymbol{y}))/\widetilde{u}_0(v(\boldsymbol{y}))$. For given functions $\widetilde{u}_0$ and $\widetilde{u}_1$, this is a function only of $v(\boldsymbol{y})$ which implies condition 1 of Theorem 8.2.8. Conversely, assume $v(\boldsymbol{y})$ is a sufficient statistic and let (p_0, p_1) be arbitrary positive a priori probabilities.[7] Choose $\phi(\boldsymbol{y}) = \mathsf{p}_{\boldsymbol{Y}}(\boldsymbol{y})$. Then to demonstrate (8.62), we must show that $\mathsf{p}_{\boldsymbol{Y}|X}(\boldsymbol{y}|x)/\mathsf{p}_{\boldsymbol{Y}}(\boldsymbol{y})$ is a funtion of $\boldsymbol{y}$ only through v:

$$\frac{\mathsf{p}_{\boldsymbol{Y}|X}(\boldsymbol{y} \mid x)}{\mathsf{p}_{\boldsymbol{Y}}(\boldsymbol{y})} = \frac{\mathsf{p}_{X|\boldsymbol{Y}}(x \mid y)}{p_x} = \frac{\mathsf{p}_{X|V}\big(x \mid v(\boldsymbol{y})\big)}{p_x},$$

where we have used (8.55) in the last step. This final quantity depends on $\boldsymbol{y}$ only through $v(\boldsymbol{y})$, completing the proof. □

[7] Classical statisticians can take (p_0, p_1) to be arbitrary positive numbers summing to 1 and take $\mathsf{p}_{\boldsymbol{Y}}(\boldsymbol{y}) = p_0\mathsf{p}_{\boldsymbol{Y}|X}(\boldsymbol{y}|0) + p_1\mathsf{p}_{\boldsymbol{Y}|X}(\boldsymbol{y}|1)$.

A sufficient statistic can be found from (8.62) in much the same way as from the likelihood ratio; namely, calculate the likelihood and isolate the part depending on x. The procedure is often trickier than with the likelihood ratio, where the irrelevant terms simply cancel. With some experience, however, one can work directly with the likelihood function to put it in the form of (8.62). As the proof here shows, however, the formalism of a complete probability model with a priori probabilities can simplify proofs concerning properties that do not depend on the a priori probabilities

Note that the particular choice of $\phi(\boldsymbol{y})$ used in the proof is only one example, which has the benefit of illustrating the probability structure of the relationship. One can use any choice of (p_0, p_1), and can also multiply $\phi(\boldsymbol{y})$ by any positive function of $v(\boldsymbol{y})$ and divide $\widetilde{u}_x\big(v(\boldsymbol{y})\big)$ by the same function to maintain the relationship in (8.62).

Continuous observations

Theorem 8.2.8 and the second part of Theorem 8.2.9 extend easily to continuous observations by simply replacing PMFs with PDFs where needed. There is an annoying special case here where, even though $\boldsymbol{Y}$ has a density, a sufficient statistic V might be discrete or mixed. An example of this is seen in Exercise 8.15. We assume that V has a density when $\boldsymbol{Y}$ does for notational convenience, but the discrete or mixed case follows with minor notational changes.

The first part of Theorem 8.2.9 however, becomes problematic in the sense that the conditional distribution of $\boldsymbol{Y}$ conditional on $\boldsymbol{V}$ is typically neither a PMF nor a PDF. The following example looks at a very simple case of this.

Example 8.2.10 (The simplest Gaussian vector detection problem) Let Z_1, Z_2 be IID Gaussian rv s, $\mathcal{N}(0, \sigma^2)$. Let $Y_1 = Z_1 + X$ and $Y_2 = Z_2 + X$, where X is independent of Z_1, Z_2 and takes the value ± 1. This is a special case of Example 8.2.4 with dimension $n = 2$ with $\boldsymbol{b} = (1, 1)^{\mathsf{T}}$ and $\boldsymbol{a} = (-1, -1)^{\mathsf{T}}$. As shown there, $v(\boldsymbol{y}) = 2(y_1 + y_2)$ is a sufficient statistic. The likelihood functions are constant over concentric circles around $(1, 1)$ and $(-1, -1)$, as illustrated in Figure 8.4.

Consider the conditional density $\mathsf{f}_{\boldsymbol{Y}|V}(\boldsymbol{y}|v)$. In the (y_1, y_2) plane, the condition $v = 2(y_1 + y_2)$ represents a straight line of slope -1 in the (y_1, y_2) plane, hitting the vertical axis at $y_2 = v/2$. Thus $\mathsf{f}_{\boldsymbol{Y}|V}(\boldsymbol{y} \mid v)$ (to the extent it has meaning) is impulsive on this line and zero elsewhere.

To see this more clearly, note that (Y_1, Y_2, V) are three rv s that do not have a joint density since each one can be represented as a function of the other two. Thus the usual rules for manipulating joint and conditional densities as if they were PMFs do not work. If we convert to a different basis, with the new basis vectors on the old main diagonals, then $\boldsymbol{Y}$ is deterministic in one basis direction and Gaussian in the other.

Returning to the general case, we could replace (8.61) by a large variety of special cases, and apparently it can be replaced by measure-theoretic Radon–Nikodym derivatives, but it is better to simply realize that (8.61) is not very insightful even in the discrete case and there is no point struggling to recreate it for the continuous case.[8] Exercise 8.5 develops a substitute for (8.61) in the continuous case, but we will make no use of it.

The following theorem summarizes the conditions we have verified for sufficient statistics in the continuous observation case. The theorem simply combines Theorems 8.2.8 and 8.2.9.

Theorem 8.2.11 *Let $V = v(\boldsymbol{Y})$ be a function of $\boldsymbol{Y}$ for a binary hypothesis X with a continuous observation $\boldsymbol{Y}$. The following (for all sample values $\boldsymbol{y}$) are equivalent conditions for $v(\boldsymbol{y})$ to be a sufficient statistic:*

1. *A function u exists such that $\Lambda(\boldsymbol{y}) = u\big(v(\boldsymbol{y})\big)$.*
2. *For any given positive a priori probabilities, the a posteriori probabilities satisfy*

$$\mathsf{p}_{X|\boldsymbol{Y}}(x \mid \boldsymbol{y}) = \mathsf{p}_{X|V}\big(x \mid v(\boldsymbol{y})\big). \tag{8.63}$$

3. *The likelihood ratio of $\boldsymbol{y}$ is the same as that of $v(\boldsymbol{y})$, i.e., if V has a density,*

$$\Lambda(\boldsymbol{y}) = \frac{\mathsf{f}_{V|X}\big(v(\boldsymbol{y}) \mid 1\big)}{\mathsf{f}_{V|X}\big(v(\boldsymbol{y}) \mid 0\big)}. \tag{8.64}$$

4. *Functions $\phi(\boldsymbol{y})$ and $\widetilde{u}_x\big(v(\boldsymbol{y})\big)$ exist such that*

$$\mathsf{f}_{\boldsymbol{Y}|X}(\boldsymbol{y} \mid x) = \phi(\boldsymbol{y})\,\widetilde{u}_x\big(v(\boldsymbol{y})\big). \tag{8.65}$$

8.3 Binary detection with a minimum-cost criterion

In many binary detection situations there are unequal positive costs, say C_0 and C_1, associated with a detection error given $X = 0$ and $X = 1$. For example, one kind of error in a medical prognosis could lead to serious illness and the other to an unneeded operation. A *minimum-cost* decision is defined as a test that minimizes the expected cost over the two types of errors with given a priori probabilities. As shown in Exercise 8.13, this is also a threshold test with the threshold $\eta = (p_0C_0)/(p_1C_1)$. The medical prognosis example above, however, illustrates that, although a cost criterion allows errors to be weighted according to seriousness, it provides no clue about evaluating seriousness, i.e., the cost of a life versus an operation (that might itself be life-threatening).

A more general version of this minimum-cost criterion assigns cost $C_{\ell k} \geq 0$ for $\ell, k \in \{0,1\}$ to the decision $\hat{x} = k$ when $X = \ell$. This only makes sense when $C_{01} > C_{00}$ and $C_{10} > C_{11}$, i.e., when it is more costly to make an error than not. With a little thought, it can be seen that for binary detection, this only complicates the notation. That is, visualize having a fixed cost of $p_0C_{00} + p_1C_{11}$ in the absense of detection errors. There is then an additional cost of $C_{01} - C_{00}$ when $X = 0$ and an error is made. Similarly, there is an additional cost of $C_{10} - C_{11}$ when $X = 1$ and an error is made. This is then a threshold test with threshold $\eta = p_0(C_{01} - C_{00})/p_1(C_{10} - C_{11})$.

[8] The major drawback of the lack of a well-behaved conditional density of Y given V is that the symmetric form of the Markov property is difficult to interpret without going more deeply into measure theory than appropriate or desirable here.

So far, we have looked at the minimum-cost rule, the MAP rule, and the ML rule. All of them are threshold tests where the decision is based on whether the likelihood ratio is above or below a threshold. For all of them, (8.12)–(8.16) also determine the probability of error conditional on $X = 0$ and $X = 1$.

There are other binary hypothesis testing problems in which threshold tests are in a sense inappropriate. In particular, the cost of an error under one or the other hypothesis could be highly dependent on the observation $\boldsymbol{y}$. For example, with the medical prognosis referred to above, one sample observation might indicate that the chance of disease is small but the possibility of death if untreated is very high. Another observation might indicate the same chance of disease, but little danger.

When we look at a threshold test where the cost depends on the observation, we see that the threshold $\eta = (p_0 C_0)/(p_1 C_1)$ depends on the observation $\boldsymbol{y}$ (since C_1 and C_0 depend on $\boldsymbol{y}$). Thus a minimum-cost test would compare the likelihood ratio to a data-dependent threshold, which is not what is meant by a threshold test.

This distinction between threshold tests and comparing likelihood ratios to data-dependent thresholds is particularly important for sufficient statistics. A sufficient statistic is a function of the observation $\boldsymbol{y}$ from which the likelihood ratio can be found. This does not necessarily allow a comparison with a threshold dependent on $\boldsymbol{y}$. In other words, in such situations, a sufficient statistic is not necessarily sufficient to make minimum-cost decisions.

The decision criteria above have been based on the assumption of a priori probabilities (although some of the results, such as Fisher–Neyman factorization, are independent of those probabilities). The next section describes a sense in which threshold tests are optimal even when no a priori probabilities are assumed.

8.4 The error curve and the Neyman–Pearson rule

In this section, we look at situations in which there is no need to assign a priori probabilities to binary hypotheses. In this case, any given detection rule gives rise to a pair of error probabilities, one conditional on $X = 0$ and the other on $X = 1$. It is always possible to make one of these error probabilities small at the expense of the other; an extreme example is to choose $\hat{x}(\boldsymbol{y}) = 0$ for all $\boldsymbol{y}$. Then the error probability conditional on $X = 0$ is 0 but that conditional on $X = 1$ is 1. Our objective here, however, is to make both error probabilities small, and, to be more specific, to minimize one error probability given a constraint on how large the other can be.

As a way of visualizing this problem, consider a plot using the two error probabilities as axes. Conceptually, each possible detection rule can then be considered as a point on this plot. Essentially, these pairs contain a lower left envelope, and the pairs on this envelope are 'optimal' in the sense that each pair above the envelope can 'be improved' by moving to the envelope with one or both error probabilities decreased at no expense to the other.

This lower left envelope (after being carefully defined) is called the *error curve*. Given any particular detection rule, there will be some point on this lower envelope for which each error probability is less than or equal to that for the given rule. Stated slightly differently, given any upper limit on one of the error probabilities, there is a point on this envelope that minimizes the other error probability. A decision rule that minimizes one error probability given a limit on the other is called a *Neyman–Pearson* rule [19]. Essentially then, the Neyman–Pearson rule corresponds to the points on the error curve.

In what follows, we define the error curve carefully and show that the set of points on it consist of a slight generalization of the class of threshold tests. We defined threshold tests earlier, using a deterministic tie-breaking rule if the observation $\boldsymbol{y}$ lies exactly on the threshold. The generalization required here is to use a randomized tie-breaking rule for observations that lie on the threshold.

Initially we assume that the observation $\boldsymbol{Y}$ is an n-**rv** with a positive joint density under each hypothesis. Later we also allow $\boldsymbol{Y}$ to be discrete or arbitrary .

Any test, i.e., any deterministic rule for selecting a binary hypothesis from a sample value $\boldsymbol{y}$ of $\boldsymbol{Y}$, can be viewed as a function mapping each possible observation $\boldsymbol{y}$ to 0 or 1. If we define A as the set of n-vectors $\boldsymbol{y}$ that are mapped to hypothesis 1 for a given test, then the test can be labeled by the corresponding set A.

Given a test A (i.e., a test labeled by the set A) and given $X = 0$, an error is made whenever $\boldsymbol{y} \in A$, i.e., whenever $\boldsymbol{y}$ is mapped into hypothesis 1. Similarly, an error is made if $X = 1$ and $\boldsymbol{y} \in A^c$. Thus the error probabilities, given $X = 0$ and $X = 1$, respectively are

$$\Pr\{\boldsymbol{Y} \in A \mid X = 0\}, \qquad \Pr\left\{\boldsymbol{Y} \in A^c \mid X = 1\right\}.$$

Note that these conditional error probabilities depend only on the test A and not on the (here undefined) a priori probabilities. We will abbreviate these error probabilities as

$$q_0(A) = \Pr\{\boldsymbol{Y} \in A \mid X = 0\}, \qquad q_1(A) = \Pr\left\{\boldsymbol{Y} \in A^c \mid X = 1\right\}.$$

If A is a threshold test, with threshold η, the set A is given by

$$A = \left\{\boldsymbol{y} : \frac{\mathsf{f}_{\boldsymbol{Y}|X}(\boldsymbol{y} \mid 1)}{\mathsf{f}_{\boldsymbol{Y}|X}(\boldsymbol{y} \mid 0)} \geq \eta\right\}.$$

Since threshold tests play a very special role here, we abuse the notation by using η in place of A to refer to a threshold test at η. That is, $q_0(\eta)$ is shorthand for $\Pr\left\{e_\eta \mid X = 0\right\}$ and $q_1(\eta)$ for $\Pr\left\{e_\eta \mid X = 1\right\}$. We can now characterize the relationship between threshold tests and other tests. The following lemma is illustrated in Figure 8.5.

Lemma 8.4.1 *Consider a two-dimensional plot in which the pair $\left(q_0(A), q_1(A)\right)$ is plotted for each A. Then for each theshold test η, $0 \leq \eta < \infty$, and each arbitrary test A, the point $\left(q_0(A), q_1(A)\right)$ lies in the closed half plane above and to the right of a straight line of slope $-\eta$ passing through the point $\left(q_0(\eta), q_1(\eta)\right)$.*

Proof In proving the lemma, we will use Theorem 8.1.2, demonstrating[9] the optimality of a threshold test for the MAP problem. As the proof here shows, that optimality for the MAP problem (which assumes a priori probabilities) implies some properties

[9] Recall that Exercise 8.12 generalizes Theorem 8.1.2 to essentially arbitrary likelihood distributions.

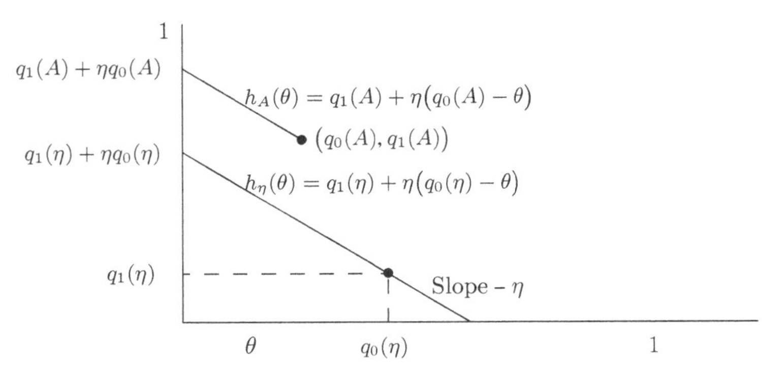

Figure 8.5 Illustration of Lemma 8.4.1.

relating $q_\ell(A)$ and $q_\ell(\eta)$ for $\ell = 0, 1$. These quantities are defined independently of the a priori probabilities, so the properties relating them are valid in the absence of a priori probabilities.

For any given η, consider the a priori probabilities (p_0, p_1) for which $\eta = p_0/p_1$. The overall error probability for test A using these a priori probabilities is then

$$\Pr\{e(A)\} \;=\; p_0q_0(A) + p_1q_1(A) \;=\; p_1\big[q_1(A) + \eta q_0(A)\big].$$

Similarly, the overall error probability for the threshold test η using the same a priori probabilities is

$$\Pr\{e(\eta)\} \;=\; p_0q_0(\eta) + p_1q_1(\eta) \;=\; p_1\left[q_1(\eta) + \eta q_0(\eta)\right].$$

This latter error probability is the MAP error probability for the given p_0, p_1, and is thus the minimum overall error probability (for the given p_0, p_1) over all tests. Thus

$$q_1(\eta) + \eta q_0(\eta) \;\le\; q_1(A) + \eta q_0(A).$$

As shown in the figure, $q_1(\eta) + \eta q_0(\eta)$ is the point at which the straight line $h_\eta(\theta) = q_1(\eta) + \eta\big(q_0(\eta) - \theta\big)$ of slope $-\eta$ through $\big(q_0(\eta), q_1(\eta)\big)$ crosses the ordinate axis. Similarly, $q_1(A) + \eta q_0(A)$ is the point at which the straight line $h_A(\theta) = q_1(A) + \eta\big(q_0(A) - \theta\big)$ through $\big(q_0(A), q_1(A)\big)$ of slope $-\eta$ crosses the ordinate axis. Thus all points on the second line, including $\big(q_0(A), q_1(A)\big)$ lie in the closed half plane above and to the right of all points on the first, completing the proof. □

The straight line of slope $-\eta$ through the point $\big(q_0(\eta), q_1(\eta)\big)$ has the equation $h_\eta(\theta) = q_1(\eta) + \eta(q_0(\eta) - \theta)$. Since the lemma is valid for all η, $0 \le \eta < \infty$, the point $\big(q_0(A), q_1(A)\big)$ for an arbitrary test A lies above and to the right of the straight line $h_\eta(\theta)$ for each η, $0 \le \eta < \infty$. The upper envelope of this family of straight lines is called the error curve, $u(\theta)$, defined by

$$u(\theta) \;=\; \sup_{0\le\eta<\infty} h_\eta(\theta) \;=\; \sup_{0\le\eta<\infty} q_1(\eta) + \eta(q_0(\eta) - \theta). \tag{8.66}$$

The lemma then asserts that for every test A (including threshold tests), we have $u(q_0(A)) \le q_1(A)$. Also, since every threshold test lies on one of these straight lines,

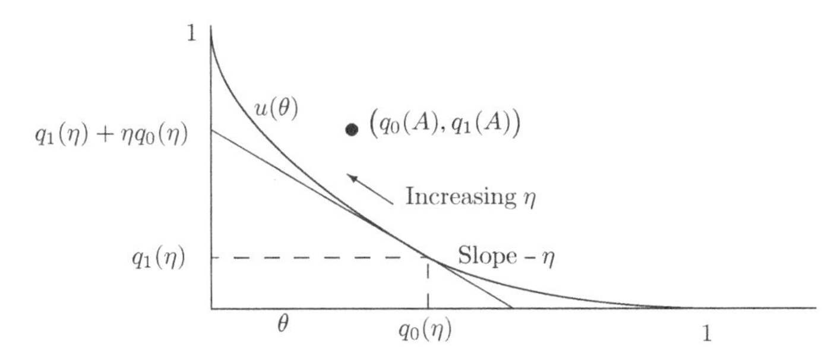

Figure 8.6 Illustration of the error curve $u(\theta)$ (see (8.66)). Note that $u(\theta)$ is convex, lies on or above its tangents, and on or below all tests. It can also be seen, either directly from the curve above or from the definition of a threshold test, that $q_1(\eta)$ is non-decreasing in η and $q_0(\eta)$ is non-increasing.

and therefore on or below the curve $u(\theta)$, we see that the pair $\big(q_0(\eta), q_1(\eta)\big)$ for each η must lie on the curve $u(\theta)$. Finally, since each of these straight lines forms a tangent of $u(\theta)$ and lies on or below $u(\theta)$, the function $u(\theta)$ is convex.[10] Figure 8.6 illustrates the error curve.

The error curve essentially gives us a tradeoff between the probability of error given $X = 0$ and that given $X = 1$. Threshold tests, since they lie on the error curve, provide optimal points for this tradeoff. Unfortunately, as we see in an example shortly, not all points on the error curve correspond to threshold tests. We will see later, however, that by generalizing threshold tests to randomized threshold tests, we can reach all points on the error curve.

Before proceeding to this example, note that Lemma 8.4.1 and the definition of the error curve apply to a broader set of models than discussed so far. First, the lemma still holds if $\mathsf{f}_{Y|X}(y \mid \ell)$ is zero over an arbitrary set of y for one or both hypotheses ℓ. The likelihood ratio $\Lambda(y)$ is infinite where $\mathsf{f}_{Y|X}(y \mid 1) > 0$ and $\mathsf{f}_{Y|X}(y \mid 0) = 0$. This means that $\Lambda(Y)$, conditional on $X = 1$, is defective, but this does not affect the proof of the lemma (see Exercise 8.15 for a fuller explanation of the effect of zero densities).

In addition, it can be seen that the lemma also holds if Y is an n-tuple of discrete rv s or if Y is a mixture of discrete and continuous components (such as being the sum of a discrete and continuous rv). With the help of Exercise 8.12, it can be seen that what is needed is for $\Lambda(Y)$ to be a rv conditional on $X = 0$ and a possibly-defective rv conditional on $X = 1$. We now summarize the results so far in a theorem.

[10] A region $\mathcal{A}$ of a vector space is said to be *convex* if, for every $x_1, x_2 \in \mathcal{A}$, $\lambda x_1 + (1-\lambda)x_2$ is also in $\mathcal{A}$ for all $\lambda \in (0, 1)$. Examples include the entire vector space and the space of probability vectors. A function $f(x)$ defined on a convex region $\mathcal{A}$ is convex if for every $x_1, x_2 \in \mathcal{A}$, $f(\lambda x_1 + (1-\lambda)x_2) \le \lambda f(x_1) + (1-\lambda)f(x_2)$ for all $\lambda \in (0, 1)$. Examples for one-dimensional regions are functions with non-negative second derivatives, and, more generally, functions lying on or above all their tangents. The latter allows for step discontinuities in the first derivative, which we soon see is required here.

Theorem 8.4.2 *Consider a binary hypothesis testing problem in which the likelihood ratio $\Lambda(\boldsymbol{Y})$ is a rv conditional on $X = 0$ and a possibly-defective rv conditional on $X = 1$. Then the error curve is convex, all threshold tests lie on the error curve, and all other tests lie on or above the error curve.*

The following example now shows that not all points on the error curve need correspond to threshold tests.

Example 8.4.3 A particularly simple example of a detection problem uses a discrete observation that has only two sample values, 0 and 1. Assume that

$$\mathsf{p}_{Y|X}(0 \mid 0) = \mathsf{p}_{Y|X}(1 \mid 1) = \frac{2}{3}, \qquad \mathsf{p}_{Y|X}(0 \mid 1) = \mathsf{p}_{Y|X}(1 \mid 0) = \frac{1}{3}.$$

In the communication context, this corresponds to a single use of a binary symmetric channel with crossover probability 1/3. The threshold test in (8.11), using PMFs in place of densities, is then

$$\Lambda(y) = \frac{\mathsf{p}_{Y|X}(y \mid 1)}{\mathsf{p}_{Y|X}(y \mid 0)} \underset{\hat{x}(y)=0}{\overset{\hat{x}(y)=1}{\gtrless}} \eta. \tag{8.67}$$

The only possible values for y are 1 and 0, and we see that $\Lambda(1) = 2$ and $\Lambda(0) = 1/2$. Under ML, i.e., with $\eta = 1$, we would choose $\hat{x}(y) = y$. For a threshold test with $\eta \le 1/2$, however, (8.67) says $\hat{x}(y) = 1$ for both $y = 1$ and $y = 0$, i.e., the decision is independent of the observation. We can understand this intuitively in the MAP case since $\eta \le 1/2$ means that the a priori probability $p_0 \le 1/3$ is so small that the observation cannot overcome this initial bias. In the same way, for $\eta > 2$, $\hat{x}(y) = 0$ for both $y = 1$ and $y = 0$. Summarizing,

$$\hat{x}(y) = \begin{cases} 1 & \text{for} \quad \eta \le 1/2, \\ y & \text{for} \quad 1/2 < \eta \le 2, \\ 0 & \text{for} \quad 2 < \eta. \end{cases} \tag{8.68}$$

For $\eta \le 1/2$, we have $\hat{x}(y) = 1$, so, for both $y = 0$ and $y = 1$, an error is made for $X = 0$ but not $X = 1$. Thus $q_0(\eta) = 1$ and $q_1(\eta) = 0$ for $\eta \le 1/2$. In the same way, the error probabilities for all η are given by

$$q_0(\eta) = \begin{cases} 1 & \text{for} \quad \eta \le 1/2, \\ 1/3 & \text{for} \quad 1/2 < \eta \le 2, \\ 0 & \text{for} \quad 2 < \eta, \end{cases} \qquad q_1(\eta) = \begin{cases} 0 & \text{for} \quad \eta \le 1/2, \\ 1/3 & \text{for} \quad 1/2 < \eta \le 2, \\ 1 & \text{for} \quad 2 < \eta. \end{cases} \tag{8.69}$$

We see that $q_0(\eta)$ and $q_1(\eta)$ are discontinuous functions of η, the first jumping down at $\eta = 1/2$ and $\eta = 2$, and the second jumping up. The error curve for this example is generated as the upper envelope of the family of straight lines $\{h_\eta(\theta);\ 0 \le \eta < \infty\}$,

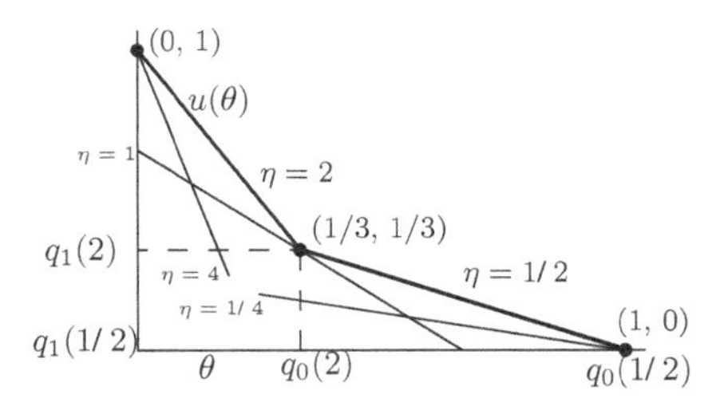

Figure 8.7 Illustration of the error curve $u(\theta)$ for Example 8.4.3. The three possible error pairs for threshold tests, $(q_0(\eta), q_1(\eta))$ are the indicated dots. That is, $(q_0(\eta), q_1(\eta)) = (1, 0)$ for $\eta \leq 1/2$. This changes to $(1/3, 1/3)$ for $1/2 < \eta \leq 2$ and to $(0, 1)$ for $\eta > 2$. The error curve (see (8.66)) for points to the right of $(1/3, 1/3)$ is maximized by the straight line of slope $-1/2$ through $(1, 0)$. Similarly, the error curve for points to the left of $(1/3, 1/3)$ is maximized by the straight line of slope -2 ($\eta = 2$) through $(1/3, 1/3)$. One can visualize the tangent lines as an inverted see-saw, first see-sawing around (0,1), then around $(1/3, 1/3)$, and finally around $(1, 0)$.

where $h_\eta(\theta)$ has slope $-\eta$ and passes through the point $(q_0(\eta), q_1(\eta))$. The equation for these lines, using (8.69), is as follows:

$$q_1(\eta) + \eta(q_0(\eta) - \theta) = \begin{cases} 0 + \eta(1 - \theta) & \text{for} \quad \eta \leq 1/2, \\ \frac{1}{3} + \eta(\frac{1}{3} - \theta) & \text{for} \quad 1/2 < \eta \leq 2, \\ 1 + \eta(0 - \theta) & \text{for} \quad 2 < \eta. \end{cases}$$

The straight lines for $\eta \leq 1/2$ pass through the point $(1, 0)$ and $\sup_{\eta \leq 1/2}[q_0(\eta) + \eta q_1(\eta)]$ for each $\theta \leq 1$ is achieved at $\eta = 1/2$. This is illustrated in Figure 8.7. Similarly, the straight lines for $1/2 < \eta \leq 2$ all pass through the point $(1/3, 1/3)$. For each $\theta < 1/3$, $u(\theta) = \sup_\eta \left[q_1(\eta) + \eta(q_0(\eta) - \theta)\right]$ is on the line of slope -2 through the point (1/3, 1/3). The three possible values for $(q_0(\eta), q_1(\eta))$ are shown by dots and the supremum of the tangents is shown by the piecewise linear curve.

Let us look more carefully at the tangent of slope $-1/2$ through the points $(1, 0)$ and (1/3, 1/3). This corresponds to the MAP test at $\eta = 1/2$, i.e., $p_0 = 1/3$. As seen in (8.68), this MAP test selects $\hat{x}(y) = 1$ for each y. This selection for $y = 0$ is a don't-care choice for the MAP test with $\eta = 1/2$ since $\Lambda(Y{=}0) = 1/2$. If the test selected $\hat{x}(1) = 0$ for $\eta = 1/2$ instead, the MAP error probability would not change, but the error probability given $X = 0$ would decrease to 1/3 and that for $X = 1$ would increase to 1/3.

It is not hard to verify (since there are only four tests, i.e., deterministic rules, for mapping a binary variable to another binary variable) that no test A can lie on an interior point of the straight line between (1/3, 1/3) and $(1, 0)$. However, if we use a randomized rule, mapping $y = 0$ to $\hat{x} = 0$ with probability ϕ and to $\hat{x} = 1$ with probability $1 - \phi$ (along with always mapping 1 to 1), then all points on the straight line from (1/3, 1/3) to $(1, 0)$ are achieved as ϕ goes from 0 to 1. In other words, a don't-care choice for MAP becomes an important choice in the tradeoff between q_0 and q_1.

In the same way, all points on the straight line from (0, 1) to (1/3, 1/3) can be achieved by a randomized rule that maps $y = 1$ to $\hat{x} = 0$ with probability ϕ and to $\hat{x} = 1$ with probability $1 - \phi$ (along with always mapping 0 to 0).

In the general case, the error curve will contain straight line segments whenever the CDF $\mathsf{F}_{\Lambda(Y)|X}(\eta|0)$ contains discontinuities. Such discontinuities always occur for discrete observations and sometimes for continuous observations (see Exercise 8.15). To see the need for these straight line segments, assume that $\mathsf{F}_{\Lambda(Y)|X}(\eta|0)$ has a discontinuity of size $\beta > 0$ at a given point $\eta \in (0, \infty)$. Then $\Pr\{\Lambda(\boldsymbol{Y}) = \eta \mid X{=}0\} = \beta$ and the MAP test at η has a don't-care region of probability β given $X = 0$. This means that if the MAP test is changed to resolve the don't-care case in favor of $X = 0$, then the error probability q_0 is decreased by β and the error probability q_1 is increased by $\beta\eta$.

Expressing this in terms of equations, the error probabilities, given $X = 0$ and $X = 1$, with a threshold test at η have been denoted as

$$q_0(\eta) = \Pr\{\Lambda(Y) \geq \eta \mid X = 0\}, \qquad q_1(\eta) = \Pr\{\Lambda(Y) < \eta \mid X = 1\}.$$

Modifying the threshold test to choose $\hat{x} = 0$ in the don't-care cases for MAP, the resulting error probabilities are denoted as $\widetilde{q}_0(\eta)$ and $\widetilde{q}_1(\eta)$, where

$$\widetilde{q}_0(\eta) = \Pr\{\Lambda(Y) > \eta \mid X = 0\}, \qquad \widetilde{q}_1(\eta) = \Pr\{\Lambda(Y) \leq \eta \mid X = 1\}. \tag{8.70}$$

What we just showed is that if $\Pr\{\Lambda(Y) = \eta \mid X = 0\} = \beta$, then

$$\widetilde{q}_0(\eta) = q_0(\eta) - \beta, \qquad \widetilde{q}_1(\eta) = q_1(\eta) + \eta\beta.$$

Lemma 8.4.1 is easily seen to be valid whichever way the MAP don't-care cases are resolved, and thus both $(q_0(\eta), q_1(\eta))$ and $(\widetilde{q}_0(\eta), \widetilde{q}_1(\eta))$ lie on the error curve. Since all tests lie above and to the right of the straight line of slope $-\eta$ through these points, the error curve has a straight line segment between these points. As explained in the example, points on the straight line between $(\widetilde{q}_0(\eta), \widetilde{q}_1(\eta))$ and $(q_0(\eta), q_1(\eta))$ can be realized by resolving these don't-care cases in a random fashion.

Definition 8.4.4 *A* ***randomized threshold rule*** *at η is a rule that detects $\hat{x} = 0$ for $\Lambda(\boldsymbol{y}) < \eta$, $\hat{x} = 1$ for $\Lambda(\boldsymbol{y}) > \eta$ and detects $\hat{x} = 0$ with some given probability ϕ, $0 < \phi < 1$ and $\hat{x} = 1$ with probability $1 - \phi$ for $\Lambda(\boldsymbol{y}) = \eta$.*

As in the example, if there is a straight line segment of slope $-\eta$ on the error curve, then a randomized threshold rule at η achieves each point on the straight line from $(\widetilde{q}_0(\eta), \widetilde{q}_1(\eta))$ to $(q_0(\eta), q_1(\eta))$ as ϕ goes from 0 to 1.

8.4.1 The Neyman–Pearson detection rule

Definition 8.4.5 *A* ***Neyman–Pearson rule*** *is a binary threshold rule (perhaps randomized) that, given the constraint that* $\Pr\{e \mid X{=}0\} \leq \theta$, *satisfies* $\Pr\{e \mid X{=}1\} \leq u(\theta)$, *where $u(\theta)$ is given in (8.66).*

Given the error curve, $u(\theta)$, we have essentially seen how to construct a Neyman–Pearson rule. The point $(\theta, u(\theta))$ either corresponds to a threshold test at one or more values of η, or it lies on a straight line of slope η between two threshold tests. In the first case, the threshold test at η is a Neyman–Pearson rule, and in the second case it is a randomized threshold rule, choosing $\hat{x} = 1$ for $\Lambda(\boldsymbol{y}) > \eta$, choosing $\hat{x} = 0$ for $\Lambda(\boldsymbol{y}) < \eta$, and choosing randomly when $\Lambda(\boldsymbol{y}) = \eta$.

This is summarized in the following theorem; the proof clarifies a couple of points that might have been confusing above.

Theorem 8.4.6 *Consider a binary hypothesis testing problem in which the likelihood ratio $\Lambda(Y)$ is a rv conditional on $X = 0$ and a possibly-defective rv conditional on $X = 1$. Then for any detection rule and any θ, $0 < \theta < 1$, the constraint that* $\Pr\{e \mid X = 0\} \leq \theta$ *implies that* $\Pr\{e \mid X = 1\} \geq u(\theta)$, *where $u(\theta)$ is the error curve. Furthermore,* $\Pr\{e \mid X = 1\} = u(\theta)$ *if the Neyman–Pearson rule for θ is used.*

Proof For any given η, the threshold test at η makes an error conditional on $X = 0$ if $\Lambda \geq \eta$. The threshold rule modified to map don't-care cases to 0 makes an error if $\Lambda > \eta$, i.e.,

$$\widetilde{q}_0(\eta) = 1 - \mathsf{F}_{\Lambda|X}(\eta|0), \qquad q_0(\eta) = 1 - \mathsf{F}_{\Lambda|X}(\eta|0) - \Pr\{\Lambda{=}\eta \mid X{=}0\}.$$

Since $\mathsf{F}_{\Lambda|X}(\eta|0)$ is a distribution function, it can have discontinuities, but if it has a discontinuity at η the discontinuity has size $\Pr\{\Lambda{=}\eta \mid X{=}0\}$ and must lie between $1 - q_0(\eta)$ and $1 - \widetilde{q}_0(\eta)$. This means that for any $\theta \in (0, 1)$, there is either an η for which $\theta = q_0(\eta)$ or there is an η for which $\widetilde{q}_0(\eta) \leq \theta < q_0(\eta)$. Thus either a threshold test at η can be used or a randomized threshold rule at η can be used for a Neyman–Pearson rule.

Theorem 8.4.2 shows that any deterministic test has error probabilities lying on or above the error curve. The question remaining is whether an arbitrary randomized rule can lie below the error curve. A randomized rule is a convex combination of a set of deterministic rules (i.e., it uses each of the deterministic tests with a probability that adds to 1 over the set of tests). However, a convex combination of points each above the error curve must also be above the error curve. □

8.4.2 The min–max detection rule

A final interesting example of a randomized threshold rule is the min–max rule. This is a rule that minimizes the maximum of the two error probabilities, $q_0(A)$ and $q_1(A)$. Now if the pair $(q_0(A), q_1(A))$ does not lie on the error curve, then the maximum of the two (or both together if they are equal) can be reduced. We have seen that all points on the error curve correspond either to threshold tests or to randomized threshold rules, so for any pair $(q_0(A), q_1(A))$ there is either a threshold test or randomized threshold rule that is at least as good in the min–max sense.

We can see geometrically that the error probabilities for a min–max rule must lie at the intersection of the error curve with the 45° line through the origin. Since this point is on the error curve, it can be achieved either by a threshold rule or a randomized threshold rule.

8.5 Finitely many hypotheses

Consider hypothesis testing problems with $\mathsf{M} \geq 2$ hypotheses. In a Bayesian setting, X is then a rv with the possible values $0, 1, \ldots, \mathsf{M}-1$ and a priori probabilities $p_0, \ldots, p_{\mathsf{M}-1}$.

Assume that, for each ℓ, there is a cost C_ℓ of making an incorrect decision, i.e., the cost is C_ℓ if $X = \ell$ and $\hat{x} \neq \ell$. We assume that this cost depends on neither the observation $\boldsymbol{y}$ nor the particular incorrect decision. A *minimum-cost test* is a decision rule that minimizes the expected cost of a decision.

In this section, we analyze minimum-cost tests for $\mathsf{M} \geq 2$ and compare them with $\mathsf{M} = 2$ as analyzed in Section 8.3. If $C_\ell = 1$ for all ℓ, then a minimum-cost test is a MAP test, so MAP is a special case of minimum cost. For an observed sample value $\boldsymbol{y}$ of the observation rv $\boldsymbol{Y}$, we can compare the expected costs of different decisions,

$$\mathsf{E}\left[\text{cost of } \hat{x}{=}k \mid \boldsymbol{Y}{=}\boldsymbol{y}\right] = \sum_{\ell \neq k} C_\ell \, \mathsf{p}_{X|\boldsymbol{Y}}(\ell \mid \boldsymbol{y}) \tag{8.71}$$

$$= \sum_{\ell} C_\ell \, \mathsf{p}_{X|\boldsymbol{Y}}(j \mid \boldsymbol{y}) - C_k \, \mathsf{p}_{X|\boldsymbol{Y}}(k \mid \boldsymbol{y}). \tag{8.72}$$

Since the sum on the right-hand side of (8.72) is common to all k, the expected cost of making decision k is minimized by choosing the hypothesis for which an error would be most costly, i.e.,

$$\hat{x}(\boldsymbol{y}) = \arg\max_k C_k \, \mathsf{p}_{X|\boldsymbol{Y}}(k \mid \boldsymbol{y}). \tag{8.73}$$

This conforms with intuition, which would weight both the a posteriori probability of the hypothesis and the cost of an error. It improves intuition by specifing the weighting. Assuming that the observation $\boldsymbol{y}$ has a positive probability density conditional on each hypothesis, we have

$$\mathsf{p}_{X|\boldsymbol{Y}}(k \mid \boldsymbol{y}) = \frac{p_k \mathsf{f}_{\boldsymbol{Y}|k}(\boldsymbol{y})}{\mathsf{f}_{\boldsymbol{Y}}(\boldsymbol{y})}, \qquad \hat{x}(\boldsymbol{y}) = \arg\max_k C_k \, p_k \mathsf{f}_{\boldsymbol{Y}|X}(\boldsymbol{y} \mid k). \tag{8.74}$$

Defining the likelihood ratio $\Lambda_{\ell k}(\boldsymbol{y}) = \mathsf{f}_{\boldsymbol{Y}|X}(\boldsymbol{y} \mid l)/\mathsf{f}_{\boldsymbol{Y}|X}(\boldsymbol{y} \mid k)$, the maximization in (8.74) can be viewed as a set of binary threshold comparisons, i.e., for all ℓ, k with $\ell > k$

$$\Lambda_{\ell k}(\boldsymbol{y}) = \frac{\mathsf{f}_{\boldsymbol{Y}|X}(\boldsymbol{y} \mid l)}{\mathsf{f}_{\boldsymbol{Y}|X}(\boldsymbol{y} \mid k)} \; \mathop{\gtrless}_{\hat{x}(y) \neq \ell}^{\hat{x}(y) \neq k} \; \frac{C_k p_k}{C_\ell p_\ell} = \eta_{\ell k}. \tag{8.75}$$

The meaning of (8.75) is that a threshold test between ℓ and k rejects either ℓ or k (while using the same arbitrary tie-breaking rule as in the binary case). After eliminating either ℓ or k as the minimum-cost decision, the remaining contender must still be tested against the other possibilities. If these binary choices are made sequentially, each time comparing two yet unrejected hypotheses, then each binary test eliminates one contender and $\mathsf{M} - 1$ binary tests are needed to establish the minimum-cost choice among the M hypotheses.

This procedure is essentially the same as an athletic tournament in which each binary game eliminates one team, with the ultimate survivor winning the tournament.

The meaning of the above tournament when some of the likelihoods are zero requires some interpretation. We assume as usual that the region for which all the likelihoods are 0 is ignored. Then if a likelihood ratio is 0 or ∞, the 0 likelihood hypothesis is eliminated. If the likelihood ratio is undefined (both likelihoods are 0), then it makes no difference which hypothesis is eliminated, since the other will eventually be eliminated

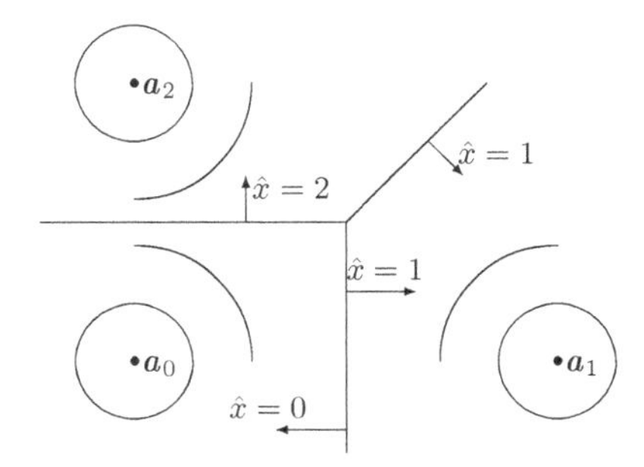

Figure 8.8 Decision regions for a 3-ary alphabet of vector signals in IID Gaussian noise. For ML detection, the decision regions are Voronoi regions, i.e., regions separated by perpendicular bisectors between the signal points. For the MAP case with unequal a priori probabilities, or the minimum-cost case, these perpendicular bisectors are shifted while remaining perpendicular to the straight lines between the signal points

by an hypothesis with positive likelihood. In essence, the tournament above, combined with common sense, finds a minimum-cost hypothesis.

Example 8.5.1 Consider the same communication situation as in Figure 8.4, but assume the source alphabet is $\{0, 1, \ldots, \mathsf{M}-1\}$. The letter to be transmitted, say ℓ, is mapped into a signal $\boldsymbol{a}_\ell = (a_{\ell 1}, a_{\ell 2}, \ldots, a_{\ell n})$. The noise in each dimension is IID Gaussian, $\mathcal{N}(0, \sigma^2)$. The LLR evaluated in (8.41), applied to any given pair, say $\ell > k$, of hypotheses is given by

$$\mathrm{LLR}_{\ell k}(\boldsymbol{y}) \quad = \quad \frac{(\boldsymbol{a}_\ell - \boldsymbol{a}_k)^{\mathsf{T}}}{\sigma^2}\left(\boldsymbol{y} - \frac{\boldsymbol{a}_\ell + \boldsymbol{a}_k}{2}\right) \quad \mathop{\gtrless}_{\hat{x}(\boldsymbol{y}) \neq \ell}^{\hat{x}(\boldsymbol{y}) \neq k} \quad \ln(\eta_{\ell k}). \tag{8.76}$$

The threshold $\eta_{\ell k}$ (as defined in (8.75)) is 1 for the ML case, p_k/p_ℓ for the MAP case, and $p_k C_k/(p_\ell C_\ell)$ for the minimum-cost case.

The geometric interpretation of this, in the space of observed vectors $\boldsymbol{y}$, is shown for the ML case in Figure 8.8. The decision threshold between each pair of hypotheses is the perpendicular bisector of the line joining the two signals. Note that $(\boldsymbol{a}_\ell - \boldsymbol{a}_k)^{\mathsf{T}}\boldsymbol{y}$ is a sufficient statistic for the binary test between ℓ and k. Thus if the dimension n of the observed vectors $\boldsymbol{y}$ is greater than $\mathsf{M}-1$, we can reduce the problem to $\mathsf{M}-1$ dimensions by transforming to a coordinate basis in which, for each j, $1 \le j \le m-1$, $\boldsymbol{a}_j - \boldsymbol{a}_0$ is a linear combination of $\mathsf{M}-1$ (or perhaps fewer) basis vectors. Using the theorem of irrelevance, the components of $\boldsymbol{y}$ in all other directions can be ignored.

Even after the simplification of representing an additive Gaussian noise M-ary detection problem in the appropriate $\mathsf{M}-1$ or fewer dimensions, calculating the probability of error for each hypothesis can be messy. For example, in Figure 8.8, $\Pr\{e \mid X{=}2\}$ is the probability that the noise, added to $\boldsymbol{a}_2$, carries the observation $\boldsymbol{y}$ outside of the region where $\hat{x} = 2$.

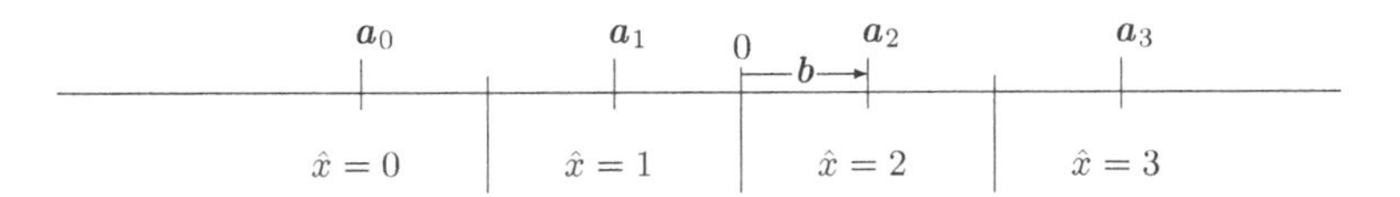

Figure 8.9 Decision regions for a 4-ary alphabet of colinear vector signals in IID Gaussian noise. The LLRs depend on $\boldsymbol{y}$ only through the one-dimensional sufficient statistic $\boldsymbol{a}^\mathsf{T}\boldsymbol{y}$. For ML detection, the decision regions are intervals over $\boldsymbol{a}^\mathsf{T}\boldsymbol{y}$ bounded by the points half way between the signal points.

This can be evaluated numerically using a two-dimensional integral over the given constraint region. In typical problems of this type, however, the boundaries of the constraint region are several standard deviations away from $\boldsymbol{a}_2$ and the union bound is often sufficient to provide a good upper bound to the error probability. That is, the error event, conditional on $X = k$, is the union of the events that the individual binary thresholds are crossed. Thus, using (8.76),

$$\Pr\{e \mid X{=}k\} \le \sum_{\ell \neq k} \Pr\left\{ \frac{(\boldsymbol{a}_\ell - \boldsymbol{a}_k)^\mathsf{T}}{\sigma^2} \left(\boldsymbol{y} - \frac{\boldsymbol{a}_\ell + \boldsymbol{a}_k}{2} \ge \ln \eta_{\ell k} \right) \right\}. \tag{8.77}$$

Using (8.44) to evaluate the terms on the right-hand side,

$$\Pr\{e \mid X{=}k\} \le \sum_{\ell \neq k} Q\left(\frac{\sigma \ln(\eta_{\ell k})}{\|\boldsymbol{a}_\ell - \boldsymbol{a}_k\|} + \frac{\|\boldsymbol{a}_\ell - \boldsymbol{a}_k\|}{2\sigma} \right). \tag{8.78}$$

Example 8.5.2 As a yet further simplified version of Example 8.5.1, suppose that $\mathsf{M} = 4$ and the four signal points are $-3\boldsymbol{b},\ -\boldsymbol{b},\ \boldsymbol{b},\ 3\boldsymbol{b}$ for some arbitrary vector $\boldsymbol{b}$. Then $\boldsymbol{a}_\ell = (2\ell - 3)\boldsymbol{b}$ for each signal $\boldsymbol{a}_\ell$, $0 \le \ell \le 3$. From (8.76), the LLRs and threshold tests are

$$\mathrm{LLR}_{\ell k}(\boldsymbol{y}) \;=\; \frac{2(\ell - k)\boldsymbol{b}^\mathsf{T}\big(\boldsymbol{y} - (\ell + k - 3)\boldsymbol{b}\big)}{\sigma^2} \;\mathop{\gtrless}_{\hat{x}(\boldsymbol{y}) \neq \ell}^{\hat{x}(\boldsymbol{y}) \neq k}\; \ln(\eta_{\ell k}). \tag{8.79}$$

As one might expect, each LLR depends on $\boldsymbol{y}$ only through $\boldsymbol{b}^\mathsf{T}\boldsymbol{y}$, so in this case $\boldsymbol{b}^\mathsf{T}\boldsymbol{y}$ is a sufficient statistic for all of the LLRs. If the test is ML, then $\ln(\eta_{\ell k}) = 0$ for all ℓ, k and the decision regions (in terms of $\boldsymbol{b}^\mathsf{T}\boldsymbol{y}$) are illustrated in Figure 8.9.

The error probability (for ML detection) can be worked out exactly for this example by inspection. For the outer two signal points, an error can be made only by noise exceeding $\|\boldsymbol{b}\|$ in the direction toward the inner points. For the two inner points, an error is made if the noise exceeds $\|\boldsymbol{b}\|$ in either direction. Thus

$$\Pr\{e \mid X = k\} = \begin{cases} Q\left(\frac{\|\boldsymbol{b}\|}{\sigma}\right) & \text{for } k = 0, 3, \\ 2Q\left(\frac{\|\boldsymbol{b}\|}{\sigma}\right) & \text{for } k = 1, 2. \end{cases}$$

8.5.1 Sufficient statistics with $\mathsf{M} \geq 2$ hypotheses

We have just shown that minimum-cost, MAP, and ML decisions can be made between $\mathsf{M} \geq 2$ hypotheses by using a tournament between binary threshold tests. Each of these binary tests might be simplified by the use of a sufficient statistic for that binary test. In Example 8.5.2, the same sufficient statistic worked for each binary test, and this section shows how to handle more general cases. What we want is a single function (often multi-dimensional) of the observation $\boldsymbol{y}$ that works as a sufficient statistic for all the binary tests.

Definition 8.5.3 *For hypothesis testing with $\mathsf{M} \geq 2$ hypotheses, $\{0, \ldots, \mathsf{M}{-}1\}$, a **sufficient statistic** is any function $v(\boldsymbol{y})$ of the observation $\boldsymbol{y}$ from which each binary likelihood ratio can be calculated, i.e., for which a set of functions $u_{\ell k}$; $0 \leq k < \ell \leq \mathsf{M} - 1$ exists such that $\Lambda_{\ell k}(\boldsymbol{y}) = u_{\ell k}(v(\boldsymbol{y}))$.*

As in the case of binary hypotheses, there are several equivalent conditions that could be used as a definition, and each provides its own insights. For a large number of hypotheses, this original definition is somewhat awkward because of the large number of binary comparisons replacing what is essentially the choice of a maximum term. The definition involves a set of $\mathsf{M}(\mathsf{M}-1)/2$ functions, $u_{\ell k}(v(\boldsymbol{y})$, but these functions are highly redundant since the corresponding set of likelihood ratios in the definition are composed from only M likelihoods. The equivalent definitions we now establish are virtually the same as in the binary case, but do not suffer this increase in complexity.

We state the following theorem in terms of continuous observations (thus again omitting the first part of Theorem 8.2.9, which could be included for the discrete case).

Theorem 8.5.4 *Let $V = v(\boldsymbol{Y})$ be a function of $\boldsymbol{Y}$ for an hypothesis rv X with M sample values and with a continuous observation $\boldsymbol{Y}$. The following (for all sample values $\boldsymbol{y}$) are equivalent conditions for $v(\boldsymbol{y})$ to be a sufficient statistic:*

1. *A set of functions $u_{\ell k}(v)$ exists such that $\Lambda_{\ell k}(\boldsymbol{y}) = u_{\ell k}\big(v(\boldsymbol{y})\big)$ for each $0 \leq \ell < k \leq \mathsf{M} - 1$.*
2. *For any given positive a priori probabilities, the a posteriori probabilities satisfy*

$$\mathsf{p}_{X|\boldsymbol{Y}}(k \mid \boldsymbol{y}) = \mathsf{p}_{X|V}\big(k \mid v(\boldsymbol{y})\big) \qquad \text{for } 0 \leq k \leq \mathsf{M} - 1. \tag{8.80}$$

3. *The likelihood ratio of $\boldsymbol{y}$ is the same as that of $v(\boldsymbol{y})$, i.e., if V has a density,*

$$\Lambda_{\ell k}(\boldsymbol{y}) = \frac{\mathsf{f}_{V|X}\big(v(\boldsymbol{y}) \mid \ell\big)}{\mathsf{f}_{V|X}\big(v(\boldsymbol{y}) \mid k\big)}. \tag{8.81}$$

4. *Functions $\phi(\boldsymbol{y})$ and $\widetilde{u}_k\big(v(\boldsymbol{y})\big)$ exist for $0 \leq k \leq \mathsf{M} - 1$ such that*

$$\mathsf{f}_{\boldsymbol{Y}|X}(\boldsymbol{y} \mid k) = \phi(\boldsymbol{y})\, \widetilde{u}_k\big(v(\boldsymbol{y})\big). \tag{8.82}$$

Proof The proof is very similar to those for Theorems 8.2.8 and 8.2.9, so identical details are omitted. We show that condition 1 implies condition 2 implies condition 3 implies condition 1, and then that condition 2 implies condition 4 implies condition 1. Starting with Bayes' law followed by the use of condition 1,

$$\begin{aligned} \mathsf{p}_{X|\boldsymbol{Y}}(k \mid \boldsymbol{y}) &= \frac{p_k \mathsf{f}_{\boldsymbol{Y}|X}(\boldsymbol{y} \mid k))}{\sum_{\ell=0}^{\mathsf{M}-1} p_\ell \mathsf{f}_{\boldsymbol{Y}|X}(\boldsymbol{y} \mid \ell))} \\ &= \frac{p_k}{\sum_{\ell=0}^{\mathsf{M}-1} p_\ell \Lambda_{\ell k}(\boldsymbol{y})} = \frac{p_k}{\sum_{\ell=0}^{\mathsf{M}-1} p_\ell u_{\ell k}\big(v(\boldsymbol{y})\big)}, \end{aligned}$$

where $u_{kk}(v)$ is taken to be 1 and $u_{\ell k}\big(v(\boldsymbol{y})\big) = \big[u_{k\ell}\big(v(\boldsymbol{y})\big)\big]^{-1}$ for $\ell > k$. This shows that $\mathsf{p}_{X|\boldsymbol{Y}}(0 \mid \boldsymbol{y})$ is a function of $\boldsymbol{y}$ only through $v(\boldsymbol{y})$, i.e., that $\mathsf{p}_{X|\boldsymbol{Y}}(0 \mid \boldsymbol{y})$ is the same for all $\boldsymbol{y}$ with a common value of $v(\boldsymbol{y})$. Thus $\mathsf{p}_{X|\boldsymbol{Y}}(\ell \mid \boldsymbol{y})$ is the conditional probability of $X = \ell$ given $V = v(\boldsymbol{y})$. This establishes (8.80).

Next we show that condition 2 implies condition 3. Taking the ratio of (8.80) for $X = \ell$ to that for $X = k$,

$$\frac{\mathsf{p}_{X|\boldsymbol{Y}}(\ell \mid \boldsymbol{y})}{\mathsf{p}_{X|\boldsymbol{Y}}(k \mid \boldsymbol{y})} = \frac{\mathsf{p}_{X|V}\big(\ell \mid v(\boldsymbol{y})\big)}{\mathsf{p}_{X|V}\big(k \mid v(\boldsymbol{y})\big)}.$$

Applying Bayes' law to each numerator and denominator above and canceling $\mathsf{f}_{\boldsymbol{Y}}(\boldsymbol{y})$ from both terms on the left, $\mathsf{f}_V\big(v(\boldsymbol{y})\big)$ from both terms on the right, and p_ℓ and p_k throughout, we get (8.81). The argument is essentially the same if V has a PMF or a mixed distribution.

Going from condition 3 to condition 1 is obvious, since the right-hand side of (8.81) is a function of $\boldsymbol{y}$ only through $v(\boldsymbol{y})$.

To prove condition 4 from condition 2, assume that $v(\boldsymbol{y})$ satisfies (8.80) for given $\{p_0, \ldots, p_{\mathsf{M}-1}\}$. Choose $\phi(\boldsymbol{y}) = \mathsf{f}_{\boldsymbol{Y}}(\boldsymbol{y})$. Then to satisfy (8.82), we must have

$$\widetilde{u}_\ell\big(v(\boldsymbol{y})\big) = \frac{\mathsf{f}_{\boldsymbol{Y}|X}(\boldsymbol{y} \mid \ell)}{\mathsf{f}_{\boldsymbol{Y}}(\boldsymbol{y})} = \frac{\mathsf{p}_{X|\boldsymbol{Y}}(\ell \mid y)}{p_\ell} = \frac{\mathsf{p}_{X|V}\big(\ell \mid v(\boldsymbol{y})\big)}{p_\ell}.$$

This indeed depends on $\boldsymbol{y}$ only through $v(\boldsymbol{y})$.

Finally, assume $v(\boldsymbol{y})$ satisfies (8.82). Taking the ratio of (8.82) for $x = \ell$ to that for $x = k$, the term $\phi(\boldsymbol{y})$ cancels. Thus condition 1 is satisfied by choosing $u_{\ell k} = \widetilde{u}_\ell / \widetilde{u}_k$. □

Example 8.5.5 Consider the four colinear signals in Gaussian noise of Example 8.5.2 again. Assuming that the observation is n-dimensional, we have

$$\begin{aligned} \mathsf{f}_{\boldsymbol{Y}|X}(\boldsymbol{y} \mid \ell) &= (2\pi\sigma^2)^{n/2} \exp \sum_{j=1}^{n} \frac{-(y_j - (2\ell - 3)b_j)^2}{2\sigma^2} \\ &= \left[(2\pi\sigma^2)^{n/2} \exp \sum_{j=1}^{n} \frac{-y_j^2}{2\sigma^2}\right] \left[\exp \frac{2(2\ell-3)\boldsymbol{b}^\mathsf{T}\boldsymbol{y} - (2\ell-3)^2 \boldsymbol{b}^\mathsf{T}\boldsymbol{b}}{2\sigma^2}\right]. \end{aligned}$$

The first factor above is a function of $\boldsymbol{y}$ and does not depend on ℓ (it is simply the density of the n-dimensional noise) and the second term is a function of both ℓ and $\boldsymbol{b}^\mathsf{T}\boldsymbol{y}$. Thus these terms can be chosen as $\phi(\boldsymbol{y})$ and $\widetilde{u}_\ell(v(\boldsymbol{y}))$ in (8.82). Note that this choice is just as natural as the choice of $\mathsf{f}_{\boldsymbol{Y}}(\boldsymbol{y})$ for $\phi(\boldsymbol{y})$ used in the proof of Theorem 8.5.4.

8.5.2 More general minimum-cost tests

Section 8.3 discussed minimum-cost decisions for binary hypotheses. In one formulation (the one followed earlier in this section), C_ℓ is taken to be the cost of an error when $X = \ell$. In the other, more general, formulation, $C_{\ell,k}$ is taken to be the cost for $X = \ell$ and $\hat{x} = k$. In this subsection we consider that more general formulation for $\mathsf{M} \geq 2$ hypotheses. In the binary case, the more general formulation simply complicated the notation. Here it has a more fundamental effect.

For this more general set of costs, the expected cost of deciding on a given hypothesis, k, $0 \leq k \leq \mathsf{M}-1$ is given by

$$\mathsf{E}\left[\text{cost of } \hat{x}{=}k \mid \boldsymbol{Y}{=}\boldsymbol{y}\right] = \sum_{\ell=0}^{\mathsf{M}-1} C_{\ell,k}\, \mathsf{p}_{X|\boldsymbol{Y}}(\ell \mid \boldsymbol{y}).$$

The minimum-cost decision minimizes this sum for each observation $\boldsymbol{y}$, i.e.,

$$\hat{x}(\boldsymbol{y}) = \arg\min_k \sum_{\ell=0}^{\mathsf{M}-1} C_{\ell,k}\, \mathsf{p}_{X|\boldsymbol{Y}}(\ell \mid \boldsymbol{y}). \tag{8.83}$$

We use the usual convention of choosing the largest numbered decision in the case of a tie.

This decision cannot be put in the form of a tournament of binary threshold decisions, although it could be viewed as a tournament of more complicated binary decisions, where the binary decision between k and j, say, would involve $C_{\ell,k}$ and $C_{\ell,j}$ for all ℓ.

This decision can be made in terms of a sufficient statistic, $v(\boldsymbol{y})$, for the observation $\boldsymbol{y}$. To see this, we simply replace $\mathsf{p}_{X|\boldsymbol{Y}}(\ell \mid \boldsymbol{y})$ in (8.83) by $\mathsf{p}_{X|V}\big(\ell \mid v(\boldsymbol{y})\big)$.

This substitution is justified by condition 2 of Theorem 8.5.4. There does not seem to be much more of general interest that can be said of these more general minimum-cost decisions.

8.6 Summary

The mathematical problem studied in this chapter is very simple. A probability space contains a discrete rv X with a finite set of sample values (hypotheses) and contains a rv $\boldsymbol{Y}$ whose sample values are called observations. We are to choose a function $\hat{x}(\boldsymbol{y})$ that maps each observation $\boldsymbol{y}$ into an hypothesis. This function is called a decision, a detected value, or a choice of hypothesis, and there are a number of possible criteria for choosing this function.

The simplest criterion is the MAP criterion, where for each $\boldsymbol{y}$, $\hat{x}(\boldsymbol{y})$ maximizes $\Pr\{X = \hat{x}(\boldsymbol{y}) \mid \boldsymbol{Y} = \boldsymbol{y}\}$. In the case of a tie, the largest such $\hat{x}(\boldsymbol{y})$ is arbitrarily chosen. This criterion (almost obviously) also maximizes the overall probability of correct choice, averaged over $\boldsymbol{Y}$.

A slightly more general criterion is the minimum-cost criterion, where for each hypothesis k, there is a cost C_k of choosing incorrectly (choosing $\hat{x}(\boldsymbol{y}) \neq k$) when $X = k$. The minimum-cost decision chooses $\hat{x}(\boldsymbol{y})$ to minimize expected cost (with the same tie-breaking rule as before). MAP is thus a special case of minimum-cost for which $C_k = 1$ for each k.

If X is binary, then minimum-cost and MAP detection are threshold tests where the likelihood ratio $\Lambda(\boldsymbol{y}) = \mathsf{f}_{\boldsymbol{Y}|X}(\boldsymbol{y} \mid 1)/\mathsf{f}_{\boldsymbol{Y}|X}(\boldsymbol{y} \mid 0)$ is compared with a numerical threshold η. For minimum-cost, $\eta = C_0 p_0 / C_1 p_1$ where (p_0, p_1) are the a priori probabilities of X.

Both threshold tests and the corresponding expected costs or error probabilities can often be simplified by a function $v(\boldsymbol{y})$ called a sufficient statistic and defined by the property that $\Lambda(\boldsymbol{y})$ is a function of $v(\boldsymbol{y})$. There are a number of equivalent definitions of sufficient statistics for binary hypotheses specified by Theorems 8.2.8, 8.2.9, and 8.2.11. Several examples, including detection of a binary signal in various forms of Gaussian noise and detection of the rate of a Poisson process, illustrate the usefulness of sufficient statistics.

There are a number of both practical and theoretical reasons for having decision criteria that do not depend on a priori probabilities. One such criterion is the ML criterion, which is the same as the MAP criterion with $p_0 = p_1$. A more fundamental criterion is the Neyman–Pearson criterion, which minimizes $\Pr\{e \mid X = 1\}$ subject to a given upper bound on $\Pr\{e \mid X = 0\}$. As the bound on $\Pr\{e \mid X = 0\}$ is varied from 0 to 1, the minimum value of $\Pr\{e \mid X = 1\}$ traces out a curve called the error curve. The points on the error curve are the threshold tests as η varies from 0 to ∞, combined with randomized rules to break ties in the threshold rule, thus providing flexibility in proportioning between error types.

For hypothesis testing with $\mathsf{M} \geq 2$ hypotheses, MAP tests and minimum-cost tests can be realized by a tournament of threshold tests. This provides a straightforward generalization of the results for binary hypothesis testing, and is often useful in understanding the regions of observations that are mapped into each of the hypotheses. It is often more straightforward to view MAP and minimum-cost tests simply as optimizations over the set of hypotheses. The equivalent definitions of sufficient statistics generalize for $\mathsf{M} \geq 2$, and some of those definitions are useful in viewing the tests as straightforward optimizations. The Neyman–Pearson rule can also be generalized for $\mathsf{M} \geq 2$. It then minimizes the error probability of one hypothesis given upper bounds on all the others.

8.7 Exercises

Exercise 8.1 In this exercise, we evaluate $\Pr\{e_\eta \mid X = \boldsymbol{a}\}$ and $\Pr\{e_\eta \mid X = \boldsymbol{b}\}$ for binary detection from vector signals in Gaussian noise directly from (8.40) and (8.41).

(a) By using (8.40) for each sample value $\boldsymbol{y}$ of $\boldsymbol{Y}$, show that

$$\mathsf{E}\,[\mathrm{LLR}(\boldsymbol{Y}) \mid X{=}\boldsymbol{a}] = \frac{-(\boldsymbol{b}-\boldsymbol{a})^{\mathsf{T}}(\boldsymbol{b}-\boldsymbol{a})}{2\sigma^2}.$$

Hint: Note that, given $X = \boldsymbol{a}$, $\boldsymbol{Y} = \boldsymbol{a} + \boldsymbol{Z}$.

(b) Defining $\gamma = \|\boldsymbol{b} - \boldsymbol{a}\|/(2\sigma)$, show that

$$\mathsf{E}\left[\mathrm{LLR}(\boldsymbol{Y}) \mid X{=}\boldsymbol{a}\right] = -2\gamma^2.$$

(c) Show that

$$\mathsf{VAR}\left[\mathrm{LLR}(\boldsymbol{Y}) \mid X{=}\boldsymbol{a}\right] = 4\gamma^2.$$

Hint: Note that the fluctuation of $\mathrm{LLR}(\boldsymbol{Y})$ conditional on $X = \boldsymbol{a}$ is $(1/\sigma^2)(\boldsymbol{b} - \boldsymbol{a})^{\mathsf{T}}\boldsymbol{Z}$.

(d) Show that, conditional on $X = \boldsymbol{a}$, $\mathrm{LLR}(\boldsymbol{Y}) \sim \mathrm{N}(-2\gamma^2, 4\gamma^2)$. Show that, conditional on $X = \boldsymbol{a}$, $\mathrm{LLR}(\boldsymbol{Y})/2\gamma \sim \mathrm{N}(-\gamma, 1)$.

(e) Show that the first half of (8.44) is valid, i.e., that

$$\Pr\left\{e_\eta \mid X{=}\boldsymbol{a}\right\} = \Pr\{\mathrm{LLR}(\boldsymbol{Y}) \geq \ln\eta \mid X{=}\boldsymbol{a}\} = Q\left(\frac{\ln\eta}{2\gamma} + \gamma\right).$$

(f) By essentially repeating (a)–(e), show that the second half of (8.44) is valid, i.e., that

$$\Pr\left\{e_\eta \mid X{=}\boldsymbol{b}\right\} = Q\left(\frac{-\ln\eta}{2\gamma} + \gamma\right).$$

Exercise 8.2 (Generalization of Exercise 8.1) **(a)** Let $U = (\boldsymbol{b} - a)^{\mathsf{T}}K_Z^{-1}\boldsymbol{Y}$. Find $\mathsf{E}\left[U \mid X{=}\boldsymbol{a}\right]$ and $\mathsf{E}\left[U \mid X{=}\boldsymbol{b}\right]$.

(b) Find the conditional variance of U conditional on $X = \boldsymbol{a}$. Hint: See the hint in Exercise 8.1(c).

(c) Give the threshold test in terms of the sample value u of U, and evaluate $\Pr\left\{e_\eta \mid X{=}\boldsymbol{a}\right\}$ and $\Pr\left\{e_\eta \mid X{=}\boldsymbol{b}\right\}$ from this and (b). Show that your answer agrees with (8.53).

(d) Explain what happens if $[K_{\mathbf{Z}}]$ is singular. Hint: You must look at two separate cases, depending on the vector $\boldsymbol{b} - \boldsymbol{a}$.

Exercise 8.3 **(a)** Let $\boldsymbol{Y}$ be a discrete observation **rv** for a binary detection problem, let $\boldsymbol{y}$ be the observed sample value. Let $v = v(\boldsymbol{y})$ be a sufficient statistic and let V be the corresponding rv. Show that $\Lambda(\boldsymbol{y})$ is equal to $\mathsf{p}_{V|X}(v(\boldsymbol{y}) \mid \boldsymbol{b})/\mathsf{p}_{V|X}(v(\boldsymbol{y}) \mid \boldsymbol{a})$. In other words, show that the likelihood ratio of a sufficient statistic is the same as the likelihood ratio of the original observation.

(b) Show that this also holds for the ratio of probability densities if V is a continuous rv or **rv**.

Exercise 8.4 **(a)** Show that if $v(\boldsymbol{y})$ is a sufficient statistic according to condition 1 of Theorem 8.2.8, then

$$p_{X|\boldsymbol{Y}V}\left(x \mid \boldsymbol{y}, v(\boldsymbol{y})\right) = p_{X|\boldsymbol{Y}}(x \mid \boldsymbol{y}).$$

(b) Consider the subspace of events conditional on $V(\boldsymbol{y}) = v$ for a given v. Show that for $\boldsymbol{y}$ such that $v(\boldsymbol{y}) = v$,

$$p_{X|\boldsymbol{Y}V}\left(x \mid \boldsymbol{y}, v(\boldsymbol{y})\right) = p_{X|V}(x \mid v).$$

(c) Explain why this argument is valid whether $\boldsymbol{Y}$ is a discrete or continuous **rv** and whether V is discrete, continuous or part discrete and part continuous.

Exercise 8.5 (**a**) Let $\boldsymbol{Y}$ be a discrete observation **rv** and let $v(\boldsymbol{y})$ be a function of the sample values of $\boldsymbol{Y}$. Show that

$$\mathsf{p}_{\boldsymbol{Y}|VX}\big(\boldsymbol{y} \mid v(\boldsymbol{y}), x\big) = \frac{\mathsf{p}_{\boldsymbol{Y}|X}(\boldsymbol{y} \mid x)}{\mathsf{p}_{V|X}\big(v(\boldsymbol{y}) \mid x\big)}.$$

(**b**) Using Theorem 8.2.8, show that the above fraction is independent of X if and only if $v(\boldsymbol{y})$ is a sufficient statistic.

(**c**) Now assume that $\boldsymbol{Y}$ is a continuous observation **rv**, that $v(\boldsymbol{y})$ is a given function, and that $V = v(\boldsymbol{Y})$ has a probability density. Define

$$\mathsf{f}_{\boldsymbol{Y}|VX}(\boldsymbol{y} \mid v(\boldsymbol{y}), x) = \frac{\mathsf{f}_{\boldsymbol{Y}|X}(\boldsymbol{y} \mid x)}{\mathsf{f}_{V|X}\big(v(\boldsymbol{y}) \mid x\big)}.$$

One can interpret this as a strange kind of probability density on a conditional sample space, but it is more straightforward to regard it simply as a fraction. Show that $v(\boldsymbol{y})$ is a sufficient statistic if and only if this fraction is independent of x. Hint: Model your derivation on that in (b), modifying (b) as necessary to do this.

Exercise 8.6 (**a**) Consider Example 8.2.5, and let $\boldsymbol{Z} = [A]\boldsymbol{W}$, where $\boldsymbol{W} \sim \mathrm{N}(0, I)$ is normalized IID Gaussian and $[A]$ is non-singular. The observation **rv** $\boldsymbol{Y}$ is $\boldsymbol{a} + \boldsymbol{Z}$ given $X = \boldsymbol{a}$ and is $\boldsymbol{b} + \boldsymbol{Z}$ given $X = \boldsymbol{b}$. Suppose the observed sample value $\boldsymbol{y}$ is transformed into $\boldsymbol{v} = [A^{-1}]\boldsymbol{y}$. Explain why $\boldsymbol{v}$ is a sufficient statistic for this detection problem (and thus why MAP detection based on $\boldsymbol{v}$ must yield the same decision as that based on $\boldsymbol{y}$).

(**b**) Consider the detection problem where $\boldsymbol{V} = [A^{-1}]\boldsymbol{a} + \boldsymbol{W}$ given $X = \boldsymbol{a}$ and $[A^{-1}]\boldsymbol{b} + \boldsymbol{W}$ given $X = \boldsymbol{b}$. Find LLR($\boldsymbol{v}$) for a sample value $\boldsymbol{v}$ of $\boldsymbol{V}$. Show that this is the same as the LLR for a sample value $\boldsymbol{y} = [A]\boldsymbol{v}$ of $\boldsymbol{Y}$.

(**c**) Find $\Pr\{e \mid X{=}\boldsymbol{a}\}$ and $\Pr\{e \mid X{=}\boldsymbol{b}\}$ for the detection problem in (b) by using the results of Example 8.2.4. Show that your answer agrees with (8.53). Note: The methodology here is to transform the observed sample value to make the noise IID; this approach is often both useful and insightful.

Exercise 8.7 Binary frequency shift keying (FSK) with incoherent reception can be modeled in terms of a four-dimensional observation vector $\boldsymbol{Y} = (Y_1, Y_2, Y_3, Y_4)^{\mathsf{T}}$, where $\boldsymbol{Y} = \boldsymbol{U} + \boldsymbol{Z}$. The L-rv $\boldsymbol{Z} \sim \mathrm{N}(0, \sigma^2[I])$ is independent of X. Under $X = 0$, $\boldsymbol{U} = (a\cos\phi, a\sin\phi, 0, 0)^{\mathsf{T}}$, whereas under $X = 1$, $\boldsymbol{U} = (0, 0, a\cos\phi, a\sin\phi)^{\mathsf{T}}$. The rv ϕ is uniformly distributed between 0 and 2π and is independent of X and $\boldsymbol{Z}$. The a priori probabilities are $p_0 = p_1 = 1/2$.

(**a**) Convince yourself from the circular symmetry of the situation that the ML receiver calculates the sample values v_0 and v_1 of $V_0 = Y_1^2 + Y_2^2$ and $V_1 = Y_3^2 + Y_4^2$ and chooses $\hat{x} = 0$ if $v_0 \geq v_1$ and chooses $\hat{x} = 1$ otherwise.

(**b**) Find $\Pr\{V_1 > v_1 \mid X{=}0\}$ as a function of $v_1 > 0$.

(**c**) Show that

$$p_{Y_1, Y_2|X,\phi}(y_1, y_2 \mid 0, 0) = \frac{1}{2\pi\sigma^2} \exp\left[\frac{-y_1^2 - y_2^2 + 2y_1 a - a^2}{2\sigma^2}\right].$$

(d) Show that

$$\Pr\{V_1 > V_0 \mid X{=}0, \phi{=}0\} = \int \mathsf{f}_{Y_1,Y_2|X,\phi}(y_1, y_2 \mid 0, 0) \Pr\left\{V_1 > y_1^2 + y_2^2\right\} dy_1 dy_2.$$

Show that this is equal to $(1/2)\exp(-a^2/(4\sigma^2))$.

(e) Explain why this is the probability of error (i.e., why the event $V_1 > V_0$ is independent of ϕ), and why $\Pr\{e \mid X{=}0\} = \Pr\{e \mid X{=}1\}$.

Exercise 8.8 Binary FSK on a Rayleigh fading channel can be modeled in terms of a four-dimensional observation vector $\boldsymbol{Y} = (Y_1, Y_2, Y_3, Y_4)^{\mathsf{T}}$, where $\boldsymbol{Y} = \boldsymbol{U} + \boldsymbol{Z}$. The L-rv $\boldsymbol{Z} \sim \mathrm{N}(0, \sigma^2[I])$ is independent of $\boldsymbol{U}$. Under $X = 0$, $\boldsymbol{X} = (U_1, U_2, 0, 0)^{\mathsf{T}}$, whereas under $X = 1$, $\boldsymbol{U} = (0, 0, U_3, U_4)^{\mathsf{T}}$. The rv s $U_i \sim \mathrm{N}(0, a^2)$ are IID. The a priori probabilities are $p_0 = p_1 = 1/2$.

(a) Convince yourself from the circular symmetry of the situation that the ML receiver calculates sample values v_0 and v_1 for $V_0 = Y_1^2 + Y_2^2$ and $V_1 = Y_3^2 + Y_4^2$ and chooses $\hat{x} = 0$ if $v_0 > v_1$ and chooses $\hat{x} = 1$ otherwise.

(b) Find $\mathsf{f}_{V_0|X}(v_0 \mid 0)$ and find $\mathsf{f}_{V_1|X}(v_1 \mid 0)$.

(c) Let $W = V_0 - V_1$ and find $\mathsf{f}_W(w \mid X{=}0)$.

(d) Show that $\Pr\{e \mid X{=}0\} = [2 + a^2/\sigma^2)]^{-1}$. Explain why this is also the unconditional probability of an incorrect decision.

Exercise 8.9 A disease has two strains, 0 and 1, which occur with a priori probabilities p_0 and $p_1 = 1 - p_0$ respectively.

(a) Initially, a rather noisy test was developed to find which strain is present for patients with the disease. The output of the test is the sample value y_1 of a rv Y_1. Given strain 0 ($X = 0$), $Y_1 = 5 + Z_1$, and given strain 1 ($X = 1$), $Y_1 = 1 + Z_1$. The measurement noise Z_1 is independent of X and is Gaussian, $Z_1 \sim \mathrm{N}(0, \sigma^2)$. Give the MAP decision rule, i.e., determine the set of observations y_1 for which the decision is $\hat{x} = 1$. Give $\Pr\{e \mid X{=}0\}$ and $\Pr\{e \mid X{=}1\}$ in terms of the function $Q(x)$.

(b) A budding medical researcher determines that the test is making too many errors. A new measurement procedure is devised with two observation rv s Y_1 and Y_2. Y_1 is the same as in (a). Y_2, under hypothesis 0, is given by $Y_2 = 5 + Z_1 + Z_2$, and, under hypothesis 1, is given by $Y_2 = 1 + Z_1 + Z_2$. Assume that Z_2 is independent of both Z_1 and X, and that $Z_2 \sim \mathrm{N}(0, \sigma^2)$. Find the MAP decision rule for $\hat{x}$ in terms of the joint observation (y_1, y_2), and find $\Pr\{e \mid X{=}0\}$ and $\Pr\{e \mid X{=}1\}$. Hint: Find $\mathsf{f}_{Y_2|Y_1,X}(y_2 \mid y_1, 0)$ and $\mathsf{f}_{Y_2|Y_1,X}(y_2 \mid y_1, 1)$.

(c) Explain in laymen's terms why the medical researcher should learn more about probability.

(d) Now suppose that Z_2, in (b), is uniformly distributed between 0 and 1 rather than being Gaussian. We are still given that Z_2 is independent of both Z_1 and X. Find the MAP decision rule for $\hat{x}$ in terms of the joint observation (y_1, y_2) and find $\Pr(e \mid X{=}0)$ and $\Pr(e \mid X{=}1)$.

(e) Finally, suppose that Z_1 is also uniformly distributed between 0 and 1. Again find the MAP decision rule and error probabilities.

Exercise 8.10 **(a)** Consider a binary hypothesis testing problem, and denote the hypotheses as $X=1$ and $X=-1$. Let $\boldsymbol{a} = (a_1, a_2, \ldots, a_n)^{\mathsf{T}}$ be an arbitrary real n-vector and let the observation be a sample value $\boldsymbol{y}$ of the **rv** $\boldsymbol{Y} = X\boldsymbol{a}+\boldsymbol{Z}$, where $Z \sim \mathrm{N}\ (0, \sigma^2[I_n])$ and $[I_n]$ is the $n \times n$ identity matrix. Assume that Z and X are independent. Find the ML decision rule and find the probabilities of error $\Pr(e \mid X{=}0)$ and $\Pr(e \mid X{=}1)$ in terms of the function $Q(x)$.

(b) Now suppose a third hypothesis, $X = 0$, is added to the situation of (a). Again the observation **rv** is $\boldsymbol{Y} = X\boldsymbol{a} + \boldsymbol{Z}$, but here X can take on values $-1, 0$, or $+1$. Find a one-dimensional sufficient statistic for this problem (i.e., a one-dimensional function of y from which the likelihood ratios

$$\Lambda_1(\boldsymbol{y}) = \frac{p_{Y|X}(y \mid 1)}{p_{Y|X}(y \mid 0)} \quad \text{and} \quad \Lambda_{-1}(\boldsymbol{y}) = \frac{p_{Y|X}(y \mid -1)}{p_{Y|X}(y \mid 0)}$$

can be calculated).

(c) Find the ML decision rule for the situation in (b) and find the probabilities of error, $\Pr(e \mid X{=}x)$ for $x = -1, 0, +1$.

(d) Now suppose that $Z_1, \ldots, Z_n$ in (a) are IID and each is uniformly distributed over the interval -2 to $+2$. Also assume that $\boldsymbol{a} = (1, 1, \ldots, 1)^{\mathsf{T}}$. Find the ML decision rule for this situation.

Exercise 8.11 A sales executive hears that one of his salespeople is routing half of his incoming sales to a competitor. In particular, arriving sales are known to be Poisson at rate one per hour. According to the report (which we view as hypothesis $X = 1$), each second arrival is routed to the competition; thus under hypothesis 1 the interarrival density for successful sales is $f(y|X{=}1) = ye^{-y}; \ y \geq 0$. The alternative hypothesis $(X = 0)$ is that the rumor is false and the interarrival density for successful sales is $f(y|X{=}0) = e^{-y}; \ y \geq 0$. Assume that, a priori, the hypotheses are equally likely. The executive, a recent student of stochastic processes, explores various alternatives for choosing between the hypotheses; he can only observe the times of successful sales, however.

(a) Starting with a successful sale at time 0, let S_i be the arrival time of the ith subsequent successful sale. The executive observes $S_1, S_2, \ldots, S_n (n \geq 1)$ and chooses the maximum a posteriori probability hypothesis given this observation. Find the joint probability density $f(S_1, S_2, \ldots, S_n|X{=}1)$ and $f(S_1, \ldots, S_n|X{=}0)$ and give the decision rule.

(b) This is the same as (a) except that the system is in steady state at time 0 (rather than starting with a successful sale). Find the density of S_1 (the time of the first arrival after time 0) conditional on $X = 0$ and on $X = 1$. What is the decision rule now after observing $S_1, \ldots, S_n$.

(c) This is the same as (b), except rather than observing n successful sales, the successful sales up to some given time t are observed. Find the probability, under each hypothesis, that the first successful sale occurs in $(s_1, s_1 + \Delta]$, the second in $(s_2, s_2 + \Delta], \ldots,$ and the last in $(s_{N(t)}, s_{N(t)} + \Delta]$ (assume Δ very small). What is the decision rule now?

Exercise 8.12 This exercise generalizes the MAP rule to cases where neither a PMF nor a PDF exists. To view the problem in its simplest context, assume a binary decision, a one-dimensional observation ($y \in \mathbb{R}^1$), and fixed a priori probabilities p_0, p_1. An arbitrary test, denoted test A, can be defined by the set A of observations that are mapped into decision $\hat{x} = 1$. Using such a test, the overall probability of correct decision is given by (8.6) as

$$\Pr\left\{\hat{X}_A(Y) = X\right\} = p_0 \Pr\left\{Y \in A^{\mathrm{C}} \mid X = 0\right\} + p_1 \Pr\{Y \in A \mid X = 1\}. \tag{8.84}$$

The maximum of this probability over all tests A is then

$$\sup_A \Pr\left\{\hat{X}_A(Y) = X\right\} = \sup_A \left[p_0 \Pr\left\{Y \in A^{\mathrm{C}} \mid X = 0\right\} + p_1 \Pr\{Y \in A \mid X = 1\}\right]. \tag{8.85}$$

We first consider this supremum over all A consisting of a finite union of disjoint intervals. Then (using measure theory), we show that the supremum over all measurable sets A is the same as that over finite unions of disjoint intervals.

(a) If A is a union of k intervals, show that A^{C} is a union of at most $k+1$ disjoint intervals. Intervals can be open or closed on each end and can be bounded by $\pm\infty$.

(b) Let $\mathtt{I}$ be the partition of $\mathbb{R}$ created by the intervals of both A and of A^c. Let $\mathtt{I}_j$ be the jth interval in this partition. Show that

$$\Pr\{\hat{x}_A(Y) = X\} \le \sum_j \max\left[p_0 \Pr\{Y \in \mathtt{I}_j \mid X = 0\},\ p_1 \Pr\{Y \in \mathtt{I}_j \mid X = 1\}\right].$$

Hint: Break (8.84) into intervals and apply the MAP principle on an interval basis.

(c) The expression on the right in (b) is a function of the partition but is otherwise independent of A. It corresponds to a test where Y is first quantized to a finite set of intervals, and the MAP test is then applied to this discrete problem. We denote this test as MAP($\mathtt{I}$). Let the partition $\mathtt{I}'$ be a refinement of $\mathtt{I}$ in the sense that each interval of $\mathtt{I}'$ is contained in some interval of $\mathtt{I}$. Show that

$$\Pr\left\{\hat{x}_{\mathrm{MAP}(\mathcal{I})}(Y) = X\right\} \le \Pr\left\{\hat{x}_{\mathrm{MAP}(\mathcal{I}')}(Y) = X\right\}.$$

(d) Show that for any two finite partitions of $\mathbb{R}$ into intervals, there is a third finite partition that is a refinement of each of them.

(e) Consider a sequence $\{A_j, j \ge 1\}$ (each is a finite union of intervals) that approaches the supremum (over finite unions of intervals) of (8.85). Demonstrate a corresponding sequence of successive refinements of partitions $\{\mathtt{I}_j, j \ge 1\}$ for which $\Pr\{\hat{x}_{\mathrm{MAP}(\mathcal{I}_j)}(Y) = X\}$ approaches the same limit.

Note what this exercise has shown so far: if the likelihoods have no PDF or PMF, there is no basis for a MAP test on an individual observation y. However, by quantizing the observations sufficiently finely, the quantized MAP rule has an overall probability of being correct that is as close as desired to the optimum over all rules where A is a finite (but arbitrarily large) union of intervals. The arbitrarily fine quantization means that the decisions are arbitrarily close to pointwise decisions, and the use of quantization means that infinitely fine resolution is not required for the observations.

(f) Next suppose the supremum in (8.85) is taken over all measureable sets $A \in \mathbb{R}$. Show that, given any measurable set A and any $\epsilon > 0$, there is a finite union of intervals A' such that $\Pr\{A \neq A' \mid X = \ell\} \leq \epsilon$ both for $\ell = 0$ and $\ell = 1$. Show from this that the supremum in (8.85) is the same whether taken over measurable sets or finite unions of intervals.

Exercise 8.13 For the minimum-cost hypothesis testing problem of Section 8.3, assume that there is a cost C_0 of choosing $X = 1$ when $X = 0$ is correct, and a cost C_1 of choosing $X = 0$ when $X = 1$ is correct. Show that a threshold test minimizes the expected cost using the threshold $\eta = (C_0 p_0)/(C_1 p_1)$.

Exercise 8.14 **(a)** For given θ, $0 < \theta \leq 1$, let η^* achieve the supremum $\sup_{0 \leq \eta < \infty} q_1(\eta) + \eta(q_0(\eta) - \theta)$. Show that $\eta^* \leq 1/\theta$. Hint: Think in terms of Lemma 8.4.1 applied to a very simple test.

(b) Show that the magnitude of the slope of the error curve $u(\theta)$ at θ is at most $1/\theta$.

Exercise 8.15 Consider a binary hypothesis testing problem where X is 0 or 1 and a one-dimensional observation Y is given by $Y = X + U$, where U is uniformly distributed over $[-1, 1]$ and is independent of X.

(a) Find $\mathsf{f}_{Y|X}(y \mid 0)$, $\mathsf{f}_{Y|X}(y \mid 1)$ and the likelihood ratio $\Lambda(y)$.

(b) Find the threshold test at η for each η, $0 < \eta < \infty$ and evaluate the conditional error probabilities, $q_0(\eta)$ and $q_1(\eta)$.

(c) Find the error curve $u(\theta)$ and explain carefully how $u(0)$ and $u(1/2)$ are found. Hint: $u(0) = 1/2$.

(d) Find a *discrete* sufficient statistic $v(y)$ for this problem that has three sample values.

(e) Describe a decision rule for which the error probability under each hypothesis is 1/4. You need not use a randomized rule, but you need to handle the don't-care cases under the threshold test carefully.

9 Random walks, large deviations, and martingales

9.1 Introduction

Definition 9.1.1 *Let $\{X_i;\ i \geq 1\}$ be a sequence of independent identically distributed (IID) random variables (rv s), and let $S_n = X_1 + X_2 + \cdots + X_n$. The integer-time stochastic process $\{S_n;\ n \geq 1\}$ is called a* ***random walk****, or, more precisely, the* ***one-dimensional random walk*** *based on $\{X_i;\ i \geq 1\}$.*

For any given n, S_n is simply a sum of IID rv s, but here the behavior of the entire random walk *process*, $\{S_n;\ n \geq 1\}$, is of interest. Thus, for a given real number $\alpha > 0$, we might want to find the probability that the sequence $\{S_n;\ n \geq 1\}$ contains any term for which $S_n \geq \alpha$ (i.e., that a threshold at α is ever crossed) or to find the distribution of the smallest n for which $S_n \geq \alpha$.

We know that S_n/n essentially tends to $\mathsf{E}[X] = \overline{X}$ as $n \to \infty$. Thus if $\overline{X} < 0$, S_n will tend to drift downward and if $\overline{X} > 0$, S_n will tend to drift upward. This means that the results to be obtained will depend critically on whether $\overline{X} < 0$, $\overline{X} > 0$, or $\overline{X} = 0$. Since results for $\overline{X} > 0$ can easily be found from results for $\overline{X} < 0$ by considering $\{-S_n;\ n \geq 1\}$, we usually focus on either the case $\overline{X} < 0$ or the case $\overline{X} = 0$. To avoid trivialities we shall always assume in addition that X is non-deterministic.

In the case $\overline{X} < 0$, the random walk drifts downward, with random excursions around the mean, and those random excursions might result in crossing a threshold at some positive value α. If such a positive threshold is crossed, it usually occurs before n becomes very large.

As one might expect, both the results and the techniques have a very different flavor for a zero-mean random walk, i.e., a walk with $\overline{X} = 0$. In this case S_n/n essentially tends to 0 and we will see that the random walk typically wanders around in a rather aimless fashion. With increasing n, σ_{S_n} increases as $\sqrt{n}$; this behavior is often called diffusion.

We can visualize a zero-mean random walk as a zero-mean gambling game where S_n represents our winnings at time n. A common fallacy in such a game is to imagine that if we have a run of bad luck, then, since $\mathsf{E}[S_n] = 0$ for all n, a run of good luck should occur very soon to return our winnings to 0. In fact, given a loss, $S_k = s_k < 0$ at time k, then $\mathsf{E}[S_n \mid S_k = s_k] = s_k$ for all $n \geq k$. The fallacy arises because $\mathsf{E}\left[S_n/n \mid S_k = s_k\right] = s_k/n$, which approaches 0 as $n \to \infty$. One's intuition often confuses the sample average S_n/n with the sample value S_n.

The following three subsections discuss three special cases of random walks. The first two, simple random walks and integer random walks, will be useful throughout as examples, since they can be easily visualized and analyzed. The third special case is that of renewal processes, which we have already studied, and which will provide additional insight into the general study of random walks.

After this, Section 9.2 shows how a major application area, G/G/1 queues, can be viewed in terms of random walks. This section also illustrates why questions related to threshold crossings are so important in random walks.

Section 9.3 then returns to study arbitrary random walks. Many of the interesting problems concerning threshold crossings involve large deviation behavior of the underlying sums, S_n, of IID rv s. This section develops some important large deviation results about S_n. Section 9.4 then combines these results with a renewed study of stopping rules. This leads to a powerful generalization of Wald's equality known as Wald's identity; this will help us to view a random walk as an actual process rather than a collection of individual rv s.

The remainder of the chapter is devoted to a rather general type of stochastic processes called martingales. The topic of martingales is both a subject of interest in its own right and also a tool that provides additional insight into random walks, laws of large numbers, and other basic topics in probability and stochastic processes.

9.1.1 Simple random walks

Suppose $X_1, X_2, \ldots$ are IID binary rv s, each taking on the value 1 with probability p and -1 with probability $q = 1 - p$. Letting $S_n = X_1 + \cdots + X_n$, the sequence of sums $\{S_n;\ n \geq 1\}$, is called a *simple random walk*. Note that S_n is the difference between positive and negative occurrences in the first n trials, and thus a simple random walk is little more than a notational variation on a Bernoulli process. For the Bernoulli process, X takes on the values 1 and 0, whereas for a simple random walk X takes on the values 1 and -1. For the random walk, if $X_m = 1$ for m out of n trials, then $S_n = 2m - n$, and

$$\Pr\{S_n = 2m - n\} = \frac{n!}{m!(n-m)!}\, p^m(1-p)^{n-m}. \tag{9.1}$$

This distribution allows us to answer questions about S_n for any given n, but it is not very helpful in answering such questions as the following: for any given integer $k > 0$, what is the probability that the sequence $S_1, S_2, \ldots$ ever reaches or exceeds k? This probability can be expressed as[1] $\Pr\left\{\bigcup_{n=1}^{\infty}\{S_n \geq k\}\right\}$ and is referred to as the probability that the random walk *crosses a threshold* at k. Exercise 9.1 demonstrates the surprisingly simple result that, for a simple random walk with $p \leq 1/2$, this threshold crossing probability is

$$\Pr\left\{\bigcup_{n=1}^{\infty}\{S_n \geq k\}\right\} = \left(\frac{p}{1-p}\right)^k. \tag{9.2}$$

[1] This same probability is sometimes expressed as $\Pr\left\{\sup_{n\geq 1} S_n \geq k\right\}$, but there is a slight difference for general random walks, since the event $\{\sup_{n\geq 1} S_n \geq k\}$ can include sample sequences that approach k but never quite meet it. It is simpler to avoid this unimportant issue by not using the supremum notation to refer to threshold crossings.

Sections 9.3 and 9.4 treat this same question for general random walks, but the results are less simple. They also treat questions such as the overshoot given a threshold crossing, the time at which the threshold is crossed given that it is crossed, and the probability of crossing such a positive threshold before crossing any given negative threshold.

9.1.2 Integer-valued random walks

Suppose next that $X_1, X_2, \dots$ are arbitrary IID integer-valued rv s. We can again ask for the probability that such an integer-valued random walk crosses a threshold at k, i.e., that the event $\bigcup_{n=1}^{\infty}\{S_n \geq k\}$ occurs, but the question is considerably harder than for the special case of simple random walks. Since this random walk takes on only integer values, it can be represented as a Markov chain with the set of integers forming the state space. A simple random walk can also be represented as a Markov chain, but has the special property of being a birth–death chain.

For the Markov chain representation of the general integer-valued random walk, threshold crossing problems become first-passage time problems. These problems can be attacked by the Markov chain tools we already know, but the special structure of the random walk provides new approaches and simplifications that will be explained in Sections 9.3 and 9.4.

9.1.3 Renewal processes as special cases of random walks

If $X_1, X_2, \dots$ are IID positive rv s, then $\{S_n;\ n \geq 1\}$ is both a special case of a random walk and also the sequence of arrival epochs of a renewal counting process, $\{N(t);\ t > 0\}$. In this special case, the sequence $\{S_n;\ n \geq 1\}$ must eventually cross a threshold at any given positive value α. Thus the question of interest here is not *whether* a threshold is crossed but rather *when* it is crossed and with what overshoot. These are familiar questions from the study of renewal theory, where $N(\alpha)$ was effectively defined as the largest n for which $S_n \leq \alpha$ and thus $N(\alpha) + 1$ is the smallest n for which $S_n > \alpha$, i.e.,

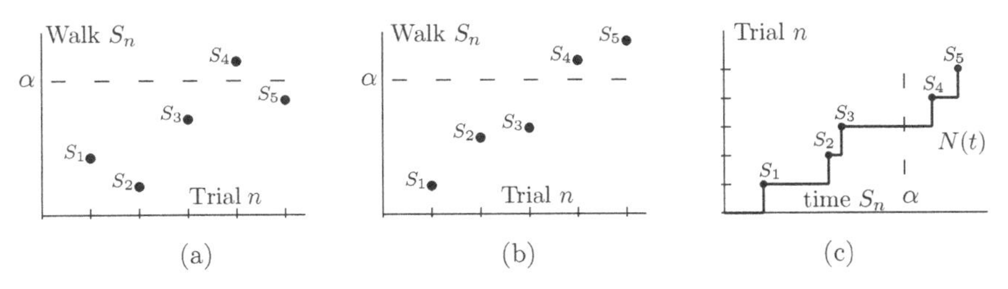

Figure 9.1 The sample function in (a) illustrates a random walk $S_1, S_2, \dots$ with arbitrary (positive and negative) step sizes $\{X_i;\ i \geq 1\}$. The sample function in (b) illustrates a random walk, $S_1, S_2, \dots$ with only positive step sizes $\{X_i > 0;\ i \geq 1\}$. Thus, $S_1, S_2, \dots$ in (b) are renewal points in a renewal process. Note that the axes in (b) are reversed from the usual depiction of a renewal process. The usual depiction, illustrated in (c) for the same sample points, also shows the corresponding counting process. The step sizes in (b) become the time increments between arrivals in (c). The random walks in (a) and (b) each illustrate a threshold at α, which in each case is crossed on trial 4 with an overshoot $S_4 - \alpha$.

the smallest n for which the threshold at α is strictly exceeded. When there is a positive overshoot, it is given by $S_{N(\alpha)+1} - \alpha$, and is familiar as the residual life at α.

Figure 9.1 illustrates the difference between positive random walks, i.e., renewal processes, and arbitrary random walks. Note that the renewal process in Figure 9.1(b) is illustrated with the axes reversed from the conventional renewal process representation. We usually view each renewal epoch as a time (epoch) and view $N(\alpha)$ as the number of renewals up to time α, whereas with random walks, we usually view the number of trials as a discrete-time variable and view the sum of rv s as some kind of amplitude or cost. There is no mathematical difference between these viewpoints, and each viewpoint is often helpful.

9.2 The queueing delay in a G/G/1 queue

Before analyzing random walks in general, we introduce an important problem area that is often best viewed in terms of random walks. This section will show how to represent the queueing delay in a G/G/1 queue as a threshold crossing problem in a random walk.

Consider a G/G/1 queue with first come, first served (FCFS) service. We shall associate the probability that a customer must wait more than some given time α in the queue with the probability that a certain random walk crosses a threshold at α. Let $X_1, X_2, \ldots$ be the interarrival times of a G/G/1 queueing system; thus these variables are IID with an arbitrary cumulative distribution function (CDF) $\mathsf{F}_X(x) = \Pr\{X_i \le x\}$. Assume that arrival 0 enters an empty system at time 0, and thus $S_n = X_1 + X_2 + \cdots + X_n$ is the epoch of the nth arrival after time 0. Let $Y_0, Y_1, \ldots$ be the service times of the successive customers. These are independent of $\{X_i;\, i \ge 1\}$ and are IID with some given CDF $\mathsf{F}_Y(y)$. Figure 9.2 shows the arrivals and departures for an illustrative sample path of the process and illustrates the queueing delay for each arrival.

Let W_n be the queueing delay for the nth customer, $n \ge 1$. The system time for customer n is then defined as the queueing delay W_n plus the service time Y_n. As illustrated in Figure 9.2, customer $n \ge 1$ arrives X_n time units after the beginning of customer $(n-1)$'s system time. If $X_n < W_{n-1} + Y_{n-1}$, i.e., if customer n arrives before the end of customer $(n-1)$'s system time, then customer n must wait in the queue until n finishes service (in the figure, for example, customer 2 arrives while customer 1 is still in the queue). Thus

$$W_n = W_{n-1} + Y_{n-1} - X_n \qquad \text{if } X_n \le W_{n-1} + Y_{n-1}. \tag{9.3}$$

On the other hand, if $X_n > W_{n-1} + Y_{n-1}$, then customer $n-1$ (and all earlier customers) have departed when n arrives. Thus n starts service immediately and $W_n = 0$. This is the case for customer 3 in the figure. These two cases can be combined in the single equation

$$W_n = \max[W_{n-1} + Y_{n-1} - X_n,\ 0] \quad \text{for } n \ge 1, \qquad W_0 = 0. \tag{9.4}$$

Since Y_{n-1} and X_n are coupled together in this equation for each n, it is convenient to define $U_n = Y_{n-1} - X_n$. Note that $\{U_n;\, n \ge 1\}$ is a sequence of IID rv s. From (9.4), $W_n = \max[W_{n-1} + U_n,\ 0]$, and iterating on this equation,

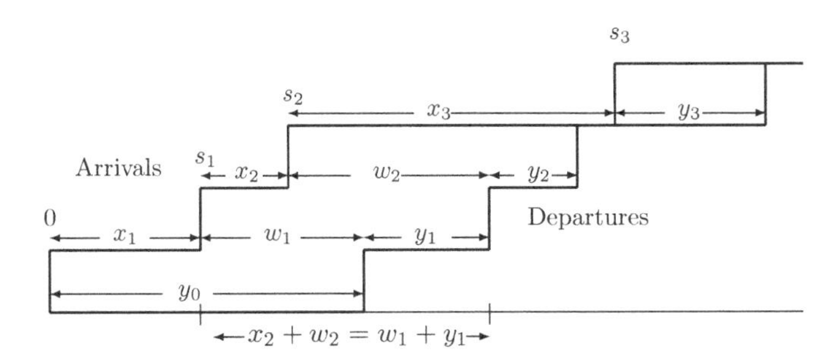

Figure 9.2 Sample path of arrivals and departures from a G/G/1 queue. Customer 0 arrives at time 0 and enters service immediately. Customer 1 arrives at time $s_1 = x_1$. For the case shown above, customer 0 has not yet departed, i.e., $x_1 < y_0$, so customer 1's time in queue is $w_1 = y_0 - x_1$. As illustrated, customer 1's system time (queueing time plus service time) is $w_1 + y_1$.

Customer 2 arrives at $s_2 = x_1 + x_2$. For the case shown above, this is before customer 1 departs at $y_0 + y_1$. Thus, customer 2's wait in queue is $w_2 = y_0 + y_1 - x_1 - x_2$. As illustrated above, $x_2 + w_2$ is also equal to customer 1's system time, so $w_2 = w_1 + y_1 - x_2$. Customer 3 arrives when the system is empty, so enters service immediately with no wait in queue, i.e., $w_3 = 0$.

$$\begin{aligned}
W_n &= \max[\max[W_{n-2}+U_{n-1},\ 0]+U_n,\ 0] \\
&= \max[(W_{n-2} + U_{n-1} + U_n),\ U_n,\ 0] \\
&= \max[(W_{n-3}+U_{n-2}+U_{n-1}+U_n),\ (U_{n-1}+U_n),\ U_n,\ 0] \\
&= \cdots \\
&= \max[(U_1+U_2+\cdots+U_n),\ (U_2+U_3+\cdots+U_n),\ldots,(U_{n-1}+U_n),\ U_n,\ 0]. \qquad (9.5)
\end{aligned}$$

If we look at the terms in the maximization in (9.5) working back from the end, we see that they can be viewed as the terms in a random walk starting at time n and working backwards. To be more specific, the first term in that backward walk is denoted as $Z_1^n = U_n$. The next term is denoted $Z_2^n = U_n + U_{n-1}$. Similarly, the ith term is

$$Z_i^n = \sum_{j=0}^{i-1} U_{n-j}.$$

With this terminology, (9.5) becomes

$$W_n = \max[0, Z_1^n, Z_2^n, \ldots, Z_n^n]. \qquad (9.6)$$

Note that the terms in $\{Z_i^n;\ 1 \le i \le n\}$ are the first n terms of a random walk, but it is not the random walk based on $U_1, U_2, \ldots$, but rather the random walk going backward, starting with U_n. Note also that W_{n+1}, for example, is the maximum of a different set of variables, i.e., it is the walk going backward from U_{n+1}. Fortunately, this does not matter for the analysis since the ordered variables $(U_n, U_{n-1} \ldots, U_1)$ are statistically identical to $(U_1, \ldots, U_n)$. The probability that the wait is greater than or equal to a given value α is

$$\Pr\{W_n \ge \alpha\} = \Pr\left\{\max[0, Z_1^n, Z_2^n, \ldots, Z_n^n] \ge \alpha\right\}. \qquad (9.7)$$

This says that, for the nth customer, $\Pr\{W_n \geq \alpha\}$ is equal to the probability that a random walk $\{Z_i^n;\ 1 \leq i \leq n\}$ based on the IID rv s U_i crosses a threshold at α by the nth trial.

In the same way, $\Pr\{W_{n+1} \geq \alpha\}$ is the probability that a random walk based on the IID rv s U_i crosses a threshold at α by trial $n+1$. We see from this that

$$\cdots \leq \Pr\{W_n \geq \alpha\} \leq \Pr\{W_{n+1} \geq \alpha\} \leq \cdots . \tag{9.8}$$

Since this sequence of probabilities is non-decreasing and bounded by 1, it must have a limit as $n \to \infty$, and this limit is denoted $\Pr\{W \geq \alpha\}$. Mathematically,[2] this limit is the probability that a random walk based on $\{U_i;\ i \geq 1\}$ ever crosses a threshold at α. Physically, this limit is the probability that the queueing delay is at least α for any given very-large-numbered customer (i.e., for customer n when the influence of a busy period starting n customers earlier has died out). These results are summarized in the following theorem.

Theorem 9.2.1 *Let $\{X_i;\ i \geq 1\}$ be the IID interarrival intervals of a G/G/1 queue, let $\{Y_i;\ i \geq 0\}$ be the IID service times, and assume that the system is empty at time 0 when customer 0 arrives. Let W_n be the queueing delay for the nth customer. Let $U_n = Y_{n-1} - X_n$ for $n \geq 1$ and let $Z_i^n = U_n + U_{n-1} + \cdots + U_{n-i+1}$ for $1 \leq i \leq n$. Then for every $n \geq 1$, $W_n = \max[0, Z_1^n, Z_2^n, \ldots, Z_n^n]$. Also, for each $\alpha > 0$, $\Pr\{W_n \geq \alpha\}$ is the probability that the random walk based on $\{U_i;\ i \geq 1\}$ crosses a threshold at α on or before the nth trial. Finally, $\Pr\{W \geq \alpha\} = \lim_{n\to\infty} \Pr\{W_n \geq \alpha\}$ is equal to the probability that the random walk based on $\{U_i;\ i \geq 1\}$ ever crosses a threshold at α.*

It was not required in establishing the above theorem, but we can understand this maximization better by realizing that if the maximization is achieved at $U_i + U_{i+1} + \cdots + U_n$, then a busy period must start with the arrival of customer $i-1$ and continue at least through the service of customer n. To see this intuitively, note that the analysis above starts with the arrival of customer 0 to an empty system at time 0, but the choice of 0 time and customer number 0 has nothing to do with the analysis, and thus the analysis is valid for any arrival to an empty system. Choosing the largest customer number before n that starts a busy period must then give the correct queueing delay, and thus maximize (9.5). Exercise 9.2 provides further insight into this maximization.

Note that the theorem specifies the CDF of W_n for each n, but says nothing about the joint distribution of successive queueing delays. These are not the same as the distribution of successive terms in a random walk because of the reversal of terms above.

We shall find relatively simple bounds and approximations to the probability that a random walk crosses a positive threshold in Section 9.3 and 9.4. These can be applied, via Theorem 9.2.1, to the distribution of queueing delay for the G/G/1 queue (and thus also for the M/G/1 and M/M/1 queues).

[2] More precisely, the sequence of queueing delays $W_1, W_2 \ldots$ converges in distribution to W, i.e., $\lim_n \mathsf{F}_{W_n}(w) = \mathsf{F}_W(w)$ for each w. We refer to W as the queueing delay in steady state.

9.3 Threshold crossing probabilities in random walks

Let $\{X_i;\ i \geq 1\}$ be a sequence of IID rv s, each with the CDF $\mathsf{F}_X(x)$, and let $\{S_n;\ n \geq 1\}$ be a random walk with $S_n = X_1 + \cdots + X_n$. The major objective of this section and the next is to develop results about $\Pr\left\{\bigcup_{n=1}^{\infty}\{S_n \geq \alpha\}\right\}$, i.e., the probability that the random walk $\{S_n;\ n \geq 1\}$ ever crosses a given threshold at some $\alpha > 0$. We assume throughout that $\mathsf{E}[X] < 0$ and that $\Pr\{X > 0\} > 0$. We focus on events known as large deviations where α is sufficiently large that $\Pr\left\{\bigcup_{n=1}^{\infty}\{S_n \geq \alpha\}\right\}$ is too small for the central limit theorem (CLT) to provide a good approximation. In this section, we expand on the treatment of the Chernoff bound in Section 1.6.3 to understand the behavior of $\Pr\{S_n \geq \alpha\}$. Then in Section 9.4, we develop a result called the Wald identity and use it along with the Chernoff bound to further study $\Pr\left\{\bigcup_{n=1}^{\infty}\{S_n \geq \alpha\}\right\}$.

Let $I(X)$ be the interval of r over which the moment generating function (MGF) $\mathsf{g}(r) = \mathsf{E}\left[e^{rX}\right]$ is finite. As seen by example in Section 1.5.5, $I(X)$ might be finite or infinite at each end, and if finite, might be open or closed at that end. Let r_- and r_+ be the lower and the upper end respectively of $I(X)$. Thus r_- and r_+ each might be finite or infinite and $\mathsf{g}(r_-)$ and $\mathsf{g}(r_+)$ each might be finite or infinite. We assume throughout that $r_- < 0 < r_+$, which implies that X has finite moments of all orders. Under the conditions above (including $\overline{X} < 0$) we will see that $\Pr\left\{\bigcup_{n=1}^{\infty}\{S_n \geq \alpha\}\right\}$ is bounded by an exponentially decreasing function of α for $\alpha > 0$. This will also be generalized to a bound on the probability that a threshold at $\alpha > 0$ is crossed before a threshold at some $\beta < 0$ is crossed. We will show that these bounds are exponentially tight in a sense to be specified.

9.3.1 The Chernoff bound

The Chernoff bound was derived and discussed in Section 1.6.3. It was shown in (1.64) that for any $r \geq 0$, $r \in I(X)$,

$$\Pr\{S_n \geq na\} \leq \exp\left(n[\gamma(r) - ra]\right), \tag{9.9}$$

where $\gamma(r) = \ln \mathsf{g}(r)$ is the semi-invariant MGF of X. The tightest bound of this form is found by optimizing over r, i.e., by

$$\Pr\{S_n \geq na\} \leq \exp[n\mu(a)], \qquad \text{where } \mu(a) = \inf_{r \geq 0;\, r \in I(X)} \gamma(r) - ra. \tag{9.10}$$

Lemma 1.6.1 showed that $\mu(a) < 0$ for all $a > \overline{X}$. This in turn implies that $\Pr\{S_n \geq na\} \to 0$ at least exponentially in n for all $a > \overline{X}$.

Optimizing (9.9) over r is relatively straightforward, since (see Exercise 1.26) $\gamma''(r) > 0$ for $0 < r < r_+$. Figure 9.3 illustrates the optimization and gives a graphical construction for finding $\mu(a)$ in the range of a for which $a = \gamma'(r)$ for some $r \in (0, r_+)$. The optimized bound in this range can then be expressed in the parametric form

$$\Pr\left\{S_n \geq n\gamma'(r)\right\} \leq \exp[n(\gamma(r) - r\gamma'(r)]. \tag{9.11}$$

Since $\gamma'(r)$ is an increasing function of r from $\overline{X}$ at $r = 0$ to $\sup_{r<r_+} \gamma'(r)$, this specifies the optimized Chernoff bound for all a in the range $\overline{X} < a < \sup_{r<r_+} \gamma'(r)$.

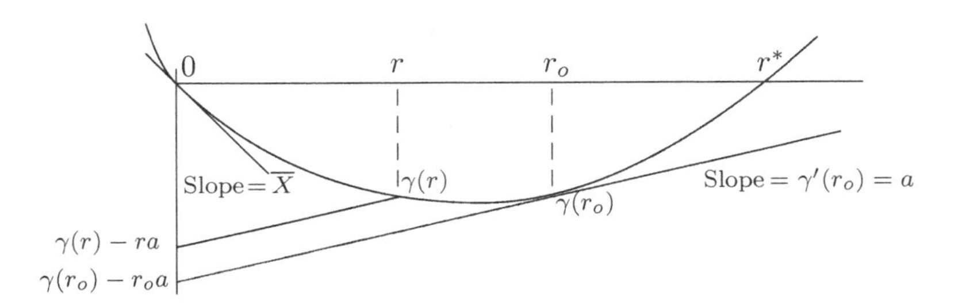

Figure 9.3 Graphical minimization of $\gamma(r) - ra$. Since $\gamma'(0) = \overline{X}$ and $\overline{X} < 0$, $\gamma(r)$ must have a negative slope at $r = 0$. Since $\Pr\{X > 0\} > 0$ by assumption also, X cannot be deterministic and $\gamma''(r) > 0$ over the interval where $\gamma(r)$ exists. Except in a very special case to be discussed in Figure 9.4, $\gamma(r)$ grows without bound as $r \to r_+$. To minimize $\gamma(r) - ra$ over r, note that $\gamma(r) - ar$ is the vertical axis intercept of a line of slope a through the point $(r, \gamma(r))$. The minimum occurs when the line of slope a is tangent to the $\gamma(r)$ curve.

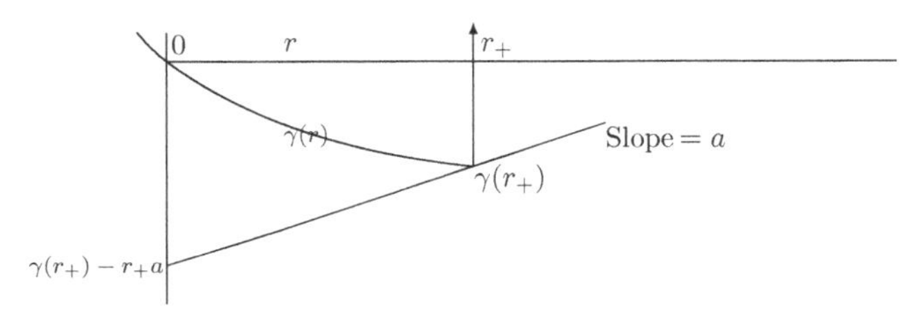

Figure 9.4 Graphical minimization of $\gamma(r) - ra$ for the case where $\gamma(r_+) < 0$. For $r \leq r_+$, $\gamma(r) - ra$ is the vertical axis intercept of a line of slope a through $(r, \gamma(r))$. If $\gamma'(r) = a$ for some $r < r_+$, the minimum occurs at that r. For $a \geq \lim_{r \to r_+} \gamma'(r)$, however (as shown in the figure), the minimum occurs at $r = r_+$.

For many rv s, $\gamma'(r)$ increases without bound with increasing $r < r_+$, and for such rv s, (9.10) and (9.11) are equivalent. For completeness, however, we should also understand the optimization if $\sup_{r<r_+} \gamma'(r)$ is finite and we are interested in $a > \sup_{r<r_+} \gamma'(r)$. There are two types of such situations, one with $r_+ = \infty$ and the other with $r_+ < \infty$.

Exercise 9.7 shows that if $r_+ = \infty$ and $\sup_{r<\infty} \gamma'(r) < \infty$, then $\mu(a)$ as defined in (9.10) is $-\infty$ for $a > \sup_{r<r_+} \gamma'(r)$. Thus $\Pr\{S_n \geq na\} = 0$ in this case. This case occurs, for example, if X is bounded by $X \leq b$ and a exceeds b. Since $S_n \leq nb$ in this case, it is not surprising that $\Pr\{S_n \geq na\} = 0$.

Exercise 9.8 shows that if $r_+ < \infty$ and $\sup_{r<r_+} \gamma'(r) < \infty$, then $\gamma(r_+) < \infty$. It also shows examples in which this occurs. Figure 9.4 extends the graphical minimization of Figure 9.3 to this case. It is almost obvious from the figure that $\mu(a) = \gamma(r_+) - r_+a$ in this case if $a > \sup_{r<r_+} \gamma'(r)$.

We can thus summarize the optimization of the Chernoff bound in the following theorem.

Theorem 9.3.1 *The optimized exponent $\mu(a)$ in (9.10) for arbitrary $a > \overline{X}$ in the Chernoff bound is given by*

$$\mu(a) = \begin{cases} \gamma(r) - r\gamma'(r) & \text{for } a \in (0, \sup_{r<r_+} \gamma'(r)), \text{ and } \gamma'(r) = a, \\ -\infty & \text{for } a > \sup_{r<r_+} \gamma'(r), \text{ and } r_+ = \infty, \\ \gamma(r_+) - r_+ a & \text{for } a > \sup_{r<r_+} \gamma'(r), \text{ and } r_+ < \infty. \end{cases} \tag{9.12}$$

In principle, we could now bound the probability of threshold crossing, $\Pr\left\{\bigcup_{n=1}^{\infty}\{S_n \geq \alpha\}\right\}$, by using the union bound over n and then bounding each term by (9.12). This would be quite tedious since, for a fixed α, the a in (9.12) such that $\alpha = an$, would vary with n. Instead, we pause to develop the concept of tilted probabilities. We will use these tilted probabilities in three ways: first, to get a better understanding of the Chernoff bound, second, to prove that the optimized Chernoff bound is exponentially tight in the limit of large n, and, third, to prove the Wald identity in the next section. The Wald identity in turn will provide an upper bound on $\Pr\left\{\bigcup_{n=1}^{\infty}\{S_n \geq \alpha\}\right\}$ that is much simpler than the above use of the union bound.

9.3.2 Tilted probabilities

As above, let $\{X_n;\ n \geq 1\}$ be IID and let $I(X)$ be the interval over which the MGF of X is finite. We will assume that X is discrete with the probability mass function (PMF) $\mathsf{p}_X(x)$, but the case in which X is continuous or arbitrary is essentially handled by replacing PMFs by probability density functions (PDFs) or CDFs. For any fixed $r \in I(X)$, define a new PMF (called a tilted PMF) on X by

$$\mathsf{q}_{X,r}(x) = \mathsf{p}_X(x)\exp[rx - \gamma(r)]. \tag{9.13}$$

Note that $\mathsf{q}_{X,r}(x) \geq 0$ for all sample values x and $\sum_x \mathsf{q}_{X,r}(x) = \sum_X \mathsf{p}_X(x)e^{rx}/\mathsf{E}\left[e^{rx}\right] = 1$, so this is a valid probability assignment.

Imagine a random walk that sums the IID rv s $X_1, X_2, \ldots$, but uses this tilted probability assignment on $X_1, X_2, \ldots$. We then have the same mapping from sample points of the underlying sample space to sample values of rv s, but we are dealing with a new probability measure, i.e., we have changed the probability model, and thus changed the probabilities in the random walk. We define this tilted probability measure so that the tilted versions of $X_1, X_2, \ldots$ are IID in this new measure.

For notational convenience, $X_1, X_2, \ldots$ will denote these rv s in both the original and the tilted probability measure. The expectations and other moments will all change, so for $r \in I(X)$, r is used as a subscript to denote the moments and MGFs in the new probability measure. The mean of X in the new, tilted, probability measure is thus denoted as $\mathsf{E}_r[X]$. Using (9.13),

$$\begin{aligned} \mathsf{E}_r[X] = \sum_x x\mathsf{q}_{X,r}(x) &= \sum_x x\mathsf{p}_X(x)\exp[rx - \gamma(r)] \\ &= \frac{1}{\mathsf{g}(r)} \sum_x \frac{d}{dr}\mathsf{p}_X(x)\exp[rx] \\ &= \frac{\mathsf{g}'(r)}{\mathsf{g}(r)} = \gamma'(r). \end{aligned} \tag{9.14}$$

Higher moments of X under the tilted measure can be calculated in the same way, but, more elegantly, the MGF of X under the tilted measure can be seen to be $\mathsf{E}_r[\exp(tX)] = \mathsf{g}(t+r)/\mathsf{g}(r)$. For the semi-invariant MGF then,

$$\gamma_r(t) = \gamma(t+r) - \gamma(r).$$

Using (9.13), using the tilted PMF on each component of an n-tuple $\boldsymbol{X}^n = (X_1, X_2, \ldots, X_n)$, we get

$$\mathsf{q}_{\boldsymbol{X}^n,\,r}(x_1, \ldots, x_n) = \mathsf{p}_{\boldsymbol{X}^n}(x_1, \ldots, x_n) \exp\left(\sum_{i=1}^{n} [rx_i - \gamma(r)] \right). \tag{9.15}$$

We now relate the PMF of the sum, $\sum_{i=1}^{n} X_i$, in the original probability measure to that in the tilted probability measure. From (9.15), note that for every n-tuple $(x_1, \ldots, x_n)$ for which $\sum_{i=1}^{n} x_i = s_n$ (for any given s_n), we have

$$\mathsf{q}_{\boldsymbol{X}^n,r}(x_1, \ldots, x_n) = \mathsf{p}_{\boldsymbol{X}^n}(x_1, \ldots, x_n) \exp[rs_n - n\gamma(r)].$$

Summing over all $\boldsymbol{x}^n$ for which $\sum_{i=1}^{n} x_i = s_n$, we then get a remarkable relation between the PMF for S_n in the original and the tilted probability measure:

$$\mathsf{q}_{S_n,r}(s_n) = \mathsf{p}_{S_n}(s_n) \exp[rs_n - n\gamma(r)]. \tag{9.16}$$

This equation is a key to large deviation theory applied to sums of IID rv s. The tilted measure of S_n, for positive r, increases the probability of large values of S_n and decreases that of small values. Since S_n is an IID sum under the tilted measure, however, we can use the laws of large numbers and the CLT around the tilted mean to get good estimates and bounds on the behavior of S_n far from the mean for the original measure.

Definition 9.3.2 *Let $\{a_n;\, n \geq 1\}$ be a sequence of numbers and for each n, let b_n be a bound on a_n, i.e., $a_n \leq b_n$. This sequence of bounds is **exponentially tight** if the following limits exist with equality as shown.*

$$\lim_{n \to \infty} \frac{\ln a_n}{n} = \lim_{n \to \infty} \frac{\ln b_n}{n}.$$

An equivalent statement is that the sequence of bounds (usually simply called a bound) is exponentially tight if, for any $\epsilon > 0$, $a_n \geq b_n e^{-\epsilon n}$ for all sufficiently large n. In the usual situation, such as that of the Chernoff bound, the bound has the form $a_n \leq e^{-\beta n}$ for all n, and the bound is exponentially tight if β cannot be enlarged without violating the bound for sufficiently large n.

It is not quite clear from the definition why exponential tightness is an important concept. For a bound such as $\Pr\{S_n \geq na\} \leq \exp[n\mu(a)]$ for example, the bound might be exponentially tight but be off by orders of magnitude for all n because of a neglected coefficient. For many, or perhaps most, applications for which a small probability of exceeding some level is desired, n is a parameter to be chosen subject to an associated cost. A tight exponent in such a bound then gives a good estimate of the rate at which $\Pr\{S_n \geq na\}$ decreases with increasing n and thus of the corresponding tradeoff between cost and probability. If the bound has the form $a_n \leq c_n e^{-\beta n}$, where c_n is, say, a polynomial in n, then the coefficient c_n, while perhaps important, is less important than the factor $e^{-\beta n}$ for sufficiently large n, since $(\ln[c_n e^{-\beta n}])/n$ approaches $-\beta$ as $n \to \infty$.

The following theorem applies the weak law of large numbers (WLLN) to $\sum_{i=1}^{n} X_n$ under the tilted measure to essentially show that the Chernoff bound for the original probability measure is exponentially tight.

Theorem 9.3.3 *Let $\{X_n;\ n \geq 1\}$ be an IID sequence of rv s. Let $r_- < 0$ and $r_+ > 0$ be the endpoints of the interval where $\gamma(r) = \ln(\mathsf{E}\left[e^{rX}\right])$ is finite. Let $S_n = \sum_{i=1}^{n} X_i$ for each $n \geq 1$. Then for each $r \in (0, r_+)$, and each $\epsilon > 0$, there is an n_o such that for all $n \geq n_o$,*

$$\Pr\left\{S_n \geq n\gamma'(r)\right\} \;\geq\; \exp[n(\gamma(r) - r\gamma'(r) - r\epsilon)]. \tag{9.17}$$

Discussion Note that $\Pr\left\{S_n \geq n\gamma'(r)\right\}$ is the probability of an event in the original probability measure. If we look at that event in the tilted measure for the given r, it becomes the (tilted) probability that S_n is greater than or equal to the tilted mean. From the WLLN, S_n is with high probability close to $n\mathsf{E}_r[X] = \gamma'(r)$ in the tilted measure. The theorem simply uses ϵ's carefully to convert this to the original probability measure.

Proof The WLLN, (1.75), says that $\lim_{n\to\infty} \Pr\left\{|S_n/n - \overline{X}| > \epsilon\right\} = 0$ for each $\epsilon > 0$, or, equivalently, $\lim_{n\to\infty} \Pr\left\{|S_n/n - \overline{X}| \leq \epsilon\right\} = 1$. The meaning of the limit on n is that for any $\epsilon > 0$ and $\delta > 0$, there is an n_o large enough that

$$\Pr\left\{\left|\frac{S_n}{n} - \overline{X}\right| \leq \epsilon\right\} \geq 1 - \delta \qquad \text{for all } n \geq n_o. \tag{9.18}$$

The WLLN can be applied to the tilted measure as well as the original measure. In the tilted measure, (9.14) shows that the mean of X is $\gamma'(r)$. Thus, assuming that X is discrete for notional convenience, and assuming $n \geq n_o$ for the n_o required for the given ϵ and δ, we can rewrite (9.18), using the tilted measure, as

$$1 - \delta \;\leq\; \sum_{(\gamma'(r)-\epsilon)n \,\leq\, s_n \,\leq\, (\gamma'(r)+\epsilon)n} \mathsf{q}_{S_n,r}(s_n) \tag{9.19}$$

$$= \sum_{(\gamma'(r)-\epsilon)n \,\leq\, s_n \,\leq\, (\gamma'(r)+\epsilon)n} \mathsf{p}_{S_n}(s_n)\exp[rs_n - n\gamma(r)] \tag{9.20}$$

$$\leq \sum_{(\gamma'(r)-\epsilon)n \,\leq\, s_n \,\leq\, (\gamma'(r)+\epsilon)n} \mathsf{p}_{S_n}(s_n)\exp[n(r\gamma'(r) + r\epsilon - \gamma(r))] \tag{9.21}$$

In (9.20), $\mathsf{q}_{S_n,r}(s_n)$ is expressed via the original probabilities; then (9.21) upper bounds s_n in the exponent by the upper bound in the summation. The resulting sum can then be conveniently upper bounded by extending the sum to arbitrarily large s_n, yielding

$$1 - \delta \leq \sum_{(\gamma'(r)-\epsilon)n \,\leq\, s_n} \mathsf{p}_{S_n}(s_n)\exp[n(r\gamma'(r) + r\epsilon - \gamma(r))] \tag{9.22}$$

$$= \exp\left[n(r\gamma'(r) + r\epsilon - \gamma(r))\right] \Pr\left\{S_n \geq n(\gamma'(r) - \epsilon)\right\}. \tag{9.23}$$

$$\Pr\left\{S_n \geq n(\gamma'(r) - \epsilon)\right\} \geq (1 - \delta)\exp\left[-n(r\gamma'(r) + r\epsilon - \gamma(r))\right], \tag{9.24}$$

where (9.24) is a rearrangement of (9.23). It can be seen from Figure 9.3 that this is essentially equivalent to (9.17). Exercise 9.9 shows how the ϵ on the left and the δ on the right can be absorbed into the ϵ on the right. □

The structure of the above proof can also be used to derive a number of other results. The general approach is to use the tilted probability measure to focus on the tail of the distribution, then to use some known probability result on the tilted distribution, and finally to transfer the result back to the original distribution. The manipulation of ϵ's and δ's above becomes automatic after a little practice.

Note that this theorem applies only to the first case in (9.12), i.e., to values of $a \in (0, \sup_{r<r_+} \gamma'(r))$. For the second case in (9.12), where $r_+ = \infty$ and $a > \sup_{r<r_+} \gamma'(r))$, there is no need of tightness, since the upper bound is 0. The theorem is extended by a truncation argument in Exercise 9.10 to show that the third case in (9.12), where $r_+ < \infty$ and $a > \sup_{r<r_+} \gamma'(r)$, is also exponentially tight.

Theorems 9.3.1 and 9.3.3 are important large deviation results since they establish the true exponential decay in n for $\Pr\{S_n \geq na\}$, where S_n is a sum of n IID rv s that have an MGF. Since these results are somewhat abstract, it will be helpful to understand the result in a less abstract but less general context. This was done for the binomial distribution in Section 1.6.3 and the following section generalizes this to the multinomial distribution, i.e., to rv s that are discrete with a finite number of possible sample values.

9.3.3 Large deviations and compositions

Let $\{a_j;\ 1 \leq j \leq \mathsf{M}\}$ be the set of sample values for X and let $\boldsymbol{x} = (x_1, \ldots, x_n)$ be an n-tuple of these sample values. The *composition* or *type* of $\boldsymbol{x}$ is defined to be the M-dimensional vector $\widetilde{\boldsymbol{p}} = (\widetilde{p}_1, \widetilde{p}_2, \ldots, \widetilde{p}_\mathsf{M})$, where, for each j, $n\widetilde{p}_j$ is the number of occurrences of a_j in $\boldsymbol{x}$. Thus the set of possible sample sequences for $\boldsymbol{x}$ is partitioned into compositions, and each composition $\widetilde{\boldsymbol{p}}$ consists of the set of n-tuples that have the given number $n\widetilde{p}_j$ of occurrences of a_j for each j. The number of n-tuples that have a given composition $\widetilde{\boldsymbol{p}} = (\widetilde{p}_1 \ldots, \widetilde{p}_\mathsf{M})$ is then the multinomial $n!/[(\widetilde{p}_1 n)!(\widetilde{p}_1 n)! \cdots (\widetilde{p}_\mathsf{M} n)!]$. We assume throughout this section that $n\widetilde{p}_j > 0$ for each j; the extension to zero values for one or more $n\widetilde{p}_j$ is essentially accomplished simply by omitting those components, but requires frequent discussion of distracting special cases. The Stirling bounds can be used on the multinomial in the same way as on the binomial in Exercise 1.9 to get

$$\frac{n!}{\prod_j (n\widetilde{p}_j)!} < \frac{\sqrt{2\pi n}}{\prod_j \sqrt{2\pi n\widetilde{p}_j}} \exp\left[-n \sum_j \widetilde{p}_j \ln \widetilde{p}_j\right], \tag{9.25}$$

$$\frac{n!}{\prod_j (n\widetilde{p}_j)!} > \frac{\sqrt{2\pi n}}{\prod_j \sqrt{2\pi n\widetilde{p}_j}} \exp\left[-n \sum_j \widetilde{p}_j \ln \widetilde{p}_j\right] \exp\left[\frac{-1}{12n} \sum_j (1/\widetilde{p}_j)\right]. \tag{9.26}$$

The ratio of the upper bound to the lower bound clearly approaches 1 with increasing n, uniformly in $\widetilde{p}_1, \ldots, \widetilde{p}_\mathsf{M}$ over any region where each component is bounded away from 0. The bound in (9.25) is thus asymptotically tight. The rate, $\sum_j -\widetilde{p}_j \ln \widetilde{p}_j$, at which the number of n-tuples in a composition $\widetilde{\boldsymbol{p}}$ increases with n is called the entropy of $\widetilde{\boldsymbol{p}}$.

Similarly, a discrete rv with PMF $\boldsymbol{p}$ has entropy $\sum_j -p_j \ln p_j$. Entropy is a central concept in information theory; see, for example, Chapter 2 of [10] for a fuller discussion.

Now suppose that $\boldsymbol{X} = (X_1, \dots, X_n)$ is an n-tuple of discrete IID rv s with the set of possible values $\{a_1, \dots, a_{\mathsf{M}}\}$ and the PMF $\boldsymbol{p} = (p_1, \dots, p_{\mathsf{M}})$. We assume $p_j > 0$ for each j throughout. If $\boldsymbol{x}$ is a particular n-tuple of composition $\widetilde{p}_1, \dots, \widetilde{p}_{\mathsf{M}}$, then $\Pr\{\boldsymbol{X} = \boldsymbol{x}\} = p_1^{n\widetilde{p}_1} p_2^{n\widetilde{p}_2} \cdots p_{\mathsf{M}}^{n\widetilde{p}_{\mathsf{M}}}$. Multiplying this by the number of n-tuples of composition $\widetilde{\boldsymbol{p}}$, we get

$$\begin{aligned}\Pr\{\boldsymbol{X} \in \widetilde{\boldsymbol{p}}\} &= \frac{n!}{\prod_j (n\widetilde{p}_j)!} \prod_j \exp(n\widetilde{p}_j \ln p_j) \\ &< \frac{\sqrt{2\pi n}}{\prod_j \sqrt{2\pi n \widetilde{p}_j}} \exp\left[-n \sum_j \widetilde{p}_j \ln \frac{\widetilde{p}_j}{p_j}\right]. \end{aligned} \tag{9.27}$$

This is asymptotically tight since (9.25) is asymptotically tight. The Kullback–Leibler divergence (divergence for short or relative entropy) between two probability vectors $\widetilde{\boldsymbol{p}}$ and $\boldsymbol{p}$ is defined to be

$$D(\widetilde{\boldsymbol{p}} \| \boldsymbol{p}) = \sum_j \widetilde{p}_j \ln \frac{\widetilde{p}_j}{p_j}. \tag{9.28}$$

Substituting this into (9.27)

$$\Pr\{\boldsymbol{X} \in \widetilde{\boldsymbol{p}}\} < \frac{\sqrt{2\pi n}}{\prod_j \sqrt{2\pi n \widetilde{p}_j}} \exp\left[-nD(\widetilde{\boldsymbol{p}} \| \boldsymbol{p})\right]. \tag{9.29}$$

As seen in Exercise 9.12, $D(\widetilde{\boldsymbol{p}} \| \boldsymbol{p}) \geq 0$ with strict inequality except for $\boldsymbol{p} = \widetilde{\boldsymbol{p}}$. It is also shown that $D(\widetilde{\boldsymbol{p}} \| \boldsymbol{p})$ is convex in $\widetilde{\boldsymbol{p}}$ over the region where $\widetilde{\boldsymbol{p}}$ is a probability vector.

Under the assumption that $n\widetilde{p}_j$ is a positive integer for each j, $1 \leq j \leq \mathsf{M}$, the coefficient $\sqrt{2\pi n}/\prod_j \sqrt{2\pi n \widetilde{p}_j}$ is upper bounded by 1 (and in fact by $\pi^{-(\mathsf{M}-1)/2}$). This can be seen by starting with the binary case and using induction on M. Thus, going back to arbitrary $\boldsymbol{p}$, we have the simpler relation,

$$\Pr\{\boldsymbol{X} \in \widetilde{\boldsymbol{p}}\} \leq \exp\left[-nD(\widetilde{\boldsymbol{p}} \| \boldsymbol{p})\right]. \tag{9.30}$$

Since the coefficient $\sqrt{2\pi n}/\prod_j \sqrt{2\pi n \widetilde{p}_j}$ is non-exponential in n and (9.29) is asymptotically tight, we see that (9.30) is exponentially tight. Note the difference between being asymptotically tight and exponentially tight; (9.29) (as seen by the upper and lower bounds in (9.25) and (9.26)) is loose only in the factor $\exp\left[(-1/12n)\sum_j (1/\widetilde{p}_j)\right]$ which approaches 1 as $n \to \infty$. On the other hand, (9.30) is also loose in the factor $\sqrt{2\pi n}/\prod_j \sqrt{2\pi n \widetilde{p}_j}$; this goes to 0 as a power of n, but not exponentially in n. Exponential tightness allows us to ignore terms that might be significant, but less significant asymptotically than the exponential terms.

The exponential tightness of (9.30) is probably the best way to see the intuitive significance of divergence. It signifies how far $\widetilde{\boldsymbol{p}}$ is from $\boldsymbol{p}$ by giving the exponent in the probability that an n-tuple drawn according to $\boldsymbol{p}$ 'looks like' an n-tuple of composition $\widetilde{\boldsymbol{p}}$.

Next we look at $\Pr\{S_n \geq na\}$ in terms of compositions. Recall that $S_n = X_1 + \cdots + X_n$, where the X_n are IID with sample values $a_1, \dots, a_{\mathsf{M}}$ and probabilities $p_1, \dots, p_{\mathsf{M}}$. The event $\{S_n \geq na\}$ consists of all those n-tuples belonging to compositions $\widetilde{\boldsymbol{p}}$ for which $\sum_j \widetilde{p}_j a_j \geq a$. Thus, we have

$$\Pr\{S_n \geq na\} = \sum_{\widetilde{\boldsymbol{p}}:\, n\widetilde{p}_j \in \mathbb{Z},\, \sum_j \widetilde{p}_j a_j \geq a} \Pr\{\boldsymbol{X} \in \widetilde{\boldsymbol{p}}\} \tag{9.31}$$

$$\leq \sum_{\widetilde{\boldsymbol{p}}:\, n\widetilde{p}_j \in \mathbb{Z},\, \sum_j \widetilde{p}_j a_j \geq a} \exp[-nD(\widetilde{\boldsymbol{p}} \| \boldsymbol{p})] \tag{9.32}$$

$$\leq n^{\mathsf{M}+1} \sup_{\widetilde{\boldsymbol{p}}:\, \sum_j \widetilde{p}_j a_j \geq a} \exp[-nD(\widetilde{\boldsymbol{p}} \| \boldsymbol{p})]. \tag{9.33}$$

In the final expression, the number of compositions in the sum has been upper bounded by the number of all compositions of n-tuples, which in turn is bounded by recognizing that $n\widetilde{p}_j$ is an integer between 0 and M for each j. Each term in the sum has then been upper bounded by the supremum without an integer constraint on $n\widetilde{p}_j$. This result, and its extension to constraints other that $\sum \widetilde{p}_j a_j \leq a$, is called Sanov's theorem.

The supremum in (9.33) could be calculated by minimizing $D(\widetilde{\boldsymbol{p}} \| \boldsymbol{p})$ over $\widetilde{\boldsymbol{p}}$ using Lagrange multipliers for the constraints $\sum_j \widetilde{p}_j = 1$ and $\sum_j \widetilde{p}_j a_j \geq a$. The following more elegant approach to this minimization uses tilted probabilities, tilted from $\boldsymbol{p}$ to $\boldsymbol{q}_r$ where r satisfies $\gamma'(r) = a$. Thus $q_{j,r} = p_j \exp[ra_j - \gamma(r)]$ and the mean of S_n, according to the tilted distribution, is na. The motivation for this is that we used these tilted probabilities earlier to evaluate $\Pr\{S_n \geq na\}$ and the tilting emphasized the probabilities of sample n-tuples close to the threshold $s_n \approx na$. Here we do the same thing in terms of compositions rather than individual n-tuples. Writing $\boldsymbol{p}$ in terms of $\boldsymbol{q}_r$, we have

$$D(\widetilde{\boldsymbol{p}} \| \boldsymbol{p}) = \sum_j \widetilde{p}_j \ln \left[\frac{\widetilde{p}_j}{q_{j,r} e^{-ra_j + \gamma(r)}} \right]$$

$$= -\gamma(r) + \sum_j \widetilde{p}_j r a_j + \sum_j \widetilde{p}_j \ln \left[\frac{\widetilde{p}_j}{q_{j,r}} \right]$$

$$= -\gamma(r) + \sum_j \widetilde{p}_j r a_j + D(\widetilde{\boldsymbol{p}} \| \boldsymbol{q}_r)$$

$$\geq -\gamma(r) + ra \qquad \text{for } \widetilde{\boldsymbol{p}} \text{ such that } \sum_j \widetilde{p}_j a_j \geq a. \tag{9.34}$$

The final expression is valid since the divergence is non-negative (see Exercise 9.12). Substituting (9.34) into (9.33) for each term over which the supremum is taken, we surprisingly have

$$\Pr\{S_n \geq na\} \leq n^{\mathsf{M}+1} \exp\big(n[\gamma(r) - ra]\big), \qquad \gamma'(r) = a.$$

This is the Chernoff bound, except for the additional factor of $n^{\mathsf{M}+1}$, so it seems we have not accomplished much beyond deriving an old result in a less general and somewhat weaker form. What is new is that we can now see that (9.33) is satisfied with equality for $\widetilde{\boldsymbol{p}} = \boldsymbol{q}_r$ and thus the supremum in (9.33) is achieved by $\widetilde{\boldsymbol{p}} = \boldsymbol{q}_r$. In other words, subject to the integer constraint on $\boldsymbol{q}_r n$, the composition most likely to give rise to $S_n \geq na$ is the tilted probability $\boldsymbol{q}_r$ using the r for which $\sum_j q_{j,r} a_j = a$.

To summarize the results of this section, we started by showing that the divergence $D(\widetilde{p}\|p)$ has a fundamental interpretation as $\lim_n(-1/n)\ln\Pr\{X \in \widetilde{p}\}$ where X is an n-rv drawn according to p, i.e., D is the exponential rate of decrease in the probability that p 'looks like' $\widetilde{p}$. As indicated by the notation, $D(\widetilde{p}\|p)$ depends only on $\widetilde{p}$ and p and not on the values $a = (a_1, \ldots, a_M)$ taken on by X. As indicated by (9.31)–(9.33), $\Pr\{S_n \geq na\}$ is then the probability of the union of compositions bounded by the linear constraint $a^{\mathsf{T}}\widetilde{p} \geq a$. The dominant such compositions are those close to q_r. An important use of divergences is to bound or estimate more general unions of compositions.

9.3.4 Back to threshold crossings

Consider again the probability that a random walk crosses a positive threshold α, i.e., $\Pr\left\{\bigcup_n\{S_n \geq \alpha\}\right\}$. Assume that $\overline{X} < 0$ and that the endpoints r_- and r_+ of the interval where $\mathsf{g}(r)$ exists satisfy $r_- < 0 < r_+$. We can use the optimized Chernoff bound to look at each event in this union. For values of n such that $\alpha/n = \gamma'(r)$ for some r in $(0, r_+)$, the first part of (9.12) says that

$$\Pr\{S_n \geq \alpha\} \leq \exp\{n[\gamma(r) - r\gamma'(r)]\} \qquad \text{for } r \text{ such that } \gamma'(r) = \alpha/n.$$

Since $\gamma'(r) = \alpha/n$ in this equation, we can replace n in the exponent with $\alpha/\gamma'(r)$, getting

$$\Pr\{S_n \geq \alpha\} \leq \exp\left\{\alpha\left[\frac{\gamma(r)}{\gamma'(r)} - r\right]\right\}, \qquad \text{where } \gamma'(r) = \alpha/n. \tag{9.35}$$

We have seen that $\gamma'(r)$ is continuous and increasing in r for $r \in (0, r_+)$, and we assume temporarily that $\lim_{r\to r_+}\gamma'(r) = \infty$ so that this equation has a solution for $\gamma'(r) = \alpha/n$ for each $\alpha > 0$ and $n \geq 1$.

A graphic interpretation of this equation is given in Figure 9.5. Note that the exponent in α, namely $[\gamma(r)/\gamma'(r)] - r$, is the negative of the horizontal axis intercept of the tangent of slope $\gamma'(r) = \alpha/n$ to the curve $\gamma(r)$ in Figure 9.5.

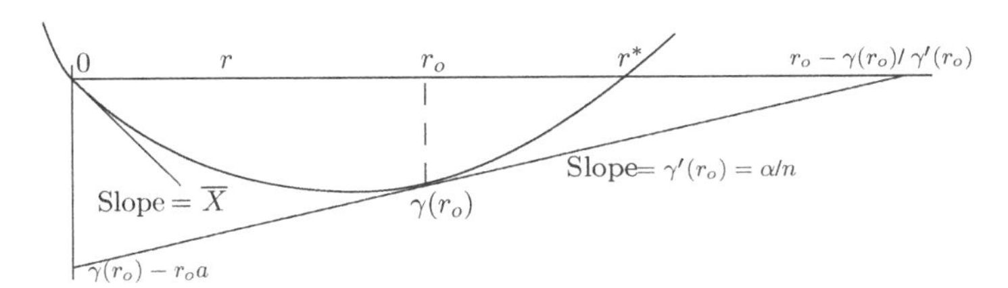

Figure 9.5 The exponent in α for $\Pr\{S_n \geq \alpha\}$, minimized over r. The minimization is the same as that in Figure 9.3, but $\gamma(r_o)/\gamma'(r_o) - r_o$ is the negative of the horizontal axis intercept of the line tangent to $\gamma(r_o)$ at $r_o = r$.

For fixed $\alpha > 0$ (still assuming that $\gamma'(r)$ is unbounded), we see that the slope α/n is small for large n and this horizontal intercept is large. As n is decreased, the slope increases, the point of tangency, r_o, increases, and the horizontal intercept decreases. When r_o increases to the point labeled r^* in the figure, namely the $r > 0$ at which $\gamma(r) = 0$, then the horizontal intercept also reaches r^*. When n decreases even further,

r_o becomes larger than r^*, $\gamma(r_0) > 0$, and the horizontal intercept starts to increase again.

Since the horizontal intercept for each n exceeds r^*, we see that (9.35) can be replaced with the looser and simpler upper bound,

$$\Pr\{S_n \geq \alpha\} \leq \exp(-r^*\alpha) \qquad \text{for arbitrary } \alpha > 0, \text{ all } n \geq 1, \tag{9.36}$$

where r^* is the positive root of $\gamma(r)$.

This bound is quite loose for many values of n, but since the bound in (9.35) is exponentially tight, the bound in (9.36) is correspondingly tight for $n \approx \alpha/\gamma'(r^*)$. Our primary interest here is in the probability of threshold crossing, $\Pr\left\{\bigcup_n\{S_n \geq \alpha\}\right\}$. The union bound shows that this is upper bounded by a sequence of terms each upper bounded by $e^{-r^*\alpha}$. At the same time, $\Pr\left\{\bigcup_n\{S_n \geq \alpha\}\right\}$ is lower bounded by $\Pr\{S_n \geq \alpha\}$ for each choice of n. For $n \approx \alpha/\gamma'(r^*)$, the exponent of this lower bound is essentially $-r^*\alpha$.

We thus conclude, without going through all the ϵ's and δ's, that

$$\lim_{\alpha\to\infty} \frac{1}{\alpha} \ln \Pr\left\{\bigcup_n \{S_n \geq \alpha\}\right\} = -r^*. \tag{9.37}$$

We now recall that (9.36) and (9.37) were derived under the assumption that $\sup_{r>0} \gamma'(r) = \infty$. If $\gamma'(r)$ is bounded, we must consider the second and third alternatives in (9.12). The second alternative consists of cases where $r_+ = \infty$ and shows that $\Pr\{S_n \geq \alpha\} = 0$ for $\alpha/n > \sup_r \gamma'(r)$. For these cases, (9.36) is still valid. These cases occur when X is bounded by some b, and the threshold α cannot be reached until n exceeds α/b. Since $\gamma(r)$ is unbounded in this case, the positive root, r^*, of $\gamma(r)$ still exists, and there are values of $n \approx \alpha/\gamma'(r^*)$. Thus (9.37) again holds in this case.

The third alternative in (9.12) is viewed graphically in Figure 9.4. In this case, if we redefine r^* as the supremum of values of $r > 0$ such that $\gamma(r) < 0$, we again have (9.36). In this case, the horizontal intercept in Figure 9.4 lies at $r^* - n\gamma(r^*)/\alpha$. We can establish (9.37) in this case by choosing n/α to be arbitrarily small and then increasing n and α together for exponential tightness.

The following theorem summarizes these results.

Theorem 9.3.4 *Let $\{X_n;\ n \geq 1\}$ be IID and let $S_n = X_1 + \cdots + X_n$ for each $n \geq 1$. Let 0 be in the interior of the interval where $\gamma(r) = \ln \mathsf{E}\left[e^{rX}\right]$ exists and let $r^* = \sup\{r > 0 : \gamma(r) < 0\}$. Then (9.37) holds and (9.36) is valid for all $n \geq 1$ and $\alpha > 0$.*

In the next section, we develop Wald's identity, which allows us to show (under almost the same conditions) that the probability of threshold crossing, namely $\Pr\left\{\bigcup_{n=1}^{\infty}\{S_n \geq \alpha\}\right\}$, is upper bounded by $\exp(-r^*\alpha)$. This is a much simpler result than (9.36), since it bounds the entire union of events over n using the same bound as used in (9.36) for each term. Thus the main purpose of Theorem 9.3.4 is that it establishes the exponential tightness of the bound.

The other purpose of this section has been to provide information on *when* a threshold is crossed as well as *whether* it is crossed. These bounds are also useful in a wide variety of situations involving large deviations other than those regarding threshold crossings.

9.4 Thresholds, stopping rules, and Wald's identity

In this section, we focus on random walks with both a positive and a negative threshold and ask questions about which threshold is crossed first and when that first threshold crossing occurs. Let J be the time at which the first of the two thresholds is crossed. To be explicit, Lemma 9.4.1 demonstrates the almost obvious fact that if X is not deterministically 0, then a threshold must be crossed eventually with probability 1. In other words, we show that J is a rv rather than a possibly-defective rv.

Figure 9.6 illustrates two sample paths and how they cross thresholds, say at $\alpha > 0$ and $\beta < 0$. More specifically, the random walk first crosses a threshold at trial n if $\beta < S_i < \alpha$ for $1 \leq i < n$ and either $S_n \geq \alpha$ or $S_n \leq \beta$. For now we make no assumptions about the mean or MGF of each X_i.

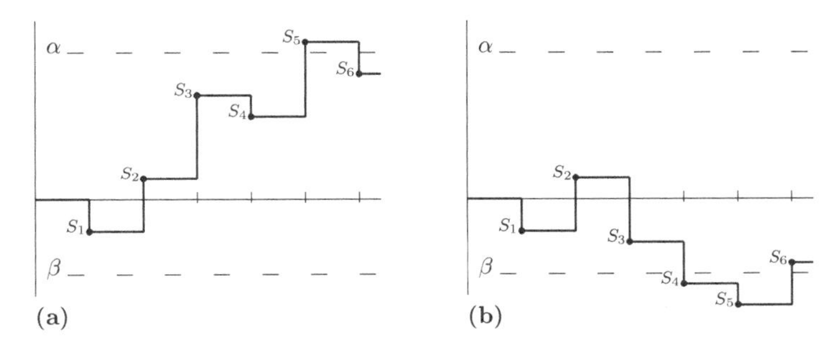

Figure 9.6 Two sample paths of a random walk with two thresholds. In (a), the threshold at α is crossed at $J = 5$. In (b), the threshold at β is crossed at $J = 4$.

Lemma 9.4.1 *Let $\{X_i; i \geq 1\}$ be IID rv s, not identically 0. For each $n \geq 1$, let $S_n = X_1 + \cdots + X_n$. Let $\alpha > 0$ and $\beta < 0$ be arbitrary, and let J be the smallest n for which either $S_n \geq \alpha$ or $S_n \leq \beta$. Then J is a rv (i.e., $\lim_{m\to\infty} \Pr\{J \geq m\} = 0$) and $r = 0$ is in the interior of the interval where the MGF of J, $\mathsf{E}\left[e^{rJ}\right]$ exists.*

Proof Since X is not identically 0, there is some n for which either $\Pr\{S_n \leq -\alpha + \beta\} > 0$ or for which $\Pr\{S_n \geq \alpha - \beta\} > 0$. For any such n, define ϵ by

$$\epsilon = \max[\Pr\{S_n \leq -\alpha + \beta\}, \ \Pr\{S_n \geq \alpha - \beta\}].$$

For any integer $k \geq 1$, given that $J > n(k-1)$, and given any value of $S_{n(k-1)}$ in (β, α), a threshold will be crossed by time nk with probability at least ϵ. Thus,

$$\Pr\{J > nk \mid J > n(k-1)\} \leq 1 - \epsilon.$$

Iterating on k,

$$\Pr\{J > nk\} \leq (1 - \epsilon)^k.$$

This shows that J is finite with probability 1 (WP1) and that $\Pr\{J \geq j\}$ goes to 0 at least geometrically in j. It follows that the MGF $\mathsf{g}_J(r)$ of J is finite in a region around $r = 0$, and that J has moments of all orders. □

The event $J = n$ (i.e., the event that a threshold is first crossed at trial n) is a function of $X_1, X_2, \dots, X_n$. Thus, in the notation of Section 5.5, J is a stopping trial for the sequence $\{X_n;\ n \geq 1\}$. In the following subsection, we derive Wald's identity for the two-threshold problem at hand. As will be seen, Wald's identity is closely related to Wald's equality. The Wald identity will be derived under more general conditions in Section 9.8.1.

9.4.1 Wald's identity for two thresholds

Theorem 9.4.2 (Wald's identity for two thresholds) *Let $\{X_i;\ i \geq 1\}$ be IID and let $\gamma(r) = \ln\{\mathsf{E}\left[e^{rX}\right]\}$. Let $I(X)$ be the interval of r over which $\gamma(r)$ exists. For each $n \geq 1$, let $S_n = X_1 + \cdots + X_n$. Let $\alpha > 0$ and $\beta < 0$ be arbitrary, and let J be the smallest n for which either $S_n \geq \alpha$ or $S_n \leq \beta$. Then for each $r \in I(X)$,*

$$\mathsf{E}\left[\exp(rS_J - J\gamma(r))\right] = 1. \tag{9.38}$$

Discussion The following proof uses the tilted probability distributions of Section 9.3.2. The theorem and proof are valid without the assumptions that $\mathsf{E}[X] < 0$ and $0 \in I(X)$.

Proof Assume that each X_i is discrete with the common PMF $\mathsf{p}_X(x)$. For the non-discrete case, the PMFs and their sums can be replaced by CDFs and Stieltjes integrals, thus complicating the technical details but not introducing any new ideas.

For any given $r \in I(X)$, we use the tilted PMF $\mathsf{q}_{X,r}(x)$ given in (9.13) as

$$\mathsf{q}_{X,\,r}(x) = \mathsf{p}_X(x)\exp[rx - \gamma(r)].$$

Using this tilted measure for each X_i in the n-tuple $\boldsymbol{X}^n = (X_1, X_2, \dots, X_n)$, and taking the X_i to be independent in the tilted probability measure, we have

$$\begin{aligned}\mathsf{q}_{\boldsymbol{X}^n,\,r}(\boldsymbol{x}^n) = \prod_{i=1}^{n} \mathsf{q}_{X_i,\,r}(x_i) \;&=\; \prod_{i=1}^{n} \mathsf{p}_{X_i}(x_i)\exp[rx_i - \gamma(r)] \\ &= \mathsf{p}_{\boldsymbol{X}^n}(\boldsymbol{x}^n)\exp[rs_n - n\gamma(r)], \qquad \text{where } s_n = \sum_{i=1}^{n} x_i.\end{aligned}$$

Now let $\mathcal{T}_n$ be the set of n-tuples $X_1, \dots, X_n$ such that $\beta < S_i < \alpha$ for $1 \leq i < n$ and either $S_n \geq \alpha$ or $S_n \leq \beta$. That is, $\mathcal{T}_n$ is the set of $\boldsymbol{x}^n$ for which the stopping trial J has the sample value n. The PMF for the stopping trial J in the tilted measure is then

$$\begin{aligned}\mathsf{q}_{J,\,r}(n) = \sum_{\boldsymbol{x}^n \in \mathcal{T}_n} \mathsf{q}_{\boldsymbol{X}^n,\,r}(\boldsymbol{x}^n) \;&=\; \sum_{\boldsymbol{x}^n \in \mathcal{T}_n} \mathsf{p}_{\boldsymbol{X}^n}(\boldsymbol{x}^n)\exp[rs_n - n\gamma(r)] \\ &= \mathsf{E}\left[\exp[rS_n - n\gamma(r) \mid J{=}n\right] \Pr\{J = n\},\end{aligned} \tag{9.39}$$

where the expectation refers to the original measure and the summation over $\mathcal{T}$ corresponds to $J = n$. Lemma 9.4.1 applies to the tilted PMF on this random walk as well as to the original PMF, and thus the sum of $\mathsf{q}_{J,\,r}(n)$ over n is 1. Summing the expression on the right-hand side of (9.39) over n yields $\mathsf{E}\left[\exp(rS_J - J\gamma(r))\right]$, completing the proof. □

After understanding the details of this proof, one sees that it is essentially just the statement that J is a non-defective stopping rule in the tilted probability space.

We next give a number of examples of how Wald's identity can be used.

9.4.2 The relationship of Wald's identity to Wald's equality

The trial J at which a threshold is crossed in Wald's identity is a stopping trial in the terminology of Chapter 5. For r in the interior of $I(X)$, the derivative of (9.38) with respect to r is

$$\mathsf{E}\left[(S_J - J\gamma'(r)) \exp\{rS_J - J\gamma(r)\}\right] = 0.$$

Assuming that 0 is in the interior of $I(X)$, we can set $r = 0$. Recalling that $\gamma(0) = 0$ and $\gamma'(0) = \overline{X}$, this becomes Wald's equality as established in Theorem 5.5.3:

$$\mathsf{E}[S_J] = \mathsf{E}[J]\,\overline{X}. \tag{9.40}$$

This is somewhat less general than Wald's equality as stated in Theorem 5.5.3, since we assume here that 0 is in the interior of $I(X)$, and we also assume two thresholds (which automatically satisfies the constraint in Wald's equality that $\mathsf{E}[J] < \infty$).

9.4.3 Zero-mean random walks

Wald's equality provides no information about $\mathsf{E}[J]$ when $\overline{X} = 0$, but Wald's identity does provide some useful information since the second derivative of (9.38) with respect to r is

$$\mathsf{E}\left[\left(S_J - J\gamma'(r)\right)^2 - J\gamma''(r)\right] \exp\{rS_J - J\gamma(r)\} = 0.$$

At $r = 0$, this is

$$\mathsf{E}\left[S_J^2 - 2JS_J\overline{X} + J^2\overline{X}^2 - J\,\sigma_X^2\right] = 0. \tag{9.41}$$

This equation is often difficult to use because of the cross term between S_J and J, but its main application comes in the case where $\overline{X} = 0$. In this case, (9.41) simplifies to

$$\mathsf{E}\left[S_J^2\right] = \sigma_X^2\,\mathsf{E}[J]\,. \tag{9.42}$$

Example 9.4.3 (The simple random walk with zero mean) Consider the simple random walk of Section 9.1.1 with $\Pr\{X{=}1\} = \Pr\{X{=}-1\} = 1/2$, and assume that $\alpha > 0$ and $\beta < 0$ are integers. Since S_n takes on only integer values and changes only by ± 1, it takes on the value α or β before exceeding either of these values. Thus S_J is either α or β. Let q_α denote $\Pr\{S_J = \alpha\}$. The expected value of S_J is then $\alpha q_\alpha + \beta(1 - q_\alpha)$. From Wald's equality, $\mathsf{E}[S_J] = 0$, so

$$q_\alpha = \frac{-\beta}{\alpha - \beta}; \qquad 1 - q_\alpha = \frac{\alpha}{\alpha - \beta}. \tag{9.43}$$

Using (9.42) with $\sigma_X^2 = 1$ and evaluating $\mathsf{E}\left[S_J^2\right]$,

$$\mathsf{E}[J] = \mathsf{E}\left[S_J^2\right] = \alpha^2 q_\alpha + \beta^2(1 - q_\alpha). \tag{9.44}$$

Substituting (9.43) into this and simplifying,

$$\mathsf{E}[J] = -\beta\alpha. \tag{9.45}$$

As a sanity check, note that if α and β are each multiplied by some large constant k, then $\mathsf{E}[J]$ increases by k^2. Since $\sigma^2_{S_n} = n$, we would expect S_n to fluctuate with increasing n, with typical values growing as $\sqrt{n}$, and thus it is reasonable that the expected time to reach a threshold increases with the product of the distances to the thresholds.

We also notice that if β is decreased toward $-\infty$, while holding α constant, then $q_\alpha \to 1$ and $\mathsf{E}[J] \to \infty$. This helps explain Example 5.5.4 where one plays a coin-tossing game, stopping when finally ahead. This shows that if the coin tosser has a finite capital β, i.e., stops on crossing either a positive threshold at 1 or a negative threshold at $-\beta$, then the coin tosser wins a small amount with high probability and loses a large amount with small probability.

For more general random walks with $\overline{X} = 0$, there is usually an overshoot when the threshold is crossed. If the magnitudes of α and β are large relative to the range of X, however, it is often reasonable to ignore the overshoots. Repeating the analysis of the simple random walk as an approximation, and including the value of σ^2_X, we get the approximation $\mathsf{E}[J] \approx -\beta\alpha/\sigma^2_X$.

9.4.4 Exponential bounds on the probability of threshold crossing

We next apply Wald's identity to complete the analysis of crossing a threshold at $\alpha > 0$ when $\overline{X} < 0$.

Corollary 9.4.4 *Under the conditions of Theorem 9.4.2, assume that $\overline{X} < 0$ and that $r^* > 0$ exists such that $\gamma(r^*) = 0$. Then*

$$\Pr\{S_J \geq \alpha\} \leq \exp(-r^*\alpha). \tag{9.46}$$

Proof Wald's identity, with $r = r^*$, reduces to $\mathsf{E}\left[\exp(r^*S_J)\right] = 1$. We can express this as

$$\Pr\{S_J \geq \alpha\}\,\mathsf{E}\left[\exp(r^*S_J) \mid S_J \geq \alpha\right] + \Pr\{S_J \leq \beta\}\,\mathsf{E}\left[\exp(r^*S_J) \mid S_J \leq \beta\right] = 1. \tag{9.47}$$

Since the second term on the left is non-negative,

$$\Pr\{S_J \geq \alpha\}\,\mathsf{E}\left[\exp(r^*S_J) \mid S_J \geq \alpha\right] \leq 1. \tag{9.48}$$

Given that $S_J \geq \alpha$, we see that $\exp(r^*S_J) \geq \exp(r^*\alpha)$. Thus

$$\Pr\{S_J \geq \alpha\}\exp(r^*\alpha) \leq 1, \tag{9.49}$$

which is equivalent to (9.46). □

This bound is valid for all $\beta < 0$ and thus it is clear intuitively that it is also valid in the absence of a lower threshold. When there is no lower threshold, however, the stopping rule becomes defective and the proof of Theorem 9.4.2 no longer holds. Exercise 9.14 verifies that (9.49) is still valid in the absence of a lower threshold.

We see from this that the case of a single threshold is little more than a special case of the two-threshold problem, but as seen in the zero-mean simple random walk, having a second threshold is often valuable in further understanding the single-threshold case.

Corollary 9.4.4 is also valid in the special case illustrated in Figure 9.4, where $\gamma(r) < 0$ for all $r \in (0, r_+]$ and we have defined r^* to be r_+. This is shown in Exercise 9.15.

The Chernoff bound in (9.36) shows that $\Pr\{S_n \geq \alpha\} \leq \exp(-r^*\alpha)$ for each n; (9.46) is typically much tighter, since it shows that $\exp(-r^*\alpha)$ is an upper bound on the probability of the union of these terms. As discussed earlier, however, the Chernoff bound also shows that the result is exponentially tight and it provides some intuition about the result.

When Corollary 9.4.4 is applied to the G/G/1 queue in Theorem 9.2.1, (9.46) is referred to as the *Kingman bound.*

Corollary 9.4.5 (Kingman bound) *Let $\{X_i;\ i \geq 1\}$ and $\{Y_i;\ i \geq 0\}$ be the interarrival intervals and service times of a G/G/1 queue that is empty at time 0 when customer 0 arrives. Let $\{U_i = Y_{i-1} - X_i;\ i \geq 1\}$, and let $\gamma(r) = \ln\{\mathsf{E}\left[e^{rU}\right]\}$ be the semi-invariant MGF of each U_i. Assume that $\gamma(r)$ has a root at $r^* > 0$. Then W_n, the queueing delay of the nth arrival, and W, the steady-state queueing delay, satisfy*

$$\Pr\{W_n \geq \alpha\} \leq \Pr\{W \geq \alpha\} \leq \exp(-r^*\alpha) \qquad \textit{for all } \alpha > 0. \tag{9.50}$$

For a random walk with $\overline{X} > 0$, the exceptional circumstance is $\Pr\{S_J \leq \beta\}$. This can be analyzed by changing the sign of X and β and using the results for a negative expected value. These exponential bounds are not valid for $\overline{X} = 0$, and we will not analyze that case here other than the result in (9.42).

Note that the simple bound on the probability of crossing the upper threshold in (9.46) (and thus also the Kingman bound) is an upper bound (rather than an equality) because, first, the effect of the lower threshold was eliminated (see (9.48)), and, second, the overshoot was bounded by 0 (see (9.49)). The effect of the second threshold can be taken into account by recognizing that $\Pr\{S_J \leq \beta\} = 1 - \Pr\{S_J \geq \alpha\}$. Then (9.47) can be solved, getting

$$\Pr\{S_J \geq \alpha\} = \frac{1 - \mathsf{E}\left[\exp(r^*S_J) \mid S_J{\leq}\beta\right]}{\mathsf{E}\left[\exp(r^*S_J) \mid S_J{\geq}\alpha\right] - \mathsf{E}\left[\exp(r^*S_J) \mid S_J{\leq}\beta\right]}. \tag{9.51}$$

Solving for the terms on the right-hand side of (9.51) usually requires analyzing the overshoot upon crossing a barrier, and this is often difficult (see Chapter 12 of [9]), for example). For the case of the simple random walk, overshoots do not occur, since the random walk changes only in unit steps. Thus, for α and β integers, we have $\mathsf{E}\left[\exp(r^*S_J) \mid S_J{\leq}\beta\right] = \exp(r^*\beta)$ and $\mathsf{E}\left[\exp(r^*S_J) \mid S_J{\geq}\alpha\right] = \exp(r^*\alpha)$. Substituting this in (9.51) yields the exact solution for the simple random walk:

$$\Pr\{S_J \geq \alpha\} = \frac{\exp(-r^*\alpha)[1 - \exp(r^*\beta)]}{1 - \exp[-r^*(\alpha - \beta)]}. \tag{9.52}$$

Solving the equation $\gamma(r^*) = 0$ for the simple random walk with probabilities p and q yields $r^* = \ln(q/p)$. This is also valid if X takes on the three values -1, 0, and $+1$ with $p = \Pr\{X = 1\}$, $q = \Pr\{X = -1\}$, and $1 - p - q = \Pr\{X = 0\}$. It can be seen that if α and $-\beta$ are large positive integers, then the simple bound of (9.46) is almost exact for this example.

Equation (9.52) is sometimes used as an approximation for (9.51) for general random walks. Unfortunately, for many applications, the overshoots are more significant than the effect of the opposite threshold. Thus (9.52) is only negligibly better than (9.46) as an approximation, and has the further disadvantage of not being a bound.

9.5 Binary hypotheses with IID observations

The objective of this section is to understand how to make a decision between two hypotheses, $X = 0$ or $X = 1$, on the basis of a sequence of observations $Y_1, Y_2, \ldots,$ with the property that $Y_1, Y_2, \ldots,$ are IID conditional on $X = 0$ and also IID conditional on $X = 1$. In Section 9.5.1 we will analyze the large deviation aspects of binary detection with a large but fixed number of observations. Section 9.5.2 follows this with an analysis of binary detection when the number of observations is variable, and in particular when the observer can choose when to make a decision based on the observations already made. We will see that this choice often reduces to a problem of threshold crossing in a random walk.

9.5.1 Binary hypotheses with a fixed number of observations

Consider the binary hypothesis testing problem of Section 8.2 in which X is a binary hypothesis with a priori probabilities p_0 and p_1. The observation $Y_1, Y_2, \ldots$ conditional on $X = 0$, is a sequence of IID rv s, each with the probability density $\mathsf{f}_{Y|X}(y \mid 0)$. Conditional on $X = 1$, the observations are IID, each with density $\mathsf{f}_{Y|X}(y \mid 1)$. For any given number n of sample observations, $y_1, \ldots, y_n$, the likelihood ratio is

$$\Lambda_n(\boldsymbol{y}) = \prod_{i=1}^{n} \frac{\mathsf{f}_{Y|X}(y_i \mid 1)}{\mathsf{f}_{Y|X}(y_i \mid 0)}.$$

The log-likelihood ratio (LLR), $s_n(\boldsymbol{y}) = \ln \Lambda(\boldsymbol{y})$ is then

$$s_n = \sum_{i=1}^{n} z_i, \qquad \text{where } z_i = \ln \frac{\mathsf{f}_{Y|X}(y_i \mid 1)}{\mathsf{f}_{Y|X}(y_i \mid 0)}. \tag{9.53}$$

The MAP test gives the maximum a posteriori probability of correct decision based on the n observations, $y_1, \ldots, y_n$. Specifically, it is the following threshold test, where the threshold η is given by $\eta = p_0/p_1$:

$$s_n \begin{cases} \geq \ln \eta; & \text{select } \hat{x}{=}1, \\ < \ln \eta; & \text{select } \hat{x}{=}0. \end{cases} \tag{9.54}$$

The Chernoff bound can be used to provide an exponentially tight bound on the probability of error, given $X = 0$, resulting from a threshold test. That is, conditional on

$X = 0$, S_n is a sum of n IID rv s $Z_1, \dots, Z_n$ whose sample values are given by (9.53) and whose PDF (conditional on $X = 0$) is determined by $\mathsf{f}_{Y|X}(y|0)$. Given $X = 0$, an error is made using the threshold test in (9.54) if $S_n \geq \eta$, i.e.,

$$\Pr\{e_\eta \mid X = 0\} = \Pr\{S_n \geq \ln \eta \mid X{=}0\}. \tag{9.55}$$

To upper bound this by the Chernoff bound, we use the semi-invariant MGF $\gamma_0(r)$ of Z given $X = 0$:

$$\begin{aligned} \gamma_0(r) &= \ln \int_y \mathsf{f}_{Y|X}(y \mid 0) \exp\left\{ r \left[\ln \frac{\mathsf{f}_{Y|X}(y \mid 1)}{\mathsf{f}_{Y|X}(y \mid 0)} \right] \right\} dy \\ &= \ln \int_y [\mathsf{f}_{Y|X}(y \mid 0)]^{1-r} [\mathsf{f}_{Y|X}(y \mid 1)]^r \, dy. \end{aligned} \tag{9.56}$$

The optimized Chernoff bound (conditional on $X = 0$) is then

$$\Pr\{e_\eta \mid X{=}0\} \leq \exp\left\{ n \min_{r \geq 0} [\gamma_0(r) - ra] \right\}, \qquad \text{where } a = \frac{1}{n} \ln \eta. \tag{9.57}$$

We can find the Chernoff bound for $\Pr\{e_\eta \mid X{=}1\} = \Pr\{S_n < \eta \mid X = 1\}$ in the same way. The rv $Z = \ln \big(f(y|1)/f(y|0)\big)$ conditional on $X = 1$ has a PDF arising from $f(y|1)$, so the semi-invariant MGF for Z conditional on $X = 1$ is given by

$$\gamma_1(r) = \ln \int_y [\mathsf{f}_{Y|X}(y \mid 0)]^{-r} [\mathsf{f}_{Y|X}(y \mid 1)]^{1+r} \, dy. \tag{9.58}$$

Since we are bounding the lower tail of S_n conditional on $X = 1$, the Chernoff bound is now optimized over $r \leq 0$, i.e., the optimized Chernoff bound (conditional on $X = 1$) is

$$\Pr\{e_\eta \mid X{=}1\} \leq \exp\left\{ n \min_{r \leq 0} [\gamma_1(r) - ra] \right\}, \qquad \text{where } a = \frac{1}{n} \ln \eta. \tag{9.59}$$

If we now compare $\gamma_0(r)$ from (9.56) with $\gamma_1(r)$ from (9.58), we find, surprisingly, that $\gamma_1(r) = \gamma_0(r + 1)$. If we substitute this into (9.59) and then change $r + 1$ throughout to r, this becomes

$$\Pr\{e_\eta \mid X{=}1\} \leq \exp\left\{ n \min_{r \leq 1} [\gamma_0(r) + (1{-}r)a] \right\}, \qquad \text{where } a = \frac{1}{n} \ln \eta. \tag{9.60}$$

If the minimization in (9.59) (for a given threshold η) occurs for $r \in (0, 1)$, then it can be seen by comparison of (9.59) and (9.60) that the minimization of (9.60) occurs at the same value of r. Figure 9.7 illustrates these optimizations of (9.57) and (9.60) together for a common threshold η.

Recall from Sections 8.3 and 8.4 that binary threshold tests are optimal not only for MAP detection but also for minimum-cost detection and (subject to a tie-breaking randomization) for the Neyman–Pearson rule.

As illustrated in Figure 9.7, the slope $\gamma_0'(r) = (1/n) \ln \eta$ at which the optimized bound occurs increases with η, varying from $\gamma'(0) = \mathsf{E}[Z \mid X{=}1]$ at $r = 0$ to $\gamma'(1) = \mathsf{E}[Z \mid X{=}0]$ at $r = 1$. The tradeoff between the two error exponents is seen to vary as the two ends of an inverted see-saw. One could, in principle, achieve a still

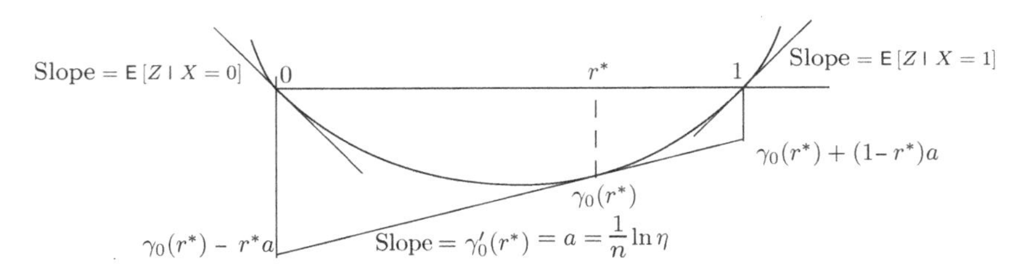

Figure 9.7 Graphical description of the optimization over r for (9.57) and (9.60). Note that since $\gamma_1(r) = \gamma_0(r+1)$ and $\gamma_1(0) = 0$, we must have $\gamma_0(1) = 0$; this can also be seen directly from (9.56). Also, since $\gamma''(r) > 0$ for $0 < r < 1$, we see that $\gamma'(0) < 0$ and $\gamma'(1) > 0$. Thus $\mathsf{E}[Z|X=0] < 0$ and the optimization for (9.57) is the same as that in Figure 9.3. Similarly, $\mathsf{E}[Z|X{=}1] > 0$ and the optimization for (9.60) uses the same principle. Assuming that $\gamma_0'(r) = a$ for some $r^* \in (0,1)$, both optimizations occur at that same value of r^*.

larger magnitude of exponent for $\Pr\{e_\eta \mid X{=}0\}$ by using $r > 1$, but this would be somewhat silly since $\Pr\{e_\eta \mid X{=}1\}$ would then be very close to 1 and it would usually make more sense to simply decide on $X = 0$ without looking at the observed data at all.

We can view the tradeoff of exponents above as a large deviation form of the Neyman–Pearson rule. The Neyman–Pearson rule is specified for a given number n of observations, and thus does not ask how the rule should be modified for different n. Let $e(n)$ denote the error event as a function of n and let us fix the allowable exponent to the error probability $\Pr\{e(n) \mid X{=}0\}$ to be exponential in n and then choose a test to minimize the exponent for $\Pr\{e(n) \mid X{=}1\}$. Since these bounds are exponentially tight, this gives us an appropriate large deviation result, focusing on behavior for large n. We can summarize these results in the following theorem.

Theorem 9.5.1 *Consider two hypotheses, $X = 0$ and $X = 1$. Conditional on each hypothesis, the observation is an IID n-rv $\mathbf{Y} = (Y_1, \ldots, Y_n)$ with conditional density $\mathsf{f}_{Y|X}(y|\ell)$ for $\ell = 0, 1$. Assume that 0 and 1 are in the interior of the interval where $\gamma_0(r) = \ln \int \mathsf{f}_{Y|X}^{1-r}(y|0)\mathsf{f}_{Y|X}^{r}(y|1)\,dy$ exists. Then for all integer $n \geq 1$ and all $r \in (0,1)$, the following bounds can be met simultaneously by using a threshold test at $\eta = e^{n\gamma_0'(r)}$:*

$$\Pr\{e(n) \mid X{=}0\} \leq \exp\left\{n[\gamma_0(r) - r\gamma_0'(r)]\right\}; \tag{9.61}$$

$$\Pr\{e(n) \mid X{=}1\} \leq \exp\left\{n[\gamma_0(r) + (1-r)\gamma_0'(r)]\right\}. \tag{9.62}$$

Furthermore, these bounds are exponentially tight in the sense that for a given $r \in (0,1)$ and any given $\epsilon > 0$, there is an n_o large enough so that for all $n > n_o$ and all tests, either

$$\Pr\{e(n) \mid X{=}0\} \geq \exp\left\{n[\gamma_0(r) - r\gamma_0'(r) - \epsilon]\right\} \tag{9.63}$$

or

$$\Pr\{e(n) \mid X{=}1\} \geq \exp\left\{n[\gamma_0(r) + (1-r)\gamma_0'(r) - \epsilon]\right\} \tag{9.64}$$

Proof Note that (9.61) and (9.62) are simply the result of the minimization over r in (9.57) and (9.60). These are Chernoff bounds, and from Theorem 9.3.3 are exponentially tight for fixed r, thus satisfying both (9.63) and (9.64) using threshold tests at $\eta = e^{n\gamma_0'(r)}$. Finally, since the tests here are threshold tests, and the threshold test for each n lies on the error curve, Theorem 8.4.2 implies that one or the other must be valid for arbitrary tests. □

Maximum likelihood (ML) tests provide an important special case of the theorem. We can regard ML as a threshold test with $\eta = 1$ (and a priori probabilities can be viewed as either equal or unknown). With $\eta = 1$, $\gamma_0'(r) = 0$ at the optimal r. The following error probabilities are then exponentially tight in the sense of the theorem:

$$\Pr\{e(n) \mid X = 0\} \le \exp\left[n \min_r \gamma_0(r)\right]; \qquad \Pr\{e(n) \mid X = 1\} \le \exp\left[n \min_r \gamma_0(r)\right]. \tag{9.65}$$

Another special case is the MAP test with given a priori probabilities, p_0, p_1. The threshold $\eta = p_0/p_1$ does not vary with increasing n. Theorem 9.5.1 is clearly applicable, but $\gamma_0'(r) = (1/n)\ln(\eta)$ and approaches 0 with increasing n.

In the limit as $n \to \infty$, the exponent $\gamma_0(r) - r\gamma_0'(r)$ approaches $\min_r \gamma_0(r)$ for any fixed η. Thus the simpler ML bound of (9.65) can be used in this case for large n. What is happening here is that the effect of η becomes unimportant in the exponent for very large n. In effect, there is so much information from the observations that the a priori probabilities are unimportant.

The opposite extreme from the ML case in Theorem 9.5.1 is that in which an error under one hypothesis, say $X = 1$, is so much more serious than an error under $X = 0$ that we want to maximize the exponential rate at which $\Pr\{e(n) \mid X{=}1\}$ approaches 0 with increasing n subject only to the constraint that $\Pr\{e(n) \mid X{=}0\}$ approaches 0, perhaps subexponentially, as $n \to \infty$.

The solution to this extreme case is known as Stein's lemma and is almost obvious from the geometry of Figure 9.7. By letting the point of tangency in the figure approach 0 with increasing n, the exponent for $\Pr\{e(n) \mid X{=}1\}$ approaches $\mathsf{E}[Z \mid X{=}0]$ and the exponent for $\Pr\{e(n) \mid X = 0\}$ approaches 0. Note that $\mathsf{E}[Z \mid X{=}0]$ is given by

$$\mathsf{E}[Z \mid X{=}0] = \int \mathsf{f}(y|0) \ln \frac{\mathsf{f}(y|1)}{\mathsf{f}(y|0)}\, dy = -D(\mathsf{f}(y|0)||\mathsf{f}(y|1)), \tag{9.66}$$

where the divergence $D\big(\mathsf{f}(y|0)\|\mathsf{f}(y|1)\big)$ is the continuous rv version of the discrete divergence in (9.28).

Corollary 9.5.2 (Stein's lemma) *For the conditions of Theorem 9.5.1, there is a sequence of tests for $n \ge 1$ for which*

$$\lim_{n\to\infty} (1/n) \ln \Pr\{e(n) \mid X{=}1\} = -D(\mathsf{f}(y|0)||\mathsf{f}(y|1)) \tag{9.67}$$

and

$$\lim_{n\to\infty} \Pr\{e \mid X = 0\} = 0. \tag{9.68}$$

Given (9.68), the result in (9.67) is exponentially tight.

A guided proof, showing how to take the limit as $r \to 0$ and $n \to \infty$ is given in Exercise 9.20.

Divergence has a somewhat broader interpretation here than as the exponent of a composition in Section 9.3.3. In both cases, it is the exponent for the probability that a sequence from one distribution 'looks like' a sequence from another distribution, but in the composition case, it is the exponent for the probability of an exact match, whereas here it is the exponent for the probability of a set of sequences that is large enough to have a probability close to 1 for the opposite hypothesis.

9.5.2 Sequential decisions for binary hypotheses

Common sense tells us that it should be valuable to make additional observations when the current observations do not lead to a clear choice. With such a possibility, there are three possible choices at the end of each observation: choose $\hat{x} = 1$, choose $\hat{x} = 0$, or continue with additional observations. We will analyze such a setup by viewing it as a threshold crossing problem. We establish two thresholds, $\alpha > 0$ and $\beta < 0$. We then look at a pair of random walks. The first random walk is the sequence of LLRs given that $X = 0$ and the second is the sequence of LLRs given that $X = 1$.

The observer sees the successive values of the random walk, but does not know whether it is the random walk conditional on $X = 0$ or on $X = 1$. If the random walk first crosses threshold α, however, the decision $\hat{x} = 1$ is made. Conversely, if the threshold β is first crossed, the decision $\hat{x} = 0$ is made.

Let us first analyze the random walk conditional on $X = 0$. The random walk is then the sequence of LLRs $\{S_n; n \geq 1\}$, where $S_n = \sum_{i=1}^{n} Z_i$ and $Z_i = \ln\big(\mathsf{f}(y_i|1)/\mathsf{f}(y_i|0)\big)$. Because of the conditioning on $X = 0$, Y_i has the PDF $\mathsf{f}_{Y|X}(y_i|0)$. The stopping rule is to stop when the random walk crosses either a threshold[3] at $\alpha > 0$ or a threshold at $\beta < 0$

Since the decision $\hat{x} = 1$ is made if $S_J \geq \alpha$ and $\hat{x} = 0$ is made if $S_J \leq \beta$, we see that for the random walk conditional on $X = 0$, an error is made if $S_J \geq \alpha$. We denote the error event as $e(J)$. Using the probability distribution for $X = 0$, we apply (9.46), along with $r^* = 1$, to get

$$\Pr\{e(J) \mid X{=}0\} \;=\; \Pr\{S_J \geq \alpha \mid X{=}0\} \;\leq\; \exp(-\alpha). \tag{9.69}$$

Given $X = 1$, $S_n = \sum_{i=1}^{n} Z_i$ is also a random walk, but the probability measure is different. That is, $Z_i = \ln\big(\mathsf{f}(y_i|1)/\mathsf{f}(y_i|0)\big)$, where y_i now has the PDF $\mathsf{f}_{Y|X}(y_i|1)$. The same stopping rule must be used, since the decision to stop at n can be based only on $Z_1, \ldots, Z_n$ and not on knowledge of X.

For the random walk conditional on $X = 1$, an error is made if the threshold at β is crossed before that at α. Either by recognizing the fundamental symmetry between $X = 0$ and $X = 1$ or by repeating the analysis leading to (9.46) for a lower bound rather than an upper bound, we get

$$\Pr\{e(J) \mid X{=}1\} \;=\; \Pr\{S_J \leq \beta \mid X{=}1\} \;\leq\; \exp(\beta). \tag{9.70}$$

[3] It is unfortunate that the word 'threshold' has a universally accepted meaning for random walks (i.e., the meaning we are using here), and the word 'threshold test' has a universally accepted meaning for hypothesis testing. Threshold tests for hypothesis testing, as used in the previous section and in Chapter 8, are non-sequential. The test here, however, is sequential, at each epoch making a ternary choice between continuing or stopping with $H = 0$, or stopping with $H = 1$.

It is rather surprising that these bounds do not depend at all on the likelihoods involved in the decision and that they are so simple. It turns out that the distribution of J conditional on $X = 0$ and $X = 1$ does involve the likelihoods. Also the exact values of $\Pr\{S_J \le \beta \mid X{=}0\}$ and $\Pr\{S_J \le \beta \mid X{=}1\}$ depend on the likelihoods.

The error probabilities can be made as small as desired by increasing the magnitudes of α and β, but there is a cost involved in increasing these magnitudes. The cost in increasing α is essentially to increase the number of observations required when $X = 1$. From Wald's equality,

$$\mathsf{E}[J \mid X{=}1] \;=\; \frac{\mathsf{E}[S_J \mid X{=}1]}{\mathsf{E}[Z \mid X{=}1]} \;\approx\; \frac{\alpha + \mathsf{E}[\text{overshoot} \mid S_J \ge \alpha]}{\mathsf{E}[Z \mid X{=}1]}.$$

In the approximation, we have ignored the possibility of S_J crossing the threshold at α conditional on $X = 1$ since this is a very-small-probability event when α and β have large magnitudes. Thus we see that the expected number of observations (given $X = 1$) is essentially linear in $|\beta|$. Similarly,

$$\mathsf{E}[J \mid X{=}0] \;=\; \frac{\mathsf{E}[S_J \mid X{=}0]}{\mathsf{E}[Z \mid X{=}0]} \;\approx\; \frac{\beta + \mathsf{E}[\text{overshoot} \mid S_J \le \beta]}{\mathsf{E}[Z \mid X{=}0]}.$$

This gives us another interpretation of divergence, since we recall that $\mathsf{E}[Z \mid X{=}1] = D(\mathsf{f}(y|1)\|\mathsf{f}(y|0))$ and $\mathsf{E}[Z \mid X{=}0] = -D(\mathsf{f}(y|0)\|\mathsf{f}(y|1))$. More specifically, the ratio of the expected number of observations when $X = 1$ to the exponent of error probability when $X = 0$ is $D(\mathsf{f}(y|1)\|\mathsf{f}(y|0))$. Similarly, the ratio of the expected number of observations when $X = 0$ to the exponent of error probability when $X = 1$ is $D(\mathsf{f}(y|0)\|\mathsf{f}(y|1))$. In other words, the tradeoff between one error probability and the opposite number of observations is given by the corresponding divergence. Thus the two divergences are the most significant quantities involved in sequential decisions.

We next ask what has been gained quantitatively by using the sequential decision procedure here. Suppose we compare the sequential procedure to a fixed-length test with $n = \alpha/\mathsf{E}[Z \mid X{=}1]$. Referring to Figure 9.7, we see that if we choose the slope $a = \gamma_0'(1) = \mathsf{E}[Z \mid X{=}1]$, then the (exponentially tight) Chernoff bound on $\Pr\{e \mid X{=}0\}$ is given by $e^{-\alpha}$, but the exponent on $\Pr\{e \mid X{=}1\}$ is 0. In other words, by using a sequential test as described here, we simultaneously get the error exponent for $X = 0$ that a fixed test would provide if we gave up entirely on an error exponent for $X = 1$, and vice-versa.[4]

A final question to be asked is whether any substantial improvement on this sequential decision procedure would result from letting the thresholds at α and β vary with the number of observations. Assuming that we are concerned only with the expected number of observations, the answer is no. We will not carry this argument out here, but it consists of using the Chernoff bound as a function of the number of observations. This shows that there is a typical number of observations at which most errors occur, and changes in the thresholds elsewhere can increase the error probability, but not substantially decrease it.

[4] In the communication context, decision rules are used to detect sequentially transmitted data. The use of a sequential decision rule usually requires feedback from receiver to transmitter, and also requires a variable rate of transmission. Thus the substantial reductions in error probability are accompanied by substantial system complexity.

9.6 Martingales

Definition 9.6.1 *A **martingale** is an integer-time stochastic process $\{Z_n;\ n \geq 1\}$ with the properties that* $\mathsf{E}[|Z_n|] < \infty$ *for all $n \geq 1$ and*

$$\mathsf{E}\left[Z_n \mid Z_{n-1}, Z_{n-2}, \ldots, Z_1\right] = Z_{n-1}; \qquad \textit{for all } n \geq 2. \tag{9.71}$$

The name martingale comes from gambling terminology where martingales refer to gambling strategies in which the amount to be bet is determined by the past history of winning or losing. If one visualizes Z_n as representing the gambler's fortune at the end of the nth play, the definition above means, first, that the game is fair (in the sense that the expected increase in fortune from play $n-1$ to n is zero), and, second, that the expected fortune on the nth play depends on the past only through the fortune on play $n-1$.

The important part of the definition of a martingale, and what distinguishes martingales from other kinds of processes, is the form of dependence in (9.71). However, the restriction that $\mathsf{E}[|Z_n|] < \infty$ is also important, particularly since martingales are so abstract and general that one often loses the insight to understand intuitively when this restriction is important. Students are advised to ignore this restriction when first looking at something that might be a martingale, and to check later after acquiring some understanding.

There are two interpretations of (9.71); the first and most straightforward is to view it as shorthand for $\mathsf{E}\left[Z_n \mid Z_{n-1}{=}z_{n-1},\ Z_{n-2}{=}z_{n-2},\ \ldots,\ Z_1{=}z_1\right] = z_{n-1}$ for all possible sample values $z_1, z_2, \ldots, z_{n-1}$. The second is that $\mathsf{E}\left[Z_n \mid Z_{n-1}{=}z_{n-1},\ \ldots,\ Z_1{=}z_1\right]$ is a function of the sample values $z_1, \ldots, z_{n-1}$ and thus $\mathsf{E}\left[Z_n \mid Z_{n-1}, \ldots, Z_1\right]$ is a rv which is a function of the rv s $Z_1, \ldots, Z_{n-1}$ (and, for a martingale, a function only of Z_{n-1}). Students are encouraged to take the first viewpoint initially and to write out the expanded type of expression in cases of confusion. The second viewpoint, however, is very powerful, and, with experience, is the more useful viewpoint.

It is important to understand the difference between martingales and Markov chains. For the Markov chain $\{X_n;\ n \geq 1\}$, each rv X_n is conditioned on the past only through X_{n-1}, whereas for the martingale $\{Z_n;\ n \geq 1\}$, it is only the expected value of Z_n that is conditioned on the past only through Z_{n-1}. The rv Z_n itself, conditioned on Z_{n-1}, can also be dependent on all the earlier Z_i's. It is very surprising that so many results can be developed using such a weak form of conditioning.

In what follows, we give a number of important examples of martingales, then develop some results about martingales, and then discuss those results in the context of the examples.

9.6.1 Simple examples of martingales

Example 9.6.2 (Random walks) One example of a martingale is a zero-mean random walk, since if $Z_n = X_1 + X_2 + \cdots + X_n$, where the X_i are IID and zero mean, then

$$\mathsf{E}\left[Z_n \mid Z_{n-1}, \ldots, Z_1\right] = \mathsf{E}\left[X_n + Z_{n-1} \mid Z_{n-1}, \ldots, Z_1\right] \tag{9.72}$$

$$= \mathsf{E}[X_n] + Z_{n-1} \ = \ Z_{n-1}. \tag{9.73}$$

Extending this example, suppose that $\{X_i;\ i \geq 1\}$ is an arbitrary sequence of IID rv s with mean $\overline{X}$ and let $\widetilde{X}_i = X_i - \overline{X}$. Then $\{S_n;\ n \geq 1\}$ is a random walk with $S_n = X_1 + \cdots + X_n$ and $\{Z_n;\ n \geq 1\}$ is a martingale with $Z_n = \widetilde{X}_1 + \cdots + \widetilde{X}_n$. The random walk and the martingale are simply related by $Z_n = S_n - n\overline{X}$, and thus general results about martingales can easily be applied to arbitrary random walks.

Example 9.6.3 (Sums of dependent zero-mean rv s) Let $\{X_i;\ i \geq 1\}$ be a sequence of dependent rv s satisfying $\mathsf{E}\left[X_i \mid X_{i-1}, \ldots, X_1\right] = 0$ for all $i > 0$. Then $\{Z_n;\ n \geq 1\}$, where $Z_n = X_1 + \cdots + X_n$, is a zero-mean martingale. This can be seen by induction as follows:

$$\begin{aligned}\mathsf{E}\left[Z_n \mid Z_{n-1}, \ldots, Z_1\right] &= \mathsf{E}\left[X_n + Z_{n-1} \mid Z_{n-1}, \ldots, Z_1\right] \\ &= \mathsf{E}\left[X_n \mid X_{n-1}, \ldots, X_1\right] + \mathsf{E}\left[Z_{n-1} \mid Z_{n-1}, \ldots, Z_1\right] = Z_{n-1}.\end{aligned}$$

This is a more general example than it appears, since given any martingale $\{Z_n;\ n \geq 1\}$, we can define $X_n = Z_n - Z_{n-1}$ for $n \geq 2$ with $X_1 = Z_1$. Then $\mathsf{E}\left[X_n \mid X_{n-1}, \ldots, X_1\right] = 0$ for $n \geq 2$. If the martingale is zero mean (i.e., if $\mathsf{E}[Z_1] = 0$), then $\mathsf{E}[X_1] = 0$ also. This means that results for zero-mean martingales can be translated into results about dependent conditionally zero-mean rv s and vice-versa.

Example 9.6.4 (Product-form martingales) Another example is a product of unit mean IID rv s. Thus if $Z_n = X_1 X_2 \cdots X_n$, we have

$$\begin{aligned}\mathsf{E}\left[Z_n \mid Z_{n-1}, \ldots, Z_1\right] &= \mathsf{E}\left[X_n Z_{n-1} \mid Z_{n-1}, \ldots, Z_1\right] \\ &= \mathsf{E}[X_n]\,\mathsf{E}\left[Z_{n-1} \mid Z_{n-1}, \ldots, Z_1\right] \qquad (9.74) \\ &= \mathsf{E}[X_n]\,\mathsf{E}\left[Z_{n-1} \mid Z_{n-1}\right] = Z_{n-1}. \qquad (9.75)\end{aligned}$$

A particularly simple case of this product example is where $X_n = 2$ with probability $1/2$ and $X_n = 0$ with probability $1/2$. Then for each $n \geq 1$,

$$\Pr\left\{Z_n = 2^n\right\} = 2^{-n}; \qquad \Pr\{Z_n = 0\} = 1 - 2^{-n}; \qquad \mathsf{E}[Z_n] = 1. \qquad (9.76)$$

Thus $\lim_{n\to\infty} Z_n = 0$ WP1, but $\mathsf{E}[Z_n] = 1$ for all n and $\lim_{n\to\infty} \mathsf{E}[Z_n] = 1$. This is an important example to keep in mind when trying to understand why proofs about martingales are necessary and non-trivial. This type of phenomenon will be clarified somewhat by Lemma 9.8.4 when we discuss stopped martingales in Section 9.8.

An important example of a product-form martingale is as follows: let $\{X_i;\ i \geq 1\}$ be an IID sequence, and let $\{S_n = X_1 + \cdots + X_n;\ n \geq 1\}$ be a random walk. Assume that the semi-invariant MGF $\gamma(r) = \ln\{\mathsf{E}\left[\exp(rX)\right]\}$ exists for some given r. For each $n \geq 1$, let Z_n be defined as

$$\begin{aligned}Z_n &= \exp\{rS_n - n\gamma(r)\} \qquad (9.77) \\ &= \exp\{rX_n - \gamma(r)\}\exp\{rS_{n-1} - (n-1)\gamma(r)\} \\ &= \exp\{rX_n - \gamma(r)\}Z_{n-1}. \qquad (9.78)\end{aligned}$$

Taking the conditional expectation of this,

$$\begin{aligned} \mathsf{E}\left[Z_n \mid Z_{n-1}, \ldots, Z_1\right] &= \mathsf{E}\left[\exp(rX_n - \gamma(r))\right] \mathsf{E}\left[Z_{n-1} \mid Z_{n-1}, \ldots, Z_1\right] \\ &= Z_{n-1}, \end{aligned} \tag{9.79}$$

where we have used the fact that $\mathsf{E}\left[\exp(rX_n)\right] = \exp(\gamma(r))$. Thus we see that $\{Z_n;\ n \geq 1\}$ is a martingale of the product form.

9.6.2 Scaled branching processes

A final simple example of a martingale is a 'scaled down' version of a branching process $\{X_n;\ n \geq 0\}$. Recall from Section 6.7 that, for each n, X_n is defined as the aggregate number of elements in generation n. Each element i of generation n, $1 \leq i \leq X_n$ has a number of offspring $Y_{i,n}$ which collectively constitute generation $n+1$, i.e., $X_{n+1} = \sum_{i=1}^{X_n} Y_{i,n}$. The rv s $Y_{i,n}$ are IID over both i and n.

Let $\overline{Y} = \mathsf{E}\left[Y_{i,n}\right]$ be the mean number of offspring of each element of the population. Then $\mathsf{E}\left[X_n \mid X_{n-1}\right] = \overline{Y}X_{n-1}$, which resembles a martingale except for the factor of $\overline{Y}$. We can convert this branching process into a martingale by scaling it. That is, define $Z_n = X_n/\overline{Y}^n$. It follows that

$$\mathsf{E}\left[Z_n \mid Z_{n-1}, \ldots, Z_1\right] = \mathsf{E}\left[\frac{X_n}{\overline{Y}^n} \mid X_{n-1}, \ldots, X_1\right] = \frac{\overline{Y}X_{n-1}}{\overline{Y}^n} = Z_{n-1}. \tag{9.80}$$

Thus $\{Z_n;\ n \geq 1\}$ is a martingale. We will see the surprising result later that this implies that Z_n converges WP1 to a limiting rv as $n \to \infty$.

9.6.3 Partial isolation of past and future in martingales

Recall that for a Markov chain, the states at all times greater than a given n are independent of the states at all times less than n, conditional on the state at time n. The following lemma shows that at least a small part of this independence of past and future applies to martingales.

Lemma 9.6.5 *Let $\{Z_n;\ n \geq 1\}$ be a martingale. Then for any $n > i \geq 1$,*

$$\mathsf{E}\left[Z_n \mid Z_i, Z_{i-1}, \ldots, Z_1\right] = Z_i. \tag{9.81}$$

Proof For $n = i+1$, $\mathsf{E}\left[Z_{i+1} \mid Z_i, \ldots, Z_1\right] = Z_i$ by the definition of a martingale. Similarly, for $n = i+2$,

$$\mathsf{E}\left[Z_{i+2} \mid Z_{i+1}, \ldots, Z_1\right] = Z_{i+1}.$$

The expectation on the left is a rv that is a function of $Z_1, \ldots, Z_{i+1}$, and that function, according to the equation, is Z_{i+1}. If we take the expectation of this (i.e., of Z_{i+1}) conditional on $Z_i, \ldots, Z_1$, we get

$$\mathsf{E}\left[Z_{i+2}|Z_i,\ldots,Z_1\right] = \mathsf{E}\left[Z_{i+1} \mid Z_i,\ldots,Z_1\right] \;=\; Z_i. \tag{9.82}$$

For $n = i + 3$, (9.82), with i incremented, shows us that the rv $\mathsf{E}\left[Z_{i+3} \mid Z_{i+1},\ldots,Z_1\right]$ (which is, in general, a function of $Z_1,\ldots,Z_{i+1}$) is equal to Z_{i+1}. Taking the conditional expectation of this rv over Z_{i+1} conditional on $Z_i,\ldots,Z_1$, we get

$$\mathsf{E}\left[Z_{i+3} \mid Z_i,\ldots,Z_1\right] = Z_i.$$

This argument can be applied successively to any $n > i$. □

This lemma is particularly important for $i = 1$, where it says that $\mathsf{E}[Z_n \mid Z_1] = Z_1$. The left-hand side of this is a rv which is a function (in fact the identity function) of Z_1. Thus, by taking the expected value of each side, we see that

$$\mathsf{E}[Z_n] = \mathsf{E}[Z_1] \qquad \text{for all } n > 1. \tag{9.83}$$

9.7 Submartingales and supermartingales

Submartingales and supermartingales are simple generalizations of martingales that provide many useful results for very little additional work. We will subsequently derive the Kolmogorov submartingale inequality, which is a powerful generalization of the Markov inequality. We use this both to give a simple proof of the strong law of large numbers (SLLN) and also to better understand threshold crossing problems for random walks.

Definition 9.7.1 *A* ***submartingale*** *is an integer-time stochastic process* $\{Z_n;\, n \geq 1\}$ *that satisfies the relations*

$$\mathsf{E}[|Z_n|] < \infty; \quad \mathsf{E}\left[Z_n \mid Z_{n-1}, Z_{n-2},\ldots,Z_1\right] \geq Z_{n-1}; \quad n \geq 1. \tag{9.84}$$

A ***supermartingale*** *is an integer-time stochastic process* $\{Z_n;\, n \geq 1\}$ *that satisfies the relations*

$$\mathsf{E}[|Z_n|] < \infty; \quad \mathsf{E}\left[Z_n \mid Z_{n-1}, Z_{n-2},\ldots,Z_1\right] \leq Z_{n-1}; \quad n \geq 1. \tag{9.85}$$

In terms of our gambling analogy, a submartingale corresponds to a game that is at least fair, i.e., where the expected fortune of the gambler either increases or remains the same. A *supermartingale* is a process with the opposite type of inequality. The prefixes *sub* and *super* are the opposites of what common sense would dictate, but these definitions are too standard to change.

Since a martingale satisfies both (9.84) and (9.85) with equality, a martingale is both a submartingale and a supermartingale. Note that if $\{Z_n;\, n \geq 1\}$ is a submartingale, then $\{-Z_n;\, n \geq 1\}$ is a supermartingale, and conversely. Thus, some of the results to follow are stated only for submartingales, with the understanding that they can be applied to supermartingales by changing signs as above.

Lemma 9.6.5, with the equality replaced by inequality, also applies to submartingales and supermartingales. That is, if $\{Z_n;\, n \geq 1\}$ is a submartingale, then

$$\mathsf{E}\left[Z_n \mid Z_i, Z_{i-1},\ldots,Z_1\right] \geq Z_i, \quad 1 \leq i < n, \tag{9.86}$$

and if $\{Z_n;\ n \geq 1\}$ is a supermartingale, then

$$\mathsf{E}\left[Z_n \mid Z_i, Z_{i-1}, \dots, Z_1\right] \leq Z_i, \quad 1 \leq i < n. \tag{9.87}$$

Equations (9.86) and (9.87) are verified in the same way as Lemma 9.6.5 (see Exercise 9.24). Similarly, the appropriate generalization of (9.83) is that if $\{Z_n;\ n \geq 1\}$ is a submartingale, then

$$\mathsf{E}\left[Z_n\right] \geq \mathsf{E}\left[Z_i\right] \qquad \text{for all } i,\ 1 \leq i < n \tag{9.88}$$

and if $\{Z_n;\ n \geq 1\}$ is a supermartingale, then

$$\mathsf{E}\left[Z_n\right] \leq \mathsf{E}\left[Z_i\right] \qquad \text{for all } i,\ 1 \leq i < n. \tag{9.89}$$

A random walk $\{S_n;\ n \geq 1\}$ with $S_n = X_1 + \cdots + X_n$ is a submartingale, martingale, or supermartingale respectively for $\overline{X} \geq 0$, $\overline{X} = 0$, or $\overline{X} \leq 0$. Also, if X has a semi-invariant MGF $\gamma(r)$ for some given r, and if Z_n is defined as $Z_n = \exp(rS_n)$, then the process $\{Z_n;\ n \geq 1\}$ is a submartingale, martingale, or supermartingale respectively for $\gamma(r) \geq 0$, $\gamma(r) = 0$, or $\gamma(r) \leq 0$. The next example gives an important way in which martingales and submartingales are related.

Example 9.7.2 (Convex functions of martingales) Figure 9.8 illustrates the graph of a convex function h from $\mathbb{R}$ to $\mathbb{R}$. Recall that h is convex if

$$\mu h(x_1) + (1-\mu)h(x_2) \geq h(\mu x_1 + (1-\mu)x_2)$$

for every real x_1, x_2 and every $\mu, 0 < \mu < 1$.

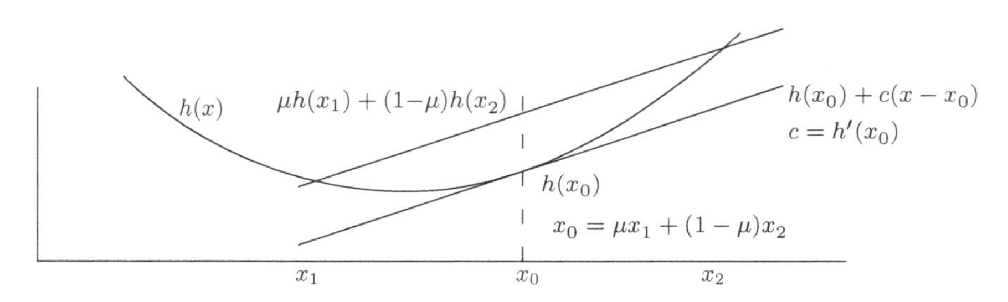

Figure 9.8 Convex function $h(x)$: every chord of $h(x)$ lies on or above the curve. Equivalently, for each x_0, there is a c such that, for all x, $h(x_0) + c(x - x_0) \leq h(x)$, i.e., all tangents of h lie on or below the curve.. If h is differentiable at x_0, then c is the derivative of h at x_0.

Geometrically, this says that every chord of h lies on or above h. By lowering each chord while holding its slope constant, we see that an equivalent condition is that each tangent to h lies on or below h. If $h(x)$ has a derivative at x_0, then the slope c of the tangent at x_0 is the value of that derivative and $h(x_0) + c(x - x_0)$ is the equation, as a function of x, for the tangent line at x_0. If $h(x)$ has a discontinuous slope at x_0, then there might be many choices for c; for example, $h(x) = |x|$ is convex, and for $x_0 = 0$, one could choose any c in the range -1 to $+1$ to form a tangent.

A simple condition that implies convexity is a non-negative second derivative everywhere. This is not a necessary condition, however, and functions (such as $|x|$) are convex even when the first derivative does not exist everywhere.

Lemma 9.7.3 (Jensen's inequality) *If h is a convex function from $\mathbb{R}$ to $\mathbb{R}$ and X is a rv with $\mathsf{E}[|X|] < \infty$, then*

$$h(\mathsf{E}[X]) \leq \mathsf{E}[h(X)]. \tag{9.90}$$

Proof Let $x_0 = \mathsf{E}[X]$ and choose c so that $h(x_0) + c(x - x_0) \leq h(x)$ for all x. Using the rv X in place of x and taking expected values of both sides, we get (9.90). □

Note that for any particular event A, this same argument applies to X conditional on A, so that $h(\mathsf{E}[X \mid A]) \leq \mathsf{E}[h(X) \mid A]$. Jensen's inequality is very widely used; it is a minor miracle that we have not required it previously.

Theorem 9.7.4 *Assume that h is a convex function from $\mathbb{R}$ to $\mathbb{R}$, that $\{Z_n;\ n \geq 1\}$ is a martingale and that $\mathsf{E}[|h(Z_n)|] < \infty$ for all n. Then $\{h(Z_n);\ n \geq 1\}$ is a submartingale.*

Proof For any choice of $z_1, \ldots, z_{n-1}$, we can use Jensen's inequality with the conditioning event to get

$$\mathsf{E}\left[h(Z_n)|Z_{n-1}{=}z_{n-1}, \ldots, Z_1{=}z_1\right] \geq h\left(\mathsf{E}\left[Z_n \mid Z_{n-1}{=}z_{n-1}, \ldots, Z_1{=}z_1\right]\right) = h(z_{n-1}). \tag{9.91}$$

For any choice of numbers $h_1, \ldots, h_{n-1}$ in the range of the function h, let $z_1, \ldots, z_{n-1}$ be arbitrary numbers satisfying $h(z_1){=}h_1, \ldots, h(z_{n-1}){=}h_{n-1}$. For each such choice, (9.91) holds, so that

$$\begin{aligned}\mathsf{E}\left[h(Z_n) \mid h(Z_{n-1}){=}h_{n-1}, \ldots, h(Z_1){=}h_1\right] &\geq h\left(\mathsf{E}\left[Z_n \mid h(Z_{n-1}){=}h_{n-1}, \ldots, h(Z_1){=}h_1\right]\right)\\ &= h(z_{n-1}) = h_{n-1},\end{aligned} \tag{9.92}$$

completing the proof. □

Some examples of this result, applied to a martingale $\{Z_n;\ n \geq 1\}$, are as follows:

$$\{|Z_n|;\ n \geq 1\} \ \text{ is a submartingale} \tag{9.93}$$

$$\{Z_n^2;\ n \geq 1\} \text{ is a submartingale if } \mathsf{E}\left[Z_n^2\right] < \infty;\ n \geq 1 \tag{9.94}$$

$$\{\exp(rZ_n);\ n \geq 1\} \ \text{ is a submartingale if } \mathsf{E}\left[\exp(rZ_n)\right] < \infty;\ n \geq 1. \tag{9.95}$$

A function of a real variable $h(x)$ is defined to be concave if $-h(x)$ is convex. It then follows from Theorem 9.7.4 that if h is concave and $\{Z_n;\ n \geq 1\}$ is a martingale, then $\{h(Z_n);\ n \geq 1\}$ is a supermartingale (assuming that $\mathsf{E}[|h(Z_n)|] < \infty$). For example, if $\{Z_n;\ n \geq 1\}$ is a positive martingale and $\mathsf{E}[|\ln(Z_n)|] < \infty$, then $\{\ln(Z_n);\ n \geq 1\}$ is a supermartingale.

9.8 Stopped processes and stopping trials

The definition of stopping trials in Section 5.5 applies to arbitrary integer-time processes $\{Z_n;\ n \geq 1\}$ as well as to IID sequences. Recall that J is a stopping trial for a sequence $\{Z_n;\ n \geq 1\}$ of rv s if $\mathbb{I}_{J=n}$ is a function of $Z_1, \ldots, Z_n$ and if J is a rv.

If $\mathbb{I}_{J=n}$ is a function of $Z_1, \ldots, Z_n$ and J is a defective rv, then J is called a defective stopping trial. For some of the results to follow, it is unimportant whether J is a rv or a defective rv (i.e., whether or not the process stops with probability 1). If it is not specified whether J is a rv or a defective rv, we refer to the stopping trial as a *possibly-defective stopping trial*; we consider J to take on the value ∞ if the process does not stop.

Definition 9.8.1 *A* ***stopped process*** $\{Z_n^*;\ n \geq 1\}$ *for a possibly-defective stopping trial* J *relative to a process* $\{Z_n;\ n \geq 1\}$ *is the process for which* $Z_n^* = Z_n$ *for* $n \leq J$ *and* $Z_n^* = Z_J$ *for* $n > J$.

As an example, suppose Z_n models the fortune of a gambler at the completion of the nth trial of some game, and suppose the gambler then modifies the game by deciding to stop gambling under some given circumstances (i.e., at the stopping trial). Thus, after stopping, the fortune remains constant, so the stopped process models the gambler's fortune in time, including the effect of the stopping trial.

As another example, consider a random walk with a positive and a negative threshold, and consider the process to stop after reaching or crossing a threshold. The stopped process then stays at that point after the threshold crossing as an artifice to simplify analysis. The use of stopped processes is similar to the artifice that we employed in Section 4.5 for first-passage times in Markov chains; recall that we added an artificial trapping state after the desired passage to simplify analysis.

We next show that the possibly-defective stopped process of a martingale is itself a martingale; the intuitive reason is that, before stopping, the stopped process is the same as the martingale, and, after stopping, $Z_n^* = Z_{n-1}^*$. The following theorem establishes this and the corresponding results for submartingales and supermartingales.

Theorem 9.8.2 *Given a stochastic process* $\{Z_n;\ n \geq 1\}$ *and a possibly-defective stopping trial* J *for the process, the stopped process* $\{Z_n^*;\ n \geq 1\}$ *is a submartingale if* $\{Z_n;\ n \geq 1\}$ *is a submartingale, is a martingale if* $\{Z_n;\ n \geq 1\}$ *is a martingale, and is a supermartingale if* $\{Z_n;\ n \geq 1\}$ *is a supermartingale.*

Proof First we show that, for all three cases, the stopped process satisfies $\mathsf{E}\left[|Z_n^*|\right] < \infty$ for any given $n \geq 1$. Conditional on $J = i$ for some $i < n$, we have $Z_n^* = Z_i$, so

$$\mathsf{E}\left[|Z_n^*| \mid J = i\right] = \mathsf{E}\left[|Z_i| \mid J = i\right] < \infty \quad \text{for each } i < n \text{ such that } \Pr\{J = i\} > 0.$$

The reason for this is that if $\mathsf{E}\left[|Z_i| \mid J = i\right] = \infty$ and $\Pr\{J = i\} > 0$, then $\mathsf{E}\left[|Z_i|\right] = \infty$, contrary to the assumption that $\{Z_n;\ n \geq 1\}$ is a martingale, submartingale, or supermartingale. Similarly, for $J \geq n$, we have $Z_n^* = Z_n$ so

$$\mathsf{E}\left[|Z_n^*| \mid J \geq n\right] = \mathsf{E}\left[|Z_n| \mid J \geq n\right] < \infty \quad \text{if } \Pr\{J \geq n\} > 0.$$

Averaging,

$$\mathsf{E}\left[|Z_n^*|\right] = \sum_{i=1}^{n-1} \mathsf{E}\left[|Z_n^*| \mid J{=}i\right] \Pr\{J{=}i\} + \mathsf{E}\left[|Z_n^*| \mid J \geq n\right] \Pr\{J \geq n\} < \infty.$$

Next assume that $\{Z_n;\ n \geq 1\}$ is a submartingale. For any given $n > 1$, consider an arbitrary initial sample sequence $(Z_1 = z_1, Z_2 = z_2, \ldots, Z_{n-1} = z_{n-1})$. Note that z_1 specifies whether or not $J = 1$. Similarly, (z_1, z_2) specifies whether or not $J = 2$, and so forth up to $(z_1, \ldots, z_{n-1})$, which specifies whether or not $J = n - 1$. Thus $(z_1, \ldots, z_{n-1})$ specifies the sample value of J for $J \leq n - 1$ and specifies that $J \geq n$ otherwise.

For $(z_1, \ldots, z_{n-1})$ such that $\mathbb{I}_{J \geq n} = 0$, we have $z_n^* = z_{n-1}^*$. For all such sample values,

$$\mathsf{E}\left[Z_n^* | Z_{n-1}^* {=} z_{n-1}^*, \ldots, Z_1^* {=} z_1^*\right] = z_{n-1}^*. \tag{9.96}$$

For the remaining case, where $(z_1, \ldots, z_{n-1})$ is such that $\mathbb{I}_{J \geq n} = 1$, we have $z_n^* = z_n$. Thus

$$\mathsf{E}\left[Z_n^* | Z_{n-1}^* {=} z_{n-1}^*, \ldots, Z_1^* {=} z_1^*\right] \geq z_{n-1}^*. \tag{9.97}$$

The same argument works for martingales and supermartingales by replacing the inequality in (9.97) by equality for the martingale case and the opposite inequality for the supermartingale case. □

Theorem 9.8.3 *Given a stochastic process $\{Z_n;\ n \geq 1\}$ and a possibly-defective stopping trial J for the process, the stopped process $\{Z_n^*;\ n \geq 1\}$ satisfies the following conditions for all $n \geq 1$ if $\{Z_n;\ n \geq 1\}$ is a submartingale, martingale, or supermartingale respectively:*

$$\mathsf{E}[Z_1] \leq \mathsf{E}\left[Z_n^*\right] \leq \mathsf{E}[Z_n] \qquad (\textit{submartingale}); \tag{9.98}$$

$$\mathsf{E}[Z_1] = \mathsf{E}\left[Z_n^*\right] = \mathsf{E}[Z_n] \qquad (\textit{martingale}); \tag{9.99}$$

$$\mathsf{E}[Z_1] \geq \mathsf{E}\left[Z_n^*\right] \geq \mathsf{E}[Z_n] \qquad (\textit{supermartingale}). \tag{9.100}$$

Proof First assume that $\{Z_n;\ n \geq 1\}$ is a submartingale. Theorem 9.8.2 shows that $\{Z_n^*;\ n \geq 1\}$ is a submartingale, so from (9.88), $\mathsf{E}\left[Z_1^*\right] \leq \mathsf{E}\left[Z_n^*\right]$ for all $n \geq 1$. Since $Z_1 = Z_1^*$, this establishes the first half of (9.98). For the second half, choose any $m \leq n$ and let $z_1, z_2, \ldots, z_m$ be any initial sample segment such that $J = m$. Then $z_n^* = z_m$, so

$$\mathsf{E}\left[Z_n^* | Z_m{=}z_m, \ldots, Z_1{=}z_1\right] = z_m.$$

On the other hand, since $\{Z_n;\ n \geq 1\}$ is a submartingale,

$$\mathsf{E}\left[Z_n | Z_m{=}z_m, \ldots, Z_1{=}z_1\right] \geq z_m.$$

Since this is true for all $z_1, \ldots, z_m$ for which $J = m$, $\mathsf{E}[Z_n \mid J = m] \geq \mathsf{E}\left[Z_n^* | J = m\right]$. This is true for all $m \leq n$, and is clearly valid with equality for $m > n$. Averaging over J then yields the second half of (9.98).

Finally, if $\{Z_n;\ n \geq 1\}$ is a supermartingale, then $\{-Z_n;\ n \geq 1\}$ is a submartingale, verifying (9.100). Since a martingale is both a submartingale and supermartingale, (9.99) follows and the proof is complete. □

Consider a (non-defective) stopping trial J for a martingale $\{Z_n;\, n \geq 1\}$. Since the stopped process is also a martingale, we have

$$\mathsf{E}\left[Z_n^*\right] = \mathsf{E}\left[Z_1^*\right] = \mathsf{E}\left[Z_1\right];\; n \geq 1. \tag{9.101}$$

Since $Z_n^* = Z_J$ for all $n \geq J$ and since J is finite WP1, we see that $\lim_{n\to\infty} Z_n^* = Z_J$ WP1. Surprisingly, $\mathsf{E}[Z_J]$ is not necessarily equal to $\lim_{n\to\infty} \mathsf{E}\left[Z_n^*\right] = \mathsf{E}[Z_1]$. The binary product martingale in (9.76) gives an example of inequality. Taking the stopping trial J to be the smallest n for which $Z_n = 0$, we have $Z_J = 0$ WP1, and thus $\mathsf{E}[Z_J] = 0$. But $Z_n^* = Z_n$ for all n, and $\mathsf{E}\left[Z_n^*\right] = 1$ for all n. The problem here is that, given that the process has not stopped by time n, Z_n and Z_n^* each have the value 2^n. Fortunately, in most situations, this type of bizarre behavior does not occur and $\mathsf{E}[Z_J] = \mathsf{E}[Z_1]$. To get a better understanding of when $\mathsf{E}[Z_J] = \mathsf{E}[Z_1]$, note that for any n, we have

$$\mathsf{E}\left[Z_n^*\right] = \sum_{i=1}^{n} \mathsf{E}\left[Z_n^* \mid J = i\right] \Pr\{J = i\} + \mathsf{E}\left[Z_n^* \mid J > n\right] \Pr\{J > n\} \tag{9.102}$$

$$= \sum_{i=1}^{n} \mathsf{E}\left[Z_J \mid J = i\right] \Pr\{J = i\} + \mathsf{E}\left[Z_n \mid J > n\right] \Pr\{J > n\}. \tag{9.103}$$

The left-hand side of (9.103) is $\mathsf{E}[Z_1]$ for all n. If the final term on the right-hand side converges to 0 as $n \to \infty$, then the sum must converge to $\mathsf{E}[Z_1]$. If $\mathsf{E}[|Z_J|] < \infty$, then the sum also converges to $\mathsf{E}[Z_J]$. Without the condition $\mathsf{E}[|Z_J|] < \infty$, the sum might consist of alternating terms which converge, but whose absolute values do not converge, in which case $\mathsf{E}[Z_J]$ does not exist (see Exercise 9.27 for an example). Thus we have established the following lemma.

Lemma 9.8.4 *Let J be a stopping trial for a martingale $\{Z_n;\, n \geq 1\}$. Then $\mathsf{E}[Z_J] = \mathsf{E}[Z_1]$ if and only if*

$$\lim_{n\to\infty} \mathsf{E}\left[Z_n \mid J > n\right] \Pr\{J > n\} = 0 \qquad \textit{and} \quad \mathsf{E}[|Z_J|] < \infty. \tag{9.104}$$

9.8.1 The Wald identity

Recall the generating function product martingale of (9.77) in which $\{Z_n = \exp[rS_n - n\gamma(r)];\, n \geq 1\}$ is a martingale defined in terms of the random walk $\{S_n = X_1 + \cdots + X_n;\, n \geq 1\}$. From (9.99), we have $\mathsf{E}[Z_n] = \mathsf{E}[Z_1]$, and since $\mathsf{E}[Z_1] = \mathsf{E}\left[\exp\{rX_1 - \gamma(r\}\right] = 1$, we have $\mathsf{E}[Z_n] = 1$ for all n. Also, for any possibly-defective stopping trial J, we have $\mathsf{E}\left[Z_n^*\right] = \mathsf{E}[Z_1] = 1$. This is the first part of the following theorem, and the second part follows from Lemma 9.8.4.

Theorem 9.8.5 (Wald's identity) *If J is a possibly-defective stopping trial for the martingale of (9.77), then the stopped process $\{Z_n^*;\, n \geq 1\}$ satisfies*

$$\mathsf{E}\left[Z_n^*\right] = 1 \qquad \textit{for all } n \geq 1. \tag{9.105}$$

If J is a non-defective stopping trial, and if (9.104) holds, then

$$\mathsf{E}[Z_J] = \mathsf{E}\left[\exp\{rS_J - J\gamma(r)\}\right] = 1. \tag{9.106}$$

If there are two thresholds, one at $\alpha > 0$, and the other at $\beta < 0$, and the stopping rule is to stop when either threshold is crossed, then (9.106) is just the Wald identity for two thresholds, (9.38). If there is only one threshold, with $\alpha > 0$ and $\mathsf{E}[X] < 0$, then J is defective but (9.105) still holds. If we choose $r = r^*$ (the positive root of $\gamma(r)$), then $Z_n = \exp(rS_n - r\gamma(r))$ becomes $Z_n = \exp(r^*S_n)$. In this case, (9.105) shows that $\Pr\{S_J \geq \alpha\} \leq e^{-r^*\alpha}$, with no need for the limiting argument used to derive (9.49).

Theorem 9.8.5 can also be used for other stopping rules. For example, for some given integer n, let J_{n+} be the smallest integer $i \geq n$ for which $S_i \geq \alpha$ or $S_i \leq \beta$. Then, in the limit $\beta \to -\infty$, $\Pr\{S_{J_{n+}} \geq \alpha\} = \Pr\{\cup_{i=n}^{\infty}(S_i \geq \alpha)\}$. Assuming $\overline{X} < 0$, we can find an upper bound to $\Pr\{S_{J_{n+}} \geq \alpha\}$ for any $r > 0$ and $\gamma(r) \leq 0$ (i.e., for $0 < r \leq r^*$) by the following steps

$$1 = \mathsf{E}\left[\exp\{rS_{J_{n+}} - J_{n+}\gamma(r)\}\right] \geq \Pr\{S_{J_{n+}} \geq \alpha\} \exp[r\alpha - n\gamma(r)]$$
$$\Pr\{S_{J_{n+}} \geq \alpha\} \leq \exp[-r\alpha + n\gamma(r)], \qquad 0 \leq r \leq r^*. \tag{9.107}$$

9.9 The Kolmogorov inequalities

We now use the previous theorems to establish Kolmogorov's submartingale inequality, which is a major strengthening of the Markov inequality. Just as the Markov inequality in Section 1.8 was used to derive the Chebychev inequality, the Chernoff bound, and the WLLN, the Kolmogorov submartingale inequality will be used to develop a number of related inequalities and then to prove the SLLN and the martingale convergence theorem. The SLLN here assumes only a second moment rather than the fourth moment assumed in Section 5.2. Perhaps more important than the increased generality of the SLLN here is that the proof suggests a number of more general results not requiring IID rv s.

The Kolmogorov submartingale inequality will follow easily from looking at a particular class of stopping rules for non-negative submartingales. Each such stopping rule is characterized by a threshold a and a trial number m and is denoted by $J(m,a)$. Stopping occurs either at the first crossing of the threshold at a or at trial m, whichever comes first.

Since stopping occurs by trial m at the latest, we see that $Z_{J(m,a)} \geq a$ if and only if the threshold is crossed at some trial $n \leq m$. In terms of events, this means that

$$\{Z_{J(m,a)} \geq a\} = \left\{\max_{1\leq n\leq m} Z_n \geq a\right\}. \tag{9.108}$$

Applying the Markov inequality to $Z_{J(m,a)}$, we then get

$$\Pr\left\{\max_{1\leq n\leq m} Z_n \geq a\right\} = \Pr\{Z_{J(m,a)} \geq a\} \leq \frac{\mathsf{E}\left[Z_{J(m,a)}\right]}{a}. \tag{9.109}$$

It is only a small step from this to Kolmogorovs's submartingale inequality.

Theorem 9.9.1 (Kolmogorov's submartingale inequality) *Let $\{Z_n; n \geq 1\}$ be a non-negative submartingale. Then for any positive integer m and any $a > 0$,*

$$\Pr\left\{\max_{1\le n\le m} Z_n \ge a\right\} \le \frac{\mathsf{E}[Z_m]}{a}. \tag{9.110}$$

Proof Using first the inequality (9.109),

$$\Pr\left\{\max_{1\le n\le m} Z_n \ge a\right\} \le \frac{\mathsf{E}\left[Z_{J(m,a)}\right]}{a} = \frac{\mathsf{E}\left[Z_m^*\right]}{a}.$$

The equality above follows because the process must stop by trial m. Finally, from (9.98), $\mathsf{E}\left[Z_m^*\right] \le \mathsf{E}[Z_m]$, yielding (9.110). □

The following simple corollary shows that (9.110) has a limiting form for non-negative martingales.

Corollary 9.9.2 (Non-negative martingale inequality) *Let $\{Z_n;\ n \ge 1\}$ be a non-negative martingale. Then*

$$\Pr\left\{\sup_{n\ge 1} Z_n \ge a\right\} \le \frac{\mathsf{E}[Z_1]}{a} \qquad \textit{for all } a > 0. \tag{9.111}$$

Proof For a martingale, $\mathsf{E}[Z_m] = \mathsf{E}[Z_1]$. Thus, from (9.110), $\Pr\left\{\max_{1\le i\le m} Z_i \ge a\right\} \le \mathsf{E}[Z_1]/a$ for all $m > 1$. Passing to the limit $m \to \infty$ essentially yields (9.111). Exercise 9.28 illustrates why the limiting operation is a little tricky, and then shows that it is valid. □

The next three corollaries are straightforward consequences of the Kolmogorov submartingale inequality. Their proofs are given in Exercise 9.31. The first bears the same relationship to the Kolmogorov submartingale inequality as the Chebychev inequality does to the Markov inequality. When applied to martingales, it is sometimes called the Kolmogorov martingale inequality.

Corollary 9.9.3 *Let $\{Z_n;\ n \ge 1\}$ be a martingale with $\mathsf{E}\left[Z_n^2\right] < \infty$ for all $n \ge 1$. Then*

$$\Pr\left\{\max_{1\le n\le m} |Z_n| \ge b\right\} \le \frac{\mathsf{E}\left[Z_m^2\right]}{b^2}; \quad \textit{for all integer } m \ge 2, \textit{all } b > 0. \tag{9.112}$$

Corollary 9.9.4 (Kolmogorov's random walk inequality) *Let $\{S_n;\ n \ge 1\}$ be a random walk with $S_n = X_1 + \cdots + X_n$, where $\{X_i;\ i \ge i\}$ is a set of IID rv s with mean $\overline{X}$ and variance σ^2. Then for any positive integer m and any $\epsilon > 0$,*

$$\Pr\left\{\max_{1\le n\le m} |S_n - n\overline{X}| \ge m\epsilon\right\} \le \frac{\sigma^2}{m\epsilon^2}. \tag{9.113}$$

Recall that the simplest form of the WLLN was given in (1.74) as $\Pr\left\{|S_m/m - \overline{X}| \ge \epsilon\right\} \le \sigma^2/(m\epsilon^2)$. This is strengthened in (9.113) to upper bound the probability that any of the first m terms deviate from the mean by more than $m\epsilon$. It is this strengthening that will allow us to prove the SLLN assuming only a finite variance.

The following corollary provides a tight exponential bound to the probability of crossing a threshold before a given number of trials.

Corollary 9.9.5 *Let $\{S_n;\ n \geq 1\}$ be a random walk, $S_n = X_1 + \cdots + X_n$, where each X_i has mean $\overline{X} < 0$ and semi-invariant moment generating function $\gamma(r)$. For any $r > 0$ such that $0 < \gamma(r) < \infty$ (i.e., for $r > r^*$), and for any $a > 0$*

$$\Pr\left\{\max_{1\leq i\leq n} S_i \geq \alpha\right\} \leq \exp\{-r\alpha + n\gamma(r)\}. \tag{9.114}$$

Proof For $r > r^*$, $\{\exp(rS_n);\ n \geq 1\}$ is a submartingale. Taking $a = \exp(r\alpha)$ in (9.110), we get (9.114). □

The following theorem about supermartingales is, in a sense, the dual of the Kolmogorov submartingale inequality. Note, however, that it applies to the terms $n \geq m$ in the supermartingale rather than $n \leq m$.

Theorem 9.9.6 *Let $\{Z_n;\ n \geq 1\}$ be a non-negative supermartingale. Then for any positive integer m and any $a > 0$,*

$$\Pr\left\{\bigcup_{i\geq m} \{Z_i \geq a\}\right\} \leq \frac{\mathsf{E}[Z_m]}{a}. \tag{9.115}$$

Proof For given $m \geq 1$ and $a > 0$, let J be a possibly-defective stopping trial defined as the smallest $i \geq m$ for which $Z_i \geq a$. Let $\{Z_n^*;\ n \geq 1\}$ be the corresponding stopped process, which is also non-negative and is a supermartingale from Theorem 9.8.2. For any $k > m$, note that $Z_k^* \geq a$ if and only if $\max_{m\leq i\leq k} Z_i \geq a$. Thus

$$\Pr\left\{\max_{m\leq i\leq k} Z_i \geq a\right\} = \Pr\left\{Z_k^* \geq a\right\} \leq \frac{\mathsf{E}\left[Z_k^*\right]}{a}.$$

Since $\{Z_n^*;\ n \geq 1\}$ is a supermartingale, (9.89) shows that $\mathsf{E}\left[Z_k^*\right] \leq \mathsf{E}\left[Z_m^*\right]$. On the other hand, $Z_m^* = Z_m$ since the process cannot stop before epoch m. Thus $\Pr\left\{\max_{m\leq i\leq k} Z_i \geq a\right\}$ is at most $\mathsf{E}[Z_m]/a$. Since k is arbitrary, we can pass to the limit, getting (9.115) and completing the proof. □

9.9.1 The SLLN

We now use the Kolmogorov submartingale inequality to prove the SLLN assuming only a second moment rather than the fourth moment assumed in Section 5.2.1. Perhaps more important than requiring only a second moment is that the proof suggests a number of SLLN type results not requiring IID rv s. The SLLN is also true assuming only a first absolute moment, but the truncation argument we used for the WLLN in Theorem 1.7.4 is inadequate here.

Theorem 9.9.7 (SLLN) *Let $\{X_i;\ i \geq 1\}$ be a sequence of IID rv s with zero mean and standard deviation $\sigma < \infty$. Let $S_n = X_1 + \cdots + X_n$ for each $n \geq 1$. Then*

$$\Pr\left\{\lim_{n\to\infty} \frac{S_n}{n} = 0\right\} = 1. \tag{9.116}$$

Discussion The theorem generalizes trivially to rv s with a non-zero mean. We simply let the X of the theorem be the fluctuation of the original rv. The statement then becomes

$$\Pr\left\{ \lim_{n\to\infty} \frac{S_n}{n} - \overline{X} = 0 \right\} = 1. \tag{9.117}$$

Proof Let $Z_n = S_n^2$ for each n. Then $\{Z_n;\, n \geq 1\}$ is a submartingale. Let $\epsilon > 0$ be fixed and for each $\ell \geq 1$, consider the stopping rule $J(2^\ell, \epsilon 2^{2\ell})$, here abbreviated as J_ℓ. From the Kolmogorov submartingale inequality,

$$\Pr\left\{ \max_{1\leq n\leq 2^\ell} Z_n \geq \epsilon 2^{2\ell} \right\} \leq \frac{\mathsf{E}\left[Z_{J_\ell}\right]}{\epsilon 2^{2\ell}} \leq \frac{\mathsf{E}\left[Z_{2^\ell}\right]}{\epsilon 2^{2\ell}}. \tag{9.118}$$

For each ℓ, $Z_{2^\ell}/(\epsilon 2^{2\ell})$ is a rv of mean $2^\ell \sigma^2/(\epsilon 2^{2\ell}) = 2^{-\ell}\sigma^2/\epsilon$. Thus

$$\sum_{\ell\geq 1} \mathsf{E}\left[\frac{Z_{J_\ell}}{\epsilon 2^{2\ell}}\right] \leq \sum_{\ell\geq 1} 2^{-\ell}\sigma^2/\epsilon = \sigma^2/\epsilon < \infty.$$

It then follows from Lemma 5.2.1 (the lemma on convergence WP1 used to prove the SLLN in Chapter 5) that

$$\Pr\left\{ \omega : \lim_{\ell\to\infty} \frac{Z_{J_\ell}(\omega)}{\epsilon 2^{2\ell}} = 0 \right\} = 1.$$

This means that there is a set of sample points, say Ω_1, of probability 1 such that, for each $\omega \in \Omega_1$ and each $\epsilon > 0$ (and thus in particular the ϵ chosen above) there is an $m(\omega, \epsilon)$ such that $Z_{J_\ell}(\omega)/2^{2\ell} < \epsilon$ for all $\ell \geq m(\omega, \epsilon)$. From (9.108),

$$\max_{1\leq n\leq 2^\ell} \frac{Z_n(\omega)}{2^{2\ell}} < \epsilon \qquad \text{for } \ell \geq m(\omega,\epsilon),\ \omega \in \Omega_1. \tag{9.119}$$

Now (9.119) remains valid if the maximum is restricted to $2^{\ell-1} < n < 2^\ell$, and, in this region, $Z_n/n^2 < Z_n/2^{2(\ell-1)} = 4Z_n/2^{2\ell}$. Thus

$$\max_{2^{\ell-1}<n\leq 2^\ell} \frac{Z_n(\omega)}{n^2} < 4\epsilon \qquad \text{for } \ell \geq m(\omega,\epsilon),\ \omega \in \Omega_1. \tag{9.120}$$

This means that $Z_n(\omega)/n^2 < 4\epsilon$ for all $n > 2^{m(\omega,\epsilon)-1}$. Since this is true for all $\epsilon > 0$, it means that $\{Z_n(\omega)/n^2;\, n \geq 1\}$ converges to 0, and thus $\{Z_n/n^2;\, n \geq 1\}$ converges to 0 WP1. Since $Z_n = S_n^2$, we can take the square root of Z_n/n^2 to see that $\{S_n/n;\, n \geq 1\}$ converges to 0 WP1. □

9.9.2 The martingale convergence theorem

Another famous result that follows from the Kolmogorov submartingale inequality is the martingale convergence theorem. This states that if a martingale $\{Z_n;\, n \geq 1\}$ has the property that there is some finite M such that $\mathsf{E}\left[|Z_n|\right] \leq M$ for all n, then $\lim_{n\to\infty} Z_n$ exists (and is finite) WP1. This is a powerful theorem, but the limitation that $\mathsf{E}\left[|Z_n|\right] \leq M$ for all n is far more than a technical restriction; for example, it is not satisfied by a zero-mean random walk. We prove the theorem with the additional restriction that there is some finite M such that $\mathsf{E}\left[Z_n^2\right] \leq M$ for all n.

Theorem 9.9.8 (Martingale convergence theorem) *Let $\{Z_n;\ n \geq 1\}$ be a martingale and assume that there is some finite M such that $\mathsf{E}\left[Z_n^2\right] \leq M$ for all n. Then there is a rv Z such that, for all sample sequences except a set of probability 0, $\lim_{n\to\infty} Z_n = Z$.*

Proof From Theorem 9.7.4 and the assumption that $\mathsf{E}\left[Z_n^2\right] \leq M$, $\{Z_n^2;\ n \geq 1\}$ is a submartingale. Thus, from (9.88), $\mathsf{E}\left[Z_n^2\right]$ is non-decreasing in n, and since $\mathsf{E}\left[Z_n^2\right]$ is bounded, $\lim_{n\to\infty} \mathsf{E}\left[Z_n^2\right] = M'$ for some $M' \leq M$. For any integer k, the process $\{Y_n = Z_{k+n} - Z_k;\ n \geq 1\}$ is a zero-mean martingale (see Exercise 9.42). Thus from Corollary 9.9.3,

$$\Pr\left\{\max_{1\leq n\leq m} |Z_{k+n} - Z_k| \geq b\right\} \leq \mathsf{E}\left[(Z_{k+m} - Z_k)^2\right]/b^2. \tag{9.121}$$

Next, observe that $\mathsf{E}\left[Z_{k+m}Z_k \mid Z_k = z_k, Z_{k-1} = z_{k-1}, \dots, Z_1 = z_1\right] = z_k^2$, and therefore, $\mathsf{E}\left[Z_{k+m}Z_k\right] = \mathsf{E}\left[Z_k^2\right]$. Thus $\mathsf{E}\left[(Z_{k+m} - Z_k)^2\right] = \mathsf{E}\left[Z_{k+m}^2\right] - \mathsf{E}\left[Z_k^2\right] \leq M' - \mathsf{E}\left[Z_k^2\right]$. Since this is independent of m, we can pass to the limit, obtaining

$$\Pr\left\{\sup_{n\geq 1} |Z_{k+n} - Z_k| \geq b\right\} \leq \frac{M' - \mathsf{E}\left[Z_k^2\right]}{b^2}. \tag{9.122}$$

Since $\lim_{k\to\infty} \mathsf{E}\left[Z_k^2\right] = M'$, we then have, for all $b > 0$,

$$\lim_{k\to\infty} \Pr\left\{\sup_{n\geq 1} |Z_{k+n} - Z_k| \geq b\right\} = 0. \tag{9.123}$$

This means that WP1 a sample sequence of $\{Z_n;\ n \geq 1\}$ is a Cauchy sequence, and thus approaches a limit, concluding the proof. □

Example 9.9.9 (Limits of orthonormal expansions of random processes) Assume that $\{X_n;\ n \geq 1\}$ is a sequence of zero-mean independent rv s with finite variances, $\mathsf{E}\left[X_n^2\right] = \sigma_n^2 < \infty$. Let $\{\phi_n(t);\ n \geq 1, t \in \mathbb{R}\}$ be a sequence of orthonormal functions (see Section 3.6.3). Suppose that a random process is defined as an orthonormal expansion, $X(t) = \lim_{\ell\to\infty} X^\ell(t)$, where $X^\ell(t) = \sum_{n=1}^{\ell} X_n\phi_n(t)$.

We want to use the martingale convergence theorem to show that this limit exists WP1 for each t. For a given t, we see that $\{X^\ell(t);\ \ell \geq 1\}$ is simply a sequence of rv s. Since the underlying rv s X_n are independent and zero mean, the sums $X^\ell(t) = \sum_{n=1}^{\ell} X_n\phi_n(t)$ form a martingale (assuming each $\phi_n(t)$ is finite).

If we now make the further assumption that $\sum_{n=1}^{\infty} \sigma_n^2 |\phi_n(t)|^2$ is finite for each t, then the partial sums from 1 to ℓ are bounded independent of ℓ and the martingale convergence theorem applies. Thus for each t, the limit $\lim_{\ell\to\infty} X^\ell(t)$ exists WP1.

It should be clear that this same approach can be used in many situations to show the limit of a sequence of rv s is itself a rv WP1.

Example 9.9.10 (Branching processes) The martingale convergence theorem can also be interpreted relatively easily for branching processes. For a branching process

$\{X_n;\ n \geq 1\}$, where $\overline{Y}$ is the expected number of offspring of an individual, $\{X_n/\overline{Y}^n;\ n \geq 1\}$ is a martingale that satisfies the above conditions. If $\overline{Y} \leq 1$, the branching process dies out WP1, so $X_n/\overline{Y}^n$ approaches 0 WP1. For $\overline{Y} > 1$, however, the branching process dies out with some probability less than 1 and approaches ∞ otherwise. Thus, the limiting rv Z is 0 with the probability that the process ultimately dies out, and is positive otherwise. In the latter case, for large n, the interpretation is that when the population is very large, a law of large numbers effect controls its growth in each successive generation, so that $X_n/X_{n-1} \approx \overline{Y}$ for large X_{n-1}. The distribution of Z conditional on $Z > 0$ essentially depends on the random number of trials until X_n becomes large.

9.10 A simple model for investments

Consider a range of possible investments, numbered 1 to ℓ, available to an investor. Each investment fluctuates in value over time, and the investor is faced with the problem of buying and selling those investments over time so as to maintain a portfolio that is desirable in some sense. We introduce a particularly simple model of such portfolio management in this section, partly for the purpose of illustrating some features of stochastic processes and partly as a theoretical introduction to an important application area.[5]

This is an area in which mathematics has sometimes run amok and where it is difficult to thread a path between common sense, common practice, and mathematical models. The analysis here is not intended to address these practical concerns but only to introduce a theory that might be useful to those who also take the time and effort to fully address the practical side of the area.

We consider a discrete-time model, where the time unit could be microseconds (for program trading) or days or months (for more casual investors). Each available investment k, $1 \leq k \leq \ell$ is characterized by a non-negative rv $X(k)$ called the price-relative. This is the ratio of the price of the kth investment at the end of an epoch (a time unit) to that at the beginning of that epoch. As the kth investment evolves over time, we take $X_n(k)$ to be this ratio at epoch n. For each k, we assume that the sequence of price-relatives $\{X_n(k);\ n \geq 1\}$ is a sequence of IID rv s with the distribution of $X(k)$. The quantity $X_n(k) - 1$ is known as the return, per unit invested, over the nth epoch. The return is often given in percent and is in many ways the natural way to measure how successful an investment is. It is more convenient here, however, to use price-relatives in our analysis.

The assumption that $\{X_n(k);\ n \geq 1\}$ is IID is often unrealistic in practice, and in fact much of the work by investment analysts is devoted to understanding how the statistics

[5] Some people view investing in stocks as simply a form of gambling with no redeeming social value. However, venture capitalists play an important role in the formation of new companies, thus providing jobs to many people. Vulture capitalists, on the other hand, often destroy companies for short-term personal gain. For the most part, however, when one investor buys a stock, another investor sells it, and the act of buying bids up the price of that stock and thus indirectly helps the corresponding company and its employees. We leave the social aspects of investing as part of the application area, and concentrate here on the stochastic processes involved.

of these rv s change over time. Our purpose here, however, is to understand the principles involved in allocating resources among investments over time, and for this, the simplicity of the IID assumption is desirable.

If \$1 is put into investment k at the beginning of epoch 1 and left to fluctuate over n epochs with no further purchases or sales, then the wealth of the investor in that investment at time n is the rv $W_n(k) = \prod_{m=1}^{n} X_m(k)$ (in dollars). We want to use the laws of large numbers to analyze this later, so it is convenient to define the rv $L_n(k)$, denoted as log-wealth, to be the log of $W_n(k)$. Thus

$$L_n(k) = \sum_{m=1}^{n} \ln X_m(k).$$

Thus the log-wealth in k, per dollar invested, i.e., $L_n(k)$, is the sum of n IID rv s.

For the set of all ℓ investments, we assume there is a joint distribution on the random vector $\boldsymbol{X} = \big(X(1), \dots, X(\ell)\big)^{\mathsf{T}}$ and that this distribution is known to the investor. There is no need to assume that these investments are independent of each other. We do assume, however, that the time sequence of rvs $\{\boldsymbol{X}_n;\ n \geq 1\}$ is an IID sequence. Note that each rv $\boldsymbol{X}_n$ is a vector of ℓ rv s, $X_n(1), \dots, X_n(\ell)$; the subscript refers to time and the parentheses to the particular investment. The IID assumption is over time and not over investments.

We want to consider what happens if there is a constant fractional allocation, $\boldsymbol{\lambda} = \big(\lambda(1), \dots, \lambda(\ell)\big)^{\mathsf{T}}$ between investments such that, at the beginning of each epoch n, a fraction $\lambda(k)$ of the investor's wealth is put into investment k. This means that as the values of the individual investments shift, the portfolio is rebalanced at the end of each epoch to maintain the allocation $\boldsymbol{\lambda}$ for the beginning of the next epoch. We assume there is no commission on the buying and selling required by this rebalancing, so the investor's wealth at each time n changes only according to the allocation $\boldsymbol{\lambda}$ and the joint distribution of $\boldsymbol{X}$.

Finally assume that $\sum_k \lambda(k) = 1$ and $\lambda(k) \geq 0$ for each k. The latter constraint, $\lambda(k) \geq 0$, rules out short sales (i.e., selling stock that is not owned) and the former, $\sum \lambda(k) = 1$, rules out investing on margin (i.e., borrowing money to invest). It also implies that the investor is always fully invested. We will assume that the ℓth investment is 'cash,' i.e., an investment where $X(\ell) = 1$ with probability 1, so that a cash investment incorporates what is usually meant by being not fully invested.

Let the investor's wealth at the beginning of epoch 1 be denoted as W_0. Then, generalizing the notation above for single investments, we denote the wealth at the end of epoch 1 for the allocation $\boldsymbol{\lambda}$ as the rv $W_1(\boldsymbol{\lambda}) = W_0 \sum_k \lambda(k) X_1(k)$. The wealth at the end of epoch 2 is found by taking the wealth $W_1(\boldsymbol{\lambda})$ at the end of epoch 1, rebalancing it with the fractions $\lambda(1), \dots, \lambda(\ell)$, and investing that rebalanced amount over epoch 2, i.e.,

$$W_2(\boldsymbol{\lambda}) = W_0 \left[\sum_k \lambda(k) X_1(k)\right] \left[\sum_k \lambda(k) X_2(k)\right].$$

The wealth $W_n(\boldsymbol{\lambda})$ at the end of epoch n is similarly W_0 times a product of n IID rv s. Since W_0 is just an extra parameter, we set it to 1, giving us

$$W_n(\boldsymbol{\lambda}) = \prod_{m=1}^{n} \Big[\sum_k \lambda(k) X_m(k) \Big]. \tag{9.124}$$

The log-wealth at the end of epoch n is the rv $L_n(\boldsymbol{\lambda}) = \ln W_n(\boldsymbol{\lambda})$ given by

$$L_n(\boldsymbol{\lambda}) = \sum_{m=1}^{n} \ln \left[\sum_k \lambda(k) X_m(k) \right]. \tag{9.125}$$

Note that $L_n(\boldsymbol{\lambda})$ is the sum of n IID rv s and we can apply the laws of large numbers (given the existence of a mean). As discussed shortly, this means that the wealth of the investor grows (or shrinks) exponentially at a rate given by $\mathsf{E}\big[\ln \sum_k \lambda(k) X(k)\big]$. Before being more precise about this, we consider an example.

Example 9.10.1 (Triple or nothing) Suppose there is only one possible investment other than cash, so we omit the parentheses denoting investments. Assume that $X = 3$ or $X = 0$, each with probability 1/2. This appears to be a very profitable investment, since the expected growth in one time unit from this investment alone is 1.5 (i.e., an expected return of 50% per epoch). If we constantly choose $\lambda = 1$, i.e., nothing in cash, we see from (9.124) that W_n is 3^n with probability 2^{-n} and 0 with probability $1 - 2^{-n}$. Thus $\mathsf{E}[W_n] = (3/2)^n$, but $\lim_{n\to\infty} W_n = 0$ WP1. In the same way, $L_n = n\ln(3)$ with probability 2^{-n} and $L_n = -\infty$ otherwise. We call this the go-for-broke strategy, since $\mathsf{E}[W_n]$ is very large, but $\lim_{n-\infty} W_n = 0$ WP1.

The solution to this dilemma occurs to most gamblers who like to continue gambling. They keep a fraction λ of their wealth in the given investment and put $1-\lambda$ in cash. They can then continue to invest a fraction λ of their wealth at each epoch, while maintaining a fraction $1 - \lambda$ in cash. Using (9.124) and (9.125) to find the expected wealth and log wealth at the end of epoch 1, we get

$$\mathsf{E}[W_1(\lambda)] = \frac{1}{2}(1 - \lambda + 3\lambda) + \frac{1}{2}(1 - \lambda) \;=\; 1 + \frac{\lambda}{2}, \tag{9.126}$$

$$\mathsf{E}[L_1(\lambda)] = \frac{1}{2}\ln(1 - \lambda + 3\lambda) + \frac{1}{2}\ln(1 - \lambda) \;=\; \frac{1}{2}\ln(1 + \lambda - 2\lambda^2). \tag{9.127}$$

The maximum of $\mathsf{E}[L_1(\lambda)]$ over λ is easily seen to be $(1/2)\ln(9/8) = 0.059$, achieved at $\lambda = 1/4$. Thus $L_n(1/4)$ is a sum of n IID rv s each of mean $(1/2)\ln(9/8)$. By the law of large numbers, $(1/n)L_n(1/4)$ is close to $(1/2)\ln(9/8)$ with high probability for large n. Since $W_n(1/4) = e^{L_n(1/4)}$, this suggests that $W_n(1/4)$ might be close to $\exp[(n/2)\ln(9/8)]$ with high probability. This is strange, however, since it is also easy to see that $\mathsf{E}\big[W_n(1/4)\big] = (9/8)^n$.

The explanation of this somewhat paradoxical behavior is that in a typical set of n epochs, the fraction of epochs in which $X_n = 3$ is close to $n/2$. Note that if there are exactly $n/2$ winning epochs, then $W_n(\lambda) = (9/8)^{n/2}$. The much larger value for $\mathsf{E}[W_n(\lambda)]$ comes from those very rare but profitable sequences in which the fraction of winning epochs is much larger than $n/2$.

We see that $W_n(\lambda)$ is one of those peculiar rv s for which the expected value and typical values are far apart. However, it should be clear that by keeping 3/4 of the wealth in cash, the investment has changed from a lottery type investment to a safe investment which grows exponentially with a return of 6.07% per epoch.

Exercise 9.35 describes and analyzes a similar game called the double or quarter game.

9.10.1 Portfolios with constant fractional allocations

We now return to the general case with ℓ investments, but still consider only situations where the same $\boldsymbol{\lambda}$ is used at each epoch n. Using (9.125),

$$L_n(\boldsymbol{\lambda}) = \sum_{m=1}^{n} Y_m(\boldsymbol{\lambda}),$$

where $Y_m(\boldsymbol{\lambda})$ is the log of the combined price relative over investments at epoch m, i.e.,

$$Y_m(\boldsymbol{\lambda}) = \ln\left(\sum_k \lambda(k) X_m(k)\right). \tag{9.128}$$

In order to use either the SLLN or the WLLN in analyzing $L_n(\boldsymbol{\lambda})$, we must assume that $\mathsf{E}\left[|Y(\boldsymbol{\lambda})|\right] < \infty$. Note from (9.127) that $\lim_{\lambda\to 1}\mathsf{E}\left[Y(\lambda)\right] = -\infty$ for Example 9.10.1. Thus the law of large numbers is not justified there for $\lambda = 1$. This is not surprising, since the investor then goes broke WP1. This problem can be avoided in general by a positive cash allocation, i.e., $\lambda(\ell) > 0$. For the opposite extreme, ruling out situations where $\mathsf{E}\left[Y(\boldsymbol{\lambda})\right] = +\infty$, does not appear to be a significant restriction.

With the assumption that $\mathsf{E}\left[|Y(\boldsymbol{\lambda})|\right] < \infty$, the SLLN applies, i.e.,

$$\lim_{n\to\infty}\frac{1}{n}L_n(\boldsymbol{\lambda}) = \mathsf{E}\left[Y(\boldsymbol{\lambda})\right] \qquad \text{WP1}.$$

Since $W_n(\boldsymbol{\lambda}) = \exp\big(L_n(\boldsymbol{\lambda})\big)$, this says that

$$W_n(\boldsymbol{\lambda}) = \exp\left[n\left(\frac{1}{n}L_n(\boldsymbol{\lambda})\right)\right], \quad \text{where } \lim_{n\to\infty}\frac{1}{n}L_n(\boldsymbol{\lambda}) = \mathsf{E}\left[Y(\boldsymbol{\lambda})\right] \quad \text{WP1}. \tag{9.129}$$

In other words, the sequence of wealths, using the constant allocation $\boldsymbol{\lambda}$, is exponential in n, with a sequence of rates $(1/n)L_n(\boldsymbol{\lambda})$ approaching $\mathsf{E}\left[Y(\boldsymbol{\lambda})\right]$ WP1.

This can be expressed more compactly by using $W_n(\boldsymbol{\lambda}) = \exp\big(L_n(\boldsymbol{\lambda})\big)$ in Lemma 5.3.3 to see that

$$\lim_{n\to\infty}\left[W_n(\boldsymbol{\lambda})\right]^{1/n} = \exp\big(\mathsf{E}\left[Y(\boldsymbol{\lambda})\right]\big) \qquad \text{WP1}. \tag{9.130}$$

Both (9.129) and (9.130) are precise ways of saying that $W_n(\boldsymbol{\lambda})$ is asymptotically roughly approximated by $\exp[n\mathsf{E}\left[Y(\boldsymbol{\lambda})\right]]$. The following theorem, which is proved in Exercise 9.37, provides another way of saying the same thing.

Theorem 9.10.2 *Assume that* $\mathsf{E}\left[|Y(\boldsymbol{\lambda})|\right] < \infty$. *Then for every* $\epsilon, \delta > 0$, *there exists an* n_o *such that*

$$\Pr\left\{\bigcap_{n\geq n_o}\left\{\exp\left[n\big(\mathsf{E}[Y(\boldsymbol{\lambda})]-\epsilon\big)\right]\leq W_n(\boldsymbol{\lambda})\leq\exp\left[n\big(\mathsf{E}[Y(\boldsymbol{\lambda})]+\epsilon\big)\right]\right\}\right\}>1-\delta. \tag{9.131}$$

This says that, with arbitrarily high probability (i.e., $1-\delta$), $W_n(\boldsymbol{\lambda})$ lies between exponential bounds of rates $\mathsf{E}[Y(\boldsymbol{\lambda})]-\epsilon$ and $\mathsf{E}[Y(\boldsymbol{\lambda})]+\epsilon$ for all sufficiently large n.

The following corollary provides a final attempt to make this a little less abstract.

Corollary 9.10.3 *Let $\boldsymbol{\lambda}_1$ and $\boldsymbol{\lambda}_2$ be alternative constant allocations and assume $\mathsf{E}[Y(\boldsymbol{\lambda}_1)]>\mathsf{E}[Y(\boldsymbol{\lambda}_2)]$. Then for every $\delta>0$, there is an n_o such that*

$$\Pr\{W_n(\boldsymbol{\lambda}_1)>W_n(\boldsymbol{\lambda}_2)\text{ for all }n\geq n_o\}>1-\delta. \tag{9.132}$$

This follows by applying the theorem to both $\boldsymbol{\lambda}_1$ and $\boldsymbol{\lambda}_2$ and choosing ϵ to be half the difference between $\mathsf{E}[Y(\boldsymbol{\lambda}_1)]$ and $\mathsf{E}[Y(\boldsymbol{\lambda}_2)]$

Let us continue to consider only portfolios with a constant allocation $\boldsymbol{\lambda}$. It then makes sense to maximize the exponential growth rate $\mathsf{E}[Y(\boldsymbol{\lambda})]$. Remember that this is not the maximum rate at which $\mathsf{E}[W_n]$ can grow, but rather the maximum rate in the sense of (9.129).

We will find in the next subsection that understanding this maximization problem is also the key to understanding arbitrary portfolios rather than portfolios with constant $\boldsymbol{\lambda}$.

The maximization of $\mathsf{E}[Y(\boldsymbol{\lambda})]$ over $\boldsymbol{\lambda}$ is an example of maximizing a concave function over a convex region. To spell this out, a region R of $\mathbb{R}^\ell$ is said to be convex if $\eta\boldsymbol{\lambda}_1+(1-\eta)\boldsymbol{\lambda}_2\in\mathrm{R}$ for all $\boldsymbol{\lambda}_1,\boldsymbol{\lambda}_2\in\mathrm{R}$ and $\eta\in(0,1)$. Geometrically, this says that all points on the straight line segment between $\boldsymbol{\lambda}_1$ and $\boldsymbol{\lambda}_2$ are also in R. For the case here, the maximization of $\mathsf{E}[Y(\boldsymbol{\lambda})]$ is over the region $\mathrm{R}\in\mathbb{R}^\ell$ for which $\sum_k\lambda(k)=1$ and $\lambda(k)\geq 0$ for $1\leq k\leq\ell$. It is easy to verify that this R is a convex region.

A function h from a convex region $\mathrm{R}\in\mathbb{R}^\ell$ to $\mathbb{R}$ is said to be concave over a convex region R if

$$h(\eta\boldsymbol{\lambda}_1+(1-\eta)\boldsymbol{\lambda}_2)\geq\eta h(\boldsymbol{\lambda}_1)+(1-\eta)h(\boldsymbol{\lambda}_2)\qquad\text{for all }\boldsymbol{\lambda}_1,\boldsymbol{\lambda}_2\in\mathrm{R},\ \eta\in(0,1). \tag{9.133}$$

To see this geometrically, consider the straight line segment from any $\boldsymbol{\lambda}_1\in\mathrm{R}$ to $\boldsymbol{\lambda}_2\in\mathrm{R}$. The function is concave if for each such segment, the function lies on or above the straight line function from $h(\boldsymbol{\lambda}_1)$ to $h(\boldsymbol{\lambda}_2)$. Concave functions are particularly easy to maximize, since if we start at any point and continually move in a direction to increase the function, we eventually approach the maximum value, i.e., the function does not have multiple peaks and the function is maximized at a point where no local increase is possible.

A function h is convex over a convex region if the inequality in (9.133) is reversed. Thus h is convex if and only if $-h$ is concave. All the properties of convex functions translate immediately to concave functions and vice-versa. For example, concave functions are easy to maximize and convex functions are easy to minimize.

To see that the function $\mathsf{E}[Y(\boldsymbol{\lambda})]$ is concave over the specified region of $\boldsymbol{\lambda}$, we start by the observation that the logarithm function, $\ln(t)$, has a negative second derivative for $t > 0$ and is thus concave for $t > 0$. Using this, the following equation shows that $Y(\boldsymbol{\lambda})$ is concave (i.e., it is concave for each of its sample values):

$$Y(\eta\boldsymbol{\lambda}_1 + (1-\eta)\boldsymbol{\lambda}_2) = \ln\left(\sum_k \big(\eta\lambda_1(k) + (1-\eta)\lambda_2(k)\big)X(k)\right)$$
$$\geq \eta \ln\left(\sum_k \lambda_1(k)X(k)\right) + (1-\eta)\ln\left(\sum_k \lambda_2(k)X(k)\right).$$

Finally, taking the expectation of each side, we see that $\mathsf{E}[Y(\boldsymbol{\lambda})]$ is concave over the specified region of $\boldsymbol{\lambda}$. The argument above was slightly careless about cases where $X(k)$ can take on the value 0 with positive probability. In this case, however, we can consider the region of $\mathbb{R}^\ell$ where $\mathsf{E}[Y(\boldsymbol{\lambda})] > -\infty$. This region is convex and $\mathsf{E}[Y(\boldsymbol{\lambda})]$ is concave over it.

There is a large literature about maximizing concave functions over convex regions, and there are necessary and sufficient conditions, called the Kuhn–Tucker conditions, for the maximum. These conditions are particularly simple for the case here in which the convex region is the probability simplex, i.e., the region in which $\sum_k \lambda(k) = 1$ and each $\lambda(k) \geq 0$.

The necessary and sufficient conditions for maximizing $\boldsymbol{\lambda}$ in this case are that, for some constant c, the partial derivatives satisfy

$$\frac{\partial}{\partial\lambda(j)}\mathsf{E}[Y(\boldsymbol{\lambda})] \begin{cases} = c & \text{for } \lambda(j) > 0, \\ \leq c & \text{for } \lambda(j) = 0. \end{cases} \tag{9.134}$$

Rather than proving this standard result, we give a convincing intuitive argument. If, for a given $\boldsymbol{\lambda}$, there are investments i and j for which $\partial\mathsf{E}[Y(\boldsymbol{\lambda})]/\partial\lambda(i) > \partial\mathsf{E}[Y(\boldsymbol{\lambda})]/\partial\lambda(j)$, then we can try to increase $\mathsf{E}[Y(\boldsymbol{\lambda})]$ by making an incremental increase in $\lambda(i)$ and an equal incremental decrease in $\lambda(j)$ (thus maintaining $\sum_k \lambda(k) = 1$). This can always be done if $\lambda(j) > 0$, thus explaining why the partial derivatives must be equal for $\lambda(j) > 0$ in (9.134). This incremental change cannot be done if $\lambda(j) = 0$, indicating why inequality might hold in (9.134) when $\lambda(j) = 0$.

We now proceed to look at these partial derivatives. For any $\boldsymbol{\lambda}$ for which $\mathsf{E}[Y(\boldsymbol{\lambda})] > -\infty$, the partial derivative with respect to each $\lambda(j)$ is given by

$$\frac{\partial}{\partial\lambda(j)}\mathsf{E}[Y(\boldsymbol{\lambda})] = \mathsf{E}\left[\frac{\partial}{\partial\lambda(j)}\ln\sum_k \lambda(k)X(k)\right]$$
$$= \mathsf{E}\left[\frac{X(j)}{\sum_k \lambda(k)X(k)}\right]. \tag{9.135}$$

Note that the denominator here must be positive for possible sample values of $X(1), \ldots, X(\ell)$, since otherwise $\mathsf{E}[Y(\boldsymbol{\lambda})]$ would not exist. Substituting this in (9.134) leads to the following theorem.

Theorem 9.10.4 *Let* $\mathrm{R} = \{\boldsymbol{\lambda} : \sum_k \lambda_k = 1$ and $\lambda(k) \geq 0$ for $1 \leq k \leq \ell$ $\}$. *The function* $\mathsf{E}[Y(\boldsymbol{\lambda})] = \mathsf{E}\left[\ln\left(\sum_k \lambda(k)X(k)\right)\right]$ *is maximized over* $\boldsymbol{\lambda} \in \mathrm{R}$ *if and only if* $\boldsymbol{\lambda}$ *satisfies the conditions:*[6]

$$\mathsf{E}\left[\frac{X(j)}{\sum_k \lambda(k)X(k)}\right] \begin{cases} = 1 & \text{for } \lambda(j) > 0, \\ \leq 1 & \text{for } \lambda(j) = 0. \end{cases} \tag{9.136}$$

Proof After substituting (9.135) into (9.134), we must show that the constant c in (9.134) has the value 1. Multiplying both sides of (9.135) by $\lambda(j)$ and summing over j, we get

$$\sum_j \lambda(j)\frac{\partial}{\partial\lambda(j)}\mathsf{E}[Y(\boldsymbol{\lambda})] = \sum_j \lambda(j)\mathsf{E}\left[\frac{X(j)}{\sum_k \lambda(k)X(k)}\right] = 1.$$

Substituting (9.134) into the left-hand side, we see that $c = 1$. □

An interesting interpretation of (9.136) arises if we multiply both sides by $\lambda(j)$. This says that the expected fraction of wealth in investment j at the end of an epoch is $\lambda(j)$. In other words, the allocation $\boldsymbol{\lambda}^*$ is characterized by the property that the expected fraction of wealth in each investment j at the end of an epoch is the same as that at the beginning of the epoch. We make use of this property in the next section.

It is possible for the maximizing $\boldsymbol{\lambda}$ to be non-unique, and we denote any maximizing choice as $\boldsymbol{\lambda}^*$. The maximum value of $\mathsf{E}[Y(\boldsymbol{\lambda})]$ is of course unique, and will be denoted $\mathsf{E}\left[Y(\boldsymbol{\lambda}^*)\right]$. This quantity is important since it is the maximum exponential growth rate in the sense of (9.129). We will soon see that this is true not only for constant $\boldsymbol{\lambda}$ but also all time-varying $\boldsymbol{\lambda}$.

Before looking at time-varying $\boldsymbol{\lambda}$, there is a very simple but important question that should be answered. Under what circumstances is $\mathsf{E}\left[Y(\boldsymbol{\lambda}^*)\right] > 0$? Since cash is one of the investments and has zero growth-rate, $\mathsf{E}\left[Y(\boldsymbol{\lambda}^*)\right] \geq 0$ in all cases. If cash alone maximizes $\mathsf{E}[Y(\boldsymbol{\lambda})]$, then we can take $\lambda^*(\ell) = 1$. Since the cash investment is deterministic with $X(\ell) = 1$, we then have $\sum_k \lambda^*(k)X(k) = 1$. It then follows from (9.136) that $\mathsf{E}\left[X(j)\right] \leq 1$ for all j. In summary, if $\boldsymbol{\lambda}^*(\ell) = 1$, then $\mathsf{E}\left[X(j)\right] \leq 1$ for all j. This in turn implies that if $\mathsf{E}\left[X(j)\right] > 1$ for at least one j, then cash alone cannot maximize $\mathsf{E}[L(\boldsymbol{\lambda})]$. Another way to see this is that an incremental move from cash alone toward any j such that $\mathsf{E}\left[X(j)\right] > 1$ increases $\mathsf{E}[Y(\boldsymbol{\lambda})]$ to a positive value. This proves half the following theorem.

Theorem 9.10.5 *Assuming that cash is an available investment, the exponential growth rate* $\mathsf{E}\left[Y(\boldsymbol{\lambda}^*)\right]$ *is positive if and only if* $\mathsf{E}\left[X(j)\right] > 1$ *for at least one investment* j.

[6] This theorem can be generalized to include short sales and margin buying. A short position in investment j would be represented by $\lambda(j) < 0$ and buying on margin would be represented by $\lambda(\ell) < 0$. Letting α be a constraint on the fraction of wealth in any short position, the constraints $\lambda(j) > 0$ and $\lambda(j) = 0$ in (9.136) become $\lambda(j) > -\alpha$ and $\lambda(j) = -\alpha$ for $j < \ell$. Letting β be a constraint on the fraction of wealth on margin, the constraints for cash become $\lambda(\ell) > -\beta$ and $\lambda(\ell) = -\beta$. In practice, short selling and margin buying are far more complex, since α and β can be changed dynamically outside the investor's control.

Proof We have seen that $\boldsymbol{\lambda}^*(\ell) = 1$ implies that $\mathsf{E}\left[X(j)\right] \leq 1$ for all j. We must show that $\mathsf{E}\left[X(j)\right] \leq 1$ for all j implies that $\mathsf{E}\left[Y(\boldsymbol{\lambda})\right]$ is maximized by $\boldsymbol{\lambda}(\ell) = 1$, or, equivalently, that $\mathsf{E}\left[Y(\boldsymbol{\lambda})\right] \leq 0$ for all $\boldsymbol{\lambda}$. This follows from the concavity of the log function:

$$\mathsf{E}\left[\ln \sum_j \lambda(j) X(j)\right] \leq \ln\left[\sum_j \lambda(j) \mathsf{E}\left[X(j)\right]\right] \leq 0.$$

□

As is often the case in finding conditions for a maximum, the conditions are more valuable for finding other results than for actually computing the maximum. In the situation here, the maximization conditions in (9.136) allow us to show that this maximum over constant $\boldsymbol{\lambda}$ also applies to time-varying $\boldsymbol{\lambda}$.

In order to motivate why varying $\boldsymbol{\lambda}$ from one epoch to another might be important, it will be helpful to look at a particular such strategy for Example 9.10.1. Suppose the investor splits the initial wealth into two equal piles. The first is then fully invested on successive epochs in the triple-or-nothing investment and the second uses the constant allocation $\lambda^* = 1/4$. We then have

$$\mathsf{E}\left[W_n\right] = \frac{1}{2} \times \left(\frac{3}{2}\right)^n + \frac{1}{2} \times \left(\frac{9}{8}\right)^n; \qquad \mathsf{E}\left[L_n\right] \approx \frac{n}{2} \ln(9/8) - \ln 2, \tag{9.137}$$

where we have approximated $\mathsf{E}\left[L_n\right]$ by neglecting the negligible contribution from the go-for-broke pile. By using this investment strategy, which is not a constant allocation strategy, we do essentially as well in maximizing $\mathsf{E}\left[W_n\right]$ as the go-for-broke strategy and essentially as well as the constant allocation λ^* in maximizing $\mathsf{E}\left[L_n\right]$. We postpone further discussion of such strategies that essentially maximise both $\mathsf{E}[W_n]$ and $\mathsf{E}[L_n]$ until later.

Another example of time-varying $\boldsymbol{\lambda}$ is the buy-and-hold strategy advocated by successful investors such as Warren Buffett. The reasons for such strategies concern many practical issues that are not included in the simple investing model here, most particularly the assumption that the **rv** s $\boldsymbol{X}_1, \boldsymbol{X}_2, \ldots,$ are IID with known distributions.

The next subsection proves an optimality result that applies to all investment strategies. It achieves essentially the same thing as (9.129). It finds the maximum rate at which a portfolio can increase exponentially, subject to any given positive probability of substantially achieving that rate. The bottom line is that the maximum exponential rate in this sense is $\mathsf{E}\left[Y(\boldsymbol{\lambda}^*)\right]$, the same as that for a constant allocation in Theorem 9.10.4.

9.10.2 Portfolios with time-varying allocations

According to the simple investment model here, the only choice for the investor is that of allocating investments at the beginning of each epoch. In this section, those allocations can vary from epoch to epoch and the allocation $\boldsymbol{\lambda}_m$ at epoch m can depend on all results prior to m, i.e., $\boldsymbol{\lambda}$ can be a function of $\boldsymbol{X}_1, \ldots, \boldsymbol{X}_{m-1}$. As an example of these time-varying allocations, Exercise 9.38 shows how to calculate $\boldsymbol{\lambda}_n$ at each epoch n for the buy-and-hold example and the two-pile example above. The point of Exercise 9.38 is not that there is any need to find these allocations, but simply to point out that they are special cases of the time-varying allocations in this section.

The analysis to follow is based on comparing an arbitrary time-varying strategy $\boldsymbol{\lambda}_1, \boldsymbol{\lambda}_2, \dots,$ to the constant $\boldsymbol{\lambda}^*$ that maximizes $\mathsf{E}[Y(\boldsymbol{\lambda})]$ as given by Theorem 9.10.4. Denote the wealth at the end of epoch n using the arbitrary allocation as W_n and that using $\boldsymbol{\lambda}^*$ as W_n^*.

Define $Z_n = W_n/W_n^*$ as the ratio of wealth using the arbitrary strategy to wealth using the optimal fixed $\boldsymbol{\lambda}^*$. Note that the stochastic process involved is $\{\boldsymbol{X}_n;\ n \geq 1\}$. The rv s W_n and W_n^* both exist in this same probability model using the same sequence of sample values $\{\boldsymbol{X}_n(w);\ n \geq 1\}$ for the individual investments. As usual, $W_0 = W_0^* = 1$.

Theorem 9.10.6 *The sequence $\{Z_n;\ n \geq 1\}$ of rv s with $Z_n = W_n/W_n^*$ is a supermartingale.*

Proof We can express W_n as

$$W_n = W_{n-1} \sum_j \lambda_n(j) X_n(j).$$

Expressing W_n^* in the same way,

$$\begin{aligned} Z_n = \frac{W_n}{W_n^*} &= \frac{W_{n-1} \sum_k \lambda_n(k) X_n(k)}{W_{n-1}^* \sum_k \lambda^*(k) X_n(k)} \\ &= Z_{n-1} \sum_j \lambda_n(j) \frac{X_n(j)}{\sum_k \lambda^*(k) X_n(k)}. \end{aligned} \tag{9.138}$$

Taking the expectation of each side conditional on $Z_1, \dots, Z_{n-1}$,

$$\begin{aligned} \mathsf{E}\left[Z_n \mid Z_1, \dots, Z_{n-1}\right] &= Z_{n-1} \sum_j \lambda_n(j) \mathsf{E}\left[\frac{X_n(j)}{\sum_k \lambda^*(k) X_n(k)}\right] \\ &= Z_{n-1} \sum_j \lambda_n(j) \frac{\partial}{\partial \lambda(j)} \mathsf{E}\left[Y(\boldsymbol{\lambda}^*)\right] \qquad (9.139) \\ &\leq Z_{n-1}, \qquad (9.140) \end{aligned}$$

where we have used (9.135) in (9.139) and (9.136) in (9.140). Note that the possible dependence of $\lambda_n(j)$ on $\boldsymbol{X}_1, \dots, \boldsymbol{X}_{n-1}$ does not affect this final inequality. Finally, $\mathsf{E}[|Z_n|]$ is finite for all n. In fact, since Z_n is non-negative, the argument above shows that $\mathsf{E}[|Z_n|] \leq 1$ for all n. □

We can now use this supermartingale property to show that there is little advantage to be gained by the flexibility of using time-varying allocations. Combining the non-negative supermartingale inequality of (9.115) with the fact that $\mathsf{E}[Z_n] \leq 1$, we see that for any $a > 0$,

$$\Pr\left\{\bigcup_{n \geq 1} \{Z_n \geq a\}\right\} \leq \frac{1}{a}.$$

Since $Z_n = W_n/W_n^*$, this becomes

$$\Pr\left\{\bigcup_{n\geq 1} \{W_n \geq aW_n^*\}\right\} \leq \frac{1}{a}. \tag{9.141}$$

This says that an arbitrary strategy cannot be much better than the constant allocation $\boldsymbol{\lambda}$ that maximizes the rate in (9.129). The above factor a looks important until we remember that the wealth of the optimal fixed strategy is growing exponentially with n, so a constant factor is negligible. We can use $\boldsymbol{\lambda}^*$ in (9.131) and combine this with (9.141). By combining the factor a in (9.141) with the ϵ and δ in (9.131), we have proven the following theorem.

Theorem 9.10.7 *For any $\epsilon > 0$ and $\delta > 0$, there is an n_o such that for any time-varying allocation, the wealth satisfies*

$$\Pr\left\{\bigcup_{n\geq n_o} W_n \geq \exp\left[n\left(\mathsf{E}\left[Y(\boldsymbol{\lambda}^*)\right] + \epsilon\right)\right]\right\} \leq \delta. \tag{9.142}$$

Note that Theorem 9.10.2 shows that the constant allocation strategy $\boldsymbol{\lambda}^*$ has an arbitrarily high probability of achieving a growth rate arbitrarily close to $\mathsf{E}\left[Y(\boldsymbol{\lambda}^*)\right]$. This theorem says that no other strategy has more than a vanishing probability of growing at a greater rate.

It is interesting to observe that strict inequality in (9.140) can occur only if $\lambda_n(j) > 0$ and $\lambda^*(j) = 0$ for some n and j. For all strategies that avoid investments for which $\lambda^*(j) = 0$, the stochastic process $\{Z_n;\ n \geq 1\}$ is a martingale. This implies that $\mathsf{E}\left[Z_n\right] = 1$ for all n.

One might be excused for thinking that if $\mathsf{E}\left[Z_n\right] = 1$ for all n, where $Z_n = W_n/W_n^*$, then the investment strategy leading to W_n must not be too bad, but it turns out that the martingale Z_n is very misleading. For example, the pure triple-or-nothing strategy of Example 9.10.1 is easily seen to satisfy $\mathsf{E}\left[Z_n\right] = 1$.

It also follows that the martingale convergence theorem applies if $\{Z_n;\ n \geq 1\}$ is a martingale. Exercise 9.39 provides some insight into this, but again, the process $\{Z_n;\ n \geq 1\}$ is a disappointment.

We may now ask whether this highly simplified theory provides any insight into investing. The assumption that $\{\boldsymbol{X}_n;\ n \geq 1\}$ is IID and known to the investor is obviously oversimplified and loses much of the character of investing. However, aside from insider information, information about investments is openly accessible, so price-relatives can at least be estimated. If all the available information is used in estimating the price-relatives, then it would appear that the assumption of independence from epoch to epoch is reasonable, at least as a first approximation. These estimated price-relatives would be time-varying, but as seen in Exercise 1.40, the WLLN applies to many situations of independent but non-identically distributed rv s.

Many people object to the long-term emphasis in this theory. It is possible, at the cost of more complexity, to approximate $L_n(\boldsymbol{\lambda})$ by the CLT. This provides an approximation to the speed of convergence with n in Theorem 9.10.2 but also lets us approximate $L_n(\boldsymbol{\lambda})$ as Gaussian, and thus consider tradeoffs between the mean and variance of L_n when choosing $\boldsymbol{\lambda}$.

A further generalization would be to model the evolution of the price-relatives by Markov chains. The log wealth could then be regarded in terms of Markov reward theory and also in terms of Markov modulated random walks (as developed in the next section), but we do not pursue these generalizations further here. In summary, we see that there are a number of ways to extend the simple theory here, but that they would require a great deal of input from practical applications. We thus conclude this section while it is still simple and elegant.

9.11 Markov modulated random walks

Frequently it is useful to generalize random walks to allow some dependence between the variables being summed. The particular form of dependence here is the same as the Markov reward processes of Section 4.5. The treatment in Section 4.5 discussed only expected rewards, whereas the treatment here focuses on the rv s themselves. Let $\{Y_m;\ m \geq 0\}$ be a sequence of (possibly dependent) rv s, and let

$$\{S_n;\ n \geq 1\}, \qquad \text{where } S_n = \sum_{m=0}^{n-1} Y_m. \tag{9.143}$$

Let $\{X_n;\ n \geq 0\}$ be a Markov chain, and assume that each Y_n is a function of X_n and X_{n+1}. Conditional on X_n and X_{n+1}, Y_n is independent of $Y_{n-1}, \ldots, Y_1$, and of X_i for all $i \neq n, n-1$. Assume that Y_n, conditional on X_n and X_{n+1}, has a CDF $\mathsf{F}_{ij}(y) = \Pr\{Y_n \leq y \mid X_n = i, X_{n+1} = j\}$. Thus each rv Y_n depends only on the associated transition in the Markov chain, and this dependence is the same for all n.

The process $\{S_n;\ n \geq 1\}$ is called a *Markov modulated random walk*. If each Y_m is positive, it can be viewed as the sequence of epochs in a semi-Markov process. In the general case, Y_m is associated with the transition in the Markov chain from trial m to trial $m+1$, and S_n is the aggregate reward up to but not including time n. Let $\overline{Y}_{ij}$ denote $\mathsf{E}\left[Y_n \mid X_n = i, X_{n+1} = j\right]$ and $\overline{Y}_i$ denote $\mathsf{E}[Y_n \mid X_n = i]$. Let $\{P_{ij}\}$ be the set of transition probabilities for the Markov chain, so $\overline{Y}_i = \sum_j P_{ij}\overline{Y}_{ij}$. We may think of the process $\{Y_n;\ n \geq 0\}$ as evolving along with the Markov chain. The distributions of the variables Y_n are associated with the transitions from X_n to X_{n+1}, but the Y_n are otherwise independent random variables.

In order to define a martingale related to the process $\{S_n;\ n \geq 1\}$, we must subtract the mean reward from $\{S_n\}$ and must also compensate for the effect of the state of the Markov chain. The appropriate compensation factor turns out to be the relative-gain vector defined in Section 4.5.

For simplicity, consider only finite-state ergodic Markov chains with M states. Let $\pi = (\pi_1, \ldots, \pi_{\mathsf{M}})$ be the steady-state probability vector for the chain, let $\overline{\boldsymbol{Y}} = (\overline{Y}_1, \ldots, \overline{Y}_{\mathsf{M}})^{\mathsf{T}}$ be the vector of expected rewards, let $g = \pi\overline{\boldsymbol{Y}}$ be the steady-state gain per unit time, and let $\boldsymbol{w} = (w_1, \ldots, w_{\mathsf{M}})^{\mathsf{T}}$ be the relative-gain vector. From Theorem 4.5.4, $\boldsymbol{w}$ is the unique solution to

$$\boldsymbol{w} + g\boldsymbol{e} = \overline{\boldsymbol{Y}} + [P]\boldsymbol{w}, \qquad w_1 = 0. \tag{9.144}$$

We assume a fixed starting state $X_0 = k$. As we now show, the process $\{Z_n;\ n \geq 1\}$ given by

$$Z_n = S_n - ng + w_{X_n} - w_k, \quad n \geq 1, \tag{9.145}$$

is a martingale. First condition on a given state, $X_{n-1} = i$.

$$\mathsf{E}\left[Z_n \mid Z_{n-1}, Z_{n-2}, \ldots, Z_1, X_{n-1} = i\right]. \tag{9.146}$$

Since $S_n = S_{n-1} + Y_{n-1}$, we can express Z_n as

$$Z_n = Z_{n-1} + Y_{n-1} - g + w_{X_n} - w_{X_{n-1}}. \tag{9.147}$$

Since $\mathsf{E}\left[Y_{n-1} \mid X_{n-1} = i\right] = \overline{Y}_i$ and $\mathsf{E}\left[w_{X_n} \mid X_{n-1} = i\right] = \sum_j P_{ij} w_j$, we have

$$\mathsf{E}\left[Z_n \mid Z_{n-1}, Z_{n-2}, \ldots, Z_1, X_{n-1} = i\right] = Z_{n-1} + \overline{Y}_i - g + \sum_j P_{ij} w_j - w_i. \tag{9.148}$$

From (9.144) the final four terms in (9.148) sum to 0, so

$$\mathsf{E}\left[Z_n \mid Z_{n-1}, \ldots, Z_1, X_{n-1} = i\right] = Z_{n-1}. \tag{9.149}$$

Since this is valid for all choices of X_{n-1}, we have $\mathsf{E}\left[Z_n \mid Z_{n-1}, \ldots, Z_1\right] = Z_{n-1}$. Since the expected values of all the reward variables $\overline{Y}_i$ exist, we see that $\mathsf{E}\left[|Y_n|\right] < \infty$, so that $\mathsf{E}\left[|Z_n|\right] < \infty$ also. This verifies that $\{Z_n;\ n \geq 1\}$ is a martingale. It can be verified similarly that $\mathsf{E}\left[Z_1\right] = 0$, so $\mathsf{E}\left[Z_n\right] = 0$ for all $n \geq 1$.

In showing that $\{Z_n;\ n \geq 1\}$ is a martingale, we actually showed something a little stronger. That is, (9.149) is conditioned on X_{n-1} as well as $Z_{n-1}, \ldots, Z_1$. In the same way, it follows that for all $n > 1$

$$\mathsf{E}\left[Z_n \mid Z_{n-1}, X_{n-1}, Z_{n-2}, X_{n-2}, \ldots, Z_1, X_1\right] = Z_{n-1}. \tag{9.150}$$

In terms of the gambling analogy, this says that $\{Z_n;\ n \geq 1\}$ is fair for each possible past sequence of states. A martingale $\{Z_n;\ n \geq 1\}$ with this property (i.e., satisfying (9.150)) is said to be a *martingale relative to the joint process* $\{Z_n, X_n;\ n \geq 1\}$. We will use this martingale later to discuss threshold crossing problems for Markov modulated random walks. We shall see that the added property of being a martingale relative to $\{Z_n, X_n\}$ gives us added flexibility in defining stopping times.

As an added bonus to this example, note that if $\{X_n;\ n \geq 0\}$ is taken as the embedded chain of a Markov process (or semi-Markov process), and if Y_n is taken as the time interval from transition n to $n+1$, then S_n becomes the epoch of the nth transition in the process.

9.11.1 Generating functions for Markov random walks

Consider the same Markov chain and reward variables as in the previous example, again with $X_0 = k$, and assume that for each pair of states, i, j, the MGF

$$\mathsf{g}_{ij}(r) = \mathsf{E}\left[\exp(rY_n) \mid X_n = i, X_{n+1} = j\right] \tag{9.151}$$

exists over some open interval (r_-, r_+) containing 0. Let $[\Gamma(r)]$ be the matrix with terms $P_{ij}\mathsf{g}_{ij}(r)$. Since $[\Gamma(r)]$ is an irreducible non-negative matrix, Theorem 4.4.2 shows that $[\Gamma(r)]$ has a largest real eigenvalue, $\rho(r) > 0$, and an associated positive right eigenvector, $\boldsymbol{\nu}(r) = (\nu_1(r), \ldots, \nu_{\mathsf{M}}(r))^{\mathsf{T}}$ that is unique within a scale factor. We now show that the process $\{M_n(r);\ n \geq 1\}$ defined by

$$M_n(r) = \frac{\exp(rS_n)\nu_{X_n}(r)}{\rho(r)^n \nu_k(r)} \tag{9.152}$$

is a product type martingale for each $r \in (r_-, r_+)$. Since $S_n = S_{n-1} + Y_{n-1}$, we can express $M_n(r)$ as

$$M_n(r) = M_{n-1}(r)\,\frac{\exp(rY_{n-1})\,\nu_{X_n}(r)}{\rho(r)\nu_{X_{n-1}}(r)}. \tag{9.153}$$

The expected value of the ratio of $M_n(r)/M_{n-1}(r)$ in (9.153), conditional on $X_{n-1} = i$, is

$$\mathsf{E}\left[\frac{\exp(rY_{n-1})\nu_{X_n}(r)}{\rho(r)\nu_i(r)}\,\middle|\,X_{n-1}{=}i\right] = \frac{\sum_j P_{ij}\mathsf{g}_{ij}(r)\nu_j(r)}{\rho(r)\nu_i(r)} = 1. \tag{9.154}$$

where, in the last step, we have used the fact that $\nu(r)$ is an eigenvector of $[\Gamma(r)]$. Thus, $\mathsf{E}\left[M_n(r) \mid M_{n-1}(r), \ldots, M_1(r), X_{n-1} = i\right] = M_{n-1}(r)$. Since this is true for all choices of i, the condition on $X_{n-1} = i$ can be removed and $\{M_n(r);\ n \geq 1\}$ is a martingale. Also, for $n > 1$,

$$\mathsf{E}\left[M_n(r) \mid M_{n-1}(r), X_{n-1}, \ldots, M_1(r), X_1\right] = M_{n-1}(r), \tag{9.155}$$

so that $\{M_n(r);\ n \geq 1\}$ is also a martingale relative to the joint process $\{M_n(r), X_n;\ n \geq 1\}$.

It can be verified by the same argument as in (9.154) that $\mathsf{E}[M_1(r)] = 1$. It then follows that $\mathsf{E}[M_n(r)] = 1$ for all $n \geq 1$.

One of the uses of this martingale is to provide exponential upper bounds, similar to (9.9), to the probabilities of threshold crossings for Markov modulated random walks. Define

$$\widetilde{M}_n(r) = \frac{\exp(rS_n)\min_j(\nu_j(r))}{\rho(r)^n \nu_k(r)}. \tag{9.156}$$

Then $\widetilde{M}_n(r) \leq M_n(r)$, so $\mathsf{E}[\widetilde{M}_n(r)] \leq 1$. For any $\mu > 0$, the Markov inequality can be applied to $\widetilde{M}_n(r)$ to get

$$\Pr\left\{\widetilde{M}_n(r) \geq \mu\right\} \leq \frac{1}{\mu}\mathsf{E}\left[\widetilde{M}_n(r)\right] \leq \frac{1}{\mu}. \tag{9.157}$$

For any α, and for any $r \in [0, r_+)$, we can choose $\mu = \exp(r\alpha)\rho(r)^{-n}\min_j(\nu_j(r))/\nu_k(r)$. Combining (9.156) and (9.157),

$$\Pr\{S_n \geq \alpha\} \leq \rho(r)^n \exp(-r\alpha)\nu_k(r)/\min_j(\nu_j(r)). \tag{9.158}$$

This can be optimized over r to get the tightest bound in the same way as (9.9).

9.11.2 Stopping trials for martingales relative to a process

A martingale $\{Z_n;\ n \geq 1\}$ relative to a joint process $\{Z_n, X_n;\ n \geq 1\}$ was defined as a martingale for which (9.150) is satisfied, i.e., $\mathsf{E}\left[Z_n \mid Z_{n-1}, X_{n-1}, \ldots, Z_1, X_1\right] = Z_{n-1}$. In the same way, we can define a *submartingale or supermartingale* $\{Z_n;\ n \geq 1\}$ *relative to a joint process* $\{Z_n, X_n;\ n \geq 1\}$ as a submartingale or supermartingale satisfying (9.150) with the $=$ sign replaced by $\geq$ or $\leq$ respectively. The purpose of this added complication is to make it easier to define useful stopping rules.

As generalized in Definition 5.5.6, a generalized stopping trial J for a sequence of pairs of rv s $(Z_1, X_1), (Z_2, X_2) \ldots$, is a positive integer-valued rv such that, for each $n \geq 1$, $\mathbb{I}_{\{J=n\}}$ is a function of $Z_1, X_1, Z_2, X_2, \ldots, Z_n, X_n$.

Theorems 9.8.2, 9.8.3 and Lemma 9.8.4 all carry over to martingales (submartingales or supermartingales) relative to a joint process. These theorems are stated more precisely in Exercises 9.29–9.34. To summarize them here, assume that $\{Z_n;\ n \geq 1\}$ is a martingale (submartingale or supermartingale) relative to a joint process $\{Z_n, X_n;\ n \geq 1\}$ and assume that J is a stopping trial for $\{Z_n;\ n \geq 1\}$ relative to $\{Z_n, X_n;\ n \leq 1\}$. Then the stopped process is a martingale (submartingale or supermartingale) respectively, (9.98)–(9.100) are satisfied, and, for a martingale, $\mathsf{E}[Z_J] = \mathsf{E}[Z_1]$ is satisfied if and only if (9.104) is satisfied.

9.11.3 Markov modulated random walks with thresholds

We have now developed two martingales for Markov modulated random walks, both conditioned on a fixed initial state $X_0 = k$. The first, given in (9.145), is $\{Z_n = S_n - ng + w_{X_n} - w_k;\ n \geq 1\}$. Recall that $\mathsf{E}[Z_n] = 0$ for all $n \geq 1$ for this martingale. Given two thresholds, $\alpha > 0$ and $\beta < 0$, define J as the smallest n for which $S_n \geq \alpha$ or $S_n \leq \beta$. The indicator function $\mathbb{I}_{J=n}$ of $\{J = n\}$, is 1 if and only if $\beta < S_i < \alpha$ for $1 \leq i \leq n-1$ and either $S_n \geq \alpha$ or $S_n \leq \beta$. Since $S_i = Z_i + ig - w_{X_i} + w_k$, S_i is a function of Z_i and X_i, so the stopping trial is a function of both Z_i and X_i for $1 \leq i \leq n$. It follows that J is a stopping trial for $\{Z_n;\ n \geq 1\}$ relative to $\{Z_n, X_n;\ n \geq 1\}$. From Lemma 9.8.4, we can assert that $\mathsf{E}[Z_J] = \mathsf{E}[Z_1] = 0$ if (9.104) is satisfied, i.e., if $\lim_{n\to\infty} \mathsf{E}[Z_n \mid J > n] \Pr\{J > n\} = 0$ is satisfied. Using the same argument as in Lemma 9.4.1, we can see that $\Pr\{J > n\}$ goes to 0 at least geometrically in n. Conditional on $J > n$, $\beta < S_n < \alpha$, so S_n is bounded independent of n. Also w_{X_n} is bounded, since the chain is finite state, and ng is linear in n. Thus $\mathsf{E}[Z_n \mid J > n]$ varies at most linearly with n, so (9.104) is satisfied, and

$$0 = \mathsf{E}[Z_J] = \mathsf{E}[S_J] - \mathsf{E}[J]\,g + \mathsf{E}\left[w_{X_n}\right] - w_k. \tag{9.159}$$

Recall that Wald's equality for random walks is $\mathsf{E}[S_J] = \mathsf{E}[J]\,\overline{X}$. For Markov modulated random walks, this is modified, as shown in (9.159), by the relative-gain vector terms.

The same arguments can be applied to the generating function martingale of (9.152). Again, let J be the smallest n for which $S_n \geq \alpha$ or $S_n \leq \beta$. As before, S_i is a function of $M_i(r)$ and X_i, so $\mathbb{I}_{J=n}$ is a function of $M_i(r)$ and X_i for $1 \leq i \leq n-1$. It follows that J is a stopping trial for $\{M_n(r);\ n \geq 1\}$ relative to $\{M_n(r), X_n;\ n \geq 1\}$. Next we need the following lemma.

Lemma 9.11.1 *For the martingale $\{M_n(r);\ n \geq 1\}$ relative to $\{M_n(r), X_n;\ n \geq 1\}$ defined in (9.152), where $\{X_n;\ n \geq 0\}$ is a finite-state Markov chain, and for the above stopping trial J,*

$$\lim_{n\to\infty} \mathsf{E}[M_n(r) \mid J > n] \Pr\{J > n\} = 0. \tag{9.160}$$

Proof Note that there is a $\delta > 0$ such that for all states i, j, and all $n > 1$ such that $\Pr\{J = n, X_{n-1} = i, X_n = j\} > 0$,

$$\mathsf{E}\left[\exp(rS_n \mid J = n, X_{n-1} = i, X_n = j\right] \geq \delta. \tag{9.161}$$

Since the stopped process, $\{M_n^*(r);\ n \geq 1\}$, is a martingale, we have for each m,

$$1 = \mathsf{E}\left[M_m^*(r)\right] \geq \sum_{n=1}^{m} \frac{\mathsf{E}\left[\exp(rS_n)\nu_{X_n}(r) \mid J = n\right]}{\rho(r)^n \nu_k(r)}. \tag{9.162}$$

From (9.161), we see that there is some $\delta' > 0$ such that

$$\mathsf{E}\left[\exp(rS_n)\nu_{X_n}(r)\right] / \nu_k(r) \mid J = n] \geq \delta'$$

for all n such that $\Pr\{J = n\} > 0$. Thus (9.162) is bounded by

$$1 \geq \delta' \sum_{n \leq m} \rho(r)^n \Pr\{J = n\}\,.$$

Since this is valid for all m, it follows by the argument in the proof of Theorem 9.4.2 that $\lim_{n\to\infty} \rho(r)^n \Pr\{J > n\} = 0$. This, along with (9.161), establishes (9.160), completing the proof. □

From Lemma 9.8.4, we have the desired result:

$$\mathsf{E}[M_J(r)] = \mathsf{E}\left[\frac{\exp(rS_J)\nu_{X_J}(r)}{[\rho(r)]^J \nu_k(r)}\right] = 1, \qquad r_- < r < r_+. \tag{9.163}$$

This is the extension of the Wald identity to Markov modulated random walks, and is used in the same way as the Wald identity. As shown in Exercise 9.41, the derivative of (9.163), evaluated at $r = 0$, is the same as (9.159).

9.12 Summary

Each term in a random walk $\{S_n;\ n \geq 1\}$, where $S_n = X_1 + \cdots + X_n$, is a sum of IID rv s, and thus the study of random walks is closely related to that of sums of IID rv s. The focus in random walks, however, as in most of the processes we have studied, is more in the relationship between the terms (such as which term first crosses a threshold) than in the individual terms. We started by showing that random walks are a generalization of renewal processes and are central to studying the queueing delay for G/G/1 queues.

A major focus of the chapter was on evaluating, bounding, and approximating the probabilities of very unlikely events, a topic known as large deviation theory. We started by studying the Chernoff bound to $\Pr\{S_n \geq \alpha\}$ for $\alpha > 0$ and $\mathsf{E}[X] < 0$. One of the insights gained here was that if a threshold at $\alpha >> 0$ is crossed, it is likely to be crossed at a trial $n \approx \alpha/\gamma'(r^*)$, where r^* is the positive root of $\gamma(r)$. The overall probability of crossing a threshold at α was next elegantly upper bounded by $\exp(-r^*\alpha)$. This bound resulted from Wald's identity, which is essentially the statement that appropriately tilted versions of the underlying probabilities are in fact rv s themselves. For questions of typical behavior of a random walk, the mean and variance of a rv are the major quantities of interest, but when interested in atypically large deviations, r^* is a major parameter of interest.

We next introduced martingales, submartingales, and supermartingales. These are sometimes regarded as somewhat exotic topics in mathematics, but in fact they are very useful in a large variety of relatively simple processes. For example, all of the random walk issues of earlier sections can be treated as a special case of martingales, and martingales can be used to model both sums and products of random variables. We also showed how Markov modulated random walks can be treated as martingales.

Stopping trials, as first introduced in Chapter 5, were then applied to martingales. We defined a stopped process $\{Z_n^*;\ n \geq 1\}$ to be the same as the original process $\{Z_n;\ n \geq 1\}$ up to the stopping point, and then constant thereafter. Theorems 9.8.2 and 9.8.3 showed that the stopped process has the same form (martingale, submartingale, or supermartingale) as the original process, and that the expected values $\mathsf{E}\left[Z_n^*\right]$ are between $\mathsf{E}[Z_1]$ and $\mathsf{E}[Z_n]$. We also looked at $\mathsf{E}[Z_J]$ and found that it is equal to $\mathsf{E}[Z_1]$ if and only if (9.104) is satisfied. The Wald identity can be viewed as $\mathsf{E}[Z_J] = \mathsf{E}[Z_1] = 1$ for the Wald martingale, $Z_n = \exp\{rS_n - n\gamma(r)\}$. We then found a similar identity for Markov modulated random walks.

The Kolmogorov inequalities were next developed. They are analogs of the Markov inequality and Chebyshev inequality, except they bound initial segments of submartingales and martingales rather than single rv s. They were used to prove, first, the SLLN with only a second moment and, second, the martingale convergence theorem.

A very simple model for allocating investments was developed both as a further example of the use of martingales and as a topic of intrinsic interest. A good reference for the mathematical development of this topic is [5].

A standard reference on random walks, and particularly on the analysis of overshoots is [9]. Dembo and Zeitouni [6] develop large deviation theory in a much more general and detailed way than the introduction here. The classic reference on martingales is [7], but [4] and [22] are more accessible.

9.13 Exercises

Exercise 9.1 Consider the simple random walk $\{S_n;\ n \geq 1\}$ of Section 9.1.1 with $S_n = X_1 + \cdots + X_n$ and $\Pr\{X_i = 1\} = p$; $\Pr\{X_i = -1\} = 1 - p$; assume that $p \leq 1/2$.

(a) Show that $\Pr\{\bigcup_{n\geq 1}\{S_n \geq k\}\} = [\Pr\{\bigcup_{n\geq 1}\{S_n \geq 1\}\}]^k$ for any positive integer k. Hint: Given that the random walk ever reaches the value 1, consider a new random walk starting at that time and explore the probability that the new walk ever reaches a value 1 greater than its starting point.

(b) Find a quadratic equation for $y = \Pr\left\{\bigcup_{n\geq 1}\{S_n \geq 1\}\right\}$. Hint: Explore each of the two possibilities immediately after the first trial.

(c) For $p < 1/2$, show that the two roots of this quadratic equation are $p/(1-p)$ and 1. Argue that $\Pr\{\bigcup_{n\geq 1} \{S_n \geq 1\}\}$ cannot be 1 and thus must be $p/(1-p)$.

(d) For $p = 1/2$, show that the quadratic equation in (c) has a double root at 1, and thus $\Pr\{\bigcup_{n\geq 1} \{S_n \geq 1\}\} = 1$. Note: This is the very peculiar case explained in Section 5.5.1.

(e) Let r^* be the unique positive root of $\mathsf{g}(r) = 1$ where $\mathsf{g}(r) = \mathsf{E}\left[e^{rX}\right]$. Show that $p/(1-p) = \exp(-r^*)$.

Exercise 9.2 Consider a G/G/1 queue with IID arrivals $\{X_i;\ i \geq 1\}$, IID FCFS service times $\{Y_i;\ i \geq 0\}$, and an initial arrival to an empty system at time 0. Define $U_i = Y_{i-1} - X_i$ for $i \geq 1$. Consider a sample path where $(u_1, \ldots, u_6) = (1, -2,\ 2, -1,\ 3, -2)$.

(a) Let $Z_i^6 = U_6 + U_{6-1} + \cdots + U_{6-i+1}$. Find the queueing delay for customer 6 as the maximum of the 'backward' random walk with elements $0, Z_1^6, Z_2^6, \ldots, Z_6^6$; sketch this random walk.

(b) Find the queueing delay for customers 1 to 5.

(c) Which customers start a busy period (i.e., arrive when the queue and server are both empty)? Verify that if Z_i^6 maximizes the random walk in (a), then a busy period starts with arrival $6 - i$.

(d) Now consider a forward random walk $V_n = U_1 + \cdots + U_n$. Sketch this walk for the sample path above and show that the queueing delay for each customer is the difference between two appropriately chosen values of this walk.

Exercise 9.3 A G/G/1 queue has a deterministic service time of 2 and interarrival times that are 3 with probability $p < 1/2$ and 1 with probability $1 - p$.

(a) Find the distribution of W_1, the wait in queue of the first arrival after the beginning of a busy period.

(b) Find the distribution of W_∞, the steady-state wait in queue.

(c) Repeat (a) and (b) assuming the service times and interarrival times are exponentially distributed with rates μ and λ respectively.

Exercise 9.4 Let $U = V_1 + \cdots + V_n$, where $V_1, \ldots, V_n$ are IID rv s with the MGF $g_V(s)$. Show that $g_U(s) = [g_V(s)]^n$. Hint: You should be able to do this simply in a couple of lines.

Exercise 9.5 Let $\{a_n;\ n \geq 1\}$ and $\{\alpha_n;\ n \geq 1\}$ be sequences of numbers. For some b, assume that $a_n \leq \alpha_n e^{-bn}$ for all $n \geq 1$. For each of the following choices for α_n and arbitrary b, determine whether the bound is exponentially tight. Note: You may assume $b \geq 0$ if you wish, but it really does not matter.

(a) $\alpha_n = \sum_{j=1}^{k} c_j n^{-j}$.

(b) $\alpha_n = \exp(-\sqrt{n})$.

(c) $\alpha_n = e^{-n^2}$.

What you are intended to learn from this is that an exponentially tight bound is not necessarily a reasonable approximation for large n. It is intended for sequences that are essentially exponentially decreasing in n and where one is satisfied with knowing the exponent without concern for the non-exponential coefficients.

Exercise 9.6 Define $\gamma(r)$ as $\ln[\mathsf{g}(r)]$, where $\mathsf{g}(r) = \mathsf{E}\big[\exp(rX)\big]$. Assume that X is discrete with possible outcomes $\{a_i;\ i \geq 1\}$, let p_i denote $\Pr\{X = a_i\}$, and assume that $\mathsf{g}(r)$ exists in some open interval (r_-, r_+) containing $r = 0$. For any given r, $r_- < r < r_+$, define a rv X_r with the same set of possible outcomes $\{a_i;\ i \geq 1\}$ as X, but with a PMF $q_i = \Pr\{X_r = a_i\} = p_i \exp[a_i r - \gamma(r)]$. Note that X_r is not a function of X, and is not even to be viewed as in the same probability space as X; it is of interest simply because of the behavior of its defined probability mass function. It is called a tilted rv relative to X, and this exercise, along with Exercise 9.11, will justify our interest in it.

(a) Verify that $\sum_i q_i = 1$.

(b) Verify that $\mathsf{E}[X_r] = \sum_i a_i q_i$ is equal to $\gamma'(r)$.

(c) Verify that $\mathsf{VAR}[X_r] = \sum_i a_i^2 q_i - (\mathsf{E}[X_r])^2$ is equal to $\gamma''(r)$.

(d) Argue that $\gamma''(r) \geq 0$ for all r such that $\mathsf{g}(r)$ exists, and that $\gamma''(r) > 0$ if $\gamma''(0) > 0$.

(e) Give a similar definition of X_r for a rv X with a density, and modify (a)–(d) accordingly.

Exercise 9.7 **(a)** Suppose Z is uniformly distributed over $[-b, +b]$. Find $\mathsf{g}_Z(r), \mathsf{g}'_Z(r)$, and $\gamma'(r)$ as a function of r.

(b) Show that the interval over which $\mathsf{g}_Z(r)$ exists is the real line, i.e., $r_+ = \infty$ and $r_- = -\infty$. Show that $\lim_{r\to\infty} \gamma'(r) = b$.

(c) Show that if $a > b$, then $\inf_{r \in I(X)} \gamma(r) - ra = -\infty$, so that the optimized Chernoff bound in (9.10) says that $\Pr\{S_n > na\} = 0$. Explain without any mathematics why $\Pr\{S_n > na\} = 0$ must be valid. Explain (with as little mathematical verbiage as possible) why an infimum rather than a minimum must be used in (9.10).

(d) Show that for an arbitrary rv X if $r_+ = \infty$ and $\lim_{r\to\infty} \gamma'(r) = b$, then (9.10) says that $\Pr\{S_n > na\} = 0$ for $a > b$.

Exercise 9.8 Note that the MGF of the non-negative exponential rv with density e^{-x} is $(1 - r)^{-1}$ for $r < r_+ = 1$. It can be seen from this that, $\mathsf{g}(r_+)$ does not exist (i.e., it is infinite) and both $\lim_{r\to r_+} \mathsf{g}(r) = \infty$ and $\lim_{r\to r_+} \mathsf{g}'(r) = \infty$, where the limit is over $r < r_+$. In this exercise, you are first to assume an arbitrary rv X for which $r_+ < \infty$ and $\mathsf{g}(r_+) = \infty$ and show that both $\lim_{r\to r_+} \mathsf{g}(r) = \infty$ and $\lim_{r\to r_+} \mathsf{g}'(r) = \infty$. You will then use this to show that if $\sup_{r<r_+} \mathsf{g}'(r) < \infty$, then $\gamma(r_+) < \infty$ and $\mu(a)$ in (9.9) is given by $\mu(a) = \gamma(r_+) - r_+ a$ for $a > \sup_{r<r_+} \gamma'(r)$.

(a) For X such that $r_+ < \infty$ and $\mathsf{g}(r_+) = \infty$, explain why

$$\lim_{A\to\infty} \int_0^A e^{xr_+}\, d\mathsf{F}(x) = \infty.$$

(b) Show that for any $\epsilon > 0$ and any $A > 0$,

$$\mathsf{g}(r_+ - \epsilon) \geq e^{-\epsilon A} \int_0^A e^{xr_+}\, d\mathsf{F}(x).$$

(c) Choose $A = 1/\epsilon$ and show that

$$\lim_{\epsilon\to 0} \mathsf{g}(r_+ - \epsilon) = \infty.$$

(d) Show that $\lim_{\epsilon\to 0} \mathsf{g}'(r_+ - \epsilon) = \infty$.

(e) Use (a)–(d) to show that if $r_+ < \infty$ and $\sup_{r<r_+} \mathsf{g}'(r) < \infty$, then $\mathsf{g}(r_+) < \infty$.

(f) Show that if $r_+ < \infty$ and $\sup_{r<r_+} \mathsf{g}'(r) < \infty$, then $\mu(a) = \gamma(r_+) - r_+a$ for $a > \sup_{r<r_+} \gamma'(r)$.

Exercise 9.9 (Details in proof of Theorem 9.3.3) **(a)** Show that the two appearances of ϵ in (9.24) can be replaced with two independent arbitrary positive quantities ϵ_1 and ϵ_2, getting

$$\Pr\{S_n \geq n(\gamma'(r) - \epsilon_1)\} \geq (1-\delta)\exp[-n(r\gamma'(r) + r\epsilon_2 - \gamma(r))].$$

Show that if this equation is valid for ϵ_1 and ϵ_2, then it is valid for all larger values of ϵ_1 and ϵ_2. Hint: Note that the left-hand side of (9.24) is increasing in ϵ and the right-hand side is decreasing.

(b) Show that by increasing the n_o such that this equation is valid for all $n \geq n_o$, the factor of $(1-\delta)$ can be eliminated above.

(c) For any $r \in (0, r_+)$, let δ_1 be an arbitrary number in $(0, r_+ - r)$, let $r_1 = r + \delta_1$, and let $\epsilon_1 = \gamma'(r_1) - \gamma'(r)$. Show that there is an m such that for all $n \geq m$,

$$\Pr\{S_n \geq n\gamma'(r)\} \geq \exp\left\{-n\left[(r+\delta_1)\gamma'(r+\delta_1) + (r+\delta_1)\epsilon_2 - \gamma(r+\delta_1)\right]\right\}. \quad (9.164)$$

Using the continuity of γ and its derivatives, show that for any $\epsilon > 0$, there is a $\delta_1 > 0$ so that the right-hand side of (9.164) is greater than or equal to $\exp[-n(\gamma'(r) - r\gamma(r) + \epsilon)]$.

Exercise 9.10 In this exercise, we show that the optimized Chernoff bound is tight for the third case in (9.12) as well as the first case. That is, we show that if $r_+ < \infty$ and $a > \sup_{r<r_+} \gamma'(r)$, then, for any $\epsilon > 0$, $\Pr\{S_n \geq na\} \geq \exp\{n[\gamma(r_+) - r_+a - \epsilon]\}$ for all large enough n.

(a) Let Y_i be the truncated version of X_i, truncated for some given b to $Y_i = X_i$ for $X_i \leq b$ and $Y_i = b$ otherwise. Let $W_n = Y_1 + \cdots + Y_n$. Show that $\Pr\{S_n \geq na\} \geq \Pr\{W_n \geq na\}$.

(b) Let $\mathsf{g}_b(r)$ be the MGF of Y. Show that $\mathsf{g}_b(r) < \infty$ and that $\mathsf{g}_b(r)$ is non-decreasing in b for all $r < \infty$.

(c) Show that $\lim_{b\to\infty} \mathsf{g}_b(r) = \infty$ for all $r > r_+$ and that $\lim_{b\to\infty} \mathsf{g}_b(r) = \mathsf{g}(r)$ for all $r \leq r_+$.

(d) Let $\gamma_b(r) = \ln \mathsf{g}_b(r)$. Show that $\gamma_b(r) < \infty$ for all $r < \infty$. Also show that $\lim_{b\to\infty} \gamma_b(r) = \infty$ for $r > r_+$ and $\lim_{b\to\infty} \gamma_b(r) = \gamma(r)$ for $r \leq r_+$. Hint: Use (b) and (c).

(e) Let $\gamma_b'(r) = (\partial/\partial r)\gamma_b(r)$ and let $\delta > 0$ be arbitrary. Show that for all large enough b, $\gamma_b'(r_+ + \delta) > a$. Hint: First show that $\gamma_b'(r_+ + \delta) \geq [\gamma_b(r_+ + \delta) - \gamma(r_+)]/\delta$.

(f) Show that the optimized Chernoff bound for $\Pr\{W_n \geq na\}$ is exponentially tight for the values of b in (e). Show that the optimizing r is less than $r_+ + \delta$.

(g) Show that for any $\epsilon > 0$ and all sufficiently large b,

$$\gamma_b(r) - ra \geq \gamma(r) - ra - \epsilon \geq \gamma(r_+) - r_+a - \epsilon \qquad \text{for } 0 < r \leq r_+.$$

Hint: Show that the convergence of $\gamma_b(r)$ to $\gamma(r)$ is uniform in r for $0 < r < r_+$.

(h) Show that for arbitrary $\epsilon > 0$, there exists $\delta > 0$ and b_o such that

$$\gamma_b(r) - ra \geq \gamma(r_+) - r_+a - \epsilon \qquad \text{for } r_+ < r < r_+ + \delta \text{ and } b \geq b_o.$$

(i) Note that if we put together (g), (h) and (d), and use the δ of (h) in (d), then we have shown that the optimized exponent in the Chernoff bound for $\Pr\{W_n \geq na\}$ satisfies $\mu_b(a) \geq \gamma(r_+) - r_+a - \epsilon$ for sufficiently large b. Show that this means that $\Pr\{S_n \geq na\} \geq \gamma(r_+ - r_+a - 2\epsilon$ for sufficiently large n.

Exercise 9.11 Assume that X is discrete, with possible values $\{a_i;\ i \geq 1)$ and probabilities $\Pr\{X = a_i\} = p_i$. Let X_r be the corresponding tilted rv as defined in Exercise 9.6. Let $S_n = X_1 + \cdots + X_n$ be the sum of n IID rv s with the distribution of X, and let $S_{n,r} = X_{1,r} + \cdots + X_{n,r}$ be the sum of n IID tilted rv s with the distribution of X_r. Assume that $\overline{X} < 0$ and that $r > 0$ is such that $\gamma(r)$ exists.

(a) Show that $\Pr\{S_{n,r}{=}s\} = \Pr\{S_n{=}s\}\exp[sr - n\gamma(r)]$. Hint: Read the text.

(b) Find the mean and variance of $S_{n,r}$ in terms of $\gamma(r)$.

(c) Define $a = \gamma'(r)$ and $\sigma_r^2 = \gamma''(r)$. Show that $\Pr\left\{|S_{n,r} - na| \leq \sqrt{2n}\sigma_r\right\} > 1/2$. Use this to show that

$$\Pr\left\{|\ S_n - na\ | \leq \sqrt{2n}\,\sigma_r\right\} > (1/2)\exp[-r(an + \sqrt{2n}\,\sigma_r) + n\gamma(r)].$$

(d) Use this to show that for any ϵ and for all sufficiently large n,

$$\Pr\left\{S_n \geq n(\gamma'(r) - \epsilon)\right\} > \frac{1}{2}\exp[-rn(\gamma'(r) + \epsilon) + n\gamma(r)].$$

Exercise 9.12 **(a)** Let $\boldsymbol{p} = (p_1, \ldots, p_{\mathsf{M}})$ and $\widetilde{\boldsymbol{p}} = (\widetilde{p}_1, \ldots, \widetilde{p}_{\mathsf{M}})$ be strictly positive probability vectors. Show that the divergence $D(\widetilde{\boldsymbol{p}}\|\boldsymbol{p}) \geq 0$ is 0 if $\widetilde{\boldsymbol{p}} = \boldsymbol{p}$.

(b) Let $\boldsymbol{u} = u_1, \ldots, u_{\mathsf{M}}$ satisfy $\sum_j u_j = 0$ and show that $\widetilde{\boldsymbol{p}} + \epsilon\boldsymbol{u}$ is a probability vector over a sufficiently small region of ϵ around 0. Show that

$$\frac{\partial}{\partial\epsilon}D(\widetilde{\boldsymbol{p}} + \epsilon\boldsymbol{u}\|\boldsymbol{p})|_{\epsilon=0} = 0.$$

(c) Show that

$$\frac{\partial^2}{\partial\epsilon^2}D(\widetilde{\boldsymbol{p}} + \epsilon\boldsymbol{u}\|\boldsymbol{p}) \geq 0,$$

where $\widetilde{\boldsymbol{p}} + \epsilon\boldsymbol{u}$ is a probability vector with non-zero components. Explain why this implies that D is convex in $\widetilde{\boldsymbol{p}}$ and non-negative over the region of probability vectors.

Exercise 9.13 Consider a random walk $\{S_n;\ n \geq 1\}$, where $S_n = X_1 + \cdots + X_n$ and $\{X_i;\ i \geq 1\}$ is a sequence of IID exponential rv s with the PDF $\mathsf{f}(x) = \lambda e^{-\lambda x}$ for $x \geq 0$. In other words, the random walk is the sequence of arrival epochs in a Poisson process.

(a) Show that for $\lambda a > 1$, the optimized Chernoff bound for $\Pr\{S_n \geq na\}$ is given by

$$\Pr\{S_n \geq na\} \leq (a\lambda)^n e^{-n(a\lambda - 1)}.$$

(b) Show that the exact value of $\Pr\{S_n \geq na\}$ is given by

$$\Pr\{S_n \geq na\} = \sum_{i=0}^{n-1} \frac{(na\lambda)^i\, e^{-na\lambda}}{i!}.$$

(c) By upper and lower bounding the quantity on the right above, show that

$$\frac{(na\lambda)^n\, e^{-na\lambda}}{n!\, a\lambda} \leq \Pr\{S_n \geq na\} \leq \frac{(na\lambda)^n e^{-na\lambda}}{n!(a\lambda - 1)}.$$

Hint: Use the term at $i = n - 1$ for the lower bound and note that the term on the right can be bounded by a geometric series starting at $i = n - 1$.

(d) Use the Stirling bounds on $n!$ to show that

$$\frac{(a\lambda)^n\, e^{-n(a\lambda-1)}}{\sqrt{2\pi n}\, a\lambda \exp(1/12n)} \;\le\; \Pr\{S_n \ge na\} \;\le\; \frac{(a\lambda)^n e^{-n(a\lambda-1)}}{\sqrt{2\pi n}\,(a\lambda - 1)}.$$

The point of this exercise is to demonstrate that the Chernoff bound is not only exponentially tight for this example but captures the important factors in the behavior of $\Pr\{S_n \ge na\}$.

Exercise 9.14 Consider a random walk with thresholds $\alpha > 0$, $\beta < 0$. We wish to find $\Pr\{S_J \ge \alpha\}$ in the absence of a lower threshold. Use the upper bound in (9.46) for the probability that the random walk crosses α before β.

(a) Given that the random walk crosses β first, find an upper bound to the probability that α is crossed before a yet lower threshold at 2β is crossed.

(b) Given that 2β is crossed before α, upper bound the probability that α is crossed before a threshold at 3β. Extending this argument to successively lower thresholds, find an upper bound to each successive term, and find an upper bound on the overall probability that α is crossed. By observing that β is arbitrary, show that (9.46) is valid with no lower threshold.

Exercise 9.15 This exercise verifies that Corollary 9.4.4 holds in the situation where $\gamma(r_+) < 0$ (see Figure 9.4). We have defined $r^* = r_+$ in this case.

(a) Use the Wald identity at $r = r_+$ to show that

$$\Pr\{S_J \ge \alpha\}\, \mathsf{E}\left[\exp(r_+ S_J - J\gamma(r_+)) \mid S_J \ge \alpha\right] \le 1.$$

Hint: Look at the first part of the proof of Corollary 9.4.4.

(b) Show that

$$\Pr\{S_J \ge \alpha\} \exp[r_+\alpha - \gamma(r_+)] \le 1.$$

(c) Show that

$$\Pr\{S_J \ge \alpha\} \le \exp[-r_+\alpha + \gamma(r_+)] \qquad \text{and} \qquad \Pr\{S_J \ge \alpha\} \le \exp[-r^*\alpha].$$

Note that the first bound above is a little stronger than the second, and note that lower bounding J by 1 is not necessarily a very weak bound, since this is the case where if α is crossed, it tends to be crossed for small J.

Exercise 9.16 **(a)** Use Wald's equality to show that if $\overline{X} = 0$, then $\mathsf{E}[S_J] = 0$, where J is the time of threshold crossing with one threshold at $\alpha > 0$ and another at $\beta < 0$.

(b) Obtain an expression for $\Pr\{S_J \ge \alpha\}$. Your expression should involve the expected value of S_J conditional on crossing the individual thresholds (you need not try to calculate these expected values).

(c) Evaluate your expression for the case of a simple random walk.

(d) Evaluate your expression when X has an exponential density, $\mathsf{f}_X(x) = a_1 e^{-\lambda x}$ for $x \ge 0$ and $\mathsf{f}_X(x) = a_2 e^{\mu x}$ for $x < 0$ and where a_1 and a_2 are chosen so that $\overline{X} = 0$.

Exercise 9.17 A random walk $\{S_n;\ n \geq 1\}$, with $S_n = \sum_{i=1}^{n} X_i$, has the following probability density for X_i:

$$\mathsf{f}_X(x) = \begin{cases} \dfrac{e^{-x}}{e - e^{-1}} & -1 \leq x \leq 1, \\ = 0 & \text{elsewhere.} \end{cases}$$

(a) Find the values of r for which $\mathsf{g}(r) = \mathsf{E}\left[\exp(rX)\right] = 1$.

(b) Let P_α be the probability that the random walk ever crosses a threshold at α for some $\alpha > 0$. Find an upper bound to P_α of the form $P_\alpha \leq e^{-\alpha A}$, where A is a constant that does not depend on α; evaluate A.

(c) Find a lower bound to P_α of the form $P_\alpha \geq Be^{-\alpha A}$, where A is the same as in (b) and B is a constant that does not depend on α. Hint: Keep it simple – you are not expected to find an elaborate bound. Also recall that $\mathsf{E}\left[e^{r^* S_J}\right] = 1$, where J is a stopping trial for the random walk and $\mathsf{g}(r^*) = 1$.

Exercise 9.18 Let $\{X_n;\ n \geq 1\}$ be a sequence of IID integer-valued rv s with the PMF $\mathsf{p}_X(k) = Q_k$. Assume that $Q_k > 0$ for $|k| \leq 10$ and $Q_k = 0$ for $|k| > 10$. Let $\{S_n;\ n \geq 1\}$ be a random walk with $S_n = X_1 + \cdots + X_n$. Let $\alpha > 0$ and $\beta < 0$ be integer-valued thresholds, let J be the smallest value of n for which either $S_n \geq \alpha$ or $S_n \leq \beta$. Let $\{S_n^*;\ n \geq 1\}$ be the stopped random walk; i.e., $S_n^* = S_n$ for $n \leq J$ and $S_n^* = S_J$ for $n > J$. Let $\pi_i^* = \Pr\{S_J = i\}$.

(a) Consider a Markov chain in which this stopped random walk is run repeatedly until the point of stopping. That is, the Markov chain transition probabilities are given by $P_{ij} = Q_{j-i}$ for $\beta < i < \alpha$ and $P_{i0} = 1$ for $i \leq \beta$ and $i \geq \alpha$. All other transition probabilities are 0 and the set of states is the set of integers $[-9 + \beta, 9 + \alpha]$. Show that this Markov chain is ergodic.

(b) Let $\{\pi_i\}$ be the set of steady-state probabilities for this Markov chain. Find the set of probabilities $\{\pi_i^*\}$ for the stopping states of the stopped random walk in terms of $\{\pi_i\}$.

(c) Find $\mathsf{E}[S_J]$ and $\mathsf{E}[J]$ in terms of $\{\pi_i\}$.

Exercise 9.19 **(a)** Conditional on $X = 0$ for the hypothesis testing problem, consider the rv s $Z_i = \ln[\mathsf{f}(Y_i|X{=}1)/\mathsf{f}(Y_i|X{=}0)]$. Show that r^*, the positive solution to $\mathsf{g}(r) = 1$, where $\mathsf{g}(r) = \mathsf{E}\left[\exp(rZ_i)\right]$, is given by $r^* = 1$.

(b) Assuming that Y is a discrete rv (under each hypothesis), show that the tilted rv Z_r with $r = 1$ has the PMF $\mathsf{p}_Y(y|X{=}1)$.

Exercise 9.20 (Proof of Stein's lemma) **(a)** Use a power series expansion around $r = 0$ to show that $\gamma_0(r) - r\gamma'(r) = -r^2\gamma_0''(0)/2 + o(r^2)$.

(b) Choose $r = n^{-1/3}$ in (9.61) and (9.62). Show that (9.61) can then be rewritten as

$$\begin{aligned} \Pr\{e(n) \mid X{=}0\} &\leq \exp\left\{n[-r^2\gamma_0''(0)/2 + o(r^2)]\right\} \\ &= \exp\left\{-n^{1/3}\gamma_0''(0)/2 + o(n^{1/3})]\right\}. \end{aligned}$$

Thus $\Pr\{e(n) \mid X = 0\}$ approaches 0 with increasing n exponentially in $n^{1/3}$.

(c) Using the same power series expansion for r in (9.62), show that

$$\Pr\{e(n) \mid X{=}1\} \le \exp\left\{n[r^2\gamma_0''(0)/2 + \gamma_0'(0) + o(r)]\right\}$$
$$= \exp\left\{n\gamma_0'(0) + o(n^{2/3})]\right\}.$$

Thus $\lim_n(1/n)\ln\Pr\{e(n) \mid X{=}1\} = \gamma_0'(0) = \mathsf{E}[Z \mid X = 0]$.

(d) Explain why the bound in (9.62) is exponentially tight given that $\Pr\{e(n) \mid X = 0\} \to 0$ as $n \to \infty$.

Exercise 9.21 (The secretary problem or marriage problem) This illustrates a very different type of sequential decision problem from those of Section 9.5. Alice is looking for a spouse and dates a set of n suitors sequentially, one per week. For simplicity, assume that Alice must make her decision to accept a suitor for marriage immediately after dating that suitor; she cannot come back later to accept a formerly rejected suitor. Her decision must be based only on whether the current suitor is more suitable than all previous suitors. Mathematically, we can view the dating as continuing for all n weeks, but the choice at week m is a stopping rule. Assume that each suitor's suitability is represented by a real number and that all n numbers are different. Alice does not observe the suitability numbers, but only observes at each time m whether suitor m has the highest suitability so far. The suitors are randomly permuted before Alice dates them.

(a) A reasonable algorithm for Alice is to reject the first k suitors (for some k to be optimized later) and then to choose the first suitor $m > k$ that is more suitable than all the previous $m-1$ suitors. Find the probability q_k that Alice chooses the most suitable of all n when she uses this algorithm for given k. Hint 1: What is the probability (as a function of m) that the most suitable of all n is dated on week m. Hint 2: Given that the most suitable of all suitors is dated at week m, what is the probability that Alice chooses that suitor? For $m > k$, show how this involves the location of the most suitable choice from 1 to $m-1$, conditional on m's overall optimality. Note: If no suitor is selected by the algorithm, it makes no difference whether suitor n is chosen at the end or no suitor is chosen.

(b) Approximating $\sum_{i=1}^{j} 1/i$ by $\ln j$, show that for n and k large

$$q_k \approx \frac{k}{n}\ln(n/k).$$

Ignoring the constraint that n and k are integers, show that the right-hand side above is maximized at $k/n = e$ and that $\max_k q_k \approx 1/e$.

(c) (Optional) Show that the algorithm of (a), optimized over k, is optimal over all algorithms (given the constraints of the problem). Hint: Let p_m be the maximum probability of choosing the optimal suitor given that no choice has been made before time m. Show that

$$p_m = \max\left[\frac{1}{n} + \frac{m-1}{m}p_{m+1},\ p_{m+1}\right];$$

part of the problem here is understanding exactly what p_m means.

(d) Explain why this is a poor model for choosing a spouse (or for making a best choice in a wide variety of similar problems). Caution: It is not enough to explain why this is not closely related to these real problems, You should also explain why this gives very little insight into such real problems.

Exercise 9.22 **(a)** Suppose $\{Z_n;\ n \geq 1\}$ is a martingale. Verify (9.83); i.e., $\mathsf{E}[Z_n] = \mathsf{E}[Z_1]$ for $n > 1$.

(b) If $\{Z_n;\ n \geq 1\}$ is a submartingale, verify (9.88), and if a supermartingale, verify (9.89).

Exercise 9.23 Suppose $\{Z_n;\ n \geq 1\}$ is a martingale. Show that

$$\mathsf{E}\left[Z_m \mid Z_{n_i}, Z_{n_{i-1}}, \ldots, Z_{n_1}\right] = Z_{n_i} \text{ for all } 0 < n_1 < n_2 < \cdots < n_i < m.$$

Exercise 9.24 **(a)** Assume that $\{Z_n;\ n \geq 1\}$ is a submartingale. Show that

$$\mathsf{E}\left[Z_m \mid Z_n, Z_{n-1}, \ldots, Z_1\right] \geq Z_n \text{ for all } n < m.$$

(b) Show that

$$\mathsf{E}\left[Z_m \mid Z_{n_i}, Z_{n_{i-1}}, \ldots, Z_{n_1}\right] \geq Z_{n_i} \text{ for all } 0 < n_1 < n_2 < \ldots < n_i < m.$$

(c) Assume now that $\{Z_n;\ n \geq 1\}$ is a supermartingale. Show that (a) and (b) still hold with $\geq$ replaced by $\leq$.

Exercise 9.25 Let $\{Z_n = \exp[rS_n - n\gamma(r)];\ n \geq 1\}$ be the generating function martingale of (9.77), where $S_n = X_1 + \cdots + X_n$ and $X_1, \ldots X_n$ are IID with mean $\overline{X} < 0$. Let J be the possibly-defective stopping trial for which the process stops after crossing a threshold at $\alpha > 0$ (there is no negative threshold). Show that $\exp[r^*\alpha]$ is an upper bound to the probability of threshold crossing by considering the stopped process $\{Z_n^*;\ n \geq 1\}$.

The purpose of this exercise is to illustrate that the stopped process can yield useful upper bounds even when the stopping trial is defective.

Exercise 9.26 This exercise uses a martingale to find the expected number of successive trials $\mathsf{E}[J]$ until some fixed pattern, $a_1, a_2, \ldots, a_k$, of successive binary digits occurs within a sequence of IID binary rv s $X_1, X_2, \ldots$ (see Example 4.5.1 and Exercise 5.35 for alternative approaches). We take the stopping time J to be the smallest n for which $(X_{n-k+1}, \ldots, X_n) = (a_1, \ldots, a_k)$. A mythical casino and sequence of gamblers who follow a prescribed strategy will be used to determine $\mathsf{E}[J]$. The outcomes of the plays (trials), $\{X_n;\ n \geq 1\}$ at the casino is a binary IID sequence for which $\Pr\{X_n = i\} = p_i$ for $i \in \{0, 1\}$.

If a gambler places a bet s on 1 at play n, the return is s/p_1 if $X_n = 1$ and 0 otherwise. With a bet s on 0, the return is s/p_0 if $X_n = 0$ and 0 otherwise; i.e., the game is fair.

(a) Assume an arbitrary choice of bets on 0 and 1 by the various gamblers on the various trials. Let Y_n be the net gain of the casino on trial n. Show that $\mathsf{E}[Y_n] = 0$. Let $Z_n = Y_1 + Y_2 + \cdots + Y_n$ be the aggregate gain of the casino over n trials. Show that for any given pattern of bets, $\{Z_n;\ n \geq 1\}$ is a martingale.

(b) In order to determine $\mathsf{E}[J]$ for a given pattern $a_1, a_2, \ldots, a_k$, we program our gamblers to bet as follows:

(i) Gambler 1 has an initial capital of 1 which is bet on a_1 at trial 1. If $X_1 = a_1$, the capital grows to $1/p_{a_1}$, all of which is bet on a_2 at trial 2. If $X_2 = a_2$, the capital grows to $1/(p_{a_1}p_{a_2})$, all of which is bet on a_3 at trial 3. Gambler 1 continues in this way until either losing at some trial (in which case he leaves with no money) or winning on k successive trials (in which case he leaves with $1/[p_{a_1}\cdots p_{a_k}]$).

(ii) Gambler ℓ, for each $\ell > 1$, follows the same strategy, but starts at trial ℓ. Note that if the pattern $a_1, \dots, a_k$ appears for the first time at trials $n-k+1, n-k+2, \dots, n$, i.e., if $J = n$, then gambler $n-k+1$ leaves at time n with capital $1/[p(a_1)\cdots p(a_k)]$ and gamblers $j < n-k+1$ have all lost their capital. We will come back later to investigate the capital at time n for gamblers $n-k+2$ to n.

First consider the string a_1=0, a_2=1 with $k = 2$. Find the sample values of Z_1, Z_2, Z_3 for the sample sequence $X_1 = 1, X_2 = 0, X_3 = 1, \dots$. Note that gamblers 1 and 3 have lost their capital, but gambler 2 now has capital $1/p_0p_1$. Show that the sample value of the stopping time for this case is $J = 3$. Given an arbitrary sample value $n \geq 2$ for J, show that $Z_n = n - 1/p_0p_1$.

(c) Find $\mathsf{E}[Z_J]$ from (a). Use this plus (b) to find $\mathsf{E}[J]$. Hint: This uses the special form of the solution in (b), not the Wald equality.

(d) Repeat (b) and (c) using the string $(a_1, \dots, a_k) = (1, 1)$ and initially assuming $(X_1X_2X_3) = (011)$. Be careful about gambler 3 for $J = 3$. Show that $\mathsf{E}[J] = 1/p_1^2 + 1/p_1$.

(e) Repeat (b) and (c) for $(a_1, \dots, a_k) = (1, 1, 1, 0, 1, 1)$.

(f) Consider an arbitrary binary string $a_1, \dots, a_k$ and condition on $J = n$ for some $n \geq k$. Show that the sample capital of gambler ℓ is then equal to

- 0 for $\ell < n - k$;
- $1/[p_{a_1}p_{a_2}\cdots p_{a_k}]$ for $\ell = n - k + 1$;
- $1/[p_{a_1}p_{a_2}\cdots p_{a_i}]$ for $\ell = n - i + 1,\ 1 \leq i < k$ if $(a_1, \dots, a_i) = (a_{k-i+1}, \dots, a_k)$;
- 0 for $\ell = n - i + 1,\ 1 \leq i < k$ if $(a_1, \dots, a_i) \neq (a_{k-i+1}, \dots, a_k)$.

Verify that this general formula agrees with (c), (d), and (e).

(g) For a given binary string $a_1, \dots, a_k$, and each j, $1 \leq j \leq k$ let $\mathbb{I}_j = 1$ for $(a_1, \dots, a_j) = (a_{k-j+1}, \dots, a_k)$ and let $\mathbb{I}_j = 0$ otherwise. Show that

$$\mathsf{E}[J] = \sum_{i=1}^{k} \frac{\mathbb{I}_i}{\prod_{m=1}^{i} p_{a_m}}.$$

Note that this is the same as the final result in Exercise 4.28. The argument is shorter here, but more motivated and insightful there. Both approaches are useful and lead to simple generalizations.

Exercise 9.27 **(a)** This exercise shows why the condition $\mathsf{E}[|Z_J|] < \infty$ is required in Lemma 9.8.4. Let $Z_1 = -2$ and, for $n \geq 1$, let $Z_{n+1} = Z_n[1 + X_n(3n+1)/(n+1)]$, where $X_1, X_2, \dots$ are IID and take on the values $+1$ and -1 with probability $1/2$ each. Show that $\{Z_n;\ n \geq 1\}$ is a martingale.

(b) Consider the stopping trial J such that J is the smallest value of $n > 1$ for which Z_n and Z_{n-1} have the same sign. Show that, conditional on $n < J$, $Z_n = (-2)^n/n$ and, conditional on $n = J$, $Z_J = -(-2)^n(2n-1)/(n^2 - n)$.

(c) Show that $\mathsf{E}[|Z_J|]$ is infinite, so that $\mathsf{E}[Z_J]$ does not exist according to the definition of expectation, and show that $\lim_{n\to\infty} \mathsf{E}[Z_n|J > n]\Pr\{J > n\} = 0$.

Exercise 9.28 This exercise shows why the supremum of a martingale can behave markedly differently from the maximum of an arbitrarily large number of the variables. More precisely, it shows that $\Pr\{\sup_{n\geq 1} Z_n \geq a\}$ can be unequal to $\Pr\{\bigcup_{n\geq 1}\{Z_n \geq a\}\}$.

(a) Consider a martingale where Z_n can take on only the values 2^{-n-1} and $1-2^{-n-1}$, each with probability 1/2. Given that Z_n, conditional on Z_{n-1}, is independent of $Z_1,\dots,Z_{n-2}$, find $\Pr\{Z_n|Z_{n-1}\}$ for each n so that the martingale condition is satisfied.

(b) Show that $\Pr\{\sup_{n\geq 1} Z_n \geq 1\} = 1/2$ and show that $\Pr\{\bigcup_n\{Z_n \geq 1\}\} = 0$.

(c) Show that for every $\epsilon > 0$, $\Pr\{\sup_{n\geq 1} Z_n \geq a\} \leq \overline{Z}_1/(a-\epsilon)$. Hint: Use the relationship between $\Pr\{\sup_{n\geq 1} Z_n \geq a\}$ and $\Pr\{\bigcup_n\{Z_n \geq a\}\}$ while getting around the issue in (b).

(d) Use (c) to establish (9.111).

Exercise 9.29 Show that Theorem 9.7.4 is also valid for martingales relative to a joint process. That is, show that if h is a convex function of a real variable and if $\{Z_n;\ n\geq 1\}$ is a martingale relative to a joint process $\{Z_n, X_n;\ n \geq 1\}$, then $\{h(Z_n);\ n \geq 1\}$ is a submartingale relative to $\{h(Z_n), X_n;\ n \geq 1\}$.

Exercise 9.30 Show that if $\{Z_n;\ n \geq 1\}$ is a martingale (submartingale or supermartingale) relative to a joint process $\{Z_n, X_n;\ n \geq 1\}$ and if J is a stopping trial for $\{Z_n;\ n \geq 1\}$ relative to $\{Z_n, X_n;\ n \geq 1\}$, then the stopped process is a martingale (submartingale or supermartingale) respectively relative to the joint process.

Exercise 9.31 Prove Corollaries 9.9.3–9.9.5, i.e., prove the following three statements.

(a) Let $\{Z_n;\ n \geq 1\}$ be a martingale with $\mathsf{E}\left[Z_n^2\right] < \infty$ for all $n \geq 1$. Then

$$\Pr\left\{\max_{1\leq n\leq m} |Z_n| \geq b\right\} \leq \frac{\mathsf{E}\left[Z_m^2\right]}{b^2};\quad \text{for all integer } m \geq 2, \text{ and all } b > 0.$$

Hint: First show that $\{Z_n^2;\ n \geq 1\}$ is a submartingale.

(b) (Kolmogorov's random walk inequality) Let $\{S_n;\ n \geq 1\}$ be a random walk with $S_n = X_1 + \cdots + X_n$, where $\{X_i;\ i \geq i\}$ is a set of IID rv s with mean $\overline{X}$ and variance σ^2. Then for any positive integer m and any $\epsilon > 0$,

$$\Pr\left\{\max_{1\leq n\leq m} |S_n - n\overline{X}| \geq m\epsilon\right\} \leq \frac{\sigma^2}{m\epsilon^2}.$$

Hint: First show that $\{S_n - n\overline{X};\ n \geq 1\}$ is a martingale.

(c) Let $\{S_n;\ n \geq 1\}$ be a random walk, $S_n = X_1 + \cdots + X_n$, where each X_i has mean $\overline{X} < 0$ and semi-invariant MGF $\gamma(r)$. For any $r > 0$ such that $0 < \gamma(r) < \infty$, and for any $\alpha > 0$,

$$\Pr\left\{\max_{1\leq i\leq n} S_i \geq \alpha\right\} \leq \exp\{-r\alpha + n\gamma(r)\}.$$

Hint: First show that $\{e^{rS_n};\ n \geq 1\}$ is a submartingale.

Exercise 9.32 **(a)** Let $\{Z_n;\ n \geq 1\}$ be the scaled branching process of Section 9.6.2 and assume that Y, the number of offspring of each element, has finite mean $\overline{Y}$ and finite variance σ^2. Assume that the population at time 0 is 1. Show that

$$\mathsf{E}\left[Z_n^2\right] = 1 + \sigma^2\left[\overline{Y}^{-2} + \overline{Y}^{-3} + \cdots + \overline{Y}^{-n-1}\right].$$

(b) Assume that $\overline{Y} > 1$ and find $\lim_{n\to\infty} \mathsf{E}\left[Z_n^2\right]$. Show from this that the conditions for the martingale convergence theorem, Theorem 9.9.8, are satisfied.

Note: This condition is not satisfied if $\overline{Y} \leq 1$. The general martingale convergence theorem requires only a bounded first absolute moment, so it holds for $\overline{Y} \leq 1$

Exercise 9.33 Show that if $\{Z_n;\ n \geq 1\}$ is a martingale (submartingale or supermartingale) relative to a joint process $\{Z_n, X_n;\ n \geq 1\}$ and if J is a stopping trial for $\{Z_n;\ n \geq 1\}$ relative to $\{Z_n, X_n;\ n \geq 1\}$, then the stopped process satisfies (9.98), (9.99), or (9.100) respectively.

Exercise 9.34 Show that if $\{Z_n;\ n \geq 1\}$ is a martingale relative to a joint process $\{Z_n, X_n;\ n \geq 1\}$ and if J is a stopping trial for $\{Z_n;\ n \geq 1\}$ relative to $\{Z_n, X_n;\ n \geq 1\}$, then $\mathsf{E}[Z_J] = \mathsf{E}[Z_1]$ if and only if (9.104) is satisfied.

Exercise 9.35 (The double or quarter game) Consider an investment example similar to that of Example 9.10.1 in which there is only one investment other than cash. The ratio X of that investment value at the end of an epoch to that at the beginning is either 2 or 1/4, each with equal probability. Thus $\Pr\{X = 2\} = 1/2$ and $\Pr\{X = 1/4\} = 1/2$. Successive ratios, $X_1, X_2, \ldots,$ are IID.

(a) In (a)–(c), assume a fixed allocation strategy where a fraction λ is kept in the double or quarter investment and $1 - \lambda$ is kept in cash. Find the expected wealth, $W_n(\lambda)$ and the expected log wealth $\mathsf{E}[L_n(\lambda)]$ as a function of a constant $\lambda \in [0, 1]$ and $n \geq 1$. Assume unit initial wealth.

(b) For $\lambda = 1$, find the PMF for $W_4(1)$ and give a brief explanation of why $\mathsf{E}[W_n(1)]$ is growing exponentially with n but $\mathsf{E}[L_n(1)]$ is decreasing linearly toward $-\infty$.

(c) Using the same approach as in Example 9.10.1, find the value of λ^* that maximizes $\mathsf{E}[L_n(\lambda)]$. Show that your solution satisfies the optimality conditions in (9.136).

(d) Find the range of λ over which $\mathsf{E}[L_n(\lambda)]$ is positive.

(e) Find the PMF of the rv Z_n/Z_{n-1} as given in (9.138) for any given λ_n.

Exercise 9.36 (Kelly's horse-racing game) An interesting special case of this simple theory of investment is the horse racing game due to J. Kelly and described in more detail in [5]. There are $\ell - 1$ horses in a race and each horse j wins with probability $p(j)$. One and only one horse wins, and if j wins, the gambler receives $r(j)$ for each dollar bet on j and nothing for the bets on other horses. In other words, the price-relative, $X(j)$ is $r(j)$ with probability $p(j)$ and 0 otherwise for $1 \leq j \leq \ell - 1$. For cash, $X(\ell) = 1$.

The gambler's allocation of wealth on the race is $\lambda(j)$ on each horse j with $\lambda(\ell)$ kept in cash. As usual, $\sum_j \lambda(j) = 1$ and $\lambda(j) \geq 0$ for $1 \leq j \leq \ell$. Note that $X(1), \ldots, X(\ell - 1)$ are highly dependent, since only one is non-zero in any sample race.

(a) For any given allocation $\boldsymbol{\lambda}$ find the expected wealth and the expected log wealth at the end of a race for a unit initial wealth.

(b) Assume that a statistically identical sequence of races is run, i.e., $\boldsymbol{X}_1, \boldsymbol{X}_2, \ldots,$ are IID, where each $\boldsymbol{X}_n = \left(X_n(1), \ldots, X_n(\ell)\right)^{\mathsf{T}}$. Assuming a constant allocation on each race and unit initial wealth, find the expected log wealth $\mathsf{E}[L_n(\boldsymbol{\lambda})]$ at the end of the nth race and express it as $n\mathsf{E}[Y(\boldsymbol{\lambda})]$.

(c) Let $\boldsymbol{\lambda}^*$ maximize $\mathsf{E}[Y(\boldsymbol{\lambda})]$. Use the necessary and sufficient condition (9.136) on $\boldsymbol{\lambda}^*$ for horse j, $1 \le j < \ell$ to show that $\lambda^*(j)$ can be expressed in the following two equivalent ways; each uniquely specifies each $\lambda^*(j)$ in terms of $\lambda^*(\ell)$.

$$\lambda^*(j) \ge p(j) - \frac{\lambda^*(\ell)}{r(j)}; \qquad \text{with equality if } \lambda^*(j) > 0. \tag{9.165}$$

$$\lambda^*(j) = \max\left\{p(j) - \frac{\lambda^*(\ell)}{r(j)},\ 0\right\}. \tag{9.166}$$

Solving for $\lambda^*(\ell)$ (which in turn specifies the other components of $\boldsymbol{\lambda}^*$) breaks into 3 special cases which are treated below in (d), (e), and (f) respectively. The first case, in (d), shows that if $\sum_{j<\ell} 1/r(j) < 1$, then $\lambda^*(\ell) = 0$. The second case, in (e), shows that if $\sum_{j<\ell} 1/r(j) > 1$, then $\lambda^*(\ell) > \min_j(p(j)r(j))$, with the specific value specified by the unique solution to (9.168). The third case, in (f), shows that if $\sum_{j<\ell} 1/r(j) = 1$, then $\lambda^*(\ell)$ is nonunique, and its set of possible values occupy the range $[0, \min_j\big(p(j)r(j)\big)]$.

(d) Sum (9.165) over j to show that if $\lambda^*(\ell) > 0$, then $\sum_{j<\ell} 1/r(j) \ge 1$. Note that the logical obverse of this is that $\sum_{j<\ell} 1/r(j) < 1$ implies that $\lambda^*(\ell) = 0$ and thus that $\lambda^*(j) = p(j)$ for each horse j.

(e) In (c), $\lambda^*(j)$ for each $j < \ell$ was specified in terms of $\lambda^*(\ell)$; here you are to use the necessary and sufficient condition (9.136) on cash to specify $\lambda^*(\ell)$. More specifically, you are to show that $\lambda^*(\ell)$ satisfies each of the following two equivalent inequalities:

$$\sum_{j<\ell} \frac{p(j)}{\lambda^*(j)r(j) + \lambda^*(\ell)} \le 1; \qquad \text{with equality if } \lambda^*(\ell) > 0. \tag{9.167}$$

$$\sum_{j<\ell} \frac{p(j)}{\max\left[p(j)r(j),\ \lambda^*(\ell)\right]} \le 1; \qquad \text{with equality if } \lambda^*(\ell) > 0. \tag{9.168}$$

Show from (9.168) that if $\lambda^*(\ell) \le p(j)r(j)$ for each j, then $\sum_j 1/r(j) \le 1$. Point out that the logical obverse of this is that if $\sum_j 1/r(j) > 1$, then $\lambda^*(\ell) > \min_j\big(p(j)r(j)\big)$. Explain why (9.168) has a unique solution for $\lambda^*(\ell)$ in this case. Note that $\lambda^*(j) = 0$ for each j such that $p(j)r(j) < \lambda^*(\ell)$.

(f) Now consider the case in which $\sum_{j<\ell} 1/r(j) = 1$. Show that (9.168) is satisfied with equality for each choice of $\lambda^*(\ell)$, $0 \le \lambda^*(\ell) \le \min_{j<\ell}\big(p(j)r(j)\big)$.

(g) Consider the special case of a race with only two horses. Let $p(1) = p(2) = 1/2$. Assume that $r(1)$ and $r(2)$ are large enough to satisfy $1/r(1) + 1/r(2) < 1$; thus no cash allotment is used in maximizing $\mathsf{E}[Y(\boldsymbol{\lambda})]$. With $\lambda(3) = 0$, we have

$$\mathsf{E}[Y(\boldsymbol{\lambda})] = \frac{1}{2}\ln[\lambda(1)r(1)] + \frac{1}{2}\ln[\lambda(2)r(2)] = \frac{1}{2}\ln[\lambda(1)r(1)\big(1 - \lambda(1)\big)r(2)]. \tag{9.169}$$

Use this equation to give an intuitive explanation for why $\lambda^*(1) = 1/2$, independent of $r(1)$ and $r(2)$).

(h) Again consider the special case of two horses with $p(1) = p(2) = 1/2$, but let $r(1) = 3$ and $r(2) = 3/2$. Show that $\boldsymbol{\lambda}^*$ is nonunique with $(1/2, 1/2, 0)$ and $(1/4, 0, 3/4)$ as possible values. Show that if $r(2) > 3/2$, then the first solution above is unique, and if $r(2) < 3/2$, then the second solution is unique, assuming $p(1) = 1/2$ and $r(1) = 3$ throughout. Note: The case $3/2 < r(2) < 2$, provides an example where an investment in used to maximize log-w! ealth even though $\mathsf{E}[X(2)] = p(2)r(2) < 1$, i.e., horse 2 is a poor investment, but is preferable to cash in this case as a hedge against horse 1 losing.

(i) For the case where $\sum_{j<\ell} 1/r(j) = 1$, define $q(j) = 1/r(j)$ as a PMF on $\{1, \ldots, \ell - 1\}$. Show that $\mathsf{E}\left[Y(\boldsymbol{\lambda}^*)\right] = D(\mathsf{p} \parallel \mathsf{q})$ for the conditions of part f). Note: To interpret this, we could view a horse race where each horse j has probability $q(j)$ of winning the reward $r(j) = 1/q(j)$ as a 'fair game'. Our gambler, who knows that the true probabilities are $\{p(j);\ 1 \le j < \ell\}$, then has 'inside information' if $p(j) \ne q(j)$, and can establish a positive rate of return equal to $D(\mathsf{p} \parallel \mathsf{q})$.

Exercise 9.37 (Proof of Theorem 9.10.2) **(a)** Let the function $h_\epsilon(u)$ have the value 0 for $-\epsilon < \mathsf{E}[Y(\boldsymbol{\lambda})] < \epsilon$ and the value 1 elsewhere. Show that $\lim_{n\to\infty} h_\epsilon(L_n(\boldsymbol{\lambda})) = 0$ WP1.

(b) Show that for all $\epsilon > 0, \delta > 0$, there is an n_o such that

$$\Pr\left\{\bigcup\nolimits_{n\ge n_o} \left\{\left|\frac{1}{n}L_n(\boldsymbol{\lambda}) - \mathsf{E}[Y(\boldsymbol{\lambda})]\right| > \epsilon\right\}\right\} \le \delta$$

(c) Show that the probability of the complementary event is

$$\Pr\left\{\bigcap\nolimits_{n\ge n_o} \left\{\mathsf{E}[Y(\boldsymbol{\lambda})] - \epsilon \le \frac{1}{n}L_n(\boldsymbol{\lambda}) \le \mathsf{E}[Y(\boldsymbol{\lambda})] + \epsilon\right\}\right\} > 1 - \delta$$

(d) Show that

$$\Pr\left\{\bigcap\nolimits_{n\ge n_o} \left\{\exp\left[n\big(\mathsf{E}[Y(\boldsymbol{\lambda})] - \epsilon\big)\right] \le W_n(\boldsymbol{\lambda}) \le \exp\left[n\big(\mathsf{E}[Y(\boldsymbol{\lambda})] + \epsilon\big)\right]\right\}\right\} > 1 - \delta.$$

Exercise 9.38 In this exercise, you are to express the buy-and-hold strategy and the two-pile strategy of (9.137) as special cases of a time-varying allocation.

(a) Suppose the buy-and-hold strategy starts with an allocation $\boldsymbol{\lambda}_1$. Let $W_n(k)$ be the investor's wealth in investment k at time n. Show that

$$W_n(k) = \lambda_1(k) \prod_{m=1}^{n} X_m(k).$$

(b) Show that $\lambda_n(k)$ can be expressed in each of the following ways:

$$\lambda_n(k) = \frac{W_{n-1}(k)}{\sum_j W_{n-1}(j)} = \frac{\lambda_{n-1}(k)X_{n-1}(k)}{\sum_j \lambda_{n-1}(j)X_{n-1}(j)}.$$

(c) For the two pile strategy of (9.137), verify (9.137) and find an expression for $\lambda_n(k)$.

Exercise 9.39 **(a)** Consider the martingale $\{Z_n;\ n \ge 1\}$ where $Z_n = W_n/W_n^*$ and W_n is the wealth at time n for the pure triple or nothing strategy of Example 9.10.1 and W_n^* is the wealth using the constant allocation $\boldsymbol{\lambda}^*$. Find the PMF of Z where $Z = \lim_{n\to\infty} Z_n$ WP1.

(b) Now consider the two-pile strategy of (9.137) and find the PMF of the limiting rv Z for this strategy.

(c) Now consider a 'tempt-the-fates' strategy where one uses the pure triple or nothing strategy for the first three epochs and uses the constant allocation $\boldsymbol{\lambda}^*$ thereafter. Again find the PMF of the limiting rv Z.

Exercise 9.40 Consider the Markov modulated random walk in the figure below. The rv s Y_n in this example take on only a single value for each transition, that value being 1 for all transitions from state 1, 10 for all transitions from state 2, and 0 otherwise. $\epsilon > 0$ is a very small number, say $\epsilon < 10^{-6}$.

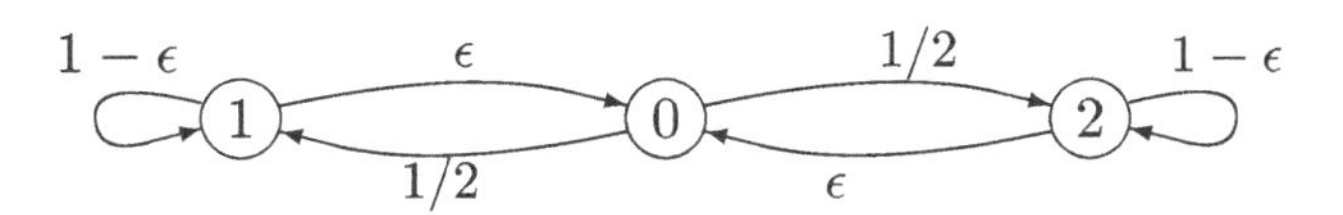

(a) Show that the steady-state gain per transition is $5.5/(1+\epsilon)$. Show that the relative-gain vector is $\boldsymbol{w} = (0, (\epsilon - 4.5)/[\epsilon(1+\epsilon)], (10\epsilon + 4.5)/[\epsilon(1+\epsilon)])$.

(b) Let $S_n = Y_0 + Y_1 + \cdots + Y_{n-1}$ and take the starting state X_0 to be 0. Let J be the smallest value of n for which $S_n \geq 100$. Find $\Pr\{J = 11\}$ and $\Pr\{J = 101\}$. Find an estimate of $\mathsf{E}[J]$ that is exact in the limit $\epsilon \to 0$.

(c) Show that $\Pr\{X_J = 1\} = (1 - 45\epsilon + o(\epsilon))/2$ and that $\Pr\{X_J = 2\} = (1 + 45\epsilon + o(\epsilon))/2$. Verify, to first order in ϵ, that (9.159) is satisfied.

Exercise 9.41 Show that (9.159) results from taking the derivative of (9.163) and evaluating it at $r = 0$.

Exercise 9.42 Let $\{Z_n;\ n \geq 1\}$ be a martingale, and for some integer m, let $Y_n = Z_{n+m} - Z_m$.

(a) Show that $\mathsf{E}\left[Y_n \mid Z_{n+m-1} = z_{n+m-1}, Z_{n+m-2} = z_{n+m-2}, \ldots, Z_m = z_m, \ldots, Z_1 = z_1\right] = z_{n+m-1} - z_m$.

(b) Show that $\mathsf{E}\left[Y_n \mid Y_{n-1} = y_{n-1}, \ldots, Y_1 = y_1\right] = y_{n-1}$.

(c) Show that $\mathsf{E}[|Y_n|] < \infty$. Note that (b) and (c) show that $\{Y_n;\ n \geq 1\}$ is a martingale.

Exercise 9.43 (Continuation of continuous-time branching) This exercise views the continuous-time branching process of Exercise 7.15 as a stopped random walk. Recall that the process was specified there as a Markov process such that for each state $j, j \geq 0$, the transition rate to $j+1$ is $j\lambda$ and to $j-1$ is $j\mu$. There are no other transitions, and in particular, there are no transitions out of state 0, so that the Markov process is reducible. Recall that the embedded Markov chain is the same as the embedded chain of an M/M/1 queue except that there is no transition from state 0 to state 1.

(a) To model the possible extinction of the population, convert the embedded Markov chain above to a stopped random walk, $\{S_n;\ n \geq 0\}$. The stopped random walk starts at $S_0 = 0$ and stops on reaching a threshold at -1. Before stopping, it moves up by 1 with probability $\lambda/(\lambda+\mu)$ and downward by 1 with probability $\mu/(\lambda+\mu)$ at each step. Give the (very simple) relationship between the state X_n of the Markov chain and the state S_n of the stopped random walk for each $n \geq 0$.

(b) Find the probability that the population eventually dies out as a function of λ and μ. Be sure to consider all three cases $\lambda > \mu$, $\lambda < \mu$, and $\lambda = \mu$.

10 Estimation

10.1 Introduction

Estimation, as considered here, involves a probabilistic experiment with two random vectors (rv s) $\boldsymbol{X} = (X_1, \ldots, X_n)^{\mathsf{T}}$ and $\boldsymbol{Y} = (Y_1, \ldots, Y_m)^{\mathsf{T}}$. The experiment is performed, but only the resulting sample value $\boldsymbol{y}$ of the rv $\boldsymbol{Y}$ is observed. The observer then 'estimates' the sample value $\boldsymbol{x}$ of $\boldsymbol{X}$ from the observation $\boldsymbol{y}$. For simplicity, we assume throughout (except where explicitly assumed otherwise) that the combined rv $(X_1, \ldots, X_n, Y_1, \ldots, Y_m)^{\mathsf{T}}$ has a finite non-singular covariance matrix and finite joint probability density. We denote the estimate of $\boldsymbol{X}$, as a function of the observed sample value $\boldsymbol{y}$, by $\hat{\boldsymbol{x}}(\boldsymbol{y})$. For a given observation $\boldsymbol{y}$, one could choose the estimate to be any desired function $\hat{\boldsymbol{x}}(\boldsymbol{y})$ of the observation. For example, $\hat{\boldsymbol{x}}(\boldsymbol{y})$ could be the conditional mean, median, or mode of $\boldsymbol{X}$ conditional on $\boldsymbol{Y} = \boldsymbol{y}$. It could also be some function of $\boldsymbol{y}$ that depends only on the likelihood $\mathsf{f}_{\boldsymbol{Y}|\boldsymbol{X}}(\boldsymbol{y}|\boldsymbol{x})$ and not on any assumed a priori distribution for $\boldsymbol{X}$.

There are many real-world estimation problems for which it is reasonable to model the conditional density $\mathsf{f}_{\boldsymbol{Y}|\boldsymbol{X}}(\boldsymbol{y}|\boldsymbol{x})$ but less reasonable to model the a priori density of the unknown $\boldsymbol{X}$ to be estimated. Classical statisticians have many estimation techniques in which $\boldsymbol{X}$ is regarded simply as a parameter. For the same reasons as explained in Chapter 8 on detection, we will continue to view $\boldsymbol{X}$ as a rv, at least for notational purposes, in these situations. Thus classical estimation procedures become those that are based on the likelihoods $\mathsf{f}_{\boldsymbol{Y}|\boldsymbol{X}}(\boldsymbol{y}|\boldsymbol{x})$ alone and not on the a priori probabilities. We mention these procedures in several places, and discuss them more in Section 10.8.

Estimation problems occur in an amazing variety of situations, often referred to as measurement, prediction, or recovery problems. For example, in communication systems, the timing and the phase of the transmitted signals must be *recovered* at the receiver. Often it is necessary to *measure* the channel, and finally, for analog data, the receiver often must *estimate* the transmitted signal at finely spaced sample times. In fact, it is often helpful in the general case to refer to the quantity to be estimated as the signal. In control systems, the state of the system must be *estimated* in order to generate appropriate control signals. In statistics, one tries to *estimate* parameters for some probabilistic system model from trial sample values. In any experimental science, one is always concerned with *measuring* quantities in the presence of noise and experimental error.

The problem of estimation is very similar to that of detection. With detection, we must decide between a discrete set of alternatives on the basis of an observation, whereas here we make a selection from a vector continuum of choices. Although this does not appear to be a fundamental difference, it leads to a surprising set of differences in approach. In many typical detection problems, the cost of different kinds of errors is secondary and we are concerned primarily with maximizing the probability of correct choice. In typical estimation problems, with a continuous a posteriori distribution, the probability of selecting the exact correct value is zero, and we are concerned with making the error small according to some given criterion.

A fundamental approach to estimation is to use a cost function, $C(\boldsymbol{x},\hat{\boldsymbol{x}})$ to quantify the cost associated with an estimate $\hat{\boldsymbol{x}}$ when the actual sample value of $\boldsymbol{X}$ is $\boldsymbol{x}$. This cost function, $C(\boldsymbol{x},\hat{\boldsymbol{x}})$ is analogous to the cost $C_{\ell k}$, defined in Section 8.3, of making decision k when ℓ is the correct hypothesis. The *minimum-cost criterion* or *Bayes criterion*, both for detection and estimation is to choose $\hat{\boldsymbol{x}}(\boldsymbol{y})$ to minimize the expected cost, conditional on the sample observation $\boldsymbol{Y} = \boldsymbol{y}$. Specifically, given the observation $\boldsymbol{y}$, choose $\hat{\boldsymbol{x}}(\boldsymbol{y})$ to minimize

$$\int C\left[\boldsymbol{x},\hat{\boldsymbol{x}}(\boldsymbol{y})\right]\, \mathsf{f}_{\boldsymbol{X}|\boldsymbol{Y}}(\boldsymbol{x} \mid \boldsymbol{y})\, d\boldsymbol{x} = \mathsf{E}\left[C[\boldsymbol{X},\hat{\boldsymbol{X}}(\boldsymbol{Y})] \mid \boldsymbol{Y}{=}\boldsymbol{y}\right].$$

Denoting the minimum-cost estimate for each $\boldsymbol{y}$ as $\hat{\boldsymbol{x}}_{\mathrm{mc}}(\boldsymbol{y})$, we then have

$$\hat{\boldsymbol{x}}_{\mathrm{mc}}(\boldsymbol{y}) \;=\; \arg\min_{\hat{\boldsymbol{x}}} \int C\left[\boldsymbol{x},\hat{\boldsymbol{x}}\right] \mathsf{f}_{\boldsymbol{X}|\boldsymbol{Y}}(\boldsymbol{x} \mid \boldsymbol{y})d\boldsymbol{x} \;=\; \arg\min_{\hat{\boldsymbol{x}}} \mathsf{E}\left[C[\boldsymbol{X},\hat{\boldsymbol{x}}] \mid \boldsymbol{Y}{=}\boldsymbol{y}\right]. \tag{10.1}$$

The notation $\arg\min_u f(u)$ means the value of u that minimizes $f(u)$ (if the minimum is not unique, any minimizing u can be chosen). This choice of estimate minimizes the expected cost for each observation $\boldsymbol{y}$. Thus it also minimizes the expected cost when averaged over $\boldsymbol{Y}$, i.e., it minimizes

$$\mathsf{E}\left[C[\boldsymbol{X},\hat{\boldsymbol{X}}(\boldsymbol{Y})]\right] \;=\; \int \mathsf{E}\left[C[\boldsymbol{X},\hat{\boldsymbol{X}}(\boldsymbol{Y})] \mid \boldsymbol{Y}{=}\boldsymbol{y}\right] \mathsf{f}_{\boldsymbol{Y}}(\boldsymbol{y})d\boldsymbol{y}$$

over possible estimation rules $\hat{\boldsymbol{X}}(\boldsymbol{Y})$. To interpret $\hat{\boldsymbol{X}}(\boldsymbol{Y})$, note that the probability model maps each sample point $\omega \in \Omega$ into a sample value $\boldsymbol{y} = \boldsymbol{Y}(\omega)$. The estimation rule maps that $\boldsymbol{y}$ into $\hat{\boldsymbol{x}}(\boldsymbol{y})$. The rv $\hat{\boldsymbol{X}}(\boldsymbol{Y})$ is the function of $\boldsymbol{Y}$ that maps each sample value $\boldsymbol{y}$ of $\boldsymbol{Y}$ into the corresponding $\hat{\boldsymbol{x}}(\boldsymbol{y})$.

10.1.1 The squared-cost function

In practice, the most important cost function is the squared-cost:

$$C[\boldsymbol{x},\, \hat{\boldsymbol{x}})] \;=\; \sum_{j=1}^{n} \left[x_j - \hat{x}_j\right]^2 . \tag{10.2}$$

This cost function will often be described in the following ways,

$$\sum_j \left[x_j - \hat{x}_j\right]^2 \;=\; (\boldsymbol{x} - \hat{\boldsymbol{x}})^{\mathsf{T}}(\boldsymbol{x} - \hat{\boldsymbol{x}}) \;=\; \|\boldsymbol{x} - \hat{\boldsymbol{x}}\|^2.$$

The estimate that minimizes $\mathsf{E}[C[\boldsymbol{X}, \hat{\boldsymbol{X}}(\boldsymbol{Y})]]$ for the squared-cost function is called the *minimum mean-squared-error estimate (MMSE estimate)*. It is also often called the Bayes least-squares estimate. In order to minimize $\mathsf{E}[\sum_j (X_j - \hat{x}_j(\boldsymbol{y}))^2 \mid \boldsymbol{Y}=\boldsymbol{y}]$ over $\hat{\boldsymbol{x}}(\boldsymbol{y})$, it is sufficient to choose $\hat{x}_j(\boldsymbol{y})$, for each j, to minimize $\mathsf{E}\left[(X_j - \hat{x}_j(\boldsymbol{y}))^2 \mid \boldsymbol{Y}=\boldsymbol{y}\right]$. This is simply the conditional second moment of X_j around $\hat{x}_j(\boldsymbol{y})$, given $\boldsymbol{Y} = \boldsymbol{y}$. This is minimized by choosing $\hat{x}_j(\boldsymbol{y})$ to be the conditional mean of X_j given $\boldsymbol{Y} = \boldsymbol{y}$.

Simple though this result is, it is a central result of estimation theory, and we state it as a theorem.

Theorem 10.1.1 *The MMSE estimate $\hat{\boldsymbol{x}}_{\text{MMSE}}(\boldsymbol{y})$ of a signal $\boldsymbol{X} = \boldsymbol{x}$ as a function of an observation $\boldsymbol{Y} = \boldsymbol{y}$, is given by*

$$\hat{\boldsymbol{x}}_{\text{MMSE}}(\boldsymbol{y}) \;=\; \mathsf{E}\left[\boldsymbol{X} \mid \boldsymbol{Y}=\boldsymbol{y}\right] \;=\; \int \boldsymbol{x}\, \mathsf{f}_{\boldsymbol{X}|\boldsymbol{Y}}(\boldsymbol{x} \mid \boldsymbol{y})\, d\boldsymbol{x}. \tag{10.3}$$

The rv $\hat{\boldsymbol{X}}_{\text{MMSE}}(\boldsymbol{Y})$ is then $\mathsf{E}[\boldsymbol{X} \mid \boldsymbol{Y}]$. Define the *estimation error* $\boldsymbol{\xi}$ for a given estimation rule as

$$\boldsymbol{\xi} = \boldsymbol{X} - \hat{\boldsymbol{X}}(\boldsymbol{Y}).$$

Since $\hat{\boldsymbol{x}}_{\text{MMSE}}(\boldsymbol{y})$ is the conditional mean of $\boldsymbol{X}$ given $\boldsymbol{Y} = \boldsymbol{y}$, the estimation error $\boldsymbol{\xi}_{\text{MMSE}}$ for the MMSE estimate satisfies $\mathsf{E}\left[\boldsymbol{\xi}_{\text{MMSE}} \mid \boldsymbol{Y}=\boldsymbol{y}\right] = 0$ for all $\boldsymbol{y}$, so $\mathsf{E}\left[\boldsymbol{\xi}_{\text{MMSE}}\right] = 0$ also. The covariance matrix of $\boldsymbol{\xi}_{\text{MMSE}}$ is then given by

$$[K_{\boldsymbol{\xi}_{\text{MMSE}}}] = \mathsf{E}\left[\boldsymbol{\xi}_{\text{MMSE}}\, \boldsymbol{\xi}^{\mathsf{T}}_{\text{MMSE}}\right] = \mathsf{E}\left[\big(\boldsymbol{X} - \hat{\boldsymbol{X}}(\boldsymbol{Y})\big)\big(\boldsymbol{X} - \hat{\boldsymbol{X}}(\boldsymbol{Y})\big)^{\mathsf{T}}\right]. \tag{10.4}$$

This can be simplified by recalling that $\mathsf{E}\left[\boldsymbol{\xi}_{\text{MMSE}} \mid \boldsymbol{Y}=\boldsymbol{y}\right] = 0$ for all $\boldsymbol{y}$. Thus multiplying this by $\hat{\boldsymbol{x}}_{\text{MMSE}}(\boldsymbol{y})$, we get

$$\mathsf{E}\left[\boldsymbol{\xi}_{\text{MMSE}} \mid \boldsymbol{Y}=\boldsymbol{y}\right] \hat{\boldsymbol{x}}^{\mathsf{T}}_{\text{MMSE}}(\boldsymbol{y}) = \mathsf{E}\left[\boldsymbol{\xi}_{\text{MMSE}}\, \hat{\boldsymbol{X}}^{\mathsf{T}}_{\text{MMSE}}(\boldsymbol{Y}) \mid \boldsymbol{Y}=\boldsymbol{y}\right] = 0.$$

Averaging over $\boldsymbol{y}$, we get the important relation that

$$\mathsf{E}\left[\boldsymbol{\xi}_{\text{MMSE}}\, \hat{\boldsymbol{X}}^{\mathsf{T}}_{\text{MMSE}}(\boldsymbol{Y})\right] = 0. \tag{10.5}$$

If we substitute this into (10.4), rewritten as $[K_{\boldsymbol{\xi}_{\text{MMSE}}}] = \mathsf{E}[\boldsymbol{\xi}_{\text{MMSE}} \big(\boldsymbol{X} - \hat{\boldsymbol{X}}(\boldsymbol{Y})\big)^{\mathsf{T}}]$, we get a useful simplified expression for the covariance of the estimation error for MMSE estimation:

$$[K_{\boldsymbol{\xi}_{\text{MMSE}}}] = \mathsf{E}\left[\boldsymbol{\xi}_{\text{MMSE}}\, \boldsymbol{X}^{\mathsf{T}}\right]. \tag{10.6}$$

10.1.2 Other cost functions

Subsequent sections of this chapter focus primarily on the squared-cost function. First, however, we briefly discuss several other cost functions. One is the absolute value cost function, where $C(\boldsymbol{x}, \hat{\boldsymbol{x}}) = \sum_n |x_n - \hat{x}_n|$; this expected cost is minimized, for each $\boldsymbol{y}$, by choosing $\hat{x}_n(\boldsymbol{y})$ to be the conditional median of $\mathsf{f}_{X_n|\boldsymbol{Y}}(x_n \mid \boldsymbol{y})$ (see Exercise 1.11(c)). The absolute value cost function places less weight on large estimation errors and more on small estimation errors than the squared-cost function. The relative merits of the conditional median relative to the conditional mean for estimation are very similar to their merits for typical values discussed in Section 1.5.3. The squared-cost function is emphasized here, primarily because of the conceptual richness and computational ease of working with means and variances rather than medians and absolute errors.

Another cost function is the maximum-tolerable-error cost function. Here, for some given number ϵ, the cost is 1 if the magnitude of any component of the error exceeds ϵ and is 0 otherwise. That is, $C(\boldsymbol{x}, \hat{\boldsymbol{x}}) = 1$ if $|x_i - \hat{x}_i| > \epsilon$ for any i and $C(\boldsymbol{x}, \hat{\boldsymbol{x}}) = 0$ otherwise. For an observation $\boldsymbol{y}$, the maximum-tolerable-error estimate is that value $\hat{\boldsymbol{x}}(\boldsymbol{y})$ that maximizes the conditional probability that the signal $\boldsymbol{X}$ lies in an n-dimensional cube with sides of length 2ϵ centered on $\hat{\boldsymbol{x}}(\boldsymbol{y})$. For ϵ very small, this is approximated by choosing $\hat{\boldsymbol{x}}(\boldsymbol{y})$ to be that $\boldsymbol{x}$ which maximizes $\mathsf{f}_{\boldsymbol{X}|\boldsymbol{Y}}(\boldsymbol{x}|\boldsymbol{y})$, i.e., one chooses the mode, for given $\boldsymbol{y}$ of $\mathsf{f}_{\boldsymbol{X}|\boldsymbol{Y}}(\boldsymbol{x}|\boldsymbol{y})$. This estimation rule is called the *maximum a posteriori probability (MAP) estimate*. One awkward property of MAP estimation is that in many cases, the ith component of $\hat{\boldsymbol{x}}_{\mathrm{MAP}}(\boldsymbol{y})$ differs from $(\hat{x}_i)_{\mathrm{MAP}}(\boldsymbol{y})$ (see Exercise 1.11). Fortunately, in many of the most typical types of applications, the mean, the median, and the mode of the conditional distribution are all the same, so one often chooses between these estimates on the basis of conceptual or computational convenience.

In studying detection, we focused on the MAP rule, whereas here we focus on MMSE. The reason is that in most detection problems, the probability of error is quite small, and, as we saw, the effect of different costs for different kinds of errors is simply to change the thresholds. Here, if we assume that $\mathsf{f}_{\boldsymbol{X}|\boldsymbol{Y}}(\boldsymbol{x} \mid \boldsymbol{y})$ is finite for each $\boldsymbol{x}$, the probability of choosing the actual sample value $\boldsymbol{x}$ exactly is equal to zero. Thus we must be satisfied with an approximate choice, and the question of cost cannot be avoided. Thus, in principle, attempting to maximize the probability of correct choice is foolish (since that probability will be zero anyway). The reason that the MAP rule is often used in estimation is, first, analytical convenience for many problems, and, second, some confidence, for the particular problem at hand, that the conditional mode is a reasonable choice.

The MAP and MMSE rules for estimation depend on a complete probabilistic model in which the a priori distibution of the signal $\boldsymbol{X}$ is specified. In many situations, estimates are desired that do not depend on an a priori distribution. The most common estimate with no dependence on $\mathsf{f}_{\boldsymbol{X}}(\boldsymbol{x})$ is called the *maximum likelihood (ML) estimate*; the ML estimate $\hat{\boldsymbol{x}}_{\mathrm{ML}}(\boldsymbol{y})$ is defined as an $\boldsymbol{x}$ that maximizes $\mathsf{f}_{\boldsymbol{Y}|\boldsymbol{X}}(\boldsymbol{y} \mid \boldsymbol{x})$ for the given $\boldsymbol{y}$. We can essentially view the ML estimate as the limit of MAP estimates as the a priori density on $\boldsymbol{X}$ becomes closer and closer to uniform. For example, we could model $\boldsymbol{X} \sim \mathrm{N}(0, \sigma^2 I)$ in the limit as $\sigma^2 \to \infty$. This limiting density does not exist, since the density approaches 0 everywhere, but the ML estimate typically approaches a limit. All of the previous comments about MAP estimates clearly carry over to ML.

10.2 MMSE estimation for Gaussian random vectors

This section explores the frequently-occurring problem of MMSE estimation of a Gaussian n-rv $\boldsymbol{X}$ from the observation of a Gaussian m-rv $\boldsymbol{Y}$, where $\boldsymbol{X}$ and $\boldsymbol{Y}$ are jointly Gaussian i.e., where $X_1, \ldots, X_n, Y_1, \ldots, Y_m$ are jointly-Gaussian rv s. We make the simplifying assumption that $\boldsymbol{X}$ and $\boldsymbol{Y}$ are jointly non-singular, i.e., that the covariance matrix of $X_1, \ldots, X_n, Y_1, \ldots, Y_m$ is non-singular.[1] Initially, we also assume that the rv s are all zero mean. As seen explicitly later, a non-zero-mean **rv** can be viewed (as usual) as a zero-mean fluctuation plus a mean.

For $\boldsymbol{X}$ and $\boldsymbol{Y}$ jointly Gaussian and zero mean, we saw in (3.42) that $\boldsymbol{X}$ can be represented in general as

$$\boldsymbol{X} = [G]\boldsymbol{Y} + \boldsymbol{V}, \tag{10.7}$$

where $\boldsymbol{V}$ is a zero-mean Gaussian n-**rv** that is statistically independent of $\boldsymbol{Y}$. The matrix $[G]$ and the covariance of $\boldsymbol{V}$ are given in (3.44) and (3.45) to be

$$[G] = [K_{\boldsymbol{X}\cdot\boldsymbol{Y}} K_{\boldsymbol{Y}}^{-1}]; \qquad [K_{\boldsymbol{V}}] = [K_{\boldsymbol{X}}] - [K_{\boldsymbol{X}\cdot\boldsymbol{Y}} K_{\boldsymbol{Y}}^{-1} K_{\boldsymbol{X}\cdot\boldsymbol{Y}}^{\mathsf{T}}], \tag{10.8}$$

where the cross-covariance $[K_{\boldsymbol{X}\cdot\boldsymbol{Y}}]$ is defined as $\mathsf{E}\left[\boldsymbol{X}\boldsymbol{Y}^{\mathsf{T}}\right]$.

Since $\boldsymbol{V}$ and $\boldsymbol{Y}$ are independent, (10.7) implies that the distribution of $\boldsymbol{X}$, conditional on $\boldsymbol{Y} = \boldsymbol{y}$, is Gaussian with mean $[G]\,y$ and covariance $[K_{\boldsymbol{V}}]$. The MMSE estimate is then the mean of this conditional distribution, i.e.,

$$\hat{\boldsymbol{x}}(\boldsymbol{y}) \;=\; [G]\boldsymbol{y} \;=\; [K_{\boldsymbol{X}\cdot\boldsymbol{Y}} K_{\boldsymbol{Y}}^{-1}]\boldsymbol{y}. \tag{10.9}$$

Given $\boldsymbol{Y} = \boldsymbol{y}$, then, $\boldsymbol{X} = \hat{\boldsymbol{x}}(\boldsymbol{y}) + \boldsymbol{V}$ and $\boldsymbol{\xi} = \boldsymbol{X} - \hat{\boldsymbol{x}}(\boldsymbol{y}) = \boldsymbol{V}$. Since $\boldsymbol{\xi} = \boldsymbol{V}$ for all $\boldsymbol{y}$, we see that $\boldsymbol{\xi}$ is Gaussian (i.e., has jointly Gaussian components) and is independent of $\boldsymbol{Y}$. Its variance is given by

$$[K_{\boldsymbol{\xi}}] \;=\; [K_{\boldsymbol{V}}] \;=\; [K_{\boldsymbol{X}}] - [K_{\boldsymbol{X}\cdot\boldsymbol{Y}} K_{\boldsymbol{Y}}^{-1} K_{\boldsymbol{X}\cdot\boldsymbol{Y}}^{\mathsf{T}}]. \tag{10.10}$$

The estimate in (10.9) is linear in the observed sample value $\boldsymbol{y}$. Also, since $\boldsymbol{Y}$ and $\boldsymbol{V}$ are independent **rv** s, the estimation error is statistically independent of the observation $\boldsymbol{Y}$. These are major simplifications that arise because the variables are jointly Gaussian.

More generally, let $\boldsymbol{X}$ and $\boldsymbol{Y}$ be jointly-Gaussian **rv** s with arbitrary means and define the fluctuations, $\widetilde{\boldsymbol{X}} = \boldsymbol{X} - \mathsf{E}[\boldsymbol{X}]$ and $\widetilde{\boldsymbol{Y}} = \boldsymbol{Y} - \mathsf{E}[\boldsymbol{Y}]$. The sample value $\boldsymbol{y}$ of $\boldsymbol{Y}$ corresponds to the sample value $\widetilde{\boldsymbol{y}} = \boldsymbol{y} - \mathsf{E}[\boldsymbol{Y}]$ of $\widetilde{\boldsymbol{Y}}$. Since $\widetilde{\boldsymbol{X}}$ and $\widetilde{\boldsymbol{Y}}$ are zero-mean and jointly Gaussian, the MMSE estimate of $\boldsymbol{X} - \mathsf{E}[\boldsymbol{X}]$ conditional on $\widetilde{Y} = \widetilde{y}$, can be found from (10.9), i.e.,

$$\hat{\boldsymbol{x}}(\boldsymbol{y}) - \mathsf{E}[\boldsymbol{X}] = [K_{\boldsymbol{X}\cdot\boldsymbol{Y}} K_{\boldsymbol{Y}}^{-1}](\boldsymbol{y} - \mathsf{E}[\boldsymbol{Y}]). \tag{10.11}$$

We have used the fact that the covariance matrix $[K_{\boldsymbol{Y}}]$ is the same as $[K_{\widetilde{\boldsymbol{Y}}}]$ and similarly $[K_{\boldsymbol{X}\cdot\boldsymbol{Y}}] = [K_{\widetilde{\boldsymbol{X}}\cdot\widetilde{\boldsymbol{Y}}}]$. Rewriting (10.11),

$$\hat{\boldsymbol{x}}(\boldsymbol{y}) = \boldsymbol{b} + [K_{\boldsymbol{X}\cdot\boldsymbol{Y}} K_{\boldsymbol{Y}}^{-1}]\boldsymbol{y}, \quad \text{where} \quad \boldsymbol{b} = \mathsf{E}[\boldsymbol{X}] - [K_{\boldsymbol{X}\cdot\boldsymbol{Y}} K_{\boldsymbol{Y}}^{-1}]\mathsf{E}[\boldsymbol{Y}]. \tag{10.12}$$

The estimate in (10.12) is a constant $\boldsymbol{b}$ plus a linear function[2] of $\boldsymbol{y}$, i.e., it is of first degree in $\boldsymbol{y}$. Note that the constant is 0 whenever $\mathsf{E}[\boldsymbol{X}] = 0$ and $\mathsf{E}[\boldsymbol{Y}] = 0$, so that the estimate is then linear in $\boldsymbol{y}$.

[1] Recall from Section 3.4.3 that Gaussian **rv** s with singular covariance matrices are usually best treated by removing the linearly dependent rv s and treating them as deterministic functions of the other rv s.

[2] We call a function $\hat{\boldsymbol{x}}(\boldsymbol{y})$ linear in $\boldsymbol{y}$ if $\hat{\boldsymbol{x}}(\boldsymbol{y}) = [A]\boldsymbol{y}$ for some matrix $[A]$. We call it first degree in $\boldsymbol{y}$ if $\hat{\boldsymbol{x}}(\boldsymbol{y}) = \boldsymbol{b} + [A]\boldsymbol{y}$, but others extend the meaning of linearity to include such functions.

The error covariance in estimating $\boldsymbol{X}$ from $\boldsymbol{Y}$ is clearly the same as that in estimating $\widetilde{\boldsymbol{X}}$ from $\widetilde{\boldsymbol{Y}}$. Thus the error covariance in this generalization to non-zero means is still given by (10.10). These results are summarized in the following theorem.

Theorem 10.2.1 *If $\boldsymbol{X}$ and $\boldsymbol{Y}$ are jointly non-singular and jointly Gaussian, the MMSE estimate of $\boldsymbol{X}$ from the observation $\boldsymbol{Y} = \boldsymbol{y}$ is given by (10.12) which, for the zero-mean case, is (10.9). The covariance of the estimation error is given by (10.10), which holds both for each $\boldsymbol{y}$ and for the average over $\boldsymbol{Y}$.*

This theorem essentially summarizes MMSE estimation of a Gaussian n-**rv** from a Gaussian m-**rv**. The following examples provide some insight into these matrix equations by looking at scalar cases. These examples are then extended to introduce MMSE recursive estimation and Kalman estimation for a scalar random variable (rv). We then take a short break from Gaussian problems to look at linear-least-squares-error (LLSE) estimation. In Section 10.4 we return to consider both MMSE and LLSE estimation for the vector case.

Example 10.2.2 (Scalar signal plus scalar noise) The simplest imaginable Gaussian estimation problem is to estimate X from the observation of $Y = X + Z$, where X and Z are zero-mean independent Gaussian rv s with variances σ_X^2 and σ_Z^2 (i.e., $X \sim \mathrm{N}\,(0, \sigma_X^2)$ and $Z \sim \mathrm{N}\,(0, \sigma_Z^2)$). Since X and Y are zero mean and jointly Gaussian, X, conditional on an observation $Y = y$, is given by (10.9) and (10.10) as $\mathrm{N}\,([K_{X \cdot Y} K_Y^{-1}]y, [K_X] - [K_{X \cdot Y} K_Y^{-1} K_{X \cdot Y}^{\mathsf{T}}])$. Since $Y = X + Z$, we have $[K_{X \cdot Y}] = \sigma_X^2$ and $[K_Y] = \sigma_X^2 + \sigma_Z^2$. From (10.9), the MMSE estimate is

$$\hat{x}(y) \;=\; \frac{[K_{X \cdot Y}]}{[K_Y]}\,y \;=\; \frac{\sigma_X^2}{\sigma_X^2 + \sigma_Z^2}\,y. \tag{10.13}$$

From the symmetry between X and Z, we can immediately express the MMSE of Z given $Y = y$ as

$$\hat{z}(y) \;=\; \frac{\sigma_Z^2}{\sigma_X^2 + \sigma_Z^2}\,y. \tag{10.14}$$

This also shows that $\hat{x}(y) + \hat{z}(y) = y$. This is not surprising since $Y = X + Z$, but it is interesting to see that MMSE estimates preserve this property. Note also that the MMSE estimate simply splits the observation between signal and noise according to their variances. From an intuitive standpoint, recall that both X and Z are zero mean, so if one has a larger variance than the other, it is reasonable to attribute the major part of Y to the variable with the larger variance, as is done in (10.13) and (10.14). The estimation error for X, namely $\xi = X - \widehat{X}(Y) = -Z + \widehat{Z}(Y)$ conditional on $Y = y$, is $\mathrm{N}\,(0, [K_X] - [K_{X \cdot Y} K_Y^{-1} K_{X \cdot Y}^{\mathsf{T}}])$. This is not a function of y, implying that ξ is statistically independent of Y; Exercise 10.2 helps develop some intuition about this independence. The variance of the conditional estimation error given $Y{=}y$, i.e., $\mathsf{E}\left[\xi^2 \mid Y{=}y\right]$, then also does not depend on y and is equal to the conditional variance averaged over Y. Thus, for all y

$$\sigma_\xi^2 = \mathsf{E}\left[\left(X - \widehat{X}(Y)\right)^2 \mid Y{=}y\right] = \sigma_X^2 - \frac{[K_{X\cdot Y}^2]}{[K_Y]}$$
$$= \sigma_X^2 - \frac{\sigma_X^4}{\sigma_X^2 + \sigma_Z^2} = \frac{\sigma_X^2\sigma_Z^2}{\sigma_X^2 + \sigma_Z^2}.$$

Often the most convenient way to express this is

$$\frac{1}{\sigma_\xi^2} = \frac{1}{\sigma_X^2} + \frac{1}{\sigma_Z^2}. \tag{10.15}$$

This final form shows that the variance of the estimation error is smaller than the original variance of X and of Z, and that this variance is increasing both with σ_X^2 and σ_Z^2.

Example 10.2.3 (Attenuated scalar signal plus scalar noise) We now generalize the previous example to $Y = hX + Z$, where h is a scale factor and X and Z are zero mean, independent, and Gaussian. The conditional probability $\mathsf{f}_{X|Y}(x|y)$ for given y is again $\mathrm{N}\left([K_{X\cdot Y}K_Y^{-1}]y,\ [K_X] - [K_{X\cdot Y}K_Y^{-1}K_{X\cdot Y}^{\mathsf{T}}]\right)$. Now $[K_{X\cdot Y}] = h\sigma_X^2$ and $\sigma_Y^2 = h^2\sigma_X^2 + \sigma_Z^2$, so (10.9) and (10.10) become

$$\hat{x}(y) = \frac{h\sigma_X^2}{h^2\sigma_X^2 + \sigma_Z^2}\,y; \qquad \sigma_\xi^2 = \sigma_X^2 - \frac{h^2\sigma_X^4}{h^2\sigma_X^2 + \sigma_Z^2} = \frac{\sigma_X^2\sigma_Z^2}{h^2\sigma_X^2 + \sigma_Z^2}. \tag{10.16}$$

This variance is best expressed as

$$\frac{1}{\sigma_\xi^2} = \frac{1}{\sigma_X^2} + \frac{h^2}{\sigma_Z^2}. \tag{10.17}$$

This follows directly from (10.16). It is insightful to also view the observation as $Y/h = X + Z/h$. Thus the variance of this scaled noise is σ_Z^2/h^2, which shows that (10.16) and (10.17) follow directly from (10.13) and (10.15).

As a slight extension, suppose $Y = hX + Z$, where X has a mean $\overline{X}$ (i.e., $X \sim \mathrm{N}(\overline{X}, \sigma_X^2)$) and Z is again zero mean. Then $\mathsf{E}[Y] = h\overline{X}$. Using (10.16) for the fluctuations in X and Y,

$$\hat{x}(y) = \overline{X} + \frac{h\sigma_X^2[y - h\overline{X}]}{h^2\sigma_X^2 + \sigma_Z^2}; \tag{10.18}$$

$$\sigma_\xi^2 = \frac{\sigma_X^2\sigma_Z^2}{h^2\sigma_X^2 + \sigma_Z^2}, \qquad \frac{1}{\sigma_\xi^2} = \frac{1}{\sigma_X^2} + \frac{h^2}{\sigma_Z^2}. \tag{10.19}$$

10.2.1 Scalar iterative estimation

Suppose that we make multiple noisy observations, $y_1, y_2, \ldots$ of a single rv $X \sim \mathrm{N}(\overline{X}, \sigma_X^2)$. After the nth observation, for each n, we want to estimate X based on

$y_1, \ldots, y_n$. We shall see that this can be done iteratively, using the estimate based on $y_1, \ldots, y_{n-1}$ to help form the estimate based on $y_1, \ldots, y_n$. The observation rv s are related to X by

$$Y_n = h_n X + Z_n,$$

where $\{Z_n;\ n \geq 1\}$ and X are independent Gaussian rv s, $Z_n \sim \mathrm{N}\ (0, \sigma^2_{Z_n})$. Let $\boldsymbol{Y}^n_m$ denote the observation rv s $Y_m, \ldots, Y_n$, so that $\boldsymbol{Y}^n$ denotes the first n observation variables. Similarly, let $\boldsymbol{y}^n_1$ denote the corresponding sample values $y_1, \ldots, y_n$. Let $\hat{x}(\boldsymbol{y}^n_1)$ be the MMSE estimate of X based on the observation of $\boldsymbol{Y}^n_1 = \boldsymbol{y}^n_1$ and let $\sigma^2_{\xi_n}$ be the conditional variance of the estimation error, $\xi_n = X - \hat{x}(\boldsymbol{y}^n_1)$.

For $n = 1$, we solved this problem in (10.18) and (10.19); given $Y_1 = y_1$,

$$\hat{x}(y_1) = \overline{X} + \frac{h_1 \sigma^2_X [y_1 - h_1 \overline{X}]}{h_1^2 \sigma^2_X + \sigma^2_{Z_1}}; \tag{10.20}$$

$$\sigma^2_{\xi_1} = \frac{\sigma^2_X \sigma^2_{Z_1}}{h_1^2 \sigma^2_X + \sigma^2_{Z_1}}, \qquad \frac{1}{\sigma^2_{\xi_1}} = \frac{1}{\sigma^2_X} + \frac{h_1^2}{\sigma^2_{Z_1}}. \tag{10.21}$$

Conditional on $Y_1 = y_1$, X is Gaussian with mean $\hat{x}(y_1)$ and variance $\sigma^2_{\xi_1}$; it is also independent of Z_2, which is $\mathrm{N}\ (0, \sigma^2_{Z_2})$, conditional or not on $Y_1 = y_1$. Thus, in the portion of the sample space restricted to $Y_1 = y_1$, and with probabilities conditional on $Y_1 = y_1$, we want to estimate X from $Y_2 = h_2 X + Z_2$; this is an instance of (10.18), here using conditional probabilities in the space $Y_1 = y_1$. Thus we use $\hat{x}(y_1)$ in place of $\overline{X}$, $\sigma^2_{\xi_1}$ in place of σ^2_X, and h_2 in place of h. The result is

$$\hat{x}(\boldsymbol{y}^2_1) = \hat{x}(y_1) + \frac{h_2 \sigma^2_{\xi_1} [y_2 - h_2 \hat{x}(y_1)]}{h_2^2 \sigma^2_{\xi_1} + \sigma^2_{Z_2}}, \tag{10.22}$$

$$\sigma^2_{\xi_2} = \frac{\sigma^2_{\xi_1} \sigma^2_{Z_2}}{h_2^2 \sigma^2_{\xi_1} + \sigma^2_{Z_2}}, \tag{10.23}$$

$$\frac{1}{\sigma^2_{\xi_2}} = \frac{1}{\sigma^2_{\xi_1}} + \frac{h_2^2}{\sigma^2_{Z_2}}. \tag{10.24}$$

Conditional on $Y_1 = y_1$ and $Y_2 = y_2$, we see that X is Gaussian with mean $\hat{x}(\boldsymbol{y}^2_1)$ and variance $\sigma^2_{\xi_2}$. What we are doing here is first conditioning on $Y_1 {=} y_1$, and then, in that conditional space, conditioning on $Y_2 {=} y_2$. This is the same as directly conditioning on $Y_1 {=} y_1, Y_2 {=} y_2$. To see this by an equation, consider the following identity:

$$\mathsf{f}_{X|Y_1Y_2}(x \mid y_1 y_2) = \frac{\mathsf{f}_{Y_2|XY_1}(y_2 \mid x y_1)}{\mathsf{f}_{Y_2|Y_1}(y_2 \mid y_1)}\, \mathsf{f}_{X|Y_1}(x \mid y_1).$$

The expression on the right is the Bayes law form of the density of X given $Y_2 = y_2$ in the space restricted to $Y_1 = y_1$. Iterating the argument in (10.22)–(10.24) to arbitrary $n > 1$, we have

$$\hat{x}(\boldsymbol{y}^n_1) = \hat{x}(\boldsymbol{y}^{n-1}_1) + \frac{h_n \sigma^2_{\xi_{n-1}} [y_n - h_n \hat{x}(\boldsymbol{y}^{n-1}_1)]}{h_n^2 \sigma^2_{\xi_{n-1}} + \sigma^2_{Z_n}}, \tag{10.25}$$

$$\sigma^2_{\xi_n} = \frac{\sigma^2_{\xi_{n-1}}\sigma^2_{Z_n}}{h_n^2\sigma^2_{\xi_{n-1}} + \sigma^2_{Z_n}}, \tag{10.26}$$

$$\frac{1}{\sigma^2_{\xi_n}} = \frac{1}{\sigma^2_{\xi_{n-1}}} + \frac{h_n^2}{\sigma^2_{Z_n}}. \tag{10.27}$$

As shown in Exercise 10.3, $\hat{x}(\boldsymbol{y}_1^n)$ and $\sigma^2_{\xi_n}$ can also be found directly from (10.9) and (10.10), and these yield the following non-iterative expressions:

$$\hat{x}(\boldsymbol{y}_1^n) = \sigma^2_{\xi_n}\sum_{j=1}^{n}\frac{h_j y_j}{\sigma^2_{Z_j}}, \tag{10.28}$$

$$\frac{1}{\sigma^2_{\xi_n}} = \frac{1}{\sigma^2_X} + \sum_{i=1}^{n}\frac{h_i^2}{\sigma^2_{Z_i}}. \tag{10.29}$$

These non-iterative forms will be discussed in a number of contexts later, since they solve the important problem of MMSE estimation of a Gaussian rv from n independent additive Gaussian noise observations. They show that the iterative approach here does not gain anything fundamentally new (although it is insightful). The next example, however, builds on this iterative approach to yield the Kalman filter, which is fundamentally different.

10.2.2 Scalar Kalman filter

Consider modifying the iterative estimation above to a situation where the rv X to be estimated is replaced by a sequence of rv s evolving in time. In particular, assume that $X_1 \sim \mathrm{N}\,(\overline{X}_1, \sigma^2_{X_1})$ and that for each $n \geq 1$, X_{n+1} evolves from X_n as

$$X_{n+1} = \alpha_n X_n + W_n\,, \qquad W_n \sim \mathrm{N}\,(0, \sigma^2_{W_n}), \tag{10.30}$$

where $W_1, W_2, \ldots,$ are independent and α_n and $\sigma^2_{W_n}$ are known scalars[3] for each n. Noisy observations $Y_1, Y_2, \ldots$ are made satisfying

$$Y_n = h_n X_n + Z_n\,, \qquad n \geq 1, \quad Z_n \sim \mathrm{N}\,(0, \sigma^2_{Z_n}), \tag{10.31}$$

where, for each $n \geq 1$, h_n is a known scalar. Finally, assume that $\{Z_n;\ n \geq 1\}$, $\{W_n; n > 1\}$, and X_1 are all statistically independent and jointly Gaussian.

For each $n \geq 1$, we want to find the MMSE estimate of both X_n and X_{n+1} conditional on the first n observations, $\boldsymbol{Y}_1^n = \boldsymbol{y}_1^n = y_1, \ldots, y_n$. We denote those estimates as $\hat{x}_n(\boldsymbol{y}_1^n)$ and $\hat{x}_{n+1}(\boldsymbol{y}_1^n)$. We denote the errors in these estimates as $\xi_n = X_n - \widehat{X}_n(\boldsymbol{Y}_1^n)$

[3] The process $X_1, X_2, \ldots$ is known as a discrete-time Gauss–Markov process. It differs from the Markov processes we have discussed in having a continuous valued rather than discrete state space and in being perhaps non-homogeneous. It is difficult to discuss general Markov processes (other than those with discrete state spaces) without measure theory. However, if the process is also Gaussian, as the one here, we can and will treat it directly using our knowledge of Gaussian sequences.

and $\zeta_{n+1} = X_{n+1} - \widehat{X}_{n+1}(\boldsymbol{Y}_1^n)$. Conditional on $Y_1 = y_1$, the solution for $\hat{x}_1(y_1)$ and the variance $\sigma^2_{\xi_1}$ of the estimation error is given by (10.20) and (10.21),

$$\hat{x}_1(y_1) = \overline{X}_1 + \frac{h_1\sigma^2_{X_1}[y_1 - h_1\overline{X}_1]}{h_1^2\sigma^2_{X_1} + \sigma^2_{Z_1}}; \tag{10.32}$$

$$\sigma^2_{\xi_1} = \frac{\sigma^2_{X_1}\sigma^2_{Z_1}}{h_1^2\sigma^2_{X_1} + \sigma^2_{Z_1}}, \qquad \frac{1}{\sigma^2_{\xi_1}} = \frac{1}{\sigma^2_{X_1}} + \frac{h_1^2}{\sigma^2_{Z_1}}. \tag{10.33}$$

Conditional on $Y_1{=}y_1$, X_1 is Gaussian with mean $\hat{x}_1(y_1)$ and variance $\sigma^2_{\xi_1}$. Now consider $X_2 = \alpha_1 X_1{+}W_1$ conditional on $Y_1{=}y_1$. In this conditional space, X_1 is Gaussian with the mean and variance above and $W_1 \sim \mathrm{N}\,(0, \sigma^2_{W_1})$ is independent of X_1. Thus, conditional on $Y_1{=}y_1$, $X_2 = \alpha_1 X_1 + W_1$ is Gaussian with mean $\alpha_1\hat{x}_1(y_1)$ and variance $\alpha_1^2\sigma^2_{\xi_1} + \sigma^2_{W_1}$. Thus the MMSE estimate $\hat{x}_2(y_1)$ and the error variance $\sigma^2_{\zeta_2}$ for X_2 conditional on $Y_1{=}y_1$ are given by

$$\hat{x}_2(y_1) = \alpha_1\hat{x}_1(y_1), \qquad \sigma^2_{\zeta_2} = \alpha_1^2\sigma^2_{\xi_1} + \sigma^2_{W_1}. \tag{10.34}$$

Remaining in the restricted space $Y_1 = y_1$, we now estimate X_2 from the additional observation $Y_2 = y_2$, i.e., we estimate X_2 from $y_1^2 = (y_1, y_2)$. Using (10.18), with $\hat{x}_2(y_1)$ in place of $\overline{X}$ and σ_{ζ_2} in place of $\sigma^2_{X_1}$,

$$\hat{x}_2(y_1^2) = \hat{x}_2(y_1) + \frac{h_2\sigma^2_{\zeta_2}[y_2 - h_2\hat{x}_2(y_1)]}{h_2^2\sigma^2_{\zeta_2} + \sigma^2_{Z_2}}; \tag{10.35}$$

$$\sigma^2_{\xi_2} = \frac{\sigma^2_{\zeta_2}\sigma^2_{Z_2}}{h_2^2\sigma^2_{\zeta_2} + \sigma^2_{Z_2}}, \qquad \frac{1}{\sigma_{\xi_2}} = \frac{1}{\sigma^2_{\zeta_2}} + \frac{h_2^2}{\sigma^2_{Z_2}}. \tag{10.36}$$

In general, we can repeat the arguments leading to (10.34)–(10.36) for each n, leading to

$$\hat{x}_n(y_1^{n-1}) = \alpha_{n-1}\hat{x}_{n-1}(y_1^{n-1}), \tag{10.37}$$

$$\sigma^2_{\zeta_n} = \alpha^2_{n-1}\sigma^2_{\xi_{n-1}} + \sigma^2_{W_{n-1}}, \tag{10.38}$$

$$\hat{x}_n(y_1^n) = \hat{x}_n(y_1^{n-1}) + \frac{h_n\sigma^2_{\zeta_n}[y_n - h_n\hat{x}_n(y_1^{n-1})]}{h_n^2\sigma^2_{\zeta_n} + \sigma^2_{Z_n}}, \tag{10.39}$$

$$\sigma^2_{\xi_n} = \frac{\sigma^2_{\zeta_n}\sigma^2_{Z_n}}{h_n^2\sigma^2_{\zeta_n} + \sigma^2_{Z_n}}, \qquad \frac{1}{\sigma^2_{\xi_n}} = \frac{1}{\sigma^2_{\zeta_n}} + \frac{h_n^2}{\sigma^2_{Z_n}}. \tag{10.40}$$

These are the scalar Kalman filter equations. The idea is that one 'filters' the observed values $Y_1, Y_2, \ldots$ to generate the successive MMSE estimates of the 'signal' $X_1, X_2, \ldots$. The variance terms, $\sigma^2_{\zeta_n}$ and $\sigma^2_{\xi_n}$, are not functions of the observations and could be pre-computed. From (10.39), the estimates are linear in the observations and $\overline{X}_1$. Finally, it is easy to get a recursion for the error variances σ_{ξ_n} directly by substituting (10.38) into (10.40):

$$\frac{1}{\sigma^2_{\xi_n}} = \frac{1}{\alpha^2_{n-1}\sigma^2_{\xi_{n-1}} + \sigma^2_{W_{n-1}}} + \frac{h_n^2}{\sigma^2_{Z_n}}. \tag{10.41}$$

An important special case of this result occurs where h_n, α_n, $\sigma^2_{Z_n}$, and $\sigma^2_{W_n}$ are each constant in n, with $0 < \alpha < 1$. The following results about this case are derived in Exercise 10.5. Starting with an arbitrary $\overline{X}_1$ and $\sigma^2_{X_1}$, the mean $\mathsf{E}[X_n]$ is $\alpha^n\overline{X}_1$, which goes to 0 with increasing n. Similarly, $\sigma^2_{X_n}$ varies, monotonically in n, from $\sigma^2_{X_1}$ to $\sigma^2_W/(1-\alpha^2)$. For each n, $\sigma^2_{\xi_n} < \sigma^2_{X_n}$. As a function of n, $\sigma^2_{\xi_n}$ is monotonic (increasing, decreasing, or constant) in n and approaches a finite limiting value λ given by the positive root of

$$\alpha^2h^2\sigma_Z^{-2}\lambda^2 + \left[h^2\sigma_Z^{-2}\sigma_W^2 + 1 - \alpha^2\right]\lambda - \sigma_W^2 = 0. \tag{10.42}$$

Each error variance $\sigma^2_{\xi_n}$ is increased by increasing σ_W, σ_Z/h or α.

10.3 LLSE estimation

We have seen that MMSE estimation is greatly simplified in the Gaussian case, since the squared error is independent of the observation. In many non-Gaussian situations, one maintains a mean-squared cost function, but minimizes it subject to the constraint that the estimate is a first-degree function of the observation. More specifically, one wants to minimize

$$\int \mathsf{E}\left[\|\boldsymbol{X} - \hat{\boldsymbol{X}}(\boldsymbol{Y})\|^2 \mid \boldsymbol{Y}{=}\boldsymbol{y}\right] \mathsf{f}_{\boldsymbol{Y}}(\boldsymbol{y})\,d\boldsymbol{y} = \mathsf{E}\left[\|\boldsymbol{X} - \hat{\boldsymbol{X}}(\boldsymbol{Y})\|^2\right] \tag{10.43}$$

subject to the condition that $\hat{\boldsymbol{x}}(\boldsymbol{y}) = [G]\boldsymbol{y} + \boldsymbol{b}$ for some matrix $[G]$ and vector $\boldsymbol{b}$. Such an estimate is called a *LLSE estimate*. Since $\hat{\boldsymbol{x}}(\boldsymbol{y})$ is restricted to have the form $[G]\boldsymbol{y} + \boldsymbol{b}$ for some matrix $[G]$ and vector $\boldsymbol{b}$, we want to find $[G]$ and $\boldsymbol{b}$ to minimize $\mathsf{E}[\|\boldsymbol{X} - [G]\boldsymbol{Y} - \boldsymbol{b}\|^2]$. For any choice of $[G]$ and $\boldsymbol{b}$, this expectation involves only the first and second moments (including cross moments) of $\boldsymbol{X}$ and $\boldsymbol{Y}$. This means that if $\boldsymbol{X}, \boldsymbol{Y}$ have the first and second moments $\mathsf{E}[\boldsymbol{X}], \mathsf{E}[\boldsymbol{Y}], [K_{\boldsymbol{X}}], [K_{\boldsymbol{Y}}], [K_{\boldsymbol{X}\cdot\boldsymbol{Y}}]$, and if $\boldsymbol{U}, \boldsymbol{V}$ are other **rv** s with the same first and second moments, then

$$\mathsf{E}\left[\|\boldsymbol{X} - [G]\boldsymbol{Y} - \boldsymbol{b}\|^2\right] = \mathsf{E}\left[\|\boldsymbol{U} - [G]\boldsymbol{V} - \boldsymbol{b}\|^2\right]$$

for all $[G]$ and $\boldsymbol{b}$. These expressions are not only equal, but they are minimized by the same $[G]$ and $\boldsymbol{b}$, and these minimizing values are functions only of the first and second moments.

For any $\boldsymbol{X}, \boldsymbol{Y}$, the MMSE estimate minimizes (10.43) without any constraint, and the LLSE estimate minimizes (10.43) with the constraint that the estimate is of first degree. Thus the mean-squared error in the MMSE estimate must be less than or equal to the mean-squared error for the LLSE estimate. For the special case in which $\boldsymbol{X}, \boldsymbol{Y}$ are jointly Gaussian, the MMSE estimate is of first degree, and thus equal to the LLSE estimate. Thus, using (10.12) for the Gaussian case, the LLSE estimate is achieved with

$$[G] = [K_{\boldsymbol{X}\cdot\boldsymbol{Y}}K_{\boldsymbol{Y}}^{-1}]; \qquad \boldsymbol{b} = \mathsf{E}[\boldsymbol{X}] - [K_{\boldsymbol{X}\cdot\boldsymbol{Y}}K_{\boldsymbol{Y}}^{-1}]\mathsf{E}[\boldsymbol{Y}]. \tag{10.44}$$

For any non-Gaussian $\boldsymbol{X}$ and $\boldsymbol{Y}$, there are Gaussian **rv** s with the same first and second moments (see Section 3.4.2), and thus the minimizing value of $\mathsf{E}[\|\boldsymbol{X} - [G]\boldsymbol{Y} - \boldsymbol{b}\|^2]$ is the same for the non-Gaussian and Gaussian cases, and the optimizing $[G]$ and $\boldsymbol{b}$ are given in (10.44). Thus (10.12) gives the LLSE estimate for both the Gaussian and non-Gaussian cases as well as the MMSE estimate for the Gaussian case.

There is an important idea here. The mean-squared error using the LLSE estimate is the same for the non-Gaussian and the Gaussian case, assuming the same first and second moments. For the Gaussian case, the MMSE estimate is the same as the LLSE estimate, so the mean-squared error is also the same. For the non-Gaussian case, the mean-squared error for the MMSE estimate is less than or equal to that for the LLSE estimate (since the MMSE estimate is not constrained to be of first degree). Thus the mean-squared error (using a MMSE estimate) for the non-Gaussian case is less than or equal to the mean-squared error (using MMSE) for the Gaussian case. Thus, for the mean-squared error criterion, Gaussian **rv** s have the largest mean-squared estimation error over all **rv** s with the same first and second joint moments. What is more, an estimator, constructed under the assumption of Gaussian **rv** s will work just as well for non-Gaussian **rv** s with the same first and second moments. Naturally, an estimator constructed to take advantage of the non-Gaussian statistics might do even better.

The LLSE error covariance in the non-Gaussian case, $\mathsf{E}[(\boldsymbol{X}-\hat{\boldsymbol{X}}(\boldsymbol{y}))(\boldsymbol{X}-\hat{\boldsymbol{X}}(\boldsymbol{y}))^{\mathsf{T}} \mid \boldsymbol{Y}=\boldsymbol{y}]$ will typically vary with $\boldsymbol{y}$. However, when this is averaged over $\boldsymbol{y}$, $\mathsf{E}[(X-\hat{\boldsymbol{X}}(\boldsymbol{y}))(\boldsymbol{X}-\hat{\boldsymbol{X}}(\boldsymbol{y}))^{\mathsf{T}}]$ depends only on first and second moments, so this average error covariance matrix is given by (10.10). We will rederive (10.9), (10.10), and (10.11) later for LLSE estimation using the first and second moments directly. The important observation here is that we can either use first and second moments directly to find LLSE estimates, or we can use the properties of Gaussian densities to find MMSE estimates for Gaussian problems, but we get the same answer in each case. Thus, we can solve either problem and get the solution to the other as a bonus. This is summarized in the following theorem.

Theorem 10.3.1 *If $\boldsymbol{X}$ and $\boldsymbol{Y}$ are jointly non-singular rv s, the LLSE estimate of $\boldsymbol{X}$ from the observation $\boldsymbol{Y} = \boldsymbol{y}$ is given by (10.12) and, for the zero-mean case, by (10.9). The error covariance, averaged over $\boldsymbol{Y}$, i.e., $[K_{\boldsymbol{\xi}}] = \mathsf{E}[(\boldsymbol{X} - \hat{\boldsymbol{X}}(\boldsymbol{Y}))(\boldsymbol{X} - \hat{\boldsymbol{X}}(\boldsymbol{Y}))^{\mathsf{T}}]$ is given by (10.10).*

It turns out to be easy to solve the LLSE optimization directly rather than using the solution to the Gaussian MMSE problem. The answer will be the same, of course, but it is instructive to see how natural the LLSE solution is. We want to find the matrix $[G]$ and the vector $\boldsymbol{b}$ that minimize

$$\mathsf{E}\left[\|\boldsymbol{X} - [G]\boldsymbol{Y} - \boldsymbol{b}\|^2\right]. \tag{10.45}$$

To start with the simplest case, let X be zero-mean and one-dimensional and let $\boldsymbol{Y}$ be a zero-mean n-**rv** with a non-singular covariance matrix $[K_{\boldsymbol{Y}}]$. The matrix $[G]$ then simplifies to a row vector $\boldsymbol{g}^{\mathsf{T}}$. Since X and $\boldsymbol{Y}$ are zero-mean, the constant b will be 0 in the minimum of (10.45). Thus we want to find $\boldsymbol{g}$ to minimize

$$\mathsf{E}\left[(X - \boldsymbol{g}^{\mathsf{T}}Y)^2\right] \;=\; \sigma_X^2 - 2\boldsymbol{g}^{\mathsf{T}}[K_{X\cdot\boldsymbol{Y}}] + \boldsymbol{g}^{\mathsf{T}}[K_{\boldsymbol{Y}}]\boldsymbol{g}.$$

The derivative of this with respect to $\boldsymbol{g}$ is $-2[K_{X\cdot\boldsymbol{Y}}] + 2\boldsymbol{g}^{\mathsf{T}}[K_{\boldsymbol{Y}}]$ (see Exercise 10.7). Minimizing by setting this equal to 0,

$$\boldsymbol{g}^{\mathsf{T}}[K_{\boldsymbol{Y}}] = [K_{X\cdot\boldsymbol{Y}}], \qquad \boldsymbol{g}^{\mathsf{T}} = [K_{X\cdot\boldsymbol{Y}}K_{\boldsymbol{Y}}^{-1}]. \tag{10.46}$$

Thus the LLSE estimate is

$$\hat{X}(\boldsymbol{Y}) = [K_{X\cdot\boldsymbol{Y}}K_{\boldsymbol{Y}}^{-1}]\boldsymbol{Y}. \tag{10.47}$$

This is the same answer as that in (10.9) for Gaussian MMSE estimation except that X here is one-dimensional. If $\boldsymbol{X}$ is a multi-dimensional **rv**, (10.47) can be applied to each component of $\boldsymbol{X}$, and $\boldsymbol{g}^{\mathsf{T}}$ is then replaced by a matrix $[G] = [K_{\boldsymbol{X}\cdot\boldsymbol{Y}}K_{\boldsymbol{Y}}^{-1}]$. Thus $\hat{\boldsymbol{X}}(\boldsymbol{Y}) = [K_{\boldsymbol{X}\cdot\boldsymbol{Y}}K_{\boldsymbol{Y}}^{-1}]\boldsymbol{Y}$, the same as (10.9). Finally, non-zero means can be included as before, resulting in (10.12).

The derivation here is conceptually simpler than that of (10.9), but the approach in (10.9) was necessary to solve the MMSE estimation problem for Gaussian problems. Without understanding Gaussian problems, the restriction to estimates of first degree would seem somewhat artificial.

10.4 Filtered vector signal plus noise

Consider estimating $\boldsymbol{X} = (X_1, \ldots, X_n)^{\mathsf{T}}$ from an observed sample of $\boldsymbol{Y} = (Y_1, \ldots, Y_m)^{\mathsf{T}}$ where $\boldsymbol{Y} = [H]\boldsymbol{X} + \boldsymbol{Z}$. Assume that $\boldsymbol{X}$ and $\boldsymbol{Z}$ are independent zero-mean Gaussian **rv** s and $[H]$ is an arbitrary real $m \times n$ matrix. Recall that we showed in Section 3.5 that if $\boldsymbol{X}, \boldsymbol{Y}$ are zero-mean, jointly Gaussian, and jointly non-singular, they can always be expressed as $\boldsymbol{Y} = [H]\boldsymbol{X} + \boldsymbol{Z}$, where $\boldsymbol{X}$ and $\boldsymbol{Z}$ are independent, and also as $\boldsymbol{X} = [G]\boldsymbol{Y} + \boldsymbol{V}$, where $\boldsymbol{Y}$ and $\boldsymbol{V}$ are independent. We also showed that both the MMSE and the LLSE estimate of $\boldsymbol{X}$ from $\boldsymbol{Y} = \boldsymbol{y}$ are given by $\hat{\boldsymbol{x}}(\boldsymbol{y}) = [G]\boldsymbol{y}$ and that the corresponding estimation-error covariance is $[K_{\boldsymbol{\xi}}] = [K_{\boldsymbol{V}}]$.

In a sense, then, we have already solved the problem of estimating $\boldsymbol{X}$ from $\boldsymbol{Y} = \boldsymbol{y}$ where $\boldsymbol{Y} = [H]\boldsymbol{X} + \boldsymbol{Z}$. However, the solutions have been given in terms of $[K_{\boldsymbol{X}\cdot\boldsymbol{Y}}]$ and $[K_{\boldsymbol{Y}}]$. In this section, we convert these solutions to expressions involving $[H]$ and $[K_{\boldsymbol{Z}}]$ directly. The problem of estimating $\boldsymbol{X}$ from $\boldsymbol{Y} = [H]\boldsymbol{X} + \boldsymbol{Z}$ is a canonical estimation problem in both the communication and control fields. For example, we might view $\boldsymbol{X}$ as the samples, at some sufficiently high sampling rate, of the input to a communication system, taken over some interval of time. Then $[H]$ represents a linear filter (perhaps time varying) that the input passes through, and $\boldsymbol{Z}$ represents the noise that is added to the filtered signal. Focusing on $[H]$ and $[K_{\boldsymbol{Z}}]$ rather than $[K_{\boldsymbol{X}\cdot\boldsymbol{Y}}]$ and $[K_{\boldsymbol{Y}}]$ is helpful in further understanding these types of estimation problems.

We start with a special case where $[H]$ is an identity matrix, i.e., $\boldsymbol{Y} = \boldsymbol{X} + \boldsymbol{Z}$. From (10.9), $\hat{\boldsymbol{x}}(\boldsymbol{y}) = [K_{\boldsymbol{X}\cdot\boldsymbol{Y}}K_{\boldsymbol{Y}}^{-1}]\boldsymbol{y}$. Since $\boldsymbol{Y} = \boldsymbol{X} + \boldsymbol{Z}$, it follows that $[K_{\boldsymbol{X}\cdot\boldsymbol{Y}}] = [K_{\boldsymbol{X}}]$, and $[K_{\boldsymbol{Y}}] = [K_{\boldsymbol{X}}] + [K_{\boldsymbol{Z}}]$. We then have

$$\hat{\boldsymbol{x}}(\boldsymbol{y}) = [K_{\boldsymbol{X}}]\Big([K_{\boldsymbol{X}}] + [K_{\boldsymbol{Z}}]\Big)^{-1}\boldsymbol{y}. \tag{10.48}$$

Similarly, from (10.10), the covariance $[K_\xi]$ of the estimation error is

$$\begin{aligned}[K_\xi] &= [K_X] - [K_{X\cdot Y}K_Y^{-1}K_{X\cdot Y}^{\mathsf T}] \;=\; [K_X] - [K_XK_Y^{-1}K_X] \\ &= [K_XK_Y^{-1}K_Y] - [K_XK_Y^{-1}K_X] \;=\; [K_XK_Y^{-1}]\Big[[K_Y] - [K_X]\Big] \\ &= [K_XK_Y^{-1}K_Z] \;=\; [K_X]\Big([K_X] + [K_Z]\Big)^{-1}[K_Z]. \end{aligned} \tag{10.49}$$

Inverting both sides of this equation, we get the vector form of (10.15),

$$[K_\xi^{-1}] = [K_Z^{-1}] + [K_X^{-1}]. \tag{10.50}$$

Substituting (10.49) into (10.48), we get a simple alternative expression for $\hat{\boldsymbol{x}}(\boldsymbol{y})$:

$$\hat{x}(y) = [K_\xi K_Z^{-1}]y. \tag{10.51}$$

Now consider the general case $\boldsymbol{Y} = [H]\boldsymbol{X} + \boldsymbol{Z}$, where $[H]$ is an arbitrary matrix (but $\boldsymbol{X}$ and $\boldsymbol{Z}$ are still zero mean). From (10.9), $\hat{\boldsymbol{x}}(\boldsymbol{y}) = [K_{X\cdot Y}K_Y^{-1}]\boldsymbol{y}$. Since $[K_{X\cdot Y}] = [K_XH^{\mathsf T}]$ and $[K_Y] = [HK_XH^{\mathsf T}] + [K_Z]$, the MMSE estimate of $\boldsymbol{X}$ is

$$\hat{\boldsymbol{x}}(\boldsymbol{y}) = [K_XH^{\mathsf T}]\Big([HK_XH^{\mathsf T}] + [K_Z]\Big)^{-1}\boldsymbol{y}. \tag{10.52}$$

Similarly, from (10.10), $[K_\xi] = [K_X] - [K_{X\cdot Y}K_Y^{-1}K_{X\cdot Y}^{\mathsf T}]$, which is

$$[K_\xi] = [K_X] - [K_XH^{\mathsf T}]\Big([HK_XH^{\mathsf T}] + [K_Z]\Big)^{-1}[HK_X]. \tag{10.53}$$

In this class of estimation problems, the dimension of $\boldsymbol{X}$ is often much smaller than that of $\boldsymbol{Y}$ and $[K_Z]$ often has a simple structure. In these cases, it turns out (see Exercise 10.6) that the expressions in (10.50) and (10.51) can be generalized for arbitrary $[H]$ to yield the alternative solutions:

$$\hat{\boldsymbol{x}}(\boldsymbol{y}) = [K_\xi H^{\mathsf T}K_Z^{-1}]\boldsymbol{y}; \tag{10.54}$$

$$[K_\xi] = \Big([K_X^{-1}] + [H^{\mathsf T}K_Z^{-1}H]\Big)^{-1}. \tag{10.55}$$

These usually save a considerable amount of matrix manipulations[4] relative to (10.52) and (10.53). The derivation of these equations is carried out in Exercise 10.6.

For completeness, the expressions in (10.52) and (10.54) are generalized to include a non-zero mean for $\boldsymbol{X}$. Using (10.52) and (10.54) for the fluctuations, $X-\overline{X}$ and $\boldsymbol{Y}-[H]\overline{X}$,

$$\hat{\boldsymbol{x}}(\boldsymbol{y}) = \overline{X} + [K_XH^{\mathsf T}]\Big([HK_XH^{\mathsf T}] + [K_Z]\Big)^{-1}(\boldsymbol{y} - [H]\overline{X}), \tag{10.56}$$

$$\hat{\boldsymbol{x}}(\boldsymbol{y}) = \overline{X} + [K_\xi H^{\mathsf T}K_Z^{-1}](\boldsymbol{y} - [H]\overline{X}). \tag{10.57}$$

The error covariance is still given by (10.53) and (10.55). The following subsections look at several simple cases of these estimates, and then proceed to analyze iterative vector estimation and vector Kalman estimation.

[4] In this day of virtually free computer computation, it is reasonable to ask who cares about computational efficiency. There are three answers. These problems are sometimes very large, additional (or uninformed) computation can cause large round-off errors, and, most important, these new formulations can lead to additional insights, which reduce programming errors.

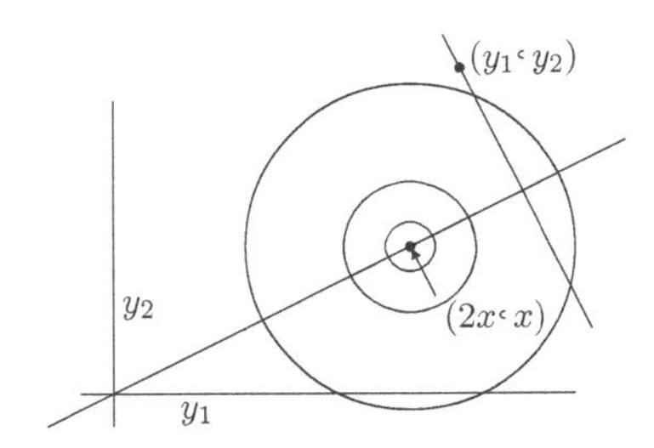

Figure 10.1 Illustration of one-dimensional signal x and two-dimensional observation $\boldsymbol{y}$ where $y_1 = 2x + z_1$ and $y_2 = x + z_2$ and (z_1, z_2) is a sample value of IID Gaussian rv s. The concentric circles are regions where $\mathsf{f}_{Y|X}(y|x)$ is constant for given x. The straight line through the origin is the locus of points $(2u, u)$ as u varies, and the other straight line is perpendicular to the first line; it is the locus of points where $2y_1 + y_2$ is constant, i.e., where the MMSE estimate of x is constant.

10.4.1 Estimate of a single random variable in IID vector noise

Let $X \sim \mathrm{N}\,(0, \sigma_X^2)$ be one-dimensional so that $\boldsymbol{Y} = \boldsymbol{h}X + \boldsymbol{Z}$. Let $\boldsymbol{Z}$ be an n-rv with independent components, each $\mathrm{N}\,(0, \sigma_{Z_i}^2)$ and independent of X. Assume $\boldsymbol{h}$ is an arbitrary n-vector. From (10.54) and (10.55),

$$\hat{x}(\boldsymbol{y}) = [K_\xi] \sum_{j=1}^{n} \frac{h_j y_j}{\sigma_{Z_j}^2}, \qquad [K_\xi^{-1}] = \frac{1}{\sigma_X^2} + \sum_{i=1}^{n} \frac{h_i^2}{\sigma_{Z_i}^2}. \tag{10.58}$$

This is the same as (10.28) and (10.29), but the derivation here adds insight into what these equations mean. In terms of the communication example, we can visualize x as being a sample value of the signal X which is 'filtered' into a sequence $(h_1 x, h_2 x, \ldots, h_n x)$. The received signal $\boldsymbol{y}$ is then $\{y_i = h_i x + z_i; 1 \le i \le n\}$. By definition, the filter 'matched' to the filter with taps $(h_1, \ldots, h_n)$ is a filter with taps $(h_n, h_{n-1}, \ldots, h_2, h_1)$. After $(y_1, y_2, \ldots, y_n)$ passes through this filter, the output sample is $h_1 y_1 + h_2 y_2 + \cdots + h_n y_n$. As seen in (10.58), $\hat{\boldsymbol{x}}(\boldsymbol{y})$ is simply a scaled version of this sum. Figure 10.1 illustrates this for $n = 2$.

10.4.2 Estimate of a single random variable in arbitrary vector noise

As a slightly more general case, let $\boldsymbol{Y} = \boldsymbol{h}X + \boldsymbol{Z}$ again, but suppose $\boldsymbol{Z}$ has an arbitrary non-singular covariance matrix $[K_{\boldsymbol{Z}}]$. We could use (10.54) and (10.55) directly, but instead we show how these formulas arise for this case. As shown in Section 3.4.2, there must be a non-singular matrix $[A]$ such that $[AA^{\mathsf{T}}] = [K_{\boldsymbol{Z}}]$. We can then express $\boldsymbol{Z}$ as $[A]\boldsymbol{W}$, where $\boldsymbol{W}$ is $\mathrm{N}\,(\boldsymbol{0}, [I])$. Let $\boldsymbol{Y}' = [A^{-1}]\boldsymbol{Y} = [A^{-1}]\boldsymbol{h}X + \boldsymbol{W}$. The MMSE of X from a sample value of $\boldsymbol{Y}'$ must be the same as that from $\boldsymbol{Y}$ since $\boldsymbol{Y}$ and $\boldsymbol{Y}'$ can be calculated from each other.

The estimation problem from $\boldsymbol{Y}'$ is the special case we solved in (10.58), where $\boldsymbol{h}$ must now be replaced with $[A^{-1}]\boldsymbol{h}$, giving the MMSE estimate

$$x(\boldsymbol{y}') = [K_\xi]\big([A^{-1}]\boldsymbol{h}\big)^{\mathsf{T}}\boldsymbol{y}' \;=\; [K_\xi]\boldsymbol{h}^{\mathsf{T}}[A^{-1}]^{\mathsf{T}}\boldsymbol{y}'.$$

Replacing y' by $[A^{-1}]y$, this becomes

$$x(y) \;=\; [K_\xi]h^{\mathsf T}[A^{-1}]^{\mathsf T}[A^{-1}]y \;=\; [K_\xi]h^{\mathsf T}[K_Z^{-1}]y.$$

The error variance can be solved in the same way, by replacing h in (10.58) by $[A^{-1}]h$ and σ_Z^2 by 1. This leads to

$$[K_\xi^{-1}] \;=\; \sigma_X^{-2} + h^{\mathsf T}[A^{-1}]^{\mathsf T}[A^{-1}]h \;=\; [K_x^{-1}] + h^{\mathsf T}[K_Z^{-1}]h.$$

These provide an independent derivation of (10.54) and (10.55) for the case where X is one-dimensional. The technique just used is called 'noise-whitening' and is often a valuable technique. What we did was to convert a problem with an arbitrary noise matrix into a problem with IID noise (called white noise). After the white noise problem was solved, the solution was converted back to the original problem.

10.4.3 Vector iterative estimation

We now make multiple noisy vector observations, $y_1, y_2, \ldots, y_n \ldots$ of a single Gaussian k-rv $X \sim \mathrm{N}\,(\overline{X}, [K_X])$. For each n, we assume the observation m-rv s $y_n \sim \mathrm{N}\,(0, [K_{Y_n}])$ have the form

$$Y_n = [H_n]X + Z_n\,; \quad Z_n \sim \mathrm{N}\,(\mathbf{0}, [K_{Z_n}]), \quad n \ge 1. \tag{10.59}$$

We assume that $X, Z_1, Z_2, \ldots$ are mutually independent and that $[H_1], [H_2], \ldots$ are known matrices. For each n, we want to find the MMSE estimate of X based on $\{y_j;\ 1 \le j \le n\}$. We will do this iteratively, using the estimate based on $y_1, \ldots, y_{n-1}$ to help in finding the estimate based on $y_1, \ldots, y_n$. Let Y^n denote the first n observation m-rv s and let y^n denote the corresponding n sample observations. Let $\hat{x}(y^n)$ and $[K_{\xi_n}]$ denote the MMSE estimate and error covariance respectively based on $y^n = y_1, y_2, \ldots, y_n$. For $n = 1$, the result is the same as that in (10.56) and (10.57), yielding

$$\hat{x}(y_1) = \overline{X} + [K_X H_1^{\mathsf T}]\Big([H_1 K_X H_1^{\mathsf T}] + [K_{Z_1}]\Big)^{-1}(y_1 - [H_1]\overline{X}), \tag{10.60}$$

$$\hat{x}(y_1) = \overline{X} + [K_{\xi_1} H_1^{\mathsf T} K_{Z_1}^{-1}](y_1 - [H_1]\overline{X}). \tag{10.61}$$

We have given these in the above two forms since each yields its own insight. From (10.53) and (10.55), the error covariance of the error, $\xi_1 = \hat{X}(Y_1) - X$, is

$$[K_{\xi_1}] = [K_X] - [K_X H_1^{\mathsf T}]\Big([H_1 K_X H_1^{\mathsf T}] + [K_{Z_1}]\Big)^{-1}[H_1 K_X], \tag{10.62}$$

$$[K_{\xi_1}^{-1}] = [K_X^{-1}] + [H_1^{\mathsf T} K_{Z_1}^{-1} H_1]. \tag{10.63}$$

Using the same argument that we used for the scalar case, we see that for each $n > 1$, the conditional mean of X, conditional on the observation y^{n-1}, is by definition $\hat{x}(y^{n-1})$ and the conditional covariance is $[K_{\xi_{n-1}}]$. Using these quantities in place of $\overline{X}$ and $[K_X]$ in (10.56) and (10.57) then yields

$$\hat{x}(y^n) = \hat{x}(y^{n-1}) + [K_{\xi_{n-1}} H_n^{\mathsf T}]\Big([H_n K_{\xi_{n-1}} H_n^{\mathsf T}] + [K_{Z_n}]\Big)^{-1}\Big(y_n - [H_n]\hat{x}(y^{n-1})\Big) \tag{10.64}$$

$$\hat{x}(y^n) = \hat{x}(y^{n-1}) + [K_{\xi_{n-1}} H_n^{\mathsf T} K_{Z_n}^{-1}](y_n - [H_n]\hat{x}(y^{n-1})). \tag{10.65}$$

From (10.53) and (10.55), the error covariance is given by

$$[K_{\xi_n}] = [K_{\xi_{n-1}}] - [K_{\xi_{n-1}}H_n^{\mathsf{T}}]\Big([H_nK_{\xi_{n-1}}H_n]^{\mathsf{T}} + [K_{Z_n}]\Big)^{-1}[H_nK_{\xi_{n-1}}], \quad (10.66)$$

$$[K_{\xi_n}^{-1}] = [K_{\xi_{n-1}}^{-1}] + [H_n^{\mathsf{T}}K_{Z_n}^{-1}H_n]. \quad (10.67)$$

10.4.4 Vector Kalman filter

Finally, consider the vector case of iterative estimation on a sequence of Gaussian k-**rv** s, $\boldsymbol{X}_1, \boldsymbol{X}_2, \ldots$, where $\boldsymbol{X}_1 \sim \mathrm{N}\,(\overline{\boldsymbol{X}}_1, [K_{\boldsymbol{X}_1}])$, and $\boldsymbol{X}_{n+1}$ evolves from $\boldsymbol{X}_n$ for $n \geq 1$ according to the equation

$$\boldsymbol{X}_{n+1} = [A_n]\boldsymbol{X}_n + \boldsymbol{W}_n\,; \quad \boldsymbol{W}_n \sim \mathrm{N}\,(0, [K_{\boldsymbol{W}_n}]), \quad n \geq 1. \quad (10.68)$$

Here, for each $n \geq 1$, $[A_n]$ is a known matrix. Noisy observations $\boldsymbol{Y}_n$ are made satisfying

$$\boldsymbol{Y}_n = [H_n]\boldsymbol{X}_n + \boldsymbol{Z}_n\,, \quad \boldsymbol{Z}_n \sim \mathrm{N}\,(0, [K_{\boldsymbol{Z}_n}]), \quad n \geq 1, \quad (10.69)$$

where for each $n \geq 1$, $[H_n]$ is a known matrix, and $[K_{\boldsymbol{Z}_n}]$ is an invertible covariance matrix.

Assume that $\boldsymbol{X}_1, \{\boldsymbol{W}_n; n \geq 1\}$ and $\{\boldsymbol{Z}_n; n \geq 1\}$ are all independent. As in the scalar case, we want to find the MMSE estimate of both $\boldsymbol{X}_n$ and $\boldsymbol{X}_{n+1}$ conditional on the sample values of $\boldsymbol{Y}_1, \boldsymbol{Y}_2, \ldots, \boldsymbol{Y}_n$. We denote these estimates as $\hat{\boldsymbol{x}}_n(\boldsymbol{y}_1^n)$ and $\hat{\boldsymbol{x}}_{n+1}(\boldsymbol{y}_1^n)$ respectively. The errors in these estimates are denoted as $\xi_n = X_n - \widehat{X}_n(\boldsymbol{Y}_1^n)$ and $\zeta_{n+1} = \boldsymbol{X}_{n+1} - \hat{\boldsymbol{X}}_{n+1}(\boldsymbol{Y}_1^n)$.

For $n = 1$, the problem is the same as in (10.60)–(10.63). Alternative forms for the estimate and error covariance are

$$\hat{\boldsymbol{x}}_1(\boldsymbol{y}_1) = \overline{\boldsymbol{X}}_1 + [K_{\boldsymbol{X}_1}H_1^{\mathsf{T}}]\Big([H_1K_{\boldsymbol{X}_1}H_1^{\mathsf{T}}] + [K_{\boldsymbol{Z}_1}]\Big)^{-1}\big(\boldsymbol{y}_1 - [H_1]\overline{\boldsymbol{X}}_1\big), \quad (10.70)$$

$$\hat{\boldsymbol{x}}_1(\boldsymbol{y}_1) = \overline{\boldsymbol{X}}_1 + [K_{\xi_1}H_1^{\mathsf{T}}K_{\boldsymbol{Z}_1}^{-1}]\big(\boldsymbol{y}_1 - [H_1]\overline{\boldsymbol{X}}_1\big); \quad (10.71)$$

$$[K_{\xi_1}] = [K_{\boldsymbol{X}_1}] - [K_{\boldsymbol{X}_1}H_1^{\mathsf{T}}]\Big([H_1K_{\boldsymbol{X}_1}H_1^{\mathsf{T}}] + [K_{\boldsymbol{Z}_1}]\Big)^{-1}[H_1K_{\boldsymbol{X}_1}], \quad (10.72)$$

$$[K_{\xi_1}^{-1}] = [K_{\boldsymbol{X}_1}^{-1}] + [H_1^{\mathsf{T}}][K_{\boldsymbol{Z}_1}^{-1}][H_1]. \quad (10.73)$$

This means that, conditional on $\boldsymbol{Y}_1{=}\boldsymbol{y}_1$, $\boldsymbol{X}_1 \sim \mathrm{N}\,(\hat{\boldsymbol{x}}_1(\boldsymbol{y}_1), [K_{\xi_1}])$. Thus, conditional on $\boldsymbol{Y}_1{=}\boldsymbol{y}_1$, $\boldsymbol{X}_2 = [A_1]\boldsymbol{X}_1 + \boldsymbol{W}_1$ is Gaussian with mean $[A_1]\hat{\boldsymbol{x}}_1(\boldsymbol{y}_1)$ and with covariance $[A_1K_{\xi_1}A_1^{\mathsf{T}}] + [K_{\boldsymbol{W}_1}]$. Thus

$$\hat{\boldsymbol{x}}_2(\boldsymbol{y}_1) = [A_1]\hat{\boldsymbol{x}}_1(\boldsymbol{y}_1), \quad [K_{\zeta_2}] = [A_1K_{\xi_1}A_1^{\mathsf{T}}] + [K_{\boldsymbol{W}_1}]. \quad (10.74)$$

Conditional on $\boldsymbol{Y}_1 = \boldsymbol{y}_1$, (10.74) gives the mean and covariance of $\boldsymbol{X}_2$. Thus, given the additional observation $\boldsymbol{Y}_2 = \boldsymbol{y}_2$, where $\boldsymbol{Y}_2 = [H_2]\boldsymbol{X}_2 + \boldsymbol{Z}_2$, we can use (10.56) and (10.57) with this mean and covariance to get the alternative forms

$$\hat{\boldsymbol{x}}_2(\boldsymbol{y}_1^2) = \hat{\boldsymbol{x}}_2(\boldsymbol{y}_1) + [K_{\zeta_2}H_2^{\mathsf{T}}]\Big([H_2K_{\zeta_2}H_2^{\mathsf{T}}] + [K_{Z_2}]\Big)^{-1}\big(\boldsymbol{y}_2 - [H_2]\hat{\boldsymbol{x}}_2(\boldsymbol{y}_1)\big), \quad (10.75)$$

$$\hat{\boldsymbol{x}}_2(\boldsymbol{y}_1^2) = \hat{\boldsymbol{x}}_2(\boldsymbol{y}_1) + [K_{\xi_2}H_2^{\mathsf{T}}K_{\boldsymbol{Z}_2}^{-1}]\big(\boldsymbol{y}_2 - [H_2]\boldsymbol{X}_2(\boldsymbol{y}_1)\big). \quad (10.76)$$

The covariance of the error is given by the alternative forms

$$[K_{\xi_2}] = [K_{\zeta_2}] - [K_{\zeta_2}H_2^{\mathsf{T}}]\Big([H_2K_{\zeta_2}H_2^{\mathsf{T}}] + [K_{Z_2}]\Big)^{-1}[H_2K_{\zeta_2}], \tag{10.77}$$

$$[K_{\xi_2}^{-1}] = [K_{\zeta_2}^{-1}] + [H_2^{\mathsf{T}}K_{Z_2}^{-1}H_2]. \tag{10.78}$$

Continuing this same argument iteratively for all $n > 1$, we obtain the Kalman filter equations:

$$\hat{x}_n(y_1^n) = \hat{x}_n(y_1^{n-1}) + [K_{\zeta_n}H_n^{\mathsf{T}}]\Big([H_nK_{\zeta_n}H_n^{\mathsf{T}}] + [K_{Z_n}]\Big)^{-1}\big(y_n - [H_n]\hat{x}_n(y_1^{n-1})\big), \tag{10.79}$$

$$\hat{x}_n(y_1^n) = \hat{x}_n(y_1^n) + [K_{\xi_n}H_n^{\mathsf{T}}K_{Z_n}^{-1}]\big(y_n - [H_n]\hat{x}_n(y_1^{n-1})\big); \tag{10.80}$$

$$[K_{\xi_n}] = [K_{\zeta_n}] - [K_{\zeta_n}H_n^{\mathsf{T}}]\Big([H_nK_{\zeta_n}H_n^{\mathsf{T}}] + [K_{Z_n}]\Big)^{-1}[H_nK_{\zeta_n}], \tag{10.81}$$

$$[K_{\xi_n}^{-1}] = [K_{\zeta_n}^{-1}] + [H_n^{\mathsf{T}}K_{Z_n}^{-1}H_n]; \tag{10.82}$$

$$\hat{x}_{n+1}(y_1^n) = [A_n]\hat{x}_n(y_1^n), \qquad [K_{\zeta_{n+1}}] = [A_nK_{\xi_n}A_n^{\mathsf{T}}] + [K_{W_n}]. \tag{10.83}$$

The alternative forms above are equivalent and differ in the size and type of matrix inversions required; these matrix inversions do not depend on the data, however, and thus can be precomputed.

10.5 Estimation for circularly-symmetric Gaussian rv s

Recall that the Gaussian rv s in Sections 10.2–10.4 are all mappings from the sample space to real valued vectors. In this section, we consider complex Gaussian rv s and in particular circularly-symmetric Gaussian rv s. Circularly-symmetric Gaussian rv s are defined in Section 3.7 and the results on conditional probability densities, which are essential for the current discussion of estimation, are developed in Section 3.7.7.

We will find that the results for conditional PDFs for the circularly-symmetric Gaussian case are so closely related to those for the real Gaussian case that the results for MMSE for the real Gaussian case carry over almost verbatim to the circularly-symmetric Gaussian case. Slightly more care is required for LLSE, but the results are still very similar.

Starting from Section 3.7.7, let $(X_1, \ldots, X_n, Y_1, \ldots, Y_m)^{\mathsf{T}}$ be a zero-mean circularly-symmetric Gaussian rv with a non-singular covariance matrix. Then $\boldsymbol{X}$ can be represented as $\boldsymbol{X} = [G]\boldsymbol{Y} + \boldsymbol{V}$, where $\boldsymbol{Y}$ and $\boldsymbol{V}$ are statistically independent and circularly-symmetric Gaussian. In other words, $\mathsf{f}_{\boldsymbol{X}|\boldsymbol{Y}}(\boldsymbol{x}|\boldsymbol{y})$ for a given $\boldsymbol{y}$ is $\mathcal{CN}\,([G]\boldsymbol{y}, [V])$. This notation means that, conditional on a given $\boldsymbol{y}$, $\boldsymbol{X}$ has a mean $[G]\boldsymbol{y}$ and is circularly-symmetric Gaussian around that mean with covariance $[K_{\boldsymbol{V}}]$. From (3.112) and (3.113), the complex matrix $[G]$ and the complex covariance $[K_{\boldsymbol{V}}]$ can be represented as

$$[G] = [K_{\boldsymbol{X}\cdot\boldsymbol{Y}}K_{\boldsymbol{Y}}^{-1}], \qquad [K_{\boldsymbol{V}}] = [K_{\boldsymbol{X}}] - [K_{\boldsymbol{X}\cdot\boldsymbol{Y}}K_{\boldsymbol{Y}}^{-1}K_{\boldsymbol{X}\cdot\boldsymbol{Y}}^{\dagger}]. \tag{10.84}$$

This is the same as (10.9) and (10.10) for real vectors except that $[\boldsymbol{K}_{\boldsymbol{X}\cdot\boldsymbol{Y}}^{\mathsf{T}}]$ there is replaced by the Hermitian transpose $[\boldsymbol{K}_{\boldsymbol{X}\cdot\boldsymbol{Y}}^{\dagger}]$ here. In addition, of course, the jointly

Gaussian conditional distribution is replaced with a circularly-symmetric Gaussian distribution.

It follows immediately, as in (10.9) and (10.10), that the MMSE estimate of $\boldsymbol{X}$ from a sample value $\boldsymbol{y}$ of $\boldsymbol{Y}$ is $\hat{\boldsymbol{x}}(\boldsymbol{y}) = [G]\boldsymbol{y}$. Note that $[G]\boldsymbol{y}$ is linear in $\boldsymbol{y}$, but the linearity is over the complex field. The estimation error $\boldsymbol{\xi}$ is independent of $\boldsymbol{Y}$ and its variance is $[K_{\boldsymbol{\xi}}] = [K_{\boldsymbol{V}}]$.

The various examples and special cases in Sections 10.2–10.4 all follow from (10.9) and (10.10), and thus can be applied directly to the circularly-symmetric case by replacing transposes with Hermitian transposes and Gaussian probability density functions (PDFs) with circularly-symmetric PDFs.

Next, let us consider LLSE estimation for complex **rv** s and particularly circularly-symmetric Gaussian **rv** s. As in the real case, we observe a sample value $\boldsymbol{y}$ of $\boldsymbol{Y}$ and want to estimate $\boldsymbol{X}$ using an estimate restricted to the form $\hat{\boldsymbol{x}}(\boldsymbol{y}) = [G]\boldsymbol{y} + \boldsymbol{b}$. We want to choose the estimate to minimize $\mathsf{E}[\|\boldsymbol{X} - \hat{\boldsymbol{X}}(\boldsymbol{Y})\|^2]$. Thus, as in Section 3.3, we want to choose $[G]$ and $\boldsymbol{b}$ to minimize $\mathsf{E}\left[\|\boldsymbol{X} - [G]\boldsymbol{Y} - \boldsymbol{b}\|^2\right]$. We will assume $\boldsymbol{X}$ and $\boldsymbol{Y}$ to be zero mean, so the minimizing value of $\boldsymbol{b}$ is 0.

As in Section 3.3, both the minimizing value of $[G]$ and the resulting error covariance are functions only of the covariances and cross-covariances of $\boldsymbol{X}$ and $\boldsymbol{Y}$ and thus will be the same for all complex **rv** s with the same combined covariance. The MMSE estimate for the zero-mean circularly-symmetric Gaussian case has been shown to be linear and thus this unconstrained estimate must be the same as the LLSE estimate. In other words, for any zero-mean complex $\boldsymbol{X}, \boldsymbol{Y}$, the LLSE estimate is $\hat{\boldsymbol{x}}(\boldsymbol{y}) = [G]\boldsymbol{y}$ and the covariance of the error is $[K_{\boldsymbol{V}}]$, where $[G]$ and $[K_{\boldsymbol{V}}]$ are given by (10.84).

We have seen in Lemma 3.7.9 that for any covariance function $[K]$ of a complex n-**rv**, there is a zero-mean circularly-symmetric Gaussian n-**rv** that has the same covariance. Thus, for any zero-mean complex n-**rv**, the LLSE estimate and the covariance of the estimation error are the same as those for the circularly-symmetric Gaussian case with the same covariance. This means that the MMSE error covariance for an arbitrary zero-mean complex n-**rv** must be less than or equal to that for the corresponding circularly-symmetric Gaussian case.

Up to this point, the results for the complex and the real case are identical (with the circularly-symmetric Gaussian playing the corresponding role to the jointly Gaussian case). This is valid whether or not the variables are circularly symmetric. This is somewhat surprising since the covariance matrices do not include the pseudo-covariance matrices, and thus do not contain all the information about second moments in this case.

The reason for the absence of the pseudo-covariance matrices here is clarified by being careful about what linearity means. What we have found in (10.84) uses linearity in the complex field, i.e., an estimate of the form $[G]\boldsymbol{y}$. This is *not* the same as the LLSE of $\boldsymbol{X}_{re}$ and $\boldsymbol{X}_{im}$ based on $\boldsymbol{Y}_{re}, \boldsymbol{Y}_{im}$. This can be seen by noting that an arbitrary linear transformation from two variables Y_{re}, Y_{im} to $\widehat{X}_{re}, \widehat{X}_{im}$ is specified by a 2×2 matrix, which contains four arbitrary real numbers. An arbitrary linear transformation from a complex variable Y to a complex variable $\widehat{X}$ is specified by a single complex variable, i.e., by two real variables.

10.6 The vector space of random variables; orthogonality

In the preceding sections, we emphasized jointly-Gaussian **rv** s (and circularly-symmetric Gaussian **rv** s). We derived MMSE estimates and error covariances and showed that the same estimates, with the same error covariances, were valid for LLSE estimation of arbitrary **rv** s. In this section, we view rv s and complex rv s respectively as elements of real and complex inner-product spaces. This will not provide many new results, but will provide additional insight into the current results, especially those comparing LLSE and MMSE for non-Gaussian rv s and complex rv s.

Viewing rv s and complex rv s as elements of vector spaces seems a little peculiar at first, but it will soon become natural. A *vector space* $\mathbb{V}$ is a set of elements, called vectors, along with two operations, addition and scalar multiplication. Under the addition operation, any two vectors, $X \in \mathbb{V}$, $Y \in \mathbb{V}$ can be added to produce another vector, denoted $X + Y$, in $\mathbb{V}$. Under scalar multiplication, any scalar α can be multiplied by any vector $X \in \mathbb{V}$ to produce another vector $\alpha X \in \mathbb{V}$.

For our purposes, there are *real* vector spaces, in which the scalars are real numbers, and *complex* vector spaces in which the scalars are complex numbers. More generally, the set of scalars can be chosen to be the elements of any given field, but the field of real numbers and the field of complex numbers are the only ones of interest here. Vector addition and scalar multiplication satisfy the following axioms for all vectors X, Y, Z and all scalars α, β:

1. Addition is commutative and associative, i.e., $X + Y = Y + X$ and $(X + Y) + Z = X + (Y + Z)$.
2. Scalar multiplication is associative, i.e., $(\alpha\beta)X = \alpha(\beta X)$.
3. Scalar multiplication by 1 satisfies $1X = X$.
4. The distributive laws hold: $\alpha(X + Y) = (\alpha X) + (\alpha Y)$; $(\alpha + \beta)X = (\alpha X) + (\beta X)$.
5. There is a zero vector $0 \in \mathbb{V}$ such that $X + 0 = X$ for all $X \in \mathbb{V}$.
6. For each $X \in \mathbb{V}$, there is a unique $-X \in \mathbb{V}$ such that $X + (-X) = 0$.

The reason for all this formalism is that vector space results can be applied to many situations other than n-tuples of real numbers. Some common examples are polynomials and other sets of functions. All that is necessary to apply all the known results about vector spaces to the given situation is to check whether the given set of elements (along with the scalar multiplication and the addition rule) satisfy the axioms above. The particular set of elements of concern to us here is the set of zero-mean rv s (or zero-mean complex rv s) defined on some given probability space. We can add zero-mean rv s in the usual way to get other zero-mean rv s, and we can scale rv s by real scalars, and it is easy to verify that all the above axioms are satisfied. The axioms are similarly satisfied for complex rv s with complex scalars.

It is important to understand that viewing a rv as a vector is very different from the **rv** s that we have been considering. A **rv** $\boldsymbol{Y} = (Y_1, Y_2, \ldots, Y_m)^{\mathsf{T}}$ is an m-tuple of rv s, and is thus a function from the sample space to the space of m-dimensional real vectors. Random vectors are used here primarily as a notational artifice to refer compactly to several rv s as a unit. They satisfy the axioms of a vector space, but are not the quantities of interest in this section. Here each rv Y_j; $1 \le j \le m$ is viewed as a vector in its own right.

An abstract vector space, as defined above, contains no notion of length or orthogonality. To achieve these notions, we must define an *inner product*, $\langle X, Y\rangle$ as an additional operation on a vector space, mapping pairs of vectors into scalars. An *inner-product space* is a vector space with an inner-product operation $\langle X, Y\rangle$ that satisfies the following axioms for all vectors X, Y, Z and all scalars α:

1. $\langle X, Y\rangle = \langle Y, X\rangle^*$ (the complex conjugate is needed only in the complex case).
2. $\alpha\langle X, Y\rangle = \langle \alpha X, Y\rangle$.
3. $\langle X + Y, Z\rangle = \langle X, Z\rangle + \langle Y, Z\rangle$.
4. $\langle X, X\rangle \ge 0$ with equality if and only if $X = 0$.

In an inner-product space, vectors X, Y are defined to be *orthogonal* if $\langle X, Y\rangle = 0$.

The *length* of a vector X is denoted by $\|X\|$ and is defined as $\|X\| = \sqrt{\langle X, X\rangle}$. Similarly, the *distance* between two vectors X and Y is $\|X - Y\| = \sqrt{\langle X - Y, X - Y\rangle}$. For the conventional real inner-product space in which a vector $\boldsymbol{x}$ is an n-tuple of real numbers, $\boldsymbol{x} = (x_1, \dots, x_n)^{\mathsf{T}}$, the inner product $\langle \boldsymbol{x}, \boldsymbol{y}\rangle$ is conventionally taken to be $\langle \boldsymbol{x}, \boldsymbol{y}\rangle = x_1 y_1 + \cdots + x_n y_n$. With this convention, length and orthogonality take on their usual geometric significance. For the real inner-product space of zero-mean rv s, the natural definition for an inner product is the covariance,

$$\langle X, Y\rangle = \mathsf{E}[XY]. \tag{10.85}$$

Note that the covariance satisfies all the axioms for an inner product above. In this inner-product space, two zero-mean rv s are orthogonal if they are uncorrelated. Also, the length of a rv is its standard deviation.

For the complex inner-product space of zero-mean complex rv s, the natural definition of inner product is

$$\langle X, Y\rangle = \mathsf{E}\left[XY^*\right]. \tag{10.86}$$

This also satisfies all the axioms for a complex inner-product space, and in particular $\langle X, X\rangle = \mathsf{E}\left[XX^*\right] \ge 0$, illustrating the need for the complex conjugate above.

Two standard inequalities of inner-product spaces are the Schwarz inequality and the triangle inequality, given respectively by

$$\langle X, Y\rangle \le \|X\|\|Y\| \quad \text{(Schwarz)}, \qquad \|X - Y\| \le \|X\| + \|Y\| \quad \text{(triangle)}. \tag{10.87}$$

These are often useful in working with variances and covariances.

We now return to the problem of finding the LLSE estimate of a zero-mean rv X from the observation of m zero-mean rv s $Y_1, \dots, Y_m$. In general, the LLSE estimate can be a constant plus a linear function, but we have seen that the constant is zero for zero-mean rv s. Thus the estimate here is a linear combination of the observed variables,

$$\widehat{X}(Y_1, \dots, Y_m) = \sum_{j=1}^{m} g_j Y_j. \tag{10.88}$$

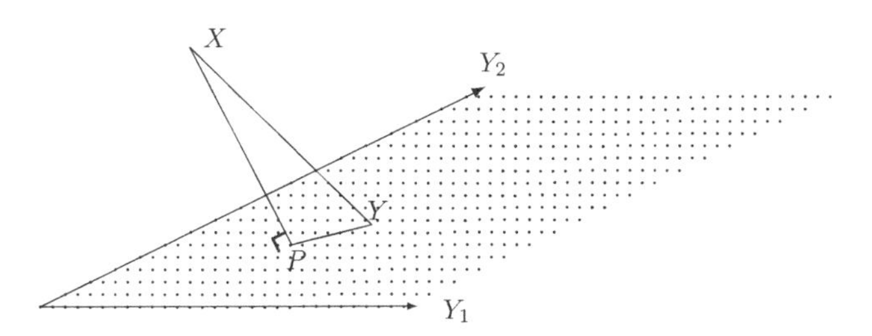

Figure 10.2 P is the projection of the vector X onto the subspace spanned by Y_1 and Y_2. That is, $X - P$ is orthogonal to all points in the subspace, and, as seen geometrically, the distance from X to P is smaller than the distance from X to any other point Y in the subspace. Note that this figure bears almost no resemblance to Figure 10.1. A vector there represented a string of sample values of rv s, whereas the vectors here are individual rv s; sample values do not appear at all.

Viewing the rv s $Y_1, \ldots, Y_m$ as vectors in the inner-product space discussed above, a linear estimate $\widehat{X}(Y_1, \ldots, Y_m)$ must then be in the subspace[5] S spanned by $Y_1, \ldots, Y_m$. The error in the estimate is the vector (i.e., the rv) $\xi = X - \widehat{X}(Y_1, \ldots, Y_m)$. Thus, in terms of this inner product space, the LLSE estimate $\widehat{X}(Y_1, \ldots, Y_m)$ is that point $P \in \mathsf{S}$ that is closest to X.

Finding the closest point in a subspace to a given vector is a fundamental and central problem for inner-product vector spaces. As indicated in Figure 10.2, the closest point is found by dropping a perpendicular from the point X to the subspace. The point where this perpendicular intersects the subspace is called the projection P of X onto the subspace. Formally, P is the *projection* of X onto S if $\langle P - X\, ,\, Y\rangle = 0$ for all $Y \in \mathsf{S}$. The following well-known linear algebra result summarizes this.

Theorem 10.6.1 (Orthogonality principle) *Let X be a vector in an inner-product space V, and let S be a finite-dimensional*[6] *subspace of V. There is a unique vector $P \in \mathsf{S}$ such that $\langle X - P, Y\rangle = 0$ for all $Y \in \mathsf{S}$. That vector P is the closest point in S to X, i.e., $\|X - P\| < \|X - Y\|$ for all $Y \in \mathsf{S}$, $Y \neq P$.*

Discussion The intuitive language and the picture suggest that this theorem is obvious geometrically, but since the theorem refers to an arbitrary inner-product space, it is certainly not obvious. At the same time, this is a central result in linear algebra and there is no point giving a proof here. In fact, for the particular inner-product spaces of interest here, the projection P is the LLSE estimate. We have essentially proven the theorem for that case and derived the LLSE estimate already, and what remains is to explain the connections.

For zero-mean rv s, we have seen that the LLSE estimate $\widehat{X}(Y_1, \ldots, Y_m)$ is a linear combination of the observation rv s $Y_1, \ldots, Y_m$ and thus it is an element of the vector

[5] A subspace of a vector space $\mathcal{V}$ is a subset of elements of $\mathcal{V}$ that constitutes a vector space in its own right. In particular, the linear combination of any two vectors in the subspace is also in the subspace; for this reason, subspaces of vector spaces are often called linear subspaces.

[6] The theorem also holds for infinite-dimensional vector spaces if they satisfy a condition called completeness.

subspace S spanned by the vectors $Y_1, \ldots, Y_m$. Being LLSE means that it is the linear combination that is the closest point in S to the vector X. Thus $\widehat{X}(Y_1, \ldots, Y_m)$ is in fact the projection P of X onto the subspace spanned by $Y_1, \ldots, Y_m$.

According to the theorem, $P = \widehat{X}(Y_1, \ldots, Y_m)$ satisfies $\langle X - P,\, Y\rangle = 0$ for all $Y \in \mathsf{S}$, A necessary and sufficient condition for this is that

$$\langle X - P,\, Y_j\rangle = 0 \qquad \text{for all } j, 1 \le j \le m.$$

Expressing P as $\sum_j g_j Y_j$ and expressing the inner products as covariances, this equation can be rewritten as

$$\mathsf{E}\left[XY_j^*\right] = \sum_k g_k \mathsf{E}\left[Y_k Y_j^*\right].$$

We have used the complex conjugate here so that this covers both the real and complex cases. If we go back to express this in the vector matrix notation of Sections 10.2–10.5, we get $\mathsf{E}\left[X\boldsymbol{Y}^*\right] = \boldsymbol{g}^{\mathsf{T}}[K_{\boldsymbol{Y}}]$, which lets us solve for $\boldsymbol{g}^{\mathsf{T}}$ as $[K_{X\cdot\boldsymbol{Y}} K_{\boldsymbol{Y}}^{-1}]$. Thus the answer is the same as in (10.46) for the real case and (10.84) for the complex case, but perhaps the similarity between the real and complex cases is more transparent than before.

Next, recall that the above discussion has been restricted to viewing *zero-mean* rv s as vectors. This is often appropriate, since rv s with a mean can usually be best viewed as a constant (the mean) plus a fluctuation. It is sometimes useful to remove this restriction and to consider the entire set of rv s in a given probability space as forming a real inner-product space. In this generalization, the inner product is again defined as $\langle X, Y\rangle = \mathsf{E}\left[XY^*\right]$, but the correlation now includes the mean values, i.e., $\mathsf{E}\left[XY^*\right] = \mathsf{E}[X]\,\mathsf{E}\left[Y^*\right] + \mathsf{E}\left[\widetilde{X}\widetilde{Y}^*\right]$.

For any rv X with a non-zero mean, the difference between X and its fluctuation $\widetilde{X}$ is a constant rv equal to $\mathsf{E}[X]$ for all sample points. This constant rv must also be interpreted as a vector in the inner-product space under consideration (since the difference of two vectors must again be a vector). Let D be the rv (i.e., vector) that maps all sample points into unity. Thus $X - \widetilde{X} = \mathsf{E}[X]\, D$ is a constant equal to $\mathsf{E}[X]$. The notion of a linear estimate involving variables with non-zero means is invariably extended to an estimate $\widehat{X}(Y_1, \ldots, Y_m) = \beta + \sum_i \alpha_i Y_i$ of first degree. Writing this in terms of the unit constant rv D,

$$\widehat{X}(Y_1, \ldots, Y_m) = \beta D + \sum_{i=1}^{n} \alpha_i Y_i. \tag{10.89}$$

In terms of vectors, the right-hand side of (10.89) is a vector within the subspace spanned by $D, Y_1, Y_2, \ldots, Y_m$. We want to choose β and $\alpha_1, \ldots, \alpha_n$ so that $P = \beta D + \sum_{j=1}^{n} \alpha_j Y_j$ is the projection of X on the subspace S spanned by $D, Y_1, \ldots, Y_m$, i.e., so that $\langle X - P, Y\rangle = 0$ for all $Y \in \mathsf{S}$. This condition is satisfied if and only if both $\langle X - P, D\rangle = 0$ and $\langle X - P, Y_j\rangle = 0$ for $1 \le j \le n$. Now for any vector Y we have $\langle Y, D\rangle = \mathsf{E}[YD] = \mathsf{E}[Y]$, so the equation $\langle X - P, D\rangle = 0$ can be rewritten as

$$\mathsf{E}[X] = \beta + \sum_{j=1}^{n} \alpha_j \mathsf{E}\left[Y_j\right]. \tag{10.90}$$

Also, $\langle X - P, Y_i \rangle = 0$ can be rewritten as

$$\mathsf{E}[XY_i] = \beta \mathsf{E}[Y_i] + \sum_{j=1}^{n} \alpha_j \mathsf{E}[Y_i Y_j]. \tag{10.91}$$

Exercise 10.9 shows that (10.89) to (10.91) are equivalent to (10.11). Note that in the world of rv s, $\widehat{X}(Y_1, \ldots, Y_m) = \beta + \sum_j \alpha_i Y_j$ is a *first-degree* function of $Y_1, \ldots, Y_m$, whereas it can also be viewed as a *linear* function of $D, Y_1, \ldots, Y_m$. In the vector space world, $\hat{\boldsymbol{X}}(Y_1, \ldots, Y_m) = \beta D + \sum_j \alpha_j Y_j$ must be viewed as a vector in the linear subspace spanned by $D, Y_1, \ldots, Y_m$. The notion of linearity here is dependent on the context.

Next consider a situation in which we want to allow a limited amount of non-linearity into an estimate. For example, suppose we consider estimating X as a linear combination of the constant rv D, the observation variables $Y_1, , Y_m$, and the squares of the observation variables $Y_1^2, \ldots, Y_m^2$. Note that a sample value y_j of Y_j also specifies the sample value y_j^2 of Y_j^2 and the sample value of D is always equal to 1. Our estimate is then

$$\hat{\boldsymbol{X}}(Y_1, \ldots, Y_m) = \beta D + \sum_{j=1}^{n} (\alpha_j Y_j + \gamma_j Y_j^2). \tag{10.92}$$

We wish to choose the scalars β, α_j, and γ_j, $1 \leq j \leq n$, to minimize the expected squared error between X and $\hat{\boldsymbol{X}}(Y_1, \ldots, Y_m)$. In terms of the inner-product space of rv s, we want to estimate X as the projection P of X onto the subspace spanned by the vectors $D, Y_1, \ldots, Y_m, Y_1^2, \ldots, Y_m^2$. It is sufficient for $X - P$ to be orthogonal to D, Y_i, and Y_i^2 for $1 \leq i \leq n$, so that $\langle X, D \rangle = \langle P, D \rangle$, $\langle X, Y_i \rangle = \langle P, Y_i \rangle$ and $\langle X, Y_i^2 \rangle = \langle P, Y_i^2 \rangle$ for $1 \leq i \leq n$. Using (10.92), the coefficients β, α_j, and γ_j must satisfy

$$\mathsf{E}[X] = \beta + \sum_{j=1}^{n} \left(\alpha_j \mathsf{E}[Y_j] + \gamma_j \mathsf{E}\left[Y_j^2\right] \right), \tag{10.93}$$

$$\mathsf{E}[XY_i] = \beta \mathsf{E}[Y_i] + \sum_{j=1}^{n} \left(\alpha_j \mathsf{E}[Y_i Y_j] + \gamma_j \mathsf{E}\left[Y_i Y_j^2\right] \right), \tag{10.94}$$

$$\mathsf{E}\left[XY_i^2\right] = \beta \mathsf{E}\left[Y_i^2\right] + \sum_{j=1}^{n} \left(\alpha_j \mathsf{E}\left[Y_i^2 Y_j\right] + \gamma_j \mathsf{E}\left[Y_i^2 Y_j^2\right] \right). \tag{10.95}$$

This example should make one even more careful about the word 'linear.' The projection P is a linear function of the vectors D, Y_i, Y_i^2, $1 \leq i \leq n$, but in the underlying probability space with rv s, Y_i^2 is a non-linear function of Y_i.

This example can be extended with cross terms, cubic terms, and so forth. Ultimately, if we consider the subspace of all rv s $g(Y_1, ..., Y_m)$, we can view the situation much more simply by viewing $\widehat{X}(Y_1, \ldots, Y_m)$ as an arbitrary function of the observation $Y_1, \ldots, Y_m$. The resulting estimate is the MMSE estimate, which, as we have seen, is the conditional mean of X, conditional on the observed sample values of $Y_1, \ldots, Y_m$.

As we progress from linear (or first-order) estimates to estimates with square terms to estimates with yet more terms, the mean-squared error must be non-increasing, since the simpler estimates are non-optimal versions of the more complex estimates with some of the terms set to zero. Viewing this geometrically, in terms of projecting from X onto a subspace S, the distance from X to the subspace must be non-increasing as the subspace is enlarged.

10.7 MAP estimation and sufficient statistics

Another approach to estimation is to look for MAP estimates, i.e., estimates that maximize the a posteriori probability of $\boldsymbol{X}$ conditional on a sample value $\boldsymbol{y}$. These techniques are very similar to those used for detection in Chapter 8. We emphasize the problem in which $\boldsymbol{Y} = [H]\boldsymbol{X} + \boldsymbol{Z}$, where $[H]$ is a given matrix and $\boldsymbol{X}$ and $\boldsymbol{Z}$ are independent of each other and are zero mean, Gaussian, and non-singular. We know that $\mathsf{f}_{\boldsymbol{X}|\boldsymbol{Y}}(\boldsymbol{x} \mid \boldsymbol{y})$ is then Gaussian for each $\boldsymbol{y}$ and thus the MAP estimate is the same as the MMSE estimate. Thus the MAP estimate is already known, but the derivation will be instructive and can be used (in more complicated form) for non-Gaussian cases, where MMSE and MAP estimates are often different.

Starting from first principles, $\boldsymbol{Y}$, conditional on a given $\boldsymbol{X} = \boldsymbol{x}$, is a Gaussian rv with covariance $[K_{\boldsymbol{Z}}]$ and mean $[H]\boldsymbol{x}$. The density, or likelihood, is then

$$\mathsf{f}_{\boldsymbol{Y}|\boldsymbol{X}}(\boldsymbol{y} \mid \boldsymbol{x}) = \frac{\exp\left(-\frac{1}{2}(\boldsymbol{y} - [H]\boldsymbol{x})^{\mathsf{T}}[K_{\boldsymbol{Z}}^{-1}](\boldsymbol{y} - [H]\boldsymbol{x})\right)}{(2\pi)^{n/2}\sqrt{\det[K_{\boldsymbol{Z}}]}}$$

$$= \frac{\exp\left(-\frac{1}{2}\left(\boldsymbol{y}^{\mathsf{T}}[K_{\boldsymbol{Z}}^{-1}]\boldsymbol{y} - ([H]\boldsymbol{x})^{\mathsf{T}}[K_{\boldsymbol{Z}}^{-1}]\boldsymbol{y} - \boldsymbol{y}^{\mathsf{T}}[K_{\boldsymbol{Z}}^{-1}H]\boldsymbol{x} + ([H]\boldsymbol{x})^{\mathsf{T}}[K_{\boldsymbol{Z}}^{-1}H]\boldsymbol{x}\right)\right)}{(2\pi)^{n/2}\sqrt{\det[K_{\boldsymbol{Z}}]}}.$$

Note that the second and third terms in the exponent above are transposes of each other and are one-dimensional; thus they must be the same. Also the first term is a function only of $\boldsymbol{y}$, so we can combine it with the denominator into some function $u(\boldsymbol{y})$, yielding

$$\mathsf{f}_{\boldsymbol{Y}|\boldsymbol{X}} = u(\boldsymbol{y})\exp\left(\boldsymbol{x}^{\mathsf{T}}[H^{\mathsf{T}}K_{\boldsymbol{Z}}^{-1}]\boldsymbol{y} - \frac{1}{2}(\boldsymbol{x}^{\mathsf{T}}[H^{\mathsf{T}}K_{\boldsymbol{Z}}^{-1}H]\boldsymbol{x}\right). \tag{10.96}$$

The *likelihood ratio* will then be taken as $\Lambda(\boldsymbol{y},\boldsymbol{x}) = \mathsf{f}_{\boldsymbol{Y}|\boldsymbol{X}}(\boldsymbol{y} \mid \boldsymbol{x})/\mathsf{f}_{\boldsymbol{Y}|\boldsymbol{X}}(\boldsymbol{y} \mid \boldsymbol{x}_o)$, where $\boldsymbol{x}_o$ is taken to be 0. Recall that for binary detection, the likelihood ratio was a function of $\boldsymbol{y}$ only, whereas here it is a function of both $\boldsymbol{y}$ and $\boldsymbol{x}$. From (10.96), with $\boldsymbol{x}_o = 0$,

$$\Lambda(\boldsymbol{y},\boldsymbol{x}) = \exp\left(\boldsymbol{x}^{\mathsf{T}}[H^{\mathsf{T}}K_{\boldsymbol{Z}}^{-1}]\boldsymbol{y} - \frac{1}{2}\boldsymbol{x}^{\mathsf{T}}[H^{\mathsf{T}}K_{\boldsymbol{Z}}^{-1}H]\boldsymbol{x}\right). \tag{10.97}$$

The ML estimate, $\hat{x}_{ML}(\boldsymbol{y})$ can be calculated from the likelihood ratio by maximizing $\Lambda(\boldsymbol{y},\boldsymbol{x})$ over $\boldsymbol{x}$ for the given observation $\boldsymbol{y}$. Similarly, the MAP estimate can be calculated by maximizing $\Lambda(\boldsymbol{y},\boldsymbol{x})\mathsf{f}_{\boldsymbol{X}}(\boldsymbol{x})$ over $\boldsymbol{x}$. Also, $\mathsf{f}_{\boldsymbol{X}|\boldsymbol{Y}}(\boldsymbol{x} \mid \boldsymbol{y})$ can be found from

$$\mathsf{f}_{\boldsymbol{X}|\boldsymbol{Y}}(\boldsymbol{x} \mid \boldsymbol{y}) = \frac{\Lambda(\boldsymbol{y},\boldsymbol{x})\mathsf{f}_{\boldsymbol{X}}(\boldsymbol{x})}{\int \Lambda(\boldsymbol{y},\boldsymbol{x}')\mathsf{f}_{\boldsymbol{X}}(\boldsymbol{x}')d\boldsymbol{x}'}. \tag{10.98}$$

The advantage of using the likelihood ratio rather than the likelihood alone is that $u(\boldsymbol{y})$ in (10.96) is eliminated. Note also that the likelihood ratio does not involve the a priori distribution on $\boldsymbol{X}$, and thus, like $\mathsf{f}_{\boldsymbol{Y}|\boldsymbol{X}}(\boldsymbol{y} \mid \boldsymbol{x})$, it can be used both where $\mathsf{f}_{\boldsymbol{X}}(\boldsymbol{x})$ is unknown or where we want to look at a range of choices for $\mathsf{f}_{\boldsymbol{X}}(\boldsymbol{x})$.

For estimation problems, as with detection problems, a *sufficient statistic* is some function of the observation $\boldsymbol{y}$ from which the likelihood ratio can be calculated for all $\boldsymbol{x}$. Thus, from (10.97), we see that $[H^{\mathsf{T}}K_{\boldsymbol{Z}}^{-1}]\boldsymbol{y}$ is a sufficient statistic for $\boldsymbol{x}$ conditional on $\boldsymbol{y}$. Note that both $[H^{\mathsf{T}}K_{\boldsymbol{Z}}^{-1}]\boldsymbol{y}$ and $\boldsymbol{x}$ have the dimension n, whereas $\boldsymbol{y}$ has dimension m. If $n < m$, as is common in communication situations, then the sufficient statistic $[H^{\mathsf{T}}K_{\boldsymbol{Z}}^{-1}]\boldsymbol{y}$ has reduced the m-dimensional observation space down to an n-dimensional space from which the estimate or decision can be made.

It is important to recognize that nothing is lost, relative to finding $\Lambda(\boldsymbol{Y})$ or $\mathsf{f}_{\boldsymbol{X}|\boldsymbol{Y}}$ if $[H^{\mathsf{T}}K_{\boldsymbol{Z}}^{-1}]\boldsymbol{y}$ is found from the observation $\boldsymbol{y}$ and then $\boldsymbol{y}$ is discarded; all of the relevant information has been extracted from $\boldsymbol{y}$. There are usually many choices of sufficient statistics. For example, the observation $\boldsymbol{y}$ itself is a sufficient statistic, and any invertible transformation of $\boldsymbol{y}$ is also sufficient. Also, any invertible transformation of $[H^{\mathsf{T}}K_{\boldsymbol{Z}}^{-1}]\boldsymbol{y}$ is a sufficient statistic. What is important about $[H^{\mathsf{T}}K_{\boldsymbol{Z}}^{-1}]\boldsymbol{y}$ is that for $n < m$, the dimensionality has been reduced from the raw observation $\boldsymbol{y}$.

The estimate in (10.54) can be viewed as first finding the sufficient statistic $\boldsymbol{s} = [H^{\mathsf{T}}K_{\boldsymbol{Z}}^{-1}]\boldsymbol{y}$, and then finding the MMSE from $\boldsymbol{s}$, i.e., $\widehat{X}(\boldsymbol{y}) = [K_{\boldsymbol{\xi}}]\boldsymbol{s}$. We now try to understand why the sufficient statistic $\boldsymbol{s}$ is operated on by $[K_{\boldsymbol{\xi}}]$ to find $\hat{\boldsymbol{x}}$. Using $\boldsymbol{s} = [H^{\mathsf{T}}K_{\boldsymbol{Z}}^{-1}]\boldsymbol{y}$ in (10.96), $\mathsf{f}_{\boldsymbol{XY}}(\boldsymbol{xy})$ can be expressed as

$$\mathsf{f}_{\boldsymbol{XY}}(\boldsymbol{xy}) = \mathsf{f}_{\boldsymbol{X}}(\boldsymbol{x})\mathsf{f}_{\boldsymbol{Y}|\boldsymbol{X}}(\boldsymbol{y} \mid \boldsymbol{x}) = u(\boldsymbol{y}) \exp\left(\boldsymbol{x}^{\mathsf{T}}\boldsymbol{s} - \frac{1}{2}\boldsymbol{x}^{\mathsf{T}}[H^{\mathsf{T}}K_{\boldsymbol{Z}}^{-1}H]\boldsymbol{x} - \frac{1}{2}\boldsymbol{x}^{\mathsf{T}}[K_{\boldsymbol{X}}^{-1}]\boldsymbol{x}\right), \tag{10.99}$$

where the denominator part of $\mathsf{f}_{\boldsymbol{X}}(\boldsymbol{x})$ has been incorporated into $u(\boldsymbol{y})$. Letting $u_1(\boldsymbol{y}) = u(\boldsymbol{y})/\mathsf{f}_{\boldsymbol{Y}}(\boldsymbol{y})$,

$$\mathsf{f}_{\boldsymbol{X}|\boldsymbol{Y}}(\boldsymbol{x} \mid \boldsymbol{y}) = u_1(\boldsymbol{y}) \exp\left(\boldsymbol{x}^{\mathsf{T}}\boldsymbol{s} - \frac{1}{2}\boldsymbol{x}^{\mathsf{T}}([H^{\mathsf{T}}K_{\boldsymbol{Z}}^{-1}H] + [K_{\boldsymbol{X}}^{-1}])\boldsymbol{x}\right). \tag{10.100}$$

Defining $[B] = [H^{\mathsf{T}}K_{\boldsymbol{Z}}^{-1}H] + [K_{\boldsymbol{X}}^{-1}]$, we can complete the square in the exponent of (10.100),

$$\mathsf{f}_{\boldsymbol{X}|\boldsymbol{Y}}(\boldsymbol{x} \mid \boldsymbol{y}) = u_1(\boldsymbol{y}) \exp\left(-\frac{1}{2}\left(\boldsymbol{x} - [B^{-1}]\boldsymbol{s}\right)^{\mathsf{T}}[B](\boldsymbol{x} - [B^{-1}]\boldsymbol{s})\right), \tag{10.101}$$

where the quadratic term in $\boldsymbol{s}$ has been incorporated into $u_1(\boldsymbol{y})$. This makes it clear that for any given $\boldsymbol{y}$, and thus any given $\boldsymbol{s} = [H^{\mathsf{T}}K_{\boldsymbol{Z}}^{-1}]\boldsymbol{y}$, we can express $\mathsf{f}_{\boldsymbol{X}|\boldsymbol{Y}}(\boldsymbol{x} \mid \boldsymbol{y})$ as $\mathrm{N}([B^{-1}]\boldsymbol{s}, [B^{-1}])$. This confirms (10.54) and (10.55), i.e., that $[B^{-1}]$ is the covariance of the estimation error, and $[B^{-1}]\boldsymbol{s}$ is the MMSE estimate. Thus this gives us nothing new, but the derivation is more insightful than the strictly algebraic derivation in Exercise 10.6.

10.8 Parameter estimation

We now focus on estimation problems in which there is no appropriate model for a priori probabilities on the quantity x to be estimated. We consider x to be a real-valued parameter, which can be viewed either as a sample value of a rv with unknown distribution or simply as an unknown value. If x is viewed as an unknown value, we do not have an overall probability space to work with – we have only a separate probability space for each value of x. We choose here to take the viewpont that X is a rv of unknown distribution, and consider only results that do not depend on any a priori distribution. By taking this choice, we can use the same notation for conditional rv s that we have used throughout, with only a little added care to remember that a priori distributions on X cannot be used in the results. This approach also allows us to compare parameter estimation to Bayesian estimation using a common notation for each problem.

Note that we will only look at estimation of a single parameter, $x \in \mathbb{R}$. This is considerably less general than the Bayesian estimation of earlier sections which considered estimation of **rv** s and complex **rv** s. As will be seen, a number of conceptual difficulties arise even with the estimation of a single parameter, and it seems preferable to understand those difficulties rather than hiding them in greater generality.

We start with a particularly simple problem. Suppose that $Y = X + Z$, where the sample value x of X is the parameter to be estimated and Z is a continuous zero-mean rv, independent of X, with the PDF $\mathsf{f}_Z(z)$. We want to estimate x from the observation of a sample value y of Y. The likelihood, which we abbreviate $\mathsf{f}(y|x)$, is then equal to $\mathsf{f}_Z(y-x)$. In other words, Y can be viewed as a rv of known fluctuation Z but unknown mean x. Statisticians view this as a problem of estimating the mean of a distribution that is otherwise known.

ML estimation is one reasonable approach to this problem, i.e., $\hat{x}_{\text{ML}}(y) = \arg\max_x f(y|x)$. Since $f(y|x) = f_Z(y-x)$, it can be seen that $\hat{x}_{\text{ML}}(y) = y + \text{mode}(Z)$. Another reasonable approach, since $\overline{Z} = 0$, is to choose $\hat{x}(y) = y$. With this choice, $\mathsf{E}[Y|X=x] = x$. For each parameter value x, then, the mean-squared error, conditional on x, is $\mathsf{E}\left[(Y-x)^2\right] = \mathsf{E}\left[Z^2\right] = \mathsf{VAR}[Z]$. If we choose the ML estimate, the mean-squared error is $\mathsf{VAR}[Z] + \text{mode}^2(Z)$. Thus if our objective is to minimize the mean-squared error, and if $\text{mode}(Z) \neq 0$, the estimate $\hat{x}(y) = y$ is preferable to ML for each value of x. Of course, if our preference is to choose the x for which $\mathsf{f}(y|x)$ is largest, then ML estimation is preferable.

The estimate $\hat{x}(y) = y$ is very similar to the MMSE estimator defined in Section 10.1.1 and shown in (10.3) to satisfy $\hat{x}_{\text{MMSE}}(y) = \mathsf{E}\left[X|Y{=}y\right]$ for each y. MMSE estimation is based on an a priori distribution for X which is unknown or undefined here. It turns out, however, that the estimator $\hat{x}(y) = y$ is the limit of MMSE estimators if, for example, we start with an a priori distribution $\mathrm{N}(0, \sigma^2)$ and consider letting σ^2 approach ∞. There is no limiting distribution as $\sigma^2 \to \infty$, but there will be a limiting MMSE estimator. To see this, note that the distribution of Y for $Y = X + Z$ becomes uniform in the limit of a uniform distribution on X. Thus in this limit, $\mathsf{E}[Y|X{=}u] = \mathsf{E}[X|Y{=}u]$ for each $u \in \mathbb{R}$.

To be able to avoid this excursion into Bayesian statistics and to treat more general parameter estimation problems, classical statisticians define the bias of an estimator $\hat{x}$ as

$$b_{\hat{x}}(x) = \mathsf{E}\left[\widehat{X}(Y) - x|X = x\right]. \tag{10.102}$$

An estimator is unbiased if $b_{\hat{x}}(x) = 0$ for all $x \in \mathbb{R}$. We have seen, for the simple problem here, that $\hat{x}(y) = y$ is unbiased, and it is not hard to convince oneself that $\hat{x}(y) = y$ is the only unbiased estimator for this case.

Recall from (10.13) that, for $Y = X + Z$ with $Z \sim \mathrm{N}(0, \sigma_Z^2)$ and $X \sim \mathrm{N}(0, \sigma_X^2)$, the MMSE estimate is given by $\hat{x}(y) = \sigma_X^2 y/(\sigma_Z^2 + \sigma_X^2)$. As we have observed, this approaches y as $\sigma_X^2 \to \infty$. For finite σ_X^2, however, this estimator is biased, reducing the magnitude of the estimate from the unbiased value. This is explained, of course, by the fact that in the Bayesian model with $X \sim \mathrm{N}(0, \sigma_X^2)$, smaller magnitudes of X are more likely than large ones, allowing the reduction of overall mean-squared error by biasing the estimate toward smaller magnitudes.

The ML estimator, according to the definition in (10.102), is also biased if the mode of Z is non-zero. However, the ML estimator is clearly not biased according to the everyday definition of bias; it treats all values of x in the same way and in fact is the special case of MAP that is specifically used to weight all values of x equally.

To summarize the above discussion, parameter estimation is used in place of Bayesian estimation when one wants to be careful not to bias the results by prior notions about the parameter. The definition of bias in (10.102) is partly successful in this regard but is sometimes misleading, so it should be used with great caution. The following examples make this even more clear.

Example 10.8.1 Suppose $Y = x^2 + Z$, where $Z \sim \mathrm{N}(0, 1)$ and x is a parameter. We can estimate x^2 from a sample value of Y, but the sample provides no clue about the sign of x. It is also clear from (10.102), by comparing any given $x > 0$ with $-x$, that no unbiased estimator exists. ML does not work any better, since $\mathsf{f}(y|x) = \mathsf{f}(y|-x)$. This also suggests that parameter estimation requires some common sense in addition to formulas.

Example 10.8.2 Suppose $Y = x^3 + Z$, where $Z \sim \mathrm{N}(0, 1)$. We can estimate x^3 as y, which as an estimator for x^3 is both unbiased and ML. The corresponding sensible estimator for x is $y^{1/3}$. This is the ML estimate, but turns out to be biased according to (10.102). An unbiased estimator appears to exist and to be slightly smaller than $y^{1/3}$ for $y > 0$, but there seems to be no reason to calculate it and no reason to believe that it is less biased than $y^{1/3}$ in any sense other than the formal definition.

The above examples and discussion are intended to make it clear that even in the very simple examples we have looked at, there is no compelling argument for using unbiased estimators if and when they exist, and no general conceptual approach to choosing one estimator over another. It is thus a pleasant surprise that there is a lower bound, called the Cramer–Rao bound, to the mean-squared estimation error. It is a function of both the parameter value and of the bias as defined in (10.102), and is quite simple for the case of unbiased estimators. Unfortunately, there are very simple estimation problems for which biased estimators are uniformly better for all x (in the mean-squared error sense) than those that are unbiased according to (10.102).

10.8.1 Fisher information and the Cramer–Rao bound

This subsection develops the concept of Fisher information and the Cramer–Rao bound for the case of a single parameter x and an observation Y that is a rv with a known conditional density $\mathsf{f}(y|x)$. We later look at vector observations where $Y_1, \ldots, Y_n$ are independent identically distributed (IID) conditional on x.

We start by defining the *score*; this is a function $v_x(y)$ of the parameter x and the sample observation y given by

$$v_x(y) = \frac{\partial \ln(\mathsf{f}(y|x))}{\partial x}. \tag{10.103}$$

Restricting our attention to cases in which $\mathsf{f}(y|x)$ is non-zero and differentiable with respect to x for all y, x,

$$v_x(y) = \frac{1}{\mathsf{f}(y|x)} \frac{\partial \mathsf{f}(y|x))}{\partial x}. \tag{10.104}$$

One interesting feature about $v_x(y)$ is that if it is zero for a given x, y, then that x is a stationary point of $\ln(\mathsf{f}(y|x))$ as a function of x for the given y. In typical cases, that x will then be the maximum of $\ln \mathsf{f}(y|x)$ over x and consequently also the maximum of $\mathsf{f}(y|x)$ for the given y. In other words, that x is the ML estimate for that y.

Let $V_x(Y)$ be the conditional rv defined by $v_x(y)$. An interesting feature of $V_x(Y)$ is that it is zero mean conditional on each x. To see this,

$$\begin{aligned} E[V_x(Y)|X{=}x] &= \int \mathsf{f}(y|x) V_x(y)\, dy = \int \mathsf{f}(y|x) \frac{1}{\mathsf{f}(y|x)} \frac{\partial \mathsf{f}(y|x)}{\partial x}\, dy \\ &= \frac{\partial \int \mathsf{f}(y|x)\, dy}{\partial x} = \frac{\partial 1}{\partial x} = 0. \end{aligned} \tag{10.105}$$

We will not attempt to be careful about conditions under which the integration and differentiation can be interchanged, nor about how to handle conditions under which the PDF goes to 0. In other words, we are not establishing a theorem here, but merely the statement that under many circumstances, not to be precisely defined, $\mathsf{E}[V_x(Y)|X{=}x] = 0$.

To get an idea about the nature of $V_x(Y)$, consider $Y = x + Z$, where $Z \sim \mathrm{N}(0, \sigma_Z^2)$. Then $\ln \mathsf{f}(y|x)$ is quadratic in $y - x$ (plus a constant). Taking the derivative,

$$V_x(Y) = \frac{Y - x}{\sigma_Z^2} \qquad \text{for } Y = x + Z \text{ and } Z \sim \mathrm{N}(0, \sigma_Z^2). \tag{10.106}$$

Thus, conditional on a given x, $V_x(Y)$ is a rv that is linear in Y around $Y = x$. Thus, conditional on a given x, $V_x(Y)$ is Gaussian with mean 0 and (since it is scaled down from Y by σ_Z^2) it has variance $1/\sigma_Z^2$. For a given sample value y of Y, $v_x(y)$ is linear in x around y and y is the ML estimate of x. If Z is non-Gaussian, the typical behavior is similar but non-linear.

The *Fisher information* $J(x)$ for a parameter estimation problem is defined as the variance (conditional on $X = x$) of the rv $V_x(Y)$, i.e.,

$$J(x) = \mathsf{VAR}\left[V_x(Y)|X = x\right]. \tag{10.107}$$

For the Gaussian example of (10.106), Y, conditional on x, has variance σ_Z^2. Since (10.106) shows that $V_x(Y)$ is scaled down from Y by σ_Z^2, we have $J(x) = 1/\sigma_Z^2$. Since better estimates of x can be made when σ_Z^2 is small, and thus $J(x)$ is large, there is some indication for this example that $J(x)$ can be viewed as a measure of how well x can be estimated.

Another indication of the relationship of $J(x)$ to how well x can be estimated comes from the following alternative representation of $J(x)$:

$$J(x) = -\mathsf{E}\left[\frac{\partial^2}{\partial x^2}\ln \mathsf{f}(Y|x)|X{=}x\right]. \tag{10.108}$$

This alternative representation can be derived by the same type of manipulation as in (10.105) and again requires the derivatives to be well behaved in a manner that we will not make precise. The interpretation of this, for a given x, is that $\mathsf{f}(y|x)$ will typically be large in the vicinity of the y, say y_x, for which x is the ML estimate. The second derivative of $\ln \mathsf{f}(y \mid x)$ in the region around y_x gives an indication of the range of x over which that ML is large. We come back to this when we talk about multiple observations, and it will make a little more sense there.

Another indication of why $J(x)$ might be viewed as measuring the 'information' about x in the observation of Y comes from the Cramer–Rao bound which follows. This is not quite a theorem, since the conditions under which it holds are quite messy and not given. It is simpler to check whether it holds for any given model of interest than to establish a general result.

Cramer–Rao bound For each x, and any estimate $\widehat{X}(Y)$,

$$\mathsf{VAR}\left[\widehat{X}(Y)|X{=}x\right] \geq \frac{\left[1 + \frac{\partial b_{\widehat{X}}(x)}{\partial x}\right]^2}{J(x)}. \tag{10.109}$$

Proof

$$\begin{aligned}
\mathsf{E}\left[V_x(Y)\widehat{X}(Y)|X{=}x\right] &= \int \mathsf{f}(y|x)V_x(y)\hat{x}(y)dy \\
&= \int \frac{\partial \mathsf{f}(y|x)}{\partial x}\hat{x}(y)dy \qquad \text{(from (10.104))} \\
&= \frac{\partial}{\partial x}\int \mathsf{f}(y|x)\hat{x}(y)dy \qquad \text{(interchange } \partial/\partial x \text{ with } \textstyle\int_y \text{)} \\
&= \frac{\partial \mathsf{E}\left[\widehat{X}(\boldsymbol{Y})|X{=}x\right]}{\partial x} = \frac{\partial[x + b_{\widehat{X}}(x)]}{\partial x} \qquad \text{(from (10.102))} \\
&= 1 + \frac{\partial b_{\widehat{X}}(x)}{\partial x}.
\end{aligned} \tag{10.110}$$

Using the Schwarz inequality, (10.87) on the covariance between $V_x(Y)$ and $\widehat{X}(Y)$ (for given x), and recalling that $V_x(Y)$ is zero mean for the given x,

$$\left(\mathsf{E}\left[V_x(Y)\widehat{X}_x(Y)|X{=}x\right]\right)^2 \le \mathsf{VAR}\left[V_x(Y)|X{=}x\right]\mathsf{VAR}\left[\widehat{X}(Y)|X{=}x\right]$$
$$= J(x)\mathsf{VAR}\left[\widehat{X}(Y)|X{=}x\right]. \tag{10.111}$$

Combining (10.110) and (10.111) yields (10.109). □

The Cramer–Rao bound leads to a simple bound on the mean-squared estimation error for any given x, i.e., $\mathsf{E}[(\widehat{X}(Y)-x)^2|X{=}x]$. This is equal to the conditional variance of $\hat{X}(Y)$ (given $X = x$) plus the square of the conditional mean of $\hat{X}(Y) - x$. Since that conditional mean is the bias, we have

$$\mathsf{E}\left[(\widehat{X}(Y)-x)^2|X{=}x\right] \ge \frac{\left[1+\frac{\partial b_{\widehat{X}}(x)}{\partial x}\right]^2}{J(x)} + \left[b_{\widehat{X}}(x)\right]^2. \tag{10.112}$$

Finally, if we restrict attention to the special case of unbiased estimates, (10.112) becomes

$$\mathsf{E}\left[(\widehat{X}(Y)-x)^2|X{=}x\right] \ge \frac{1}{J(x)}. \tag{10.113}$$

This is the usual form of the Cramer–Rao inequality. Many people mistakenly believe that unbiased estimates must be better than biased ones, and that therefore biased estimates also satisfy (10.113), but this is untrue. To see this, note that the bias can be negative, and thus $(1 + \partial b_{\hat{X}}(x)/\partial x)^2$ can be less than 1. A particularly striking example arises when we look again at the example $Y = X + Z$, where $Z \sim \mathrm{N}(0, \sigma_Z^2)$. Consider the MMSE estimator for an a priori distribution $X \sim \mathrm{N}(0, 1)$. Then, as shown in Exercise 10.15, the mean-squared error satisfies the Cramer–Rao bound with equality for all x, but the Cramer–Rao bound now varies with the bias, which varies with x. The mean-squared error for this MMSE estimator is smaller than that for the unbiased ML estimator for small x and larger for large x.

To summarize, the Fisher information exists for a wide variety (but not all) parameter estimation problems with a single parameter in $\mathbb{R}$ and a single observation in $\mathbb{R}$. For those problems, the Cramer–Rao bound exists as a function of both x and the bias $b_{\hat{X}}(x)$. The Cramer–Rao bound is simplest where the bias is 0, but unbiased estimators do not exist in all cases and are undesirable in many cases. Fortunately, however, the Cramer–Rao bound is valid both for estmators based on a priori distributions and those based on no a priori assumptions. The effect of the a priori distribution in the Cramer–Rao bound is represented by the bias, and the bound is valid for the given bias whether or not the parameter has that or any other a priori distribution.

10.8.2 Vector observations

Up until this point, we have considered only parameter estimation problems with a single real parameter and a single real observation. Generalizing to the case of vector observations, with $\boldsymbol{Y} = (Y_1, \dots, Y_m)^{\mathsf{T}}$, comes at virtually no cost but with several new insights. In particular, $V_x(\boldsymbol{Y}) = \partial \ln \mathsf{f}(\boldsymbol{y}|x)/\partial x$ is a one-dimensional rv defined in terms of the rv

$\boldsymbol{Y}$ as before. Also $V_x(\boldsymbol{Y})$ is zero mean for each x as before and the Fisher information is given both as the conditional variance of $V_x(\boldsymbol{Y})$ and as the second derivative expression in (10.108) as before. The bias is defined as in (10.102), and the Cramer–Rao bound follows as before.

Vector observations become particularly simple to work with if the observations $Y_1, \ldots, Y_m$ are IID conditional on the parameter, $X = x$. In this case, $\mathsf{f}_{\boldsymbol{Y}^m|X}(\boldsymbol{y}^m|x) = \prod_{\ell=1}^m \mathsf{f}_{Y_\ell|X}(y_\ell|x)$ and

$$\ln \mathsf{f}_{\boldsymbol{Y}^m|X}(\boldsymbol{y}^m|x) = \sum_{\ell=1}^{m} \ln \mathsf{f}_{Y_\ell|X}(y_\ell|x).$$

Taking the partial derivative with respect to x, the score for a vector observation is the sum of the scores for the individual scores, i.e.,

$$V_x(\boldsymbol{Y}^m) = \sum_{\ell=1}^{m} V_x(Y_\ell).$$

These are independent rv s, so the laws of large numbers and the central limit theorem (CLT) can be applied (all conditional on any given x).[7]

The CLT is of particular interest here, since it says that $V_x(\boldsymbol{Y}^m))/\sqrt{m}$ tends to a Gaussian distribution $\mathrm{N}\,(0, \sigma^2_{V_x(Y)}) = \mathrm{N}\,(0, J(x))$, where $J(x)$ is the Fisher information for a single observation. This is significant in two ways. First, it means that $J(x)$ provides an approximation to the *distribution* of the score for large m rather than just the variance. This in turn is useful in understanding large deviation results for parameter estimation. Second, the Cramer–Rao bound is tight in the Gaussian case, so it becomes asymptotically tight in the IID vector case. Also, the ambiguities about bias become far less important in the vector case, so the Fisher information, which is linear in the number of observations, shows that the expected mean-squared error asymptotically decreases with the number of observations as $1/(mJ(x))$.

10.8.3 Information

The Fisher information is useful in parameter estimation but provides a mnemonic name for a quantity that has only limited similarity to the everyday notion of information. Unfortunately, it also has some superficial resemblance to the information measure (sometimes called Shannon information) of information theory and communication technology. This superficial resemblance usually causes more confusion than mutual benefit.

To try to circumvent this confusion, we give a very brief and non-technical discussion of information as it arises in information theory;[8] unfortunately, information theory is a major subject which cannot be adequately developed here without a major influx of new ideas and several new chapters.

[7] See Chapters 8 and 10 of [20] for a more detailed treatment of these topics.

[8] Shannon's original monograph creating information theory [26] is highly readable and provides a beautiful example of combining engineering and mathematics to construct an elegant theory which in retrospect has created and guided modern telecommunication. Suggested texts are [12] and [5].

We are all used to voice, video, text, etc. being converted to binary strings of data before transmission over a network, processing in a computer, or storage anywhere, and we are used to referring to those bits as information. Our internet connections, for example, are referred to in megabits or gigabits per second and are viewed as information rates. Defining these as data rates rather than as information rates might be more sensible, but the word *information* makes the engineers in this field feel better. There are interesting information theoretic questions about how to convert voice, video, etc. into binary data, and this is a subject called source coding which we do not discuss here.

The question of interest here (and the question generating confusion wth Fisher information) is that of channel coding, i.e., how are these bit streams processed for transmission over the links (channels) of a network. The input to each such channel can be viewed as a stochastic process and the output as another stochastic process. The output process differs from the input because of noise and other disturbances on the channel.

In the simplest case, we can view the channel input as a discrete-time stochastic process $\{X_n;\ n \geq 1\}$ and the output as another discrete-time stochastic process $\{Y_n;\ n \geq 1\}$. Consider the case in which each Y_n, conditional on X_n, is independent of 'everything else.' For example, we might have $Y_n = X_n + Z_n$ where $\{Z_n;\ n \geq 1\}$ is IID and also independent of $\{X_n;\ n \geq 1\}$. The channel in this case is defined by the one-dimensional conditional distribution of outputs given inputs, called the channel transition probabilities and denoted as $\mathsf{f}_{Y|X}(y|x)$. The input stochastic process is determined by the bits to be transmitted and by the modulation and coding at the input.

If we look at the channel at a single discrete time, say time n, and assume there is some arbitrary PDF $\mathsf{f}_X(x)$ defined at the input, then this input plus the channel transition probabilities, $\mathsf{f}_{Y|X}(y|x)$, provide a full probabilistic characterization of the channel and its input at time n. The *mutual information* on the channel at time n is then defined to be

$$I(X;Y) = \int_{x,y} \mathsf{f}_X(x)\mathsf{f}_{Y|X}(y|x) \log_2 \frac{\mathsf{f}_{Y|X}(y|x)}{\mathsf{f}_Y(y)}\, dx\, dy, \tag{10.114}$$

where $f_Y(y) = \int_x \mathsf{f}_X(x)\mathsf{f}_{Y|X}(y|x)\, dy$.

Before describing what this mutual information has to do with the problem of transmitting binary digits from input to output, we note that the input distribution $\mathsf{f}_X(x)$ is not a characteristic of the channel and potentially can be changed depending on how we use the channel. Thus it makes sense to define the capacity of the channel (per unit time) by

$$C = \max_{\mathsf{f}_X} I(X;Y),$$

where the maximization is usually done subject to some constraints such as $\mathsf{E}\left[X^2\right] \leq \sigma^2$. A particularly well-known case of this maximization is the Gaussian noise case where $Z \sim \mathrm{N}\,(0, \sigma_Z^2)$. Then with the constraint $\mathsf{E}\left[X^2\right] \leq \sigma^2$, it turns out that $C = \frac{1}{2}\log_2(1 + \sigma^2/\sigma_Z^2)$.

The connection between this definition of mutual information, maximized over a single letter input distribution, and the problem of transmitting binary digits over the channel comes from the use of coding. A block code of rate R and block length n is defined as a mapping from the $2^{\lceil nR \rceil}$ binary strings of length $\lceil nR \rceil$ to a set of $2^{\lceil nR \rceil}$ channel input tuples. The sample value of the $\lceil nR \rceil$ binary input rv s is then mapped into the corresponding sample value $(x_1, x_2, \ldots, x_n)$ of channel inputs. These are transmitted, and the receiver, given the sample n-tuple $y_1, \ldots, y_n$, and knowing the code mapping, guesses which binary $\lceil nR \rceil$ tuple was sent. There is then an error probability $\Pr\{e\}$ that the receiver decodes incorrectly. It turns out to make very little difference whether we look at the error probability for each code word, the average over a uniform distribution of code words, or the maximum over all the code words.

The noisy channel coding theorem is a central result of information theory. This theorem says that for any $R < C$, any desired error probability $\Pr\{e\}$, and all sufficiently large block lengths, there are codes of rate R and block length n for which the error probability is less than $\Pr\{e\}$. Furthermore, for $R > C$, such codes do not exist for sufficiently small $\Pr\{e\} > 0$. In other words, mutual information, as defined in (10.114) really corresponds to reliable transmission of bits of information using coding.

We have looked only at memoryless channels above, i.e., channels for which the output at each time n depends only on the input at time n, and the theorem is much more general, although that generality will not be discussed.

The superficial connections with Fisher information are, first, that both involve log-likelihoods, second, that both have an input (or parameter) to be estimated or detected from an output (or observation), and, third, that the input distribution is somewhat artificial in both cases. For the estimation case, there are good reasons to avoid an input distribution, whereas for the communication case, avoiding an input distribution interferes with the relationship between the one-dimensional view and the coding view. This interference makes the coding theorem much more difficult and less intuitive.

10.9 Summary

The problem of estimation is very similar to that of detection. There is a probability space including an n-rv $\boldsymbol{X}$ and an m-rv $\boldsymbol{Y}$. A sample value $\boldsymbol{y}$ of $\boldsymbol{Y}$ is observed and an estimate $\boldsymbol{x}(\boldsymbol{y})$ is selected to approximate (i.e., estimate) the sample value $\boldsymbol{x}$ of $\boldsymbol{X}$. There are many approaches to defining an estimator that is optimal in some sense. A particularly important class is that of minimum-cost estimators, where a cost $C(\boldsymbol{x}, \hat{\boldsymbol{x}})$ is assigned for each sample value $\boldsymbol{x}$ and estimate $\hat{\boldsymbol{x}}$. One then chooses $\hat{\boldsymbol{x}}(\boldsymbol{y})$ to minimize the expected cost for each $\boldsymbol{y}$ and thus minimize the overall expected cost.

The most important minimimum-cost estimate is MMSE, where the cost is the squared error. The MMSE estimator is then $\hat{\boldsymbol{x}}(\boldsymbol{y}) = \mathsf{E}\left[\boldsymbol{X} | \boldsymbol{Y}{=}\boldsymbol{y}\right]$, i.e., the estimate is the conditional mean of X conditional on $\boldsymbol{Y} = \boldsymbol{y}$.

The results on conditional probabilities of Gaussian rv s in Section 3.5 were then used for cases in which the n-rv $\boldsymbol{X}$ to be estimated and the m-rv $\boldsymbol{Y}$ to be observed are zero-mean, jointly Gaussian, and non-singular. Any such case can be represented in the form $\boldsymbol{X} = [G]\boldsymbol{Y} + \boldsymbol{V}$, where $\boldsymbol{Y}$ and $\boldsymbol{V}$ are statistically independent and jointly Gaussian. The MMSE estimate for any observed sample value $\boldsymbol{y}$ is the linear function $\hat{\boldsymbol{x}}(\boldsymbol{y}) = [G]\boldsymbol{y}$. The error in the MMSE estimate is V, which surprisingly is independent of the observation $\boldsymbol{Y}$.

All estimation problems in this class can also be represented in the form $\boldsymbol{Y} = [H]\boldsymbol{X} + \boldsymbol{Z}$, where the noise $\boldsymbol{Z}$ and the signal $\boldsymbol{X}$ are statistically independent. This is the typical formulation for this class of Gaussian problems and represents cases in which the observation is a linearly 'filtered' version of the signal $\boldsymbol{X}$ added to the Gaussian noise $\boldsymbol{Z}$. There are many formulas connecting the matrices $[G]$ and $[H]$ and the covariances $[K_V]$ and $[K_Z]$ with the covariances and joint covariances of $\boldsymbol{X}$ and $\boldsymbol{Y}$.

Many special cases of these Gaussian estimation problems were considered, and there were many minor extensions, including that of non-zero mean, recursive estimation, Kalman filtering, and circularly-symmetric complex Gaussian problems. The general class of Gaussian problems is particularly important since they serve as guides even when the noise or the signal are non-Gaussian.

Non-Gaussian estimation problems are often quite messy, so it is interesting to investigate estimators that have a simple structure and do not depend on parts of the model that are difficult to justify. A major class of such estimators are the LLSE estimators. Here the estimate $\hat{x}(\boldsymbol{y})$ is restricted to have the form $[G]\boldsymbol{y} + \boldsymbol{b}$. This is the same form as the MMSE estimate for the Gaussian case with a non-zero mean. For any probability model where $\boldsymbol{X}$ and $\boldsymbol{Y}$ have finite means, covariances and cross-covariances, we showed that there is a Gaussian model with the same means and covariances. The LLSE estimator and all its error components are then the same as those for the corresponding Gaussian case. Since the error components for the MMSE estimator in any non-Gaussian case must have mean-square values that are less than or equal to those of the corresponding Gaussian case. Thus MMSE estimation is most difficult for the Gaussian case.

For any probability model, the set of zero-mean rv s in that model can be viewed as the vectors in an inner-product space where $\langle \boldsymbol{X}, \boldsymbol{Y} \rangle = \mathsf{E}\left[\boldsymbol{X}^{\mathsf{T}}\boldsymbol{Y}\right]$. This is a subspace of the inner-product space of all rv s, but it is usually easier to deal only with zero-mean rv s, serving as the fluctuations of rv s with non-zero means. This allows us to adapt all the results about inner-product spaces to rv s, and in particular allows us to use the projection theorem. We found that the projection of a rv X on the set of observed rv s is the LLSE estimate of X from $Y_1, \dots, Y_m$. This is quite different intuitively from finding $\hat{x}(\boldsymbol{y})$ from a sample value $\boldsymbol{y}$ of $\boldsymbol{Y}$ but is in fact the same thing and it is helpful to be able to go from one viewpoint to the other.

The chapter ended with a discussion of parameter estimation, i.e., estimation of a one-dimensional signal X from an observation $(Y_1, \dots, Y_m)$ where there is no a priori distribution on X. There cannot be any completely satisfying theory for this problem, since making an estimator better for one value of the parameter will make it worse for others, and no measure of importance is permitted between different values. We considered both ML estimators and unbiased estimators. The definition of estimation bias for estimators and the attendant notion of unbiased estimators is not entirely satisfying since unbiased estimators do not always exist and are not necessarily desirable. They lead to the Cramer–Rao bound, however, which provides an upper bound to mean-squared error as a function of the signal value x and the bias for that x. Perhaps the most important feature of the Cramer–Rao bound is that it provides a simple characterization of the benefit of enlarging the number m of observations, and essentially shows that the estimation error must approach 0 as $m \to \infty$.

10.10 Exercises

Exercise 10.1 (**a**) Consider the joint probability density $\mathsf{f}_{X,Z}(x,z) = e^{-z}$ for $0 \le x \le z$ and $\mathsf{f}_{X,Z}(x,z) = 0$ otherwise. Find the pair x, z of values that maximize this density. Find the marginal density $\mathsf{f}_Z(z)$ and find the value of z that maximizes this.

(**b**) Let $\mathsf{f}_{X,Z,Y}(x,z,y)$ be y^2e^{-yz} for $0 \le x \le z$, $1 \le y \le 2$ and be 0 otherwise. Conditional on an observation $Y = y$, find the joint MAP estimate of X, Z. Find $\mathsf{f}_{Z|Y}(z|y)$, the marginal density of Z conditional on $Y = y$, and find the MAP estimate of Z conditional on $Y = y$.

Exercise 10.2 Let $Y = X+Z$, where X and Z are IID and each $\mathcal{N}(0,1)$. Let $U = Z-X$.

(**a**) Explain (using the results of Chapter 3) why Y and U are jointly Gaussian and why they are statistically independent.

(**b**) Without using any matrices, write down the joint probability density of Y and U. Verify your answer from (3.22).

(**c**) Find the MMSE estimate $\hat{x}(y)$ of X conditional on a given sample value y of Y. You can derive this from first principles, or use (10.9) or Example 10.2.2.

(**d**) Show that the estimation error $\xi = X - \widehat{X}(Y)$ is equal to $-U/2$.

(**e**) Write down the probability density of U conditional on $Y = y$ and that of ξ conditional on $X = y$.

(**f**) Draw a sketch in the x, z plane of the equiprobability contours of X and Z. Explain why these are also equiprobability contours for Y and U. For some given sample value of Y, say $Y = 1$, illustrate the set of points for which $x+z = 1$. For a given point on this line, illustrate the sample value of U.

Exercise 10.3 (**a**) Let $X, Z_1, Z_2, \dots, Z_n$ be independent zero-mean Gaussian rv s with variances $\sigma_X^2, \sigma_{Z_1}^2, \dots \sigma_{Z_n}^2$ respectively. Let $Y_j = h_jX + Z_j$ for $j \ge 1$ and let $\boldsymbol{Y} = (Y_1, \dots Y_n)^{\mathsf{T}}$. Use (10.9) to show that the MMSE estimate of X conditional on $\boldsymbol{Y} = \boldsymbol{y} = (y_1, \dots, Y_n)^{\mathsf{T}}$ is given by

$$\hat{x}(\boldsymbol{y}) = \sum_{j=1}^{n} g_j y_j, \quad \text{where} \quad g_j = \frac{h_j/\sigma_{Z_j}^2}{(1/\sigma_X^2) + \sum_{i=1}^{n} h_i^2/\sigma_{Z_i}^2}. \tag{10.115}$$

Hint: Let the row vector $\boldsymbol{g}^{\mathsf{T}}$ be $[K_{X\cdot\boldsymbol{Y}}][K_{\boldsymbol{Y}}^{-1}]$ and multiply $\boldsymbol{g}^{\mathsf{T}}$ by $K_{\boldsymbol{Y}}$ to solve for $\boldsymbol{g}^{\mathsf{T}}$.

(**b**) Let $\xi = X - \widehat{X}(\boldsymbol{Y})$ and show that (10.29) is valid, i.e., that

$$1/\sigma_\xi^2 = 1/\sigma_X^2 + \sum_{i=1}^{n} \frac{h_i^2}{\sigma_{Z_i}^2}.$$

(**c**) Show that (10.28), i.e., $\hat{x}(\boldsymbol{y}) = \sigma_\xi^2 \sum_{j=1}^{n} h_j y_j/\sigma_{Z_j}^2$, is valid.

(**d**) Show that the expression in (10.29) is equivalent to the iterative expression in (10.27).

(e) Show that the expression in (10.28) is equivalent to the iterative expression in (10.25).

Exercise 10.4 Let $X, Z_1, \ldots, Z_n$ be independent, zero-mean, and Gaussian, satisfying $Y_i = h_i X + Z_i$ for $1 \le i \le n$. The point of this exercise is to show that estimation of X from $\mathbf{Y}_1^n$ is essentially no more complex than the same problem with each $h_i = 1$.

(a) Write out the MMSE estimate and variance of the estimation error from (10.28) and (10.29) where each $h_i = 1$.

(b) For arbitrary numbers $h_1, \ldots, h_n$, consider the equations $h_i Y_i = h_i X + h_i Z_i$. Let $U_i = h_i Y_i$. Write out the MMSE estimate of X as a function of $U_1, \ldots, U_n$, still using the rv s $Z_1, \ldots, Z_n$. Hint: Note that if you scale each observation U_i to U_i/h_i you have the same estimation problem as in (a).

(c) Now let $W_i = h_i Z_i$. Write out the same MMSE estimate in terms of both $U_i, \ldots, U_n$ and $W_1, \ldots, W_n$ and show that you now have (10.28) with all the h_i's back in the equation. Hint: You still have the basic equation in (a), but the variance of the Z_i's must be expressed in terms of the W_i's. Write out the error variance from (10.29) in the same way.

Exercise 10.5 **(a)** Assume that $X_1 \sim \mathrm{N}(\overline{X}_1, \sigma_{X_1}^2)$ and that for each $n \ge 1$, $X_{n+1} = \alpha X_n + W_n$, where $0 < \alpha < 1$, $W_n \sim \mathrm{N}(0, \sigma_W^2)$, and $X_1, W_1, W_2, \ldots,$ are independent. Show that for each $n \ge 1$,

$$\mathsf{E}[X_n] = \alpha^{n-1}\overline{X}_1, \qquad \sigma_{X_n}^2 = \frac{(1-\alpha^{2(n-1)})\sigma_W^2}{1-\alpha^2} + \alpha^{2(n-1)}\sigma_{X_1}^2.$$

(b) Show directly, by comparing the equation $\sigma_{X_n}^2 = \alpha^2\sigma_{X_{n-1}}^2 + \sigma_W^2$ for each two adjoining values of n, that $\sigma_{X_n}^2$ moves monotonically from $\sigma_{X_1}^2$ to $\sigma_W^2/(1-\alpha^2)$ as $n \to \infty$.

(c) Assume that sample values of $Y_1, Y_2, \ldots,$ are observed, where $Y_n = hX_n + Z_n$ and where $Z_1, Z_2, \ldots,$ are IID zero-mean Gaussian rv s with variance σ_Z^2. Assume that $Z_1, \ldots, W_1, \ldots, X_1$ are independent and assume $h \ge 0$. Rewrite the recursion for the variance of the estimation error in (10.41) for this special case. Show that if $h/\sigma_Z = 0$, then $\sigma_{\xi_n}^2 = \sigma_{X_n}^2$ for each $n \ge 1$. Hint: Compare the recursion in (b) to that for $\sigma_{\xi_n}^2$.

(d) Show from the recursion that $\sigma_{\xi_n}^2$ is a decreasing function of h/σ_Z for each $n \ge 2$. Use this to show that $\sigma_{\xi_n}^2 \le \sigma_{X_n}^2$ for each n. Explain (without equations) why this result must be true.

(e) Show that the sequence $\{\sigma_{\xi_n}^2; n \ge 1\}$ is monotonic in n. Hint: Use the same technique as in (b). From this and (d), show that $\lambda = \lim_{n\to\infty} \sigma_{\xi_n}^2$ exists. Show that the limit satisfies (10.42) (note that (10.42) must have two roots, one positive and one negative, so the limit must be the positive root).

(f) Show that for each $n \ge 1$, $\sigma_{\xi_n}^2$ is increasing in σ_W^2 and increasing in α. Note: This increase with α is surprising, since when α is close to 1, X_n changes slowly, so we would expect to be able to track X_n well. The problem is that $\lim_n \sigma_{X_n}^2 = \sigma_W^2/(1-\alpha^2)$ so the variance of the untracked X_n is increasing without limit as α approaches 1. Part (g) is somewhat messy, but resolves this issue.

(g) Show that if the recursion is expressed in terms of $\beta = \sigma_W^2/(1-\alpha^2)$ and α, then λ is decreasing in α for constant β.

Exercise 10.6 (**a**) Assume that $\boldsymbol{X}$ and $\boldsymbol{Y}$ are zero-mean, jointly Gaussian, jointly non-singular, and related by $\boldsymbol{Y} = [H]\boldsymbol{X} + \boldsymbol{Z}$ where $\boldsymbol{X}, \boldsymbol{Z}$ are independent. Show that the MMSE estimate can be expressed as

$$\hat{\boldsymbol{x}}(\boldsymbol{y}) = [K_{\xi}][H^{\mathsf{T}}][K_{\boldsymbol{Z}}^{-1}]\boldsymbol{y}.$$

Hint: Recall from (10.9) that $\hat{\boldsymbol{x}}(\boldsymbol{y}) = [G]\boldsymbol{y}$. Relate $[G]$ and $[H]$ by using (3.115).

(**b**) Show that

$$[K_{\xi}^{-1}] = [K_{\boldsymbol{X}}^{-1}] + [H^{\mathsf{T}}K_{\boldsymbol{Z}}^{-1}H].$$

Hint: A reasonable approach is to start with (10.10), i.e., $[K_{\xi}] = [K_{\boldsymbol{X}}] - [K_{\boldsymbol{X}\cdot\boldsymbol{Y}}K_{\boldsymbol{Y}}^{-1}K_{\boldsymbol{X}\cdot\boldsymbol{Y}}^{\mathsf{T}}]$. Premultiply both sides by $[K_{\boldsymbol{X}}^{-1}]$ and post-multiply both sides by $[K_{\xi}^{-1}]$. The final product of matrices here can be expressed as $[H^{\mathsf{T}}K_{\boldsymbol{Z}}^{-1}H]$ by using the hint in (a) on $[K_{\boldsymbol{Z}}^{-1}H]$.

Exercise 10.7 (**a**) Write out $\mathsf{E}\left[(X - \boldsymbol{g}^{\mathsf{T}}\boldsymbol{Y})^2\right] = \sigma_X^2 - 2[K_{X\cdot\boldsymbol{Y}}]\boldsymbol{g} + \boldsymbol{g}^{\mathsf{T}}[K_{\boldsymbol{Y}}]\boldsymbol{g}$ as a function of $\boldsymbol{g} = (g_1, g_2, \ldots, g_n)^{\mathsf{T}}$ and take the partial derivative of the function with respect to g_i for each i, $1 \le i \le n$. Show that the vector of these partial derivatives is $-2[K_{X\cdot\boldsymbol{Y}}] + 2\boldsymbol{a}^{\mathsf{T}}[K_{\boldsymbol{Y}}]$.

(**b**) Explain why the stationary point here is actually a minimum.

Exercise 10.8 For a real inner-product space, show that m vectors, $Y_1, \ldots, Y_m$ are linearly dependent if and only if the matrix of inner products, $\{\langle Y_j, Y_k\rangle;\ 1 \le j, k \le m\}$, is singular.

Exercise 10.9 Show that (10.89) to (10.91) agree with (10.11).

Exercise 10.10 Let $\boldsymbol{X} = (X_1, \ldots, X_n)^{\mathsf{T}}$ be a zero-mean complex **rv** with real and imaginary components $X_{re,j}, X_{im,j}$, $1 \le j \le n$ respectively. Express $\mathsf{E}\left[X_{re,j}X_{re,k}\right]$, $\mathsf{E}\left[X_{re,j}X_{im,k}\right]$, $\mathsf{E}\left[X_{im,j}X_{im,k}\right]$, $\mathsf{E}\left[X_{im,j}X_{re,k}\right]$ as functions of the components of $[K_{\boldsymbol{X}}]$ and $\mathsf{E}\left[\boldsymbol{X}\boldsymbol{X}^{\mathsf{T}}\right]$.

Exercise 10.11 Let $Y = Y_r + jY_{im}$ be a complex rv. For arbitrary real numbers a, b, c, d, find complex numbers α and β such that

$$\Re\left[\alpha Y + \beta Y^*\right] = aY_{re} + bY_{im},$$

$$\Im\left[\alpha Y + \beta Y^*\right] = cY_{re} + dY_{im}.$$

Exercise 10.12 (Derivation of circularly-symmetric Gaussian density) Let $\boldsymbol{X} = \boldsymbol{X}_{re} + i\boldsymbol{X}_{im}$ be a zero-mean, circularly-symmetric, n-dimensional, Gaussian complex **rv**. Let $\boldsymbol{U} = (\boldsymbol{X}_{re}^{\mathsf{T}}, \boldsymbol{X}_{im}^{\mathsf{T}})^{\mathsf{T}}$ be the corresponding $2n$-dimensional real **rv**. Let $[K_{re}] = \mathsf{E}\left[X_{re}X_{re}^{\mathsf{T}}\right]$ and $[K_{ri}] = \mathsf{E}\left[X_{re}X_{im}^{\mathsf{T}}\right]$.

(**a**) Show that

$$[K_U] = \begin{bmatrix} [K_{re}] & [K_{ri}] \\ -[K_{ri}] & [K_{re}] \end{bmatrix}.$$

(**b**) Show that

$$[K_U^{-1}] = \begin{bmatrix} B & C \\ -C & B \end{bmatrix}$$

and find the B, C for which this is true.

(c) Show that $[K_X] = 2([K_{re}] - i[K_{ri}])$.
(d) Show that $[K_X^{-1}] = \frac{1}{2}(B - iC)$.
(e) Define $\mathsf{f}_X(\boldsymbol{x}) = \mathsf{f}_U(\boldsymbol{u})$ for $\boldsymbol{u} = (\boldsymbol{x}_{re}^{\mathsf{T}}, \boldsymbol{x}_{im}^{\mathsf{T}})^{\mathsf{T}}$ and show that

$$\mathsf{f}_X(\boldsymbol{x}) = \frac{\exp\left(-\boldsymbol{x}^{\dagger}[K_X^{-1}]\boldsymbol{x}^{\dagger}\right)}{(2\pi)^n\sqrt{\det[K_U]}}.$$

(f) Show that

$$\det[K_U] = \begin{bmatrix} [K_{re}] + i[K_{ri}] & [K_{ri}] - i[K_{re}] \\ -[K_{ri}] & [K_{re}] \end{bmatrix}.$$

Hint: Recall that elementary row operations do not change the value of a determinant.
(g) Show that

$$[K_U] = \begin{bmatrix} [K_{re}] + i[K_{ri}] & 0 \\ -[K_{ri}] & [K_{re}] - i[K_{ri}] \end{bmatrix}.$$

Hint: Recall that elementary column operations do not change the value of a determinant.
(h) Show that

$$\det[K_U] = 2^{-2n}\left(\det[K_X]\right)^2$$

and from this conclude that (3.108) is valid.

Exercise 10.13 (Alternate derivation of circularly-symmetric Gaussian density) **(a)** Let X be a circularly-symmetric, zero-mean, complex Gaussian rv with covariance 1. Show that

$$\mathsf{f}_X(x) = \frac{\exp(-x^*x)}{\pi}.$$

Recall that the real part and imaginary part each have variance 1/2.
(b) Let $\boldsymbol{X}$ be an n-dimensional, circularly-symmetric, complex Gaussian zero-mean rv with $[K_X] = I_n$. Show that

$$\mathsf{f}_X(\boldsymbol{x}) = \frac{\exp(-\boldsymbol{x}^{\dagger}\boldsymbol{x})}{\pi^n}.$$

(c) Let $\boldsymbol{Y} = [H]\boldsymbol{X}$ where $[H]$ is $n \times n$ and invertible. Show that

$$\mathsf{f}_Y(\boldsymbol{y}) = \frac{\exp\left[-\boldsymbol{y}^{\dagger}([H^{-1}])^{\dagger}[H^{-1}]\boldsymbol{y}\right]}{v\pi^n},$$

where v is $d\boldsymbol{y}/d\boldsymbol{x}$, the ratio of an incremental $2n$-dimensional volume element after being transformed by $[H]$ to that before being transformed.
(d) Use this to show that that (3.108) is valid.

Exercise 10.14 **(a)** Let $Y = X^2 + Z$, where Z is a zero-mean unit variance Gaussian rv. Show that no unbiased estimate of X exists from observation of Y. Hint. Consider any $x > 0$ and compare with $-x$.

(b) Let $Y = X + Z$, where Z is uniform over $(-1, 1)$ and X is a parameter lying in $(-1,\ 1)$. Show that $\hat{x}(y) = y$ is an unbiased estimate of x. Find a biased estimate $\hat{x}_1(y)$ for which $|\hat{x}_1(y) - x| \leq |\hat{x}(y) - x|$ for all x and y with strict inequality with positive probability for all $x \in (-1, 1)$.

Exercise 10.15 **(a)** Assume that for each parameter value x, Y is Gaussian, $\mathrm{N}\ (x, \sigma_Z^2)$. Show that $v_x(y)$ as defined in (10.103) is equal to $(y - x)/\sigma_Z^2$ and show that the Fisher information is equal to $1/\sigma_Z^2$.

(b) Show that for ML estimation, the bias is 0 for all x. Show that the Cramer–Rao bound is satisfied with equality for all x in this case.

(c) Consider using the MMSE estimator for the a priori distribution $X \sim \mathrm{N}\ (0, \sigma_X^2)$. Show that the bias satisfies $b_{\hat{z}}(x) = -x\sigma_Z^2/(\sigma_Z^2 + \sigma_X^2)$.

(d) Show that the MMSE estimator in (c) satisfies the Cramer–Rao bound with equality for each x. Note that the mean-squared error, as a function of x, is smaller than that for the ML case for small x and larger for large x.

Exercise 10.16 Assume that Y is $\mathrm{N}\ (0, x)$. Show that $v_x(y)$ as defined in (10.103) is $v_x(y) = [y^2/x - 1]/(2x)$. Verify that $v_x(Y)$ is zero mean for each x. Find the Fisher information, $J(x)$.

References

[1] Bellman, R., *Dynamic Programming*. Princeton, NJ: Princeton University Press, 1957.

[2] Bertsekas, D. and J. Tsitsiklis, *An Introduction to Probability Theory*, second edn. Belmont, MA: Athena Scientific, 2008.

[3] Bertsekas, D. P., *Dynamic Programming and Optimal Control*, second edn. Belmont, MA; Athena Scientific, 2000.

[4] Bhattacharya, R. N. and E. C. Waymire, *Stochastic Processes with Applications*. New York, NY: Wiley, 1990.

[5] Cover, T. M. and J. A. Thomas, *Elements of Information Theory*, second edn. New York, NY: Wiley, 2006.

[6] Dembo, A. and O. Zeitouni, *Large Deviation Techniques and Applications*. Woods Hole, MA: Jones & Bartlett Publishers, 1993.

[7] Doob, J. L., *Stochastic Processes*. New York, NY: Wiley, 1953.

[8] Feller, W., *An Introduction to Probability Theory and Its Applications*, vol 1, third edn. New York, NY: Wiley, 1968.

[9] Feller, W., *An Introduction to Probability Theory and Its Applications*, vol 2. New York, NY: Wiley, 1966.

[10] Gallager, R. G., *Principles of Digital Communication*. Cambridge, UK: Cambridge University Press, 2008.

[11] Gallager, R. G., *Discrete Stochastic Processes*. Norwell, MA: Kluwer, 1996.

[12] Gallager, R. G., *Information Theory and Reliable Communication*. New York, NY: Wiley, 1968.

[13] Gantmacher, F. R., *The Theory of Matrices* (English translation). New York, NY: Chelsea, 1959 (2 volumes).

[14] Grimmett, G. and D. Starzaker, *Probability and Random Processes*, third edn. Oxford, UK: Oxford University Press, 2001.

[15] Harris, T. E., *The Theory of Branching Processes*. Berlin: Springer Verlag, and Englewood Cliffs, NJ: Prentice Hall, 1963.

[16] Howard, R., *Dynamic Programming and Markov Processes*. Cambridge, MA: MIT Press, 1960.

[17] Kelly, F. P., *Reversibility and Stochastic Networks*. New York, NY: Wiley, 1979.

[18] Kolmogorov, A. N., *Foundations of the Theory of Probability*. New York, NY: Chelsea, 1950 (Translation of *Grundbegriffe der Wahrscheinlchksrechnung*, Springer Berlin, 1933).

[19] Neyman, J. and E. S. Pearson, On the use and interpretation of certain test criteria for purposes of statistical inference, *Biometrica*, **20A**, 175–263, 1928.

[20] Pawitan, Y., *In All Likelihood: Statistical Modelling and Inference Using Likelihood*. Oxford, UK: Clarendon Press, 2001.

[21] Ross, S., *A First Course in Probability*, eighth edn. Englewood Cliffs, NJ: Prentice Hall, 2009.

[22] Ross, S., *Stochastic Processes*, second edn. New York, NY: Wiley, 1983.

[23] Rudin, W. *Real and Complex Analysis*, third edn. New York, NY: McGraw Hill, 1987.

[24] Rudin, W. *Principles of Mathematical Analysis*, third edn. New York, NY: McGraw Hill, 1975.

[25] Schweitzer, P. J. and A. Federgruen, The asymptotic behavior of undiscounted value iteration in Markov decision problems, *Math. of Op. Res.*, **2**, 360–381, 1967.

[26] Shannon, C. E., A mathematical theory of communication, *Bell System Technical Journal*, **27**, 379–423 and 623–656, 1948. Available on the web, http://cm.bell-labs.com/cm/ms/what/shannonday/paper.html

[27] Shiryaev, A. N., *Probability*, second edn. New York, NY: Springer, 1995.

[28] Strang, G., *Linear Algebra*, fourth edn. Wellesley, MA: Wellesley-Cambridge Press, 2009.

[29] Wald, A., *Sequential Analysis*. New York, NY: Wiley, 1947.

[30] Wolff, R. W., *Stochastic Modeling and the Theory of Queues*. Englewood Cliffs, NJ: Prentice Hall, 1989.

[31] Yates, R. D. and D. J. Goodman, *Probability and Stochastic Processes*. New York, NY: Wiley, 1999.

[32] Yates, R. D., *High Speed Round Robin Queueing Networks*, LIDS-TH-1983. Cambridge, MA: MIT, 1983.

Index

a posteriori probability, 377
a priori probability
  in decision-making, 375–376, 379
  situations not needing, 396–403
absolute value cost function, 490–491
accessible state (Markov chain), 164
age for renewal process, 255–259, 300
aperiodic class (Markov chain), 165, 306
arbitrary finite state (Markov chain), 175, 180
arbitrary random walk, 419, 423–432
arithmetic average, 36
arithmetic renewal process, 253–258
arrival process, 72–74, 74 n.3, 92–96, *see also* renewal processes
averages (time), 226–233, 328–329, 331–332, 336
axioms of probability, 8–9

backward Markov chain, 303–307, 312–317, 323
bandpass noise, 144–145
Bayes' criterion, *see* minimum-cost rule
Bayes' law, 10, 376, 392, 408, 495–496
Bellman, R., 202
Bellman dynamic programming algorithm, 190–194
Bernoulli process, *see also* binomial distribution
  analysis, 17–19
  as Markov chain, 287–288
  shrinking, 82–84, 89–90
  and SLLN, 221
  and stopping times, 235
Berry–Esseen theorem, 41
Bertsekas, D., 58, 202
Bessel's inequality, 130
binary detection with minimum-cost criterion rule, 395–396
binary Kullback–Liebler divergence, 18
binary MAP threshold rule, 379–395
binary pattern, expected wait for, 182–184, 280–281, 481–482
binomial distribution, 14, 18, 35–36, 43–44
birth–death chains, 288–289, 302–303, 305
Birth-death Markov processes, 342–344, 346–347
Blackwell's theorem, 217, 249, 253, 256, 259, 268, 299–300
Borel-Cantelli lemma, 274
bounds, *see* inequality
branching processes
  Markov chain, 309–312
  scaled, 446
Brownian motion, *see* Wiener process
Burke's theorem, 309, 347–350

Cauchy principle value, 22–23
CDF, *see* cumulative distribution function (CDF)
central limit theorem (CLT)
  description and statement, 39–44
  for conditionally IID observations, 519
  for renewal processes, 225–226
  motivation for jointly Gaussian rv s, 110
  relation to WLLN, 39–41
  Wiener process, 144
Chapman–Kolmogorov equation, 168, 170, 290, 337–341
Chebyshev inequality
  and convergence, 37
  one-sided, 67
  principles of, 32–33
  and Wiener process, 143
  and WLLN, 45
Chernoff bound
  principles of, 33–36
  as used in binary hypothis tests, 438–439
  optimization of, 423–425
  used in random walks, 431–432
circular symmetry
  and covariance matrices, 146–149
  complex Gaussian random vectors, 145–148
  for complex random vectors, 144–146
  Gaussian PDF, 150–154
  Gaussian processes, 154–155
class of states, 164
closed Jackson networks, 355–357
CLT, *see* central limit theorem (CLT)
communicating states, 164
complex rv, *see* random variable (rv); random vector
composition of random vectors, 428–431
conditional expectation, *see also* expectation
  as a rv, 26–28
  definition, 25–26

total expectation, 28
conditional probability, 9–10
continuous random variable, *see also* random variable (rv)
definition, 13, 14
expectation, 21
convergence
in distribution, 40–41, 45–48
in probability, 46–48, 70
infinite-state Markov chains, 170–175
of random variable sequences, 45–48
in mean square, 48, 70
WP1 (with probability 1), 48–51, 130–131, 217–226
convex functions of martingales, 448
countable intersection, 7
countable union, 6–7, 9
covariance function
definition, 125
and spectral density, 138–139
complex process, 146–149
Gaussian process, 125
stationary process, 127, 134–139
wide-sense stationary process, 128
covariance matrix
and Gaussian PDFs, 121–128, 150–154
convex random vectors, 146–149
definition, 108
Gaussian random vectors, 111, 113, 118–120
properties of, 114–120
Cramer–Rao bound, 515–518
cumulative distribution function (CDF)
arithmetic, 253
in central limit theorem, 40–41
definition, 13
as function of other random variables, 23–24
joint CDF, 14–15
of indicator random variable, 29
in Poisson process, 74
of renewal processes, 231–233, 263–266
use in finding expectations, 20–22
WLLN, 37–38
cycle (Markov chain), 163–164

De Morgan's law, 7
decision criteria
binary hypotheses, IID observations, 438–443
in hypothesis testing, 376–379
decision making, *see* hypothesis testing
decision rule
binary MAP threshold, 379–395
error curve, 396–402
MAP, 377–379
minimum-cost, 395–396, 404, 409, 489
min–max detection, 403
ML, 380–381
Neyman–Pearson, 397, 402–403, 439–440
randomized threshold, 402–403
decision theory (Markov), 189–202
defective rv s, 13, 104
delayed renewal process, 266–270, 300
Dembo, A., 473
detection, *see also* hypothesis testing
of antipodal signals in Gaussian noise, 381–383
of general signals in Gaussian noise, 383–385
of vector signals in Gaussian noise, 385–390
with minimum cost criterion, 395–396
discrete observation, 391–394, 400–401
discrete random variable, *see also* random variable
definition, 13–14
expectation, 20–21, 26
discrete stochastic processes, *see* Markov chain (finite or countable state); Markov process (countable state); Poisson process; renewal process
discrete-time Gauss-Markov processes, 126–127
discrete-time Gaussian processes, 125–126
distribution function, *see* cumulative distribution function
double or quarter game, 484
duration for renewal processes, 255–259
dynamic programming, 189–200

eigenvalues, 114–120, 148–153, 176–180
eigenvectors, *see* eigenvalues
elementary renewal theorem, 217, 249, 251–252, 268
embedded Markov chain (Markov processes), 324
equilibrium (renewal) process, 270
ergodic class (countable-state Markov chain)
convergence (Blackwell's theorem), 300
definition, 300
ergodic class finite-state (Markov chain)
convergence, 172–173
definition, 167
ergodic unichain, 169–170, 173–175
Erlang density, 78–79
error curve (binary hypotheses), 396–402
estimation
application areas, 488–489
complex Gaussian vectors, 505, 506
cost function, 489–491
Gaussian vectors, 491–498
in a vector space of rv s, 507–512
iterative, 494–496, 503–504
Kalman, 496–498, 504–505
LLSE, 498–500
MAP, 512–513
MMSE, 491–498
parameter, 513–521
scalar iterative signal and noise, 494–496
scalar Kalman filter, 496–498

scalar signal and noise, 493–494
vector signal, 500–505
Euler-Mascheronei constant, 60
events
as component of probability model, 2
axioms for events, 6–7
expectation (E[X]), *see also* mean
definition, 19–23
as function of other random variables, 23–26
of counting rv s in renewal processes, 249–253
and Wald's equality, 238–239
expected value, *see* expectation (E[X])
exponential bounds on thresholds, 436–438
exponential growth rate, 462–464
exponentially tight sequence, 426, 429, 438

Feller, W., 58, 271, 318
filtered processes
continuous time, 136–138
Gaussian sinc processes, 134–136
finite grain event, 3
finite sample space
assigning probabilities for, 4–5
and discrete random variable, 13
and expectation, 19–20
first-passage time
for countable-state Markov chains, 289–291
for finite-state Markov chains, 181–184
Fisher information, 515–519
Fisher–Neyman factorization, 393–395
Fourier transform, 31, 128–134, *see also* spectral density
frequency shift keying (FSK), 412–413
Frobenius theorem, 173 n.4

Gaussian hypothesis testing
binary hypothesis testing in, 381–390
for multiple hypotheses, 405–406, 408–409
G/G/1 queue
and random walks, 420–422
renewal processes, 239–242
Gaussian processes
continuous-time, 130–132
definition, 124–126
discrete-time, 125–127
Discrete-time Gauss-Markov, 126–127
Markov chain (countable state), 287–289
matrices, 114–117
MMSE estimation of, 491–498
normalized random variables in, 105–107
orthonormal expansions, 128–130
PDF, 117–124
sinc processes, 132–136
spectral density, 136–139
stationarity, 126–128
white noise, 139–142
wide-sense stationary, 128
Wiener process, 142–144
Gaussian random vector processes
circularly-symmetric complex rv estimation, 505–506
complex random vectors, 144–155
rv, 107–114
Geometry for Gaussian PDFs, 118–120

Harris, T. E., 319
Hermitian covariance matrices, 148–154
holding intervals (Markov process), 325
homogeneous Markov chain, *see* Markov chain (finite state)
horse race betting, 484–485
hypothesis testing
binary detection with minimum-cost criterion, 395–396
binary MAP, 379–395
decision criteria in, 376–379
error curve, 396–402
min–max detection rule, 403
for multiple hypotheses, 403–409
Neyman–Pearson rule, 402–403
and probability models, 375–376

IID, *see* independent identically distributed trials (IID)
IID noise, *see* white noise
increment process (Poisson), 77–78, 81–82, 142–144
independence, *see* statistical independence
independent identically distributed trials (IID), *see also* Poisson process; renewal processes
Bernouli process, 17–19
definition, 11–12
Gaussian processes, 108–109, 125
and martingales, 445–446
observation, 438–443
in real world experiments, 52–55
and tilted probabilities, 425–428
Wiener process, 144
independent increments, 77–82, 142
indicator random variable, 29
inequality
Bessel's, 130
Chebyshev, 32–33
Chernoff, 33–36, 423–425, 430–439
Cramer–Rao, 515–518
Jensen's, 449
Markov, 32
Schwartz, 508
information theory, 519–521
inherent reachability (Markov decisions), 196
inner-product space, 508–509
integer time process, *see* Markov chain (finite state)
integer-valued random walk, 419

irreducible Markov chain, 297–299, 305, 306
irreducible Markov processes, *see* Markov processes
iterative estimation, 494–496, 503–504

Jackson networks, 350–357, *see also* queues
Jensen's inequality, 449
joint probability density (Gaussian)
  definition, 117–118
  geometry, 118–120
  special case, 112–114
jointly-Gaussian random vectors, 109–111, 124–126
Jordan form, 179–180

Kalman filter, 496–498, 504–505
Kelly, F. P., 319
Kingman bound, 437–438
Kolmogorov, A. N., 5, 11, 58
Kolmogorov differential equations, 337
Kolmogorov equations, 453–458
Kolmogorov's submartingale inequality, 453–455

Laplace transform, 31, 249–251
large deviations, 428–431
last come first serve (LCFS) queues, 370–371
linear transformations, 149–150
Little's theorem, 242–245, 316
LLSE estimate, 498–500

MAP, *see* maximum a posteriori probability (MAP)
Markov chain (countable state)
  birth–death, 288–289, 302–303, 305
  branching processes, 309–312
  first-passage-time analysis of, 289–291
  irreducible, 297–299, 305, 306
  M/M/1 queue, 307–309
  recurrence of, 291–294
  renewal theory applied to, 215, 294–302
  reversible, 303–306
  round-robin service system, 312–317
Markov chain (finite state)
  classification of states, 163–167
  decision theory, 189–190, 200
  definition, 161–163
  dynamic programming algorithm, 190–194
  eigenvalues, 176–180
  matrices, 168–175
  policy-improvement algorithm, 197–200
  rewards, 180–189
  stationary decision policies, 194–200
Markov inequality, 32
Markov modulated random walks, 468–472
Markov processes
  birth–death process, 342–344, 346–347
  Chapman–Kolmogorov equation, 337–341
  definition, 324–326
  Jackson networks, 350–357
  M/M/1 queue, 326–328
  reversibility of, 344–350
  steady state distributions, 329–337
  time averages, 328–329
  uniformization, 341–342
marriage problem, 480
martingale convergence theorem, 456–458
martingales, 444
  definition, 444–447
  investment model, 458–468
  Kolmogorov equations, 453–458
  product-form, 445–446
  stopping times, 450–453
  sub- and super-, 447–449
maximum a posteriori probability (MAP)
  and estimation, 491, 512–513
  hypothesis testing, 379–395
  rule, 377–379
maximum likelihood tests (ML), 380–381, 441, 491, 512, *see also* parameter estimation
maximum-tolerable-error cost function, 491
mean, *see also* expectation
  contrast with median, 28–29
  of common rv s, 14
measure theory, 7, 7 n.8, 12, 12 n.13
median, 28–29
memoryless property (Poisson process), 75–78, 85
MGF, *see* moment generating function (MGF)
minimum mean-squared-error estimate (MMSE estimate), 490–500, 513, *see also* parameter estimation
minimum-cost rule
  definition, 395–396
  and estimation, 489
  for multiple hypotheses, 404, 409
min–max detection rule, 403
M/M/1 queue
  Markov chain (countable state), 307–309
  Markov processes, 326–328, 336–344
MMSE, *see* minimum mean-squared-error estimate (MMSE estimate)
moment generating function (MGF)
  definition, 14, 29–31
  for Markov modulated random walks, 469–471
  Gaussian random vectors, 108–109, 111
  Gaussian rv s, 106–107
  properties, 423–433

Neyman–Pearson rule, 397, 402–403, 439–440
noise
  bandpass, 144–145
  vector, 500–505
  white, 139–142
non-arithmetic renewal process, 253, 258–266

observation
  binary hypothesis testing with vector, 385–391

continuous, 394–395
discrete, 391–394, 400–401
hypothesis testing with one-dimensional, 381–385
IID, 438–443
vector, 518–519
one-dimensional random walk, *see* random walk
order statistics, 92–96
orthogonal matrix (Gaussian), 114–115
orthogonality principle, 509–512
orthonormal expansions (Gaussian), 128–130
outcomes, *see* sample points

parameter estimation, 513–521, *see also* maximum likelihood tests (ML); minimum mean-squared-error estimate (MMSE estimate)
path (Markov chain), 163–164
pathological Gaussian noise, 141
pathological Markov processes, 336–337
PDF, *see* probability density function (PDF)
periodic state (Markov chain), 165–306
PMF, *see* probability mass function (PMF)
Poisson process, *see also* Bernouli process; independent identically distributed trials (IID); renewal processes; Wiener process
arrival process in, 72–74, 74 n.3, 92–96
combining, 84–86
conditional arrival densities in, 92–96
hypothesis testing in, 390–391
Markov processes (countable-state), 306, 326–328
non-homogeneous, 89–92
order statistics, 92–96
PMF, 79–80
properties of, 75–84
subdividing, 86–89
two-dimensional, 103
policy improvement (Markov decisions), 197–200
portfolios for investment
constant fractional allocation, 461–465
time-varying allocation, 465–468
positive definite matrix, 115–117
positive recurrence
Markov processes (countable-state), 329–330, 333, 335
semi-Markov processes (countable-state), 359–360
positive semi-definite, 55
preemptive resume LCFS queues, 370–371
principle axes for Gaussian PDFs, 118–120
probability 1 (WP1)
convergence, 48–51, 130–131
and SLLN, 217–226
probability density function (PDF)
circularly-symmetric Gaussian, 150–154
definition, 13
for common continuous rv s, 14
for common rv s, 14
Gaussian, 120–124
Gaussian rv, 105–106
probability mass function (PMF)
binomial distribution, 18
definition, 13
for common discrete rv s, 14
Poisson rv, 79–80
probability models
components of, 2–4
and estimation, 488
and hypothesis testing, 375–376
martingales, 458–468
and real world randomness, 51–57
for repeated idealized experiments, 11–12
probability theory axioms, 5–9
processor sharing (round robin), 313–317
pseudo-covariance matrix, 146–148

queues
backward-moving, 315
Burke's theorem, 309, 347–350
FCFS, 239
feedback queues, 349
FIFO, 239
G/G/1, 420–422
G/G/m, 215
Jackson networks, 350–357
Little's theorem, 242–245
M/D/m/m, 373
M/G/∞, 91
M/G/1, 245
M/M/1, 88
M/M/m, 309
PASTA property, 248
Pollaczek-Khinchin theorem, 248
processor sharing, 313–317
round-robin queues, 316
tandem queues, 350
unfinished work, 246–249

random variable (rv), *see also* random vectors
complex, 13, 22, 145–146
convergence, 45–48
defective, 13, 104
definition, 12–14
Gaussian, 105–107, 145–146
indicator, 29
multiple, 14–16
observation, 381–395, 400–401, 438–443
typical value of, 28
random vectors, *see also* random variable (rv)
complex, 144–155
Gaussian, 107–114, 124–126
random walk, *see also* martingales
arbitrary, 419, 423–432

definition, 417–418
for binary hypotheses, 438–442
for sequential decisions, 442–444
G/G/1 queue, 420–422
integer, 419
Markov modulated, 468–472
renewal processes as, 419–420
simple, 418–419
with thresholds, 433–438
randomized threshold rule, 402–403
randomness
definition, 1–2
and probability models, 51–57
recurrent state (Markov chain)
definition (countable state), 291–294
definition (finite state), 164, 165
in unichain, 169
null recurrence, 288, 290, 296
positive recurrence, 288, 290, 294–296, 299
relative frequency, 5, 39, 52–56
relative-gain vector (Markov chain), 186
renewal equation, 250, 269, 278
renewal processes, *see also* arrival process; independent identically distributed trials (IID); Poisson process
arithmetic, 253–258
Bernoulli process as, 19
definition, 214–215
delayed, 266–270
Markov chain (countable state), 294–302
Markov processes, 330–335
non-arithmetic, 253, 258–266
and random walks, 419–420
number of renewals in (0, t], 249–253
rewards, 226–233, 254–266
stopping times, 233–249
Strong law for renewal processes, 219, 221–226
renewal theory, 200
repeated idealized experiments, 11–12
residual life (renewal processes), 226–228, 231–232
reversibility
Markov chain, 303–306
Markov processes, 344–350, 360–361
rewards
Markov chain, 180–189
renewal processes, 226–233, 254–266
Riemann sum, 21, 44, 62, 262–264
Ross, S., 271, 318
round-robin service and processor sharing, 312–317
Rudin, W., 58
rv, *see* random variable (rv)

sample average (in laws of large numbers), 36–39
sample points, 2
sample space, 3–9
sampled-time (Markov processes), 328–329, 336
Schwartz inequality, 508
score, 516
secretary problem, 480
semi-Markov processes (countable-state), 357–361
sequential decisions, 442–444
simple random walk, 418–419, 435–436
sinc processes (Gaussian), 132–136, 140
spectral density, *see also* Fourier transform
definition, 136–139
and white noise, 139–142
squared-cost function, 489–490
standard deviation (σ), 23–24
stationary increments, 77–82, 142
stationary policy algorithm (Markov), 194–200
stationary processes
Gaussian, 126–128, 133–134, 139–142
unichain (Markov chain), 194–200
Wiener process, 142–144
statistical independence
of random variables, 15–16
countable state Markov, 300
definition, 10–11
of Gaussian random vectors, 110–111, 120–124
in Poisson process, 86–89
of real world experiments, 55–56
steady state
equilibrium (renewal process), 270
Markov chains, 168–176, 296–299
Markov processes, 329–337
steady-state vector (Markov chain)
countable state Markov, 296–299
definition, 168–175
finite-state Markov, 168–171, 173, 299
in transient states, 181
Stein's lemma, 441
Stieltjes integral, 23, 60, 260–262
Stirling bounds, 60
stochastic matrix, 168–175
stochastic process, 16, 16 n.15, 17–19, *see also* Bernoulli process
stopped processes, 450
stopping times
martingales, 450–453, 471
renewal processes, 233–249
stopping trial, *see* stopping times
strong law for renewal processes, 217, 221–226
strong law of large numbers (SLLN), *see also* weak law of large numbers (WLLN)
and convergence WP1, 217–226
definition, 35–51
and Kolmogorov inequalities, 455–456
relative frequency in, 52
subjective probabiity, 57–58
submartingale inequality, 453–455
submartingales, 447–449
sufficient statistic

in binary hypothesis testing, 381
discrete with continuous observations, 394
for multiple hypotheses, 407–409
with discrete observations, 391–394
sums of dependent zero-mean rv s, 445
supermartingales, 447–449
symmetric complex rv estimation, 505–506
symmetry
Gaussian matrix, 114–115
as rationale for choosing probabilities, 4

theorem
Berry–Esseen, 41
Blackwell's, 217, 249, 253, 256, 259, 268, 299–300
Burke's, 309, 347–350
central limit, 18, 39–44, 110, 144, 225–226, 519
elementary renewal, 217, 249, 251–252, 268
Little's, 242–245, 316
martingale convergence, 456–458
threshold
detection rules, 379–395, 402–403
Markov modulated random walks, 471–472
and random walks, 418, 433–438
tilted probabilities, 425–428
time averages
Markov processes (countable-state), 328–329, 331–332, 336
renewal processes, 226–233
transform, *see also* moment generating functions
definition, 29–31
Fourier, 31, 128–134
Laplace, 31, 249–251
Z, 31
transient state (Markov chain), 164, 165, 291
transition probabilities (Markov chain)
periodic, 166
stochastic matrix, 168–175
trapping state (Markov chain), 181–184
Triple or nothing investments, 460–461
Tsitsiklis, J., 58
typical values of rv s, 28–29

unichain (Markov chain)
definition, 169
stationary policies, 194–200
transition matrix, 170
uniformized process (Markov), 341–342

variance
random variable, 14, 23, 25
WLLN, 36–45
vector, 110
composition, 428–431
estimation, 500–505, 507–512
Gaussian, 107–114, 124–126
jointly-Gaussian random, 124–126
noise estimation, 500–505
observations, 518–519
random, 144–155
space estimation, 507–512

Wald, Abraham, 233
Wald's equality, 217, 236–242, 246, 435
Wald's identity, 434–437
walk
Markov, 163–164, 468–472
random, 417–431
zero-mean random, 417, 435–436, 444–445, 492–494
weak law of large numbers (WLLN), *see also* strong law of large numbers (SLLN)
with finite variance, 36–39
with infinite variance, 44–45
and tilted probabilities, 427–428
for renewal processes, 223–224
white noise, 139–142
wide sense stationary (WSS)
Gaussian processes, 134–139
stochastic process, 128
Wiener process, 142–144, *see also* Poisson process
Wolff, R. W., 271, 318
WP1, *see* probability 1 (WP1)

Zeitouni, O., 473
zero-mean random walk, 417, 435–436, 444–445, 492–494
Z-transform, 31